The Study of Prosimian Behavior

Contributors

S. K. Bearder
K. J. Boskoff
P. Charles-Dominique
G. A. Doyle
G. Gray Eaton
C. M. Hladik
P. H. Klopfer
Robert D. Martin
Carsten Niemitz
Georges Pariente
Jean-Jacques Petter
Arlette Petter-Rousseaux
J. I. Pollock
Duane M. Rumbaugh
A. Schilling
Richard N. Van Horn
Alan Walker
Beverly J. Wilkerson

The Study of Prosimian Behavior

Edited by

G. A. Doyle

Primate Behaviour Research Group
University of the Witwatersrand
Johannesburg, South Africa

R. D. Martin

Department of Anthropology
University College London
London, England

ACADEMIC PRESS New York San Francisco London 1979

A Subsidiary of Harcourt Brace Jovanovich, Publishers

ACADEMIC PRESS, INC.
111 Fifth Avenue, New York, New York 10003

United Kingdom Edition published by
ACADEMIC PRESS, INC. (LONDON) LTD.
24/28 Oval Road, London NW1 7DX

Library of Congress Cataloging in Publication Data

Main entry under title:

The Study of prosimian behavior.

Includes bibliographies.
1. Primates--Behavior. 2. Mammals--Behavior.
I. Doyle, Gerald A. II. Martin, Robert D.
III. Title: Prosimian behavior.
QL737.P9S77 599'.81'05 78-9417
ISBN 0-12-222150-8

PRINTED IN THE UNITED STATES OF AMERICA

79 80 81 82 9 8 7 6 5 4 3 2 1

To the memory of

GEORGES PARIENTE

who died tragically shortly
after his own contribution had
been completed,
with the affection and respect
of all the contributors to this
volume

Contents

List of Contributors

Numbers in parentheses indicate the pages on which the authors' contributions begin.

S. K. BEARDER[1] (567), Primate Behaviour Research Group, University of the Witwatersrand, Johannesburg 2001, South Africa

K. J. BOSKOFF[2] (123), Department of Zoology, Duke University, Durham, North Carolina 27706

P. CHARLES-DOMINIQUE (247, 567), Equipe de Recherche sur les Prosimiens, Laboratoire d'Écologie Générale, Muséum National d'Histoire Naturelle, Brunoy 91800, France

G. A. DOYLE (157), Primate Behaviour Research Group, University of the Witwatersrand, Johannesburg 2001, South Africa

G. GRAY EATON (79), Oregon Regional Primate Research Center, Beaverton, Oregon 97005

C. M. HLADIK (307), Equipe de Recherche sur les Prosimiens, Laboratoire d'Écologie Générale, Muséum National d'Histoire Naturelle, Brunoy 91800, France

P. H. KLOPFER (123), Department of Zoology, Duke University, Durham, North Carolina 27706

ROBERT D. MARTIN[3] (45), Wellcome Institute of Comparative Physiology, The Zoological Society of London, Regent's Park, London, England

CARSTEN NIEMITZ (631), Anatomisches Institut, Fachbereich Medizin, Georg-August Universitat, Göttingen, West Germany

GEORGES PARIENTE[4] (411), Laboratoire d'Écologie Générale, Muséum National d'Histoire Naturelle, Brunoy 91800, France

JEAN-JACQUES PETTER (1, 247), Equipe de Recherche sur les Prosimiens, Laboratoire d'Écologie Générale, Muséum National d'Histoire Naturelle, Brunoy 91800, France

ARLETTE PETTER-ROUSSEAUX (1), Equipe de Recherche sur les Prosimiens,

[1]Present address: Wellcome Institute of Comparative Physiology, The Zoological Society of London, Regent's Park, London NW1 4RY, England.

[2]Present address: Zoological Society of San Diego, San Diego, California.

[3]Present address: Department of Anthropology, University College London, London WC1, England.

[4]Deceased.

Laboratoire d'Écologie Générale, Muséum National d'Histoire Naturelle, Brunoy 91800, France

J. I. POLLOCK[5] (359), Department of Anthropology, University College London, London WC1, England

DUANE M. RUMBAUGH (207), Department of Psychology, Georgia State University, Atlanta, Georgia 30303

A. SCHILLING (461), Equipe de Recherche sur les Prosimiens, Laboratoire d'Écologie Générale, Muséum National d'Histoire Naturelle, Brunoy 91800, France

RICHARD N. VAN HORN (79), Oregon Regional Primate Research Center, Beaverton, Oregon 97005

ALAN WALKER (543), Department of Anatomy, Harvard Medical School, Boston, Massachusetts 02115

BEVERLY J. WILKERSON (207), Department of Psychology, Georgia State University, Atlanta, Georgia 30303

[5]Present address: Department of Physiology, University of Cambridge, Cambridge, England.

Preface

"The Study of Prosimian Behavior" was first conceived in 1970 while the first editor was on a tour of the United States under the auspices of the Carnegie Corporation, and in response to many suggestions to write a book on this subject. Prosimians had only recently come into their own as objects of study because of the potential that such studies offered in understanding the course of primate evolution.

At the same time the second editor was thinking about an international seminar on prosimians to bring together all those studying these animals—to present and discuss their work with peers in a variety of disciplines. This idea was given precedence. A highly successful international seminar took place in London in 1972 ("Prosimian Biology," Martin *et al.*, 1974).[1] It was only the third work to have appeared recently devoted exclusively to the prosimian primates. Subsequent to "Prosimian Biology," Tattersall and Sussman (1975)[2] produced a valuable edited volume of contributed chapters constituting "a progress report on our knowledge of the lemurs." Charles-Dominique's (1977)[3] superb monograph, based on years of field work on five sympatric lorisid species in Central West Africa, together with Chapter 13 in this volume, destroy the myth that the lorisids of Africa and Asia "do not show the richness and variety of adaptation" characteristic of the Malagasy forms (Tattersall and Sussman, 1975). The Malagasy forms are of particular interest because in Madagascar they have not been subject to the constraints provided by competition with higher primates and, as a consequence, have undergone adaptive radiations of great variety. The lorisids are of equal interest for just the opposite reason. In the face of competition from higher primates and an even larger variety of nonprimate forms, they have successfully adapted in a variety of very different but equally complex ways.

These two major groups of prosimians must not, however, be viewed in isolation

[1]MARTIN, R. D., DOYLE, G. A., and WALKER, A. C., eds. (1974). "Prosimian Biology," Duckworth, London.

[2]TATTERSALL, I., and SUSSMAN, R. W., eds. (1975), "Lemur Biology." Plenum, New York.

[3]CHARLES-DOMINIQUE, P. (1977). "Behaviour and Ecology of Nocturnal Primates." Duckworth, London.

from one another and intelligent discussion continues on the ''when'' and ''how'' of their common ancestry (see, for example, Charles-Dominique and Martin, 1970).[4] For that matter it is equally important that the prosimians must not be viewed in isolation from the simian primates and the present volume might well be the last book, at least for some time, to be devoted exclusively to the prosimians. There is a danger of the prosimian research worker regarding himself as a ''prosimianologist'' with all the inherent consequences of insulating himself from the main stream of primatology in general.

Indeed, some primatologists have voiced disquiet recently at the growth of a discipline name for a taxonomic group with which it is primarily concerned (S. A. Altmann, personal communication; J. H. Crook, personal communication) and the resulting isolation from the still broader mainstream of evolutionary biology. Altmann notes, for instance, that primate field studies have been slow to incorporate modern ecological and sociobiological concepts and wonders if the primatologist would be better advised to interact with those whose research is more like his own conceptually rather than taxonomically. Primatology, however, is here to stay and if we heed Altmann's advice we can have the best of both worlds.

Whichever view we take there is a growing and continual need for books like the present volume. From its early conception, the idea took root that the book should be written by a group of people, each member working in a specific behavioral or behavior-related area. From time to time the scientific community needs something that will help it take stock of the progress of research in a particular area. This cannot be provided by scientific journals but only by international conferences and synoptic books. The research worker also needs the opportunity to present the fruits of a number of years of concentrated research in his own specific field of endeavor. Again, this cannot be provided by current journals with their unavoidable restrictions on space, nor in this case, can it be provided by international meetings with their equally unavoidable restrictions on time.

Reviews of recent books on prosimians have been overwhelmingly kind for a number of reasons. The main reason is that prosimians have not received their fair share of attention in the proliferation of studies on primates that has marked the last 20 odd years and yet is common cause, as Doyle and Martin (1974)[5] point out, that no full understanding of the course of primate evolution will ever be arrived at if one entire group of primates is neglected. Behavior studies, particularly field studies, are more difficult to carry out than laboratory studies, but these difficulties are by no means insurmountable as recent studies have shown. Recent books on the prosimians have gone a long way to redress both these imbalances and it would be true to say that, particularly with the publication of the present

[4]CHARLES-DOMINIQUE, P., and MARTIN, R. D. (1970). *Nature (London)* **225,** 139–144.

[5]DOYLE, G. A., and MARTIN, R. D. (1974). *In* ''Prosimian Biology'' (R. D. Martin, G. A. Doyle, and A. C. Walker, eds.), pp. 4–14. Duckworth, London.

book, behavior is no longer a neglected area although there are still a number of species that have not yet been studied at all.

Each of the contributors in the present volume has been working for many years in the area in which he has contributed a chapter. The purpose of each chapter is twofold: first, it gives each author an opportunity to present new, hitherto unpublished, data on his research and, second, it places these data within the context of a review or overview of work done in his particular area. The chapters were chosen to provide as complete a coverage as possible of the field of prosimian behavior. Some cross-referencing has been done but by and large each chapter represents a synthesis and overview in its own right. No attempt has been made at providing an overall synthesis.

So many people contribute to the production of a volume such as this and in so many ways that it would be impossible to mention them all. Special mention must be made, however, of those who helped in the practical side of its production. Dorci Carpenter-Frank, Berna Struwig, and Wendy Cullinan typed and, in many cases, retyped all the chapters. R. D. Martin translated several of the French chapters into English. Heather Brand and Tina Scheffer, Research Assistants to the senior editor, helped with bibliographical checking and in this connection special tribute must be paid to the staff of the Inter-Library Loans Department and Miss L. Van Schaardenburg. Printing and technical help with illustrations were provided by Colin Emslie and Adolph Veenstra of the Educational Technology Unit of the University of the Witwatersrand, and by Mark Hudson and Theo Bekker. Charles Darlington of the Printing Department of the University of the Witwatersrand and Frans Mashu willingly provided duplicating services. Last but not least the editors gratefully acknowledge the guidance and patience of the editorial staff of Academic Press throughout all stages of the production of this book.

G. A. Doyle
R. D. Martin

Chapter 1

Classification of the Prosimians

JEAN-JACQUES PETTER
AND ARLETTE PETTER-ROUSSEAUX

1. INTRODUCTION

Despite the interest which the prosimians attract and the numerous publications which have emerged from their study, the systematics of the group remain controversial and there are still a number of problems concerning the phylogenetic relationships of certain extant forms. The main object of this chapter is, in fact, to contribute to research on the evolution of the prosimians. Although certain criticisms will be raised, the aim is not to modify radically Simpson's (1945) classification, which is widely recognized internationally and provides a practical basis for the organization of many museum collections. It is an important point that it would be a difficult task to modify the organization of museum systems to keep pace with new publications, and nonspecialists would in any case have difficulty in finding specimens. Accordingly, the preliminary results proposed here should be regarded

ISBN 0-12-222150-8

as nothing more than a contribution to a future basis for establishing a new, stable classification of the primates.

After the first systematists had produced a variety of classifications, it was soon generally accepted, following Mivart (1873), that the Primates should be divided into two main subgroups. The broad outlines of a classification on this basis have now been adopted by the majority of workers, but there are two major points which remain problematic: the relationship of the tree shrews (Tupaiidae) to the Primates and the status of the genus *Tarsius* within the order. Further, the origins of the order Primates and the detailed relationships between the Lemuriformes and the Lorisiformes (as defined by Simpson, 1945) are still a focus for heated discussion. The specific points will be discussed first, and then the basis for overall classification of the Strepsirhini (Lemuriformes and Lorisiformes) will be considered with a view to introducing certain modifications.

2. STATUS OF THE TUPAIIDAE

Simpson's classification of the mammals (1945) included the Tupaiidae in the Primates, in line with suggestions made by Kaudern (1910), Carlsson (1922), and others (Saban, 1956, 1957) following the lead of Le Gros Clark (1934). This inclusion of tree shrews with the Primates met with some resistance, particularly from Strauss (1944). Significantly, in his systematic revision of the prosimians, in which he emphasized the value of the distinction between Strepsirhini and Haplorhini, Hill (1953) excluded the tree shrews from the order Primates. More recently, Martin (1967, 1968) has suggested that the tree shrews have no specific evolutionary relationship to the primates and that their apparent similarity to certain primates is based upon retention of ancestral placental mammalian characteristics and upon convergent adaptation for arboreality.

In addition, it has now been concluded (McKenna, 1963; van Valen, 1969) that the Oligocene fossil *Anagale* from Mongolia is probably related neither to the tree shrews nor to the primates. Thus, an entire phylogenetic edifice which, among other things, led some people to suggest that the order Primates originated in Mongolia, has collapsed.

In view of the uncertainty of the status of the tree shrews, both in terms of their evolution and from the point of view of their classification, it seems preferable at this stage to exclude them from the following discussion of the Primates.

3. STATUS OF THE TARSIIDAE

Prior to Mivart's seminal publication in 1873, Gervais (1854) had already proposed that special status should be given to the genus *Tarsius* in primate classification. This proposal was taken up by Gadow (1898) and was subsequently discussed in depth with respect to anatomical and embryological evidence by Elliot Smith (1919), Hill (1919), Pocock (1919), and Wood Jones (1919) at a special meeting of the Zoological Society of London. Finally, Simpson (1945) used a compromise solution in his classification, placing the tarsiers among the prosimians but setting them aside in their own infraorder. The first division in his classification was between the suborder Prosimii, Illiger and the suborder Anthropoidea, Mivart. [Hoffstetter (1974) has, incidentally, suggested that the latter term should be replaced by Simii, Van der Hoven 1933, in order to avoid confusion.] The suborder Prosimii was itself divided into three infraorders:

Infraorder Lemuriformes, Gregory
 Superfamily Tupaioidea, Dobson
 Superfamily Lemuroidea, Mivart
Infraorder Lorisiformes, Gregory
Infraorder Tarsiiformes, Gregory

Although this classification accorded some special status to the tarsiers, their inclusion among the prosimians has subsequently been criticized by numerous authors.

As was demonstrated by Gervais in 1854, despite their possession of a number of primitive characters which link tarsiers to the other prosimians, they exhibit several anatomical specializations (notably in placentation and in the nasal region) which indicate a relationship to monkeys, apes, and man. Such a relationship is further indicated by zoogeographical considerations (Hoffstetter, 1974) and by immunological studies (Goodman, 1975). There is an increasing tendency for the tarsiers to be regarded as descendants from a group of primates intermediate in characteristics between the ancestors of the Strepsirhini and the ancestors of the monkeys, apes, and man.

Hill (1953) took account of various criticisms of Simpson's classification of the tarsiers, by classifying the living species and various Eocene fossil forms together with the Anthropoidea. He adopted Pocock's (1918) scheme of classification of the primates into Strepsirhini (lemurs and lorises) and Haplorhini (tarsiers, monkeys, apes, and man), producing an outline which seems to provide a more acceptable reflection of the relationships of the tarsiers for many primate systematists:

Grade A: Strepsirhini, Geoffroy 1812
 Suborder: Lorisoidea, Tate Regan 1930
  Families:
   Lorisidae
   Galagidae
 Suborder: Lemuroidea, Mivart 1864
  Families:
   Lemuridae

Adapidae
Indriidae
Megaladapidae
Daubentoniidae
Plesiadapidae
Apatemyidae
Carpolestidae
Grade B: Haplorhini, Pocock 1912
Suborder: Tarsioidea, Elliot Smith 1907
Families:
Tarsiidae
Anaptomorphidae
Suborder: Pithecoidea, Pocock 1918
(containing all the remaining primate families)

4. THE ORIGIN OF THE PRIMATES

The most ancient fossil forms commonly accepted as belonging to the order Primates are from the Paleocene. However, even the most primitive species, which are apparently quite close to the insectivores in their basic characteristics, exhibit dental specializations which exclude them from direct ancestry of any modern primates.

Both Stehlin (1912) and Le Gros Clark (1934) considered some of the Paleocene fossil forms as possible ancestors for the surviving aye-aye, *Daubentonia*. But recent behavioral studies support Simpson's view (1945) that there is no reliable basis for suggesting such a direct ancestral relationship. The first fossil finds attributed to the order Primates hence represent no more than a very pale reflection of an adaptive radiation for which we have very little documentary evidence. The earliest primates were doubtless of small body size, and even if it is certain that we should look for traces of them at the beginning of the Paleocene, it is still extremely difficult to decide where they were likely to have occurred.

Most of these early fossil forms attributed to the Primates have recently been grouped together in the infraorder Plesiadapiformes (Simons, 1972; Tattersall, 1973), while Szalay (1973) has suggested establishing them in a separate suborder, Paromomyiformes. It is generally accepted that the Apatemyidae should not now be included in the Primates.

Although these Paleocene fossil forms, together with some species which survived later into the Tertiary, have no direct relationship to modern primates, discussion of their characteristics is extremely useful in defining more clearly our concepts of primate evolution.

5. FOUNDATIONS OF THE CLASSIFICATION OF THE STREPSIRHINI

As has been seen above, Simpson's classification (1945) separated the Lemuriformes clearly from the Lorisiformes. Hill (1953) had a similar clear division into two suborders (Lorisoidea and Lemuroidea), but he did note that the Cheirogaleinae (classified in the Lemuridae) exhibit certain similarities to the Lorisoidea, indicating either phylogenetic relationship or convergence. A number of earlier authors (Flower and Lydekker, 1891; Forbes, 1894), in fact, considered the Cheirogaleinae as transitional between the Lemuroidea and the Lorisoidea (as defined in Hill's classification). Petter-Rousseaux (1962) also noted resemblances between the Cheirogaleinae and the Lorisoidea in reproductive characteristics, and the whole question of the significance of such resemblances has recently been raised by Charles-Dominique and Martin (1970). These two authors analyzed the original distinctions between the Lemuroidea and Lorisoidea set out by Weber (1927). They pointed out that such distinctions had been based essentially on a comparison between *Loris* or *Perodicticus,* of the subfamily Lorisinae, with *Lemur,* of the subfamily Lemurinae. There had been virtually no consideration of the Galaginae and the Cheirogaleinae in establishing the distinctions between the Lorisoidea and the Lemuroidea, and when these two subfamilies are included in the comparison it emerges that many distinctions break down.

5.1. Analysis of the Characters Employed in the Classification of the Strepsirhini

In a recent paper, Hoffstetter (1974) has partially adopted the proposals of Szalay and Katz (1973) in suggesting a division of the order Primates into three suborders, with the Paromomyiformes treated as representatives of a special radiation distinct from those giving rise to the Strepsirhini and the Haplorhini:

Suborder: Paromomyiformes (= Plesiadapiformes)
  Infraorder: Plesiadapoidea
    Families:
      Paromomyidae (including Purgatorinae)
      Microsyopidae
      Plesiadapidae
      Carpolestidae
      Picrodontidae
Suborder: Strepsirhini
  Infraorder: Lemuriformes
    Superfamily: Adapoidea
      Family: Adapidae (including Notharctinae)

Superfamily: Lemuroidea
  Families:
    Lemuridae
    Indriidae
    Megaladapidae
Superfamily: Daubentonioidea
  Family: Daubentoniidae
Infraorder: Lorisiformes
  Superfamily: Lorisoidea
    Families:
      Lorisidae
      Cheirogaleidae?
Suborder: Haplorhini
  Infraorder: Tarsiiformes
    Superfamily: Tarsioidea
      Family: Tarsiidae (including Microchoerinae)
    Superfamily: Omomyoidea
      Family: Omomyidae (including Anaptomorphinae)
  Infraorder: Simiiformes
    Subinfraorder: Platyrrhini
    Superfamily: Ceboidea
      Families:
        Cebidae
        Xenothricidae
        Callithricidae
    Subinfraorder: Catarrhini
    Superfamily: Cercopithecoidea
      Families:
        Parapithecidae
        Cercopithecidae
    Superfamily: Hominoidea
      Families:
        Hylobatidae
        Pongidae
        Hominidae
        Oreopithecidae

This classification reflects a link between the Adapoidea and the Lemuroidea, whose common origin is perhaps indicated (as argued by Szalay and Katz, 1973) by similarities in the tympanic bulla and its relationships to the ectotympanic element. It also incorporates the suggestion that the Cheirogaleidae (now ranked as a family) should be provisionally associated with the Lorisidae, and the special status of *Daubentonia* is recognized by the creation of a special superfamily.

The main classical characters which are involved in hypotheses concerning the phylogeny of the Strepsirhini (Adapoidea + Lemuroidea + Lorisoidea), and which are specifically relevant to the status of the Cheirogaleidae and *Daubentonia,* are

1. The general form of the skull
2. The form of the lower jaw
3. The structure of the orbit
4. The structure of the bulla and its relationship to the ectotympanic element
5. The carotid circulation
6. Special features of the dentition—notably the presence or absence of a toothcomb

Emphasis on these characters has been partly dictated by the limitations of paleontological material, but it is also true that they seem to have remained relatively conservative during the evolution of the mammals. However, although such conservatism seems to be obvious in groups such as the rodents, which became specialized relatively early, it is not necessarily present in other groups such as the prosimians, whose cranial characters have remained relatively unspecialized and are probably still open to adaptive modification. For this reason, one cannot simply apply the same criteria to the study of the prosimian skull as one would, for example, to that of rodent cranial morphology.

This fact provides a fairly good explanation for the prolific discussion which has taken place with respect to prosimian systematics. Particular cranial characters have been taken by certain authors to indicate phylogenetic relationships, whereas the similarities involved may be due to convergence, regressive evolution, or simply the retention of primitive characters which do not provide evidence of recent common ancestry. A great deal of anatomical and embryological work remains to be done before such characters can be considered in sufficient detail to permit reliable conclusions to be drawn.

Any method which involves consideration of a list of the characters present in different genera and determination of relationships between successive levels of "sister taxa" can be criticized on the grounds that we do not yet fully understand the interdependence of such characters. For example, the form of the mandibular condyle is obviously associated with that of the glenoid cavity, but these characters in turn are linked to such features as dental morphology and the sites of insertion of the jaw muscles. This entire assemblage of characters, despite its impressive extent, probably carries no more weight than the most representative single character involved. Hence, compilation of a long list of associated characters does not amount to a convincing demonstration of phylogenetic relationships.

Evolution involves fine adaptation to particular ecological niches and it would seem that, unless specialization has already proceeded too far in a particular direction, anatomical features are gradually molded through mutation and natural selection to match environmental requirements. Such adaptations naturally fall into a number of large functional categories, such as feeding, locomotion, vision, olfaction, and hearing.

Progressive fine adaptation of each of these major functional categories to the possibilities presented by the ecological niche of a given species usually leads to a compromise between individual influences on anatomical adaptation. In certain cases, however, the influence of a particular character may be so predominant that the entire architecture of the skull may be fundamentally affected (as in the case of *Daubentonia*).

When the main characters, mentioned above, classically used in prosimian systematics, are viewed from this standpoint it is possible to make a number of observations in respect of each which permit better appreciation of their value.

5.1.1. The General Form of the Skull

The shape of the skull is one of the first characters to be considered in the description of a species. It is, in fact, very difficult to compare the skulls of species of greatly different body sizes and none of the recent statistical investigations (Mahé, 1972; Lessertisseur *et al.,* 1974) seems to provide a convincing solution to this problem. In addition, it is particularly difficult to interpret the tendencies present in the skulls of small-bodied forms, since the forces responsible for their evolution have in most cases been inadequate to produce marked specializations. In large-bodied species, on the other hand, complex mechanical forces associated with dental adaptations (Tattersall, 1973; Roberts and Tattersall, 1974; Cartmill, 1974) and thus with feeding behavior seem to give rise to pronounced modification of the orbit (with possible repercussions on vision), on the nasal cavity (with repercussions on olfaction), and on other features.

Tattersall and Schwartz (1974) have pointed out that it is, nevertheless, possible to make certain generalizations with respect to the facial region, but this is far more difficult for the basicranium, which is itself strongly influenced by the length of the face and by the position and orientation of the occipital condyles and the foramen magnum.

Further, various behavioral and ecological adaptations can influence the size and orientation of the orbits as, for example, in the shift from nocturnal to diurnal habits. *Avahi,* which are exclusively nocturnal, have orbits which are relatively larger than in the other, diurnal indriids. With respect to the Indriidae, Roberts and Tattersall (1974) have taken the shape of the skull as a basis for distinguishing *Avahi, Propithecus, Mesopropithecus, Archaeolemur,* and even *Hadropithecus,* on the one hand, from *Indri, Palaeopropithecus,* and *Archaeoindri,* on the other. In fact, *Indri* differs from *Propithecus* essentially in the elongation of the face and the resulting mechanical influences on the neurocranium.

Taking the skulls of *Lemur* and *Varecia* (*Lemur variagatus*), there is little difference in general proportions and Tattersall and Schwartz have no difficulty in associating these two genera with *Megaladapis.* The facial elongation found in the latter genus is considered to be responsible for numerous other differences, such as the constriction of the neurocranium around the relatively small brain, the insertion of a large posterior temporal muscle, and the presence of an enormous frontal sinus. According to Tattersall and Schwartz (1974), *Daubentonia* are also linked to the Indriidae through the shape of the skull. However, it is a difficult task to compare the skull of this genus with that of other lemur genera because extreme dental specialization has exerted an enormous influence on skull shape (Cartmill, 1974).

5.1.2. The Form of the Lower Jaw

Tattersall and Schwartz (1974) note that the form of the lower jaw is relatively constant among the prosimians, but the mandibles of the Indriidae and of the Lemuridae are clearly distinctive. Mandibular shape is closely associated with dietary specialization (the indriids all have a markedly specialized diet, involving a large proportion of leaves, whereas the lemurids tend to have more varied diets; see Hladik, Chapter 8), and it would seem to be of little value for interpreting the phylogenetic relationships between small-bodied species. The latter typically have less specialized diets and the forces influencing skull morphology are not so pronounced. Further, some dietary influences are still little understood, as is the case with the heavy dependence on gum-feeding in *Phaner furcifer* and *Galago elegantulus* (*Euoticus elegantulus*), which is probably responsible for certain modifications of the anterior part of the skull. The shape of the mandibular condyles and the structure of the jaw symphysis would also seem to be closely related to diet and hence of relatively little value in establishing evolutionary relationships.

5.1.3. The Structure of the Orbit

The structure of the medial wall of the orbit has traditionally been considered to be of importance in primatology, particularly with respect to the presence or absence of an os planum (a component of the ethmoid). Recent anatomical studies (Simons and Russell, 1960; Cartmill, 1971) have similarly diminished the phylogenetic value of this character, since the presence of an os planum is exhibited particularly by species with large, approximated orbits. Tattersall and Schwartz (1974) have summarized the following characters distinguishing the different prosimian families: The Lorisidae and Cheirogaleidae are distinguished by the large ethmoid component in the orbital wall of the former, with the palatine excluded from contact with the lachrymal, as compared with the smaller ethmoid component of the latter, associated with contact between the palatine and the lachrymal. There is resemblance between the Cheirogaleidae and the Lemuridae, and between the Indriidae and *Daubentonia,* with a portion of the palatine excluded from contact with the lachrymal. However, these distinctions are perhaps largely the consequence of other anatomical modifications, themselves the product of convergent adaptation under particular ecological conditions.

5.1.4. The Structure of the Bulla and Its Relationship to the Ectotympanic Element

The structure of the tympanic bulla, its shape, and its relationships with the ectotympanic element have recently been considered in detail by Szalay and Katz

(1973), Gingerich (1975), Hoffstetter (1974), Cartmill (1974), Groves (1974), and Saban (1975). The "lemuroid" pattern, with extension of the bulla beyond the tympanic ring resulting in its inclusion within the bulla, is found with the Adapidae, Cheirogaleidae, Lemuridae, Indriidae, and Daubentoniidae, but not with the Lorisidae. Hoffstetter (1974) considers that the presence of this character may result from convergence, since it has also been acquired independently by some nonprimate mammals. However, it is strange that this "lemuroid bulla" should be found not only with all Madagascar lemurs but also with the Eocene Adapidae.

Hoffstetter also regards regressive evolution of this character as a reasonable possibility, for the ontogenetic stages are very similar in the Lorisidae and in the Madagascar lemurs (van Kampen, 1905). Simple arrest in the development of the bulla would be adequate for appearance of the ancestral pattern in a modern form. Thus, several different interpretations are possible, and if one adds the additional possibility of convergent adaptation it is difficult indeed to draw conclusions about phylogenetic relationships on this basis. Quite apart from the fact that the interpretation of bullar structure is so uncertain, there is the additional problem that the morphological distinctions involved are not so clear-cut as has been traditionally accepted (Charles-Dominique and Martin, 1970). For example, the Cheirogaleidae exhibit a condition intermediate between that of the other Madagascar lemurs and that found in the Lorisidae, with respect to the fusion of the tympanic ring to the bulla wall. In addition, false interpretations may be drawn as a result of examining such a character in small-bodied species and comparing it directly with the condition found in large-bodied forms, since the requirements for ossification may differ drastically between the two.

5.1.5. The Carotid Circulation

The carotid circulation associated with the bulla has recently been subjected to reexamination by Saban (1963), Szalay and Katz (1973), and Cartmill (1975). All of these authors have concluded that the "lorisid" disposition, with the presence of a well-developed anterior carotid artery and atrophy of the stapedial artery, is present only in the Lorisidae and the Cheirogaleidae. The unanimous conclusion has been drawn that this disposition of the carotid circulation could not have arisen through convergence, nor through retrogressive evolution. However, even this characteristic disposition, which until recently was regarded as unique to the Lorisidae, is perhaps of only limited value for phylogenetic reconstruction. As Hoffstetter (1974) has suggested, the anterior carotid artery may be homologous with the medial enterocarotid of primitive mammals, and its retention as a primitive characteristic in Lorisidae and Cheirogaleidae would not necessarily imply close phylogenetic relationship.

5.1.6. Special Features of the Dentition

Among the various dental characters which may be considered, the differentiation of a "toothcomb" incorporating the lower incisors and canines, and the accompanying caniniform shape of the anterior lower premolar (here interpreted as $P_{\overline{2}}$) has been considered to be a common specialization of the Madagascar lemurs and the Lorisidae (Martin, 1972). The presence of the toothcomb may appear to provide a more reliable indicator of phylogenetic relationships, but there is no a priori reason why it should not have appeared through convergence in separate forms. This is all the more likely since it would seem that the "comb" is associated with the collection of gums from the surfaces of trunks and branches and that this represents an important dietary adaptation in certain species.

Although the inclusion of the lower canines in the "comb" is a unique feature in the lemurs and lorisids, similar devices have been developed in other mammalian groups. Further, the toothcomb is not identical in form in all species since, in the Indriidae, there are only four teeth instead of the usual six. [Schwartz (1974) considers that the teeth in the indriid "comb" are two incisors and two canines.] *Daubentonia,* of course, only have two teeth in the anterior part of the lower jaw. In fact, these distinctions have been used as a basis for separating the Madagascar lemur families into two groups, one containing the living Indriidae and *Daubentonia* and the other containing the remaining living lemur species.

According to Schwartz (1974), there are indications that the "comb" in indriids consists of two incisors and two canines. However, examination of the skull of a young animal (P. Charles-Dominique, personal communication) has raised certain doubts about the identification of the canine teeth. Both Peters (1866), who examined the skull of a very young animal, and Bennejeant (1936) concluded that the two specialized lower anterior teeth of *Daubentonia* are incisors which erupt at the time of birth between two pairs of deciduous incisors. However, Tattersall and Schwartz (1974) interpret the two permanent lower anterior teeth of *Daubentonia* as canines, on the basis of their position in the lower jaw and their laterally compressed form. In fact, the possibilities for modification of a dental bud according to its position and the developmental influences to which it is subjected remain to be fully examined.

According to Hoffstetter (1974), the genus *Daubentonia,* with its rodentlike dental specialization, may be considered as a derivative from an "adapoid" type lacking a dental comb. However, the retention by *Daubentonia* of numerous primitive characters indicates that no direct derivation from the Adapidae is possible. Both the specialization of the teeth of *Daubentonia* and the development of the filiform middle finger of the hand can be interpreted as adaptive characters which have emerged relatively recently.

As far as dental adaptation is concerned, *Daubentonia* could easily have developed convergent similarities to certain fossil forms, in the same way that the filiform middle finger and various cranial features are shared with some phalangerid marsupials such as *Dactylopsila* (Cartmill, 1974). Cartmill has effectively presented the case that *Daubentonia* and *Dactylopsila* have evolved to fill comparable ecological niches, in both cases in response to the absence of woodpeckers as potential competitors.

In any event, the continuously growing incisors (or canines) of *Daubentonia* could easily represent a derivation from an ancestral toothcomb. Like most other lemurs and lorisids, the aye-aye actively seeks out gums as food at certain times of the year, and it would be quite reasonable to suggest that there has been progressive refinement of an original toothcomb in the course of evolution, ultimately leading to the present rodentlike teeth (see also Martin, 1972). Development of such teeth would have permitted greater efficiency not only in the search for gums, but also in the exploitation of new dietary resources such as wood-boring larvae and very hard fruits such as the coconut (Petter, 1959, 1962a; Petter and Peyrieras, 1970). The anterior teeth in the lower jaw of *Dactylopsila,* even if they have not undergone the radical degree of transformation found with *Daubentonia,* are both enlarged and reinforced. It is quite possible that with *Daubentonia,* in contrast to *Dactylopsila,* it is precisely the prior existence of a toothcomb in an early ancestral form which has permitted the differentiation of more effective prising teeth comparable to those found in rodents.

Daubentonia provide a useful subject for discussion, since they exemplify the importance of convergence. The study carried out by Cartmill (1974) clearly demonstrates a large number of comparable anatomical traits in *Dactylopsila* and *Daubentonia.* Although the former are smaller and less extensively specialized than the latter, they exhibit a filiform manual digit (the fourth, instead of the third), claws on all digits except the hallux (permitting powerful gripping of tree surfaces), and a reinforced skull exhibiting comparable features (shortened face, globular cranium, enlargement and strengthening of the interorbital region, etc.) All of these peculiarities can be associated with the generation and dissipation of forces connected with penetration of wood by the incisor teeth. All of these features distinguishing *Daubentonia* from the other lemurs similarly distinguish *Dactylopsila* from other phalangerid marsupials.

As far as the dental morphology of the various prosimian species is concerned, particularly with respect to their order of appearance and the replacement of milk teeth, the studies of Bennejeant (1936) and of Tattersall and Schwartz (1974) have provided much interesting information. It is generally accepted that the earliest primates possessed a dental formula of $I\frac{2}{2}$, $C\frac{1}{1}$, $P\frac{4}{4}$, $M\frac{3}{3}$. The Upper Eocene fossil *Adapis parisiensis* Blainville 1849, which is one of the best known Early Tertiary primates and may be regarded as a derivative of an evolutionary sequence leading from the Early Eocene *Pelycodus* and passing through the Middle Eocene *Protoadapis,* possessed this dental formula, and its dentition is generally quite comparable with that of modern prosimians. In fact, Gingerich (1975) has emphasized that the dentition of *Hapalemur griseus* is to some extent similar to that of *Adapis parisiensis,* which was comparable in body size (hence eliminating problems of allometry). The dental features of *Hapalemur griseus* exhibit an intermediate character with respect to the various living lemur species.

The upper and lower incisors of *Adapis parisiensis* were large and spatulate, resembling those of simian primates, whereas the upper incisors of *Hapalemur* are small and close to the canines, while the lower incisors are narrow, pointed teeth incorporated into the toothcomb with the canines. Further, the upper and lower canines of *Adapis* are medium-sized, forward-sloping teeth, whereas the upper canines of *Hapalemur* are large and pointed, in contrast to the incisiform canines incorporated into the toothcomb of the lower jaw. In considering these anatomical details, it is interesting to note (as did Gingerich, 1975) that although the spatulate lower incisors of *Adapis parisiensis* are comparable to those of other *Adapis* species, the lower canine is functionally an incisor, as a unique feature of the former. Hence, the six anterior teeth of the lower jaw of *Adapis parisiensis* can be regarded as a functional unit which was perhaps adapted for prising of bark. It would be relatively easy to derive the modern lemur or loris toothcomb from this condition. In *Indri,* this toothcomb has been reduced to four teeth (two incisors and two canines, or four incisors), and in *Daubentonia* reduction has progressed still further. However, in the vast majority of living lemurs and in all living lorises the comb consists of six teeth, as in *Hapalemur,* and this provides a strong argument for a common ancestral pattern (see Martin, 1972).

The four premolar teeth of *Adapis parisiensis* exhibit progressive enlargement and increasing molariform development from front to back. The lower anterior premolar ($P_{\overline{2}}$) of *Hapalemur* is caniniform and operates in conjunction with the upper canine, but the posterior premolars in both upper and lower jaws ($P_{\overline{4}}$ and $P^{\underline{4}}$) are molariform, as in *Adapis*. The loss of one or two teeth from the premolar series is a common event in various lines of mammalian evolution (as noted by Gingerich, 1975), and the loss of $P_{\overline{1}}$ and $P^{\underline{1}}$ therefore poses no problem in the derivation of the condition in *Hapalemur* from that present in *Adapis*. The molar teeth of *Adapis parisiensis* all exhibit clear crests and small hypocones in the upper series. In the lower jaw, the last molar ($M_{\overline{3}}$) is elongated and bears a well-developed hypoconulid, as in early primates generally. In *Hapalemur,* the upper molars lack the hypocone, and the last lower molar ($M_{\overline{3}}$) lacks the hypoconulid. However, once again, Gingerich (1975) notes that the two genera exhibit such similarity in other features of their molar teeth that if they were to be found together in the same fossil deposits they would doubtless be placed in the same family and possibly in the same genus.

It is evident from the papers of Bennejeant (1936), Schwartz (1974; 1975), and Tattersall and Schwartz (1974), that comparison of the patterns of premolar eruption can throw additional light on phylogenetic relationships. For most prosimians, possessing three premolars in the upper and lower jaws, the generally accepted interpretation is that the milk teeth are $P\frac{2}{2}$, $P\frac{3}{3}$, $P\frac{4}{4}$, followed by the permanent teeth $P\frac{2}{2}$, $P\frac{3}{3}$, $P\frac{4}{4}$ (Schwartz, 1974). The Indriidae only possess two pairs of milk premolars, followed by two pairs of permanent premolars, in the upper jaw, whereas the lower jaw has four pairs of milk premolars (probably $P_{\overline{1}}$, $P_{\overline{2}}$, $P_{\overline{3}}$ and $P_{\overline{4}}$, as in early primates), which have only two pairs of permanent replacements (probably $P_{\overline{2}}$ and $P_{\overline{3}}$, according to Schwartz, 1974). However, despite the overall similarity in numbers of teeth, their eruption and development do not take place in the same sequence in all cases. Following the studies carried out by the authors cited above (particularly Schwartz, 1975), one can distinguish five groups among the living lemurs and lorises:

1. "Primitive" prosimians, including *Microcebus* and the Lorisidae (and also *Tarsius*), with a sequence of development and eruption of the milk premolars in the order $P\frac{2}{2}$, $P\frac{4}{4}$, $P\frac{3}{3}$ and of the permanent premolars in the order $P\frac{2}{2}$, $P\frac{4}{4}$, $P\frac{3}{3}$.

2. With the exception of *Lemur catta* and *Varecia,* all of the Lemuridae appear to have the following order of development and eruption: $P\frac{2}{2}$, $P\frac{3}{3}$, $P\frac{4}{4}$ for the milk premolars and $P\frac{2}{2}$, $P\frac{4}{4}$, $P\frac{3}{3}$ for the permanent premolars.

3. The Indriidae share a common sequence of development and eruption of the permanent premolars, with the posterior teeth erupting before the anterior premolars. The milk premolars probably exhibit the same sequence.

4. *Lepilemur* and *Megaladapis* exhibit a *developmental* sequence of the permanent premolars of $P\frac{2}{2}$, $P\frac{4}{4}$, $P\frac{3}{3}$, as for group (1), whereas the eruption sequence is $P\frac{4}{4}$, $P\frac{3}{3}$, $P\frac{2}{2}$, as for group (3). For the milk teeth, *Lepilemur* exhibit the sequence $P\frac{2}{2}$, $P\frac{4}{4}$, $P\frac{3}{3}$, as for group (1). These two genera (Schwartz, 1974, 1975) to some extent represent a transitional condition between the primitive condition and a more specialized condition. In addition, the similarities between *Lepilemur* and *Megaladapis* doubtless reflect a phylogenetic link between the two.

5. *Lemur catta* and *Hapalemur* exhibit a milk premolar development and eruption sequence of $P\frac{2}{2}$, $P\frac{3}{3}$, $P\frac{4}{4}$, as in group (2), and a permanent premolar development and eruption sequence of $P\frac{4}{4}$, $P\frac{3}{3}$, $P\frac{2}{2}$, as in group (3). Schwartz has pointed out that the transition from the hypothetical primitive condition of $P\frac{2}{2}$, $P\frac{4}{4}$, $P\frac{3}{3}$ to the sequence $P\frac{4}{4}$, $P\frac{3}{3}$, $P\frac{2}{2}$ could be achieved by one of three routes: first, precocious growth of $P\frac{4}{4}$; second, retardation of $P\frac{2}{2}$ (which seems the most plausible) or, third, a combination of both processes.

Of course, one can criticize the interpretation of this information, since the characters concerned are doubtless linked to other cranial features (or the consequences thereof), which are themselves already employed to establish phylogenetic relationships. However, such dental characteristics do have the advantage that they are accessible to precise formulation and, as will be seen, various conclusions based upon them seem to agree with cytogenetic data.

Overall, it can be concluded that critical examination of characters classically regarded as valuable for reconstruction of phylogenetic relationships among the Strepsirhini shows that they appear to be far less conclusive in the light of recent reanalyses. However, one can attempt to produce hypothetical schemes of the phylogenetic relationships among the Strepsirhini, though these will differ according to the phylogenetic significance attached to the various characters examined (see, in particular, Hill, 1953; Martin, 1972; Hoffstetter, 1974; Tattersall and Schwartz, 1974).

5.2. The Importance of Small-Bodied Forms

Interpretation of the evolution of the Primates is severely hindered by the paucity of small-bodied fossil forms. It is obvious that such forms were less likely to undergo fossilization than their larger-bodied relatives, and the forest environment in any case does not provide a suitable environment for fossilization. An example is provided by the lesser mouse lemur of Madagascar, *Microcebus murinus*. Numerous anatomical characters, or at least certain interpretations of them, together with cytogenetic data (Rumpler and Albignac, 1973) indicate that this lemur species has remained relatively primitive and is perhaps closest among the living forms to the ancestral primate condition. The lesser mouse lemur is a small-bodied nocturnal form which can survive forest destruction by living in scattered bushes. Specialization on particular plant foods is unnecessary, because the diet can be supplemented to a variable extent by small arthropods. Special metabolic adaptations provide the lesser mouse lemur with considerable resistance to climatic fluctuations, and nutrient storage permits survival for some time without food. An ancestral type comparable to this species would have been well-equipped to survive the marked climatic transitions of the Tertiary. A widespread ancestral form of this kind would have permitted both the development of the lemurs on Madagascar, presumably following rafting across the Mozambique Channel, and the evolutionary radiation of the Lorisidae.

In view of these considerations, and without placing too much emphasis on the present biogeographical distribution patterns which perhaps tend to lead too easily to the conclusion that the Madagascar lemurs represent an isolated cohesive unit, it would seem to be best to dispense with the division of the suborder Strepsirhini into two distinct infraorders. The subfamilies Lorisinae and Galaginae, united within the family Lorisidae, may be simply included in the Strepsirhini without indicating special separation from the various Madagascar lemur families.

6. NOTES ON THE CLASSIFICATION OF THE STREPSIRHINI

It has been shown that the various living representatives of the Strepsirhini can probably be traced back to a very early common ancestor. However, renewed study of the existing relevant fossil material is unlikely to throw much further light on this early origin, in the absence of new discoveries of remains of small-bodied forms. Anatomical and embryological characters, which still require more detailed examination, are not in themselves entirely reliable as a basis for reconstructing evolutionary relationships. However, behavioral/ecological studies and cytogenetic investigations conducted on living prosimian species have yielded a wealth of new evidence, which must be incorporated into our previous framework of knowledge in order to permit interpretation of the phylogenetic relationships within the Strepsirhini.*

*In cytogenetic studies it is now generally accepted that "Robertsonian" evolution (reciprocal translocation) represents the predominant mode of chromosomal evolution.

6.1. Classification of the Lorisidae

Fossil lorisids have been found in Miocene deposits of Africa (e.g., *Progalago*) and India, and they are undoubtedly closely allied to the ancestors of the living lorisid species.

6.1.1. Lorisinae

The most widely distributed group among the modern lorisids are the Lorisinae, containing four genera, two living in the African tropical rainforest belt (*Arctocebus* and *Perodicticus*) and the other two occurring in Asian tropical forests (*Loris* and *Nycticebus*). *Arctocebus* and *Loris,* in fact, resemble one another in numerous characters; they are both small-bodied, gracile forms with slender limbs, occupying comparable ecological niches. Their principal anatomical differences reside in the greater degree of reduction of the index finger in *Arctocebus* and in the greater elongation of the limbs in *Loris,* which also possess brachial glands and folding ear pinnae. *Perodicticus* and *Nycticebus,* on the other hand, are larger-bodied and more robust forms, though they have the same slow-climbing style of locomotion and occupy fairly similar ecological niches. These latter two genera differ from one another primarily in the presence of a short tail in *Perodicticus* (completely lost in *Nycticebus*), the possession of well-developed dorsal nuchal spines and the greater reduction of the index finger in *Perodicticus,* and the presence of brachial glands in *Nycticebus.* The systematic status of these four genera is now widely accepted, following the reviews of Hill (1953) and Fiedler (1956), with each genus being monospecific. Hill (1953) has carried out a subspecific revision.

As has been seen, the living African and Asiatic lorisines can be grouped in two pairs with one representative from each area. This can be interpreted either as a result of an early separation into two ancestral forms which have each given rise to an African and an Asian descendant (Simpson, 1967), or as the parallel evolution of two similar types in each geographical area (Fiedler, 1956; Hill, 1972).

Recent cytogenetic investigations conducted by Egozcue and Vilarasau de Egozcue (1967), Egozcue *et al.* (1966), and de Boer (1973a,b) permit more detailed examination of relationships within the Lorisinae. In fact, although *Perodicticus* and *Nycticebus* exhibit marked morphological similarities, their diploid numbers are different ($2N = 62$, versus $2N = 50$). Similarly, *Loris* and *Arctocebus,* despite their morphological resemblance, have diploid numbers of $2N = 62$ and $2N = 52$.

According to de Boer, there are clear karyological similarities between the African *Perodicticus* and the Asian *Loris,* and particularly between the African *Arctocebus* and the Asian *Nycticebus.* This applies both to the diploid numbers and to the form of the individual chromosomes. The difference between the two sets of diploid numbers ($2N = 62$ versus $2N = 52$) could be explained as the result of five centric fusion events (i.e., Robertsonian evolution). Between *Nycticebus* and *Arctocebus,* the chromosomal resemblances are so clear-cut that de Boer finds it difficult to accept that there has been independent evolution from an ancestral

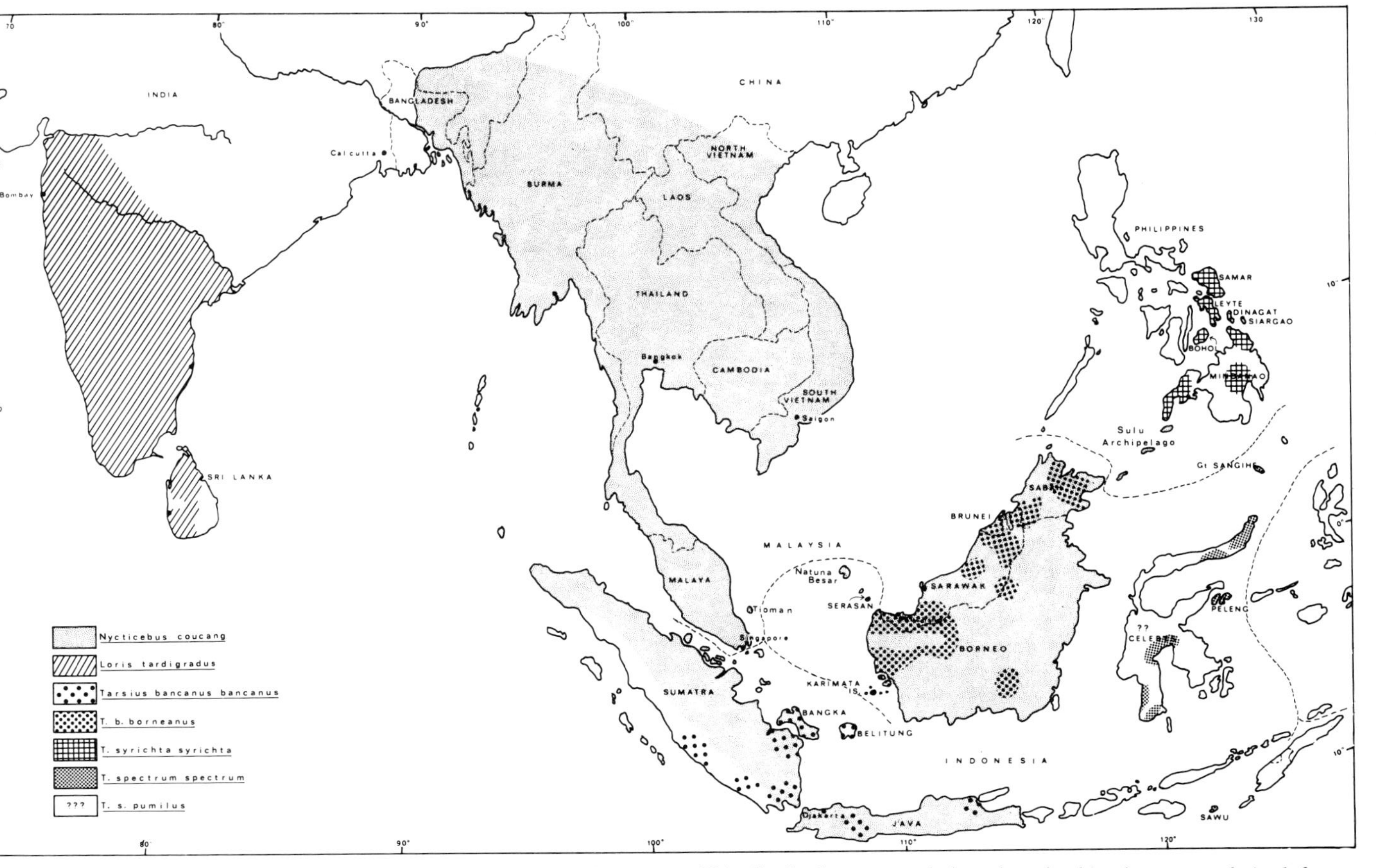

Fig. 1. Map showing the distribution of *Nycticebus*, *Loris*, and *Tarsius*. (This distribution map and the others in this chapter are derived from information from the following sources: P. Charles-Dominique, personal communication; Doyle and Bearder, 1977; Fiedler, 1956; Hill, 1953; Humbert and Cours Darne, 1965; Niemitz, personal communication; Petter, 1962b, Petter and Peyrieras, 1972; Petter *et al.*, 1972; Pringle, 1974; Vincent, 1972.)

condition of $2N = 62$. It therefore seems more likely that *Nycticebus* and *Arctocebus* are derived from a common ancestor which diverged either from the line leading to *Perodicticus* ($2N = 62$) or from that leading to *Loris* ($2N = 62$). This pattern of divergence, which was perhaps already established in the Eocene with *Pronycticebus* or in the Oligocene with *Anchomomys* (attributed to the Adapidae by Hill, 1972), doubtless required a long period of geographical separation for *Nyc-*

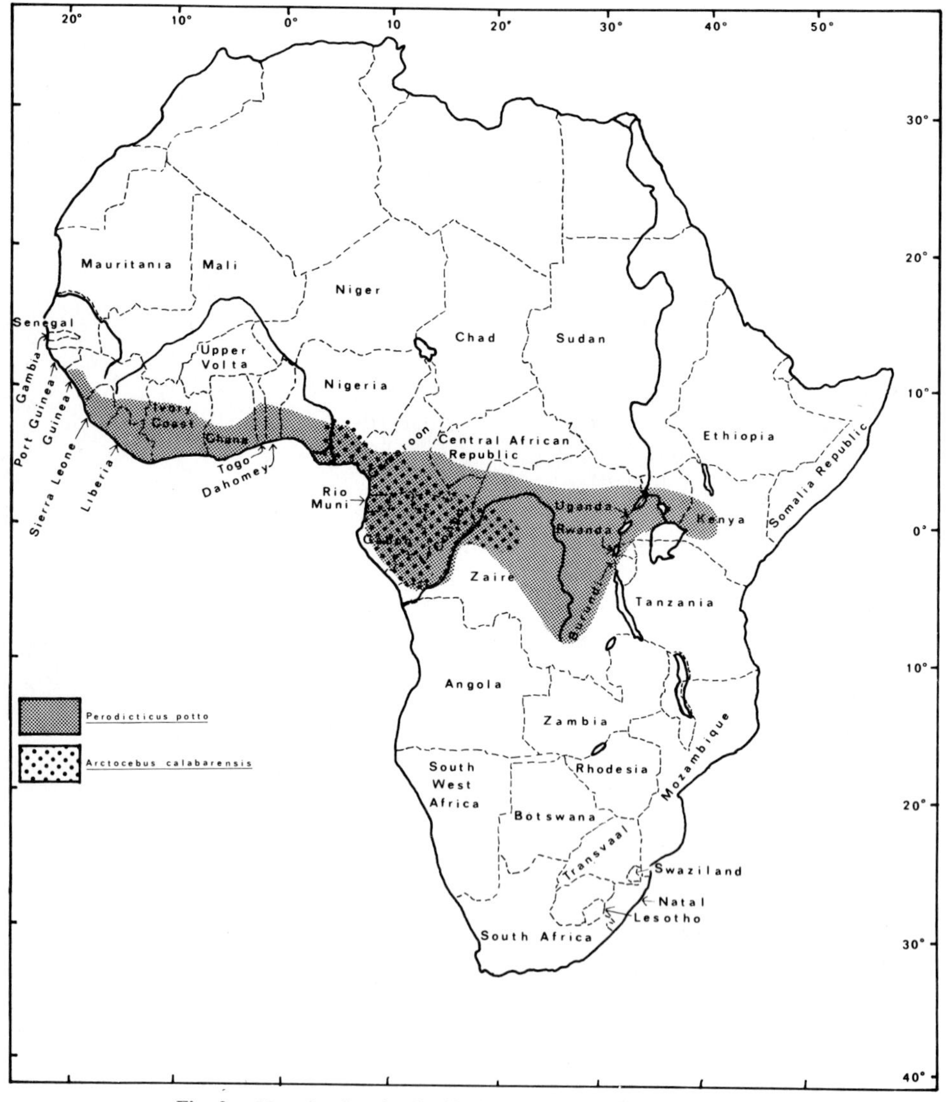

Fig. 2. Map showing the distribution of *Perodicticus* and *Arctocebus*.

ticebus and *Arctocebus* to undergo divergent evolution to produce two types paralleling those of the two species with 62 chromosomes. Figures 1 and 2 show the distribution of the Lorisines and of *Tarsius* based on the latest information available.

6.1.2. Galaginae

In contrast to the Lorisinae, the Galaginae are only represented in Africa. For a long period of time, there has been considerable confusion over their classification, and taxonomic revision of the group is still in progress. Gray (1863) distinguished five genera: *Galago* E. Geoffroy 1796, *Otolincus* Illiger 1811, *Hemigalago* Dahlbohm 1857 (already named *Galagoides* by A. Smith 1839), *Otolemur* Coquerel 1859, and *Otogale* Gray 1863 (already named *Euoticus* by Gray 1863). Schwartz (1931) regrouped the first four genera within the single genus *Galago,* which was the earliest established generic name, and maintained the last as a separate genus *Euoticus* because of dental distinctions and the peculiar possession of pointed tips on the nails (hence: "needle-clawed bushbaby"). Simpson (1945) adopted the classification proposed by Schwartz, but Hill (1953) and Fiedler (1956) maintained the distinction originally suggested by Gray in recognizing generic separation between *Galago* and *Hemigalago* (which they named *Galagoides*). Accordingly, the genus *Galago* contained three different species: *G. senegalensis, G. alleni,* and *G. crassicaudatus*. The genus *Euoticus* was regarded as containing two species: *E. elegantulus* and *E. inustus*. The latter had already been separated by Schwartz (1931) as a distinct subspecies (*G. senegalensis inustus* Schwartz 1930) because of its darker pelage coloration, and Hill (1953; following the study published by Hayman, 1937) transferred it to the genus *Euoticus* because of similar development of pointed tips on the nails.

The genus *Galagoides* accepted by Hill (1953) and Fiedler (1956) contained only one species, *G. demidovii* which, together with the lesser mouse lemur of Madagascar (*Microcebus murinus*), shares the distinction of being the smallest of the living Strepsirhini. A revision of the subspecies of *G. demidovii* was carried out by Hill (1953). However, it would seem from observations recently conducted in the field (P. Charles-Dominique, personal communication), that most of the subspecific names listed by Hill for *Galagoides demidovii* should be regarded as synonyms.

Following more recent publications (Charles-Dominique, 1971; Walker, 1974), it is now a common tendency for *Euoticus inustus* to be classified as a form closely related to *Galago senegalensis* and for the two genera *Euoticus* and *Galagoides* to be considered as synonyms of *Galago*.

S. K. Bearder (personal communication) conducted a systematic revision of forms attributed to *Galago crassicaudatus* and *G. senegalensis,* in connection with his behavioral/ecological field study of these two galagine species in South Africa:

> The classification of the two species is rendered difficult through recent isolation of many breeding populations. Patches of suitable bush or forest may be cut off from one another by barriers which need be no more than a tract of open grassland which the animals are unlikely to cross. This may well explain the large variety of subspecific characters which have been described and the

apparent differences between some of Hill's type specimens may be misleading. Buettner-Janusch (1963, 1964), for example, examined a number of *Galago crassicaudatus* from a single forest in Kenya which had the pelage coloration of at least five subspecies described by Hill, and he considers that most of the differences should be considered as minor variations of a single major population. On the other hand, he described *G. c. argentatus* as being distinctly different from *G. c. crassicaudatus,* which leads him to suggest that if they are found in the same area they may actually be separate species.

Classical treatments of this species distinguish between two main groups of races. Matschie (1905) included *crassicaudatus* and *argentatus* in a southern group of races together with *lönnbergi, umbrosus, garnetti* and *monteiri,* while *lasiotes, agisymbanus, kikuyuensis* and *panganiensis* were separated as a northern group and even given subgeneric status. Hill (1953), Schwartz (1931) and Kingdon (1971) all follow the division between a northern and a southern group of races, but *G. crassicaudatus* (Hill, 1953) or *G. panganiensis* (Kingdon, 1971) are suggested as intermediate forms between the two groups.

Cytogenetic studies carried out by Egozcue (1968, 1969, 1970) and de Boer (1972, 1973a) have permitted more precise examination of the phylogenetic relationships between the various galagine species, but much work remains to be done to clarify the actual mechanisms involved in their evolution.

On the basis of chromosomal formulae, de Boer distinguishes clearly between two groups of species within the Galaginae: (1) *G. crassicaudatus* and *G. demidovii,* with high diploid numbers ($2N$ = 62 and 58, respectively); and (2) *G. alleni* and *G. senegalensis,* with relatively low diploid numbers ($2N$ = 40 and 36, 37, or 38, respectively). One can suggest a hypothetical ancestral stage with the diploid number of 62, which has been retained in *G. crassicaudatus,* and which has been reduced in *Galago senegalensis* ($2N$ = 36, 37, or 38) by a maximum of thirteen centric fusion events.

Marked differences shown by the karyotype of *G. demidovii* have led de Boer to separate this species quite sharply from the other galagines and to propose an earlier divergence from a common ancestor (despite the somewhat reduced diploid number of $2N$ = 58) prior to that which gave rise to *G. crassicaudatus* and *G. senegalensis.* Further, de Boer considers that the karyotype of *G. alleni* ($2N$ = 40), despite its general resemblance to that of *G. senegalaensis,* differs from the latter by the proportions of different chromosome types. The cytogenetic information obtained for *G. crassicaudatus* and *G. senegalensis* confirms the conclusions drawn from morphological data. As has been pointed out by S. K. Bearder (personal communication):

Egozcue (1969) and Ying and Butler (1971) have found intraspecific variation in karyotypic structure in representatives of both *G. crassicaudatus* and *G. senegalensis* which coincided with previous subspecific ranking. Chromosome number polymorphism is reported for *G. senegalensis* even within a single subspecies, namely *G. s. braccatus* (Ying and Butler, 1971). In a cytotaxonomic study of the Lorisoidea, de Boer (1973b) concludes that the karyological variations which are encountered within these species suggest that their subspecies diverged much more than is generally assumed. He distinguishes three groups of karyotypes in *G. senegalensis,* represented by *G.s. moholi* ($2N$ = 38), *G.s. braccatus* ($2N$ = 36, 37 or 38) and *G.s. zanzibaricus* ($2N$ = 36). De Boer also reveals two different karyotypes within *G. crassicaudatus* which are clearly related to racial differences, but not in accordance with the division into northern and southern groups. He considers it unlikely that fertile hybrids could be obtained from *kikuyuensis* and *crassicaudatus*, on the one hand, and *monteiri*, on the other, due to the large differences in the structure of their chromosomes ($2N = 62$, but fundamental numbers of 90 and 76, respectively).

The striking difference between the diploid chromosome numbers of *G. senegalensis* ($2N = 38$) and *G. crassicaudatus* ($2N = 62$) was first demonstrated by Chu and Bender (1961). Furthermore, the karyotype of *G. crassicaudatus* corresponds numerically with that of *Perodicticus potto* in

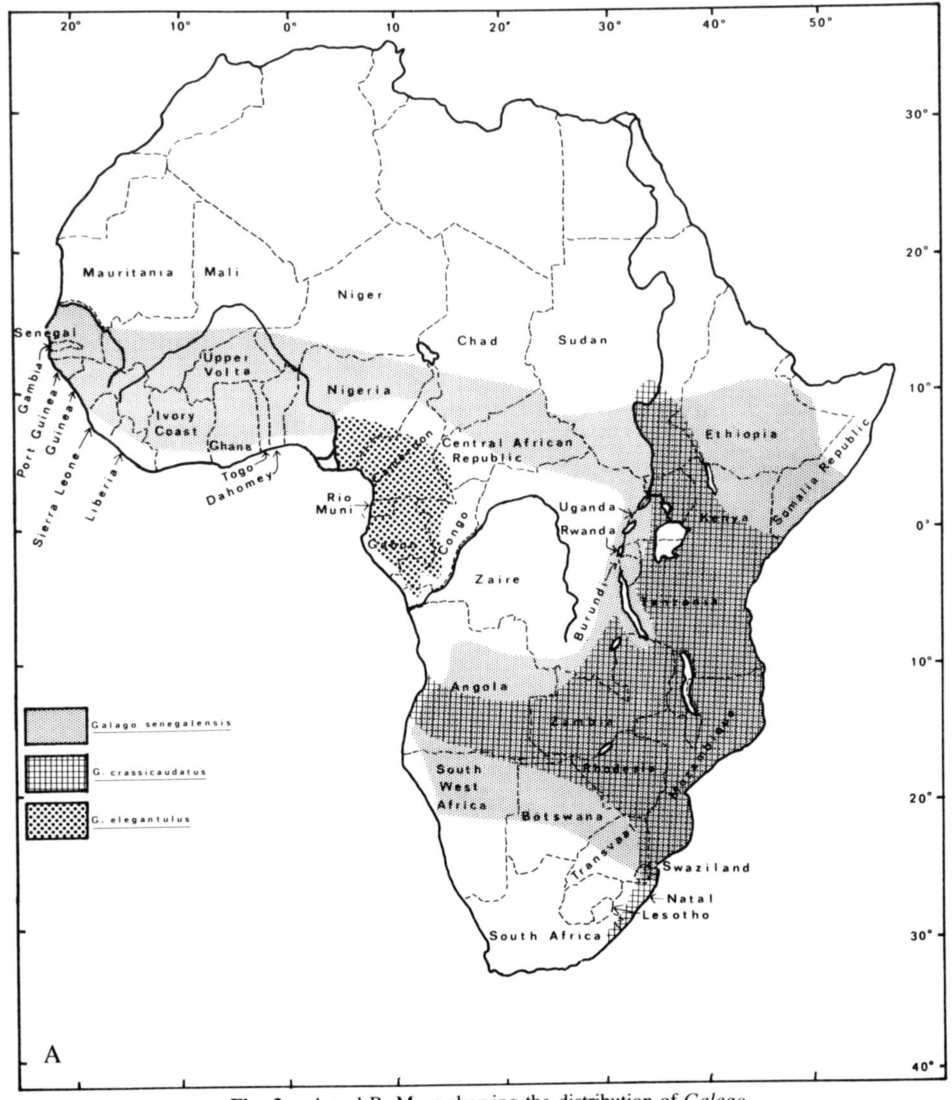

Fig. 3. A and B: Maps showing the distribution of *Galago*.

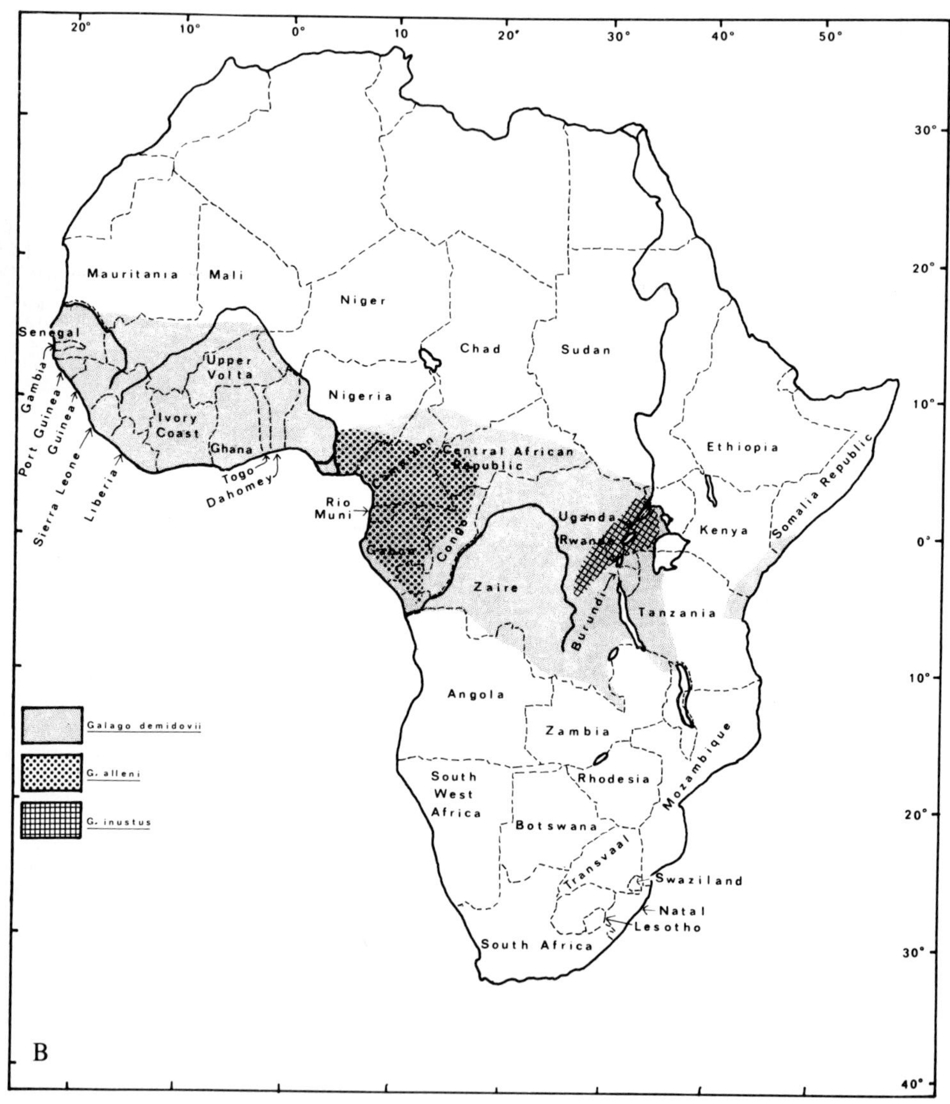

Fig. 3. (*continued*).

Africa and *Loris tardigradus* in Asia (Bender and Chu, 1963; Manna and Talukdar, 1968). This has been interpreted, in conjunction with similar fundamental numbers, as being a reflection of a true link between living representatives of the two subfamilies (Egozcue, 1969).

As a conclusion to his cytogenetic investigations, de Boer suggests that the Galaginae and the Lorisinae are probably derived from a common ancestor with 62 chromosomes. Both groups exhibit certain karyotypes with $2N = 62$, and measurements of nuclear DNA for lorisines and galagines with this number of chromosomes (*Perodicticus potto* and *Galago crassicaudatus*) show that the values are equivalent (Manfredi-Romanini *et al.*, 1972). However, the considerable differences which exist between their karyotypes probably emerged after the original separation between Lorisinae and Galaginae. As de Boer has remarked, in the Lorisinae the two species with diploid numbers of 62 exhibit relatively high fundamental numbers (96 in *Perodicticus potto* and 98 to 102 in *Loris tardigradus*), while the two species with lower diploid numbers ($2N = 52$) have even higher fundamental numbers (100 in *Arctocebus calabarensis;* 100 to 104 in *Nycticebus coucang*). In de Boer's estimation, an early division between the Lorisinae and Galaginae would seem likely in view of the clear differences in structure and behavior between the two subfamilies. Further support for this idea is provided by the fact that paleontological finds indicate that the two subfamilies have evolved independently since the Miocene. It has also been pointed out by de Boer that the results of recent comparisons of serum proteins (Goodman, 1967, 1975) are difficult to interpret. Apparently there is more difference between *Periodicticus* and *Nycticebus* than between *Nycticebus* and *Galago*. Figure 3 shows the distribution of the Galagines.

6.2. Classification of the Cheirogaleidae, Lemuridae, Lepilemuridae, Indriidae, and Daubentoniidae

These five families, whose members were included in the Infraorder Lemuriformes by Simpson (1945), are all restricted to the island of Madagascar, except for two species now found on the Comoro Islands. No fossil evidence is available to illuminate the evolution of the lemurs in Madagascar. However, there are remains of a number of recently extinct subfossil forms, including approximately fourteen species belonging to seven different genera. All of these species, which are generally larger than the surviving lemur species, formed part of the recent assemblage of Madagascar lemurs, and their extinction can be linked to the arrival of human settlers on the island (Lamberton, 1934, 1939; Walker, 1967).

The systematic status of the modern lemurs of Madagascar has remained confused for some time. Numerous authors have attempted to establish an orderly classification, notably Mivart (1873), Milne Edwards and Grandidier (1875), Leche (1896), and Pocock (1918). More recently, Schwartz (1931), followed by Hill (1953) and Petter (1962a), proposed that the group of living lemur species should be divided into three families: (1) Lemuridae (containing two subfamilies, Cheirogaleinae and Lemurinae, each comprising three genera); (2) Indriidae (comprising three genera); and (3) Daubentoniidae (containing a single genus). This classification has been reexamined and modified in a recent revision of the systematics of the Madagascar lemurs (Petter *et al.*, 1977), which takes into account new anatomical studies, recent behavioral/ecological data, and cytogenetic information. It seemed appropriate to separate, within the overall group of the "Madagascar lemurs," genera which differed in a significant number of characters and to classify them in distinct subfamilies. The various families of surviving Madagascar lemurs exhibit marked anatomical and behavioral/ecological distinctions. With few exceptions, it is difficult to identify a clear pattern of relationships between the various groups. Recent cytogenetic investigations (Buettner-Janusch, 1963; Bender and Chu, 1963; and particularly Rumpler, 1970, 1972; Rumpler and Albignac, 1970, 1972a,b, 1973) do, however, permit construction of hypotheses about the evolutionary radiation of the lemurs in conjunction with various lines of evidence from the laboratory and the field (Martin, 1972; Petter *et al.,* 1977). Nevertheless, there remain certain questions which are difficult to resolve.

6.2.1. Cheirogaleidae

By making use of the cytogenetic information obtained, and seeking hypotheses about relationships between species which require a minimum number of chromosomal rearrangements and which do not directly conflict with conclusions drawn from other areas of study, it is possible to distinguish primitive characters from those which are more advanced (derived). Bearing this distinction in mind, it would seem to be a useful step to group the Cheirogaleinae (*Cheirogaleus, Microcebus,* and *Allocebus*) and the Phanerinae (*Phaner*) in a special family, the Cheirogaleidae, to distinguish them from the Lemuridae (*Lemur, Varecia,* and *Hapalemur*). It has already been suggested (Petter, 1962a) that the Cheirogaleidae should be separated from the Lemuridae on the basis of cranial and behavioral characters. Parasitological studies (Chabaud *et al.,* 1965) have emphasized the distinction between these two families, and recent biometric studies (Mahé, 1972) and karyological investigations (Rumpler and Albignac, 1972a) have provided further support for the division between them.

The genus *Allocebus* has been provisionally left in the subfamily Cheirogaleinae (see Petter and Petter-Rousseaux, 1960). However, in the absence of a karyotype for this genus, it is difficult to reach precise conclusions about its evolutionary relationships.

The genus *Phaner* would seem to be best allocated to a distinct subfamily separate from the Cheirogaleinae, viz., the Phanerinae. Chromosomal studies on the species previously allocated to the subfamily Cheirogaleinae (Simpson, 1945) have, in fact, permitted identification of two distinct groups (Rumpler and Albignac, 1972a): (1) *Cheirogaleus* and *Microcebus,* which have identical karyotypes in terms of diploid number ($2N = 66$) and fundamental number ($NF = 66$); and (2) the genus *Phaner,* which is distinctive in its diploid number ($2N = 48$) and its fundamental number ($NF = 62$). Craniodental characters (Hill, 1953), along with other recently described characters, such as the distinctive dermatoglyphs (Rakotosamimanana and Rumpler, 1970), the presence of a specific marking gland on the ventral surface of the neck (Rumpler and Andriamiandra, 1971), and the unique behavior of *Phaner* (Petter *et al.,* 1971) all support the interpretation from the chromosomal evidence that this genus is distinct from *Microcebus* and *Cheirogaleus.*

All of the Cheirogaleinae exhibit a very primitive karyotype entirely composed of acrocentric chromosomes. Despite the considerable morphological differences be-

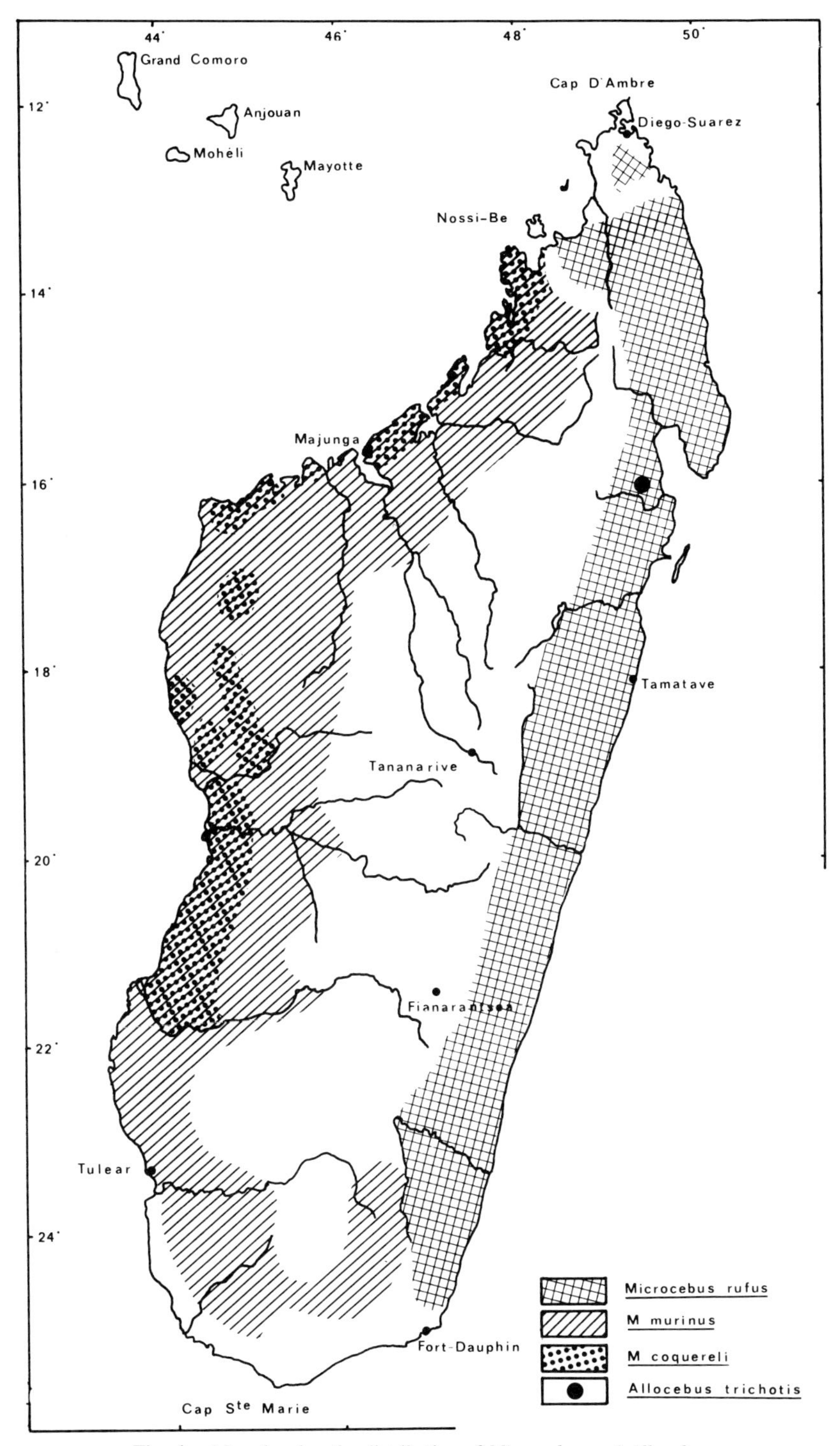

Fig. 4. Map showing the distribution of *Microcebus* and *Allocebus*.

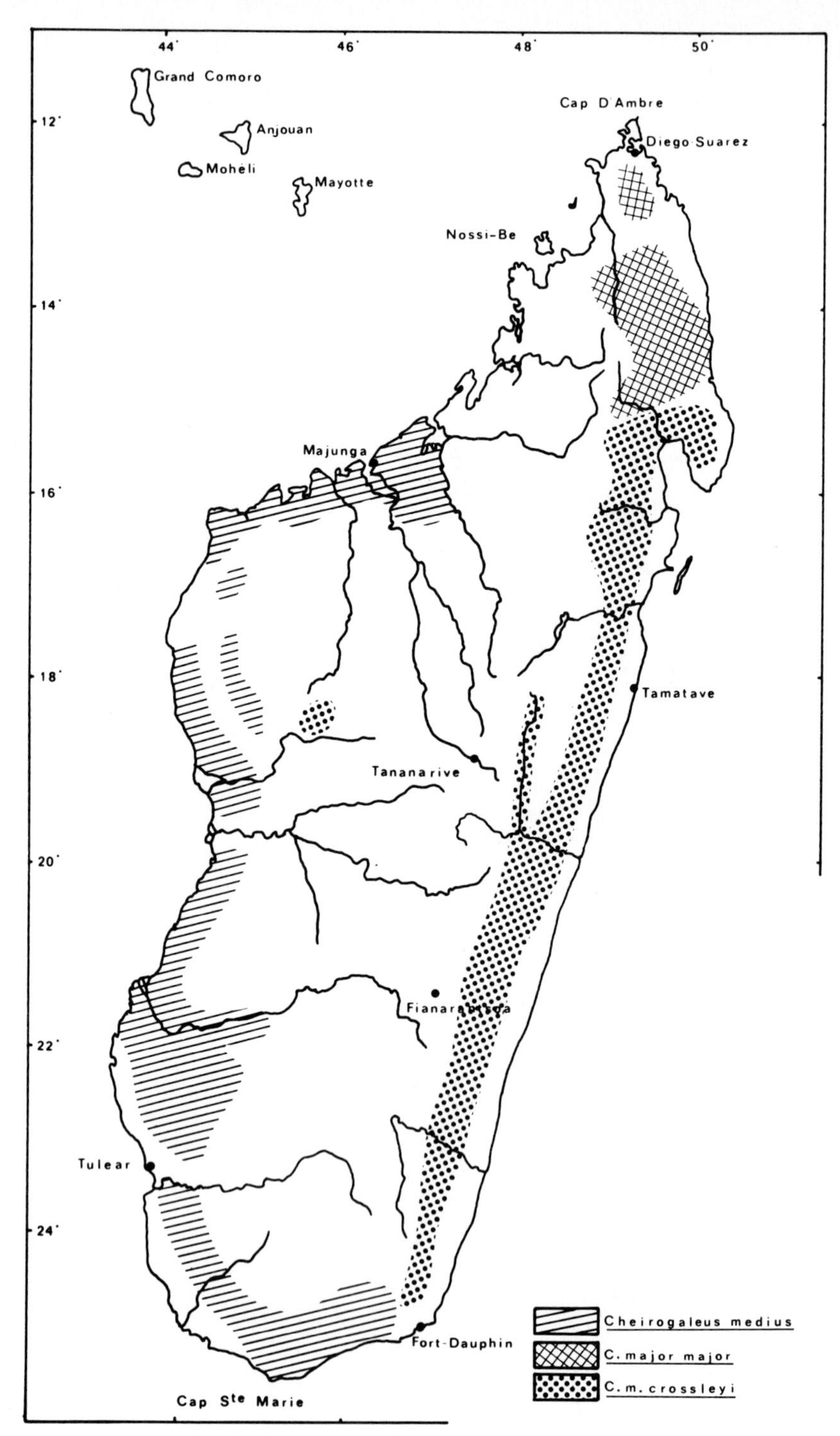

Fig. 5. Map showing the distribution of *Cheirogaleus*.

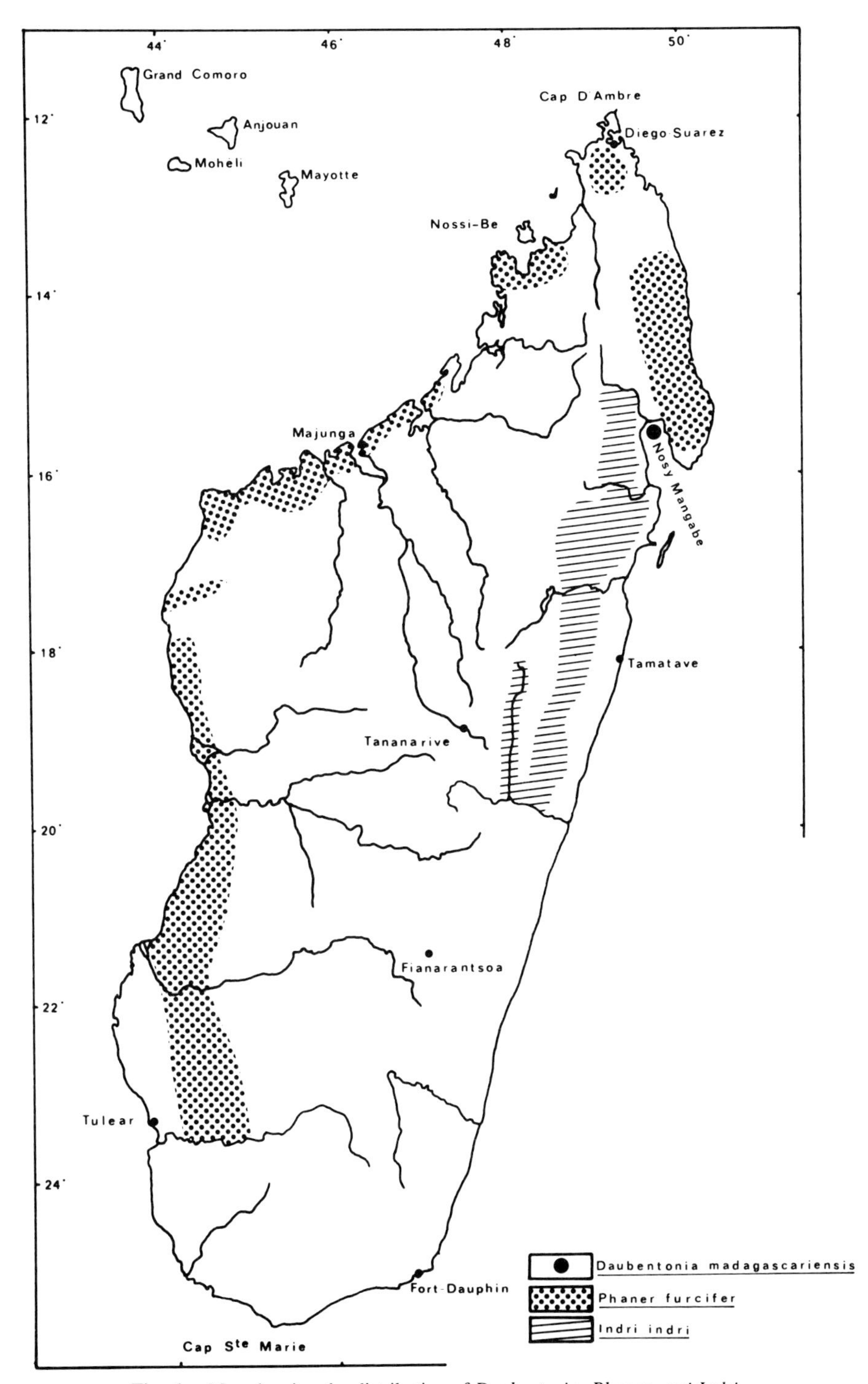

Fig. 6. Map showing the distribution of *Daubentonia, Phaner,* and *Indri*.

tween *Cheirogaleus* and *Microcebus,* the two genera exhibit the same chromosomal formula, as noted above, and it may be supposed that a common ancestor possessing that formula gave rise to the evolutionary lines leading to the Phanerinae, the Lemuridae, the Lepilemuridae, the Indriidae, and the Daubentoniidae. Accordingly, the Phanerinae could be derived directly from the condition found in the Cheirogaleinae by two successive reciprocal translocations (centric fusions). The distributions of the Cheirogaleidae are shown in Figs. 4–6.

The Lemuridae, in turn, are probably descended from an ancestor whose karyotype contained only acrocentric chromosomes, with the various evolutionary lines undergoing reduction in chromosome number through centric fusion.

6.2.2. *Lemuridae*

The three genera contained in the family Lemuridae (*Lemur, Hapalemur,* and *Varecia*) all have the same fundamental number for their chromosomes, $NF = 64$ (Rumpler and Albignac, 1970, 1972a). Within this family, it would seem to be justifiable to erect a separate genus *Varecia* for the variegated lemur (formerly *Lemur variegatus*), as was suggested by Gray (1872), Schwartz (1931), and Petter (1962a) on the basis of various anatomical, physiological, and behavioral characters. *Varecia variegata* in fact possess a series of primitive characters which render this species quite different from *Lemur* species, and it is possible that its chromosomal evolution followed a slightly different path (its karyotype can be derived directly from a primitive formula of $2N = 60$ or even from a more primitive stage). Figure 7 shows the distribution of *Varecia.*

The taxonomy of the various species of *Lemur* poses quite complex problems, particularly in view of the results of recent cytogenetic studies (Rumpler, 1972) and of hybridization experiments (Albignac *et al.,* 1971). Although the species *L. catta* and *L. rubriventer* can be distinguished with little discussion, it would now seem to be justifiable to accord specific rank to the two former subspecies of *Lemur mongoz* (*L.m. mongoz* and *L.m. coronatus*), which differ markedly in their behavior (Petter *et al.,* 1977), and to separate *Lemur macaco* from *Lemur fulvus.*

The most primitive karyotype which exists within the genus *Lemur* contains a diploid chromosome number of $2N = 60$ chromosomes, of which only two pairs are telocentric and the rest are acrocentric. This condition is found in *L. mongoz* and the majority of the subspecies of *L. fulvus.* One stage in chromosomal evolution within the genus *Lemur* is represented by the diploid number $2N = 52$ (*L.f. collaris*), probably achieved as a result of centric fusion of four pairs of acrocentric chromosomes. Two further fusion events probably gave rise to the diploid number $2N = 48$ (*L.f. albocollaris*), and two more would explain the emergence of a karyotype with $2N = 44$ (*L.m. macaco*). On lateral lines of evolution, separate from this principal lineage, there are karyotypes of $2N = 50$ for *L. rubriventer,* $2N = 48$ for *L. coronatus,* and $2N = 48$ for *L. catta.* The former two species can be derived directly from the principal lineage, but only by centric fusion of chromosomes different from those involved in the main line of evolution. *Lemur catta,* on the other hand,

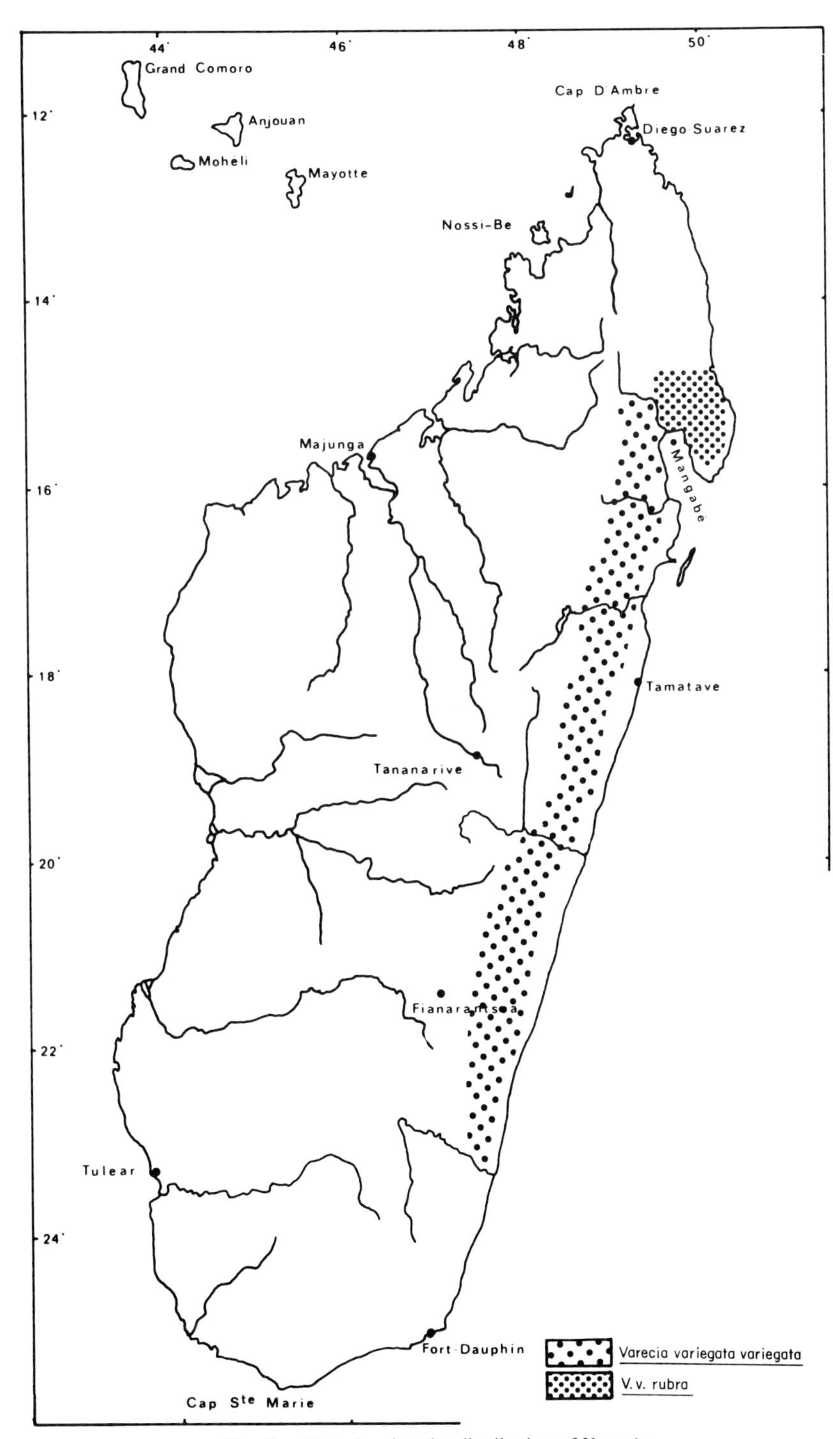

Fig. 7. Map showing the distribution of *Varecia*.

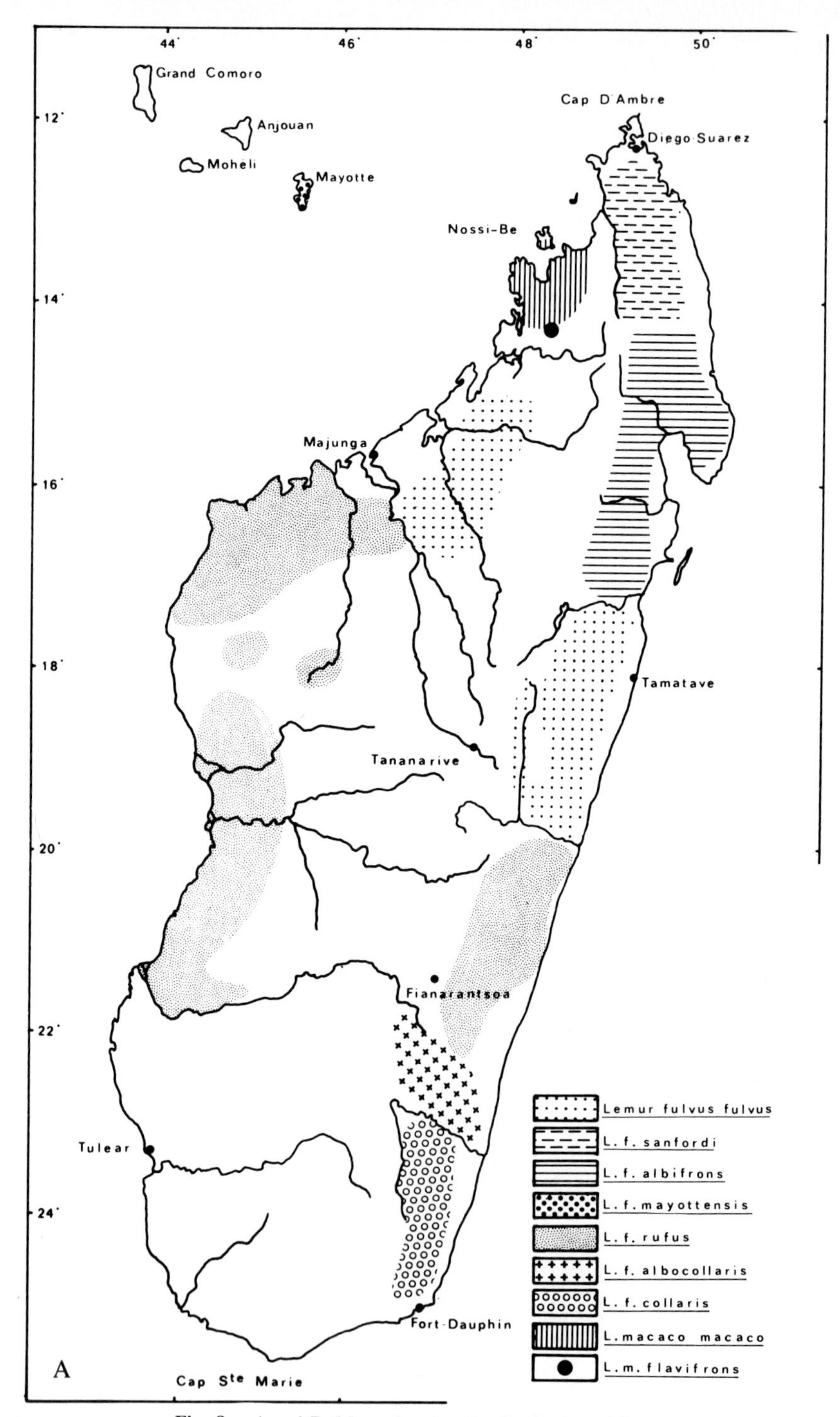

Fig. 8. A and B: Maps showing the distribution of *Lemur*.

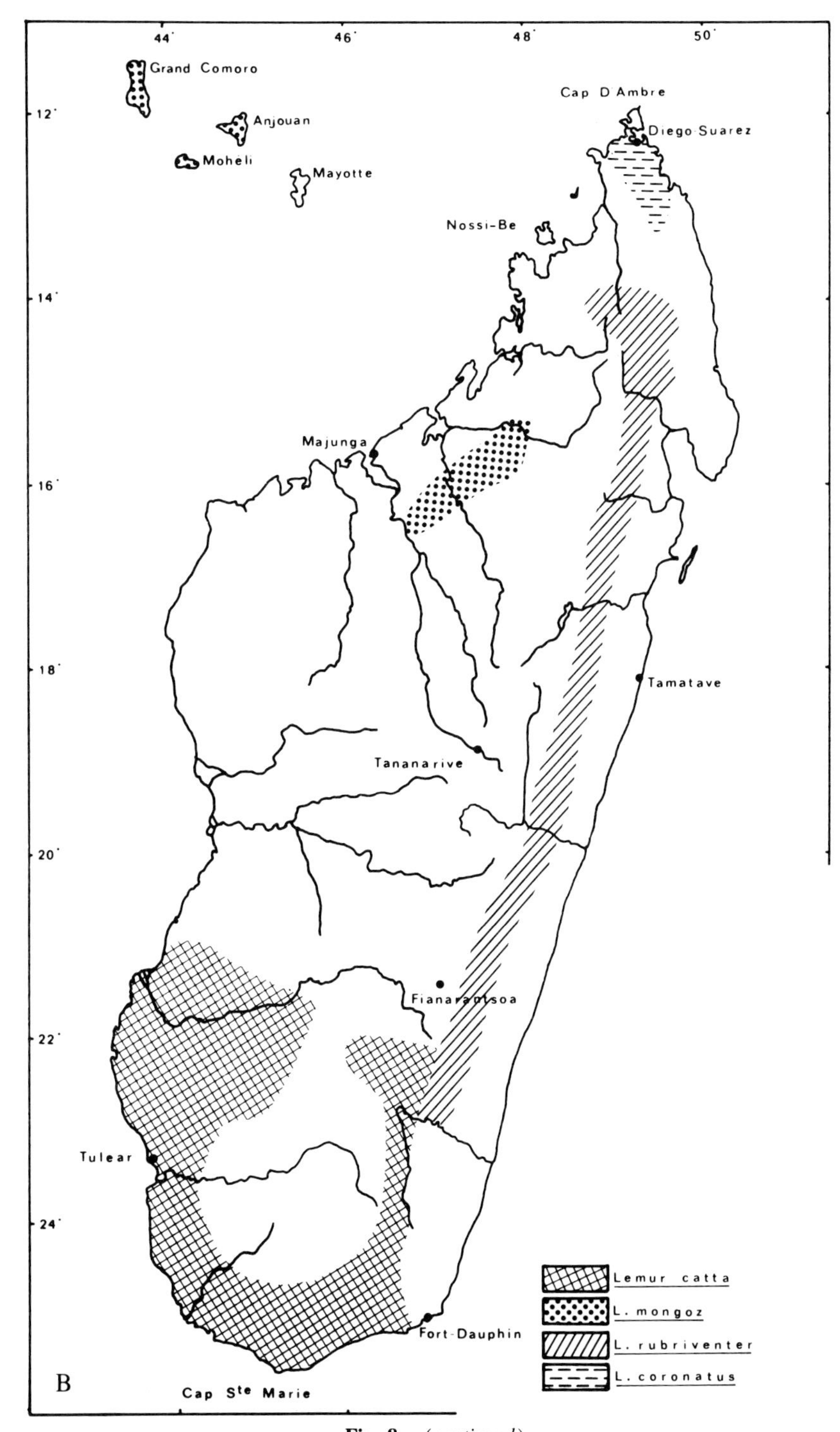

Fig. 8. *(continued).*

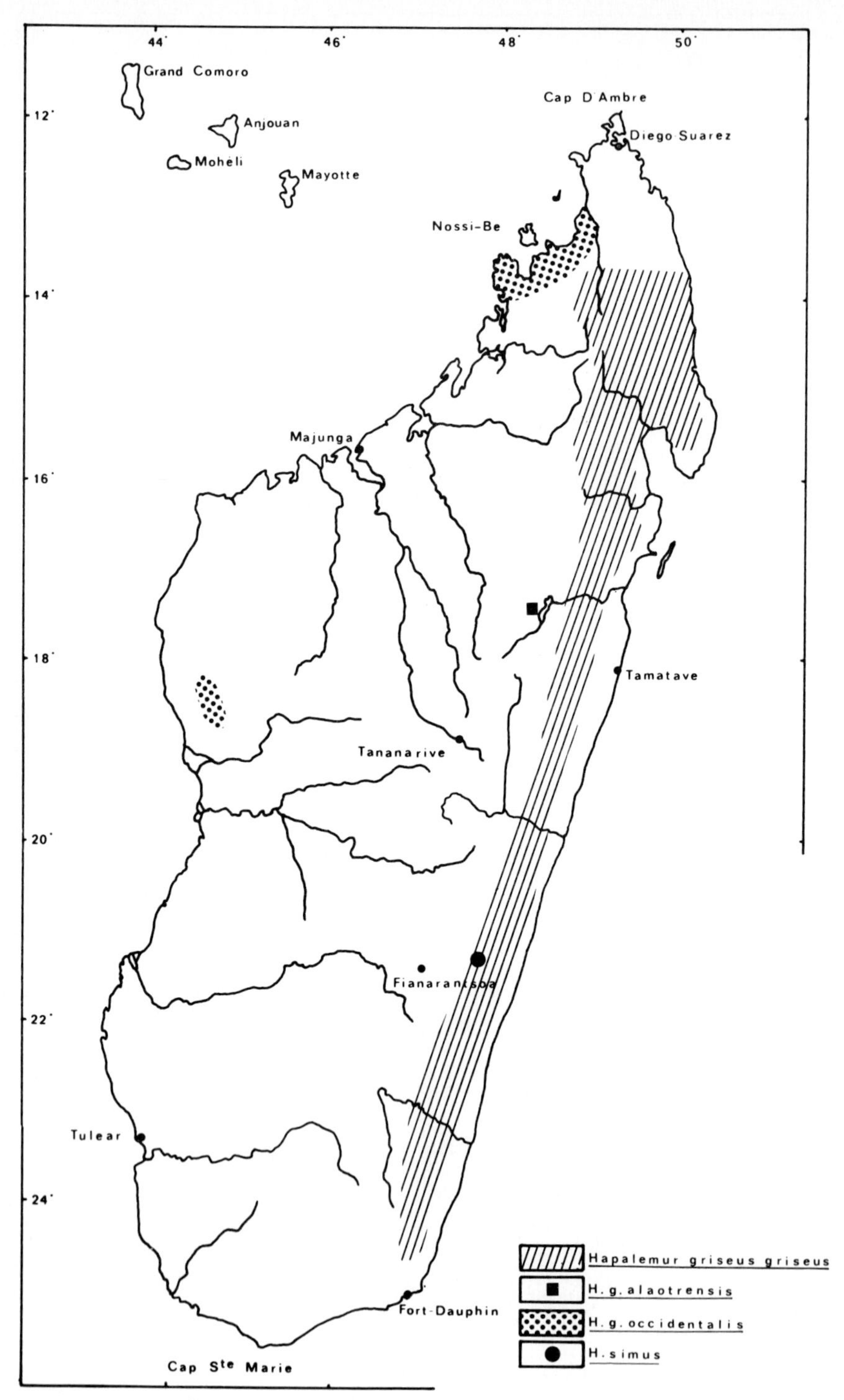

Fig. 9. Map showing the distribution of *Hapalemur*.

could only be derived from the principal lineage through two centric fusion events and two pericentric inversions. Figure 8 shows the distribution of *Lemur*.

The karyotype of *Hapalemur simus* (2*N* = 60) can be derived from the primitive karyotype of 2*N* = 60 by one pericentric inversion, while those of *Hapalemur griseus* (2*N* = 58 and 2*N* = 54) could be derived from that of *L. fulvus* (2*N* = 60) through two successive series of centric fusions. It would appear that *Lemur catta* and the *Hapalemur* species are all derived from a common ancestor and that they diverged from one another relatively recently. Figure 9 shows the distribution of *Hapalemur*.

6.2.3. Lepilemuridae

The genus *Lepilemur,* which is very different from *Lemur,* should be removed from the family Lemuridae and placed in a separate family, Lepilemuridae. Stephan and Bauchot (1965) had already proposed a separation into two tribes, Lepilemurini and Lemurini (the latter including *Lemur* and *Hapalemur*). The karyotypes of the *Lepilemur* species (Rumpler, 1972) are very distinctive, with diploid numbers varying from 2*N* = 20 to 2*N* = 38 and a range of fundamental numbers from 38–46. Craniodental characters (Hill, 1953) and other recently described characteristics of *Lepilemur* species, in particular, involving dermatoglyphs (Rumpler and Rakotosamimanana, 1972), hematology (Andriamiandra and Rumpler, 1971; Richaud *et al.,* 1971; Buettner-Janusch *et al.,* 1971), parasitology (Chabaud *et al.,* 1965), and behavior (Petter, 1962a; Charles-Dominique and Hladik, 1971), support the separation of Lepilemuridae from Lemuridae.

There can, therefore, be no doubt that the Lepilemuridae form a group which is quite distinct from the other lemurs. Their chromosomal evolution, for example, is far more complex. In addition to centric fusions, other mechanisms, such as simple translocations or pericentric inversions, are common in the evolution of *Lepilemur* karyotypes. The most primitive karyotype represented among the living *Lepilemur* species has a diploid number of 38. Systematic revision of the genus, based upon cranial characters (Petter and Petter-Rousseaux, 1960), and especially on cytogenetic analysis of numerous specimens (Rumpler and Albignac, 1977) has permitted classification into seven species, with one containing four subspecies. Figure 10 shows the distribution of *Lepilemur*.

6.2.4. Indriidae

Turning to the family Indriidae, the three genera included are clearly distinguished from one another by their karyotypes and by their fundamental numbers (Rumpler and Albignac, 1972b). The genus *Avahi* is particularly distinctive in that its chromosomes are markedly different from those of *Indri* and *Propithecus*. However, anatomical characters (Hill, 1953), marking gland morphology (Bourlière, *et al.,* 1956; Rumpler and Andriamiandra, 1971), dermatoglyphic features (Rakotosamimanana and Rumpler, 1970), and behavioral features (Petter, 1962a; Jolly, 1966; Richard, 1973) all indicate that the indriids form a cohesive group.

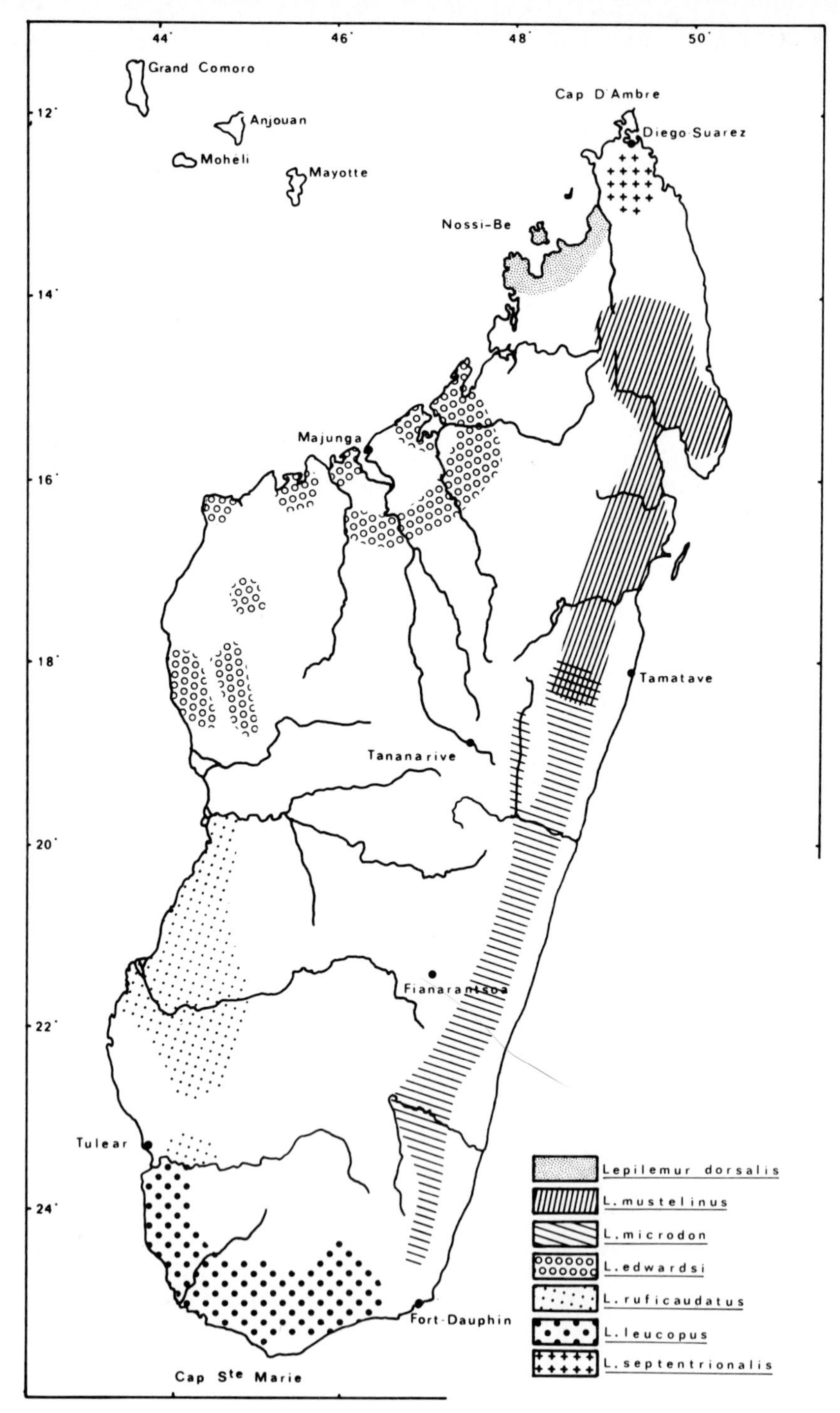

Fig. 10. Map showing the distribution of *Lepilemur*.

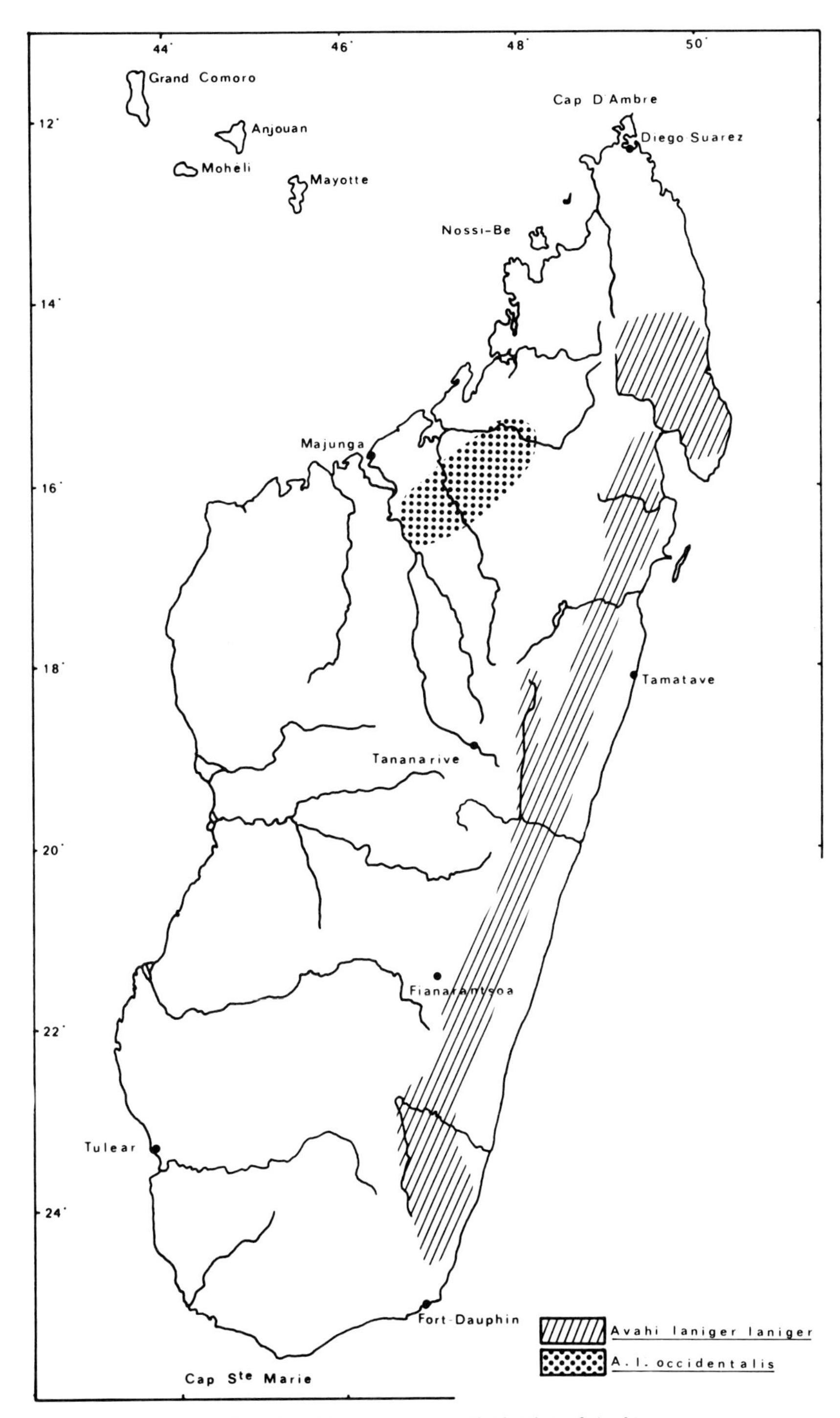

Fig. 11. Map showing the distribution of *Avahi*.

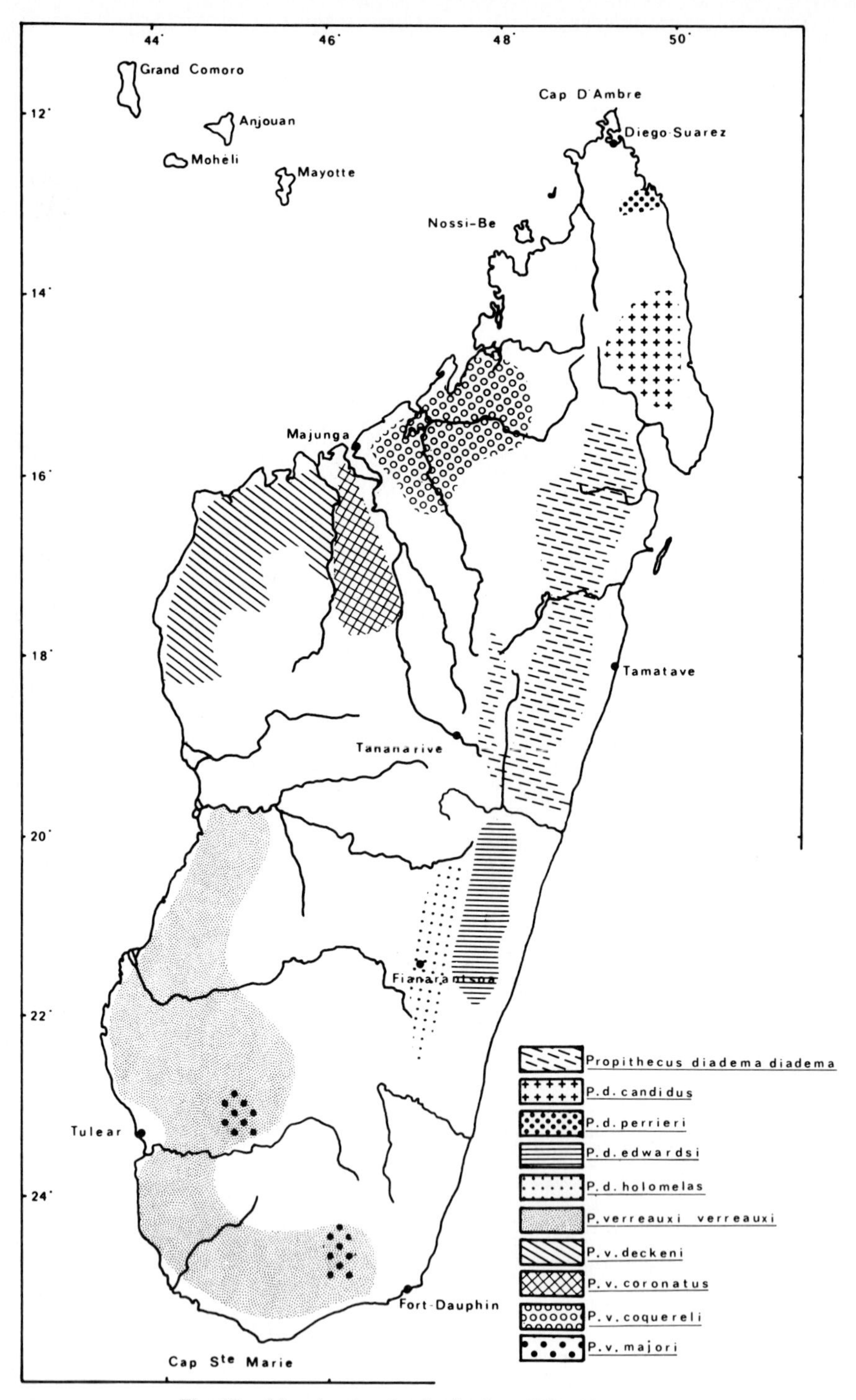

Fig. 12. Map showing the distribution of *Propithecus*.

Within the Indriidae, the karyotype of *Avahi laniger* (2*N* = 66); fundamental number, (*N* = 66), is the most primitive. It can be derived from the condition present in the Cheirogaleinae through centric fusion of two pairs of acrocentric chromosomes. Further evolution of this karyotype, probably through eight centric fusions and six pericentric inversions, could have given rise to the karyotype of *Propithecus verreauxi* (2*N* = 48); fundamental number, (*N* = 76), and eleven centric fusions and five pericentric inversions could have produced the karyotype of *Propithecus diadema* (2*N* = 42); fundamental number, (*N* = 70). The karyotype of *Indri* (2*N* = 40); fundamental number, (*N* = 74), is the most specialized of all the indriids, and it is perhaps derived from the lineage of the *Propithecus* species, diverging somewhat before *P. diadema*. The *Indri* karyotype could be derived from that of *Avahi* through twelve centric fusions and at least three pericentric inversions. The distributions of *Indri, Avahi,* and *Propithecus* are shown in Figs. 6, 11, and 12, respectively.

The family Daubentoniidae is now represented only by the species *Daubentonia madagascariensis*. The aye-aye is extremely peculiar in its anatomy (Hill, 1953), and its karyotype (2*N* = 30) and fundamental number (*N* = 54) are different from those of other Madagascar lemurs (Rumpler and Albignac, 1970). However, it has been shown that such characters do not provide sufficient grounds for postulating that the aye-aye evolved separately from other lemurs. For this reason, it seemed justifiable to suppress the distinction between Lemuroidea Mivart 1864 and Daubentonoidea Gill 1872, since this tends to exaggerate the differences between the Daubentoniidae and the other lemur families.

The karyotype of *Daubentonia,* with a diploid number of 30, is one of the most highly evolved among the lemurs. It is characterized by a large number of centric fusions and could be derived from the condition found in the Cheirogaleinae, though this (as in the case of the Lepilemuridae) would have required complex rearrangements. The distribution of *Daubentonia* is shown in Fig. 6.

As has been seen from the above discussion, the present level of cytogenetic information permits certain inferences which might be quite close to reality. The inferences are, in any case, plausible with respect to other lines of evidence. For instance, the link between the Madagascar lemurs and the African lorisids is not challenged by cytogenetic interpretation, since *Galago demidovii* could easily have developed from the karyotype condition found in the Cheirogaleinae, through two centric fusion events and the appearance of a secondary constriction.

7. PROPOSALS FOR A REVISED CLASSIFICATION OF THE EXTANT STREPSIRHINI

Order: Primates Linnaeus, 1758
Suborder: Strepsirhini E. Geoffroy, 1812
(= Lemuroidea Mivart, 1864)

	Family: Lorisidae; Subfamily: Lorisinae
Genus	*Arctocebus* (Gray, 1863)
Species	*calabarensis* (Smith, 1860) ($2N = 52$)
Subspecies	*aureus* (de Winton, 1902)
	calabarensis (Smith, 1860)
Genus	*Loris* (E. Geoffroy, 1796)
Species	*tardigradus* (Linnaeus, 1758) ($2N = 52$)
Subspecies	*grandis* (Hill and Phillips, 1932)
	lydekkerianus (Cabrera, 1908)
	malabaricus (Wroughton, 1917)
	nordicus (Hill, 1933)
	nycticeboides (Hill, 1942)
	tardigradus (Linnaeus, 1758)
Genus	*Nycticebus* (E. Geoffroy, 1812)
Species	*coucang* (Boddaert, 1785) ($2N = 62$)
Subspecies	*bancanus* (Lyon, 1906)
	bengalensis (Fischer, 1804)
	borneanus (Lyon, 1906)
	coucang (Boddaert, 1785)
	hilleri (Stone and Rehn, 1902)
	insularis (Robinson, 1917)
	natunae (Stone and Rehn, 1902)
	javanicus (E. Geoffroy, 1812)
	tenasserimensis (Elliot, 1912)
	pygmaeus (Bonhote, 1907[a])
Genus	*Perodicticus* (Bennet, 1831)
Species	*potto* (P.L.S. Muller, 1766) ($2N = 62$)
Subspecies	*edwardsi* (Bouvier, 1879)
	faustus (Thomas, 1910)
	ibeanus (Thomas, 1910)
	ju-ju (Thomas, 1910)
	potto (P.L.S. Muller, 1766)
	Family: Lorisidae; Subfamily: Galaginae
Genus	*Galago* (E. Geoffroy, 1796)
	(subgenus: *Euoticus*) (Gray, 1863)
Species	*elegantulus* (Le Conte, 1857)
Subspecies	*elegantulus* (Le Conte, 1857)
	pallidus (Gray, 1863)
	(subgenus: *Galago*)
Species	*alleni* (Waterhouse, 1837) ($2N = 40$)
Species	*crassicaudatus* (E. Geoffroy, 1812) ($2N = 62$)
Subspecies	*argentatus* (Lönnberg, 1913)
	lönnbergi (Schwartz, 1930)
	umbrosus (Thomas, 1913)

(*continued*)

	Family: Lorisidae; Subfamily: Galaginae
	garnetti (Ogilby, 1836)
	monteiri (Gray, 1836[b])
	badius (Matschie, 1905)
	crassicaudatus (E. Geoffroy, 1812)
	panganiensis (Matschie, 1906)
	lasiotis (Peters, 1876)
	agisymbanus (Coquerel, 1859)
	kikuyuensis (Lönnberg, 1912)
Species	*senegalensis* (E. Geoffroy, 1796) ($2N = 36$)
Subspecies	*albipes* (Dollman, 1909)
	braccatus (Elliot, 1907[c,d,e])
	dunni (Dollman, 1910)
	gallarum (Thomas, 1901)
	granti (Thomas and Wroughton, 1907)
	moholi (A. Smith, 1839[c,e])
	senegalensis (E. Geoffroy, 1796[e])
	sotikae (Hollister, 1920)
	zanzibaricus (Matschie, 1893[c])
Species	*inustus* (Schwartz, 1930)
	(subgenus: *Galagoides*) (Smith, 1833)
Species	*demidovii* (Fischer, 1808) ($2N = 58$)
Subspecies	*anomurus* (Pousargues, 1894)
	demidovii (Fischer, 1808)
	murinus (Murray, 1859)
	orinus (Lawrens and Washburn, 1936)
	phasma (Cabrera and Ruxton, 1926)
	poensis (Thomas, 1904)
	thomasi (Elliot, 1907)
	Family: Cheirogaleidae; Subfamily: Cheirogaleinae Gregory, 1915
Genus	*Microcebus* (E. Geoffroy, 1828)
Species	*murinus* (Miller, 1777) ($2N = 66$)
	rufus (E. Geoffroy, 1828) ($2N = 66$)
	coquereli (A. Grandidier, 1867) ($2N = 66$)
Genus	*Cheirogaleus* (E. Geoffroy, 1812)
Species	*major* (E. Geoffroy, 1812) ($2N = 66$)
Subspecies	*major* (E. Geoffroy, 1812)
	crossleyi (Grandidier, 1871)
Species	*medius* (E. Geoffroy, 1812) ($2N = 66$)
Genus	*Allocebus* (A. Petter-Rousseaux and J. J. Petter, 1967)
Species	*trichotis* (Gunther, 1875)
	Family: Cheirogaleidae; Subfamily: Phanerinae nov.
Genus	*Phaner* (Gray, 1870)
Species	*furcifer* (Blainville, 1841) ($2N = 46$)

(*continued*)

	Family: Lemuridae
Genus	*Lemur* (Linnaeus, 1758)
Species	*macaco* (Linnaeus, 1766)
Subspecies	*macaco* Linnaeus, 1766)
	flavifrons (Gray, 1867[f])
Species	*fulvus* (E. Geoffroy, 1812)
Subspecies	*fulvus* (E. Geoffroy, 1812[g]) ($2N = 60$)
	rufus (Audebert, 1800) ($2N = 60$)
	mayottensis (Schlegel, 1866) ($2N = 60$)
	albifrons (E. Geoffroy, 1796) ($2N = 60$)
	sanfordi (Archibold, 1932) ($2N = 60$)
	collaris (red beard) (E. Geoffroy, 1812) ($2N = 52$)
	albocollaris (white beard) (Rumpler 1974) ($2N = 48$)
Species	*mongoz* (Linnaeus, 1766) ($2N = 60$)
	coronatus (Gray, 1842) ($2N = 46$)
	rubriventer (I. Geoffroy, 1850) ($2N = 50$)
	catta (Linnaeus, 1758) ($2N = 58$)
Genus	*Hapalemur* (I. Geoffroy, 1851)
Species	*griseus* (Link, 1795)
Subspecies	*griseus* (Link, 1795) ($2N = 54$)
	alaotrensis (ssp. nov.)
	occidentalis (Rumpler and Albignac, 1973) ($2N = 58$)
Species	*simus* (Gray, 1870) ($2N = 60$)
Genus	*Varecia* (Gray, 1863)
Species	*variegata* (Kerr, 1792)
Subspecies	*variegata* (Kerr, 1792) ($2N = 46$)
	ruber (E. Geoffroy, 1812) ($2N = 46$)
	Family: Lepilemuridae nov.
Genus	*Lepilemur* (I. Geoffroy, 1851)
Species	*dorsalis* (Gray, 1870) ($2N = 44$)
	ruficaudatus (A. Grandidier, 1867) ($2N = 20$)
	edwardsi (F. Major, 1894) ($2N = 22$)
	leucopus (F. Major, 1894) ($2N = 26$)
	mustelinus (I. Geoffroy, 1851) ($2N = 34$)
	microdon (F. Major, 1894) ($2N = ?$)
	septentrionalis (Rumpler and Albignac, 1975)
Subspecies	*sahafarensis* (ssp. nov. Rumpler and Albignac, 1975) ($2N = 36$)
	septentrionalis (ssp. nov. Rumpler and Albignac, 1975) ($2N = 34$)
	andrafiamenensis (ssp. nov. Rumpler and Albignac, 1975) ($2N = 38$)
	ankaranensis (ssp. nov. Rumpler and Albignac, 1975) ($2N = 36$)

(*continued*)

	Family: Indriidae
Genus	*Avahi* (Gmelin, 1788) (2*N* = 66)
Species	*laniger* (Gmelin, 1788)
Subspecies	*laniger* (Gmelin, 1788)
	occidentalis (Lorentz, 1898)
Genus	*Propithecus* (Bennett, 1832) (2*N* = 48)
Species	*verreauxi* (A. Grandidier, 1867)
Subspecies	*verreauxi* (A. Grandidier, 1867)
	majori (Rothschild, 1894)
	deckeni (Peters, 1870)
	coronatus (A. Milne-Edwards, 1871)
	coquereli (A. Milne-Edwards, 1867)
Genus	*Propithecus* (Bennett, 1852) (2*N* = 42)
Species	*diadema* (Bennett, 1832)
Subspecies	*diadema* (Bennett, 1832)
	edwardsi (A. Grandidier, 1871)
	holomelas (Gunther, 1875)
	candidus (A. Grandidier)
	perrieri (Lavauden, 1931)
Genus	*Indri* (E. Geoffroy, 1796) (2*N* = 40)
Species	*indri* (Gmelin, 1788)
	Family: Daubentoniidae
Genus	*Daubentonia* (E. Geoffroy, 1795) (2*N* = 30)
Species	*madagascariensis* (Gmelin, 1788)

[a] Fiedler (1956) has given this subspecies the rank of a species on the basis of its small body size and certain dental peculiarities, but it is, nevertheless, extremely close to *N. coucang*.

[b] Subspecies with 2*N* = 62.

[c] Subspecies with 2*N* = 36.

[d] Subspecies with 2*N* = 37.

[e] Subspecies with 2*N* = 38.

[f] It is very probable that this last form (synonym: *L. nigerrimus* Sclater, 1880) is a subspecies of *L. macaco;* the color of the pelage is similar to that of *L.m. macaco*. The male is black, and the female is pale red; no lateral tufts of hair are present around the head. The geographical range of these animals would be limited to Ampasindava (N.W. of Madagascar) near the range of *L. macaco*. The only known specimens at the present are preserved skins.

[g] This subspecies includes several, as yet, poorly defined forms. Apart from the eastern form, one from Tampoketsy (now almost extinct), and another from the West have also been described. The localized form from the Comoro Islands, *L.f. mayottensis* Schlegel, 1866, is very similar and may have originated as a hybridization between *L.f. fulvus* and *L.f. rufus* imported to the island.

ACKNOWLEDGMENT

The authors particularly wish to thank Professor Doyle for the considerable amount of time spent checking the references, putting the finishing touches to the translation, and for preparing the distribution maps.

REFERENCES

Albignac, R., Rumpler, Y., and Petter, J. J. (1971). *Mammalia* **35,** 358–368.

Andriamiandra, A., and Rumpler, Y. (1971). *Bull. Acad. Malgache* **49,** 145–147.

Bender, M. A., and Chu, E.H.Y. (1963). *In* "Evolutionary and Genetic Biology of Primates" (J. Buettner-Janusch, ed.), Vol. I, pp. 261–311. Academic Press, New York.

Bennejeant, C. (1936). "Anomalies et variations dentaires chez les Primates." P. Vallier, Clermont-Ferrand.

Bourlière, F., Petter, J. J., and Petter-Rousseaux, A. (1956). *Mém. Inst. Sci. Madagascar, Sér. A* **10,** 303–304.

Buettner-Janusch, J. (1963). *In* "Evolutionary and Genetic Biology of Primates" (J. Buettner-Janusch, ed.), Vol. I, pp. 1–64. Academic Press, New York.

Buettner-Janusch, J. (1964). *Folia Primatol.* **2,** 93–110.

Buettner-Janusch, J., Washington, J. L., and Buettner-Janusch, V. (1971). *Arch. Inst. Pasteur Madagascar* **40,** 127–136.

Carlsson, A. (1922). *Acta Zool. (Stockholm)* **3,** 227–270.

Cartmill, M. (1971). *Nature (London)* **232,** 566–567.

Cartmill, M. (1974). *In* "Prosimian Biology" (R. D. Martin, G. A. Doyle, and A. C. Walker, eds.), pp. 655–670. Duckworth, London.

Cartmill, M. (1975). *In* "Phylogeny of the Primates: An Interdisciplinary Approach" (W. P. Luckett and F. S. Szalay, eds.), pp. 313–354. Plenum, New York.

Chabaud, A. G., Brygoo, E. R., and Petter, A. (1965). *Ann. Parisitol. Hum. Comp.* **40,** No. 2, 181–214.

Charles-Dominique, P. (1971). *Biol. Gabonica* **7,** 121–228.

Charles-Dominique, P., and Hladik, C. M. (1971). *Terre Vie* **1,** 3–66.

Charles-Dominique, P., and Martin, R. D. (1970). *Nature (London)* **227,** 257–260.

Chu, E.H.Y., and Bender, M. A. (1961). *Science* **133,** 1399–1405.

de Boer, L.E.M. (1972). *Genen Phaenen* **15,** 41–64.

de Boer, L.E.M. (1973a). *Genetica* **44,** 155–193.

de Boer, L.E.M. (1973b). *Genetica* **44,** 330–367.

Doyle, G. A., and Bearder, S. K. (1977). *In* "Primate Conservation" (G. H. Bourne, ed.), pp. 1–35. Academic Press, New York.

Egozcue, J. (1968). *Mamm. Chrom. Newsl.* **9,** 92–93.

Egozcue, J. (1969). *In* "Comparative Mammalian Cytogenetics" (K. Bernirschke, ed.), pp. 357–389. Springer-Verlag, Berlin and New York.

Egozcue, J. (1970). *Folia Primatol.* **12,** 236–240.

Egozcue, J., and Vilarasau de Egozcue, M. (1967). *Primates* **7,** 423–432.

Egozcue, J., Ushijima, R. N., and Vilarasau de Egozcue, M. (1966). *Mamm. Chrom. Newsl.* **22,** 204.

Elliot Smith, G. (1919). *Proc. Zool. Soc. London* pp. 465–475.

Fiedler, W. (1956). *In* "Primatologia: Handbook of Primatology" (H. D. Hofer, A. H. Schultz, and D. Starck, eds.), Vol. I, pp. 1–266. Karger, Basel.

Flower, W. H., and Lydekker, R. (1891). "An Introduction to the Study of Mammals Living and Extinct." A. and C. Black, London.

Forbes, H. O. (1894). "Handbook of the Primates," Vol. I. W. H. Allen, London.

Gadow, H. (1898). "A Classification of Vertebrates, Recent and Extinct." A. and C. Black, London.

Gervais, P. (1854). "Histoire naturelle des Mammifères." L. Curmer, Paris.

Gingerich, P. D. (1975). *In* "Lemur Biology" (I. Tattersall and R. W. Sussman, eds.), pp. 65–80. Plenum, New York.

Goodman, M. (1967). *Primates* **8,** 1–22.

Goodman, M. (1975). *In* "Phylogeny of the Primates: A multidisciplinary approach" (W. P. Luckett and F. S. Szalay, eds.), pp. 219–248. Plenum, New York.

Gray, J. E. (1863). *Proc. Zool. Soc. London* pp. 129–152.
Gray, J. E. (1872). *Proc. Zool. Soc. London* pp. 846–860.
Groves, C. P. (1974). *In* "Prosimian Biology" (R. D. Martin, G. A. Doyle, and A. C. Walker, eds.), pp. 449–474. Duckworth, London.
Hayman, R. W. (1937). *Ann. Mag. Nat. Hist.* **10,** 41–67.
Hill, J. P. (1919). *Proc. Zool. Soc. London* pp. 476–491.
Hill, W.C.O. (1953). "Primates: Comparative Anatomy and Taxonomy," Vol. I. Edinburgh Univ. Press, Edinburgh.
Hill, W.C.O. (1972). "Evolutionary Biology of the Primates." Academic Press, New York.
Hoffstetter, R. (1974). *J. Hum. Evol.* **3,** 327–350.
Humbert, H., and Cours Darne, G. (1965). *Trav. Sect. Sci. Tech., Inst. Fr. Pondichery* **6,** 1–156.
Jolly, A. (1966). "Lemur Behavior: A Madagascar Field Study." Univ. of Chicago Press, Chicago, Illinois.
Kaudern, W. (1910). *Anat. Anz.* **21–22,** 561–573.
Kingdon, J. (1971). "East African Mammals," Vol. I. Academic Press, New York.
Lamberton, C. (1934). *Mem. Acad. Malgache* **17,** 1–168.
Lamberton, C. (1939). *Mem. Acad. Malgache* **27,** 1–103.
Le Gros Clark, W. E. (1934). "Early Forerunners of Man." Baillière, London.
Leche, W. (1896). *Festschr. C. Gegenbaur* pp. 127–166.
Lessertisseur, J., Nakache, J. P., and Petit-Maire, N. (1974). *Bull. Mem. Soc. Anthropol., Paris* **9,** 269–291.
McKenna, M. C. (1963). *Proc. Int. Congr. Zool., 16th, 1963 Vol. 4,* 69–74.
Mahé, J. (1972). Doctoral Thesis, University of Paris (unpublished).
Manfredi-Romani, M. G., de Boer, L.E.M., Chiarelli, B., and Tinozzi Massari, S. (1972). *J. Hum. Evol.* **1,** 473–476.
Manna, G. K., and Talukdar, M. (1968). *Mammalia* **32,** 118–130.
Martin, R. D. (1967). Ph. D. Thesis, Oxford University (unpublished).
Martin, R. D. (1968). *Man* **3,** 377–401.
Martin, R. D. (1972). *Philos. Trans. R. Soc. London, Ser. B* **264,** 295–352.
Matschie, P. (1905). *Sitzungsber. Ges. Naturf. Freunde Berl.* **8,** 277–279.
Milne Edwards, A., and Grandidier, A. (1875). "Histoire physique, naturelle et politique de Madagascar." Imprimerie Nationale et Société d'Edition Géographique et Coloniale, Paris.
Mivart, St. G. (1873). *Proc. Zool. Soc. London* pp. 484–510.
Peters, W. H. C. (1866). *Abhandl. d. Königl. Akad. der Wiss. Berl.* pp. 79–110.
Petter, J. J. (1959). *Nat. Malgache* **11,** Nos. 1–2, 153–164.
Petter, J. J. (1962a). *Mem. Mus. Natl. Hist. Nat., Paris, Ser. A* **27,** 1–146.
Petter, J. J. (1962b). *Terre Vie* **4,** 394–416.
Petter, J. J., and Petter-Rousseaux, A. (1960). *Mammalia* **24,** 76–86.
Petter, J. J., and Peyrieras, A. (1970). *Mammalia* **34,** 167–193.
Petter, J. J., and Peyrieras, A. (1972). *J. Hum. Evol.* **1,** 379–388.
Petter, J. J., Schilling, A., and Pariente, G. (1971). *Terre Vie* **3,** 287–327.
Petter, J. J., Albignac, R., and Rumpler, Y. (1977). "Faune de Madagascar: Lémuriens." ORSTOM–CNRS, Paris.
Petter-Rousseaux, A. (1962). *Mammalia* **26,** 7–78.
Pocock, R. I. (1918). *Proc. Zool. Soc. London* pp. 19–53.
Pocock, R. I. (1919). *Proc. Zool. Soc. London* pp. 494–495.
Pringle, J. A. (1974). *Ann. Natal Mus.* **22,** 173–186.
Rakotosamimanana, B., and Rumpler, Y. (1970). *Bull Assoc. Anat.* **148,** 493–510.
Richard, A. F. (1973). Ph.D. Thesis, University of London (unpublished).
Richaud, A., Rumpler, Y., and Albignac, R. (1971). *Arch. Inst. Pasteur Madagascar* **40,** 137–144.
Roberts, D., and Tattersall, I. (1974). *Am. Mus. Novit.* **2536,** 1–9.

Rumpler, Y. (1970). *Cytogenetics* **9,** 239–244.
Rumpler, Y. (1972). *Congr. Assoc. Anat., 57th,* March 26–30, 1972.
Rumpler, Y., and Albignac, R. (1970). *Ann. Univ. Madagascar (Med. Biol.)* **12–13,** 3–11.
Rumpler, Y., and Albignac, R. (1972a). *C.R. Seances Soc. Biol. Ses Fil.* **165,** 742–747.
Rumpler, Y., and Albignac, R. (1972b). Third Conf. on Exp. Med. and Surg. in Primates, Lyon, June 21–23, 1972.
Rumpler, Y., and Albignac, R. (1973). *Am. J. Phys. Anthropol.* **38,** 261–264.
Rumpler, Y., and Albignac, R. (1977). *Am. J. Phys. Anthropol.* (in press).
Rumpler, Y., and Andriamiandra, A. (1971). *C.R. Seances Soc. Biol.* **165,** 436–442.
Rumpler, Y., and Rakotosamimanana, B. (1972). C.R. 56e Congrès des Anatomistes, Nantes 1971. (*Bull. Ass. Anat.* **154** 1127–1143).
Saban, R. (1956). *Ann. Paleontol.* **42,** 169–224.
Saban, R. (1957). *Ann. Paleontol.* **43,** 1–43.
Saban, R. (1963). *Mem. Mus. Natl. Hist. Nat., Paris. Ser A* **29,** 1–378.
Saban, R. (1975). *In* "Lemur Biology" (I. Tattersall and R. W. Sussman, eds.), pp. 83–109. Plenum, New York.
Schwartz, E. (1931). *Ann. Mag. Nat. Hist.* **10,** 41–67.
Schwartz, J. H. (1974). *Am. J. Phys. Anthropol.* **41,** 107–114.
Schwartz, J. H. (1975). *In* "Lemur Biology" (I. Tattersall and R. W. Sussman, eds.), pp. 41–63. Plenum, New York.
Simons, E. L. (1972). "Primate Evolution." Macmillan, New York.
Simons, E. L., and Russell, D. E. (1960). *Breviora* **127,** 1–14.
Simpson, G. G. (1945). *Bull. Am. Mus. Nat. Hist.* **85,** 1–350.
Simpson, G. G. (1967). *Bull. Mus. Comp. Zool.* **136,** 39–61.
Stehlin, H. G. (1912). *Abh. Schweiz. Palaeontol. Ges.* **38,** 1165–1298.
Stephan, H., and Bauchot, R. (1965). *Acta Zool. (Stokholm)* **46,** 209.
Strauss, W. L. (1944). *Am. J. Anthropol.* **4,** 243–247.
Szalay, F. S. (1973). *Folia Primatol.* **19,** 73–87.
Szalay, F. S., and Katz, C. C. (1973). *Folia Primatol.* **19,** 88–103.
Tattersall, I. (1973). *Anthropol. Pap. Am. Mus.* **52,** 1–110.
Tattersall, I., and Schwartz, J. H. (1974). *Anthropol. Pap. Am. Mus.* **52,** 139–192.
van Kampen, P. N. (1905). *Morphol. Jahrb.* **36,** 321–372.
van Valen, L. (1969). *Am. J. Phys. Anthropol.* **30,** 295–296.
Vincent, F. (1972). *Ann. Fac. Sci. Cameroun* No. 10, pp. 135–141.
Walker, A. C. (1967). *In* "Pleistocene Extinctions, the Search for a Cause" (P. S. Martin and H. E. Wright, eds.), pp. 425–432. Yale Univ. Press, New Haven, Connecticut.
Walker, A. C. (1974). *In* "Prosimian Biology" (R. D. Martin, G. A. Doyle, and A. C. Walker, eds.), pp. 435–447. Duckworth, London.
Weber, M. (1927). "Die Saügetiere." Fischer, Stuttgart.
Wood-Jones, F. (1919). *Proc. Zool. Soc. London* pp. 491–494.
Ying, K. L., and Butler, H. (1971). *Can. J. Genet. Cytol.* **13,** 793–800.

Chapter 2

Phylogenetic Aspects of Prosimian Behavior

ROBERT D. MARTIN

1. INTRODUCTION

Consideration of the evolutionary background to behavior has been an integral part of ethology since its earliest beginnings (e.g., see Tinbergen, 1951; Lorenz, 1950, 1965). Indeed, one of the main features distinguishing ethology from related subjects, such as comparative psychology, has been its emphasis on the intimate relationship between behavior and the natural environmental setting for evolution. When observing the behavior of any animal species today, we are dealing with a complex product of selection pressures which have operated under particular environmental conditions. Some of these conditions are those prevailing in the present habitat of that species, but some conditions may have operated only during a particular phase of past evolutionary history. Thus, a full understanding of the behavior exhibited by members of a particular animal group—such as the prosimians—requires both knowledge of modern habitat conditions peculiar to each species and appreciation of the broad evolutionary history of the group. We are unlikely, for example, to achieve a satisfactory framework for interpreting the evolution of primate social behavior until we have a sound basis for relating such

ISBN 0-12-222150-8

behavior not only to modern habitat conditions but also to the general evolutionary history of primates. Unfortunately, although there are numerous research workers with considerable competence in either one of these fields, very few have seriously tackled the two together. One of the main aims of this chapter, therefore, is to demonstrate the need for combining expertise in primate field studies with an adequate grounding in primate evolutionary theory.

There is now considerable interest in studying the relationships between behavior and ecology, and this may perhaps be regarded as a discipline in its own right (''behavioral ecology''). One may cite Hutchinson (1965) as one of the main contributors to the philosophical foundations of this discipline, and Wilson (1975) has recently provided an elegant synthesis of research material demonstrating many of the precise ways in which behavior and ecology are interwoven. At the level of evolution of the individual species, it is possible to conduct impressive investigations into the environmental pressures which mold behavior. In some cases, at least theoretically, it is possible to examine how such pressures may influence genetic processes. However, it is also necessary to analyze the broader evolutionary relationships between extant species, and in this respect Wilson's synthesis is notably lacking. It might be held that analysis of interspecific evolutionary relationships is too heavily dependent upon speculation to permit valid conclusions to be drawn. Yet to omit analysis is to admit failure in the attempt to reach a full synthetic theory of behavior and ecology. For example, without considering evolutionary history one cannot ''explain'' the social behavior of a particular prosimian species in ecological terms. Some aspects of its social life will undoubtedly have been retained from some past era. In short, a truly synthetic view of behavioral evolution in the prosimians requires the following:

1. Detailed field studies of individual prosimian species, involving quantitative investigation of behavior and of key ecological features.
2. Laboratory studies of functional morphology, to demonstrate the anatomical and physiological features associated with particular behavioral adaptations.
3. Recognition of biological laws governing the relationships between ecology, behavior, anatomy, and physiology.
4. A broad understanding of prosimian evolution, involving a grasp of the paleontological evidence and, in particular, the use of objective techniques for reconstructing evolutionary relationships. For such reconstructions, comprehensive coverage of the evidence, including biochemical information, is essential.

There has been considerable progress in all of these areas in recent years, and Wilson's textbook (1975) has provided a preliminary synthesis of the first three aspects. Effective integration with the broad evolutionary approach at this stage would set us well on the way to an overall synthesis. Perhaps the greatest barrier to progress at present is the absence of a universally acceptable set of objective criteria for reconstructing evolutionary relationships. This is a complex subject, which has been discussed in detail elsewhere (Martin, 1968, 1973a, 1975, 1978). Suffice it to say that there is at present a diversity of opinion about the details of prosimian evolution, and that such diversity is largely due to differences in interpretation of the evidence. Hence, if it is accepted that a broad evolutionary framework is essential for a valid interpretation of prosimian behavioral adaptation, it must also be accepted that the development of objective methods for reconstructing evolutionary relationships is a matter of priority. Without reliable phylogenetic trees to support interpretations, one cannot hope to tackle the problems of retention of behavioral adaptations to past environments (''behavioral inertia''). There are also a number of concepts, such as that of occupation of an ''adaptive zone'' by a particular group of animal species (Simpson, 1950; Jerison, 1973; Charles-Dominique, 1975), which are inherent to evolution above the species level and demand a broad phylogenetic approach.

There is, of course, a special problem which arises in discussions of behavioral evolution, particularly when applied to man. This concerns the degree to which behavior may be regarded as genetically controlled. Again, this is a complex subject which has been the center of lengthy arguments. However, many of the pitfalls can be avoided by referring to ''species-typical behavior'' rather than to ''innate behavior.'' The term ''innate'' implies that there is such firm genetic control that the behavior will develop into its typical form regardless of environmental conditions. ''Species-typical,'' on the other hand, indicates only that the behavior is characteristic of a species under a given set of environmental conditions. There must, of course, be some genetic control, if only in the organization of the central nervous system to favor particular pathways of behavioral development, but modification of environmental conditions may lead directly to behavioral modification. Accordingly, what is said in this chapter about species-typical behavior in prosimians is not particularly relevant to the vexed subject of the biological constraints on human behavior, since human beings do not now live in a single ''natural'' set of environmental conditions (see Martin, 1974, for further discussion of this point).

Another point which arises in this connection, and which is fundamental to the discussion of behavioral evolution, is that it is now widely accepted that behavioral change typically precedes morphological change in evolution (e.g., see Mayr, 1969; Hardy, 1965). It is eminently reasonable to suggest that behavior is more flexible than morphology in terms of genetic control and that changes in environmental conditions would initially give rise to shifts in behavior. The combined environmental and behavioral conditions would then, in the long term, provide an overall context favoring the evolution of appropriate morphological changes. This adds a further complexity to the analysis of behavior of any individual species, such that there are really three classes of behavior which may be distinguished:

1. Behavior adapted for past environmental conditions, which subsequently changed after that behavior had become deeply rooted in the species' repertoire (''behavioral inertia'').

2. Behavior adapted to present environmental conditions, but of sufficient antiquity to be associated with established morphological changes.

3. Behavior adapted to present environmental conditions, but of such recent origin that it may be somewhat out of step with past morphological adaptations.

Two corollaries arise from this distinction. First, any interpretation of the morphology of any living species purely in terms of its current behavior patterns must be treated with caution. Second, one can only reconstruct in the most general terms the behavior of any given fossil species. For example, to ascribe a particular diet (e.g., "insectivorous") to a single fossil species purely on the basis of molar tooth morphology would be to go beyond the confines of the evidence. Behavioral diversity within a group of closely related species, despite relative uniformity in morphology, is the rule rather than the exception among living animals.

It is important to remember that competition between animal species in a given habitat is a major influence (possibly *the* major influence) leading to specialization. Closely related species living in the same general habitat zone must undergo divergent specialization if they are to coexist. Again, behavioral divergence probably precedes morphological divergence. This applies to prosimians just as to other animal groups, and there have been two major studies to date (Charles-Dominique, 1971, 1974; Sussman, 1974) which have specifically examined intraspecific competition as a factor in behavioral evolution of prosimians. Such studies provide much new information on the ways in which prosimian behavior is attuned to specific ecological factors, and they also throw further light on the broad outlines of prosimian evolution (see Charles-Dominique, 1977).

Taking all the evidence together, one can attempt to sketch some major outlines of the evolution of prosimian behavior. Before doing so, however, it is necessary to draw a distinction between the use of behavioral data to reconstruct phylogenetic relationships and the discussion of prosimian behavioral evolution in terms of established views of prosimian phylogeny. It is the author's view that the former procedure is, as yet, unjustifiable. In the first place, there are very few features of prosimian behavior whose pattern of distribution indicates clear evolutionary ties. For example, the distinction between nocturnal and diurnal habits does not in itself indicate phylogenetic divisions. Second, the requirements for objective methods of reconstruction of phylogenetic relationships are such (Martin, 1975) that speculation about ancestral behavioral states in isolation would not provide an acceptable basis for conclusions. It is therefore preferable to set out with a preliminary outline of prosimian evolution (Fig. 1) based on the author's assessment of the current morphological and biochemical evidence. Behavioral evolution among the prosimians may then be discussed against this background. Analysis of prosimian behavior can in turn provide additional evidence to confirm or question such a phylogenetic scheme. Nevertheless, if it subsequently emerges that there are serious errors in the preliminary "evolutionary tree" based on morphological and biochemical characters, it would obviously be necessary to reinterpret the behavior. The views expressed in this chapter about likely pathways of prosimian behavioral evolution are therefore heavily dependent on the reliability of the hypothetical

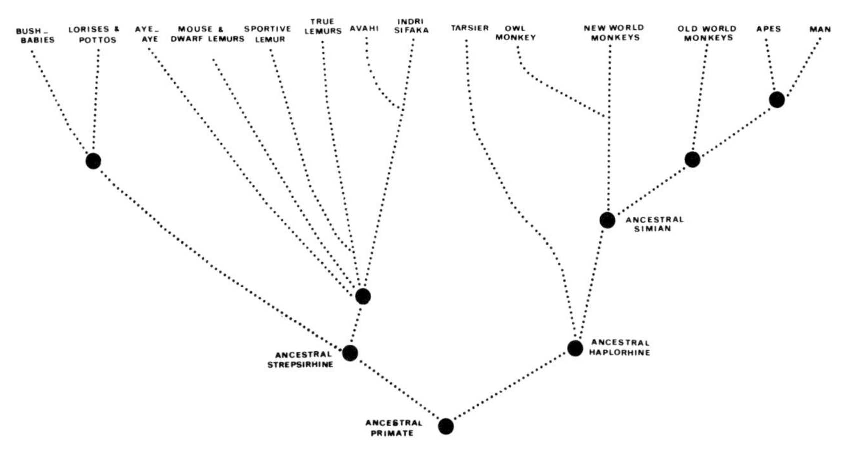

Fig. 1. A general outline of hypothetical phylogenetic relationships between living primates based on an assessment of morphological and biochemical evidence.

relationships indicated in Fig. 1. Finally, it should be noted that the author has previously attempted to trace the major lines of behavioral evolution within the Madagascar lemurs (Martin, 1972). The scope of that earlier account has now been expanded to include the lorisiform primates (lorises, pottos, and bushbabies) and the tarsiers, and several new lines of evidence are considered.

2. BODY SIZE AND SCALING EFFECTS

It might, at first sight, seem unusual to consider body size as the initial subject for discussion in a chapter on behavioral evolution in prosimians. However, both morphology and behavior are subject to considerable constraints imposed by body size. Because of certain fundamental laws which govern the relationship between numerous biological characters and body size, no analysis of the variation in characters within a group of mammals is complete without appropriate treatment of "scaling effects" (Schmidt-Nielsen, 1972; Gould, 1966, 1971). If the species within the group under investigation differ notably in body size, one should not rely upon a simple comparison of their characters as such, but take the analysis further to account for regular transformations ("scaling") required by increase or decrease in body dimensions. To take one crucial example, there is a basic law applicable to all mammals and affecting the energy consumption required for basic life processes: this is that as body size increases the basal metabolic rate per unit of body tissue decreases (Kleiber, 1932; Schmidt-Nielsen, 1970, 1972). Insofar as routine metabolic requirements typically represent the major energy expenditure in a given group of mammals, large-bodied species will be under constraints predictably different from those applying to small-bodied species. One outcome of this is that small-bodied mammal species usually need to consume food with more directly available nutrients. This certainly applies to prosimians in that small-bodied species, such as Galaginae, Lorisinae, Cheriogaleidae, Tarsiidae, generally require at least a proportion of animal food in their diets and select most of any additional dietary items from those which contain readily available carbohydrates, for example, ripe fruits and gums. Larger-bodied prosimians, such as Lepilemuridae, Lemuridae, Indriidae, on the other hand, can consume exclusively plant food which requires considerable mastication and subsequent conversion, for example, buds, leaves, and unripe fruits (and see Hladik, Chapter 8). Further, body size and utilizable energy input together determine to a large extent the size of the home range which can be covered in the search for food. Given that dietary items containing readily available nutrients are relatively scarce in a typical forest environment, whereas those containing nutrients requiring extensive conversion are relatively common, it follows that small-bodied prosimians are unlikely to form large social groups. It also follows that small-bodied prosimians are more likely to occur in secondary forest and at forest fringes, where plant productivity, and perhaps the attraction of edible prey, is higher. Large-bodied prosimians, by contrast, can live in relatively large social groups, depending upon the availability of the plant food selected, and more commonly occupy primary forest areas. Such relationships can be multiplied almost *ad infinitum,* and eventually detailed analysis of the behavior of any single prosimian species will undoubtedly require a multifactorial approach. However, the main point at this juncture is that body size must be considered as a major factor in any explanatory model.

Having established the fact that body size must be taken into account in considering prosimian behavioral evolution, we are left with the problem of determining the ways in which particular characters must be scaled up or down as body size increases or decreases. Most work to date in this area has relied on an empirical approach in which the variation of each character (e.g., brain size) with body size is examined within an animal group. One of the main texts illustrating this empirical approach is Jerison's excellent treatise (1973) on the evolution of brain size in the vertebrates. In most cases, the "allometric" variation of a particular character with body size within an animal group can be considered in terms of a standard formula:

$$A = kB^{\alpha}$$

(where A is the dimension of the character concerned, B is the body size and k, α are constants). Conversion of this formula to a logarithmic expression facilitates analysis of the allometric relationship:

$$\log A = \log k + \alpha \log B$$

This expression has the advantage that it is linear in form. When the data are plotted on logarithmic coordinates, a statistical best-fit line can be drawn, the slope of which directly indicates the constant α, while the intercept on the ordinate indicates the logarithmic value of the constant k. For instance, the standard relationship between metabolic rate and body size which has been established for mammals in general (see Schmidt-Nielsen, 1972) is:

$$A = kB^{3/4}$$

(where A stands for metabolic rate). The slope of the best-fit line on logarithmic coordinates is accordingly 0.75.

At this point, it is necessary to add a note of clarification concerning the "best-fit line" which is used with logarithmic data to identify the allometric constants α and k. A number of authors continue to use a simple *least squares regression* in this context. However, as Sacher (1970) has recently stated, this approach is inappropriate for allometric data; he recommends instead use of either the *principal axis* or the *reduced major axis*. The latter were extensively discussed by Kermack and Haldane (1950), who provided one of the most succinct statements of the basic statistical problem:

> In studies on organic correlation it is frequently desired to fit a straight line to show the general trend of the relationship between two variates. The residual scatter of the observed points about any such line is often due mainly to the biological variability of the material; and the errors of measurement may only account for a very small part of it. For such a case the conventionally used regression lines are quite unsuitable since here the terms "dependent variate" and "independent variate" have no real meaning!

Kermack and Haldane outline in their paper the statistical background to determination of the reduced major axis, and that has been used as the basis for all allometric analyses discussed in this present account. As Sacher (1970) notes, the reduced major axis provides a useful means of rapidly estimating the allometric constants when the correlation between any two variates (e.g., brain size and body size) is known to be high. In all cases considered here, the reduced major axis consistently produces higher values of α and lower values of k than are determined with the (inappropriate) linear regression technique.

This empirical approach to the relationship between quantifiable characters and body size has been used increasingly in recent years to consolidate interspecific comparisons. One of the oldest-established, and currently most popular, areas of application has been in the study of relative brain size (e.g., Stephan and Bauchot, 1965; Jerison, 1973; Gould, 1975), but it has recently been extended to the study of reproductive characters (e.g., Sacher and Staffeldt, 1974; Leutenegger, 1973; Martin, 1975), and developmental variables* and even to aspects of primate social behavior and ecology (Milton and May, 1976). However, there are a number of drawbacks in this approach which must be recognized. First, because the approach is empirical, comparisons must include reliable data on a wide range of species of differing body sizes to permit effective analysis. Complex statistical problems arise when the data are divided into subgroups for separate treatment, and the require-

*See Doyle, Chapter 5.

ments for a large sample size become even more severe when an attempt is made to demonstrate that any two groups differ significantly in the values determined for either of the allometric constants (α; k). In any case, plotting of data on logarithmic coordinates will tend to produce an approximately linear appearance even if the standard allometric formula ($A = kB^{\alpha}$) does not accurately reflect the relationships involved. One of the main problems is that it has frequently emerged that the value for α obtained is generally higher for large taxonomic groups (e.g., all prosimians considered together) than for smaller units (e.g., Lorisidae or Cheirogaleidae considered in isolation). Caution must therefore be exercised in "scaling" characters on the basis of empirical allometric formulas. However, the allometric approach has become sufficiently well established in practice to permit broad generalizations, and it is certainly preferable to use such an approach rather than to omit consideration of scaling in interspecific comparisons within a group such as the prosimians.

Apart from permitting an appreciation of the influence of "scaling effects" in evolution, allometric studies have additional value in the investigation of prosimian phylogeny. In fact, the value of the allometric exponent (α) may itself yield insights into the ways in which morphology and behavior are governed. For instance, the value of α determined with an allometric plot of neonatal weight against maternal body weight in lemurs and lorises (Leutenegger, 1973; Martin, 1975) is very close to the value for a plot of maternal metabolic rate against maternal body weight (viz., 0.75). This indicates that the size of the neonate produced may in some way be directly connected with the mother's metabolic turnover, an observation which could have important implications for the ecology of reproduction. Finally, it should be emphasized that arguments about the likely characteristics of the ancestral primate (= ancestral prosimian) depend heavily upon its approximate body size. Since we have yet to discover fossil remains of a primate species which might confidently be regarded as directly ancestral to all living primates, there is no direct evidence of the ancestral primate body size. However, the constraints of natural laws linking body size to morphological and behavioral characters would have applied to the ancestral primate just as to any living primate species. One of the implications of such constraints is that if an increase in body size occurred separately in two lines of primate evolution, many similar changes in morphology and behavior would occur more-or-less automatically as "scaling effects." This is one very significant component of the phenomenon known as parallelism. There are, for example, many morphological similarities between the better-known Eocene Adapidae (*Adapis, Notharctus, Smilodectes*) and the larger modern Madagascar lemurs, which are very similar in body size (e.g., see Gregory, 1920). Some authors (e.g., Szalay and Katz, 1973) have interpreted these similarities as reflecting retained primitive characters from a common ancestral species. For these characters to have been retained as such, the ancestral species must have been of comparable body size. However, it may equally well be argued (e.g., see Martin, 1972) that the common ancestor was a small-bodied species closer in size, morphology, and behavior to the modern bushbabies (Galaginae) or mouse and dwarf lemurs (Cheirogaleidae).

As with so many controversies about phylogenetic reconstruction, the conclusions reached depend essentially on the characters which are presumed to be primitive in the evolution of the group in question (see Martin, 1975, 1977). If characters common to the best-known adapid species and the larger lemurs are regarded as primitive for the prosimians generally, the similarities between bushbabies and mouse/dwarf lemurs appear as later specializations resulting from either parallelism or later common ancestry. Conversely, if the ancestral primate was a relatively small-bodied form, at least some of the morphological similarities between the well-documented adapids and the larger modern lemurs could be interpreted as parallel developments, with the requirements of allometric scaling providing the necessary evolutionary constraints. It must be recognized, then, that body size is an important and integral factor which must be considered in reconstructing primate phylogenetic history. In discussing the likely features of the ancestral primates, we must also take into account the probable body size range involved. If that proves to be impossible, then any characters postulated for the ancestral primates should be set on a sliding scale, allowing appropriate adjustment to the constraints of any given body size.

3. FEEDING BEHAVIOR

Because the bulk of the fossil evidence for primate evolution (and mammalian evolution generally) consists of isolated teeth and jaw fragments, a great deal of weight has traditionally been placed on dental characters in hypothetical phylogenetic reconstructions. Although broad dietary categories have been invoked in the paleontological literature for some considerable time, it is only comparatively recently that systematic attempts have been made to correlate dietary components of living primates with specific dental features. It has, for example, been customary to refer to the hypothetical ancestral placental mammals as "insectivores" (e.g., Romer, 1966; Simpson, 1950), despite the fact that the living Insectivora, which are presumed to retain relatively primitive dental features, cannot be regarded as uniformly "insectivorous." It is not even true to state that the diets of modern Insectivora are restricted to small animal prey. Obviously, any more accurate assessment of probable dietary characteristics in fossil mammal species must depend upon detailed field investigations of the *natural* diets of living species and correlations of specific dietary components with individual dental and cranial characteristics. At the same time, one must continue to bear in mind the problem of "evolutionary inertia," with morphological adaptations trailing some way behind behavioral predilections. It should not necessarily be expected that every observable dietary trait must be reflected in dentocranial morphology. Nor should it be expected that any two distantly related species which happen to have similar diets at the present time should necessarily exhibit exactly parallel dentocranial adaptations. Nevertheless, broad generalizations can be made by (1) investigating closely related species with different diets, and (2) distantly related species with similar diets. Perhaps the most spectacular illustration of this approach is provided by Cartmill's elegant comparison (1974a) of the aye-aye (*Daubentonia madagascariensis*) with other lemurs and with marsupials of the genus *Dactylopsila* which exhibit the same behavioral trait of excavating the bark of trees to locate the larvae of wood-boring insects and beetles. In this case, the behavioral adaptation is obviously long-standing, since parallel major morphological changes have occurred in both *Daubentonia* and *Dactylopsila:* development of continuously growing incisors along with appropriate reinforcement of the skull; formation of a distema between the incisors and the cheek teeth; reduction of surface area of the cheek teeth; attenuation of one digit of the hand as a probe. Cartmill has further elaborated on his dietary arguments by pointing out that *Daubentonia* and *Dactylopsila* occur in two areas of the world where woodpeckers (which also feed on wood-boring insects and beetle larvae) are conspicuously absent.

It is nonetheless still impossible to provide a summary of clear-cut evidence relating specific features of dental and cranial morphology to diets in prosimians generally. A number of different approaches have been made, with varying degrees of success, and a synthetic treatment should eventually be possible. First, there have been a number of recent papers dealing in detail with overall jaw mechanisms in prosimians and relating these to specific dental features (e.g., Mills, 1955; Every, 1974; Gingerich, 1972, 1974; Hiiemae and Kay, 1972; Kay and Hiiemae, 1974a,b). However, it is not yet clear how far the complex relationship between jaw mechanisms and dental morphology is relevant to diet as such, and it is beyond the scope of this present article to consider dental evolution in its own right. Second, a sophisticated attempt has latterly been made (Kay, 1975) to draw conclusions relating to dietary adaptations from a bivariate and multivariate statistical analysis of a number of metrical features of primate molar teeth. This analysis seems to distinguish relatively well between two broad categories described as "frugivorous" and "folivorous" (the latter for some reason overlapping with "insectivorous"), though it is not really clear which actual dental features combine to produce this statistical distinction in the multivariate plot. Further, no account is taken of certain dietary categories which have recently been confirmed for various prosimian species in field studies, for example, broad omnivorous diet in *Microcebus murinus* and *Galago demidovii,* predominant gum-feeding habit in *Phaner furcifer, Galago senegalensis,* and *G. elegantulus* (*Euoticus elegantulus*), and exclusive diet of animal prey in *Tarsius bancanus*. Hence the resolution of this multivariate statistical technique still leaves something to be desired. Finally, a number of attempts have been made to infer function and broad dietary categories for fossil and subfossil primates on the basis of detailed analyses of dentocranial features in individual species (e.g., for *Hadropithecus, Theropithecus* (*Simopithecus*), and *Australopithecus* by Jolly, 1970a,b; for *Varecia jullyi* and *V. insignis* by Seligsohn and Szalay, 1974; for *Archaeolemur* by Tattersall, 1974). These studies have produced numerous insights into specific dental and cranial features (e.g., molar cusp pattern, jaw architecture, orientation of pterygoid plates, facial morphology), but the dietary inferences have suffered from inadequate consideration of the actual diets of the living primate species used as models. In sum, all of these recent studies have broadened our perspective on the relationship between dietary components and dentocranial characters, though they do not yet permit confident identification of diet from the configuration of the jaws and teeth of any species. For example, if the molar teeth of *Phaner furcifer* and *Galago elegantulus* were known only from the fossil record, it would certainly not have been inferred from the results of any of the above studies that gums formed the bulk of the diet in those two species.

In fact, there has been one notable omission in virtually all of the discussion to date of prosimian dental evolution in relation to diet: the anterior teeth have barely been considered. Although a recent analysis of maxillary incisor width in simian primates (Hylander, 1975) has shown that frugivorous species tend to have broader incisors than folivorous species, this allometric study has not as yet been paralleled for the prosimians. With the exception of the aye-aye, where the incisors are so conspicuously developed, there has hardly been any consideration of the role of the anterior dentition in prosimian feeding behavior. Indeed, until recently, a curious situation existed whereby the upper peglike incisors of the living lemurs and lorises other than the aye-aye were regarded as vestigial and the array of the lower anterior teeth (incisors and canines) was interpreted as a special adaptation for grooming. It is now fairly widely accepted (Martin, 1972; Charles-Dominique, 1977; Szalay and Katz, 1973; Gingerich, 1975) that the peculiar "tooth-scraper" found in the lower jaw of all living lemurs and lorises, except the aye-aye, has been retained from a common ancestor possessing a "scraper" consisting of six procumbent teeth (two canines + four incisors). This hypothesis has been strengthened by the observation (Walker, 1969, 1974) that the Miocene lorisoids of East Africa also had a tooth-scraper of this kind. The basic arrangement of six teeth has been reduced by the loss of two teeth in the extant Indriidae, whereas the aye-aye retains only two continually growing teeth. Among the subfossil lemurs, the full six-tooth-scraper is present in *Megaladapis*, but *Palaeopropithecus*, *Archaeolemur*, and *Hadropithecus* have only four vertically implanted, spatulate teeth in the lower anterior set (also identifiable as a secondarily derived condition). However, it is not yet widely recognized that—as with virtually all other features of dental evolution—the basic tooth-scraper containing a total of six procumbent canines and incisors was probably developed primarily in relation to feeding behavior, not to grooming (Martin, 1972).

Interpretation of the role of the tooth-scraper has followed a tortuous path. Stein (1936) questioned the grooming function which had been proposed by previous observers, but de Lowther (1940) and Roberts (1941) subsequently put the record straight. It was then claimed by Avis (1961) that the toothcomb played a cropping role in feeding. In fact, this suggestion is doubtless applicable at least to *Lepilemur* (and, by inference, to *Megaladapis*), which have lost the upper incisors and thus resemble certain ungulates in which semiprocumbent lower anterior teeth serve a cropping function against a hardened pad formed by the upper lip over the ventral surface of the premaxillae. However, Avis overstated the case and misinterpreted some of the evidence. It was therefore hardly surprising that Buettner-Janusch and Andrew (1962), in clearly establishing the grooming function of the tooth-scraper in various lemur and galago species, went somewhat too far in ruling out *any* dietary function of this dental structure. The evidence now available clearly indicates that toothcombs in lemurs and lorises serve functions both in feeding and in grooming. Anyone who has observed lemurs (other than the aye-aye) or lorises, even to a limited extent, must have noticed the characteristic glide-and-jerk head movement associated with the passage of the tooth-scraper through the fur in grooming. One can also observe on the individual teeth in the scraper circular ''whip marks'' produced by hairs pulled between the teeth (R. D. Every, personal communication). On the other hand, field studies have now indicated a definite feeding role of the tooth-scraper in virtually all lemur and loris species concerned. In fact, Buettner-Janusch and Andrew (1962) themselves accepted that sifakas (*Propithecus verreauxi*) used the tooth-scraper as a scoop to take pulp from soft fruits in captivity. The author has also seen clear evidence in the field of *Microcebus murinus* using the tooth-scraper to scoop fruit pulp from grooved indentations made on certain favored fruits (e.g., *Uapaca* sp.). In addition, field studies have shown that *Indri indri* use the tooth-scraper to scoop fruit pulp (J. I. Pollock, personal communication) and that *Propithecus verreauxi* use the tooth-scraper to prise off pieces of bark (A. F. Richard, personal communication). With the majority of the relatively small-bodied lemur and loris species, it has now emerged that gums may form a moderate to large component of the diet. Extensive feeding on gums has been observed in *Galago elegantulus* (Charles-Dominique, 1977), *G. senegalensis* (Bearder, 1969), and *Phaner furcifer* (Petter *et al.*, 1971). With *Microcebus murinus* (Martin, 1973b), *Microcebus coquereli* (Petter *et al.*, 1971), *G. demidovii* (Charles-Dominique, 1972), *G. crassicaudatus* (Bearder, 1975), and *Perodicticus potto* (Charles-Dominique, 1977), gums have been found to comprise a less pronounced, but nevertheless significant, component of the diet. It has therefore been argued (Martin, 1972; Charles-Dominique, 1977) that the diet of the ancestors of the lemurs and lorises probably included a significant quota of gums, if their body size was relatively small. Although it is difficult under nocturnal conditions to see exactly how the tooth-scraper is involved in gum collection, it is reasonably well established (S. K. Bearder, P. Charles-Dominique, and R. D. Martin, personal observation) that in some species, at least, the lower anterior teeth are used to pierce relatively fresh gum droplets and to scoop up gum. The tongue then takes over the function of lapping up most of the semiliquid gum which is liberated. It is important to note that the tooth-scraper in these small-bodied, nocturnal prosimians is too fragile to permit actual scraping or penetration of the bark to produce a gum lick. (Some of the author's previous statements in this connection—e.g., Martin, 1972—have been somewhat misinterpreted.) These species must therefore rely on ''natural'' gum licks produced by the activities of wood-boring arthropods. One can, therefore, conclude that the emergence of the characteristic tooth-scraper in the common ancestral stock of the lemurs and lorises might have been associated with the inclusion of appreciable quantities of gum in the diet, associated with the occurrence of wood-boring arthropods. This is not to deny, however, a grooming function for the tooth-scraper throughout the evolution of the lemurs and lorises. Indeed, it is likely that (as with most small- to medium-sized mammals), the anterior teeth consistently served a grooming function in addition to their primary dietary role. It would be difficult to argue that the ancestors of the lemurs and lorises were subject to some special selection pressures favoring conversion of the lower anterior dentition into a "comb" uniquely serving a grooming function. On the other hand, it is perfectly reasonable to suggest that such modification of the lower anterior dentition for a primary dietary role (gum collection) would have been accompanied by exploitation of the tooth-scraper for grooming.

Using allometric comparisons, it is possible to take a closer look at the relative importance of (1) feeding and (2) grooming in the determination of the form of the tooth-scraper (Fig. 2). Linear dimensions of the tooth-scraper increase allometrically with increasing body size. A suitable approach, therefore, is to plot given dimensions of the tooth-scraper (basal width × central length from alveolar margin to tip = T) against some measure of body size (maximum skull length = S) on logarithmic coordinates (Fig. 3). The reduced major axis for the array of points has the formula:

$$\log T = 2.74 \log S - 3.48$$

Overall there is a very high correlation between this approximate measure of tooth-scraper surface area and skull length ($r = 0.97$). Yet departures of points above and below the reduced major axis (the "average" condition) show a clear pattern. Most species which have a larger than average tooth-scraper surface area (*Microcebus murinus, Galago senegalensis, G. elegantulus, Phaner furcifer, Perodicticus potto*) have been shown to feed on significant quantities of gum in the field. Conversely, most of those species with a smaller than average tooth-scraper surface area, such as *Loris tardigradus, Arctocebus calabarensis, Galago alleni, Cheirogaleus major, Hapalemur griseus, Lemur mongoz, Varecia variegata* (*Lemur variagatus*), have not been reported to feed on gum. It is also obvious that *Daubentonia* have a larger than average surface area of the lower anterior teeth, as would be expected from their specific use of those teeth in feeding. On this basis, one can predict that detailed field studies of *Lepilemur mustelinus* and *Avahi laniger* are likely to reveal a special feeding function of the tooth-scraper in these species, for they also have larger than average tooth-scraper surface areas. It is also notable that all of the species with larger than average tooth-scraper surface areas are nocturnal and essentially solitary during activity. There is, therefore, no correlation whatsoever between gregarious activity and tooth-scraper development. Those diurnal species which form social bands which move, groom, and feed together have tooth-scrapers of average or below average surface area. The analysis can be taken further by plotting hair length against skull length (Fig. 4) in order to see whether there is any indication that species with relatively long hair also have relatively well-

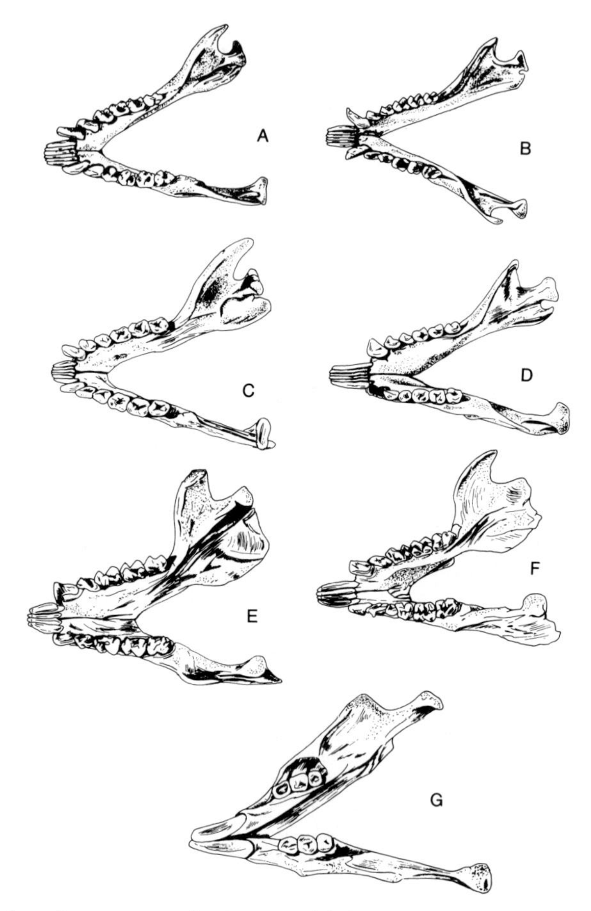

Fig. 2. Lower jaws from representative lemurs and lorises, showing variations in the form of the tooth-scraper. A, *Galago crassicaudatus;* B, *Galago elegantulus;* C, *Cheirogaleus major;* D, *Phaner furcifer;* E, *Propithecus verreauxi;* F, *Indri indri;* G, *Daubentonia madagascariensis* (drawings based on photographs in Warwick James, 1960).

developed tooth-scrapers (cf. Fig. 3). Although there is an overall tendency for hair length (H) to increase with body size (as indicated by skull length), the correlation is the lowest obtained in this study ($r = 0.89$). The reduced major axis has the formula:

$$H = 0.67\,S - 20.37$$

There is no clear-cut association either way between relative hair length and relative tooth-scraper surface area. The two species which are the most conspicuous gum feeders and also have relatively large tooth-scrapers (*Galago elegantulus* and *Phaner furcifer*) actually have shorter than average hair. This also applies to the

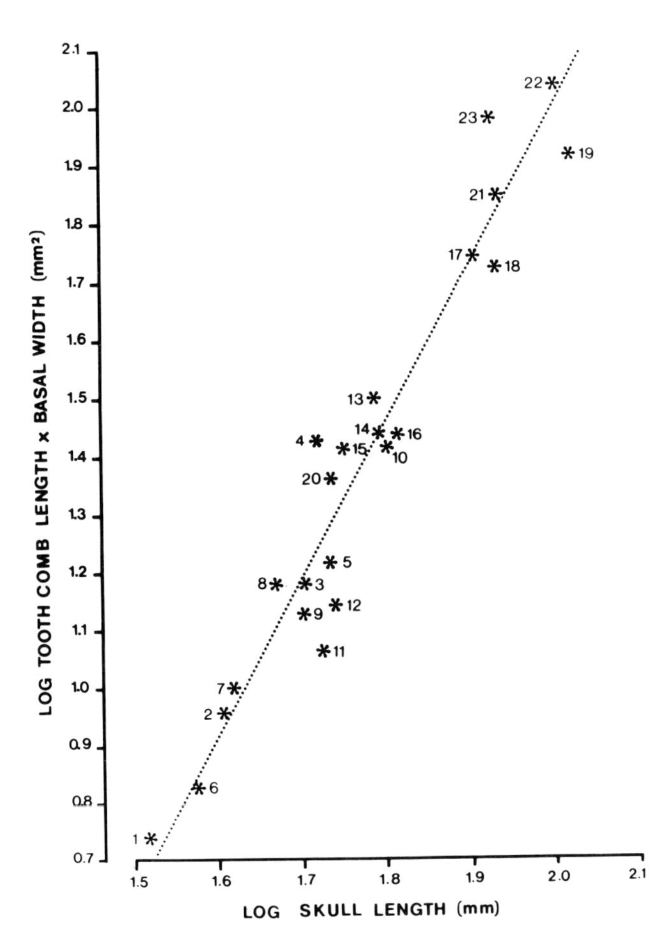

Fig. 3. "Tooth comb" surface area measure (length × basal width) plotted against skull length on logarithmic coordinates. Numbered asterisks indicate individual lemur and loris species as follows (numbers of specimens measured in parentheses): Cheirogaleidae: 1. *Microcebus murinus* (3); 2. *Cheirogaleus medius* (7); 3. *Microcebus coquereli* (1); 4. *Phaner furcifer* (3); 5. *Cheirogaleus major* (5). Galaginae: 6. *Galago demidovii* (6); 7. *Galago senegalensis* (7); 8. *Galago elegantulus* (7); 9. *Galago alleni* (7); 10. *Galago crassicaudatus* (6). Lorisinae: 11. *Loris tardigradus* (5); 12. *Arctocebus calabarensis* (8); 13. *Perodicticus potto* (7); 14. *Nycticebus coucang* (8). Lepilemuridae: 15. *Lepilemur mustelinus* (7). Lemuridae: 16. *Hapalemur griseus* (7); 17. *Hapalemur simus* (1); 18. *Lemur mongoz* (8); 19. *Varecia variegata* (7). Indriidae: 20. *Avahi laniger* (8); 21. *Propithecus verreauxi* (7); 22. *Indri indri* (7). Daubentoniidae: 23. *Daubentonia madagascariensis*. The dotted line indicates the reduced major axis.

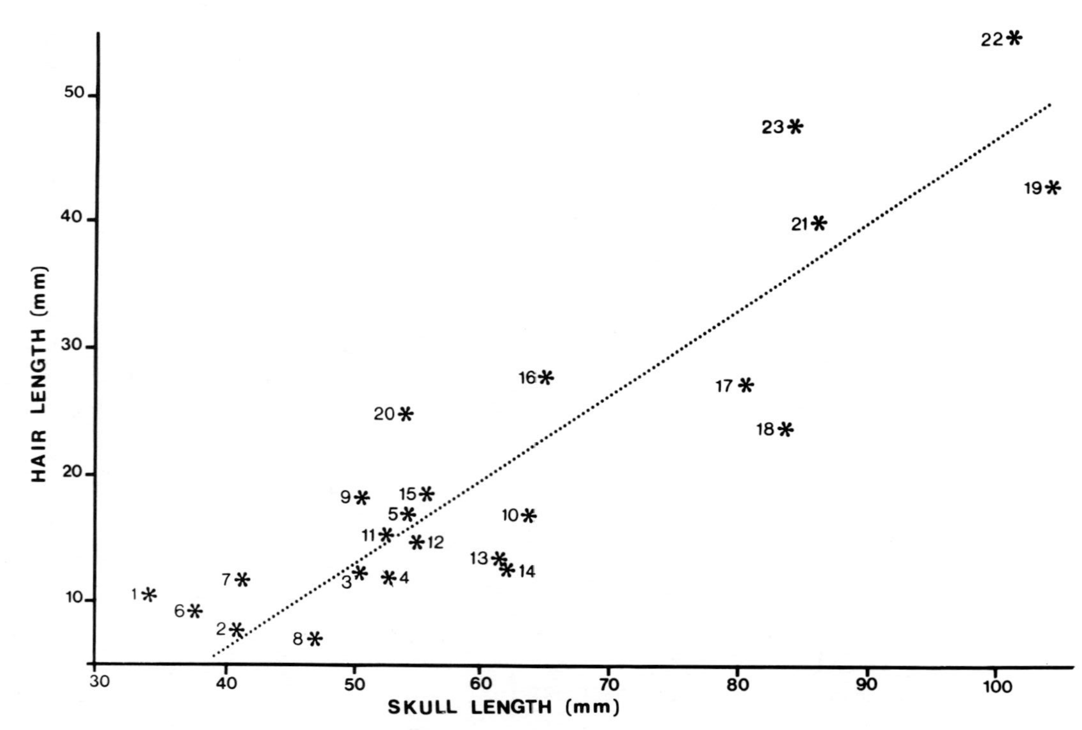

Fig. 4. Length of hair, measured on dorsal thoracic area of preserved skins, plotted against skull length. Key to numbered asterisks as for Fig. 3. The dotted line indicates the reduced major axis.

facultative gum feeder *Perodicticus potto*. Yet some species with mildly enlarged tooth-scrapers do have relatively long hair (e.g., *Microcebus murinus, Galago senegalensis*). There is also a lack of clear correlation, incidentally, between hair length and gregarious tendencies and/or frequency of social grooming. If anything, there is a weak negative correlation. One can therefore conclude that allometric analysis of the tooth-scraper indicates a clear association between gum feeding and the development of this dental device, but no obvious association with either auto-grooming or allogrooming. The evidence therefore supports the hypotheses that: (1) the primary function of the tooth-scraper is related to feeding behavior; (2) in a small-bodied common ancestor of the lemurs and lorises that function would have been involved with gum feeding; and (3) the tooth-scraper would have been used incidentally for grooming purposes. In this context, it should be noted that the six-tooth-scraper (assumed to be the primitive configuration) is found in all living lemurs and lorises below 800 gm in body weight. Above that weight, there are numerous living and subfossil lemur species exhibiting modifications of the tooth-scraper and hence a departure from the assumed ancestral condition. This itself indicates a relatively small-bodied common ancestor for the lemurs and lorises.

A simple allometric approach can also be applied to the relationship between a crude measure of cheek-teeth surface area (maximum length of cheek-tooth row × maximum width = C) and body-size. When this measure is plotted against skull length (= S) on logarithmic coordinates (Fig. 5), a surprisingly good correlation is found ($r = 0.98$), if *Daubentonia* is excluded. The cheek teeth of the aye-aye are so reduced in total surface area, undoubtedly in association with the unusual diet composed extensively of pulped larvae, that this species is quite aberrant. (The same relative reduction of cheek-teeth surface area is exhibited by *Dactylopsila* in comparison with other phalangeroid marsupials, such as *Pseudocheirus:* see Cartmill, 1974a; Fig. 3). For the remaining lemur and loris species, the formula of the reduced major axis is:

$$\text{Log } C = 2.32 \log S - 2.19$$

This, in itself, is interesting in that the slope of the line ($\alpha = 2.32$) is theoretically predictable. If it is assumed that the volume of food which can be masticated at any given time is roughly related to the surface area of the cheek teeth, and that food volume requirements largely reflect metabolic turnover, then one would expect the following relationship between cheek-teeth surface area and body size (see Section 2):

$$\log C = 0.75 \log B + k \quad (1)$$

Since skull length is, to all intents and purposes, a linear measure of body size (B), one can assume the following relationship:

$$B = k' S^3$$

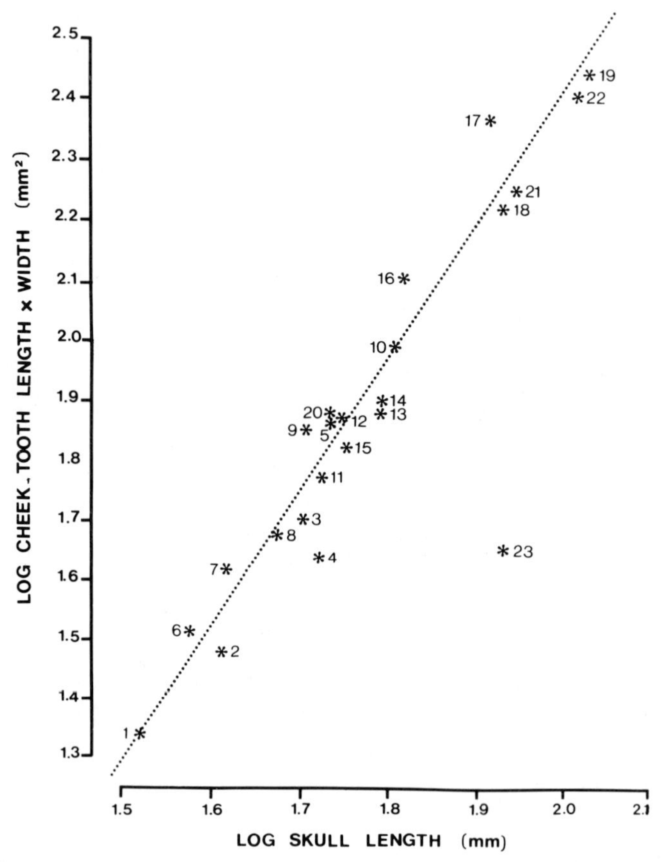

Fig. 5. Cheek–teeth surface area measure (maximum length × maximum width of tooth row) plotted against skull size on logarithmic coordinates. Key to numbered asterisks as for Fig. 3. The dotted line indicates the reduced major axis.

therefore

$$\log B = 3 \log S + \log k' \tag{2}$$

Combining Eqs. (1) and (2), one obtains the overall theoretical relationship between cheek–teeth surface area and skull length as:

$$\log C = 2.25 \log S + k''$$

This is indistinguishable from the empirically determined reduced major axis shown in Fig. 5.

It is notable from Fig. 5 that *Galago elegantulus* have a relatively smaller cheek–teeth surface area than most other galagines and that the same applies to *Phaner furcifer* in comparison with other cheirogaleids. It might, therefore, be argued that gum feeding is associated with a relative reduction in cheek–teeth surface area, as would be expected from the fact that gums typically require little or no mastication. It is also notable that the two bamboo-eating *Hapalemur* species have relatively large cheek–teeth surface areas. Of course, this simple measure of cheek–teeth area is very limited in scope, and attention must eventually be directed to the details of individual cusps and areas thereof (see Kay, 1975), but at least some general conclusions can be drawn.

Overall, it can be concluded that the surface areas of the tooth-scraper and of the cheek–teeth batteries in lemurs and lorises are very highly correlated with skull length (and hence with body size). This demonstrates the regularity of dental changes with body size, regardless of individual dietary adaptations. Over and above that, minor departures from the average condition indicated by the reduced major axis in each case provide some insight into the factors responsible for dental evolution independent of body size. In lemurs and lorises, a major factor in such evolution has clearly been the habit of gum feeding in many smaller-bodied species.

4. LOCOMOTION

The locomotor behavior of prosimians is of central importance to discussion of their evolution (see Jenkins, 1974), since it is almost universally accepted that extensive adaptation for arboreal life has been a hallmark of primate phylogeny.* Taking the ancestral primate stock as indicated in Fig. 1, it can be confidently assumed that grasping hands and feet (or at least some well-defined precursor thereof) were characteristic. Indeed, as has been pointed out elsewhere (Martin, 1968, 1972), it is the grasping foot which was of primary significance. All living primates, with the exception of man, possess a divergent hallux on each foot, operating in a pincer action against the other pedal digits. It is, therefore, reasonable to conclude that this feature can be traced back to the ancestral primates.

It is not sufficient, however, to attribute primate characteristics to arboreal life in a general sense. As Cartmill has pointed out in three perceptive discussions of this subject (1972, 1974b,c), there are numerous arboreal mammal groups which do not exhibit certain characteristic primate traits. One must, in fact, distinguish between mammals (such as tree shrews; D'Souza, 1974) which typically use broad trunk surfaces simply as an elevated counterpart to the terrestrial substrate, and those (such as most primates and arboreal marsupials) which grasp relatively fine supports which have no equivalent on the ground. Once again, the likely body size of the ancestral primate stock is relevant. If the ancestral primates were relatively small in size, then it is possible to talk of adaptation to the "fine branch niche" (Martin, 1973b; Charles-Dominique, 1977). All living prosimians, up to a body weight of

*See Walker, Chapter 12.

approximately 1 kg, typically move around on relatively fine supports (Cheirogaleidae; Lorisidae, Tarsiidae). Although some other placental mammals (e.g., some rodents; Cartmill, 1974c) occasionally make use of fine branches, they do not habitually do so. The only real parallel to the relatively small-bodied prosimians is provided by small arboreal marsupials in South America and Australia which also characteristically use fine branches. It is noteworthy that these marsupials also have grasping feet with a widely divergent hallux. Occupation of the "fine branch niche" by a relatively small-bodied ancestral primate would hence explain the emphasis on the grasping foot characteristic throughout the order Primates and at the same time provide a reason for the emphasis on vision and replacement of the primitive prehensile function of the snout by mobile, grasping hands. (Leaping between adjacent fine branches and grasping of small animal prey on nearby supports with the hands would explain the relatively large eyes, the universal possession of a postorbital bar, and the reduction of the snout and anterior teeth among living primates.)

In a previous discussion of prosimian locomotion concerned primarily with the Madagascar lemurs (Martin, 1972), it was suggested (largely following Walker, 1967; Napier and Walker, 1967) that a major characteristic of primate locomotion throughout evolution has been *hindlimb domination*. This characteristic can probably be traced back to the ancestral primates, associated as it is with the grasping action of the hallux. This is not to say that the ancestral primates were "vertical clingers and leapers" (Napier, 1967) in the strict sense; merely that arboreal locomotion involved considerable emphasis on the hindlimbs. As demonstrated previously (Hall-Craggs, 1965; Walker, 1967; Martin, 1972), one of the most striking features identifiable in the primate foot is the relative elongation of the distal segment ("load arm") of the calcaneus compared to the proximal or heel segment ("lever arm"). This feature can be illustrated by calculating for each species a "calcaneal index" (expressed as a percentage) from the ratio of the lever arm to the load arm. It has already been demonstrated (Martin, 1972) that the length of the lever arm can be used as an indicator of body size. Hence, it is possible to plot calcaneal index against lever arm in order to display the variation of the index with body size (Fig. 6).

As noted previously (Walker, 1967; Martin, 1972), on this basis one can clearly distinguish the slow-climbing Lorisinae from the typically saltatory Galaginae and Tarsiidae, with the Madagascar lemurs exhibiting a generally intermediate condition. A qualification must be applied to this observation, however. Although the galagines typically exhibit saltatory motion, Bearder's detailed field study (1975; see also Charles-Dominique and Bearder, Chapter 13) has shown that *Galago crassicaudatus* in South Africa very rarely leap in the course of arboreal movement. In some ways, it would appear that *G. crassicaudatus* have developed behavior paralleling that of the Lorisinae (relatively slow arboreal locomotion; discretion rather than versatility in movement; see Charles-Dominique, 1977). This would, therefore, seem to be an exemplary case in which current behavioral adaptations are

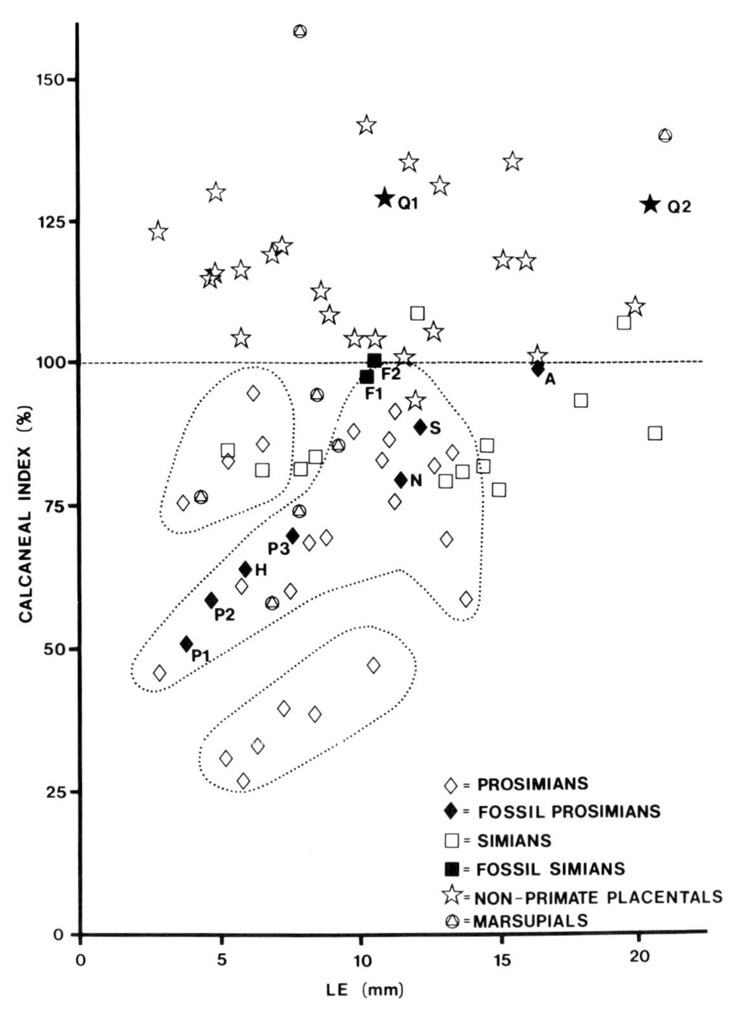

Fig. 6. Calcaneal index (%) plotted against length of heel section ("lever arm"—LE) for various primates and nonprimates. Virtually all of the primates have a calcaneal index below 100%; other placentals typically have a calcaneal index in excess of 100% (horizontal dashed line indicates demarcation). The irregular dotted contours indicate the overall ranges of variation for (top) extant Lorisinae, (middle) extant Madagascar lemurs, (bottom) extant Galaginae and *Tarsius*. Key to fossil forms (black symbols): P1, P2, P3, East African Miocene lorisids ($N = 3$); H, *Hemiacodon gracilis* ($N = 6$); N, *Notharctus osborni* ($N = 8$); S, *Smilodectes gracilis* ($N = 2$); F1, F2, Fayum primate calcanei ($N = 2$); A, *Archaeolemur edwardsi* ($N = 2$); Q1, Q2, Quercy phosphorite calcanei attributed to "*Adapis parisensis*" (CM2564) and "*Adapis magnus*" (CM2565) on size grounds ($N = 2$). (From Martin, 1972, incorporating new data.)

somewhat out of phase with certain morphological developments. If the calcaneus of *G. crassicaudatus* were known only from the fossil record, comparison with other galagines and with *Tarsius* would undoubtedly lead to the conclusion that its owner had been a typically saltatory species. One could not ask for a better demonstration of the need for caution in interpreting function in fossil species (see also Rudwick, 1964).

As a preliminary basis for comparison, Fig. 6 includes data for a number of representative nonprimate placental mammals covering the same range of body sizes. It can be seen that there is very little overlap with the prosimians, nor with the simians for that matter. As a general rule, it can be stated that—up to a body size of approximately 10 kg—living primates have a longer load arm than lever arm in the calcaneus (calcaneal index less than 100%), whereas the converse applies to living nonprimate placental mammals (calcaneal index in excess of 100%). For body sizes below 1 kg, this distinction is probably absolute. If one takes representative marsupials, on the other hand, there is a distinction between relatively small-bodied, primarily arboreal forms which typically possess grasping feet and specialized nonarboreal forms without a divergent, grasping hallux. The former overlap with the primates (points for *Didelphis, Dasyurus, Phascolarctos, Trichosurus,* and *Thylacomys* in Fig. 6), while the latter overlap with nonprimate placentals (points for *Chironectes* and *Macropus* in Fig. 6). The key to this almost certainly lies in the fact that all living primates (except *Homo*) and all living arboreal marsupials possess grasping feet with a divergent hallux. The marsupials, in fact, have been classically traced back to an arboreal common ancestor exhibiting this feature (see Martin, 1968). Relative elongation of the load arm of the calcaneus (= low calcaneal index) can be explained in terms of Morton's observation (1924) that, in any mammal species with a grasping foot, elongation of the foot primarily affects the tarsal bones, whereas in species with a nonprehensile foot, elongation affects the metatarsal bones. It can therefore be stated as a general principle for viviparous mammals (marsupials and placentals) that a low calcaneal index (less than 100%) is indicative of a grasping foot. From Fig. 6, an analysis of the species concerned indicates that calcaneal indices in excess of 100% are, by contrast, associated with nonprehensile feet.

Examination of calcanei attributed to various well-established fossil primate species (Fig. 6) shows that the characteristic low calcaneal index is also found in Eocene adapids and omomyids, Miocene lorisids, and Oligocene simians. In fact, directly associated skulls and skeletons have only been found for the North American adapids *Notharctus* and *Smilodectes,* but the low calcaneal index found with the other calcanei confirms their primate status. A certain amount of confusion has arisen, however, with respect to a number of calcanei from the Eocene Quercy phosphorites which have been attributed—exclusively on grounds of size—to *Adapis parisiensis* and *Adapis magnus*. Gregory (1920) figures a calcaneus, which he wisely referred to in inverted commas as "*Adapis parisiensis,*" showing quite different lever arm/load arm relationships to those shown by calcanei of *Pelycodus* and *Notharctus*. More recently, Decker and Szalay (1974) have figured and discussed two calcanei of very similar form from the Quercy phosphorites (Montauban collection), one attributed to *A. parisiensis* and the other to *A. magnus*. The author has located two comparable calcanei attributed to these species in the collection of the Carnegie Museum, Pittsburgh, and their calcaneal indices are plotted in Fig. 6. It can be seen that these calcanei fall right out of the range covered by all living primates (as was recognized by Decker and Szalay, 1974) and by all other early Tertiary (Eocene/Oligocene/Miocene) calcanei considered. In fact, Szalay (1975) has recently provided illustrations of calcanei attributed to *Teilhardina, ?Tetonius,* and an unidentified omomyid from the Middle Eocene of Wyoming, all of which are obviously comparable in calcaneal index to the calcaneus of *Hemiacodon*. Hence, it is obvious that either the Quercy calcanei attributed to *Adapis* have in fact been completely misidentified, or that the foot of *Adapis* was utterly different from that of contemporary notharctines and omomyids. It is, for instance, highly unlikely that the Quercy calcanei assigned to *Adapis* came from species with a grasping foot. In view of the extensive cranial similarities between adapines and notharctines, it seems more likely that the Quercy calcanei have been incorrectly identified, and it is certainly true that these calcanei are irrelevant to the main line of primate locomotor evolution. This being the case (pending any evidence to the contrary), the entire analysis of primate calcaneal evolution presented by Decker and Szalay (1974) is placed in question. In particular, the fact that a calcaneal index in excess of 100% is exhibited by certain Paleocene plesiadapoids (''paromomyiforms'') provides a further argument against regarding these forms as related to the ancestral primates, in the absence of the spurious ''connecting link'' provided by the Quercy calcanei.

In every other respect, analysis of the locomotor evidence clearly indicates that the prosimians, and probably the primates as a whole, can be traced back to an arboreal ancestor with a grasping foot exhibiting a low calcaneal index. Since all of the reliably identified early Tertiary prosimian calcanei fall into the range covered by the modern Madagascar lemurs, rather than resembling the calcanei of the Lorisinae, Galaginae, or Tarsiidae, it is among the lemurs that broad locomotor parallels to the ancestral condition should be sought. The likely body size range of the common ancestors of the lemurs and lorises is covered by the modern Cheirogaleidae, Lepilemuridae, and Lemuridae, all of which exhibit hindlimb domination in some degree for arboreal leaping. At the modal prosimian body size of 500 gm, a representative lemur species is *Cheirogaleus major,* the calcaneus of which is almost indistinguishable from those described by Walker (1970) for the Miocene lorisids of East Africa and the calcanei of the Eocene omomyid *Hemiacodon*. It should be emphasized that even among the Cheirogaleidae, which exhibit broadly uniform calcaneal morphology, there is a range of locomotor types, so that locomotor adaptation of the ancestral lemur/loris cannot be defined in greater detail on the basis of the evidence presented here. Nevertheless, it is virtually certain that the modern slow-climbing Lorisinae and the ''vertical clinging and leaping'' Galaginae and *Tarsius* have undergone divergent specialization away from an ancestral condition which possessed a potentiality for both climbing and leaping in an arboreal environment characterized by relatively fine supports.

5. THE NERVOUS SYSTEM

In a previous review of the adaptive radiation of the Madagascar lemurs (Martin, 1972), it was concluded that the lemurs and lorises could reasonably be traced back to a relatively small-bodied *nocturnal* ancestor (see also Martin, 1973b). One of the main arguments advanced to support this hypothesis was that, with very few exceptions, the lemurs (including some diurnal species) and lorises exhibit a reflecting *tapetum lucidum* behind the retina. Pirie (1959) had previously shown that the *tapetum* of *Galago crassicaudatus* incorporated riboflavin crystals and indicated ways in which the *tapetum* might assist nocturnal vision. It has since emerged (Alfieri *et al.*, 1974) that the *tapetum* in additional lemur and loris species (*Microcebus murinus, Hapalemur griseus, Perodicticus potto*) similarly incorporates riboflavin and the author has received confirmation of this for *Microcebus murinus* and *Lemur catta* from work conducted by Mr. Alan Williams (personal communication). The combined facts that the majority of living lemurs and lorises are nocturnal, that the *tapetum* is almost universal to lemurs and lorises, and that it has been found to incorporate riboflavin crystals in all species so far investigated, considerably strengthen the interpretation that the common ancestor of the lemurs and lorises was nocturnal. By contrast, neither *Tarsius* nor *Aotus* exhibit a *tapetum,* despite their being exclusively nocturnal.

A recent analysis by Charles-Dominique (1975) of the mammal and bird faunas of two separate tropical rain forest areas (one in West Africa; one in South America) has emphasized a relationship between body size and nocturnal habits in mammals. Up to a body weight of about 5 kg, above which there are severe constraints on bird flight, it is evident that diurnal life is essentially a property of birds while mammals in that weight range are typically nocturnal. Above 5 kg, mammals have very few bird competitors of comparable size, and the majority of the larger mammal species are active during the daytime. This distinction is apparent in the primates: The prosimians, which are typically relatively small in size, are predominantly nocturnal, whereas the typically larger-bodied simians are (with only one exception—*Aotus*) diurnal (Fig. 7). The modal body weight for prosimians is approximately 500 gm; that for simians is approximately ten times greater. Such relationships have presumably applied throughout the period in which birds and mammals have been evolving side-by-side, and it is therefore likely that a relatively small-bodied ancestor of the primates (or any primate subgroup) would have been nocturnal. Conversely, if any common ancestor was nocturnal, it was probably relatively small in size.

The division between the typically nocturnal prosimians and the typically diurnal simians is doubtless relevant to the question of differences in relative brain size.

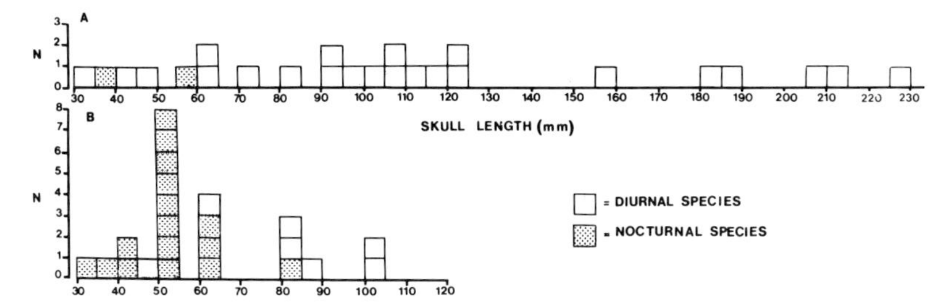

Fig. 7. Histograms showing relationship between body weight and nocturnal versus diurnal habits in living primates (weight data from Stephan *et al.*, 1970). A = simians; B = prosimians.

Thanks particularly to the efforts of Stephan and Bauchot (1965; see also Stephan, 1972; Jerison, 1973), a considerable amount of information is now available about relative brain size in primates. It has been confirmed that, for any major group of vertebrates, there is a standard allometric relationship which takes the form:

$$\log C = 0.67 \log B + b$$

where C is a measure of brain size, B a comparable measure of body size, and b is a constant. A simple interpretation of this formula is that, as body size increases in any vertebrate group, brain size increases to keep pace with the body's surface area, not its volume. Taking the data provided by Stephan *et al.* (1970), one can calculate reduced major axes on logarithmic coordinates for the following groups: (1) "basal" insectivores; (2) "progressive" insectivores; (3) prosimians; and (4) simians (Fig. 8).

It emerges that very similar slopes, all close to 0.67, are obtained for the four groups. [This had already been broadly established by Stephan (1972), but unfortunately he applied least squares regressions to the data.] It matters little to the overall result whether the tarsier is included with the lemurs and lorises (Stephan, 1972) or with the simians (Fig. 8). The important thing is that the simians, for any given body size, tend to have brains approximately three times larger than those of prosimians of comparable body size. This overall difference in relative brain size is probably attributable to the broad difference between diurnal habits in the simians and typically nocturnal habits in the prosimians. In the terminology used by Jerison (1973), one can say that the simians occupy an "adaptive zone" different from that occupied by the prosimians. If it is correct (as is widely assumed) that the ancestral placental mammals were nocturnal, and if (as the author believes) the ancestral primates were still nocturnal, then it would be reasonable to conclude that the larger average brain size of the simians is associated with a shift from nocturnal to diurnal habits. Such a shift would undoubtedly have been accompanied by an increase in average body weight in the ancestral forms concerned.

In order to present the data on relative brain size in a more directly understandable form, and to indicate the range of variation (with overlap!) found in the four

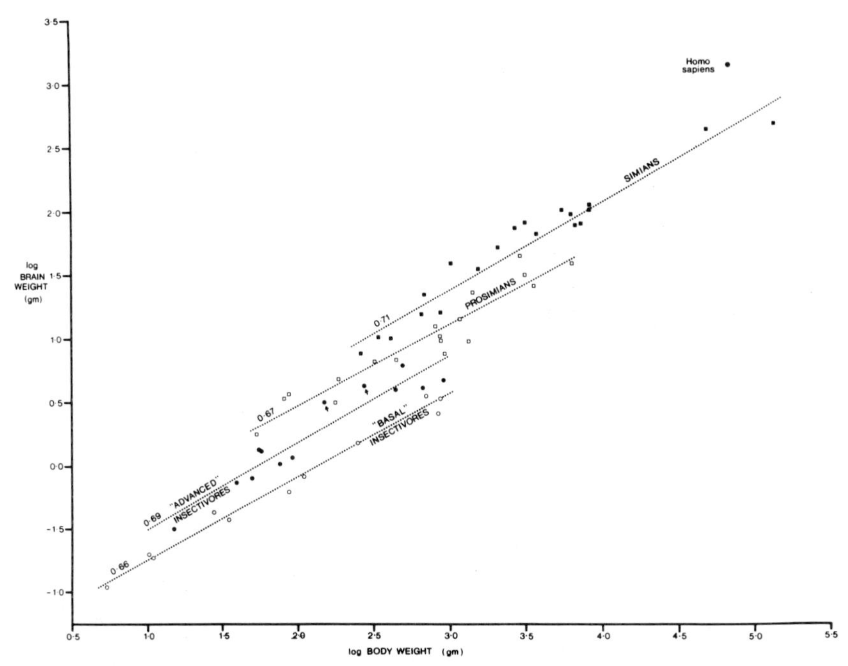

Fig. 8. Brain weight plotted against body weight on logarithmic coordinates for: "basal" insectivores; "progressive" insectivores; prosimians and simians (data from Stephan *et al.*, 1970). The reduced major axes are shown as dotted lines for each group, and the slope is indicated in each case.

mammalian groups considered, a simple transformation can be carried out. If it is assumed that the equation $\log C = 0.67 \log B + b$ represents a consistent overall relationship, one can "convert" the brain size of any given species to a "theoretical" (standardized) brain size at a standard body weight of, say, 1 kg. The following formula is used for the conversion:

$$\log S = 0.67\ (3.00 - \log B) + \log C$$

Where B = actual body weight, C = actual brain weight, and S = "standardized" brain weight. The standardized brain weights can then be plotted in the form of a histogram (Fig. 9). It is apparent that there is considerable variation within each group. Some prosimians have larger standardized brain weights than some simians, for example. However, two points are noteworthy. First, the diurnal tree shrew species (*Tupaia glis* and *Urogale everetti*) in fact overlap with essentially nocturnal prosimians, so a direct comparison of relative brain size between such tree shrews and the primates is not ecologically meaningful (see also Martin, 1968). Second, the nocturnal *Tarsius* and *Aotus* obviously occupy an intermediate position between lemurs and lorises and the simians, despite their nocturnal habits.

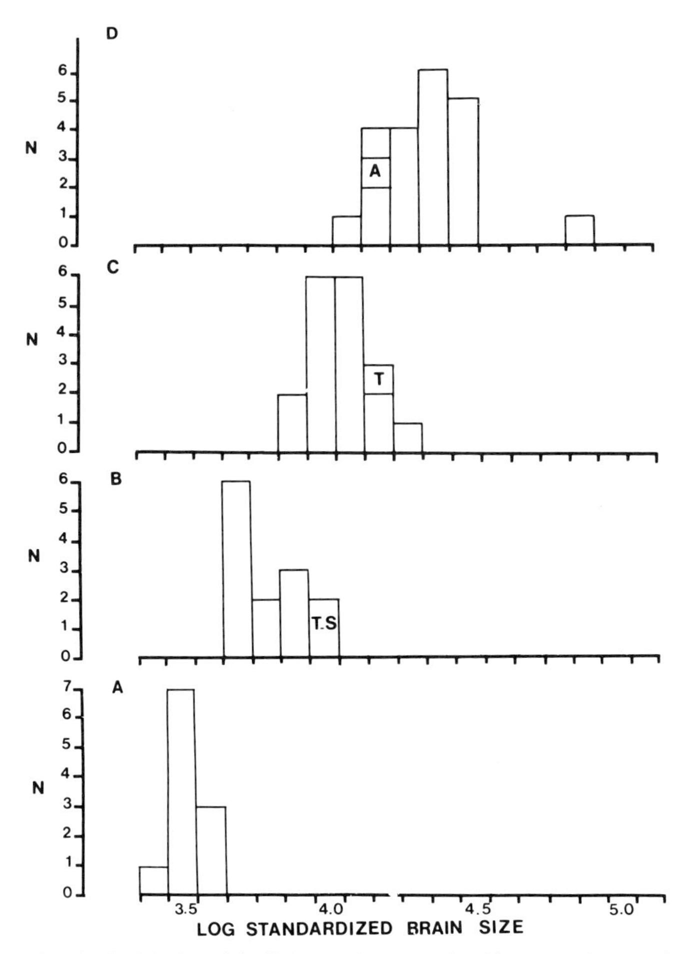

Fig. 9. "Standardized brain weights" (see text) presented as histograms for: A, "basal" insectivores; B, "progressive" insectivores; C, prosimians; D, simians. Letters on the bars of the histograms indicate species of particular interest (T.S. = tree shrews; *Tupaia glis* and *Urogale everetti*; T = *Tarsius syrichta*; A = *Aotus trivirgatus*).

It is possible to extend the analysis by considering the question of the relative size of the olfactory bulbs. There is a well-established myth in the literature (e.g., see Le Gros Clark, 1971) that the olfactory apparatus has been reduced in all primates in the course of their evolution. However, until recently there was no clear evidence that the olfactory apparatus in primates is absolutely smaller (at any given body size), rather than merely appearing inconspicuous in comparison with the predominant development of the visual apparatus and the cerebral hemispheres as a whole. One must also consider the possibility that the olfactory apparatus may have been expanded in the evolution of some other mammals, such as the insectivores (Martin, 1968, 1973a). In fact, if the volume of the olfactory bulb is plotted on logarithmic coordinates against body size (Fig. 10; see also Stephan, 1972; Jerison, 1973), it emerges that the olfactory bulbs in certain prosimian species are *not* significantly smaller than in living insectivores. The reduced major axis for all living insectivores (''basal'' and ''progressive'') has a high correlation coefficient ($r = 0.94$) indicating a consistent overall trend. The nocturnal lemurs and lorises generally fall within the range of variation of the insectivores, with the exception of *Lepilemur* and *Avahi* (both exhibiting a specialized folivorous habit, which is unusual in relatively small-bodied, nocturnal primates). The diurnal lemurs, on the other hand, have olfactory bulbs which are noticeably smaller (at any given body size) than in insectivores generally. In this, they approximate the diurnal simians. It is remarkable, however, that *Tarsius* and *Aotus* also have relatively reduced olfactory bulbs, despite their nocturnal (and at least partly insectivorous) habits, and lie at the upper limit of the simian distribution. This, along with other evidence such as the absence of a *tapetum* in *Tarsius* and *Aotus* (see also Martin, 1973b), indicates that these two primates may have secondarily returned to nocturnal habits.

The consensus of the evidence therefore indicates that nocturnal life involving at least some predation on small animals is a primitive feature for the lemurs and lorises, and possibly for the primates as a whole. The interpretation best fits the

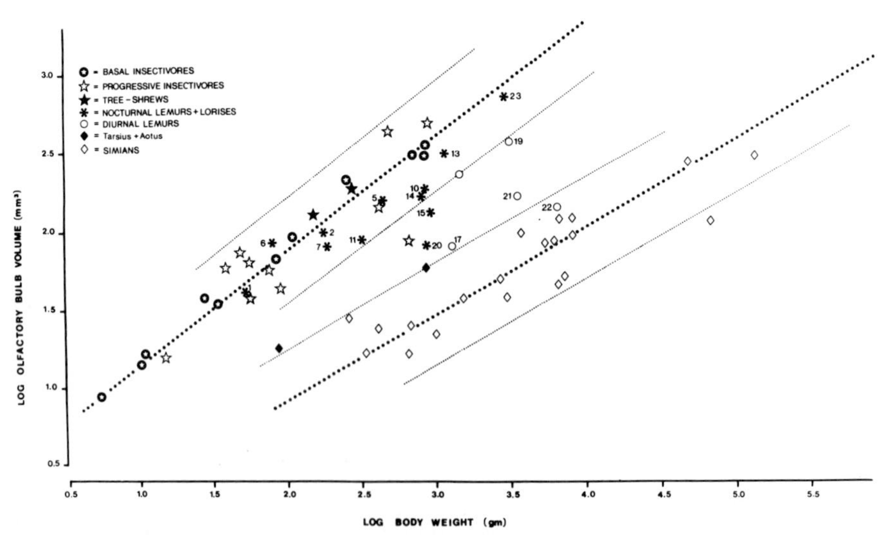

Fig. 10. Olfactory bulb volume plotted against body weight on logarithmic coordinates for: ''basal'' insectivores; ''progressive'' insectivores; prosimians; simians (data from Stephan *et al.*, 1970). The heavy dotted lines represent reduced major axes for (above) insectivores ($N = 22$) and (below) simians excluding man ($N = 20$). Light dotted lines indicate 2 standard deviations on either side of the reduced major axes (95% confidence limits).

observation that in modern lemurs and lorises with these habits, the olfactory bulbs generally show the same relative size as in the (predominantly nocturnal) insectivores and, incidentally, in representative late Mesozoic mammals (Jerison, 1973). Nocturnal habits are associated with relatively small body size, and it seems likely that the common ancestors of the lemurs and lorises (and possibly of the entire order Primates) were in the size range of 50–2500 gm. If the modal body weight of the modern nocturnal prosimians (with their relatively primitive life style) in any way reflects the ancestral primate condition, then an ancestral body weight of approximately 500 gm (comparable to that of modern *Cheirogaleus major* or *Loris tardigradus*) would be a suitably representative figure to take.

6. DISCUSSION

In examining the phylogeny of any group of animals, it is exceedingly difficult to avoid circular arguments, particularly since each statement has repercussions on all others. The foregoing account of some evolutionary aspects of body size, feeding behavior, locomotion, and the nervous system in prosimians should show how intimately these functional systems are linked. As stated at the outset, body size is of paramount importance in that many features of dentition, locomotor apparatus, and nervous system are limited by the constraints imposed by an animal's overall size. It is accordingly necessary to take scaling effects into account in discussing phylogenetic relationships between animals of differing body size. Indeed, analysis of scaling effects using allometric analysis (as illustrated in Figs. 3–10) can also provide additional information in terms of the empirically determined parameters of the allometric analysis and of deviations of individual points away from the central trend indicated by the reduced major axis. This applies to behavior (e.g., nocturnal versus diurnal habits; dietary habits) just as much as to morphology (e.g., tooth dimensions, calcaneal index, relative brain size).

Because of the intimate connection between functional systems and the circularity which can arise in discussing their evolution, the interpretations expressed in this chapter are open to challenge. The overall conclusion has been drawn that the common ancestors of the lemurs and lorises—and probably of the primates as a whole—were relatively small in body size. In terms of size and some of its morphological features the modern *Cheirogaleus major* (body weight approximately 500 gm) could well be taken as a model in the light of that conclusion. However, it has been argued (e.g., Szalay and Katz, 1973) that relatively large-bodied modern lemurs (e.g., *Lemur* spp.) provide the best model for considering the ancestry of the lemurs and lorises and, in particular, their relationship to the Eocene Adapidae. In theory, either of these two interpretations can be used to explain the major features of prosimian evolution. In practice, there are a number of clear indications that the common ancestor of the lemurs and lorises was relatively small in size.

1. The modal body weight of living prosimians is in the region of 500 gm. Species such as *Lemur catta* (approximate body weight: 1.5 kg) are somewhat unusual in exceeding this mode.

2. Nocturnal mammals are generally small-bodied. It seems likely that the ancestor of the lemurs and lorises was nocturnal; therefore it is also likely that it was small-bodied. Nocturnal ancestry for the lemurs and lorises is indicated by the almost universal possession of a reflecting *tapetum* (including diurnal species), probably incorporating riboflavin in all cases. In fact the majority of the extant prosimians are nocturnal. Analysis of the relative size of the olfactory bulb shows that, with few exceptions (notably *Tarsius*), nocturnal prosimians fall into the same range as the modern, typically nocturnal insectivores. Relative reduction of the olfactory bulb is characteristically found in diurnal lemurs and the diurnal simians. Further, diurnal lemurs have relatively larger brains than nocturnal lemurs and the typically diurnal simians have relatively larger brains than the typically nocturnal prosimians. Hence, diurnal life and the associated increase in relative brain size/decrease in relative olfactory lobe size can be regarded as a specialization on both lemurs and the simians.

3. The tooth-scraper can be interpreted as a dental modification for gum feeding and the characteristic locomotor adaptations of the primates as a group (grasping foot, prehensile hand, hindlimb domination) can be interpreted as an adaptation for movement in the "fine branch niche." Both of these developments are likely to have occurred in a relatively small-bodied ancestral form. Prosimian gum feeders are typically nocturnal and small-bodied, and fine branches most easily support an animal of limited body weight.

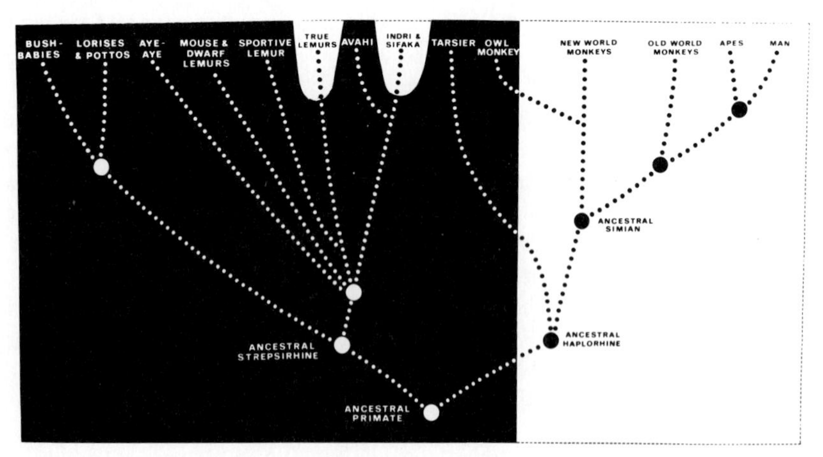

Fig. 11. The hypothetical tree shown in Fig. 1, indicating the division between nocturnal and diurnal life in living primates and suggesting how this division (in overall terms) might have influenced primate evolution.

As has been suggested previously (Martin, 1972, 1973a), the overall evidence indicates that the ancestral primates were small-bodied and nocturnal, while the prosimians largely retained these features. This may provide a basic explanation for the fact that prosimians generally exhibit a more primitive "grade" of evolution than the simians. However, it is also likely that *Tarsius* and *Aotus* are secondarily nocturnal forms derived from a diurnal ancestor which gave rise, on the one hand, to the tarsier and its relatives, and on the other to the simians. This overall interpretation is illustrated in Fig. 11.

ACKNOWLEDGMENTS

Thanks are due to Dr. June Rollinson for assistance with the measurements on lemurs and lorises used for Figs. 3–5, to my wife Anne-Elise for preparing most of the figures, to Mr. T. B. Dennett for photographic assistance, and to Miss Rachel Simper for efficient typing of the manuscript. Fossil and modern primate calcanei were examined and measured at the Peabody Museum, Yale University (thanks to Prof. E. L. Simons and Dr. G. Conroy), the Carnegie Museum, Pittsburgh (thanks to Dr. Mary Dawson and Dr. J. Schwartz), the Smithsonian Institution, Washington (thanks to Dr. C. L. Gazin), the Harvard University Museum of Comparative Zoology (thanks to Dr. A. C. Walker), and the British Museum of Natural History (thanks to Miss T. Molleson and Mrs. P. H. Napier).

Field work on lemurs and bushbabies, which has provided an essential background to the concepts developed in this chapter, has been consistently supported by the Royal Society (London).

REFERENCES

Alfieri, R., Pariente, G., and Sole, P. (1974). *Int. Soc. Clin. E.R.G. Symp.* **12,** 99–111.

Avis, V. (1961). *Am. J. Phys. Anthropol.* **19,** 55–61.

Bearder, S. K. (1969). M.Sc. Dissertation, University of the Witwatersrand, Johannesburg, South Africa (unpublished).

Bearder, S. K. (1975). Ph.D. Thesis, University of the Witwatersrand, Johannesburg, South Africa (unpublished).

Buettner-Janusch, J., and Andrew, R. J. (1962). *Am. J. Phys. Anthropol.* **20,** 127–129.

Cartmill, M. (1972). *In* "The Functional and Evolutionary Biology of Primates" (R. H. Tuttle, ed.), pp. 97–122. Aldine-Atherton, Chicago, Illinois.

Cartmill, M. (1974a). *In* "Prosimian Biology" (R. D. Martin, G. A. Doyle, and A. C. Walker, eds.), pp. 655–670. Duckworth, London.

Cartmill, M. (1974b). *Science* **184,** 436–443.

Cartmill, M. (1974c). *In* "Primate Locomotion" (F. A. Jenkins, Jr., ed.), pp. 45–83. Academic Press, New York.

Charles-Dominique, P. (1971). *Biol. Gabonica* **7,** 121–228.

Charles-Dominique, P. (1972). *Z. Tierpsychol., Beih,* **9,** 7–41.

Charles-Dominique, P. (1974). *In* "Prosimian Biology" (R. D. Martin, G. A. Doyle, and A. C. Walker, eds.), pp. 131–150. Duckworth, London.

Charles-Dominique, P. (1975). *In* "Phylogeny of the Primates: A Multidisciplinary Approach" (W. P. Luckett and F. S. Szalay, eds.), pp. 69–88. Plenum, New York.

Charles-Dominique, P. (1977). "Ecology and Behaviour of Nocturnal Prosimians." Duckworth, London.

Decker, R. L., and Szalay, F. S. (1974). *In* "Primate Locomotion" (F. A. Jenkins, Jr., ed.), pp. 261–291. Academic Press, New York.
de Lowther, F. (1940). *Zoologica (N.Y.)* **24,** 477–480.
D'Souza, F. (1974). *In* "Prosimian Biology" (R. D. Martin, G. A. Doyle, and A. C. Walker, eds.), pp. 167–182. Duckworth, London.
Every, R. D. (1974). *In* "Prosimian Biology" (R. D. Martin, G. A. Doyle, and A. C. Walker, eds.), pp. 579–619. Duckworth, London.
Gingerich, P. D. (1972). *Am. J. Phys. Anthropol.* **36,** 359–368.
Gingerich, P. D. (1974). *In* "Prosimian Biology" (R. D. Martin, G. A. Doyle, and A. C. Walker, eds.), pp. 531–541. Duckworth, London.
Gingerich, P. D. (1975). *In* "Lemur Biology" (I. Tattersall and R. W. Sussman, eds.), pp. 65–80. Plenum, New York.
Gould, S. J. (1966). *Biol. Rev. Cambridge Philos. Soc.* **41,** 587–640.
Gould, S. J. (1971). *Am. Nat.* **105,** 113–136.
Gould, S. J. (1975). *Contrib. Primatol.* **5,** 244–292.
Gregory, W. K. (1920). *Mem. Am. Mus. Nat. Hist.* [*NS.*] **3,** 49–243.
Hall-Craggs, E.C.B. (1965). *J. Anat.* **99,** 119–126.
Hardy, A. (1965). "The Living Stream." Collins, London.
Hiiemae, K., and Kay, R. F. (1972). *Nature (London)* **240,** 486–487.
Hutchinson, G. E. (1965). "The Ecological Theatre and the Evolutionary Play." Yale Univ. Press, New Haven, Connecticut.
Hylander, W. L. (1975). *Science* **189,** 1095–1098.
Jenkins, F. A., Jr., ed. (1974). "Primate Locomotion." Academic Press, New York.
Jerison, H. J. (1973). "Evolution of the Brain and Intelligence." Academic Press, New York.
Jolly, C. J. (1970a). *Man* **5,** 5–26.
Jolly, C. J. (1970b). *Man* **5,** 619–626.
Kay, R. F. (1975). *Am. J. Phys. Anthropol.* **43,** 195–215.
Kay, R. F., and Hiiemae, K. M. (1974a). *In* "Prosimian Biology" (R. D. Martin, G. A. Doyle, and A. C. Walker, eds.), pp. 501–530. Duckworth, London.
Kay, R. F., and Hiiemae, K. M. (1974b). *Am. J. Phys. Anthropol.* **40,** 227–256.
Kermack, K. A., and Haldane, J.B.S. (1950). *Biometrika* **37,** 30–41.
Kleiber, M. (1932). *Hilgardia* **6,** 315–353.
Le Gros Clark, W. E. (1971). "The Antecedents of Man." Edinburgh Univ. Press, Edinburgh.
Leutenegger, W. (1973). *Folia Primatol.* **20,** 280–293.
Lorenz, K. (1950). *Symp. Soc. Exp. Biol.* **4,** 221–254.
Lorenz, K. (1965). "Evolution and Modification of Behavior." Univ. of Chicago Press, Chicago, Illinois.
Martin, R. D. (1968). *Man* **3,** 377–401.
Martin, R. D. (1972). *Philos. Trans. R. Soc. London, Ser. B* **264,** 295–352.
Martin, R. D. (1973a). *Symp. Zool. Soc. London* **33,** 301–337.
Martin, R. D. (1973b). *In* "Comparative Ecology and Behaviour of Primates" (R. P. Michael and J. H. Crook, eds.), pp. 1–68. Academic Press, London.
Martin, R. D. (1974). *In* "The Biology of Brains" (W. B. Broughton, ed.), pp. 215–250. Blackwell, Oxford.
Martin, R. D. (1975). *In* "Phylogeny of the Primates: A Multidisciplinary Approach" (W. P. Luckett and F. S. Szalay, eds.), pp. 265–297. Plenum, New York.
Martin, R. D. (1978). *In* "Recent Advances in Primatology, Vol. 3: Evolution" (D. J. Chivers and K. A. Joysey, eds.), pp. 3–26. Academic Press, London.
Mayr, E. (1969). "Principles of Systematic Zoology." McGraw-Hill, New York.
Mills, J.R.E. (1955). *Dent. Pract.* **6,** 47–61.
Milton, K., and May, M. L. (1976). *Nature (London)* **259,** 459–462.

Morton, D. J. (1924). *Am. J. Phys. Anthropol.* **7,** 1–52.
Napier, J. R. (1967). *Am. J. Phys. Anthropol.* **27,** 333–342.
Napier, J. R., and Walker, A. C. (1967). *Folia Primatol.* **6,** 204–219.
Petter, J. J., Schilling, A., and Pariente, G. (1971). *Terre Vie* **25,** 287–327.
Pirie, A. (1959). *Nature (London)* **183,** 985.
Roberts, D. (1941). *J. Anat.* **75,** 236–238.
Romer, A. S. (1966). "Vertebrate Paleontology." Univ. of Chicago Press, Chicago, Illinois.
Rudwick, M. S. (1964). *Br. J. Philos. Sci.* **15,** 27–40.
Sacher, G. A. (1970). *In* "The Primate Brain" (C. R. Noback and W. Montagna, eds.), pp. 245–287. Appleton, New York.
Sacher, G. A., and Staffeldt, E. F. (1974). *Am. Nat.* **108,** 593–616.
Schmidt-Nielsen, K. (1970). "Animal Physiology." Prentice-Hall, Englewood Cliffs, New Jersey.
Schmidt-Nielsen, K. (1972). "How Animals Work." Cambridge Univ. Press, London and New York.
Seligsohn, D., and Szalay, F. S. (1974). *In* "Prosimian Biology" (R. D. Martin, G. A. Doyle, and A. C. Walker, eds.), pp. 543–561. Duckworth, London.
Simpson, G. G. (1950). "The Meaning of Evolution." Yale Univ. Press, New Haven, Connecticut.
Stein, R. M. (1936). *Am. Nat.* **70,** 19–28.
Stephan, H. (1972). *In* "The Functional and Evolutionary Biology of Primates" (R. Tuttle, ed.), pp. 155–174. Aldine-Atherton, Chicago, Illinois.
Stephan, H., and Bauchot, R. (1965). *Acta Zool. (Stockholm)* **46,** 1–23.
Stephan, H., Bauchot, R., and Andy, O. J. (1970). *In* "The Primate Brain" (C. R. Noback and W. Montagna, eds.), pp. 289–297. Appleton, New York.
Sussman, R. W. (1974). *In* "Prosimian Biology" (R. D. Martin, G. A. Doyle, and A. C. Walker, eds.), pp. 75–108. Duckworth, London.
Szalay, F. S. (1975). *In* "Phylogeny of the Primates: A Multidisciplinary Approach" (W. P. Luckett and F. S. Szalay, eds.), pp. 357–404. Plenum, New York.
Szalay, F. S., and Katz, C. C. (1973). *Folia Primatol.* **19,** 88–108.
Tattersall, I. (1974). *In* "Prosimian Biology" (R. D. Martin, G. A. Doyle, and A. C. Walker, eds.), pp. 563–577. Duckworth, London.
Tinbergen, N. (1951). "The Study of Instinct." Oxford Univ. Press (Clarendon), London and New York.
Walker, A. C. (1967). Ph.D. Thesis, University of London (unpublished).
Walker, A. C. (1969). *Uganda J.* **32,** 90–91.
Walker, A. C. (1970). *Am. J. Phys. Anthropol.* **33,** 249–261.
Walker, A. C. (1974). *In* "Prosimian Biology" (R. D. Martin, G. A. Doyle, and A. C. Walker, eds.), pp. 435–447. Duckworth, London.
Warwick James, W. (1960). "The Jaws and Teeth of Primates." Pitman, London.
Wilson, E. O. (1975). "Sociobiology: The New Synthesis." Belknap Press, Cambridge, Massachusetts.

Chapter 3

Reproductive Physiology and Behavior in Prosimians

RICHARD N. VAN HORN AND G. GRAY EATON

1. INTRODUCTION

What variables control sexual behavior poses a provocative issue to students of mammalian reproduction. Whether one is seeking appropriate animal models for the effective development of birth control technology or the knowledge to breed rare

ISBN 0-12-222150-8

and endangered species, the key factors are the endogenous and exogenous variables that control reproduction.

Because of the ancestral position of prosimians in primate phylogeny, a thorough knowledge of reproduction in this suborder is fundamental to the study of reproductive mechanisms within the order as a whole. In this chapter, we present a critical review of the literature on prosimian reproduction and discuss whether these data are relevant to the question of evolutionary trends in the photoperiodic and hormonal control of primate seasonality and sexual behavior. Doyle's (1974) thorough review of studies of prosimian reproduction greatly facilitated our compilation of reproductive measures. But unlike this earlier review, which was organized by topics, such as breeding season, this report presents the reproductive information within a taxonomic format. These studies varied considerably in content and thus prevented us from providing comprehensive descriptions of all prosimian species. Nevertheless, we sought to gather from them all available information about such fundamental reproductive measures as the seasonablity or aseasonality of the species, environmental correlates of breeding season, ovarian cycle length, the relation between gonadal hormones and sexual behavior, evidence of induced or spontaneous ovulation, length of gestation, litter size, postpartum estrus, courtship and copulatory behavior, and age of sexual maturation. While integrating these comparative data into a concise discussion of evolutionary trends in primate reproduction, we have emphasized our own research into the photoperiodic responses of captive *Lemur catta* (Van Horn, 1975) and into the hormonal control of sexual behavior of *Galago crassicaudatus crassicaudatus* (Eaton *et al.*, 1973b), *Macaca mulatta* (Eaton *et al.*, 1973a), *Macaca fuscata* (Eaton, 1972), and *Macaca nemestrina* (Eaton and Resko, 1974). These discussions also contain previously unpublished results from studies of sexual maturation in *G. crassicaudatus* and *L. catta* and experiments on the influence of partner preferences in female receptivity in *L. catta*.

2. TARSIIDAE

2.1. *Tarsius spectrum* (Celebesian tarsier), *T. syrichta* (Phillipines tarsier), and *T. bancanus* (Malaysian tarsier)

2.1.1. Seasonality

Zuckerman (1932) examined the seasonal distribution of pregnancies in 919 uteri collected from the Isle of Banka off the coast of Sumatra. Hubrecht collected females during all months of the year, the greatest number being collected in October and November (Zuckerman, 1932). Zuckerman (1932) compiled his analysis of seasonality in *Tarsius* from a catalogue prepared by de Lange (1921) of the embryological materials collected by Hubrecht. Although Zuckerman (1932), de Lange (1921), and van Herwerden (1925) had designated these specimens *T. spectrum,* in subsequent taxonomic analyses Hill (1953, 1955) classified them as *T. bancanus* (*contra* Asdell, 1964) since they had originated from the Isle of Banka.

The females in Hubrecht's collection showed pregnancies in every month and therefore provided no evidence of a restricted season of reproductive activity. Rather, the evidence suggested that this species breeds continuously throughout the year with little if any seasonal variation. Although van Herwerden (1925) had reported detecting a breeding season in Hubrecht's collection, Zuckerman's (1932) examination failed to corroborate this finding. In reality, van Herwerden's plot of the data across the year demonstrated a peak in October and November. However, he had plotted the absolute number rather than the percentage of pregnant females in the collection, and thus the peak probably reflects the larger samples of specimens of *T. bancanus* collected in those months.

Similarly, no convincing evidence of seasonal reproductive cycles has been reported for *T. syrichta*. Catchpole and Fulton (1943) followed the reproductive cycles of a single adult female over more than a 12-month period in New Haven, Connecticut. This female had been caught in the wild with her male infant on the island of Mindanao. In captivity, her estrous cycles were followed by vaginal smears and examinations of the external genitalia for swelling. This study verified the persistence of estrous cycles throughout most of the year with 12 cycles recorded between January, 1939, and March, 1940. The data did, however, reveal a definite period of quiescence from late August to early November. Ulmer (1963) suggests that this was not a period of seasonal quiescence but rather one of generally poor physical health since the female died 8 months later. Unfortunately, the lighting conditions under which this female was caged were not reported; thus it is not clear whether the year-round persistence of cycles occurred in constant day-length cycles of artificial light or in natural photoperiods at temperate latitudes. Another study, however, reported June births of full-term infants 2 years in a row to a female *T. syrichta* caged in natural light cycles at 40° N latitude in the Philadelphia Zoo (Ulmer, 1963). Since both infants died shortly after birth, the interval of one year cannot be explained by endogenous events associated with an extended period of nursing. This suggested that *T. syrichta* are aseasonal in tropical or constant artificial photoperiods and seasonal in temperate photoperiods. Under natural photoperiods in London, however, the discovery of sperm in the vaginal smear of a single female *T. syrichta* in August and the miscarriage of her 1-inch fetus in April suggest that parturition and conception both occur during the spring and summer months at temperate latitudes (Hill *et al.,* 1952).

Despite the reports of aseasonality by Zuckerman (1932) and Catchpole and Fulton (1943), Fogden (1974) concluded from the literature that tarsiers have a sharply defined breeding season with the onset of impregnations in October and births from January to March. This author also reported trapping a higher incidence of small individuals in March and July, a phenomenon that was interpreted to reflect the recent completion of an annual birth season. Although Fogden's (1974) data support this hypothesis, his sample is quite small with only two animals trapped in this lightweight range. These results are not, therefore, convincing evidence of a breeding season in this species. A similar problem limits the credibility of Le Gros Clark's (1924) data. On the basis of 8 specimens (5 pregnant females and 3 newborn) brought to him by native woodcutters, he concluded that tarsiers bred between October and December. However, it is unclear from his report whether specimens had been collected in all months and, consequently, the yearly profile of reproductive activity in this study remains obscure.

2.1.2. Ovarian Cycle

The ovarian cycle of tarsiers is known from studies of two specimens of *T. syrichta* captured on Mindanao. Cycles were identified in one female over a 1-year period by the examination of vaginal smears and by the external signs of estrous swelling (Catchpole and Fulton, 1943). The mean length of 12 cycles for this female was 23.5 ± 0.7 days. Vaginal smears indicated a period of cornification, which was characterized by the nearly complete loss of leukocytes that persisted no longer than 24 hours and was followed by a rapid reinvasion of leukocytes. The investigators observed no periodic appearance of erythrocytes in the vaginal smears.

Using genital swelling and changes in the male's behavior toward the female as their measures of estrus, Hill *et al.* (1952) examined cyclicity in one female *T. syrichta*. On the basis of 7 to 9 cycles of increased sexual activity, these authors reported a cycle length of approximately 28 days. They also reported the irregular occurrence of external bleeding in this female which, however, may have been due to an infection of the genital tract. No bleeding was detected at any stage of the reproductive cycle in a second female at the London Zoological Gardens, but after copulation and ejaculation, a vaginal plug formed and was extruded from the vagina a few days later (Hill *et al.*, 1952).

2.1.3. Sexual Behavior

Although neither Hill *et al.* (1952) nor Catchpole and Fulton (1943) observed copulation in their tarsiers, the former reported that an aroused male, chirping in a "birdlike" manner, chased after an estrous female, gripped her with both hands, and sniffed her genital region. Hill (1955) also stated that "copulation occurs at night and from the dorsal position."

Harrisson (1963) also observed an increase in genital examinations by the male as the female's cycle progressed toward behavioral estrus. In these examinations, the male approached the female from below. Both the male and female groomed themselves more often during the proestrous stage of the female's cycle and marked the environment with urine and feces with increasing frequency. During periods of sexual excitement, these marking behaviors by the male increased noticeably. A characteristic vocalization also occurred with increased frequency as the female approached sexual receptivity. During proestrus, both the male and female called occasionally with a "piercing-twittering 'chit-chit,' slightly higher and sharper than a squirrel's. These calls stand out because the tarsier is otherwise silent (except in distress)'' (see also Niemitz, Chapter 14). Harrisson (1963) observed one copulation in *Tarsius*. The precopulatory sequence included chases of the female by the male, bouts of vocalizations, and displays in which both the male and female suspended themselves by their hands and spread their legs. Prolonged periods of grooming intervened between these displays. When the female finally became receptive, copulation lasted for 2 minutes.

2.1.4. Gestation

Ulmer (1963) reported a gestation of about 6 months for a pregnancy in a single female. In this instance, however, the onset of pregnancy was estimated on the basis of a 6-gm increase in body weight between two weighings several months apart. The weight records provided by Hill *et al.* (1952) show that fluctuations in body weight of this magnitude occur frequently even in nonpregnant females and therefore cast some doubt on the validity of Ulmer's estimate of gestation in *T. syrichta*.

2.1.5. Litter Size

Tarsiers give birth to single offspring (Le Gros Clark, 1924; Hill, 1955). Observations by Catchpole and Fulton (1943) and Hill *et al.* (1952) on live births and a miscarriage also confirmed the absence of multiple births in *Tarsius*.

3. GALAGINAE

3.1. *Galago crassicaudatus*

3.1.1. Seasonality

In the natural habitat, Gérard (1932) observed *G. crassicaudatus* breeding during a restricted season near the end of May. Bearder (1975) observed a concentration of conceptions in the period June/July. Unlike *G. senegalensis* females, which experience a second conception during a postpartum estrus each season (Cooper, 1966; Bearder, 1969; Doyle *et al.*, 1971), *G. crassicaudatus* females are reported to conceive only once during the year in the wild. Bearder (1975) detected a single season of births in November for *G. c. umbrosus* from the northern Transvaal (N = 8 births). In Zambia, Ansell (1960) reported a birth season in August and September (Fig. 1).

Additional evidence of seasonality in *G. crassicaudatus* comes from the islands of Zanzibar and Pemba (approximately 5° to 6° S latitude), where Lumsden *et al.* (1955, cited in Haddow and Ellice, 1964) found that of 14 females only 1 was pregnant between July 13 and August 5, whereas of 13 females 7 were pregnant between August 12 and 18. In contrast, Cooper (1966) reported a bimodal breeding season among captive *G. c. pangienensis* females that were caged outdoors at the

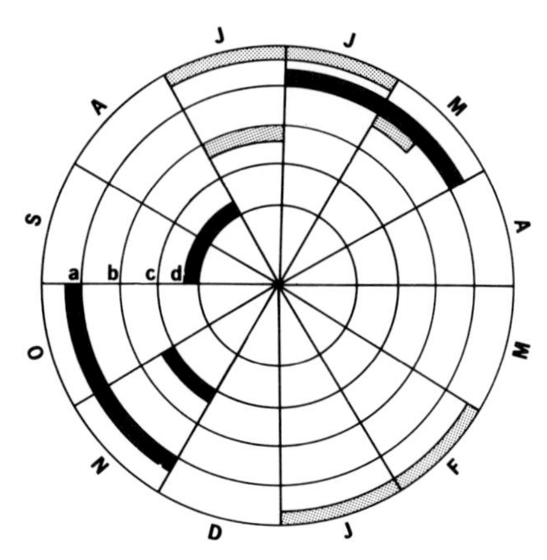

Fig. 1. Comparison of the reported seasons of matings (▨) and births (■) in *Galago crassicaudatus*. (a) San Diego Zoo (Cooper, 1966); (b) Zaire (Gérard, 1932); (c) Southern Africa (Bearder, 1975); (d) Zambia (Ansell, 1960).

San Diego Zoo. Over a 3-year period (1964–1966), these females experienced a "primary" breeding season with most conceptions concentrated in January and February. In 1965, 13 of 14 females (92.9%) experienced a "secondary" breeding season in June and July during a postpartum estrous cycle. Postpartum estrus occurred 6 to 8 weeks after delivery if the females nursed their infants and 2 to 4 weeks after delivery if the infants were taken from their mothers at birth. Cooper (1966) also reported a higher incidence of conceptions (N = 13) (92.9%) during postpartum estrous cycles (N = 14) if the infant was taken from the mother at birth. When the mother and infant remained together, however, the incidence of conceptions (N = 5) decreased to 35.7% during postpartum estrous cycles (N = 14).

In contrast to the seasonal nature of reproduction in natural or seminatural settings, captive *G. crassicaudatus* appear to breed all year round under some laboratory conditions (Buettner-Janusch, 1964). Eaton *et al.* (1973b), for example, observed that, under a regimen of constant dark-light cycles (12L:12D) of artificial illumination, reproductive cycles continued throughout the year. In contrast, Cooper's study (1966) suggests a seasonal cycle of reproduction in *G. crassicaudatus* living in natural light cycles in San Diego, California (ca. 33° N latitude); however, this study lacks appropriate statistical analysis. In this study, the greatest concentration of conceptions in the "primary" breeding season occurred during the short days of January and February, nearly half a year out of phase with the season of conceptions reported by Bearder (1975) in the Southern Hemisphere. Davis (1960) likewise recorded a birth in Washington, D.C., in July, nearly half a year out of phase with the season in the Southern Hemisphere (Fig. 1). One report that appears to contradict this photoperiodic interpretation originates from Vincent (1969), who estimated a mating season from November to June for *G. cras-*

sicaudatus on the basis of a study by Buettner-Janusch (1964). His estimate is, however, in error since Buettner-Janusch found no evidence of a breeding season in this or subsequent studies (J. Buettner-Janusch, personal communication).

3.1.2. Ovarian Cycle

Eaton *et al.* (1973b) monitored reproductive cycles in captive *G. crassicaudatus* ($N = 23$) through daily examination of vaginal smears and behavioral tests for sexual receptivity. Smears were initially classified as diestrous (string mucus and leukocytes mixed with clumped epithelial cells) or vaginal estrous (cornified epithelial cells and absence of leukocytes). During vaginal estrus, females were placed in the home cage of a male each day to detect sexual receptivity. Assays of plasma estradiol and progesterone were made from blood samples taken daily from 12 females throughout vaginal estrus and every fourth day during diestrus. All subjects were caged individually in a windowless room and were maintained on a reversed

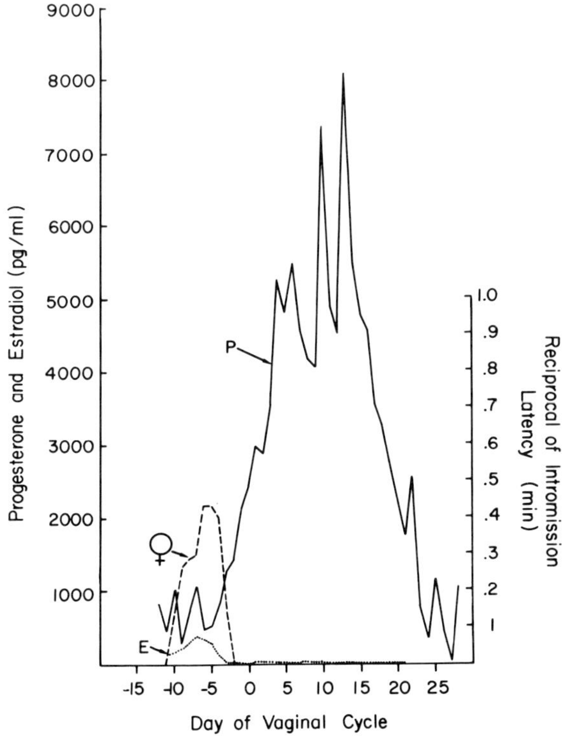

Fig. 2. Mean levels of galago (*G. crassicaudatus*) plasma progesterone (P) and estradiol (E) and an index of female sexual receptivity (reciprocal of intromission latency in minutes, ♀) across 12 females' vaginal cycles (estrogen data from 4 females). Plotted from the reappearance of leukocytes in the vaginal smears on day 0 (end of vaginal estrus, start of diestrus). (Reproduced from Eaton *et al.*, 1973b.)

light schedule (dim incandescent red light from 10.00 to 22.00 hours and white fluorescent light from 22.00 to 10.00 hours). Room termperature remained constant at 25°C.

Under these conditions, female *G. crassicaudatus* experienced a mean cycle length of 44 days (No. of cases = 10, range = 35–49 days, SD = 8.0 days). Vaginal estrus persisted for a mean duration of 12.4 days (no. of cases = 8, range = 7–24 days, SD = 3.7 days). During vaginal estrus, the external genitalia became red and swollen. In periods of sexual quiescence such as pregnancy, lactation, or diestrous phases of the cycle, a membrane covered the vaginal orifice. If vaginal smears were collected regularly, however, the vaginal orifice remained open through all stages of the reproductive cycle. Concentrations of estradiol began to rise 5 to 7 days before the females became receptive, stayed high through most of the 6-day period of behavioral estrus, and then dropped to almost nondetectable amounts 1 to 2 days before the period of receptivity ended. Progesterone began to rise as estradiol dropped and then peaked midway through the 24-day luteal phase (Fig. 2).

3.1.3. Sexual Behavior

In this same study, sexual receptivity among female *G. crassicaudatus* occurred only during vaginal estrus and appeared at a mean interval of 4.7 days (No. of cases = 28, range = 2–16 days, SD = 1.7 days) after the onset of vaginal estrus. Receptivity lasted a mean of 5.8 days (No. of cases = 28, range = 2–10 days, SD = 1.7 days) and leukocytes reappeared in the smears at a mean interval of 2.7 days (No. of cases = 28, range = 1–6 days, SD = 0.7 days) after the final day of receptivity (Fig. 2). During behavior tests, males typically approached with clucking vocalizations, sniffed the females, and then attempted to lick their genitalia. When receptive, the female crouched and diverted her tail; otherwise she withdrew and threatened. The males persisted, however, and these sequences were interspersed with mutual grooming that occurred more frequently when a female was not receptive (Table I). Typically, if the female did not threaten, the male attempted to mount and occasionally gently bit the female's back. Once the male had clasped the female's abdomen with his arms and her ankles with his feet in a dorsoventral position, he remained motionless for a few seconds and then dismounted or began to thrust. If penetration did not occur in 30–80 seconds, he would dismount and repeat the sequence until intromission was achieved. Then the rapid "searching" thrusts were replaced with slower and deeper thrusts that continued for 4–8 minutes. After the deep thrusts ended, ejaculation occurred. However, the pair maintained the intromission up to several hours during which the male thrusted 5 or 6 times about every 20 minutes. Whether he ejaculated again after these thrusts was not determined. During most tests of receptivity, however, the male–female pair was separated if the male achieved an intromission, and the authors point out that this method of testing may have extended the period of receptivity. Erythrocytes were rarely

TABLE I

Mating Behavior throughout the Female Estrous Cycle in *Galago crassicaudatus* (mean frequency/min)[a]

Behavior	Vaginal estrus	Behavioral estrus	Diestrus
Sex investigation (male)	0.97	1.46[b]	0.56[b]
Contact (male)	0.21	0.95[b]	0.18
Attempt mount (male)	0.43	0.30	0.03[b]
Clasp mount (male)	0.08	0.78[b]	0.01
Gentle bite (male)	0.04	0.22[b]	0.03
Groom (male)	0.39	0.19[b]	0.45
Groom (female)	0.27	0.11[b]	0.42

[a] Modified from Eaton *et al.* (1973b).
[b] Significant at 0.05.

observed and their occurrence in smears correlated with no particular stage of the ovarian cycle (Eaton *et al.*, 1973b).

3.1.4. Gestation

To ascertain the timing of ovulation and to determine the period of gestation, Eaton *et al.* (1973b) allowed ejaculations in 11 females on various days of behavioral estrus (Table II). Five of the six females who had received ejaculations on

TABLE II

The Effect of the Day of Mating on Conception in *Galago crassicaudatus*[a]

Female number	Day of receptivity mated	Gestation (days)
2730	2	130
2385	2	132
741	3	135
1487	3	131
2482	3	136
4585	1	No conception
740	3	No conception
733	4	No conception
3441	4	No conception
3229	10	No conception

[a]Reproduced from Eaton *et al.* (1973b).

day 2 or 3 of behavioral estrus conceived and gave birth to a single infant after a mean gestation of 132.8 days (SD = 2.6 days). Females that had received ejaculations on 1, 4, or 10 of behavioral estrus failed to conceive (Table II). Petter-Rousseaux (1962) cites a personal communication from Andrew, who reported a gestation of 130 and 135 days for two pregnancies in which both copulations and births were observed.

3.1.5. Litter Size

Among captive *G. c. crassicaudatus* at the Oregon Regional Primate Research Center, twin births have occurred in 9.3% of the pregnancies (*N* = 150 pregnancies; Table III); 30.3% of all parous females (*N* = 33) have had at least one set of twins.

In contrast, captive *G. c. argentatus* (melanistic variety, cf. Napier and Napier, 1967) in the Oregon colony have had a much higher incidence of multiple births with twins in 36.7% of all pregnancies (*N* = 30 pregnancies; Table III). In this group, three-fourths of the females (*N* = 8) had at least one twin pregnancy. However, this high incidence of twinning may reflect a sampling error rather than a subspecific difference since one female (No. 1124, Table III) and her female offspring account for 90.9% of the twinning in our colony. This observation suggests that the productivity of breeding programs in this species could be increased by systematic efforts to breed for twinning. Buettner-Janusch (1964) reported one instance of triplets and one of twins for a captive female *G. c. argentatus* (silver variety), an additional indication that this subspecies may experience multiple births more frequently than *G. c. crassicaudatus*.

3.1.6. Sexual Maturity

Buettner-Janusch (1964) reported a conception at 1 year of age in a female *G. c. crassicaudatus*. In the Oregon colony, female *G. c. crassicaudatus* (*N* = 20) first conceived at a mean age of 1.89 years (range = 0.99–4.58 years, SD = 0.85 years) and female *G. c. argentatus* (*N* = 5) first conceived at a mean age of 2.52 years (range = 1.45–3.83 years, SD = 0.97 years) (Table IV). These estimates are conservative since maturing females were not systematically paired with experienced males. On the contrary, females were often caged with inexperienced peer group males. Thus, these figures also reflect, in many cases, the sexual maturation of the males. The age at first conception could probably be lowered by pairing the prepubescent females with males of known breeding ability.

3.1.7. Postpartum Estrus

Eaton *et al.* (1973b) examined the vaginal cytology of 3 *G. crassicaudatus* females on two consecutive days after parturition and found no evidence of an immediate postpartum estrus such as that reported in *G. senegalensis* (Doyle *et al.*, 1971). Nevertheless, *G. crassicaudatus* females in the Oregon colony regularly display an estrous cycle within 1 to 2 weeks after being separated from their

TABLE III

Differences in the Incidence of Twins in *Galago c. crassicaudatus* (9.3%) and *G.c. argentatus* (36.7%)

G.c. crassicaudatus			*G.c. argentatus*		
Female number	Number of pregnancies	Number of twin pregnancies	Female number	Number of pregnancies	Number of twin pregnancies
732	1	0	1124	6	3
733	5	0	1711	5	2
734	10	0	2347	4	1
736	1	0	2348	7	1
739	11	1	2727	2	0
740	15	1	3203	3	3
741	9	1	5676	2	1
1302	6	3	6088	1	0
1304	6	0			
1487	9	1			
2186	2	0			
2385	5	0			
2482	8	2			
2532	7	2			
2730	7	1			
2743	2	0			
2745	5	0			
2747	1	0			
2749	2	0			
3175	3	1			
3278	3	0			
3434	6	0			
3441	2	1			
4528	1	0			
4867	5	0			
5300	5	0			
5576	3	0			
5673	1	0			
6133	3	0			
6149	2	0			
6465	1	0			
6558	2	0			
6588	1	0			
$N = 33$	150	14 (9.3% twins)	$N = 8$	30	11 (36.7% twins)

TABLE IV

The Age at First Conception in *G.c. crassicaudatus* and *G.c. argentatus* females.

Female number	Age at first conception in years
G.c. crassicaudatus	
1487	1.6
2186	1.0
2385	1.7
2482	1.3
2730	1.8
3175	2.5
3278	4.5
3434	1.6
3441	1.4
4528	2.7
4867	2.7
5300	1.4
5576	1.8
5673	2.9
6088	2.2
6133	1.1
6149	1.5
6465	1.5
6558	1.0
6588	1.4
$N = 20$	$\bar{X} = 1.9$
	SD = 0.85
G.c. argentatus	
1711	1.6
2347	1.4
2348	1.9
3203	3.8
5676	2.5
$N = 5$	$\bar{X} = 2.3$
	SD = 0.97

newborn or young nursing infants (Buettner-Janusch, 1964; personal observations of the authors). (See Section 3.1.1 for additional data on postpartum estrus.)

3.2. *Galago senegalensis*

3.2.1. *Seasonality*

Field studies of both northern and southern varieties of *G. senegalensis* generally support the hypothesis that females of this species experience at least two restricted periods of mating and two pregnancies per year. The data also suggest that the second conception in each bimodal season results from postpartum estrous cycles among females that had conceived some 4 months earlier in the first period of mating (Doyle, 1974). In Uganda (4° N latitude), for example, Haddow and Ellice (1964), who suspected a bimodal season among *G. senegalensis,* found 60% of their female sample ($N = 23$) lactating between February 17 and March 20 and 74% of another sample ($N = 43$) pregnant a few weeks later (May 7–June 11). Moreover, 10 of the females in the second sample (May–June) were also lactating while pregnant.

Butler (1967a) also reported indirect evidence of a bimodal breeding season in *G. s. senegalensis* females collected in the Nuba Mountains of the Sudan (10° to 12° N latitude). He reported that among 36 pregnant specimens the incidence of early embryonic stages declined gradually and that the incidence of late fetal stages increased concomitantly from December to March. This finding, which suggested that a discrete breeding season had ended in late November and early December, was corroborated by his observation (Butler, 1967b) of two captive females that displayed subsurface corpora lutea in early December and therefore had probably ovulated some 2 weeks earlier in November. Later studies of estrous cycles and births among captive females, however, indicated another period of mating in August that presumably led to births some time in November and December (Butler, 1967a). Accordingly, the pregnancies reported by him (Butler, 1967a,b) in December, January, and March could have resulted from postpartum estrous cycles after the births of infants that had been conceived in an August breeding season. However, the hypothesis of an August breeding season is based on observations of births and estrous cycles among captive animals that may not, therefore, be strictly comparable to data derived from wild populations. This distinction is especially important since the author provided no descriptions of lighting, temperature, and diet regimens among the captive females. He did, nevertheless, observe three instances of lactation and one case of ovulation among 3 wild females that had been captured in November (Butler, 1967a). These observations tend to support the hypothesis of postpartum estrous cycles in November and December, but other discrepancies in the data weaken the value of this interpretation. For example, the pregnant females collected in the field from December to March were not reported to have infants or to show signs of lactation despite the implication in these studies (Butler, 1967a,b) that these pregnancies had occurred during a postpartum breeding season. Furthermore, the gestation period (approximately 4 months) of *G. senegalensis* would permit three pregnancies per year if postpartum estrus occurred immediately after parturition (Doyle *et al.,* 1971). Because Butler (1967a) was unable to collect females from one location in all months of the year, however, his report fails to demonstrate convincingly that female *G. senegalensis* experience more than two breeding seasons per year and that a restricted August breeding season occurs in the natural habitat. Nevertheless, Butler (1967a) did report convincing evidence that females can experience estrous cycles during lactation. A lactating female that died during capture had a recently ruptured Graafian follicle and displayed histological signs of estrus in the uterus and vagina.

In the southern variety of *G. senegalensis,* Bearder (1969) observed a bimodal cycle of reproductive activity similar to that reported for the northern variety. Female *G. senegalensis moholi* delivered first in October and early November and again in late January to early February (N = 33 infants observed) (Fig. 3) with a much higher proportion of births in the second season (Doyle *et al.,* 1971). Although two births at 4-month intervals in two different females confirmed Bearder's observation of postpartum estrous cycles, his field reports of matings in September, a month before the onset of births in October (Fig. 3), contradicts the hypothesis that the second breeding season resulted solely from postpartum estrous cycles. Despite these contradictory observations from the field, Doyle *et al.* (1971) also verified postpartum estrous cycles among captive *G. senegalensis moholi.* In a sample of 12 captive females, 4 animals experienced postpartum estrous cycles. Thus, 7 of 29 births among the 12 female subjects were followed within 24 hours by postpartum cycles. Two of these cycles occurred after the death of the newborn infants, but the remainder occurred among females that successfully reared their offspring (Doyle *et al.,* 1971). In this study, however, it was not clear whether breeding was seasonally restricted. The authors suggest that the absence of seasonal cycles of temperature, photoperiods, and diet caused the breakdown of seasonality among the laboratory animals (Doyle *et al.,* 1971; Doyle, 1974).

This interpretation is corroborated by the observations of Cooper (1966) who studied the seasonal distribution of births among captive *G. senegalensis braccatus* (captured in Kenya at about 2° to 3° S latitude) in an outdoor colony at the San Diego Zoo (about 33 ° N latitude). Cooper (1966) recorded the monthly incidence of 30 births among 11 females during a 4-year period (1963–1966) and thereby pro-

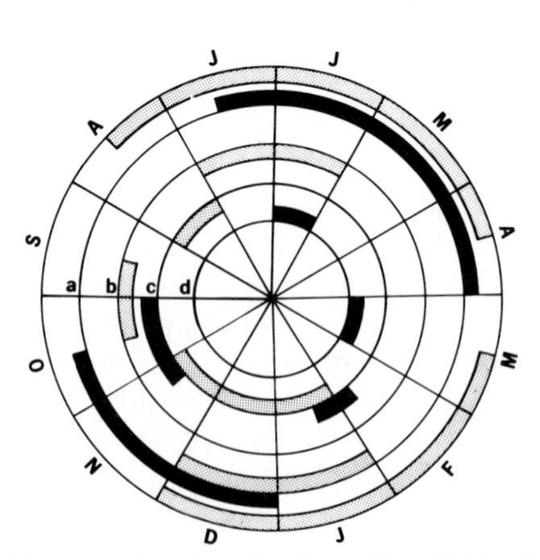

Fig. 3. Comparison of the reported seasons of matings (▩) and births (■) in *Galago senegalensis.* (a) San Diego Zoo (Cooper, 1966); (b) London Zoo (Zuckerman, 1932); (c) Southern Africa (Bearder, 1969); (d) Sudan (Butler, 1967a, b).

vided one of the more detailed accounts of seasonality and postpartum estrous cycles in *G. senegalensis*.

According to the cumulative birth records covering 4 years at the San Diego Zoo, a bimodal season with a "primary" period of mating occurred between December and mid-March (Fig. 3). During this "primary" season, each female that had failed to conceive experienced repeated estrous cycles at approximately 30-day intervals over a period of 3–4 months. Shortly after the delivery of the infants that had been conceived in this "primary" season, about half of the females (5 of 11) experienced a "secondary" breeding season. Both impregnated and unimpregnated females, however, experienced only one estrous cycle after parturition in the "secondary" season.

Unlike Doyle *et al.* (1971), who reported an immediate postpartum estrus, Cooper (1966) found postpartum estrous cycles occurring 2–4 weeks after delivery; Manley (1966a) also reported postpartum estrous cycles in three instances in which females were not separated from their newborn infants. In addition, Cooper's (1966) study also demonstrated that many females (5 of 11) did not conceive during the postpartum period. These two factors—the 2- to 4-week latency to a postpartum estrus and the absence of postpartum conceptions in many females—explain why only two pregnancies per year occur in a species whose gestation period (approximately 4 months) would accommodate three conceptions per year.

Although *G. senegalensis* females in both northern and southern latitudes appear to experience more than one conception, the available data on the timing and degree of restriction of the breeding seasons suggest a complex relation between annual photoperiod cycles and reproductive activity. For example, females in San Diego at approximately 33° N latitude experienced two restricted periods of mating each year, the onset of the "primary" breeding season occurring during the shortest days in December and January (Fig. 3). The fact that this season persisted for several years suggests that photoperiods control the onset of the "primary" season. De Lowther (1940) also reported the onset of estrous cycles in a captive *G. senegalensis moholi* at ca. 41° N latitude. Estrous cycles in this animal usually began in December and January over a 3-year period. In South Africa, on the other hand, the period of mating that contains no births (presumably the "primary" season) begins in June (Fig. 3), 6 months out of phase with the "primary" season at northern latitudes, that is, in both hemispheres the "primary" breeding season begins during the shortest days of the year. Zuckerman (1932) reported a similar finding in captive *G. senegalensis moholi* in the London Zoo where most conceptions were concentrated in December and January with a smaller peak 4 months later in April. Unlike the studies of Cooper (1966), Bearder (1969), and Zuckerman (1932), which support a photoperiodic interpretation, Butler (1967a,b) reports the onset of a bimodal season in August at 10°–12° N latitude in Uganda. Whether Butler's (1967a,b) preliminary observations of seasonality in captive animals can be validly applied to populations and whether subspecific variability (Doyle *et al.*, 1971) or reduced environmental variability can account for the breakdown of a restricted breeding season in equatorial populations must be determined by additional studies in the field and laboratory.

3.2.2. Ovarian Cycle

During seasonal periods of sexual quiescence or the diestrous phase of the ovarian cycle, the vulva of the female *G. senegalensis* is imperforate. Around the time of ovulation, however, the vaginal orifice opens widely and the labia and clitoris become red and inflated (Butler, 1967b). In the *G. senegalensis senegalensis* females studied by Butler (1967b), this swelling persisted for 24–48 hours and the epithelial lining of the vagina developed a shiny white appearance throughout estrus. Histological sections of the vagina revealed a lining epithelium 0.3 mm thick with the superficial half composed entirely of cornified epithelial cells in the estrous female. At this stage of the ovarian cycle vaginal smears contained fully cornified epithelial cells but no leukocytes. Although the period of vaginal estrus (fully cornified smears, white shiny appearance of vaginal epithelium, and swollen labia and clitoris) continued for as long as 1–7 days (de Lowther, 1940; Sauer and Sauer, 1963), behavioral estrus (sexual receptivity) persisted for only 1–3 days within this period (Butler, 1967b; Doyle *et al.*, 1971). In Butler's (1967b) study, approximately 2 weeks after the vagina had opened, leukocytes reappeared and the cornified epithelial cells gradually disappeared from vaginal smears and were replaced by noncornified epithelial cells with large nuclei. Butler (1967b) examined the ovaries of two females 14 days after the vaginal orifice had opened and found cornified epithelial cells and leukocytes in the vaginal smears of both females. The vaginal orifices of these animals had not yet closed, but their labia and clitorises were no longer red and swollen. The ovaries of each female had a subsurface corpus luteum (770 and 500 μm in diameter) and Graafian follicles up to 700 μm in diameter. In the ovary of a third female, Butler (1967b) reported a large pedunculate corpus luteum (1350 μm in diameter) 29 days after the vaginal orifice had opened. Most of the luteal cells appeared morphologically normal with only minor signs of atresia. In this female, which had received an ejaculation 26 days earlier, leukocytes had reappeared in vaginal smears 15 days after copulation. By 26 days after copulation, the vaginal smears contained leukocytes and noncornified epithelial cells. In addition to the large corpus luteum, both ovaries contained Graafian follicles up to 700 μm in diameter with large antra and no signs of atresia. A vaginal plug formed within the vaginal canal after ejaculation.

Estimates of cycle length in this species vary according to whatever facet of the reproductive cycle is being measured (Table V). Estimates of cycle length based on changes in the external genitalia of 4 females with 31 estrous cycles yielded a mean periodicity of 31.7 days with a range of 19–31 days (Manley, 1966a). De Lowther (1940) used changes in the external genitalia and sexual behavior of one female to measure estrus and estimated the cycle length at 43.5 days with a range of 43–45 days. On the basis of vaginal smears in a single female, Petter-Rousseaux (1962) reported a mean cycle length of 39.0 days with a range of 36–42 days.

TABLE V

Periodicity of Estrous Cycles in *G. senegalensis*

Measure	$\bar{X}$ period (days)	Range (days)	Sample		Species	Author
			No. females	No. cycles		
External genitalia	31.7	19–31	4	31	*G. senegalensis*	Manley, 1966a
External genitalia and sexual behavior	43.5	43–45	1	2	*G. senegalensis moholi*	de Lowther, 1940
Vaginal histology	39.0	36–42	1	3	*G. senegalensis*	Petter-Rousseaux, 1962
Sexual behavior and external genitalia	ca. 30.0	Unreported	Unreported	Unreported	*G. senegalensis braccatus*	Cooper, 1966

Doyle *et al.* (1971) reported subspecies differences in several measures of reproductive activity for *G. senegalensis,* but some of the variability in the recorded estimates of cycle length may well be due to the lack of uniformity in the attributes measured.

3.2.3. Sexual Behavior

Doyle *et al.* (1967) reported 5 cases of mating and courtship behavior in 3 *G. senegalensis moholi*. Both male and female *G. senegalensis* engaged in "urine-marking" throughout the female's ovarian cycle. Males and females marked in the same way, but males displayed a higher incidence. The urine was deposited in the hand cupped under the urethra, and the foot on the same side of the body was then wiped 1 to 4 times in the urine. Frequently, the behavior was repeated on the opposite side.

During the anestrous or diestrous phases of a female's cycle, the male characteristically approached the female from the front and then examined her genitalia. Each examination of a sexually inactive female was accompanied or followed by urine-marking by the male who then rarely continued to investigate or to mount her. The quiescent female in turn usually jumped away from the male before contact could occur. Occasionally, a brief chase followed before the male lost interest. At times, especially during late pregnancy, the female turned directly toward the male and struck his face with both hands.

In contrast, the behavior of a male toward an estrous female showed marked changes: the number of approaches increased and urine marking was sometimes 30% higher than when the female was quiescent (Doyle *et al.,* 1967). Since both changes occurred just after the vaginal membrane had opened, the authors suggested that the white vaginal discharge associated with the opening aroused the male through olfactory mechanisms.

If two or more males were present, the dominant male drove the subordinate males away from a sexually attractive female. As the dominant male approached her, she allowed him to make brief contact before leaping away. The male pursued immediately, making low "clucking" vocalizations as he did so. This sequence was repeated 3 or 4 times before the female allowed the male to mount; before mounting, the male occasionally licked the female's genitalia. The mount began with the male approaching the female from behind and grasping her lower waist with his forearms, his thighs spread widely to the sides and his feet flat on the supportive surface. Occasionally a double-foot-clasp mount was observed. The male's eyes remained half closed, his back was fully arched, and his chin was pressed into the nape of the female's neck, but he neither bit nor gripped the neck fur. During the mount, the male vocalized with a loud call that ended in a whistle.

Unlike the male, the female showed no increase in urine-marking frequency during estrus. During copulation, she crouched low and usually looked straight ahead with her eyes wide open but emitted no vocalizations. Sometimes, however, she appeared to resist the male by turning her head and attempting to bite him. The male, however, countered this attack by pressing harder on her neck with his chin. Occasionally, the female escaped from the mount and was pursued immediately by the male, which attempted to mount again. This sequence of mounts and escapes sometimes occurred 3 or 4 times in rapid succession before a successful mount was achieved.

Copulatory behavior consisted of alternating periods of chasing and short mounts (10–50 seconds) interspersed with sequences of longer mounts (up to 450 seconds) which were separated by periods of up to 30 minutes. For example, on the first day of estrus one female was mounted 22 times, with 5 mounts that lasted more than 30 seconds. On the second day of estrus, she was mounted only 3 times, but the mounts lasted from 277 to 380 seconds. During a postpartum estrus, this same female was mounted 12 times with mounts that lasted from 60 to 460 seconds. In the long periods between mounts, the male and female showed no sexual interest in each other. However, the dominant male continued to drive away subordinate males if they approached the female between mounts.

Intromission and ejaculation probably occurred only during the longer mounts when the female did not attempt to escape and when the male and female rested during the final two minutes of the mount sequence (Doyle *et al.*, 1967). In this final stage of copulation, the male maintained a relaxed hold on the female and remained motionless. After ejaculation, both male and female disengaged and groomed themselves.

In group situations, there is some evidence that the dominant male and an estrous female *G. senegalensis* formed a monogamous pair, both of which drove subordinate males away. Anestrous females, on the other hand, were permitted to approach the copulating pair and even to groom one or both of them without interrupting the copulation. Doyle *et al.* (1967) observed attempts by subordinate males to disrupt a copulation by approaching the pair. In response to such approaches, the female lunged and bit at the intruders. The authors concluded that at this stage the female is monogamous and would not mate with a subordinate male even if the dominant male were removed. If this observation could be verified in controlled experiments, it would demonstrate that in the control of sexual behavior social variables can override hormonal variables even at the prosimian level of neural organization.

3.2.4. Gestation

De Lowther (1940), who measured gestation in a single pregnancy from mating to delivery, reported a value of 120 days. Manley (1966a), however, who measured gestation in 4 pregnancies among *G. senegalensis senegalensis* from the day of ejaculation to the day before the infant was discovered with the mother, found a gestational period of 145 days (1.15 SD). Finally, Doyle *et al.* (1971) reported a mean gestation of 123.5 days (range 121–124 days) in 13 pregnancies among 7 female *G. senegalensis moholi*.

Butler (1960) reported that in early pregnancy the corpus luteum reaches a maximum diameter of 1.5–2.0 mm that decreases to 1.0 mm by the time the embryo has grown to a crown–rump (CR) length of 4 mm. By the time the fetus measures 12.0 mm CR length, that is, near the end of the first trimester of pregnancy, the corpus luteum is replaced by a corpus albicans that lacks luteal cells and is only 0.5 mm in diameter. When the fetus is 20 mm CR length, the corpus albicans is still clearly visible. Since accessory corpora lutea were not found in the ovaries of pregnant females, Butler (1960) concluded that after the first trimester pregnancy is maintained by the placenta.

3.2.5. Litter Size

The incidence of multiple births varies among the subspecies of *G. senegalensis.* In captive *G. s. braccatus* and in *G. s. senegalensis* captured in the wild, Cooper (1966) reported predominantly single births with only rare exceptions. Doyle *et al.* (1971), on the other hand, recorded 16 sets of twins and one set of triplets in 29 pregnancies among 12 captive *G. senegalensis moholi.*

3.2.6. Sexual Maturity

Both male and female *G. senegalensis* reach sexual maturity well before 1 year of age (Doyle, Chapter 5). Manley (1966b) observed estrous changes in the external genitalia of two females at 196 and 209 days of age. One of these females copulated and conceived during her third estrous cycle at 257 days of age. Manley (1966b) also observed mounting behavior in two young males at 153 and 183 days of age. These observations were confirmed by Doyle *et al.* (1967), who reported a mating between a young female (201 days of age) and a young male (326 days of age). Doyle *et al.* (1967) saw no difference between the mating behavior of young animals and that of experienced animals.

3.2.7. Postpartum Estrus

According to various studies, *G. senegalensis* experience a postpartum estrus, but in some subspecies the conditions under which it occurs and the latency between parturition and estrus differ (see discussion in Section 3.2.1 for details).

4. LORISINAE

4.1. *Loris tardigradus*

4.1.1. Seasonality

Studies of *L. tardigradus* in their natural habitat contain conflicting reports on the timing of the breeding season. Rao (1927; 1932, cited in Ramaswami and Anand Kumar, 1965) described a biphasic seasonality with estrous cycles and mating beginning in April and May. In the first breeding period, unimpregnated females experienced two estrous cycles with a periodicity of about 2 weeks. Females who failed to conceive during the April–May mating period became sexually receptive again 6 months later in October and November. In this second period, unimpregnated females again experienced only two estrous cycles.

Ramaswami and Anand Kumar (1965) also observed a biphasic breeding season in the slender loris. Over a period of 2 years, they examined the vaginal smears of 170 animals and reported vaginal cornification first in June–July and then again in September–November (Fig. 4). However, the presentation of their data on the incidence of estrous cycles is too general to provide compelling evidence for their conclusions.

In a study of a single captive female under a constant day length regimen (13L:11D), Manley (1966a) reported two estrous cycles separated by an anestrous period of 164 days. If such a biannual rhythm could be verified in studies of larger samples, the findings would suggest that in this species endogenous rhythms regulate the biphasic rhythm of seasonality. An alternative explanation of the biphasic breeding season might involve the mechanism of postpartum estrous cycles like those in the genus *Galago*. This is an appealing alternative since the length of gestation of this species ($\bar{X}$ = 167 days) approximates the 6 months between the two mating periods reported by Rao (1927) (Fig. 4). The data provided by Ramaswami and Anand Kumar (1962) conflict with this hypothesis, however, since in their study the interval between the first and second periods of mating is much shorter than the gestation period of this species (Fig. 4). Furthermore, there is no evidence that postpartum estrous cycles occur frequently enough in *L. tardigradus* to account for a second mating period each year. In fact, Ramaswami and Anand Kumar (1965) have suggested that *L. tardigradus* experience a second mating period each year only if they fail to conceive during the first mating period.

Although female *L. tardigradus* appeared to breed during restricted seasons, spermatogenesis continued throughout the year in a large sample (N = 151) of male *L. tardigradus* (Ramakrishna and Prasad, 1967).

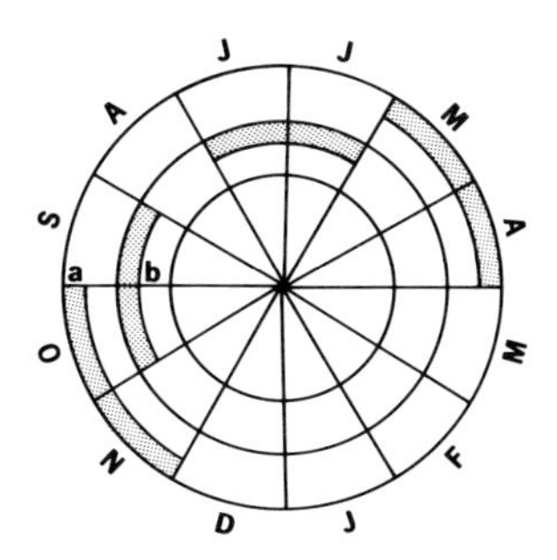

Fig. 4. Comparison of the reported season of matings (▨) in *Loris tardigradus*. (a) Rao (1927, 1932); (b) Ramaswami and Anand Kumar (1965).

4.1.2. Ovarian Cycle

No studies of vaginal cytology or sexual behavior have precisely defined the periodicity of estrous cycles in *L. tardigradus*. Manley (1966a), however, reported irregular intervals of 6, 11, 18, 40, 48, and 51 days between adjacent cycles and periods of 6 and 18 days between instances of "full" estrus. The remaining intervals occurred between periods of "lesser" estrous changes. It is unclear whether the criterion of "full" estrus was maximal turgescence of the external genitalia or the appearance of sexual receptivity.

Rao (1927) described a proestrous period of 7–10 days, during which the external genitalia became swollen and a sanguineous discharge appeared at the vaginal orifice. At the end of proestrus, the female was reported to have become receptive and to have remained in estrus for 1 week. However, Manley (1967) reported that *L. tardigradus* females copulate only once during each estrous cycle. After an ejaculation, a vaginal plug forms (Ramaswami and Anand Kumar, 1965; Manley, 1967). Ramaswami and Anand Kumar (1965) verified the enlargement of the external genitalia around the time of ovulation but did not observe a sanguineous discharge. They also established that cornified epithelial cells in vaginal smears coincide with turgescence of the external genitalia. In their study, sexual receptivity occurred only in females with fully cornified vaginal smears (i.e., complete absence of leukocytes). During anestrus, pregnancy, or lactation, on the other hand, vaginal smears consisted predominantly of leukocytes, and the ovaries contained Graafian follicles of minimal development. Females with both leukocytes and nucleated epithelial cells in their vaginal smears had large numbers of mature Graafian follicles and were not sexually receptive (Ramaswami and Anand Kumar, 1965).

To determine whether ovulation was spontaneous or induced, the authors prevented mating in a female with a fully cornified vaginal smear. An autopsy several days later revealed the early formation of a corpus luteum despite the absence of copulation.

4.1.3. Gestation

Counting from the day of mating to the day before parturition inclusive, Nicholls (1939) reported a gestation of 174 days in one pregnancy. Manley (1966a) measured two pregnancies in the same manner and obtained values of 160 and 166 days. The three values for gestation in *L. tardigradus* yield a mean gestation of 167 days ($\pm$ 7.02 SD).

4.1.4. Litter Size

In a sample of 87 gravid uteri, Ramaswami and Anand Kumar (1965) found twin fetuses in 56% of the cases and single fetuses in the remainder. The authors also suggested that the twins were biovular since two corpora lutea were found in all cases.

4.1.5. Postpartum Estrus

Manley (1966a) observed a female *L. tardigradus* copulating 5 days after the death of her newborn infant. In a subsequent pregnancy, however, the same female successfully raised her infant and experienced no estrous cycle until 7 weeks after parturition. Similarly, Ramaswami and Anand Kumar (1965) found no evidence of vaginal cornification or pregnancy in their histological examinations of 41 lactating females. One lactating female did, however, provide circumstantial evidence of a postpartum estrus since her ovaries contained one fully developed and one recently formed corpus luteum.

4.2. *Arctocebus calabarensis*

4.2.1. Seasonality

In their study of seasonality in *A. calabarensis,* Jewell and Oates (1969a,b) examined 7 adult females and 14 infants. The females were examined for signs of lactation, ovulation, or pregnancy. They estimated the birth dates of the infants by their weights; however, these estimates probably involve considerable error since data on the growth rate from birth were available for only one infant. The authors speculated that a breeding season occurred over a 7-month period with a peak in January, but potential inaccuracies in their estimates of birth dates and their failure to collect animals in all months of the year weaken the reliability of their proposal. For example, the 7 adult females were collected only in the months of March, April, May, and June.

In contrast to this study (Jewell and Oates 1969a,b), Charles-Dominique (1968) reported births (N = 24) among *A. calabarensis* in all months of the year except June, July, and August when he was absent from the field. His study covered a period of 2 years and revealed no seasonal peak of births.

4.2.2. Ovarian Cycle

Manley (1966a) followed the estrous cycles of a captive *A. calabarensis* for 6 months and reported values of 45, 36, 37, and 38 days for the periodicity of successive cycles. He did not take vaginal smears to verify the presence of cornified epithelial cells but determined estrus by signs of turgescence in the external genitalia. Copulation occurred only in the final estrous cycle of the series.

4.2.3. Gestation

Manley (1966a, 1967) reported gestations of 131, 136, and 134 days in two female *A. calabarensis* for a mean value of 133.67 days (± 2.52 SD), whereas Jewell and Oates (1969a) recorded a gestation of 131 days in a single pregnancy. The four measures yield a mean value of 133.00 days gestation (± 2.45 SD).

4.2.4. Postpartum Estrus

Three days after parturition, Manley (1966a) observed a postpartum estrus in a female that failed to nurse or otherwise care for her newborn infant. Jewell and Oates (1969a) also reported a female that delivered 2 infants at an interval of 132 days. In another female, these authors also reported pregnancy during lactation.

4.3. *Perodicticus potto*

4.3.1. Seasonality

Ioannou (1966) monitored the vaginal cytology of 2 captive females over a period of 24 months and found no seasonal variations in the incidence of reproductive cycles. Males also showed an active reproductive status throughout the year since sperm were found in the vaginas of the two females during all months of the 2-year study. Unfortunately, the lighting conditions of these animals were not specified.

Charles-Dominique (1968), on the other hand, reported a distinct season for *P. potto* in Gabon with the birth season extending from August to December and with births peaking in August and September. Males also showed a seasonal increase in testicular volume and weights during the mating season.

4.3.2. Ovarian Cycle

Petter-Rousseaux (1962), who examined the vaginal smears of a single female over the period of a year, observed no changes in the turgescence or coloration of the external genitalia and a patent vaginal orifice at all stages of the cycle. In this animal, periods of full cornification reappeared at about 50-day intervals. Manley (1966a), however, observed changes in the turgidity and coloration of the external genitalia of a female *P. potto* during estrus; the vaginal orifice opened only during the period of maximum turgescence. Measuring the intervals between periods of maximal turgescence and opening of the vaginal orifice ($N = 8$), Manley (1966a) calculated a mean cycle length of 39 days (range 34–47 days). Ioannou (1966) examined daily samples of vaginal exfoliate in 2 adult pottos for 2 years and determined mean cycle lengths of 38.8 days (S.E. ± 0.94) and 37.0 days (S.E. ± 1.59). The females remained in vaginal estrus (i.e., complete loss of leukocytes and fully cornified vaginal epithelium) for 2.2 (S.E. ± 0.20) days.

4.3.3. Gestation

Butler and Juma (1970) identified sperm in the vaginal smear of a potto on December 13, and on June 1 of the following year found a newborn infant in the female's cage. These observations indicate a gestation of 170 days and provide the most reliable measure of gestation in this species.

4.3.4. Litter Size

Among 13 pregnancies recorded in captivity, no multiple births were reported (Cowgill, 1969).

4.3.5. Postpartum Estrus

Cowgill (1969) reported an immediate postpartum estrous cycle after the death of a neonate and copulatory behavior within a day of the infant's death. Although she made no study of postpartum estrus in females that successfully reared their infants, she did observe two successive births of infants to the same female at an interval of 195 days. This observation suggests a postpartum estrous cycle within a month of delivery.

4.4. *Nycticebus coucang*

4.4.1. Seasonality

Zuckerman (1932) examined Hubrecht's collection (see de Lange, 1921) of feral *N. coucang* (*N* = 146 females) and found that pregnant females (*N* = 35) had been collected in all months of the year except July. His graph of Hubrecht's material shows a slight increase in the incidence of pregnant females during September–November. Although no statistical analysis of the data appears in the report, the frequency of pregnancies appears not to have fluctuated at significant levels throughout the year. Crandall (1964) reported births to a captive female slow loris in February, March, April, June, August, September, and December. He concluded, therefore, that this female would have been polyestrous throughout the year if pregnancy had not intervened. Manley (1966a) also reported recurrent estrous cycles throughout the year in two captive females as long as they remained unimpregnated. His females, however, were caged in artificial light under a photoperiod cycle of 13L:11D throughout the period of study.

4.4.2. Ovarian Cycle

Manley (1966a) followed the estrous cycles of two female *N. coucang*. One was examined continually for 2 years, the other for 9 months. He noted periodic changes in the external genitalia of both females with increased turgescence and reddening of the labia and clitoris at full estrous. The labium moved caudally from the clitoris during turgescence to produce an obvious lateral elongation of the vaginal opening. He also measured the intervals between successive cycles of turgescence in his females. Female 1 experienced eleven successive cycles, female 2, six successive cycles. The total of 17 cycles yielded a mean periodicity of 42.3 days (range 37–54). The same investigator reported the formation of a vaginal plug after ejaculation (Manley, 1967).

4.4.3. *Gestation*

In a single pregnancy, Manley (1967) reported a gestational period of 193 days.

4.4.4. *Litter Size*

Manley (1966a) reported the birth of a single infant. However, Crandall (1964) recorded the birth of twins in 2 of 8 pregnancies to a slow loris in captivity.

4.4.5. *Postpartum Estrus*

Manley (1966a) observed no evidence of an immediate postpartum estrus in a female that successfully reared her offspring. She did, however, experience her first postpartum estrous cycle 39 days after parturition.

5. CHEIROGALEINAE

5.1. *Microcebus murinus*

5.1.1. *Seasonality*

In their natural habitat, mouse lemurs (*M. murinus*) breed during a restricted season from August–September to March (Petter-Rousseaux, 1970; Martin, 1972a,b; Andriantsiferana *et al.*, 1974). Sexually active females, however, became quiescent when they were moved during the breeding season from long photoperiods in Madagascar to short photoperiods in northern latitudes (Petter-Rousseaux, 1970). After prolonged exposure to photoperiods at northern latitudes, female mouse lemurs resumed a seasonal cycle of reproductive activity that began in March and April, nearly 6 months out of phase with the Malagasy season (Fig. 5). Furthermore, laboratory studies of captive mouse lemurs demonstrated that long photoperiods (16L:8D) in artificially lighted rooms induced estrous cycles whereas a change from long to shorter photoperiods inhibited cycles (Petter-Rousseaux, 1970, 1972; Martin, 1972a). Experimenting with this photoperiodic response, Petter-Rousseaux (1972) returned the breeding season of *Microcebus* in Paris to a Malagasy rhythm by exposing the animals to a regimen of artificial light varying

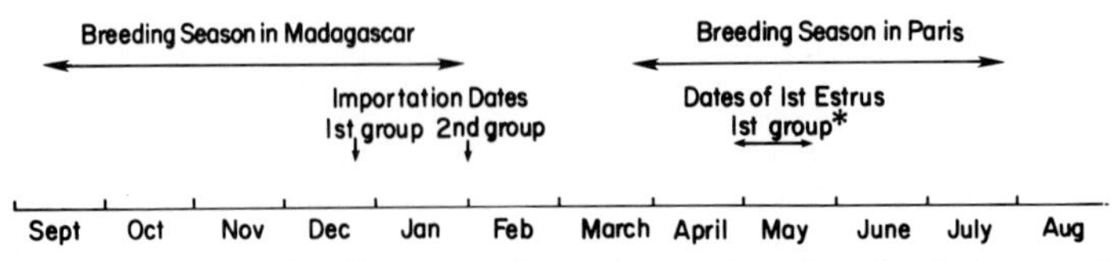

Fig. 5. Differences in the breeding season in northern and southern hemispheres under natural photoperiods in *Microcebus murinus*. *(The second group failed to enter estrus after importation to Paris). (Reproduced from Petter-Rousseaux, 1970.)

from about 11 to 13 hours of light per day, which simulated the seasonal changes of photoperiods in Madagascar (Fig. 6). This regimen regulated the reproductive cycles just as effectively as the natural light cycles in Paris, with a total magnitude of 8 hours difference between the longest and shortest days. Thus, the magnitude of the annual changes in day length does not appear to be the effective stimulus in the photoperiodic regulation of estrous cycles in the mouse lemur. Because females failed to experience estrous cycles until the following year when they were brought into longer photoperiods in Paris at the end of the Malagasy breeding season in February (Fig. 5), Petter-Rousseaux (1970) suggested that *Microcebus* females undergo a refractory period at the end of each breeding season during which they remain insensitive to photoperiod increments for several months.

Male *M. murinus* also experienced seasonal cycles of reproductive activity from July to January in Madagascar (Spühler, 1935; Martin, 1972a; Andriantsiferana *et al.*, 1974). During decreasing photoperiods, the testes gradually decreased in size until they were only about 8 mm long before partially entering the inguinal canal. During long photoperiods of the breeding season, however, the testes increased to about 19–22 mm long (Petter-Rousseaux, 1972), attained 8 times their quiescent volume, and showed extensive spermatogenic activity (Spühler, 1935).

To induce two breeding seasons per year, both Martin (1972a) and Petter-Rousseaux (1972) exposed captive *M. murinus* to compressed annual light cycles. Both investigators used an amplitude of photoperiod changes similar to the magnitude of annual light changes in Madagascar (about 2 hours). After Martin (1972a) had exposed his animals to a Madagascar regimen that had been compressed into a period of 9 months, he reported a highly synchronized set of estrous cycles and increases in testicular size when the photoperiod had increased to about 13 hours of light per day. But when Petter-Rousseaux (1972) compressed the Malagasy photoperiod to a semestrial pattern, she produced two breeding seasons each year (Fig. 7). Both male and female *M. murinus* responded to the accelerated light regimens.

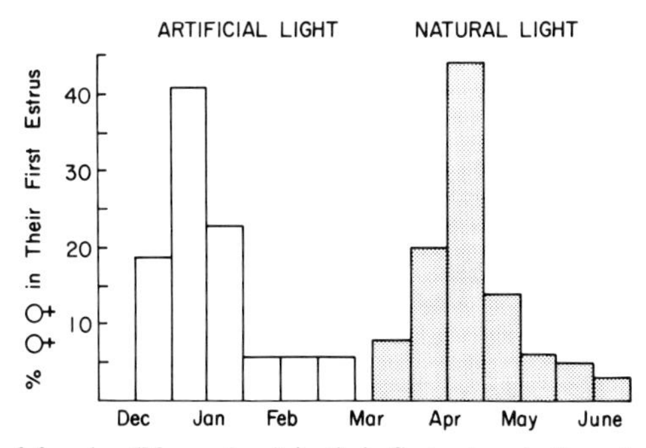

Fig. 6. Percentage of females (*M. murinus*) in their first estrus in 2-week intervals each month during natural light in Paris (▨); experiments conducted with 16L-8D beginning September 20 (□). (Reproduced from Petter-Rousseaux, 1970.)

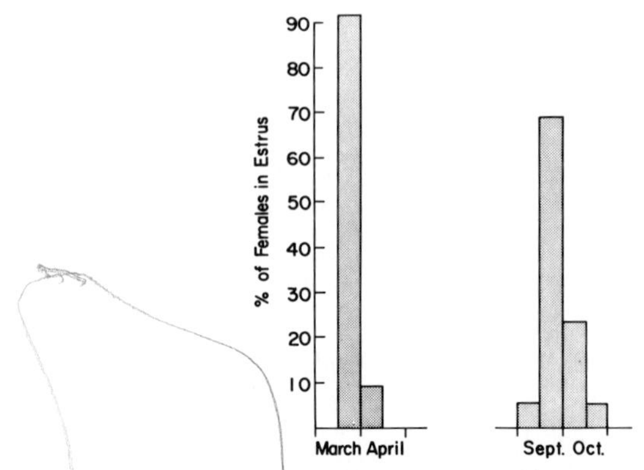

Fig. 7. The production of two breeding seasons in one year by exposing captive *M. murinus* to a semestrial photoperiod cycle. (Reproduced from Petter-Rousseaux, 1972.)

Under the semestrial light regimen, estrous cycles became more highly synchronized than they were in either the Malagasy or Parisian light regimens with 91 percent of the females ($N = 11$) cycling in the last two weeks of March in response to the first light increment and 68 percent of the females ($N = 22$) cycling in the last 2 weeks of September in response to the second light increment. However, the latency from the shortest photoperiod to the onset of estrous cycles was 2 weeks shorter than it had been under natural light cycles. Moreover, females under the semestrial light regimen tended to experience only a single estrous cycle during each breeding season, whereas females in natural photoperiods experienced 2–4 estrous cycles before becoming quiescent.

The testicles of male *M. murinus* underwent a normal range of changes in size under the semestrial light regimen, but the latency of response was slower than in natural photoperiods. Under the Parisian or Malagasy light regimen, the testes began to hypertrophy one month after the onset of increasing day length, one month earlier than under the accelerated or semestrial light regimen.

Martin (1972a,c) suggested that each female *M. murinus* in the wild bears two successive litters during the breeding season. He estimated that about 80 percent of the females conceive during September and reported a second peak of pregnancies from January to March (Martin, 1972a,b,c). Although he based this hypothesis on circumstantial evidence, it has been corroborated by laboratory studies in which two females produced two litters in one artificially produced breeding season. In this study, however, the females did not raise their first litter (R. D. Martin, personal communication). Martin (1972c) suggests that gestation (ca. 60 days) and nursing (ca. 40 days) are brief enough to accommodate two pregnancies per breeding season without conceptions during postpartum estrous cycles. However, Andriantsiferana *et al.* (1974) reported that two females whose infants survived gave birth twice in one breeding season at intervals of 71 and 75 days. Their observations suggest, therefore, that postpartum estrous cycles are at least partially responsible for the second peak of pregnancies in each breeding season.

5.1.2. *Ovarian Cycle*

During reproductive quiescence, the vulva in mouse lemurs remains imperforate. At the onset of an estrous cycle, a pink swelling forms at the base of the clitoris and persists for 4 to 6 days. A small slitlike opening appears in this swelling and then gradually enlarges to form a wide opening into the vaginal canal. During this period (2–5 days), highly colored folds surround the vaginal orifice and then gradually fade. Two to three days after the opening of the vaginal orifice, only epithelial cells with extensive cornification are seen in vaginal smears; copulation usually occurs on the third day after vaginal opening. On days 3 to 5 of vaginal opening, the number of cornified epithelial cells declines sharply and leukocytes again become numerous. The vaginal orifice then closes gradually over a period of 5–10 days, during which all epithelial cells, cornified or undifferentiated, gradually disappear and leukocytes begin to predominate in the vaginal smears. The estrous cycles studied by Petter-Rousseaux (1964) ranged from 38 to 78 days with 78% of the cycles (N = 23) falling between 45 and 55 days. The periodicity of estrous cycles remained unchanged whether the female copulated, was isolated, or remained with males (Petter-Rousseaux, 1964). The duration of the estrous cycles studied by Andriantsiferana *et al.* (1974) was also highly variable and ranged from 28 to 108 days with a mean period of about 56 days (± SD 20.00). These authors also recorded the formation of a large white vaginal plug after copulation.

5.1.3. *Gestation*

Within 12 days after copulation, the vaginal orifice of the impregnated mouse lemur closes and remains imperforate until delivery. Gestation (25 cases) lasts for about 60 days with a range of 54–68 days and a mean of about 63 days (± SD 7.06) (Petter-Rousseaux, 1964; Martin, 1972a; Andriantsiferana *et al.*, 1974). Within 12 days after delivery, the vagina again closes (Petter-Rousseaux, 1964).

5.1.4. *Litter Size*

The litter usually consists of 2 to 3 infants (Martin, 1972a), but occasionally single offspring are delivered (Petter-Rousseaux, 1964).

5.1.5. *Sexual Maturity*

Although Petter-Rousseaux (1964) reported that *M. murinus* of both sexes reach sexual maturity at 7 to 10 months of age, the tabulation of her data shows that females (N = 6) in her sample experienced their first estrous cycles between 9.63 and 29.30 months of age and that males (N = 4) showed their first cycle of testicular enlargement between the ages of 7 and 19 months of age (Table VI). Andriantsiferana *et al.* (1974) reported 3 females that gave birth at 1 year of age.

TABLE VI

Age at Sexual Maturity (*M. murinus*)[a]

Female	Age at first estrous swelling and vaginal opening (months)
67	17.5
69	18.3
109	29.3
107a	16.2
107b	14.8
127a	9.6
Male	**Age at first testicular enlargement (months)**
68	6– 7
107c	18–19
107aa	7– 8
107ab	7– 8

[a] Reproduced from Petter-Rousseaux (1964).

5.1.6. Postpartum Estrus

Despite the suggestion by Martin (1972a,c) that *M. murinus* females conceive twice within each restricted breeding season (see discussion, Section 5.1.1), neither Petter-Rousseaux (1962, 1964) nor Martin (1972a,b,c) had reported irrefutable cases of postpartum estrous cycles in this species. In two cases when females failed to raise their infants, Petter-Rousseaux (1962) observed that vaginal closure was slower and that the vulva began to undergo estrous changes shortly after delivery. Neither of these females, however, had copulated after parturition. However, fertile postpartum matings have now been observed in cases in which the litter was not reared (A. F. Glatson and R. D. Martin, personal communication). Nevertheless, if *M. murinus* females do produce two litters per year in the natural habitat, as Martin suggested (1972a,c), the postpartum estrus provides at least a plausible mechanism. In fact, Andriantsiferana *et al.* (1974) recorded conceptions in two lactating females during postpartum estrous cycles and thereby confirmed this hypothesis.

5.2. Cheirogaleus major

5.2.1. Seasonality

Petter-Rousseaux (1962, 1964) postulated that *C. major* in Madagascar have a summer breeding season that is analogous to that of *M. murinus,* with the onset of mating in October and births from November to February (Petter-Rousseaux, 1964; Doyle, 1974). However, she based this postulate on a single female and her two infants (approximately 1–2 months old); therefore, it remains open to modification by future studies in either laboratory or field. Petter-Rousseaux (1962) also reported a seasonal cycle of testicular changes in *C. major* that resembled the changes observed in *M. murinus*.

5.2.2. Ovarian Cycle

The estrous cycle in *C. major* closely resembles that of *M. murinus*. Over a period of 4–5 days, the external genitalia begin to swell, after which the vaginal orifice, which was previously imperforate, opens and remains patent for 2–3 days. During this stage of the ovarian cycle, copulation occurs. The vaginal orifice then gradually closes, but at a much slower rate than in *M. murinus*. Estrous cycles of unimpregnated females recur at approximately 30-day intervals during the breeding season (Petter-Rousseaux, 1964).

5.2.3. Sexual Behavior

The male *C. major* holds the receptive female with his hands and licks her flanks and neck without biting. During this behavior, the male also vocalizes, waves his tail, and grips the female's ankles. The receptive female usually responds to the male's advances by lying with her ventral surface on the ground. Copulatory sequences last 2 to 3 minutes and are repeated several times at approximately 10-minute intervals (Petter-Rousseaux, 1964).

5.2.4. Gestation

On the basis of one pregnancy, Petter-Rousseaux (1962, 1964) determined the gestation of *C. major* to be 70 days. Copulation was observed in this female on April 27 to 29 and the young were born on July 7.

6. INDRIIDAE

6.1. *Propithecus verreauxi*

6.1.1. Seasonality

At both the northern (approximately 12° S latitude) and southern (approximately 26° S latitude) extremes of their distribution on Madagascar, *P. verreauxi* breed in a restricted season between January and March (Petter-Rousseaux, 1964; Jolly, 1966; Richard, 1974). Although unimpregnated females of several prosimian species experience a polyestrous breeding season, *P. verreauxi* appear to ovulate only once during the breeding season. In 2 successive years, the northern subspecies (*P. v. coquereli*) had unimodal birth seasons that lasted only 21 days (Richard, 1974). In southern Madagascar, Jolly (1966) also reported a brief unimodal birth season in *P. v. verreauxi* that lasted only 10 days. The absence of subsequent birth peaks susggests that all females either conceive on their first ovulatory estrous cycle or experience only one ovulation in each breeding season whether impregnation occurs or not. These alternative explanations of the unimodal birth season can be effectively tested only in captivity, where the estrous cycles of unimpregnated females can be monitored over extended periods.

Despite the evidence of a unimodal birth season, however, Richard (1974) suggests that a single anovulatory estrous cycle precedes the receptive (and presumably ovulatory) estrous cycle of the breeding season proper in *P. verreauxi* as it does in *L. catta* (Jolly, 1966). Some 3 to 4 weeks before copulation, she observed the external genitalia of 4 of her females ($N = 9$) change to pink. Richard (1974) observed that during this period male behavior directed toward females increased.

6.1.2. Sexual Behavior

Richard (1974) observed 3 males copulate with 2 female partners. All copulations occurred while the female grasped a vertical trunk. Often the preliminary approaches by the male were repulsed by female lunges, slaps, and bites. In addition, other males often disrupted the mounts by attacking the mounted male. Both females presented to male partners by rolling up their tails. Immediately before copulation the females "doubled up in a squatting position." The male mounted by grasping the female's legs with his hands and holding the tree trunk with his feet. Copulation included a series of 1 to 6 mounts. In early mounts, only a few thrusts were observed and intromission was not positively identified. During longer mount sequences, the males thrust from 24 to 53 times and Richard's observations suggest that these longer mounts ended with ejaculation (Table VII).

6.1.3. Gestation

Petter-Rousseaux (1962) observed copulation and birth in one female *P. verreauxi*. Gestation lasted just over 5 months (162 days).

TABLE VII

Copulatory Behavior (*Propithecus verreauxi*)[a]

				Terminal mount	
Date	Male ID	Female ID	Number mounts	Duration (minutes)	Number thrusts
March 3	F	F1	6	—	24
March 4	P	F1	4	4	40
March 5	INT	FN1	1	4	53

[a] Data tabulated from Richard (1974).

7. LEMURIDAE

7.1. *Lemur catta*

7.1.1. *Seasonality*

Jolly (1966, 1967) observed four matings of *L. catta* females in the natural habitat. At the time of mating, the external genitalia of the four females had become pink and swollen, but returned to a deflated colorless condition in the two females that were sighted again on the day after mating. On the basis of these observations, Jolly (1966, 1967) suggested that ring-tailed females remain sexually receptive for only 1 day during the estrous cycle. In addition, observations of matings and changes in the external genitalia of females in four feral troops led her to infer that mating occurs in April during a brief period of no more than 2 weeks. For example, all Troop 1 females ($N = 9$) displayed changes in the external genitalia over a period of 12 days between April 16 and 27. Furthermore, Troop 2 females ($N = 6$) all had changes in the external genitalia on April 23 when mating was also observed. Additional females in Troops 3 and 4 displayed changes in the genitalia on April 21 and 22.

Three to four weeks before the onset of receptive estrous cycles, Jolly (1966, 1967) observed changes in the genitalia of 5 troop 1 females ($N = 9$). No matings, however, were seen at this time. These observations led Jolly (1966, 1967) to suggest that a set of nonreceptive estrous cycles or a period of "pseudoestrus" precedes the onset of the mating season by about 1 month. During subsequent observations, she observed a restricted period of births in late August and thus confirmed her hypotheses about the synchrony of estrous cycles, the brevity of the mating season in April, and the absence of conceptions during the "pseudoestrous" period. However, Budnitz and Dainis (1975) observed a major concentration of births in August and September with fewer births in October and November. Thus, most but not all females probably conceived during their first estrous cycle but a few individuals must have conceived during subsequent cycles. In Madagascar, the breeding season occurs during the shortest days of the year, an indication of a relation between photoperiods and breeding. Furthermore, the breeding season occurred 6 months out of phase with the season in the southern hemisphere after ring-tailed lemurs were moved from Madagascar to the northern hemisphere (45° N latitude) in Portland, Oregon (Evans and Goy, 1968; Van Horn, 1975) (Fig. 8). Captive animals at the Oregon Regional Primate Research Center have bred in the fall months (mean breeding season, late November to early December) when they were exposed to natural light cycles (Fig. 8). Animals caged indoors in constant temperature and humidity but with natural light through windows maintained a breeding season that was indistinguishable from that of animals caged outdoors and subjected to much broader fluctuations of temperature and humidity (Fig. 9).

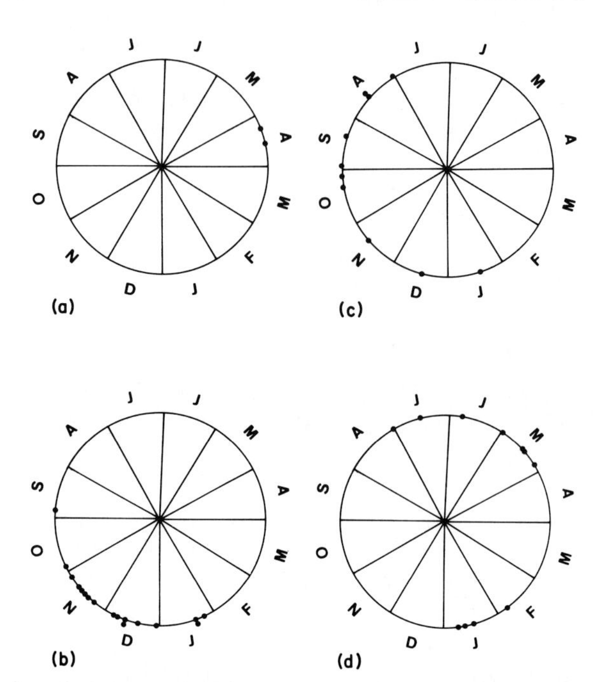

Fig. 8. Differences in the seasonal distribution of conceptions (●) in *Lemur catta* under (a) natural light in Madagascar, (b) natural light in Oregon, (c) constant day length cycle of 9L:15D, and (d) an experimental light regimen. (Modified from Van Horn, 1975.)

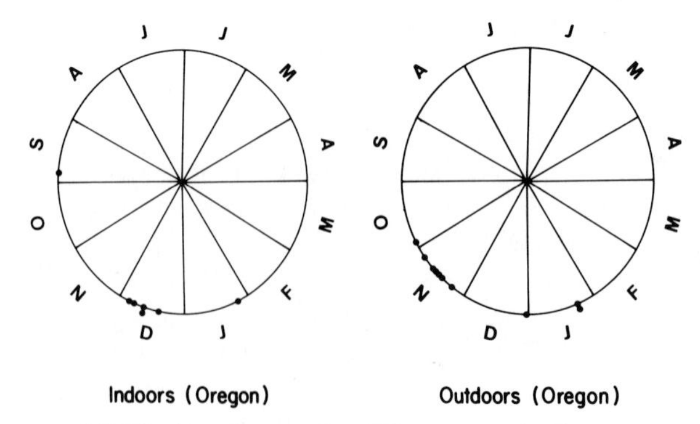

Fig. 9. The seasonal distributions of conceptions (●) among captive *Lemur catta* under natural light in indoor and outdoor cages in Oregon. (Modified from Van Horn, 1975.)

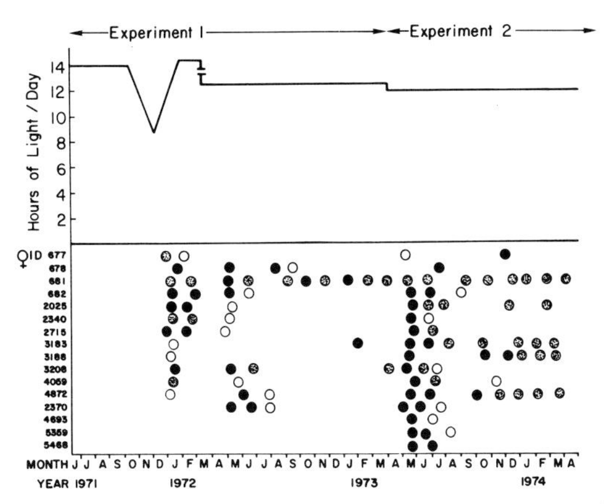

Fig. 10. The relation between the experimental light cycles and the temporal distribution of estrous cycles (1971–1974). (○) Behavioral estrus, i.e., sexual receptivity; (◎) vaginal estrus; (●) conception; (∓) undetermined rate of photoperiod decrement due to equipment failure during Experiment 1 regime. (Reproduced from Van Horn, 1975.)

Experiments with captive *L. catta* have demonstrated the inhibition of all estrous cycles by long photoperiods (14L:10D) and the reactivation of estrous cycles by a return to short photoperiods (9.0L:15D). However, in a subsequent study, 94% of quiescent females (15 of 16) also resumed estrous cycles within 60 days after a 30-minute decrease in the photoperiod from a regimen of 12.5L:11.5D to 12.0L:12.0D (Fig. 10) (Van Horn, 1975). Thus, the experimental data on captive females demonstrate the ability of both temperate and tropical photoperiods to regulate reproductive cycles in *L. catta* (Fig. 10).

7.1.2. Ovarian Cycle

Evans and Goy (1968) monitored the estrous cycles of 5 female *L. catta* by examining their vaginal smears. They classified smears as diestrous (predominately mucus and leukocytes with large nuclei in epithelial cells), proestrous (decreasing numbers of leukocytes and increase in size of noncornified epithelial cells), vaginal estrous (full cornification of 75% or more of the epithelial cells), and metestrous (dense aggregations of large noncornified epithelial cells). Using this classification, the authors reported a mean ovarian cycle length (N = 17 cycles) of 39.3 days (SE = ± 0.16 days; range = 33–45 days). Within the ovarian cycle, vaginal estrus persisted for a mean duration of 4.7 days (SE = ± 0.3 days; range = 3–7 days) and the period of leukocytic absence covered a mean interval of 9.7 days (SE = ± 0.6 days). R. N. Van Horn (personal observation) also measured the periodicity of estrous cycles (N = 40) among 16 captive females by examining their vaginal smears and observing sexual receptivity and obtained a mean value of 39.87 days (SD = ± 2.66 days; range = 32–45 days) between successive cycles of either behavioral estrus (sexual receptivity) or vaginal estrus.

To measure the duration of sexual receptivity, females in vaginal estrus ($N = 9$) were paired with breeding males every 6 hours in an observation room. When the male intromitted within a 30-minute period, the onset of receptivity was recorded and the pair was separated before ejaculation. Pairs tests continued at 6-hour intervals until receptivity ceased. In these tests, females were paired with more than one male to eliminate the influence of partner preferences on receptivity (R. N. Van Horn, personal observation). These tests yielded a mean value of 21.66 hours (SD $= \pm$ 5.76 hours; range = less than 8–44 hours) for the duration of receptivity (Van Horn and Resko, in press).

7.1.3. Sexual Behavior

The behavioral repertoire of ring-tailed lemurs includes several forms of scent marking. For example, female *L. catta* rub the perineum on vertical or horizontal surfaces to deposit urine and secretions from the vagina and perineum. The frequency of female genital marking, however, does not fluctuate significantly during either the short-term ovarian cycle or the long-term annual cycle of reproduction (Evans and Goy, 1968).

Male ring-tailed lemurs also mark with the genital area by rubbing the scrotum and perineum on branches and trunks. In addition, male *L. catta* mark with specialized scent glands on the wrist (antebrachial gland) and shoulder (brachial gland). The secretions from these glands are frequently mixed during "shoulder-rubbing" when the male rubs the wrist glands against the shoulder glands. Shoulder-rubbing may occur by itself or it may precede other displays such as "arm-marking" or "tail-marking." In arm marking, male lemurs deposit secretions from the forearm glands by dragging the wrist forcefully across trunks and branches. They may at times also deposit secretions from both the wrist and shoulder glands onto the tail by folding the tail over the shoulder and then drawing it downward between the shoulder gland and the wrist glands of the folded arms. Tail-marking occurs in both aggressive and sexual encounters and regularly precedes tail-flicking displays in which a male rapidly waves his tail in the face of another animal. In captivity, the frequency of tail displays between males decreased significantly during the nonbreeding season, whereas the frequency of arm marking did not differ significantly regardless of the time of year (Evans and Goy, 1968). In the natural habitat, the incidence of tail displays increased during the breeding season and frequently developed into "stink fights" in which two males arm-marked, tail-marked, and tail-flicked toward one another in aggressive interactions (Jolly, 1966).

In feral troops, stink fights were observed when more than one male attempted to approach a receptive female. For example, the copulatory sequences observed by Jolly (1966) involved at least two males that approached the receptive female. The males displayed toward one another until one male had driven all other males away from the female. These stink fights lasted from 5 to 25 minutes and the female then presented to the male who had successfully maintained his position near her. The female presented either by raising her tail and hindquarters or by crouching flat on a branch in a lordosis posture. The male then mounted her with his arms clasped around her waist. Copulation included 3 to 12 mounts but not all mounts included intromission and thrusting. Other males repeatedly approached the copulating pair and disrupted the mounts.

In captivity, additional details of the copulatory sequences have been observed (R. N. Van Horn, personal observation). During pair tests, males showed an increased interest in proestrous females and frequently tail-marked, tail-flicked, and flattened their ears against their head as they approached her from the front. If the female slapped, lunged, or bit at the male in response to such frontal approaches, the male usually retreated rapidly. If, however, the female did not attack the male during his frontal approaches, he usually moved behind her and continued the displays. Receptive females did not look back at the displaying male but rather presented toward him and directed their eyes to the front. In response, the male reached for the female's waist, tail-flicked, ear-flattened, and then mounted. At any point in this sequence, nevertheless, a nonreceptive female might drive the male away by looking back over her shoulder or by turning and attacking the male with slaps and bites.

Although captive ring-tailed lemurs displayed true estrous behavior with sexual receptivity confined to a brief period ($\bar{X}$ = 21.7 hours) around the time of ovulation, hormonal factors were not the sole determinants of sexual receptivity. During pair tests, 12 receptive females were tested with more than one male to determine the influence of partner preferences on sexual receptivity. To prevent ejaculations in these tests, the pair was separated immediately after an intromission had been achieved. The receptive female was then immediately tested with additional males. Finally, the female was again tested with the original male to assure that she was still receptive. Seven of the females in this study (N = 12) failed to copulate with one or more males during their period of sexual receptivity. However, females did not form monogamous pair bonds, since several females copulated with more than one male (Van Horn and Resko, in press).

7.1.4. Gestation

Copulation and birth were observed in 14 captive *L. catta*. In these matings, females were allowed to receive only one ejaculation and the presence of sperm was verified by the examination of vaginal smears. The gestational periods of these pregnancies were calculated from the day of mating to the day before birth inclusive. The data yielded a mean gestational period of 135.64 days (S.D. = ± 1.63 days; range = 134–138 days) (R. N. Van Horn, personal observation).

7.1.5. Litter Size

Female *L. catta* (N = 21) at the Oregon Regional Primate Research Center have experienced 60 term pregnancies with 11 cases of twins (18.33%). However, only 6

TABLE VIII

The Incidence of Twins in Captive *Lemur catta* (18.3%)

Female number	Number of pregnancies	Number of twin pregnancies
677	7	4
678	6	0
681	1	0
682	6	2
2025	5	0
2340	3	2
2370	5	1
2371	2	0
2715	2	1
3183	2	0
3188	3	0
3208	4	0
4059	2	0
4693	1	0
4872	3	0
5359	2	0
5468	1	0
6047	1	0
6048	1	1
6438	1	0
6460	1	0
6548	1	0
N = 22	60	11 (18.3%)

females (27.3%) in our captive population account for the twin births (Table VIII) (R. N. Van Horn, personal observation).

7.1.6. *Sexual Maturity*

In the spring months of each year, sexually immature ring-tailed lemurs were placed in large outdoor cages to form bisexual groups of 3 to 13. Under these conditions, the females (N = 9) first conceived at a mean age of 19.56 months (SD = ± 5.57 months; range = 10–31 months) (Table IX). This estimate of the age of sexual maturity may, however, be an underestimation since the females were not caged with experienced male partners and the influences of partner preferences were not controlled (R. N. Van Horn, personal observation). Furthermore, the effect of natural photoperiods on the onset of puberty, although documented in rodents (Hoffman, 1973), remains to be defined in primate species.

TABLE IX

The Age at First Conception in Captive *Lemur catta* Females

Female number	Age at first conception (years)
2340	2.6
3188	1.6
3208	1.5
4693	1.8
4872	1.8
5359	0.8
5468	1.8
6047	1.4
6048	1.4
$N = 9$	$\bar{X} = 1.6$ SD = 0.46

7.1.7. Postpartum Estrus

Although no systematic studies have examined postpartum estrous cycles in *L. catta,* casual observations in captivity verify that females regularly experience a postpartum estrus if they are separated from their infants within 2 to 3 weeks after parturition. However, the percentage of females that experience a postpartum estrus after being separated from their infant appears to decrease as the infant matures. For example, very few females experience a sudden estrous cycle if the separation occurs when the infant is 6 months of age (R. N. Van Horn, personal observation).

8. SUMMARY

8.1. Seasonality

Seven of the eleven species examined in this review showed a seasonal pattern of reproduction. All Malagasy prosimians appear to breed seasonally under natural conditions. Studies of *Tarsius bancanus* and *Nycticebus coucang,* however, showed pregnancies occurring in all months of the year. Among the African lorises (*Arctocebus calabarensis* and *Perodicticus potto*), both aseasonal (Ioannou, 1966) and seasonal (Charles-Dominique, 1968) reproduction has been reported. Despite the seasonal nature of reproduction in female lorises (*Loris tardigradus*), however, male *L. tardigradus* displayed no evidence of a seasonal cycle of testicular activity

(Ramakrishna and Prasad, 1967). Male pottos (*P. potto*) also bred in all months (Ioannou, 1966). In contrast, male mouse lemurs (*Microcebus murinus*) undergo marked seasonal changes in their reproductive tracts, as do male *Cheirogaleus major*. (Petter-Rousseaux, 1962, 1964; Martin, 1972a). Because the available data are not sufficient to enable us to determine whether males of other prosimian species do or do not experience seasonal changes in their reproductive status, there is a great need for future studies to examine the elements of reproductive physiology in the male.

In studies of seasonal species, it has often been difficult to evaluate the relation between environmental variables and the breeding season. For example, because of too few studies, no meaningful correlations can be made in *L. tardigradus, Propithecus verreauxi,* and *C. major*. On the other hand, studies of *Galago senegalensis* demonstrate significant variability in the timing of the breeding season and thereby complicate attempts to correlate the breeding season with exogenous factors (Zuckerman, 1932; Haddow and Ellice, 1964; Cooper, 1966; Butler, 1967a,b; Bearder, 1969; Doyle, 1974). Despite this variability, however, the breeding season for captive *G. senegalensis* in the northern hemisphere (33° N latitude) occurred in December and January (Cooper, 1966), nearly 6 months out of phase with the onset of the South African breeding season in June (Bearder, 1969). This reversal of the breeding season in opposite hemispheres and the onset of breeding during the shortest days of the year in both hemispheres suggest that photoperiods regulate the annual cycles of reproduction in this species. Butler's (1967a,b) studies of captive *G. senegalensis* at 10°–12° N latitude in Uganda appear to conflict with this photoperiodic explanation, however, since he reported the onset of the breeding season in August. On the other hand, he failed to specify the conditions under which his captive animals were caged. Therefore, his findings do not preclude the factor of photoperiodic regulation in this species.

The breeding season of *Galago crassicaudatus,* like that of *G. senegalensis,* begins during the shortest days of the year in both South Africa (Bearder, 1975) and San Diego (Cooper, 1966). Under a constant regimen of artificial illumination (12L:12D), however, captive *G. crassicaudatus* continue to breed throughout the year (Eaton *et al.*, 1973b). These observations demonstrate how a species could be seasonal in temperate latitudes and aseasonal in equatorial latitudes where annual photoperiod changes are minimal.

Studies of *M. murinus* and *Lemur catta* provide the most convincing evidence of photoperiodic regulation in prosimian reproduction. Both male and female *M. murinus* became sexually active during the long photoperiods of either natural or experimental light cycles (Petter-Rousseaux, 1970, 1972; Martin, 1972a). In contrast, *L. catta* became sexually active during short photoperiods and sexually quiescent during long photoperiods. Experiments have documented that reproductive cycles can be elicited in quiescent females in both temperate and tropical photoperiods. These studies, however, did not examine the effects of photoperiods on males or females in the absence of the opposite sex. Therefore, behavioral and pheromonal feedback relations between the sexes may be significant factors in the regulation of prosimian reproduction (Van Horn, 1975, also personal observations).

The effects of photoperiods on reproduction in both *M. murinus* and *L. catta* emphasize the need to examine the environmental correlates of the breeding season in other primate species. To do this, field studies should examine the distribution of mating and births throughout the year and laboratory studies should consider carefully the lighting conditions of captive animals.

8.2. Hormones and Behavior

In some ways, the reproductive physiology and behavior of prosimians resemble those of other mammals more than they do those of simians. Prosimians, for example, are sexually receptive only for a brief period during their ovarian cycle, whereas at least some simians are potentially receptive during all phases of the cycle. During both the follicular and the luteal stages of their ovarian cycles, for example, female pig-tailed macaques (*Macaca nemestrina*) released confined males and then permitted them to mount and intromit with equal frequency (Eaton and Resko, 1974).

Despite differences in the duration of receptivity, the cyclic patterns of plasma estrogen and progesterone are similar in the few simian and prosimian species that have been studied (*G. crassicaudatus:* Eaton *et al.*, 1973b; *L. catta:* Van Horn and Resko; *M. mulatta:* Hotchkiss *et al.*, 1971; *M. nemestrina:* Eaton and Resko, 1974; *Homo sapiens:* Ross *et al.*, 1970). Relatively low levels of estrogen and progesterone prevail until a few days before ovulation when follicular estrogen surges and ovulation occurs. Estrogen then subsides, and progesterone rises and remains high until the regression of the corpus luteum causes it to subside and the cycle ends (Figs. 2 and 11).

Because of the similarity in these cyclic patterns of gonadal steroids, differences in neural sensitivity to the action of these (and other) hormones rather than differences in neural–hormonal mechanisms appear to underlie the simian–prosimian differences in female sexual behavior. However, general statements on primate hormone-behavior relationships may not yet be made because only a few species of these taxonomic groups have been studied. For example, of the more than 100 species and subspecies of prosimians, only 11 have been mentioned in this review. There is also a paucity of information on the reproductive physiology and behavior of the New World monkeys and great apes. Thus, without sufficient data on hormone fluctuations and behavior, it appears that the hormone-behavior relationships of the chimpanzee (*Pan troglodytes*) and gorilla (*Gorilla gorilla*) are more like those of prosimians than of other simians because the sexual behavior of these great apes occurs with greatest frequency during the follicular phase of the female's ovarian cycle (Yerkes and Elder, 1936; Young and Orbison, 1944; Nadler, 1975). However, simple correlations between hormone titers and sexual behavior do not necessarily demonstrate cause–effect relations. Furthermore, in any study of sexual be-

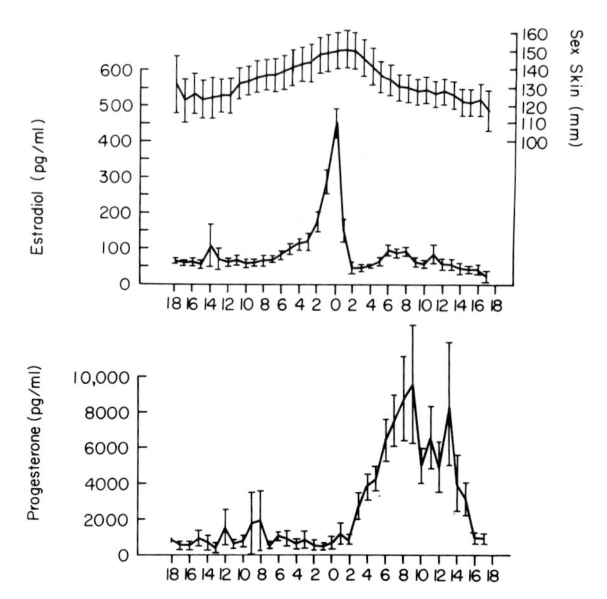

Fig. 11. Mean (± SE) plasma estradiol and progesterone concentrations from 8 female pigtailed macaques. (Cycles were aligned from the day of peak estradiol titers [day 0], and the luteal phase was defined as beginning 3 days later.) (Modified from Eaton and Resko, 1974.)

havior, it is difficult to differentiate between the effects of some hormones on female attractiveness to the male and the effects of other hormones on her receptivity to his advances. In the rhesus monkey, female attractiveness appears to be controlled by estrogen, her receptivity by testosterone (Dixson *et al.,* 1973). Questions of evolutionary trends in hormone-behavior relationships must, therefore, remain unanswered until more complete information on both simians and prosimians is available (Rowell, 1972).

ACKNOWLEDGMENTS

We thank Drs. C. H. Phoenix and A. F. Dixson for helpful criticism. We are also indebted to Margaret T. Barss for editorial comments, and to Isabel G. McDonald and Janice E. Weinstock for assistance with reference materials. Publication No. 842 of the Oregon Regional Primate Research Center, supported in part by Grants RR-00163 and HD-05969 from the National Institutes of Health, U.S.P.H.S.

REFERENCES

Andriantsiferana, R., Rarijanona, Y., and Randrianaivo, A. (1974). *Mammalia* **38,** 234–243.
Ansell, W.F.H. (1960). "Mammals of Northern Rhodesia." Govt. Printer, Lusaka.

Asdell, S. A. (1964). "Patterns of Mammalian Reproduction." Cornell Univ. Press (Comstock), Ithaca, New York.
Bearder, S. K. (1969). M.Sc. Dissertation, University of the Witwatersrand, Johannesburg, South Africa (unpublished).
Bearder, S. K. (1975). Ph.D. Thesis, University of the Witwatersrand, Johannesburg, South Africa (unpublished).
Budnitz, N., and Dainis, K. (1975). *In* "Lemur Biology" (I. Tattersall and R. W. Sussman, eds.), pp. 219–235. Plenum, New York.
Buettner-Janusch, J. (1964). *Folia Primatol.* **2,** 93–110.
Butler, H. (1960). *Proc. Zool. Soc. London* **135,** 423–430.
Butler, H. (1967a). *Folia Primatol.* **5,** 165–175.
Butler, H. (1967b). *J. Zool.* **151,** 143–162.
Butler, H., and Juma, M. B. (1970). *Lab. Primate Newsl.* **9,** 16.
Catchpole, H. R., and Fulton, J. F. (1943). *J. Mammal.* **24,** 90–93.
Charles-Dominique, P. (1968). *In* "Entretien de Chizé I" (R. Canivenc, ed.), pp. 2–5. Masson, Paris.
Cooper, R. W. (1966). "Fourth Annual Report," pp. 5–8. Inst. Comp. Biol., Zool. Soc. San Diego, San Diego, California.
Cowgill, U. M. (1969). *Folia Primatol.* **11,** 144–150.
Crandall, L. S. (1964). "The Management of Wild Animals in Captivity." Univ. of Chicago Press, Chicago, Illinois.
Davis, M. (1960). *Mammalia* **41,** 401–402.
de Lange, D. (1921). "Catalogue of the Embryological Material of Lemuridae (*Tarsius* and *Nycticebus*) and Dermoptera (*Galeopithecus*)." Collection of the Hubrecht Laboratory, Utrecht, Holland.
de Lowther, F. (1940). *Zoologica (N.Y.)* **25,** 433–459.
Dixson, A. F., Everitt, B. J., Herbert, J., Rugman, S. M., and Scruton, D. M. (1973). *Proc. Int. Congr. Primatol., 4th, 1972* Vol. 2, pp. 36–63.
Doyle, G. A. (1974). *Behav. Nonhum. Primates* **5,** 155–338.
Doyle, G. A., Pelletier, A., and Bekker, T. (1967). *Folia Primatol.* **7,** 169–197.
Doyle, G. A., Andersson, A., and Bearder, S. K. (1971). *Folia Primatol.* **14,** 15–22.
Eaton, G. G. (1972). *Horm. Behav.* **3,** 133–142.
Eaton, G. G., and Resko, J. A. (1974). *J. Comp. Physiol. Psychol.* **86,** 919–925.
Eaton, G. G., Goy, R. W., and Phoenix, C. H. (1973a). *Nature (London)* **242,** 119–120.
Eaton, G. G., Slob, A., and Resko, J. A. (1973b). *Anim. Behav.* **21,** 309–315.
Evans, C. S., and Goy, R. W. (1968). *J. Zool.* **156,** 181–197.
Fogden, M. (1974). *In* "Prosimian Biology" (R. D. Martin, G. A. Doyle, and A. C. Walker, eds.), pp. 151–165. Duckworth, London.
Gérard, P. (1932). *Arch. Biol.* **43,** 93–151.
Haddow, A. J., and Ellice, J. M. (1964). *Trans. R. Soc. Trop. Med. Hyg.* **58,** 521–538.
Harrisson, B. (1963). *Malay. Nat. J.* **17,** 218–231.
Hill, W.C.O. (1953). *Proc. Zool. Soc. London* **123,** 13–16.
Hill, W.C.O. (1955). "Primates: Comparative Anatomy and Taxonomy," Vol. II. Edinburgh Univ. Press, Edinburgh.
Hill, W.C.O., Porter, A., and Southwick, M. D. (1952). *Proc. Zool. Soc. London* **122,** 79–119.
Hoffman, J. C. (1973). *Handb. Physiol., Sec. 7: Endocrinol.* **2,** Part 1, 57–77.
Hotchkiss, J., Atkinson, L. E., and Knobil, E. (1971). *Endocrinology* **89,** 177–183.
Ioannou, J. M. (1966). *J. Reprod. Fertil.* **11,** 455–457.
Jewell, P. A., and Oates, J. F. (1969a). *J. Reprod. Fertil.,* Suppl. **6,** 23–28.
Jewell, P. A., and Oates, J. F. (1969b). *Zool. Afr.* **4,** 231–248.
Jolly, A. (1966). "Lemur Behavior: A Madagascar Field Study." Univ. of Chicago Press, Chicago, Illinois.
Jolly, A. (1967). *In* "Social Communication among Primates" (S. A. Altmann, ed.), pp. 3–14. Univ. of Chicago Press, Chicago, Illinois.

Le Gros Clark, W. E. (1924). *Proc. Zool. Soc. London* **1,** 217–223.
Lumsden, W.H.R., Ellice, J. M., and Hewitt, L. E. (1955). "East African Virus Research Institute Annual Report," No. 6, (1956), p. 9. Govt. Printer, Nairobi.
Manley, G. H. (1966a). *Symp. Zool. Soc. London* **15,** 493–509.
Manley, G. H. (1966b). *Symp. Zool. Soc. London* **17,** 11–39.
Manley, G. H. (1967). *Int. Zoo Yearb.* **1,** 80–81.
Martin, R. D. (1972a). *In* "Breeding Primates" (W.I.B. Beveridge, ed.), pp. 161–171. Karger, Basel.
Martin, R. D. (1972b). *Philos. Trans. R. Soc. London, Ser. B* **264,** 295–353.
Martin, R. D. (1972c). *Z. Tierpsychol., Beih.* **9,** 43–89.
Nadler, R. D. (1975). *Science* **189,** 813–814.
Napier, J. R., and Napier, P. H. (1967). "A Handbook of Living Primates." Academic Press, New York.
Nicholls, L. (1939). *Nature (London)* **143,** 246.
Petter-Rousseaux, A. (1962). *Mammalia* **26,** 1–88.
Petter-Rousseaux, A. (1964). *In* "Evolutionary and Genetic Biology of Primates" (J. Buettner-Janusch, ed.), Vol. 2, pp. 92–132. Academic Press, New York.
Petter-Rousseaux, A. (1970). *Ann. Biol. Anim., Biochim. Biophys.* **10,** 203–208.
Petter-Rousseaux, A. (1972). *Ann. Biol. Anim., Biochim. Biophys.* **12,** 367–375.
Ramakrishna, P. A., and Prasad, M.R.N. (1967). *Folia Primatol.* **5,** 176–189.
Ramaswami, L. S., and Anand Kumar, T. C. (1962). *Naturwissenschaften* **5,** 115–116.
Ramaswami, L. S., and Anand Kumar T. C. (1965). *Acta Zool. (Stockholm)* **46,** 257–273.
Rao, C.R.N. (1927). *J. Bombay Nat. Hist. Soc.* **32,** 206–208.
Rao, C.R.N. (1932). *J. Mysore Univ.* **6,** 140.
Richard, A. (1974). *In* "Prosimian Biology" (R. D. Martin, G. A. Doyle, and A. C. Walker, eds.), pp. 49–74. Duckworth, London.
Ross, G. T., Cargille, C. M., Lipsett, M. B., Payford, P. L., Marshall, J. R., Stroth, C. A., and Rodbard, D. (1970). *Recent Prog. Horm. Res.* **26,** 1–62.
Rowell, T. E. (1972). *Adv. Study Behav.* **4,** 69–105.
Sauer, E.G.F., and Sauer, E. M. (1963). *J. S.W. Afr. Sci. Soc.* **16,** 5–36.
Spühler, O. (1935). *Z. Zellforsch. Mikrosk. Anat.* **23,** 442–463.
Ulmer, F. A. (1963). *Zool. Gart., Leipzig* **27,** 106–121.
van Herwerden, M. (1925). *Anat. Rec.* **30,** 221–223.
Van Horn, R. N. (1975). *Folia Primatol.* **24,** 203–220.
Van Horn, R. N., and Resko, J. A. *Endocrinol.* (in press).
Vincent, F. (1969). Doctoral Thesis, University of Paris (unpublished).
Yerkes, R. M., and Elder, J. H. (1936). *Comp. Psychol. Monogr.* **13,** 1–39.
Young, W. C., and Orbison, W. D. (1944). *J. Comp. Psychol.* **44,** 492–500.
Zuckerman, S. (1932). *Proc. Zool. Soc. London* pp. 1059–1075.

Chapter 4

Maternal Behavior in Prosimians

P. H. KLOPFER AND K. J. BOSKOFF

1. INTRODUCTION

1.1. Why Study Maternal Behavior in Prosimians?

Studies of prosimian maternal behavior contact two disparate areas. On the one hand, they provide descriptions of a little-known group of primates, and thus touch

ISBN 0-12-222150-8

problems of primate evolution. On the other, they have relevance to the area of infant development. In this regard, prosimians are an advantageous group with which to generate models of the ontogenesis of social behavior.

The old and limited geographical range of extant prosimians (restricted to Madagascar for most species) suggests a close relationship among species. Most prosimian species are threatened with extinction because of rapid destruction of their habitats. Thus, limited opportunities for their study remain. As to captive animals, few large collections of prosimians exist outside of Madagascar. In short, the unique geographical and systematic position of prosimians (Buettner-Janusch, 1963) and their endangered status make studies of prosimians especially important to primatologists and others interested in the origins of the social behavior of primates (see Martin, 1972a).

The second area of contact is with the study of maternal–infant relations and the subsequent development of social relations between maturing infants and other adults. No attempt will be made to review the literature dealing with this subject in humans (e.g., Gewirtz, 1969). The animal studies are only slightly less voluminous, particularly experimental studies that utilize nonhuman primates. Much of this work has been recently reviewed (Klopfer *et al.*, 1973), or summarized in topic-oriented anthologies (e.g., Rheingold, 1963; Schrier *et al.*, 1965a,b, 1971; DeVore, 1965; Jay, 1968; Fox, 1972). It is epitomized by the reports of Hinde *et al.* (1964; Hinde and Spencer-Booth, 1971; Hinde, 1972), Preston *et al.* (1970), Castell and Wilson (1971), and Denenberg (1969), to name but a few examples. It should, however, be noted that only a very few primate species are represented by these studies.

1.2. The Problem of Species Differences and Extrapolation

There has been a tendency to extrapolate directly from particular studies to man, sometimes with politically important effects. D. Lehrman has asked how United States congressmen who opposed child-care centers would have responded had studies of motherless bonnet monkeys been presented to them rather than studies of rhesus monkeys. Bonnet macaques tolerate separation far better than do rhesus macaques (Kaufman and Rosenblum, 1969; Harlow and Harlow, 1965). This implies that evolutionary and comparative studies of behavior are of limited value if one expects, merely by extrapolation, to understand the origins, functions, or mechanisms of behavior. Such studies must rather (in the words of G. E. Hutchinson, 1965) be designed to generate the rules of the evolutionary game as it is played in the ecological theatre.

Similarities between organisms can be due to convergence (analogies) as well as to kinship (homologies). The green color of many insects and amphibians of the tropical rain forest can arise from very different sources, physical structure, on the one hand (hairs, refracting scales), or pigments, on the other. Even pigments with similar spectral reflectances may be composed of different molecules. The important distinction between analogies and homologies is largely dependent on prior knowledge of the phyletic relationship and physiology of the animals in question. If one already knows bats to be mammals and flies insects, then one can classify their wings as analogous structures. If one has no basis for deciding upon relationship, one cannot conclude that the wings may be homologies (Klopfer, 1973).

Behavior at the organismic level is especially malleable and subject to convergence. What one can do, however, is to generate "the rules of the game"; that is, one can specify both the ecological factors and evolutionary accidents that have provided the significant constraints on particular patterns of behavior. Having done this for an array of species, generalizations applicable to others would then be expected to emerge. For instance, Crook (1970) and Eisenberg *et al.* (1972) have examined the relation between habitat and social structure in various primates. Arboreal, diurnal, leaf-eating species tend toward small home ranges and troops with but a single adult male, irrespective of taxonomic relationships. The common problems of predation by cats and foraging in tree tops have encouraged the evolution of similar adaptations (Eisenberg *et al.*, 1972).

Prosimian primates are excellent subjects for studying the relationship between ecology and behavior. Though, for the most part, they inhabit a limited range, related species have nonetheless adapted to a variety of ecological conditions. Thus, one can look for constellations of correlations with which to formulate "rules of the game"—rules which serve a predictive as well as a descriptive function, relating habitat choice and behavior.

For example, it might be found that several prosimian species which are nocturnal and solitary-living also tend to traverse a relatively large home range and feed on insects rather than on leaves. After examining the correlations among these factors, these data can be used to look at species for which information is incomplete. Suppose there is a species that is available only in captivity, or is inaccessible in its natural habitat. The animal eats insects and is active during the night. Can a prediction be made, with any degree of reliability, that this species is also solitary?

Of course, as warned earlier, care must be taken in extrapolations. It is quite tempting to assume that the closely related prosimian species should display similarities in patterns of relationships between ecology and behavior. However, differences among species are important, too, and indicate ways in which animals with similar physical characteristics have evolved different mechanisms for dealing with the same problems.

In this chapter, the focus of attention is on maternal behavior—how the various prosimian species handle the production and rearing of young—and its relation to other aspects of life style. By compiling the data available on the prosimians at this time, one can see what correlations are apparent between patterns of life style and maternal care. Table I presents the data on parameters of life style in prosimians, while Table II does the same for parameters of reproductive and maternal behavior.

TABLE I

Parameters of Life Style in Prosimians

Family	Microhabitat	Diet	Activity pattern	Social organization
Tarsiidae	Secondary forest range: 0–8 m $\bar{x}$ = 2–3 m	Insectivorous	Nocturnal	Male–female pairs; with overlapping ranges
Lorisidae				
A. Lorisinae	Primary and secondary forest Range: low levels (0–5 m); High levels (5–30 m)	Omnivorous (primarily insectivorous)	Nocturnal	Solitary or male–female pairs Overlapping ranges (*Perodicticus*)
B. Galaginae	Primary and secondary forest; canopy and undergrowth; woodland, bush, scrub Range: 0–50 m	Insectivorous and frugivorous	Nocturnal	Sleeping groups (*G. senegalensis; G. crassicaudatus*) overlapping male–female ranges (*G. demidovii*)
Daubentoniidae	??? Range: 0–70 m	Insectivorous	Nocturnal	Solitary
Indriidae	Range: mostly above 13 m; occasionally to ground	Folivorous	Nocturnal (*Avahi*) Diurnal (*Indri, Propithecus*)	Family groups (2–5 members)

Cheirogaleidae				
A. Cheirogaleinae				
1. *Microcebus*	Fine branches of secondary forest (0–30 m); most activity 2–4 m	Omnivorous (insects and fruit)	Nocturnal	Overlapping ranges; males, alone or pairs; females, nesting groups
2. *Cheirogaleus*	Thick branches of low levels	Omnivorous (fruits and insects)	Nocturnal	Solitary or pairs
B. Phanerinae	Medium-sized branches, low and high	Insectivorous (and vegetable resins)	Nocturnal	Solitary or pairs
Lepilemuridae	Bush and gallery forests 3–10 m	Folivorous	Nocturnal	Overlapping ranges
Lemuridae				
A. *Hapalemur*	Lower regions of forest and bamboo thickets	Folivorous	Crepuscular	Family groups (3–6)
B. *Lemur* and *Varecia*	Gallery forests	Omnivorous (folivorous, frugivorous)	Diurnal (*L. mongoz*, nocturnal)	Troops 2–5 (*Varecia*, *L. mongoz*); 4–17 (*L. fulvus*); 16–24 (*L. catta*)

2. PATTERNS OF LIFE STYLE AND MATERNAL BEHAVIOR IN PROSIMIANS

Species of the infraorder Prosimii display many subtle and puzzling variations in maternal behavior. For example, some prosimian mothers transport their infants by mouth, some carry their babies on their bellies while lending a supportive hand, and still others carry the young on their backs with support depending entirely on the offspring's ability to cling. Similarities and differences among species are of course interesting to describe and categorize, but they become informative only when one also considers the constraints that shape a particular species' maternal behavior. Why (to follow up on the present example) are there such differences in modes of transport? Is one determinant the type of habitat through which the female must navigate, or the sort of food for which she must search; when she is active and when she must sleep; the sort of social structure of which she is a part; the state of development of her newborn infant; the course of growth and maturation which the infant takes? How are these factors, and others, related to the style of maternal care?

Descriptive data on most species are still far from complete; a final satisfactory analysis, therefore, of relations between ecological and behavioral parameters must necessarily await the filling-in of information gaps. However, at this point, the information available on the various prosimian species can be pulled together, and a search for obvious trends can be begun (see Tables I and II). The following pages are devoted to a more detailed description of the information contained in the tables, and point to indications of emergent "rules" describing the relationship between life style and maternal care.

2.1. Family: Tarsiidae

The three species comprising the tarsier family are found exclusively in southeastern Asian rain forests, principally in secondary forest growth. Tarsiers possess several characteristics which make them different from other prosimians; for example, they have two grooming claws on each foot rather than the one typical of most other prosimians; they do not have a toothcomb formed by the lower canines and incisors; and they have a diploid number of 80 chromosomes, among the highest number reported for mammals (Buettner-Janusch, 1973). These nocturnal prosimians display the vertical clinging and leaping type of locomotion. They range vertically from the ground to approximately 8 m (Fogden, 1974), with most activity occurring at 2 to 3 m (see Niemitz, Chapter 14). Tarsiers eat insects and small vertebrates. Their activity is essentially nocturnal; they become active at twilight, rest during the middle of the night, and are again foraging before daylight (Buettner-Janusch, 1973; Fogden, 1974). Niemitz (Chapter 14) observed that activity began only after complete darkness had fallen. During the day, they rest in dense undergrowth.

Tarsier social organization has been considered to be that of male–female pairs (Napier and Napier, 1967; Buettner-Janusch, 1973); however, recent field work by

TABLE II

Parameters of Reproduction and Maternal Care in Prosimians

Family	Duration of gestation	Litter size	State of infant at birth	Mode of transport by mother	Nest?
Tarsiidae	180 Days	1	Eyes open; can cling and crawl	Clings to abdomen (occasionally by mouth)	No
Lorisidae					
A. Lorisinae	130; 165; 193; 219 days	1–2	Eyes open; can cling to abdomen and branches	Clings to abdomen	No
B. Galaginae	120–145; 130–135; 180 days	1–2	Eyes open; can cling to abdomen and branches	Mouth; then by back	Yes
Daubentoniidae	?	1	?	Clings to fur—back or abdomen?	Yes
Indriidae	120–150 days	1	Eyes open; clings to mother's fur	Clings to abdomen; then to back	No
Cheirogaleidae					
A. Cheirogaleinae					
1. *Microcebus*	60 days	1–3 ($\bar{x}$ = 2)	Incapable of locomotion	Mouth	Yes
2. *Cheirogaleus*	70 days	2–3	Eyes closed	Mouth	Yes
B. Phanerinae	?	1	?	Clings to abdomen; then to back	Yes
Lepilemuridae	120–150 days	1	Eyes open; clings to branches	Mouth	Yes
Lemuridae					
A. *Hapalemur*	135–150 days	1–2	?	Clings to abdomen; mouth; back	Yes
B. *Lemur* and *Varecia*	120–135 days	1–2 (*Lemur*) 2–3 (*Varecia*)	Eyes open; can cling (*Lemur*)	Clings to abdomen; then back (*Lemur*); mouth (*Varecia*)	No (*Lemur*) Yes (*Varecia*)

Fogden (1974) suggests a system similar to that of *Microcebus murinus* and *Galago demidovii*. According to Fogden's data, *Tarsius bancanus* males and females (equal sex ratio) live in small home ranges, with relatively exclusive areas among members of the same sex, and rather extensive overlap between the sexes. He found no evidence that females share nest sites. The largest males were often seen accompanied by different females; smaller adult males also had contact with females, but may have been subordinate to the larger males for breeding purposes. Another group of immature males occupied marginal positions, with no permanent home range and no sexual activity. (See Niemitz, Chapter 14, for a further discussion of social grouping.) Tarsiers are much less vocal than other prosimians (Napier and Napier, 1967; Jolly, 1972). It has been suggested, therefore, that tarsiers are less social, but this merits further attention.

Napier and Napier (1967) and Walker (1968) report that tarsiers breed all year round, but Fogden (1974) has recently described a sharply defined breeding season which coincides with fluctuations in abundance of insects. The duration of gestation is approximately 180 days (Napier and Napier, 1967) and typically one infant is produced. Niemitz (Chapter 14) provides a description of the development of a tarsier infant. The baby is born with eyes open, and is capable not only of clinging to the mother's belly, but also of scrambling around at about 2 days (see Napier and Napier, 1967; Niemitz, Chapter 14). No nest is built. The infant is transported on the mother's belly, but Ulmer (1963, cited in Napier and Napier, 1967) reports that the mother occasionally carries an infant by mouth. The young tarsier can climb at 2 days, jump at 4 days, and displays almost the full adult repertoire of locomotor acts by 19 days.

2.2. Family: Lorisidae

2.2.1. Subfamily: Lorisinae

Among the four genera of lorisine primates little information is available on the life style in the wild of the two Asian forms, *Loris tardigradus* (slender loris) and *Nycticebus coucang* (slow loris), while quite a good deal is now known of the two African forms, *Periodicticus potto* (potto) and *Arctocebus calabarensis* (angwantibo) (see Charles-Dominique, 1971). All four species are nocturnal and primarily insectivorous.

Loris tardigradus are reported to catch and eat birds in the wild (Phillips, 1931), as well as lizards (Pocock, 1939; Sanderson, 1957) and, in captivity, to supplement a largely insectivorous diet with fruit and small vertebrates like baby mice (Bishop, 1964) and, in the wild, besides insects and fruit, they eat leaves, seeds, birds and birds' eggs, and lizards (see Napier and Napier, 1967). *Arctocebus calabarensis* eat live insects and fruit in captivity (Jewell and Oates, 1969) but Charles-Dominique (1966) reports that, in the wild, caterpillars make up 90% of their diet, the remaining 10% being made up of other insects and a small amount of vegetable matter. Bishop (1964) reports that *P. potto* are much more omnivorous, at least in captivity. Charles-Dominique (1966) reports that animal matter comprises, on an average, only 20% of the diet of this species in the wild, mainly large and slow insects, the remainder being made up of fruit and gum. Charles-Dominique (1971) notes that the larger lorisids eat the larger insects, the stomachs of a sympatric species like *Galago alleni* containing none of the insects found in the stomachs of *P. potto*. His studies of five sympatric species of lorisid in Gabon led him to conclude that food preferences are learned from the mother (and see Hladik, Chapter 8).

The lorisines have the widest distribution of all prosimians, ranging from African tropical rain forests (*P. potto* and *A. calabarensis*) to southeast Asia (*N. coucang*) including Ceylon and southern India (*L. tardigradus*). Each genus inhabits a particular vertical range of the forest. For instance, *A. calabarensis* live in the lower levels (0–5 m) of primary and secondary forest, while *P. potto* prefer the forest canopy, 10–30 m in primary, 5–15 m in secondary forest (Charles-Dominique, 1974b). Members of this subfamily have a characteristic slow-moving locomotion, climbing very slowly and quadrupedally without leaping.

The social structure of the lorisines is largely unknown; they are generally considered to live singly or in pairs (Napier and Napier, 1967). Charles-Dominique (1974a) has recently described the social organization of *P. potto*. As in other prosimians considered to be solitary, the potto system involves overlapping ranges of males and females. Large males and females are sedentary, while smaller males do not maintain fixed ranges. The duration of gestation apparently ranges from about 130 days for *A. calabarensis* to about 167 days for *L. tardigradus* and 193 days for *N. coucang* and *P. potto* (see Doyle, Chapter 5, for details on the relation of gestation to development.) Females possess two or three pairs of mammae, and typically only one infant is produced at a time. Cowgill (1974) reports an instance in which a captive *P. potto* female produced twins but then had difficulty feeding them both.

At birth, lorisines have their eyes open and are capable of clinging tightly to the fur of the mother's belly. They also can cling to branches, and are often left by the mother on the first day after their birth. These species do not build nests for the young (Martin, 1972a). The infant achieves independent locomotion at about 2 weeks (in *N. coucang*) (Napier and Napier, 1967).

2.2.2. Subfamily: Galaginae

Much more is known about the galagine relatives of the lorisines. The galagos are also nocturnal; they are found only on the African continent. *Galago elegantulus* (*Euoticus elegantulus*) and *G. inustus* inhabit the forest canopy, from as low as 3–4 m and up to 50 m (Charles-Dominique, 1974b). *Galago demidovii* live in dense canopy vegetation, ranging at 10–30 m heights in primary forest, and 0–10 m in secondary forest (Charles-Dominique, 1972, 1974b). *Galago alleni* take to the undergrowth of primary forests, living at heights of 0–2 m (Charles-Dominique, 1974b). *Galago crassicaudatus* inhabit dense parts of secondary forest, while *G. senegalensis* have a wider range of habitats, including open woodland, bush, and scrub (Bearder and Doyle, 1974). All the galagos eat insects, but their diets can vary with seasonal changes in availability of food, and in captivity they will also take fruits and vegetables (Charles-Dominique, 1971; Napier and Napier, 1967). In contrast to the slow-climbing lorisines, galagos have been classified as vertical clingers and leapers which, if they do descend to the ground, hop bipedally or quadrupedally. Specialized extremities aid both in locomotion and in prey catching.

The social organization of only a few species of galagos has been studied in the field. Charles-Dominique (1971) has studied the social system of *G. alleni* and has commented on *G. elegantulus*. *Galago senegalensis* and *G. crassicaudatus* have been observed by Bearder and Doyle (1974). Both appear to form stable sleeping groups of from 2 to 4 (*G. crassicaudatus*) or 2 to 6 (*G. senegalensis*) individuals. These associations, whose members usually sleep in nests or tree holes during the day, may be family groups involving an adult male and female and several generations of their offspring, or females with infants and males with older offspring. Solitary animals are also frequently sighted. At night, animals usually disperse to forage alone.

Charles Dominique (1972) has described the social organization of *G. demidovii* in the wild. Like *Lepilemur* and *Microcebus,* this species of galago is organized into a system of overlapping ranges. Female ranges overlap each other, and ranges of larger males overlap with those of several females. Other males display varying degrees of contact with males and females.

Gestation lengths are recorded for *G. senegalensis* as 120 days (Lowther, 1940), 121–125 days (Doyle *et al.,* 1967, 1971), and 144–146 days (Manley, 1966). The record is 130–135 days for *G. crassicaudatus* (Buettner-Janusch, 1964), and 180 days for *G. alleni* (Vincent, 1969), though 130 days may be a more likely estimate (G. A. Doyle, personal communication). Multiple births are most frequent in the former two genera, although single births are not uncommon. Doyle *et al.* (1971) report that first births for *G.s. moholi* are usually singletons, while later births are usually twins.

Galago neonates have open eyes and are able to cling tightly to the mother's belly fur. They, like the lorisines, can cling to branches and are often left there from day 1 on. They are also able to crawl clumsily at an early age: *G. senegalensis,* day 1 (Doyle, Chapter 5), *G. demidovii,* day 1 (Vincent, 1969), and *G. crassicaudatus,* day 8 (Rosenson, 1972). Maternal transportation of the infant *G. crassicaudatus* is by mouth initially, and then by means of the infant's clinging to the mother's back or abdomen (Buettner-Janusch 1964). Vincent (1969) reports that *G. demidovii* young are carried by mouth for about 4 weeks, while Rosenson (1972) found that in the laboratory *G. crassicaudatus* mothers carried their babies for 7 weeks in this fashion. The development of *Galago* infants is reviewed by Doyle (see Doyle, Chapter 5, 1974a, 1974b; Doyle *et al.* 1967, 1969, 1971); since purposes here involve only summarizing parameters of maternal care, the reader is directed to his chapter for further information on galagos.

2.3. Family: Daubentoniidae

The aye-aye (*Daubentonia madagascariensis*) is one of the most elusive and endangered of the prosimian species. Little information is available on this species, which alone comprises the family Daubentoniidae. It ranges throughout the rain forests of eastern coastal and northern Madagascar. It is an arboreal and nocturnal animal that builds sleeping nests some 40–70 m above ground in tree forks (Petter and Petter, 1967). *Daubentonia* are slow-moving animals, occasionally descending to the ground (Petter, 1967). The diet consists of insects and fruits; a uniquely slender third digit on the hand is used to probe holes for food and to drink water.

Daubentonia appear to be a solitary-living species; the only reported groups are those composed of females with their young (Petter, 1962, 1967). Virtually nothing is known regarding reproductive behavior of the aye-aye except that it apparently produces only one infant which clings to the mother's fur (Petter, 1965). The female aye-aye possesses a pair of inguinal mammae, a feature peculiar to this species and different from all other primates.

2.4. Family: Indriidae

Indriids are difficult to observe in the wild and have been virtually impossible to maintain in captivity. They are leaf-eating, arboreal creatures, and seem to exhibit a somewhat wider variety of life styles than is found in the other prosimian families.

2.4.1. *Avahi laniger*

The woolly lemur, *Avahi laniger,* is the smallest indriid and is the only nocturnal genus of the family. It is found in eastern and northwestern rainforests of Madagascar, and eats leaves, fruits, bark, and flowers (Napier and Napier, 1967). They have been reported (Petter, 1962) to form family units of two to four individuals; however, Buettner-Janusch (1973) indicates a more solitary social system with females accompanied by infants but not by adult males. Recently, Pollock (1975) found groups composed of an average of two individuals, ranging from one to four.

Mating occurs once a year. One infant is born following a 120- to 150-day gestation period (Walker, 1968). The baby is carried on the mother's belly for a while, and is then transported on her back as in other indriids. Petter-Rousseaux (1964) reports infants still clinging in this fashion at 150 days.

2.4.2. *Indri indri*

The largest indriid, *Indri indri,* unlike *Avahi,* is strictly diurnal. Known to the Malagasy as *Babakoto,* "little father" or "man of the forest," *Indri* inhabit the eastern rain forests, feeding on leaves and fruits. The social system is based on a group, composed of parents and immature offspring. According to usch (1973) sightings of up to 6 animals have been made. More recent , though, has revealed an average group size of 3, with a population de... timated at 9–15 animals per km^2 (Pollock, 1975). Petter and Peyriéras (1974) suggest that the normal group size is actually 2 to 4 individuals, but has increased as a result of human disturbance.

Little information on infant development or maternal behavior exists at present, except that one infant at a time is produced, probably at infrequent intervals of up to 3 years (Pollock, 1975). Although the gestation period was earlier estimated at 60 days (see Walker, 1968), Pollock (1975) estimates a 120- to 155-day duration from his field studies. The infant is carried ventrally, like *Propithecus,* for approximately 4 to 5 months, after which time it is carried dorsally until about 7 or 8 months of age (Pollock, 1975).

2.4.3. Propithecus verreauxi

Of the three indriid genera, most information is available on *Propithecus,* especially on *P. verreauxi.* There are more subspecies of *Propithecus* than of other indriids and they are more widely distributed. These animals have been successfully maintained in captivity, and a private reserve in Madagascar has made field studies possible (e.g., Jolly, 1966; Klopfer and Klopfer, 1970). Called "sifaka" by the natives because of their distinctive vocalizations, *Propithecus* live in all forest areas throughout Madagascar, except for the northwest corner of the island. Diurnal like *Indri,* they have a similar diet of fruits, flowers, and leaves. The social organization is again based on a family group, composed of 3–5 and up to 7 or 9 members (Jolly, 1966; Richard, 1974). Jolly has studied the composition of large numbers of *Propithecus* troops, and reports a higher proportion of males than females, though this was not true of four groups studied by Richard (1974).

Mating occurs in a restricted breeding season and synchrony of estrus has been documented by Jolly (1966) and Richard (1974). The duration of gestation has not been completely resolved; records of both 130 days (Petter-Rousseaux, 1962; Jolly, 1966; Richard, 1974) and 150 days (Petter-Rousseaux, 1964; Buettner-Janusch, 1973) exist in the literature. One infant is born to a given female, and usually only one infant a year is born to a given troop (Jolly, 1966). The neonate is born with its eyes open and a complete natal coat, and is capable of grasping the mother's fur and clinging transversely to her abdomen. The *Propithecus* mother assumes a special sitting position for her young infant, with knees up and trunk leaning back to form a sort of bowl or "cuvette" for the baby (Jolly, 1966).

Jolly (1966) provides the most complete description of early infant development in wild *Propithecus.* The neonate is not active at birth, but begins to be so at about day 3 to 4. By the age of 6 weeks (day 42), the infant crawls onto the mother's back, but does not consistently do this until 3 months of age. The age cited in the literature for this transfer from mother's ventrum to back varies from day 30 (Petter-Rousseaux, 1964; Buettner-Janusch, 1973) to 42 days (Jolly, 1966). Observations in captivity of a mother–infant pair reveal an age of about 50 days for both initial and consistent transfer (K. J. Boskoff, personal observation). Obviously there are differences between wild and captive animals, and undoubtedly also among different housing arrangements, e.g., group versus isolated situations. The *Propithecus* infant, like *Indri,* continues to ride on the mother for up to 7 months (Petter-Rousseaux, 1964; Buettner-Janusch, 1973).

"Multiple parenting" is perhaps the most significant aspect of maternal behavior documented in *Propithecus* (Jolly, 1966). The parturient mother tends to remain separated from her troop during and immediately following birth; however, all other troop members are interested in the neonate, and soon are allowed to groom both mother and baby. The infant grows up surrounded by caretaking relatives, mother, father, uncles, possibly an aunt or two in a larger troop, and older brothers and sisters. It seems essential to obtain more information on both *Avahi* and *Indri* to see whether this pattern is unique to *Propithecus* or common to the entire family.

2.5. Family: Cheirogaleidae

2.5.1. Subfamily: Cheirogaleinae

2.5.1.1. *Microcebus murinus.* Among the cheirogaleines, perhaps the best studied species is the mouse lemur, *Microcebus murinus.* Martin (1972b, 1973) reports that these arboreal creatures inhabit the dense foliage and fine branches of secondary forest bushes of Madagascar. They range vertically from 0 to 30 m, with most activity occurring between 2 to 4 m, and often descend to the ground. *Microcebus* are quadrupedal and move in a scurrying, rapid-running fashion, as well as by agile leaps (Martin, 1972b).

Microcebus, like the other members of the cheirogaleine subfamily, are strictly nocturnal, sleeping in nests during daylight hours. The diet is an omnivorous one, consisting primarily of fruits and insects. Local concentrations or population nuclei have been described by Martin (1973). Separate male and female nesting sites in bushes and tree holes are found throughout most of the year. Males usually live in pairs or alone and females in larger groups with an average of 4 females (ranging from 1 to 15) and their infants per nest. Within a given area, males and females coexist in a ratio of approximately 1:4. Martin (1973) has concluded that the female groups are stable, though females vary their sleeping sites among several different nests, sharing their beds with many of the females in the region.

During the breeding season, which is suspected to occur twice during the rainy season, females living in social contact exhibit a synchrony of estrus and mating; at this time males can be found in female nest sites. [See Martin (1973) for his description of peripheral and central males and their role in reproduction.] Gestation lasts approximately 60 days, and multiple births are most common, with an average of 2 infants per litter. The neonatal *Microcebus murinus* has its eyes closed and is relatively incapable of movement (Petter-Rousseaux, 1964; Martin, 1972b). Transportation of the infant is by mouth; the female grasps the baby's flank in her teeth, and moves one infant at a time. The newborn is capable of clinging to twigs but never clings to the mother. The eyes open by day 4, and by day 10 the infant can crawl, although it remains in the nest unless transported by the mother. By 2 months, the infant has been weaned and is virtually an adult in its physical and behavioral development. It is no longer carried and the mother is preparing for the second litter of the season (Martin, 1972b, 1973; Andriantsiferana *et al.,* 1974).

2.5.1.2. *Cheirogaleus.* The genus *Cheirogaleus* (dwarf lemur) has not been studied as extensively as has *Microcebus,* but some data are available. Also arboreal and nocturnal, *Cheirogaleus* traverse thicker branches than *Microcebus,* but in a similar horizontal fashion with brief darting movements. However *Cheirogaleus* are not as agile as their mouse lemur cousins and perform less leaping. *Cheirogaleus major* inhabit the lower levels of the rain forest and are reported to live alone or in pairs, nesting in tree holes. The diet is varied, including fruits and insects, but includes fewer insects than does the *Microcebus* diet. *Cheirogaleus medius* are similar but inhabit dry forest (Petter, 1962, 1965).

Petter-Rousseaux (1964) has described reproduction in *Cheirogaleus major* in the laboratory. She reports a 70-day gestation period. In our laboratory, however, a *Cheirogaleus medius* female gave birth to infants less than 60 days following the birth and death of a litter (K. J. Boskoff, personal observation). Two to three infants are born with eyes closed. Like *Microcebus,* the *Cheirogaleus* female carries her young by the skin of the flank or back in her teeth; however, the *Cheirogaleus* infant shows a faster rate of locomotor development, being able to crawl on day 1. By 3 weeks the infant can climb vertical limbs, begins to eat fruit, and often follows the mother out of the nest.

2.5.2. Subfamily: Phanerinae

The last member of the cheirogaleid family, the fork-marked lemur (*Phaner furcifer*) is a little-known prosimian. Also nocturnal, *Phaner* live among medium-sized branches of the Madagascan rain forest, travelling quadrupedally by climbing, springing, and running along branches. Most activity occurs at heights of 3–4 m, but *Phaner* have been seen on the ground, and up to heights of 10 m (Petter *et al.,* 1975). These animals sleep in tree holes or build spherical leaf nests (Pariente, 1974). The nests are inhabited both by *Phaner* and by sympatric *Microcebus coquereli,* but Petter *et al.* (1975) have concluded that *Phaner* are the builders since in areas inhabited by only *M. coquereli* such nests are not found. Individuals do not sleep in the same nest at all times, and often sleep in tree holes or on the tree branches (Petter *et al.,* 1975).

Phaner are reported to sleep alone or in pairs (Petter, 1962; Jolly, 1972), and Petter *et al.* (1975) report solitary activity in feeding. The diet is a specialized one, consisting of vegetable resins (gum, sap) and insect secretions.

Gestation is not on record, and Petter *et al.* (1975) report single young born in mid-November (in Madagascar). The infant is left in a tree hole at first, then carried ventrally by the mother; at a later age, the infant is transported dorsally (Petter *et al.*, 1975).

2.6. Family: Lepilemuridae

The nocturnal *Lepilemur* inhabit the lower regions (3–10 m) of bush and gallery and rain forests of Madagascar. During the day, *Lepilemur* sleep in tree hollows or nests (Martin, 1972a; Pariente, 1974). The diet is primarily a folivorous one, consisting of leaves and flowers; this diet prevents these animals from having to move extensive distances to feed.

Hladik and Charles-Dominique (1974) have recently updated knowledge about *Lepilemur* social organization, which involves population nuclei similar to those described for *Microcebus* and *Galago demidovii*. Territories of individual males and females overlap; male territories are larger than those of females, so that the range of a given male may overlap with those of several females. The larger males may be associated with five females, while other males interact with only one or two females.

Reproduction in *Lepilemur* has been described by Petter-Rousseaux (1964). After a 120–150-day gestation, one infant is delivered. The newborn has its eyes open and is fairly active. It can cling to branches and the mother often leaves her baby clinging while she feeds (Petter, 1965). When the mother rests, the infant clings to her, but transportation of the infant is by mouth. At 1 month of age, the *Lepilemur* young can climb and jump.

2.7. Family: Lemuridae

2.7.1. Subfamily: Lemurinae

2.7.1.1. *Hapalemur*. Living in lower regions of the forest and bamboo thicket of Madagascar (Petter, 1962), the two species of the *Hapalemur* genus have a diet of primarily bamboo shoots and leaves. One variety of *H. griseus* has been seen to descend to the ground, while another, larger variety, reportedly swims (Petter and Peyriéras, 1974). Earlier reports ascribed a nocturnal activity pattern to *Hapalemur* (see Walker, 1968) but more recent work has revealed a crepuscular pattern (Petter, 1965; Petter and Peyriéras, 1974). They are active in early morning, rest during most of the day, resuming activity in late afternoon until nightfall. In captivity at the Duke University Primate Facility, *Hapalemur* are also crepuscular.

The social organization of *Hapalemur* involves a family group of between 3 and 6 individuals with an adult male–female pair and their juvenile offspring (Petter, 1962; Petter and Peyriéras, 1974). Little information is available regarding reproduction. Records from the Duke University Primate Facility indicate a gestation period of 135–150 days. Both single (three instances) and twin (one instance) births have occurred; two births occurred in October, and two were in June. From field research, Petter and Peyriéras (1974) report that single young are born in January and February.

These latter authors report that the infant is carried dorsally from birth, and that they did not see any ventral transport. At the Duke University Primate Facility, however, the infant has been seen to cling and be carried ventrally, to be left in a nestbox, and to be carried in the mother's mouth (K. J. Boskoff, personal observations). Petter (1962, 1965) also reported mouth transport and depositing of the infant in a tree fork. By 3 weeks of age in captivity, the infant rides on the mother's back.

2.7.1.2. *Lemur* and *Varecia*. The six species of *Lemur* and one of *Varecia* are, with one exception, diurnal or crepuscular prosimians which inhabit the Madagascan forests. All are frugivorous and folivorous. They traverse the larger branches of the trees, locomoting quadrupedally and with great agility.

Virtually nothing is known about *Lemur rubriventer* or *L. coronatus* in the wild. A recent paper by Tattersall and Sussman (1975) has provided the first major description of *Lemur mongoz* ecology. This species, previously described as diurnal by Hill (1953), was found to be strictly nocturnal during the (limited) period of study of less than 2 months. These animals exhibited a very specialized diet of mainly nectar from flowers. They were never seen on the ground, but remained in the highest levels of 10–15 m in continuous canopy. Social organization was found to consist of family groups, with adult male and female and immature young. A subadult female which was seen to travel alone and copulate with a male from another group led the authors to suggest that offspring reaching maturity tend to become independent of the parental group. No further information regarding reproduction and maternal behavior is available at this time.

The remaining three species of *lemur*—*L. catta, L. fulvus,* and *L. macaco*—have been studied more extensively in the wild (for *L. catta:* Jolly, 1966; Klopfer, 1972; Sussman, 1972, 1974; for *L. fulvus:* Harrington, 1971, 1974, 1975; Sussman, 1972, 1974, 1975; for *L. macaco:* Petter, 1965; Jolly, 1966). *Lemur catta* inhabit the drier gallery forests of western and southern Madagascar, where they are often sympatric with *L. fulvus* (except in the southern regions, where *L. fulvus* do not live). From Sussman's (1972) field study, it is known that *L. catta* feed on twice as many plant types as do *L. fulvus,* and range twice as far during the day. *Lemur macaco* are also entirely herbivorous and frugivorous and, like *L. catta,* feed on a wide range of plant types (Jolly, 1966). In fact both Petter (1965) and Jolly (1966) note that they are very similar to *L. catta* in their general activity patterns, sleeping, waking and mode of locomotion. *Lemur catta* are also the most terrestrial of the lemurs, spending about 20% of their time on the ground in the wild (Jolly, 1966) but may spend as much as 30% of the time on the ground, unlike *L. fulvus,* for example, which rarely descend to the ground, preferring to remain in the forest canopy (Sussman, 1974).

These three lemur species form social groupings of larger size than other prosimians. *Lemur catta,* for instance, are reported by Jolly (1966) to live in troops of 12–24 animals, by Klopfer and Klopfer (1970) 16–20 animals, and by Sussman (1974) 15–20 animals. *Lemur fulvus,* on the other hand, form smaller groups of around 10 individuals (Ramanantsoa, 1975); Harrington (1974) saw two troops of 12 members each, and Sussman (1974) reported troops ranging from 4 to 17 animals with a mean of 9.5. Petter (1965) reported *L. macaco* troops of 10 to 15 animals and Jolly (1966) reported a mean troop size of 9 for the same species.

One mating season per year occurs in these lemurines. Gestation lasts for 120–135 days. Typically one infant is produced by *Lemur,* but twins are not uncommon. The infant is well-developed, with eyes open, and capable of clinging to the mother's abdomen. This sort of transport precedes the dorsal clinging position; the age of transfer from ventral to dorsal surface differs with the species.

Varecia variegata (*Lemur variegatus*) live in the dense, closed canopy of eastern Madagascar, and is thus extremely difficult to observe. However, it has been reported that this species lives in small groups of 2–5 individuals (see Napier and Napier, 1967; Klopfer and Klopfer, 1970).

Varecia, in contrast to *Lemur,* usually deliver twins or triplets. The infants do not cling to the mother, but are kept instead in a nest and transported by mouth.

3. DETAILED STUDIES OF MATERNAL CARE IN *LEMUR* AND *VARECIA*

3.1. Methods of Study

It is appropriate to describe first the methods that have been employed in studies of maternal behavior, for there is ample ground to suspect that what one sees is influenced by how one looks.

In general, methods of study may be characterized as based on naturalistic or laboratory techniques. The former may be anecdotal or may employ formal sampling procedures. The latter may be merely observational, utilizing the same procedures as in the natural setting, or may be manipulative or experimental.

3.1.1. Naturalistic Observational Techniques (in Madagascar)

These methods may be illustrated by the description of our own procedure in Madagascar during the observations of mother–infant relations in *Lemur catta.* A minimum of 100 hours of contact with the lemurs was recorded during each of three periods. The procedure was to stalk the animals from a distance that was sufficient to avoid elicitation of the "click-grunt" alarm (see Jolly, 1966). This distance varied from 3 to over 30 m, depending on cover. In fact, the animals in undisturbed areas, or where they are protected, take remarkably little cognizance of man, having been rarely frightened or fed by him in the area chosen for study. Adults and older juveniles frequently approached within arm's reach when one sat quietly. Binoculars (8 × 30 mm or 7 × 50 mm) were used for identification; distances were either paced or estimated, the estimates being periodically checked with an optical range finder. Observations began at about 0500 hours, which was some 30 minutes before sunrise, and extended to no later than 1800 hours (often only to 1700 hours). The lemurs actually continued to feed until dark (1900 to 1930 hours, depending on cloud cover), but rarely moved any distance after 1700 hours.

Upon locating a troop and taking up an acceptable vantage point, the observer selected the nearest female with young and proceeded to observe her for at least 30 minutes at a time. Particular animals could be recognized due to individual differences in appearance. If the pair were lost sight of prior to the completion of the 30-minute period, those observations were not included in the data summary. After any one female had been observed for a total of 10 hours (spread over several days), no further data from her were included. Thus, each of the 100 hours of observation of mother–infant pairs was distributed among at least ten different females. No female was represented by less than 30 minutes of continuous observation, nor by more than 10 hours of total observation. This procedure was developed to minimize bias that could result from a small and nonrandomly selected sample. At 30-second intervals during a 30-minute observation session observers recorded the following items: Was the infant asleep or awake (i.e., infant nursing or stirring, eyes open)? Was it in contact with its mother, some other animal, or was it separated? If it was separated, by what distance and for how long was it separated? Other details were recorded as opportunity provided, e.g., groups, and intratroop synchrony in behavior.

The 30-second intervals at which maternal–filial relations were sampled were indicated by a small battery-operated audiosignal marker, pocket size, with an ear-plug head set. The quality of the 30-second time signal differed from that at 60-second intervals, allowing for a convenient and foolproof timing of events.

3.1.2. Observational Techniques in the Laboratory (at Duke University Primate Facility)

The subjects of the laboratory studies were 7 females and their infants for each of the two species of lemur, *Lemur catta* and *L. fulvus,* and 3 female *Varecia variegata* and their 8 infants (see Figs. 1–3).

The observation chambers of the Duke University Primate Facility, within which the animals lived, are bare, hexagonal rooms with a volume of about 82 m^3. The ceiling tapers to a height of 5 m; the maximum distance from corner to corner across the floor is 5.3 m. Corner shelves and a single central post with several horizontal branches provide perches. Observations were made through one-way glass either

Fig. 1. Two *Lemur catta* mothers pose with their vertically clinging twin and single infants, aged 7 weeks. (Photograph: K. J. Boskoff.)

from an overhead mezzanine or at ground level (see Klopfer and Klopfer, 1970). The rooms are not fully soundproof, but only especially loud sounds distract the animals. Two groups of animals can be observed while in adjoining outdoor runs, about 150 m^2 in area. Here, the observer watches from inside a special building.

The animals were introduced into their chambers at least several weeks prior to the birth of an infant. Formal recording of data commenced upon the discovery of a newly born infant, usually within hours, always within 24 hours after birth, and continued for 215 days, by which age the young no longer maintain a special relationship with their mothers.

The animals were maintained on a light schedule with variable daylength approximating the natural seasonal pattern. Nocturnal observations were made to establish activity patterns but the maternal data were collected only during daylight hours. Observations were evenly spaced throughout those periods of the day when the animals were known to be most active. (Nocturnal activities were observed with a "starlight scope," a light amplifying and magnifying device.) These observations were supplemented by videotape records, which served to confirm that periods of presumed activity were indeed that.

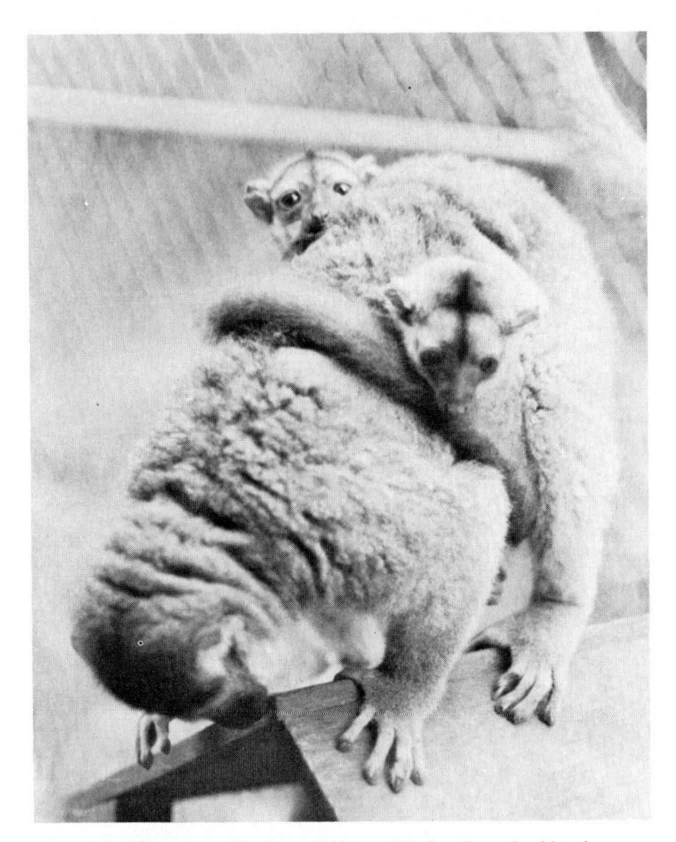

Fig. 2. A *Lemur fulvus* female sniffs the substrate while her 7-week-old twins peer out from their transverse cling which wraps them around the mother's body. (Photograph: K. J. Boskoff.)

Each observation period, scheduled for a particular and different time each day, was of 60 minutes duration. Data were compiled by noting the position or responses during 3 second periods, at half-minute intervals. Records were kept of with whom an infant was in contact, distance from the mother, and manner of behavior, particularly whether grooming or being groomed. Data were grouped into 30-minute data periods and for these the percent of all observation periods for which an infant was separated, mean and maximum distance from the mother, grooming and nursing frequency, etc., were computed. Differences were assessed by conventional nonparametric statistics.

Experimental procedures, of course, vary considerably with the question being asked. Details on data-taking procedures employed elsewhere have been summarized by Hinde (1972).

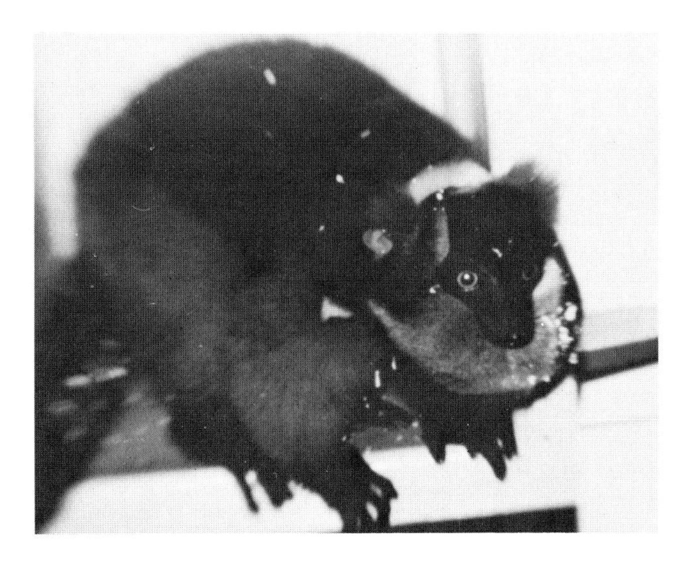

Fig. 3. A *Varecia variegata* mother retrieves her sawdust-covered week-old infant which has fallen out of the nest box. She uses the typical mouth-carry. (Photograph: K. J. Boskoff.)

3.2. Caveats

The separation of mother and infant among some species of macaques may lead to a severe reduction in the infant's activity. It may assume a rigid, seated posture, hands clasped over bowed head, clearly an unhappy, depressed-appearing creature. So similar is the aspect of separated rhesus infants to that of comparably treated human babes, that the same term has been applied to describe both: anaclitic depression (note Kaufman and Rosenblum, 1969). Unfortunately, this label is more than a purely descriptive one: it implies a particular etiology and mechanism. Once the label has been applied, however, this implication unobtrusively passes from assumed to accepted fact. Rhesus macaques are then studied for the insight they provide into human behavioral disorders. Yet, how likely is the behavior of separated rhesus and human infants to be truly homologous? Other species of macaques, assuredly evolutionarily more akin to rhesus than is either species to man, do not show the same syndrome (at least not if a surrogate parent is available). Even if one is not generally skeptical about the utility and validity of so-called homologies, these must surely be suspect when created by a label. Since descriptive studies of behavior, such as these, inevitably require the use of labels, and common labels are generally more appealing than novelties—ergo "goal box" rather than "chamber X" (note Hinde, 1966)—we must be especially wary of unjustified extrapolations (Klopfer, 1973).

A second cautionary note concerns misleading quantification. The fact that a majority of a population of animals (51%) behaves in a particular fashion may indeed be more significant than the fact that the minority behaves differently. It is equally possible that the existence of a lonely exception will prove the more significant event. A recent example is afforded by studies of maternal care in elephant seals (*Mirounga angustirostis*). One group of workers had reported that females tolerated alien young, even allowing them to suckle. In a later, more thorough study, a second group of observers rejected this conclusion, asserting that elephant seal mothers restricted their care to their own young. Quantitatively, this conclusion was correct: of 50 animals, 25 (50%) were not seen nursing alien kids. However, 18 others were seen to nurse aliens. Comparable studies with goats had indicated that fewer than 1 mother in 20 would normally accept alien kids; a figure of 36% acceptance (a minimum value, since the 25 "exclusive" mothers might have nursed aliens in the observer's absence) surely does not (by contrast) justify the assertion that elephant seal mothers are exclusive. In short, group means must be viewed in a variety of contexts rather than allowed to dictate simplistic conclusions (Klopfer and Gilbert, 1967; Le Boeuf *et al.*, 1972).

Finally, correlations are not necessarily causal explanations. Any particular habitat or social structure may allow more than one set of solutions to the challenges it poses. Thus, the vertically ranging *L. catta* mother might also solve her child-rearing problems by adopting the *V. variegata* practice of using a nest (if nest sites are available). Lemurs, however, by virtue of their ecological diversity and physical similarities, are especially well suited to comparative studies which may be expected to provide insights into the rules of the game, which, hopefully, will also lead to a discovery of causal principles.

3.3. Patterns of Maternal Care

3.3.1. Lemur catta

Shortly after birth, the infant may be seen clinging to the ventral surface of its mother, its body axis parallel to hers, its head upward. Grooming of the infant by the mother is frequent, as are grooming episodes by the juveniles, born the previous year. The mother even tolerates juveniles sharing her lap simultaneously with the infant. When the juveniles' grooming of the infant becomes too vigorous, the mother may snap at them and, for a brief time, drive them off. The males, at this stage, usually avoid the females, though occasional approaches of a male are tolerated. Grooming of the female by the male occurs, but the male's efforts to as much as sniff the infant are repulsed for the first few days of its life. Thereafter, by a week's age, the males may participate in the communal grooming of the infant.

Infants apparently shift from the ventral to a dorsal position (body axis still parallel to that of the mother) on their mother by the first week. At 2 weeks of age, some infants may even begin piggy-backing on their juvenile siblings, and by the third week independent exploration of the environment may begin. No mother was herself seen to initiate a separation until nearly the end of the fourth week. When a mother did herself move off independently of the infant while the latter was on an exploration or with a sibling, such movement promptly triggered the infant's return to her.

By the fifth week, infants regularly engage in play with the juveniles, involving chasing and wrestling. On occasion, an infant even swings from an older sibling's tail. The mother continues to serve as a base to which frequent returns are made. This is also the age at which solid foods are first taken.

The first indication of weaning, i.e., interruptions of nursing attempts which are initiated by the mother, were noted at 10 weeks, though considerable nursing was still occurring after 12 weeks, when the infants were reasonably mobile and eating solids. Curiously, one infant, at an age of 9 months 22 days, resumed nursing 12 days after the birth of a younger sibling, and continued to do so for 3 months.

With weaning, the infants become more closely bound to the group of juveniles than to the mother and adolescence begins. Figures 4 and 5 illustrate the growing independence of the infant, and describe 10 infants which were systematically observed.

3.3.2. *Lemur fulvus*

Among these animals, the infants are first carried across the mother's ventral surface with body axis perpendicular to her own. The approach of other animals is generally rebuffed throughout the first week. Thereafter, the mother, her infants cradled across her belly, may sleep close beside a male, while most other females continue to be cuffed or stared down if they approach. The male consort is occasionally allowed to sniff or groom the infant. After 2 weeks, the infants first crawl out of the maternal lap and onto their mother's back, though they may be forcefully returned to the lap by the mother. However, from this age on, the infants become more restive, craning their heads and following the movement of other animals.

It is not until after the fourth week that the infants are seen to scramble regularly onto the female's back and be allowed to remain there (still with body axis perpendicular to the mother's). However, when the mother moves or others threaten, the baby is returned to her belly. At about this time, too, the infants began making short excursions away from the mother, and making contact with other animals, whose approach is now more often tolerated.

Solid food was first seen taken after 7 weeks, though nursing continued beyond 12 weeks. Some indications of weaning (avoidance of the infant or restraints on nursing) are first evident after the seventh week. By 9 weeks, the infants are off their mothers for considerable periods, approaching and contacting the other animals of the group. Thus, while remaining closer to their mothers than young *L. catta* early in life, the infant *L. fulvus* achieves independence at about the same age. This is illustrated by Figs. 4 and 5, which present a typical picture of 10 systematically observed infants.

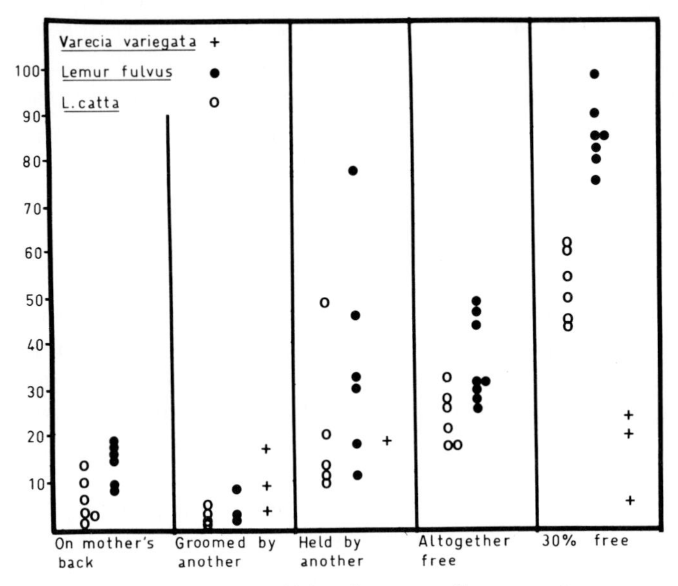

Fig. 4. Age at which various events first occurred.

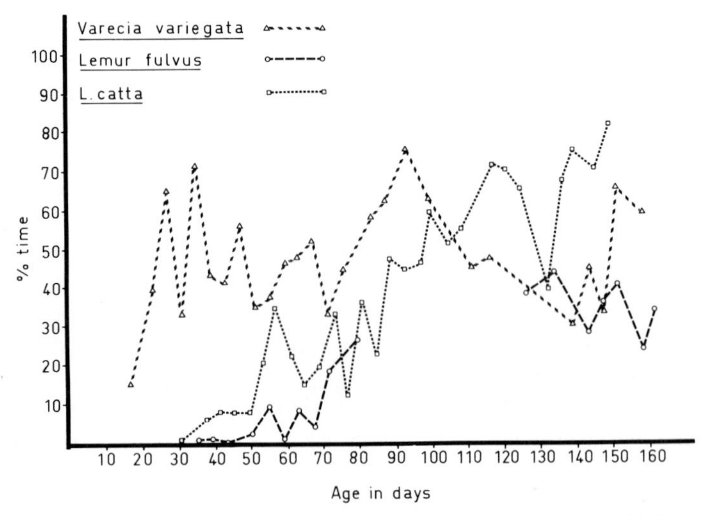

Fig. 5. Percentage of waking time during which infant is not in contact with its mother.

3.3.3. *Varecia variegata*

Varecia variegata have not been systematically observed with young in the wild, and the number of laboratory births has been small. From the first, the infants are transported maternally by mouth. The infants do cling to the mother when in contact with her, and may be dragged some distance when she begins to move out of the nest. Also, the infants may grasp at the neck of the mother as she picks them up in her mouth; however, the grasping of *V. variegata* is never that of *L. catta* or *L. fulvus*. When not being carried, they are left (often for lengthy periods and even on the day of birth) in a nest box which is used exclusively by mother and infant. Later, occasionally infants may be left unattended on the floor while the mother eats.

In observations reported in an initial study (Klopfer and Klopfer, 1970), no other animals were seen to groom the infants during the first two months. However, in a more recent study (Klopfer and Dugard, 1976), the father in the family group (mother, father and twin infants) was seen to groom the twins from the third week after birth. In the extended group (mother, father, adult male, adult female, male yearling sibling, and triplet infants), the sibling was allowed to participate in infant care; it groomed the triplets from the first week on, and often slept near them or directly in the nest box. The other animals were not allowed near the infants until about week 10. It is interesting to note, though, that in recent incidental observations of an extended family group (mother, father, adult female, twin yearling siblings, twin infants), the nonmaternal adult female was seen responding to infant distress cries with maternal vocalizations and to retrieve an infant by mouth. No clearly consistent relationship between infants and other animals emerges from these data.

There are instances, however, in which infants, by 3 weeks of age, give a "defense threat" display (mouth open, body rigid) at the approach of other animals. At about this age, infants actively begin to follow the mother and clamber onto accessible tree branches in early independent excursions. By 5 weeks of age, infants can climb to the top of the trees, and by 7 weeks they are as fully mobile and as agile as the adults. At this age, too, they were seen to chew solids, though they continued to nurse and to sleep with and be groomed by the mother. The sample size in *V. variegata,* however, has been limited to 8 infants distributed among 3 mothers, so these results are less definite than those from the other species.

One can see that these three species, *Lemur catta, L. fulvus,* and *V. variegata,* do differ radically in the duration and exclusiveness of the maternal–infant bond. What is the significance of these different patterns of care? Are the more communally reared *L. catta* infants better buffered against the trauma of parental separation? Are the "independent" *V. variegata* better able to invade and colonize new areas?

3.4. Developmental Events

When one notes the predictable differences in behavior that distinguish different species living in similar habitats, one is attending to the constraints that limit behavioral plasticity. Constraints are rarely omnipotent. Even under the most egalitarian conditions, animals of identical genotype are likely to exhibit some individual variation. One interesting question concerns the degree to which variations among adults can be systematically related to events occurring earlier in their lives. The existence of ontogenetic constraints, "as the twig is bent so grows the limb," is an article of faith for many developmental psychologists. It is at least plausible, however, to consider that the ultimate shape of the limb is not materially or irreversibly influenced by events occurring earlier in life, much as the cambium of an oak will simply grow around the wire that binds it; not that adult behavior must be catalogued in an "either-this-or-that" dichotomy. Different aspects of behavior may be differently determined, and both experiential and maturational events may feed back to influence one another. In goats, for instance, there is a specific and brief period following parturition when the female is receptive toward young; however, the female's behavior, whether maternal or aggressive, may still be determined by seemingly trivial nuances in the behavior of the young. Thus, a female goat which happens, as a result of certain experimental procedures, to be ambivalent toward neonatal kids, will react to a kid by lowering her head and presenting her horns, or by extending her neck and sniffing, according to whether the kid's initial move is a hesitant retreat or a confident step toward her. If the former, the repulsing movement of the doe becomes more intense, as does the kid's retreat, which may then be followed by a violent butt (Klopfer and Klopfer, 1968). If the latter, the sniff will likely give way to a lick, an encouragement to the kid to nuzzle the doe, which stimulates the adoption, by the doe, of a nursing posture. The entire episode may pass within a minute or two with the final outcome, adoption or rejection, hinging on the first tentative movements of the kid. The phenomenon of imprinting, as described by Lorenz (1935), Hess (1972), and others, is a comparable instance of such cascading interactions.

In order to understand the significance of differences in behavior, and their origins, it is instructive to compare *Lemur catta* and *Lemur fulvus* with respect to their life styles and pattern of utilization of particular habitats, as seen in Table III (Sussman, 1975; Budnitz and Dainis, 1975; Harrington, 1975). In addition, in the laboratory, infants of the three species were subjected to one of three sets of conditions: reared alone with the mother; reared with both parents; reared with both parents along with other adults and juveniles. In all these groups the mother was removed for a 7-day period when her infant was 150 days of age and again at 200 days. At 150 days the infant can be self-sufficient, though it still spends much of its time with its mother. Also, by this age in the wild, some of the mothers are already coming into estrus at which time they would normally lessen their hold on the infants. In the lab, the effects of the brief separation on the infant's relations with other animals and with its mother upon her return were measured, as well as its general rate of maturation.

In these three species, for which some experimental data are in hand, it appears the infant is not greatly disturbed by a week-long separation from its mother so long

TABLE III

Summary Comparison of *Lemur catta* and *Lemur fulvus rufus*[a]

	Lemur catta	*Lemur fulvus rufus*
Percentage (%) of time spent on the ground	30	0.4
Average length of day range (m)	920	125–150
Percentage (%) of feeding which takes place on the ground	28	0
Percentage (%) of feeding which takes place in the continuous canopy and emergent layer	34	82
Number of plant species eaten	24	11
Activity to rest ratio	1.44	0.79

[a] From Sussman (1972).

as another animal, even if is only the male, is available. However, a wholly isolated youngster acts as if traumatized. These animals are first hyperactive, calling constantly, then pass into a "depressive phase" not unlike that described for the macaque (Harlow and Harlow, 1965). With another animal present, however, there is scant sign of upset; the attention usually directed toward the mother is simply redirected to another. The mother's return, however, is greeted with an increase in the number and duration of contacts initiated by the infant, though the mother generally does not reciprocate this attention to the extent that she did before the separation. In human terms, the infants seemed pleased at the mother's return, but the latter could hardly care less.

No lasting result of the separation has been noted. After 1 week of resumed contact all measures had returned to their preseparation level (see Figs. 4 and 5). This result was somewhat surprising: a longer-lasting effect might be expected, at least in *L. fulvus,* due to the rather exclusive nature of the early mother–infant relationship. But the *L. catta* infants, which early in life are not nearly so sequestered, are equally flexible, so notions correlating the exclusiveness of early relations and subsequent dependency must be revised.

The generalized accounts given above should not obscure the fact that there are considerable differences from one individual to the next: the specific features of maternal care and infant development differ more in degree than in kind (Klopfer and Dugard, 1976). We are still largely ignorant of the long-term effects of these different rearing conditions. Are the more precocial, socially-reared *L. catta* better buffered against the trauma of parental separation than *L. fulvus? Lemur catta* do range more widely and actively up and down the forest's extent; hence the likelihood of infants needing assistance might be greater. The fact that *V. variegata* infants are scarcely carried at all may be related to the apparent high frequency of multiple births in this species. This obviously begs the question as to what is cause and what effect, though it does underscore the major finding to which lemur studies have led: the incredible range of behavioral adaptations and evolutionary plasticity to be found in primates, even members of an ancient, closely bound group.

3.5. Species versus Individual Differences

The data from all animals observed to date are summarized in Fig. 1. The age at which certain benchmark events occur is here displayed as a function of age, i.e., when the infant first leaves its mother, or is first held by another animal, etc. Considerable interspecific overlap is seen; only the measure which deals with a quantitative change in behavior (in proportion of its waking hours the infant is unattached to another animal) is nonoverlapping.

Sample sizes are too small as yet to allow statistical examination of the effects of group size on these measures. It is probable that the age at which the infant first accepts a surrogate mother will decrease slightly if more than one possible surrogate is available, but no other effects seem evident.

A relation between the sex of the infant and these measures was also sought. In macaques, differences in the behavior of mothers toward the male and female infants have been noted during the first three months of life (Mitchell, 1968), as well as differences between male and female infants. A comparison of mother–infant contact and grooming frequencies and the distance the infant moved from its mother at various ages was made between two matched pairs (male and female) of *Lemur catta* infants and their mothers and two of *Lemur fulvus*. Both the between sexes/within species and the between sexes/across species measures overlapped completely. There is a possibility that in *L. fulvus* the males develop independence more rapidly: by the end of the 215-day observation period, the two male *L. fulvus* were spending over 80% of their time away from their mothers, compared with less than 50% for each of the two females. This difference did not appear until after the infants were 150 days of age (5 months), and thus is hardly comparable to the differences seen in the more slowly maturing macaques before the age of 3 months.

The behavior described here is apparently more strongly related to individual characteristics, including age and prior experience, than to species identity. This does not mean that specific differences do not exist or even that they can be disregarded. Under natural conditions, most *L. catta* youngsters will be similarly reared in the same habitat. They will differ specifically as well as individually from young *L. fulvus*. What the data do suggest is that the behavior of individual lemurs is not tightly constrained but is holistically related to an interrelated series of interactions. J. Kagan has remarked upon a similar phenomenon, based on observations of a human population in Guatemala.

> These observations force the psychologist to commit himself to an interactionist view of development in a devious way. It does not seem to be the case that there is a set of fixed characteristics that lead inexorably to a particular pattern of psychological growth. Nor is the behaviorist view that the child is soft putty in the hands of experience a useful alternative. As indicated earlier, Darwin's

prejudice for a totally environmental interpretation of speciation prevented him from considering the organism's inherent contribution to its own evolution. A large number of psychological environmentalists in the United States are in a similar position. They have made the elaboration of mental capability too dependent on continual environmental input from the outside, and have failed to see the brain's own contribution to its autochthonous growth. (Kagan, 1972, unpublished manuscript, pp. 59–60)

One is tempted to conclude that while maintaining a particular pattern of tradition of behavior is essential to a species, there are several options as to which of several patterns are finally selected. The adage, "There's more than one way to skin a cat" may prove to have a wider currency than biologists anticipated.

3.6. Conclusions

From studies of the three lemur species, we see three different patterns of maternal behavior. In *L. catta,* the infant is precocial at birth and begins to be active within the first week. It clings vertically to the mother's ventral surface, and begins to ride dorsally within a few days after birth. Other animals (juvenile siblings in laboratory studies, and other maternal females in the wild; Jolly, 1966) are allowed to contact the infant almost from the beginning.

In contrast, *L. fulvus* infants are less precocial at birth, not showing much activity until about 2 or 3 weeks of age. They cling transversely to the mother's abdomen where they are carried for 4 or more weeks. At this point transportation changes to transverse clinging by the infant to the mother's back. At first only the male is allowed near the infant, and only after a week or so following birth. Later other animals are allowed to groom the infant.

The third pattern of maternal care is that of *V. variegata*. Unlike *L. catta* and *L. fulvus,* in which single births are most frequent but twinning is not uncommon, *V. variegata* invariably deliver twins or triplets. The infants are kept in a nest box, where the mother often leaves the young alone in order to feed herself. Maternal transport of the infants is by mouth; the babies never cling as do the other two species. With regard to interactions between infants and other animals, no consistent pattern has yet emerged; *V. variegata* appear to represent a mixed pattern, with contact depending on the group composition, presence or absence of siblings, relationship between the mother and other females, and so on.

Can we relate these differing patterns of maternal care to the patterns of life style of the species? And can we see consistencies in such relationships across all prosimian species? There are many factors listed below to consider for such an analysis:

1. Geographical distribution (type of forest habitat, climate, photoperiod)
2. Vertical range (low levels or canopy regions)
3. Activity cycle (nocturnal, diurnal, crepuscular)
4. Diet (insectivorous, folivorous, omnivorous)
5. Locomotion style (vertical clinger and leaper, horizontal jumper, slow-moving climber)

6. Social organization (solitary and/or pairs, population nuclei, family groups, larger more complex troops)
7. Duration of gestation (long, short)
8. Number of infants (single, multiple)
9. Number of mammae (1, 2 or 3 pairs)
10. State of infant at birth (eyes open or closed, grasping abilities, active or immobile)
11. Method of infant transport (by mouth, infant clinging to ventrum or back)
12. Infant development (rate of maturation)

Among the lemurs, for example, one can begin to attempt a correlative analysis with such factors as 8 and 9. The lemurs tend to possess 2 to 3 pairs of mammae, and multiple births do occur. *Lemur fulvus* and *L. catta* have 2 or 3 pairs, produce both single and twin births. *Varecia variegata,* on the other hand, have 3 pairs and always deliver twins or triplets (see Table IV). Considering nonlemurine species, the cheirogaleines, like *V. variegata,* also have 3 pairs and consistently produce multiple births. The lorisines and galagines show a pattern similar to *L. fulvus* and *L. catta,* in which the female has 2 or 3 pairs of mammae, and produces 1, 2, or 3 infants. The indriids and the aye-aye, on the other hand, have only one pair of mammae, and normally produce single infants.

From this simple analysis, one can proceed to consider other factors. Are there tendencies for infants of these multiple births to be more or less precocial at birth? Do they tend to be left in nests and transported by mouth or by clinging to the mother? Do they tend to belong to low-living species or canopy dwellers? Are they

TABLE IV

Frequency of Multiple Births in *Lemur* Species—Duke University Primate Facility, 1961–1974[a]

	Lemur catta	*Lemur fulvus* (including subspecies *fulvus, rufus,* and *albifrons*)	*Varecia variegata*
No. of individuals giving birth	7	24	4
Total live births	14	80	7
No. and percentage multiple births	1 (7%)	8 (10%)	6 (86%)[c]
No. and percentage of individuals with any multiple births	1 (14%)	5 (21%)[b]	4 (100%)
No. of triplets	0	0	3

[a] Courtesy D. Anderson.

[b] Three of the five *Lemur fulvus* individuals with multiple births were *Lemur fulvus albifrons;* these 3 were also 60% of our *albifrons* sample. Without the 5 *Lemur fulvus albifrons,* the percentage of *Lemur fulvus* individuals with multiple births would only be 10.5%, not 21%.

[c] There were indications that a second infant had been born on the sole occasion a single birth was recorded, but that this infant had been cannabilized prior to the arrival of attendants.

more likely to be nocturnal or diurnal? These and many such questions can be asked, and after a thorough process of factor analysis, patterns should emerge, patterns which will allow some understanding of how a group of animals, living in different habitats but faced with the same problem (i.e., how to successfully reproduce and raise the young) can respond to ecological constraints such as forest habitat, climate, availability of food, etc., to solve that problem for those particular conditions. It is to be hoped that studies of prosimians, because of their unique position, will lead to a similar understanding of other mammalian species. At this point, though, the knowledge of prosimian life style in the wild is so lacking for many species, and laboratory situations, though invaluable for allowing close scrutiny of maternal–infant relations, do not allow for the many environmental factors operative in the natural setting. Any satisfactory analysis, then, of the relations among the life style factors and patterns of maternal behavior, must remain obscure for now. However, this approach to the study of maternal behavior (and indeed, to any form of behavior under consideration) will undoubtedly prove more useful in the future, when more basic information is available for comparative analysis.

Finally, does a study of lemurs contribute to an understanding of man? Much of man's behavior is explained, justified, or excused on the assumption that it represents a biological heritage that we can no more deny than our parentage. This has been particularly true of such patterns as those labeled ''aggression,'' ''territoriality,'' and ''mother love.'' The evolutionary origins of these behavior patterns and their underlying mechanisms are understood on the basis of extrapolations from or analogies with particular species. It is only occasionally admitted that the endpoint of a particular extrapolation will depend crucially on the species selected. Given the multiplicity of mechanisms that subserve common ends, even among related species, the choice of species becomes so arbitrary as to preclude any meaningful conclusions. English robins defend individual territories; Galapagos mockingbirds may have communal territories. How can it be known which of these provides the more appropriate analog or homolog of the behavior of men?

Animal studies can enlighten the view of the human condition to the degree that they generate the rules of the game, i.e., define the functional attributes that particular ecological conditions demand. Studies of lemurs are revealing certain of these ''rules of the game.'' They thereby provide a more reliable basis for extrapolation than phyletic comparisons, and provide insights into the origins and control of human behavior which might otherwise be denied us.

ACKNOWLEDGMENTS

The authors wish to thank the following for their advice, help, and criticisms: J. Bergeron, A. Clark, C. Dewey, J. Dubbs, C. Erickson, L. Fairchild, M. Gould, J. Hailman, L. McGeorge, L. Reinherz, L. Rosenson, D. Rubenstein, and L. Vick; and G. A. Doyle for redoing Figs. 4 and 5.

P.H. Klopfer's research is supported by a grant from NIMH, MH04453.

REFERENCES

Andriantsiferana, R., Rarijaona, Y., and Randrianaivo, A. (1974). *Mammalia* **38,** 234–243.

Bearder, S. K., and Doyle, G. A. (1974). *In* "Prosimian Biology" (R. D. Martin, G. A. Doyle, and A. C. Walker, eds.), pp. 109–130. Duckworth, London.

Bishop, A. (1964). *In* "Evolutionary and Genetic Biology of the Primates" (J. Buettner-Janusch, ed.), Vol. 2, pp. 133–325. Academic Press, New York.

Budnitz, N., and Dainis, K. (1975). *In* "Lemur Biology" (I. Tattersall and R. W. Sussman, eds.), pp. 219–235. Plenum, New York.

Buettner-Janusch, J. (1963). "The Origins of Man." Wiley, New York.

Buettner-Janusch, J. (1964). *Folia Primatol.* **2,** 93–110.

Buettner-Janusch, J. (1973). "Physical Anthropology: A Perspective." Wiley, New York.

Castell, R., and Wilson, C. (1971). *Behaviour* **39,** 202–211.

Charles-Dominique, P. (1966). *Biol. Gabonica* **2,** 347–353.

Charles-Dominique, P. (1971). *Biol. Gabonica* **7,** 121–228.

Charles-Dominique, P. (1972). *Z. Tierpsychol, Beih.* **9,** 7–41.

Charles-Dominique, P. (1974a). *Mammalia* **38,** 355–379.

Charles-Dominique, P. (1974b). *In* "Prosimian Biology" (R. D. Martin, G. A. Doyle, and A. C. Walker, eds.), pp. 131–150. Duckworth, London.

Cowgill, U. M. (1974). *In* "Prosimian Biology" (R. D. Martin, G. A. Doyle, and A. C. Walker, eds.), pp. 261–272. Duckworth, London.

Crook, J. H., ed. (1970). "Social Behavior in Birds and Mammals." Academic Press, New York.

Denenberg, V. (1969). *In* "The Behaviour of Domestic Animals" (E.S.E. Hafez, ed.), pp. 95–130. Baillière, London.

DeVore, I., ed. (1965). "Primate Behavior: Field Studies of Monkeys and Apes." Holt, New York.

Doyle, G. A. (1974a). *In* "Prosimian Biology" (R. D. Martin, G. A. Doyle, and A. C. Walker, eds.), pp. 213–231. Duckworth, London.

Doyle, G. A. (1974b). *Behav. Nonhum. Primates* **5,** 155–353.

Doyle, G. A., Pelletier, A., and Bekker, T. (1967). *Folia Primatol.* **7,** 169–197.

Doyle, G. A., Andersson, A., and Bearder, S. K. (1969). *Folia Primatol.* **11,** 215–238.

Doyle, G. A., Andersson, A., and Bearder, S. K. (1971). *Folia Primatol.* **14,** 15–22.

Eisenberg, J. F., Muckenhirn, N. A., and Rudran, R. (1972). *Science* **176,** 863–874.

Fogden, M.P.L. (1974). *In* "Prosimian Biology" (R. D. Martin, G. A. Doyle, and A. C. Walker, eds.), pp. 151–165. Duckworth, London.

Fox, M. W. (1972). *Behaviour* **41,** 298–313.

Gewirtz, J. L. (1969). *In* "Handbook of Socialization Theory and Research" (D. A. Goslin, ed.), pp. 57–212. Rand McNally, Chicago, Illinois.

Harlow, H. F., and Harlow, M. K. (1965). *Behav. Nonhum. Primates* **2,** 287–334.

Harrington, J. (1971). Ph.D. Thesis, Duke University, Durham, North Carolina (unpublished).

Harrington, J. (1974). *In* "Prosimian Biology" (R. D. Martin, G. A. Doyle, and A. C. Walker, eds.), pp. 331–346. Duckworth, London.

Harrington, J. (1975). *In* "Lemur Biology" (I. Tattersall and R. W. Sussman, eds.), pp. 259–279. Plenum, New York.

Hess, E. (1972). *Ann. N. Y. Acad. Sci.* **193,** 124–136.

Hill, W.C.O. (1953). "Primates: Comparative Anatomy and Taxonomy," Vol. I. Strepsirrhini, Univ. Press, Edinburgh.

Hinde, R. A. (1972). *J. Psychosom. Res.* **16,** 227.

Hinde, R. A. (1966). "Animal Behaviour." McGraw-Hill, New York.

Hinde, R. A., and Spencer-Booth, Y. (1971). *Anim. Behav.* **19,** 165–173.

Hinde, R. A., Rowell, T. E., and Spencer-Booth, Y. (1964). *Proc. Zool. Soc. London* **143,** 609–649.

Hladik, C. M., and Charles-Dominique, P. (1974). *In* "Prosimian Biology" (R. D. Martin, G. A. Doyle, and A. C. Walker, eds.), pp. 23–37. Duckworth, London.

Hutchinson, G. E. (1965). "The Ecological Theatre and the Evolutionary Play." Yale Univ. Press, New Haven, Connecticut.

Jay, P., ed. (1968). "Primates: Studies in Adaptation and Variability." Holt, New York.

Jewell, P. A., and Oates, J. F. (1969). *Zool. Afr.* **4,** 231–248.

Jolly, A. (1966). "Lemur Behavior: A Madagascar Field Study." Univ. of Chicago Press, Chicago, Illinois.

Jolly, A. (1972). "The Evolution of Primate Behavior." Macmillan, New York.

Kagan, J. (1972). Privately distributed manuscript.

Kaufman, I. C., and Rosenblum, L. A. (1969). *Ann. N. Y. Acad. Sci.* **159,** 681–695.

Klopfer, P. H. (1972). *Z. Tierpsychol.* **30,** 277–296.

Klopfer, P. H. (1973). *Ann. N. Y. Acad. Sci.* **223,** 113–119.

Klopfer, P. H., and Dugard, J. (1976). *Z. Tierpsychol.* **40,** 210–220.

Klopfer, P. H., and Gilbert, B. K. (1967). *Z. Tierpsychol.* **23,** 757–760.

Klopfer, P. H., and Klopfer, M. S. (1968). *Z. Tierpsychol.* **24,** 862–866.

Klopfer, P. H., and Klopfer, M. S. (1970). *Z. Tierpsychol.* **27,** 984–996.

Klopfer, P. H., McGeorge, L., and Barnett, R. J. (1973). "Module in Biology." Addison-Wesley, Reading, Massachusetts.

Le Boeuf, B. J., Whiting, R. J., and Gantt, R. F. (1972). *Behaviour* **43,** 121–156.

Lorenz, K. Z. (1935). *J. Ornithol.* **83,** 137–213, 324–413.

Lowther, F. de L. (1940). *Zoologica* **25,** (*N.Y.*) 433–459.

Manley, G. H. (1966). *Symp. Zool. Soc. London* **15,** 493–509.

Martin, R. D. (1972a). *Philos. Trans. R. Soc. London, Ser. B* **264,** 295–352.

Martin, R. D. (1972b). *Z. Tierpsychol. Beih.* **9,** 43–89.

Martin, R. D. (1973). *In* "Comparative Ecology and Behaviour of Primates" (R. P. Michael and J. H. Crook, eds.), pp. 1–68. Academic Press, London.

Mitchell, C. (1968). *Child Dev.* **39,** 611–620.

Napier, J. R., and Napier, P. H. (1967). "A Handbook of Living Primates." Academic Press, New York.

Pariente, G. (1974). *In* "Prosimian Biology" (R. D. Martin, G. A. Doyle, and A. C. Walker, eds.), pp. 183–198. Duckworth, London.

Petter, J. J. (1962). *Ann. N. Y. Acad. Sci.* **102,** 267–281.

Petter, J. J. (1965). *In* "Primate Behavior: Field Studies of Monkeys and Apes" (I. DeVore, ed.), pp. 292–319. Holt, New York.

Petter, J. J. (1967). "The aye-aye of Madagascar." Society for French American Cultural Services and Educational Aid, New York.

Petter, J. J., and Petter, A. (1967). *In* "Social Communication Among Primates" (S. A. Altmann, ed.), pp. 195–205. Univ. of Chicago Press, Chicago, Illinois.

Petter, J. J., and Peyriéras, A. (1974). *In* "Prosimian Biology" (R. D. Martin, G. A. Doyle, and A. C. Walker, eds.), pp. 39–48. Duckworth, London.

Petter, J. J., Schilling, A., and Pariente, G. (1975). *In* "Lemur Biology" (I. Tattersall and R. W. Sussman, eds.), pp. 209–218. Plenum, New York.

Petter-Rousseaux, A. (1962). *Mammalia* **26,** Suppl. 1, 1–88.

Petter-Rousseaux, A. (1964). *In* "Evolutionary and Genetic Biology of Primates" (J. Buettner-Janusch, ed.), Vol. 2, pp. 91–132. Academic Press, New York.

Phillips, W.W.A. (1931). *Spolia Zeylan.* **16,** 205–208.

Pocock, R. I. (1939). "The Fauna of British India," Vol. I. Taylor & Francis, London.

Pollock, J. I. (1975). *In* "Lemur Biology" (I. Tattersall and R. W. Sussman, eds.), pp. 287–311. Plenum, New York.

Preston, D. C., Baker, R. P., and Seay, B. (1970). *Dev. Psychol.* **3,** 298–306.
Ramanantsoa, G. A. (1975). *Defenders Wildl. News* **50,** 148–149.
Rheingold, H., ed. (1963). ''Maternal Behavior in Mammals.'' Wiley, New York.
Richard, A. (1974). *In* ''Prosimian Biology'' (R. D. Martin, G. A. Doyle, and A. C. Walker, eds.), pp. 49–74. Duckworth, London.
Rosenson, L. M. (1972). *Anim. Behav.* **20,** 677–688.
Sanderson, I. T. (1957). ''The Monkey Kingdom.'' Doubleday, Garden City, New York.
Schrier, A. M., Harlow, H. F., and Stollnitz, K. F., eds. (1965a). ''Behavior of Non-Human Primates,'' Vol. 1. Academic Press, New York.
Schrier, A. M., Harlow, H. F., and Stollnitz, F., eds. (1965b). ''Behavior of Non-Human Primates,'' Vol. 2. Academic Press, New York.
Schrier, A. M., Harlow, H. F., and Stollnitz, F., eds. (1971). ''Behavior of Non-Human Primates,'' Vol. 3. Academic Press, New York.
Sussman, R. W. (1972). Ph.D. Thesis, Duke University, Durham, North Carolina, (unpublished).
Sussman, R. W. (1974). *In* ''Prosimian Biology'' (R. D. Martin, G. A. Doyle, and A. C. Walker, eds.), pp. 75–108. Duckworth, London.
Sussman, R. W. (1975). *In* ''Lemur Biology'' (I. Tattersall and R. W. Sussman, eds.), pp. 237–258. Plenum, New York.
Tattersall, I., and Sussman, R. W. (1975). Anthropol. Pap. Am. Mus. Nat. Hist. **52,** 193–216.
Vincent, F. (1969). Thèse de Doctorat d'Etat, Sciences Naturelles, Fac. Sci., Université de Paris.
Walker, E. P. (1968). ''Mammals of the World,'' 2nd ed. Johns Hopkins Press, Baltimore, Maryland.

Chapter 5

Development of Behavior in Prosimians with Special Reference to the Lesser Bushbaby, *Galago senegalensis moholi*

G. A. DOYLE

ISBN 0-12-222150-8

1. INTRODUCTION

Campbell and Thompson (1968) point out that there are two kinds of developmental studies of animals: those concerned with normative description of behavior in relation to age, and predictive studies concerned with the effects produced by experimental treatment applied at different ages. They suggest quite rightly, as does Hinde (1966), that a descriptive account of the way in which behavior changes with age under normal conditions must precede an analysis of the factors which influence these changes.

The literature, however, reveals that most studies of development in primates are of the predictive kind and are concerned only with specific aspects of development, for example, social behavior (see Hinde, 1971, for an excellent review and detailed list of references). Most of the very few studies of normative development in primates have also been devoted largely to specific aspects of behavior, e.g. maternal–infant behavior, social behavior. Very few studies of normative development have been quantitative and few have been concerned with the entire period of development from birth to adulthood. Hinde (1966) notes that detailed descriptions of the normal development of behavior in animals, over even a very small part of the life span, are rare. He lists a few such studies of which only one is concerned with primates, that of Hines (1942), who made a detailed study of development related directly to morphological or structural changes in osteology and myology, for example, reflex behavior, posture, progression, with little interest in more complex behavior, like social behavior or intelligence, in a number of Rhesus macaques some of which were studied from birth until one year of age.

Since Hinde's (1966) brief review at least two more important and detailed studies concerned with normative behavioral development over a wide range of behaviors in simian primates have appeared. These are Hinde and Spencer-Booth's (1967) study of *Macaca mulatta* over a period of $2\frac{1}{2}$ years and Rosenblum's (1968) study of *Saimiri sciureus* over the first 10 months of life.

Schultz (1969) draws attention to the basic importance of information regarding the profound behavioral changes that occur with age. He notes that changes with age occur throughout life and are most dramatic in the early stages of development proceeding with generally decreasing tempo until the relatively stable conditions of adulthood. Very few of the evolutionary specializations which characterize the adult of each primate species are fully present until fairly late in development and full understanding of these characteristics is precluded until their ontogeny has been studied in detail. Development may be defined as a process involving all the changes that occur during progress toward the adult state. It implies not only the orderly and gradual emergence of adult patterns of behavior, but also the gradual decrease and eventual disappearance of a number of important evolutionary specializations characteristic of early and transitory stages of development. The implicit end product is the crystallization of all physical and behavioral attributes into the adult norm.

It follows that the acquisition of adult patterns of behavior is not determined solely by their first appearance but, more importantly, by the attainment of typical patterns of frequency characteristic of the adult. Primates begin to eat and groom very early in infancy displaying the full adult pattern. For some time, however, the frequency of such patterns will not be typically adult. After weaning the frequency of eating, for example, may well exceed that of adults and may only drop to the adult norm on the attainment of full adult weight. Attainment of the typical adult frequency of social grooming may have to wait until full social integration which, in many social primates, may follow sometime after adulthood. Successful breeding, on the other hand, usually considered an important and reliable index of the attainment of adult status, may occur while the individual is still a juvenile, in terms of other indices like size and weight (Schultz, 1969) or, in animals with a restricted breeding season, like *Lemur catta* (Jolly, 1966) which mate during one week of the year, it may follow sometime after the attainment of adulthood.

The full development of behavior to adulthood needs to be studied in many selected primate species if the appearance and nature of all behavioral specializations are to be fully understood. While the study of specific aspects of development is important, it is only the study of the full range of behavioral development which will yield the information necessary for determining the attainment of adulthood.

In a recent review of studies of prosimian behavior (Doyle, 1974), no detailed studies, and certainly no quantitative studies, of the development of behavior in any prosimian could be found; Ehrlich's (1974) more recent study of development in *Galago crassicaudatus* and *Nycticebus coucang* is confined to the first 10 weeks of life. The aim of this chapter is, first, to satisfy this need for normative studies, with one species of prosimian, the lesser bushbaby (*Galago senegalensis moholi*), by presenting both descriptive and quantitative data from the laboratory and, second, for purposes of comparison, to review the literature on normative development in other species of prosimian in the laboratory and in the field.

2. METHOD

2.1. Facilities

Since 1964 the bushbabies have been housed in the Primate Behaviour Laboratory of the University of the Witwatersrand under conditions designed to approximate their natural habitat and to allow as free a range of naturalistic behaviors as seminatural laboratory conditions allow, consistent with the need to observe behavior under conditions of minimal interference.

A detailed description of the Primate Behaviour Laboratory and the routine methods of care and maintenance have been fully presented in a previous paper (Doyle and Bekker, 1967) and only a brief description will be given here. The bushbabies are confined to cages approximately 5 × 6 × 7 feet, in groups of one adult male, one or two adult females, and their young. The cages are equipped with branching systems, ledges, and suitably placed glass-sided nest boxes. The floors are covered with a layer of wood shavings. Temperatures fluctuate between 15° and 30°C, which is well within the seasonal range characteristic of their natural habitat. Duplication of the normal diurnal variation in light is achieved by automatic time switches. Red light, between 1200 and 2300 hours, under which animals are awake and active, alternates with white light for the remaining 13 hours. Animals are fed daily with finely chopped-up fruits and vegetables in season, mealworms, a commercially available food supplement in porridge form containing balanced proportions of proteins, vitamins, minerals and salts, and milk supplemented with a vitamin B-complex syrup. At regular intervals locusts and other insects are added to the diet. The cages are thoroughly cleaned and food is placed on the ledges between 0900 and 1200 hours.

Observations are made through large one-way mirrors separating the bushbaby cages from an observation room which remains in darkness. The observation room is equipped with tape recorders and instruments for recording both the frequency and duration of events (an Esterline-Angus 20-channel events recorder and two batteries each of 24 electronically operated seconds timers/events counters).

Under these conditions the animals have bred freely and at the time of writing over 60 infants have been born of which approximately 75% have survived to adulthood. Both the low mortality rate of infants and adults and the nature of occasional deaths suggest that the conditions under which the animals are kept are optimal.

2.2. Subjects

The subjects of the quantitative observations forming the second part of the present study were two pairs of laboratory born twins and two single laboratory born infants, all born to wild-caught mothers. The two pairs of twins were O (female) and P (female), born to Pr, and E (female) and F (male), born to Vi. The two single infants were A (male) and C (female), both born to Pr. Each had been one of a pair of twins, the other of each pair having died in early adolescence.

Each pair of twins was observed together and each singleton was observed alone. The six subjects, therefore, were the subject of four independent sets of observations. Throughout the period of observation all animals remained in "family" groups with their parents, with the exception of O and P, where the father was absent until they were approximately 14 weeks of age. In the case of E and F a second female was also present throughout the period of the observation. These six infants are referred to as the primary subjects. In addition other infants were born to the mothers of the primary subjects such that throughout much of the period of observation all primary subjects were in close association with infants younger than themselves and, except for A, with infants older than themselves as well.

The descriptive data forming the first part of the present study were based not only on the observations of the above 6 primary subjects but on approximately 35 other infants as well, all living under similar conditions, and which were observed regularly in the course of other studies or periodically, particularly during the first few weeks of life. The observations on which the quantitative data are based were carried out over a period of 2 years. Much of the descriptive data was extracted from our records gathered over a period of approximately 5 years.

2.3. Procedure

All observations were made by the author. Each of the 6 primary subjects was observed daily for a 50-minute period over an unbroken span of 53 or 54 weeks between 1400 and 1700 hours (Period 2). This choice of time was determined on the basis of field and laboratory observations (Bearder, 1969; Pinto *et al.*, 1974) that general activity during this time, following a period of heightened activity for the first few hours of the night, is even and regular before rising again to a less noticeable peak for the last few hours before dawn. O and P were also observed the first 25 minutes of the waking day (Period 1) from birth for 189 successive days and E and F, under the same conditions, for the first 147 days. In addition A and C were observed at weekly intervals during the last 50 minutes of the waking day (Period 3) for the first 14 weeks of life. As a further check other 50-minute periods of the waking day were sampled from time to time for all infants.

2.3.1. Descriptive Data

The descriptive data, recorded either by hand or on a tape recorder, are based on all observations of both primary and secondary subjects made during Periods 1, 2, and 3, a total of nearly 1800 50-minute observations.

2.3.2. Quantitative Data

The quantitative data are based entirely on 1498 Period 2 observations of primary subjects only. These data were recorded on the instruments mentioned above and were entered on data sheets following each observation. The data sheets also provided for the recording of descriptive data which could not be quantified. The quantitative data, both activity counts and the time scores, were averaged for each sex and are illustrated graphically against baselines for adults of both sexes.

2.3.3. Adult Normative Data

Normative data for adults, against which data from the 6 primary subjects are compared, were extracted from records representing approximately 7000 50-minute Period 2 animal observation periods over a 6-year span of time and are based on 6 wild-caught animals, 2 males and 4 females, all of which were adjusted to laboratory conditions at the time that observations were made. To illustrate how these

TABLE I

Breakdown for Individuals of Either Sex of the Adult Baseline Data for *Galago senegalensis* against which Quantitative Behavioral Data from the Six Primary Infant Subjects Are Compared

	Male					Female				
	Subject	No. obs.	Range	M[a]	M̄[b]	Subject	No. obs.	Range	M[a]	M̄[b]
Time active (%)	VO	102	0–100	40.23	44.1	IN	118	0–99	44.44	44.50
	JA	106	0–100	47.79		VI	114	0–92	36.64	
						AN	143	0–93	57.88	
						PR	116	0–100	35.79	
Jumps	VO	102	0–324	74	104	IN	128	0–438	152	115
	JA	129	0–250	129		VI	160	0–262	89	
						AN	143	0–406	148	
						PR	137	0–394	77	
Contact urine-washes	VO	111	0–9	2.24	2.46			NIL		
	JA	118	0–13	2.67						
Total urine washes	VO	111	0–16	6.07	5.60	IN	126	0–5	0.86	0.90
	JA	118	0–9	5.16		VI	157	0–4	0.98	
						AN	141	0–4	0.89	
						PR	139	0–9	0.85	

Eats	VO	99	0–14	2.99	3.19	IN	127	0–18	4.76	2.70
	JA	116	0–12	3.36		VI	163	0–10	1.77	
						AN	144	0–12	2.97	
						PR	138	0–12	1.72	
Drinks	VO	99	0–14	2.67	2.86	IN	127	0–14	1.42	2.94
	JA	116	0–26	3.03		VI	162	0–10	2.91	
						AN	144	0–14	3.72	
						PR	138	0–16	3.56	
Self-grooms	VO	95	0–20	4.86	5.24	IN	79	0–18	4.67	5.58
	JA	73	0–12	5.73		VI	139	0–20	5.58	
						AN	141	0–20	6.82	
						PR	133	0–20	5.13	
Allo-grooms	VO	95	0–16	2.32	3.19	IN	79	0–20	6.23	3.95
	JA	73	0–18	4.33		VI	139	0–18	4.53	
						AN	141	0–15	3.36	
						PR	133	0–16	2.62	

[a] M = average of number of observations.
[b] $\bar{M}$ = average of M.

norms were arrived at Table I shows the range and mean for each subject in the adult sample and the number of observations on which these were based for all behaviors under consideration in Section 4.

These adult means, derived as they are from a very large sample of observations, nevertheless cannot be said to represent a completely accurate picture of normal adult behavior because of the large variation both within and between observation periods and within and between animals. While these adult normative data are based on Period 2 observations under normal conditions, it is of interest to note that most of these norms change radically under what may be called ''new cage conditions,'' i.e., when animals are introduced into a new cage or after the home cage has been thoroughly cleaned out.

In 50-minute observations under new cage conditions the same animals are much more active than normal, the males more so than the females. Briefly the number of 50-minute observations under new cage conditions, for each behavior pattern considered, ranged from 3 to 7. Males are active 100% of the time and females slightly less. Mean number of jumps for males was about 360 and for females 249. Mean contact urine-washing in males rose to 8.40 and total urine-washing to 21.0 while in females total urine-washing rose to 5.14. There is little noticeable change in frequency of eating and drinking, a noticeable drop in the frequency of self-grooming and no allogrooming at all takes place. A similar pattern of behavior is typical of Period 1 observations as well, but to a somewhat lesser extent. Period 2 observations, for both infants and adults, are more typical of general behavior during the greater part of the active period.

3. DESCRIPTIVE OBSERVATIONS ON DEVELOPMENT

3.1. Development at Birth

The infant *Galago senegalensis* is born covered with fur of a uniform gray color and with its eyes fully open. At the first opportunity, within half an hour after birth, if the mother stops grooming it, it is able to crawl awkwardly on all fours, belly on the substrate, almost invariably toward the mother, turn on its back, kick with its legs, and grip with its hands, pulling itself into the suckling position in search of the nipple. It turns over on its hands and feet again when its mother leaves. It may also climb on her back clinging tightly with hands and feet to the mother's fur, occasionally being dragged out of the nest box in the process before releasing its grip (Doyle *et al.,* 1967).

The above is probably true of all the lorisids. Charles-Dominique (1977) reports that the newborn *Galago demidovii* is covered with a sparse coat of fur and that the eyes are partially open. The infant's hands are capable of grasping an object; if suspended from a branch it will hang from its hands alone. It can crawl into the nest, snuggle under the mother, and locate a nipple. The same author reports that the eyes of the infant *Perodicticus potto* and *Arctocebus calabarensis* are also open at birth.

Infants of both species are born on branches and are covered with a thick coat of fur. After being cleaned both infants climb slowly on the mother, grip her fur, and take up a position on her belly without the mother providing manual assistance of any kind (Charles-Dominique, 1977). For the first few days, at least in *A. calabarensis,* the infant accompanies the mother in this way in her movements about the environment. Charles-Dominique (1966) notes that by 17 hours the infant *A. calabarensis* already displays well-coordinated movements. The infant *G.s. zanzibaricus* is likewise born with eyes open; its body is covered with ash-gray fur and it is able to cling to objects (Gucwinska and Gucwinski, 1968).

In most other lemurine and indriid infants the eyes are open at birth and they are able to cling with hands and feet to the mother's fur and move around actively. This has been reported for *Propithecus verreauxi, Lemur catta, L. macaco, Lepilemur mustelinus,* as well as for *Daubentonia* (Petter-Rousseaux, 1964; Petter, 1965), for *Indri indri* (J. I. Pollock, personal communication), and *Avahi laniger* (R. D. Martin, personal communication). Jolly (1966) reports, however, that the infant *P. verreauxi* only becomes really active at 3 to 4 days after birth, its activity prior to that time presumably being limited to gripping the mother tightly. *Hapalemur griseus* and *Varecia variegata* (*Lemur variegatus*) infants, on the other hand, are reported to be relatively helpless at birth, are unable to cling to the mother, and are left for long periods of time either on a suitable platform like a tree fork, in the case of *H. griseus,* or in a prepared nest in the case of *V. variegata* (Petter-Rousseaux, 1964; Petter, 1965).

The cheirogaleine infants, at least *Microcebus murinus* and *Cheirogaleus major,* are born with eyes closed. The *M. murinus* infant can cling at birth (though not usually to the mother), the fur is through, some of the teeth are through (R. D. Martin, personal communication), and it is capable of getting into position to suckle (Petter-Rousseaux, 1964; Martin, 1972a). Martin (1972a) notes too that it is able to support its entire weight suspended from one foot at birth. Petter-Rousseaux (1964) reports that the *C. major* infant is born relatively helpless but can crawl haltingly. Le Gros Clark (1924) reports that the *Tarsius* infant is born with eyes open and is able to cling tightly to a vertical surface.

The early ability of the infant to cling to the mother is probably of adaptive significance in that it enables the infant to suckle with marginal help from the mother, as Blackwell and Menzies (1968) note for *Perodicticus potto.* It is probably necessary too for infants to move around actively from as early an age as possible both to ensure location of the nipple on their own and because, as Doyle *et al.* (1967, 1969) show in *Galago senegalensis,* the more active the infant is the more it is noticed by the mother.

3.2. Early Postnatal Development

From a very early age infant prosimians display less reliance on their mothers and greater ability to move about their environments than do infants of simian primates.

This is probably necessary as a defence against predators. At 1 day of age the infant *Galago senegalensis* is able to cling safely to a branch where it may be left temporarily by its mother while being moved from one nest box to another. Its grip is so tight that it cannot be dislodged by an over-playful adolescent. Although it spends a great deal of time sleeping, resting, or suckling it has become increasingly active during its wakeful periods. The infant *G. demidovii* may also be left clinging alone to a branch for short periods from about 3 to 7 days of age. Charles-Dominique (1977) notes that it remains completely immobile, practically immune to discovery.

The ability to cling from an early age either to the mother's fur or to a branch is probably characteristic of the lorisids, but almost certainly of the lorisines. Anderson (1971) reports the infant *Perodicticus potto* clinging securely underneath a branch at 1 day of age and both Manley (1966b) and Hill (1937) report the same for the infant *Nycticebus coucang*. The same phenomenon appears in the infant *Arctocebus calabarensis* and probably also from as early as 1 day of age since Jewell and Oates (1969a) report that the infant is never seen with the mother at night in the wild. Charles-Dominique (1977) also reports on this phenomenon of "infant parking" in *P. potto* and *A. calabarensis* in both the laboratory and in the field beginning at 3 to 8 days of age. He notes that infants are parked on a branch at dusk where they remain suspended until retrieved at dawn, although this behavior is not consistent in that infants may be parked one night and carried the next. Ehrlich (1974) reports that both *Galago crassicaudatus* and *N. coucang* infants on the first day of life are able to grasp a pencil with either hands or feet and maintain their hold when the pencil is lifted. The strength of this response gradually diminishes although the process is significantly more rapid in *G. crassicaudatus* than in *N. coucang*. In respect of the hand the response disappeared entirely in *G. crassicaudatus* by 17 days but not until 36 days in *N. coucang*. In respect of the feet the response disappeared entirely in *G. crassicaudatus* at 29 days but not until 67 and 99 days in two *N. coucang* infants and even later in a third.

The *Lepilemur mustelinus* infant climbs around unassisted at about 3 days (Petter-Rousseaux, 1964; Petter, 1965) and is very active at 3 to 4 days (Jolly, 1966). The *Lemur catta* infant is also very active by this age, moving around on the mother and on other females at 3 to 4 days (Petter-Rousseaux, 1964; Petter, 1965). The *Varecia variegata* infant, on the other hand, less precocious than most lemurines, only begins to crawl on its belly toward its mother at 4 days (Petter-Rousseaux, 1964; Klopfer and Klopfer, 1970).

The cheirogaleines are also relatively slow to develop initially. *Microcebus murinus* infants open their eyes at 2 to 4 days and crawl on the mother's back at 8 days (Petter-Rousseaux, 1962, 1964; Martin, 1972a). *Cheirogaleus major* infants open their eyes at between 2 days (Petter-Rousseaux, 1964) and 6 days (A. Schilling, personal communication) and at 5 days they cannot yet walk or climb (Petter-Rousseaux, 1964). Niemitz (1974) reports that an infant *Tarsius bancanus,* probably 2 to 3 days when first acquired, clung tightly to its mother's body ventroventrally with arms halfway round her body at the height of the lower ribs.

3.3. Early Vocalizations

Andrew (1964) notes that an example of the development of adult social activities out of the early mother/infant relationship is seen in the development of adult vocalizations out of the distress call of the infant which follows loss of physical contact. In the more social prosimians, like *Lemur fulvus* and *L. catta,* distress on loss of contact is more acute, is accompanied by more intense vocalization and persists more markedly into adulthood, than in less social prosimians, like *Galago crassicaudatus,* for instance (Andrew, 1964). Petter-Rousseaux (1964) reports that the *Cheirogaleus major* infant emits piercing whistles in distress and soft cries when licked by the mother. She also reports that the *Varecia variegata* infant emits a shrill cry in response to the mother. *Galago senegalensis* infants have been heard uttering high-pitched distress squeaks as early as a few hours after birth when left on the nest box ledge after being dragged out by the mother (Doyle *et al.,* 1967). They have been heard uttering soft notes when suckling, and they emit distress signals when left alone or removed for weighing, all before 1 day of age.

At 7 days the *Galago senegalensis* infant has been heard to emit a high-pitched threat signal, and Blackwell (1969) reports hearing the defense cry in *G. demidovii* at 10 days of age. Charles-Dominique (1977) reports that from a very early age the infant *G. demidovii, G. alleni, G. elegantulus* (*Euoticus elegantulus*), *Perodicticus potto,* and *Arctocebus calabarensis* respond with high-pitched cries to the mother's contact call and will also spontaneously vocalize its distress when separated from the mother. Bearder (1975) reports squeaks and a maternal call at 3 days and clicks and crackles at 9 days in infant *G. crassicaudatus* in the wild. Ehrlich (1974) reports that both *G. crassicaudatus* and *Nycticebus coucang* infants emit a variety of sounds during the first weeks of life, largely in response to human handling and separation from the mother. The most commonly heard of these sounds is the click heard in both species from the first day of life. In *N. coucang* the whistle and twitter, two calls characteristic of adults, were heard on the sixth and seventh day, respectively, and in *G. crassicaudatus* the "adult group call" appeared on the 58th day.

Niemitz (1974) reports a threat response, accompanied by a high-pitched whistle, in an infant *Tarsius bancanus* by 3 days of age and "greeting sounds" on the approach of a human caretaker in another infant of the same species at about 10 days. The same infant uttered sounds of distress at the same age on being left alone. At about 10 days too one infant uttered continuous high-pitched squeaks, not audible beyond 3 to 4 m, while grasping the fur of a foster mother with both hands. Niemitz (1974) reports that many different types of vocalization were recorded in one infant *Tarsius bancanus* until it died at about 50 days of age.

From the earliest age, all prosimian infants studied are reported to vocalize their distress when alarmed or when contact with the mother is broken (Petter-Rousseaux, 1964; Andrew, 1964). Some are reported to emit a softer call when contact with the mother is reestablished or when suckling (Andrew, 1964). Vocal communication between mother and infant is discussed in greater detail elsewhere (Doyle, 1974).

3.4. Infant Transport and Retrieval

The way in which the *Galago senegalensis* infant is carried by the mother in the mouth, the occasions on which infants are carried, and the decline in frequency of oral transportation over the period of infancy have been described in detail by Doyle *et al.* (1969). In that study, in which four pairs of twins were observed daily for an unbroken period of 14 weeks, it was reported that no carrying behavior was seen after 56 days, except on very rare occasions, although unsuccessful attempts to carry continued long after this age.

In the laboratory, carrying behavior in *Galago senegalensis* is devoted initially almost entirely to changing infants from one nest site to another, particularly while still young and confined to the nest but, occasionally, even up to 35 days. After infants emerge from the nest, carrying is devoted almost entirely to retrieving them back to the nest. Bearder (1969) reports that, in the wild, carrying behavior, from the time infants are venturing from the nest to the age of 28 days, is also devoted almost entirely to retrieval, except where the mother's help is needed in difficult crossings from one tree to another. Up to just over 3 weeks of age an infant that falls or is otherwise in difficulty cries and is immediately retrieved back to the nest box or a branch. By about 25 days it does not require its mother's assistance and can get back by itself.

Bearder (1975) first reported the *G. crassicaudatus* infant in the wild being carried in the mouth at 3 days and that from 8 days it may be carried clinging to the mother's back. He reports that if there are twins one is carried in the mouth and the other clings to the mother's back. Davis (1960) reports too that an infant is first carried in the mouth and later clings to the mother's back, and both he and Buettner-Janusch (1964) report that in an alarm situation the infant is carried clinging with hands and feet to the mother's abdomen. Buettner-Janusch (1964) reports seeing twins carried, one clinging to the ventral surface and the other clinging to the dorsal surface of the mother, but that, as Bearder (1975) and Davis (1960) reported, an infant is carried in the mouth for the first 7 days and only after this on the abdomen or back. In the wild Bearder (1975) saw no carrying behavior after the infants were 70 days of age, but 6-month-old wild-caught twins clung to their mother with all fours when first released in the laboratory. Charles-Dominique (1977) notes that, although the *Galago elegantulus* infant is carried in the mouth like other galagines, including *G. demidovii* and *G. alleni,* it may climb on the mother's back, grip the dorsal fur, and be carried in this fashion.

Unlike the galagines, *Perodicticus potto* and *Arctocebus calabarensis* infants are first transported by the mother, clinging tightly to her ventral surface and later riding on her back (Charles-Dominique, 1977). This is probably true of the Asian lorisines as well. When the infant *P. potto* and *A. calabarensis* are older they let go of the mother and follow her or they may move from the spot where they have been parked (see Section 3.2 above) without waiting to be retrieved.

Lemurine and indriid infants are first transported clinging to the ventral surface of the mother and later riding or clinging to the dorsal surface. *Varecia variegata, Lepilemur mustelinus,* and *Hapalemur griseus* are exceptions and, like the cheirogaleines, are carried in the mother's mouth (Petter-Rousseaux, 1965; Petter, 1965; Basilewsky, 1965; Klopfer and Klopfer, 1970; Charles-Dominique and Hladik, 1971), at least initially in the case of *H. griseus* since both J. J. Petter (personal communication) and R. D. Martin (personal communication) report that later on the infant is carried clinging to the mother's fur at night.

As early as day 3 the *Lemur catta* infant may ride on its mother's back and by 7 to 9 days it is frequently on its mother's back (Petter, 1965; Jolly, 1966; Klopfer and Klopfer, 1970). The *L. macaco* infant does not ride on its mother's back till 15 days (Basilewsky, 1965) and Klopfer and Klopfer (1970) report twin *L. fulvus* first venturing on to the mother's back at 16 and 19 days, but that they were forcibly removed until 32 days. The *Propithecus verreauxi* infant first rides on its mother's back at 30 to 42 days (Petter-Rousseaux, 1964; Jolly, 1966; Klopfer and Klopfer, 1970), but the *Indri* infant is already spending more and more time on its mother's back by 14 days.

Retrieval in the lemurines and indriids probably takes the form of the mother providing subtle body movement cues indicating to the infant that it must climb on her back, as has been noted for *Saimiri sciureus,* or in the form of the mother touching the infant with the hand as in *Macaca radiata* and *M. nemestrina* (Rosenblum, 1968), and as Petter (1965) noted for *Indri*.

Some of the information relevant to this section is included in Table IV (see Section 6.1).

3.5. Locomotion and Manipulation

The *Galago senegalensis* infant walks quadrupedally with its belly raised from the substrate and stands supported by its hands on a vertical surface within a few days after birth. By 4 days it climbs over other animals resting in the nest box. By 6 days it climbs unsteadily toward its mother in the nest box, and by 8 days it can take tiny bipedal hops. By about 14 days it is capable of fast quadrupedal running on branches. By 21 days it is jumping actively, and before 6 weeks it may jump nearly 1 m and its locomotor ability is quite elaborate (Bearder, 1969). Up to 18 days it is likely to remain clinging tightly when left out on a branch but by 19 days it will climb rapidly to the top branches. As early as 12 days the infant may follow its mother out of the nest box and for the next 2 weeks or so thereafter it will usually follow her back if not retrieved. By 28 days it is likely to follow the mother even to the extent of crossing open spaces through the trees or on the ground and taking short jumps in the process (Bearder, 1969). By 46 days it is still following the mother, taking short jumps when she does and generally investigating whatever arouses her curiosity.

Gucwinska and Gucwinski (1968) report that *G.s. zanzibaricus* are still very awkward at the age of 10 days and unable to leap. The *Galago demidovii* infant is taking short jumps and clinging to the mesh wall of the cage by 9 days (Vincent, 1969), still walking unsteadily at 10 to 18 days (Vincent, 1969; Blackwell, 1969; Brand, 1977), very agile by 23 days (Brand, 1977), and jumping as much as 30 cm by 30 to 33 days (Woodhull and Woodhull, 1969; Blackwell, 1969). Ehrlich (1974) reports *Galago crassicaudatus* infants moving forward on all fours by 8 days, climbing up and down by 14 days, but not jumping until 25 days. Bearder (1975) reports that the *G. crassicaudatus* infant begins following its mother around at 25 days.

The cheirogaleines, although they may reach adulthood sooner than most other prosimians, are not as precocious in their early developmental stages. The *Microcebus murinus* infant is still crawling with its belly on the substrate and cannot yet walk at 10 days; not until 15 days is it moving around actively and climbing easily (Petter-Rousseaux, 1964; Martin, 1972a). At 20 days it is walking well and by 21 to 25 days it is running and jumping (Petter-Rousseaux, 1964; Petter, 1965; Martin, 1972a). The infant *Cheirogaleus major* does not raise the upper part of its body until 15 days and by 21 days it is still not walking steadily; not until 25 days is its locomotor ability sufficiently well developed to enable it to follow its mother (Petter-Rousseaux, 1964).

The infant *Lepilemur mustelinus* develops relatively quickly and by 30 days it climbs swiftly and agilely and jumps short distances; by 75 days it may jump as much as 75 cm, compared to the larger and slower developing *Propithecus verreauxi* which is only jumping 50 cm at 90 days (Petter-Rousseaux, 1964). The *Varecia variegata* infant, although relatively undeveloped in very early infancy, exhibits a high degree of precocity in later infancy and by 19 days it is actively following its mother and climbing around the environment; at 50 days it is as agile as an adult (Petter-Rousseaux, 1964; Klopfer and Klopfer, 1970).

In the other lemurines, indriids, and lorisines, which do not remain initially in a nest, early locomotion is more a matter of being transported clinging to the mother.

The *Lemur catta* infant appears to develop at a faster pace than other lemurines and indriids. At 4 days it will catch on to branches while clinging to its mother (Jolly, 1966) and at 30 days, although the full adult grip and prehensive pattern is present, it still does not walk properly and loses its balance easily (Bishop, 1964; Petter-Rousseaux, 1964).

Anderson (1971) reports that the infant *Perodicticus potto* walks alone on a branch at 21 days and at about 45 days of age actively climbs about the cage. Charles-Dominique (1966) reports that the *Arctocebus calabarensis* infant does not display well-coordinated movements until 17 days and that the *Nycticebus coucang* infant is still unsteady on its feet at 14 days. Ehrlich (1974), however, reports that the *N. coucang* infant is able to move forward on all fours by 6 days and may climb up a vertical slope at 5 days and down at 7 days.

By 2 days at least the infant *Tarsius bancanus* is capable of slow quadrupedal climbing and takes its first jump at 4 days, from which time on it prefers jumping to quadrupedal locomotion (Niemitz, 1974). A second infant of the same species, at about 7 days, could jump up to 40 cm from horizontal to vertical, horizontal to horizontal, and vertical to horizontal, displaying coordination of the extremities in the process (Niemitz, 1974). At this age too the infant displayed extremely fast and well-coordinated quadrupedal walking and climbing up a vertical surface. By about 19 days the infant displayed the full adult locomotor range (Niemitz, 1974).

3.6. Play

The young of all or most mammals and particularly primates probably play a good deal. Dolhinow and Bishop (1970) suggest that play constitutes preparation for the rigors of adult life and without which the young primate would be ill-prepared to compete successfully or even survive in an adult world. Play, however, does not appear to persist in any marked degree into adulthood in any primate species, humans being a notable exception. In the lesser bushbaby, for example, there is no evidence of what could be interpreted as play behavior beyond late adolescence. At most it can be said that adults tolerate the playfulness of infants and juveniles but have never been observed to play with them.

The *Galago senegalensis* infant begins to play at a very early age and twins have been observed play-wrestling and playing with nest material at 6 days of age. By 20 days a great deal of time is devoted to play, much of it with adolescents, and by 42 days, in both the laboratory and the field, play involving peers, juveniles, adults, and inanimate objects has become quite elaborate (Bearder, 1969).

Galago demidovii infants are playing boisterously with peers by 17 days of age (Blackwell, 1969) and Bearder (1975) reports seeing *G. crassicaudatus* infants playing together and with their mother at 32 days and playing with an adult male at 126 days. Ehrlich (1974) reported *G. crassicaudatus* initiating rough and tumble play by about the fifth week. In *Nycticebus coucang,* on the other hand, social play was not observed until after 10 weeks of age. Charles-Dominique (1971) reports that a young *G. alleni* kept as a pet spent much of its time in locomotor play.

All infant prosimians studied have been reported to play (Doyle, 1974); probably with those that spend early infancy in the maternal nest, particularly in those species where twins are common, like *Galago senegalensis,* play begins at an earlier age than in those species where early infancy is spent clinging to the mother. Jolly (1966) reports, for example, that in *Lemur* play only begins when the infant leaves the mother, and Klopfer and Klopfer (1970) report that in *Lemur catta,* perhaps the most precocious of the lemurines, play does not begin until about 31 days. Niemitz (1974) also reports that play occurred in two infant *Tarsius bancanus* later than in other prosimians. He describes three kinds of play clearly elaborated by about 33 days in one infant. The first kind appears to be directed at satisfying a tactile need;

the second bears obvious resemblance to fighting play; and the third concerns play with inanimate objects. As Doyle (1974) noted in other prosimians, he reported no sexual elements in play in *Tarsius bancanus*.

3.7. Threat, Defense, and Reactions to Alarm

Galago senegalensis infants exhibit the typical open-mouthed defensive threat and appropriate posture as early as 7 days of age. At 12 days the infants' reaction to the adult alarm call is to huddle quietly in a corner of the nest box. Up to about 37 days of age infants, if in the nest box, pay no attention to alarm calls, and if outside they go immediately to the nest box. After this age and typically around 50 days of age the reaction of infants to the adult alarm call is to repeatedly go in and out of the nest box, or to go to the nest box, put their heads out, look around, come out cautiously, and remain alert while the alarm call persists. Only later will they typically go to the top branches as adults do.

Niemitz (1974) reports an infant *Tarsius bancanus* at about 2 days of age displaying a threatening gesture typical of an adult, though weaker, when approached too fast by a caretaker. On the following day the same gesture was given in a comparable situation but this time it was stronger and accompanied by a high-pitched whistle. Blackwell (1969) reports the defense posture and defense cry in the *Galago demidovii* infant at 10 days of age. The threat display is reported at about 20 days in the infant *Varecia variegata* (Petter-Rousseaux, 1964; Klopfer and Klopfer, 1970) and, together with accompanying vocalizations, is probably present from an early age in all prosimians.

3.8. Achievement of Independence

Although much of the above has been concerned implicitly, if not explicitly, with the achievement of independence from the mother, two important factors in this process should be considered separately. The first is the increase in physical distance between infant and mother, leading to the infant spending more time with its peers, other adults, and finally to its spending no more time with its own mother, if at all, than with other conspecifics. The second is the actual weaning process or the decrease in the reliance on the mother for nourishment to the point where the mother ceases to be a source of nourishment.

Where convenient, some of the information relevant to this section has been extracted for inclusion in Table IV (see Section 6.1).

3.8.1. Separation from the Mother

By the time the infant primate is physically exploring its environment and initiating regular contact with peers the mother, or the maternal nest, is still regarded as a home base from which the infant ventures into its environment, and to which it regularly returns (Rheingold and Eckerman, 1970). The secure home base seems to be a prerequisite to exploration and social development in all primates. Rheingold and Eckerman (1970) note too that in both humans and other higher primates the mother actively and implicitly encourages separation. The same process may be seen in prosimians. At first the mother is solicitous and protective. Gradually she becomes more tolerant both of the approach of conspecifics and of the infant leaving her for short periods over small distances. This is followed by her encouraging the infant to leave her more often and for longer and, finally, to her intolerance of the infant's presence. This process appears similar in all prosimians studied although features of this separation process may differ between species particularly in respect of such things as duration and exclusiveness of the mother/infant bond (Klopfer and Klopfer, 1970).

Emergence from the nest is probably an important transitional period in the development of nesting prosimians since it represents the beginning of exploration of the environment and independence from the mother. By this time infants have become extremely active and the development of the locomotor and sensory apparatus facilitates proper exploration of the environment, the location of food, and consequent reduction in the need for mother's milk (Doyle *et al.,* 1969).

The *Galago senegalensis* infant is reported to leave the nest box in the laboratory for the first time at between 10 and 14 days (Sauer and Sauer, 1963; Kolar, 1965; Manley, 1966b; Gucwinska and Gucwinski, 1968), and it is at this age that its appearance is first reported in the wild (Bearder, 1969). In the laboratory it may, however, get out of the nest box of its own accord as early as 7 days, but it immediately gets back in (Doyle *et al.,* 1969). By 9 days it will remain outside the nest box if not retrieved by the mother and by 11 days it is getting out continuously, by which time Doyle *et al.* (1969) report a large drop in the amount of time during the waking hours that the infant spends nursing/suckling. Before the age at which the infant *G. senegalensis* first gets out of its nest box and begins to explore its environment, its activities can be said to be virtually directed to and by the mother. Whatever activities it is engaged in when the mother enters the nest box, it will stop and move toward her to be groomed and suckled. By the time it leaves the nest box to explore the environment it begins occasionally to ignore the mother's presence and will often continue whatever activity it is involved in in her presence. Already by 15 days the mother too for the first time may ignore an infant she would normally retrieve and by 21 days the frequency of ignoring infants has increased.

By 29 days *Galago senegalensis* infants are interacting more with adolescents than with their mother and at 33 days the mother may push an infant away from the food tray. Infants show little inclination to remain in the nest box with the mother by 35 days, and at about 38 days the mother frequently ignores her infants when they approach her. By 47 days this negative attitude has markedly increased. Bearder (1969) reports that from 28 to 42 days infants in the wild spend most of the night away from the mother and after 42 days they leave the sleeping site at sunset of their own accord and explore independently over an increasingly wide range.

The infant *Galago crassicaudatus* has emerged from the nest by 21 days in the wild (Bearder, 1975). It begins associating with other adults at 32 days, and moves from one tree to another without its mother's help for the first time at 40 days (Bearder, 1975). Ehrlich (1974) notes that in the laboratory mothers actively begin avoiding their infants by 35 days of age. In the wild the infant *G. crassicaudatus* follows other conspecifics at 50 days, crosses open spaces on the ground at 80 days and at 120 days it may lead the group during movement through the trees or make independent forays (Bearder, 1975). The male moves off independently or migrates to a new home range at 300 days, although the female may continue to follow its mother up to 426 days (Bearder, 1975).

Little is reported of this stage in development in other nesting prosimians. *Microcebus murinus* infants do not usually emerge of their own accord from the nest until day 21 (Martin, 1972a). *Varecia variegata* infants emerge from the nest by at least 19 days (Petter-Rousseaux, 1964) and Klopfer and Klopfer (1970) note that the mother *V. variegata* actively begins avoiding its infants as early as 21 days. The *Galago demidovii* infant first emerges from the nest at 11 to 16 days (Blackwell, 1969; Vincent, 1969; Brand, 1977) and both Woodhull and Woodhull (1969) and Charles-Dominique (1977) report that it is widely exploring its environment at 2 to 3 weeks. Charles-Dominique (1977) reports that, as soon as they are able, young *G. demidovii* will follow the mother about and the female will continue to do so even when adult, remaining in the matriarchal system for years. He notes that the same is true of the young *G. alleni* female. The young male *G. demidovii,* on the other hand, may soon begin to follow other females and at 5 to 7 months it may follow an established male such that their home ranges come to coincide. When puberty occurs by 8 to 10 months it leaves to adopt a nomadic existence.

The developmental landmark of "first emerging from the nest" has no real counterpart in those prosimians in which infants are not confined to a nest but which cling to the mother and remain in close physical contact. The analogy in these species would probably be the time when infants first leave the mother's body to explore the environment.

Although in the lorisines infants are left clinging alone on a branch from an early age, Ehrlich (1974) reminds us that at this age this in no way represents active avoidance on the part of the mother and that in *Nycticebus coucang* the mother returns periodically for brief checks. She notes, in fact, that the *N. coucang* infant remains dependent on its mother for longer than does the *Galago crassicaudatus* infant. Charles-Dominique (1977) reports that at 3 to 4 months the young *Perodicticus potto* and *Arctocebus calabarensis* infants are still following the mother either riding on her back or walking behind her. By 6 months of age the young *P. potto* male is moving around on its own and no longer sleeping with her. The young female continues sleeping with the mother until about 8 months of age.

In *Lemur catta* the first sign of independence appears at about 13 days, when the infant moves off on to the backs of siblings (Petter, 1965; Jolly, 1966; Klopfer and

Klopfer, 1970). It makes its first tentative exploration of the environment at 18 days; by 30 days it is frequently and independently exploring its environment and between 27 and 37 days the mother moves off independently of her infant for the first time (Klopfer and Klopfer, 1970; Klopfer, 1972). In the first 30 days it never sleeps away from its mother, although by this time it may be spending most of its waking time with its peers (Klopfer and Klopfer, 1970; Klopfer, 1972).

In *Lemur fulvus* the duration of mother/infant contact is longer than in *L. catta*. An infant *L. fulvus* studied by Klopfer (1972) made its first real excursion from its mother on day 46 and climbed on the male's back. Klopfer (1972) notes that this initial separation may be delayed if mother and infant are separated from other conspecifics. By 55 days it may move as much as 2 m from its mother and by 60 days it is spending more time away from its mother with its peers (Klopfer and Klopfer, 1970; Klopfer, 1972). By 100 to 110 days it is more part of the juvenile group and, like *L. catta*, it is making maximum use of the available space (Klopfer and Klopfer, 1970).

Klopfer and Klopfer (1970) report that the *Propithecus verreauxi* infant first leaves its mother at 45 days to explore its environment, but Jolly (1966) reports that it does not lose contact with its mother until 60 days; before this time, even though it may crawl on to a branch, it maintains contact with its mother with one foot. By 90 days it is spending more time with its peers (Petter-Rousseaux, 1964).

3.8.2. Weaning

In *Galago senegalensis* there is a noticeable drop in the amount of time spent suckling at 14 days, which coincides with the time when infants first leave the nest box of their own accord (Doyle *et al.*, 1969). Up to the age of about 21 days most suckling takes place inside the nest box, and after this time most suckling occurs out of the nest box where infants have already located food and drink. At 45 days there is another fairly drastic drop in time spent suckling. From this time on, very little maternal behavior of any kind is seen, and quite often infants attempting to suckle are ignored or discouraged. Weaning is probably complete by 10 to 11 weeks although, since infants continue sleeping with their mothers for some time yet, at least in the laboratory, some suckling during the day may take place (Doyle *et al.*, 1969).

This is probably the pattern for most galagines. Buettner-Janusch (1964) reports that weaning in *Galago crassicaudatus* starts at 21 days and Bearder (1975) reports seeing no suckling behavior in the same species in the wild after 10 weeks though Ehrlich (1974) notes that in the laboratory suckling had not ceased by 10 weeks. Vincent (1969) reports that lactation in *G. demidovii* lasts for 60 days, though Charles-Dominique (1977) reports weaning by day 45 and Brand (1977) reports that the infant is fully independent of the mother by 37 to 39 days.

In the lorisines the data do not appear to be as clear-cut. Blackwell and Menzies (1968) report the end of lactation in *Perodicticus potto* in two cases at 5 weeks and 7 weeks. Ehrlich (1974) reports that *Nycticebus coucang* infants were still suckling at 10 weeks and Manley (1966b) reports that they may occasionally suckle as late as 4 months. P. Charles-Dominique (personal communication) notes that though the young *P. potto* and *A. calabarensis* continue sleeping with the mother until relatively late, 6 to 8 months in the case of the former, it cannot be construed that suckling continues up to this age.

Weaning seems to be complete earlier in the cheirogaleines than in any other prosimians (see Table IV and Section 6.1). The prosimians that suckle their infants longest are probably the lemurines and indriids. Klopfer and Klopfer (1970) report the first signs of weaning in *Lemur catta* at 69 days and in *L. fulvus* at 50 days. Incidence and frequency of suckling are less easy to determine in those prosimians that carry their infants around with them and probably the most reliable sign of the completion of weaning in these species is when the infant is fully independent of the mother for transportation.

3.8.3. Sexual Maturity

The age at which animals first mate successfully is often regarded as a good index of the attainment of adult status. In the laboratory, however, *Galago senegalensis* mate successfully well before reaching adult weight, another important criterion of adult status. This is particularly true of females. In our laboratory female twins both conceived at about 120 days and each produced live singletons within a few days of one another at approximately 245 days. A third mated at 168 days and gave birth at 291 days. These females weighed between 115 and 140 gm at conception, all well below the minimum range of the adult female (see Section 5 below). More typical of the general trend in our laboratory, however, is to conceive for the first time at about 200 days. Subspecific variations of this and other respects have been noted by Doyle *et al.* (1971).

The male *Galago senegalensis* comes into sexual maturity much later than the female. The first successful mating in a male in our colony occurred at 326 days, and Andersson (1969) notes that not until about a year does the male give the full adult aggressive pattern and emit the male call, although our data (see Section 4.3 below) indicate that the frequency of contact urine-washing, which is typically associated with sexual interest and behavior in the male, is reached well before this time. Manley (1966b) noted sexual activity in the male *G. senegalensis* at 139, 153, and 183 days but reported that the genitalia were still immature.

Reports on the onset of sexual maturity in other prosimians are scanty and may not always be reliable, based as they usually are on very small samples. In addition R. D. Martin (personal communication,) notes that sexual maturity may be accelerated under laboratory conditions where seasonal changes can be artificially manipulated.

Most lemurines and indriids studied are reported to be sexually mature and adult in size at 18 months, but owing to their very restricted annual breeding season they only mate for the first time at $2\frac{1}{2}$ years (Petter-Rousseaux, 1964; Jolly, 1966).

4. QUANTITATIVE OBSERVATIONS ON THE DEVELOPMENT OF SOME ADULT BEHAVIORS

4.1. Ingestive Behavior

The earliest occasion on which an infant *Galago senegalensis* was observed eating was at 17 days, in the case of a female, and 20 days in the case of a male. Although both sexes were regularly observed licking, handling, and smelling solid food as early as 17 days, it was not until 22 days that both sexes were seen eating with any regularity. A female infant was first seen drinking from the container at 16 days, and by 19 days both sexes were drinking regularly. Bearder (1969) observed infants taking solid food in the wild at 21 days, and notes that gum is eaten before insects. In the laboratory infants were first observed catching and eating insects at 28 days.

The increase in frequency in eating and drinking behavior, as a function of age, is plotted in Figs. 1 and 2, respectively, for both sexes. Frequency refers to the number of times animals were seen to take food in the mouth, either directly or by means of the hands, and to lap milk from the container.

Figure 1 shows that male and female infants have begun to eat as often as adults by between 8 and 15 weeks, which coincides with the time when weaning is

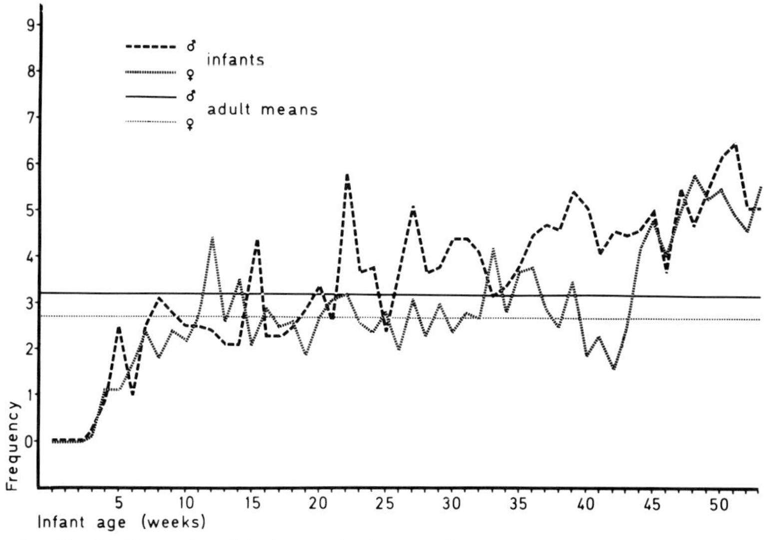

Fig. 1. Graphic illustration of the increase in the mean frequency of eating as a function of age in the 2 male and 4 female primary subjects. Each point on the graph represents the weekly mean of fourteen 50-minute observations in the case of the 2 male subjects and twenty-eight 50-minute observations in the case of the 4 female subjects.

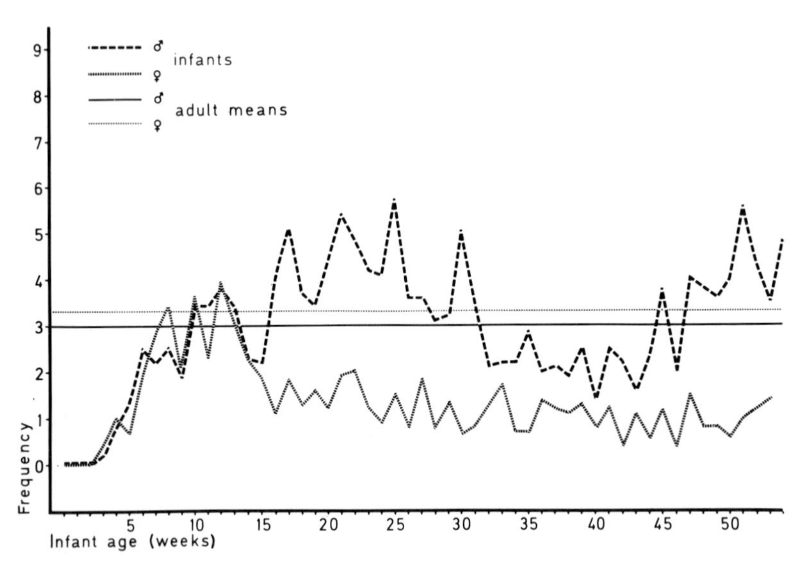

Fig. 2. Graphic illustration of the increase in the mean frequency of drinking as a function of age in the 2 male and 4 female primary subjects. Each point on the graphs was derived in the same manner as those for Fig. 1.

complete at 10 to 11 weeks. From 21 weeks onward male infants steadily eat more often than adults and particularly from 34 weeks onward when they eat almost twice as frequently as adults. Female infants continue eating at approximately the same frequency as adults up to about 43 weeks, after which they eat much more frequently.

Although frequency of eating is not necessarily a reliable indicator of amount eaten, it is probably highly correlated with amount eaten. Figure 11 (see Section 5 below) shows, for instance, that both male and female infants still weigh considerably less than adults at 58 weeks. This would suggest that infants of both sexes need to continue eating more than adults long after this age.

Figure 2 shows that drinking behavior follows much the same pattern as eating behavior for the first 10 or 11 weeks, after which the pattern changes quite distinctly. Male infants continue drinking at more or less the same frequency as adults, whereas females begin and continue drinking considerably less often.

It could be that the pooled data for drinking and eating would yield a more reliable picture of the development of ingestive behavior than the two considered separately. The liquid food available to the animals (see Section 2.1 above) is such that some animals might prefer a liquid to a solid diet. Table I shows, for instance, that an adult female which ate far more often than two of the others, drank far less often than the same two. It should also be borne in mind that the porridge is semiliquid in form and could have satisfied the need for liquid. This factor could explain the low drinking frequency of female infants, which showed a stronger preference for porridge than wild-caught adults.

The data for eating and drinking pooled together would show that from about 10 weeks female infants take food slightly less frequently than adults until about 43 weeks and thereafter at about the same frequency, suggesting that from this time their need for sustenance is about the same as that for adults. For male infants, on the other hand, the pooled data would show a consistently higher frequency of eating and drinking from about 16 weeks onward, suggesting a greater need for food than the adult. These deductions concerning the greater need for sustenance on the part of the male, compared with the female, based on frequency of eating and drinking and weight gain, are supported by an examination of the difference in activity levels between the growing sexes. Figures 7 and 8 (see Section 4.4 below) show that from the age of about 6 weeks male infants are consistently and considerably more active than adult males, but that from about the same age female infants are no more active than adults.

The above data suggest that *Galago senegalensis* have not reached full adulthood at one year.

The cheirogaleines appear to take solid food at about the same time as the galagines, while the lorisines and lemurines are much slower.

Le Gros Clark (1924) and Hill (1955) report that *Tarsius* accept live food at 21 days. Niemitz (1974) reports an infant *Tarsius bancanus* taking mealworm halves from a caretaker and eating them at about 5 days. At about 15 days it accepted and ate live cricket larvae but showed no prey-catching behavior. A second infant of the same species caught and ate live mealworms at about 12 days and 4 days later was catching insects and eating them adult fashion. It was not until about 35 days, however, that the same infant was observed drinking on its own for the first time.

4.2. Grooming

Twin *Galago senegalensis* infants were first observed both self-grooming and allogrooming on day 2. Neither activity was seen again until a male and female infant were both seen grooming each other at 5 days and two female infants were seen self-grooming at 7 days. Not until 9 and 13 days, respectively, however, can self-grooming and allogrooming (involving both other infants and adults, mainly the mother) be said to have become a regular part of their behavioral repertoire.

The increase in frequency in self-grooming and allogrooming, as a function of age, is plotted in Figs. 3 and 4, respectively, for both sexes. Both figures reveal a much earlier appearance of these behaviors than ingestive behaviors. In both cases adult frequency is reached at 5 to 6 weeks. In neither case is there any observable difference between the sexes, at least until the animals are nearly a year old. Figure 3 shows that from about 22 weeks of age both male and female infants self-groom considerably more frequently than adults, but that by 45 weeks they have settled into the adult pattern of frequency.

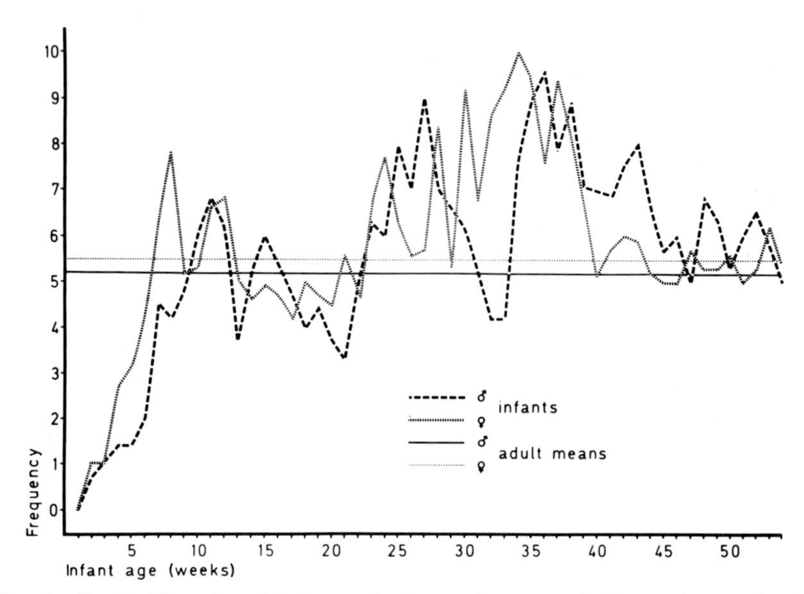

Fig. 3. Graphic illustration of the increase in the mean frequency of self-grooming as a function of age in the 2 male and 4 female primary subjects. Each point on the graphs was derived in the same manner as those for Fig. 1.

Figure 4 shows that from 8 to 38 weeks both males and females allogroom considerably more than adults and that for most of this period, from 8 to 28 weeks, males appear to allogroom more than females. The typical adult pattern of frequency is reached at 40 weeks by which time also females consistently allogroom more frequently than males. In gathering these data no attempt was made to record the status and sex of the animal being groomed.

There is general agreement on the importance of allogrooming in the primates, including the prosimians, as a means of satisfying dependency needs, in establishing and maintaining group cohesion and integrity and reducing tension and aggression, and in initiating and maintaining both social and sexual contact between individuals [see Terry (1970) for a review of this behavior in primates generally and see Doyle (1974) for a review of this behavior in prosimians]. As such one would expect the frequency of allogrooming, in particular, to level off at the adult frequency at an age consistent with full independence and integration into the adult group.

Grooming behavior in prosimians is well described in the literature, particularly in the more social species like some species of *Lemur* (see Doyle, 1974), but there are very few data on the ontogeny of this form of behavior in any prosimian species and none, as far as the author is aware, on the social prosimians. Vincent (1969) reports seeing infant *Galago demidovii* self-groom and allogroom at 15 days and Bearder (1975) saw *G. crassicaudatus* infants allogroom in the wild at 32 days.

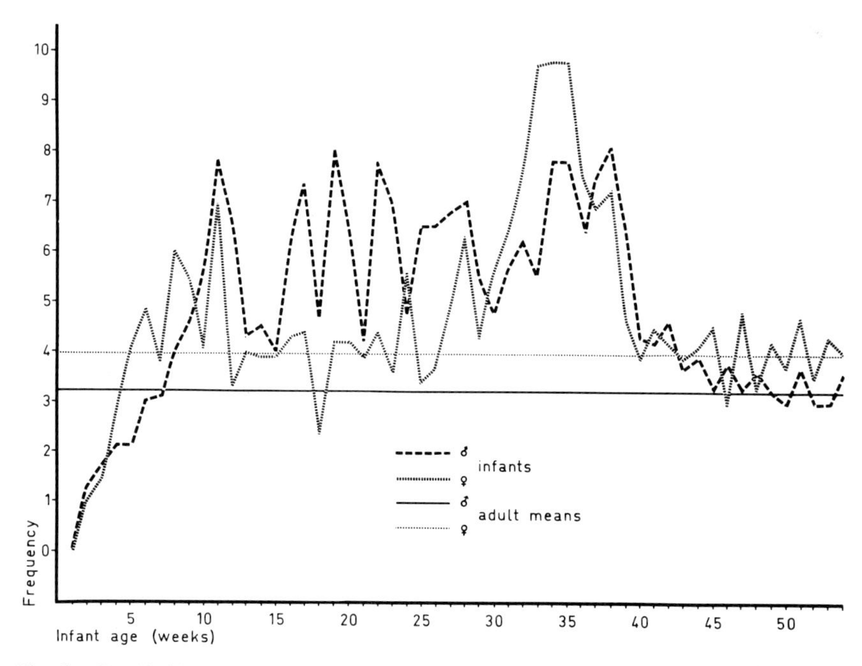

Fig. 4. Graphic illustration of the increase in the mean frequency of allogrooming as a function of age in the 2 male and 4 female primary subjects. Each point on the graphs was derived in the same manner as those for Fig. 1.

Perodicticus potto infants allogroom at 9 days, present the arm for grooming, and self-groom at 16 days (Anderson, 1971). Infant *Microcebus murinus* are reported allogrooming at 21 days (Petter-Rousseaux, 1964; Petter, 1965; Martin, 1972a). Niemitz (1974) reports on the development of self-grooming in two infant *Tarsius bancanus*. One infant, at 2 to 4 days of age, licked its lips, forearms, hands, and feet after feeding. At 7 to 8 days its grooming was much improved and at 11 to 12 days it used its grooming claw properly for the first time. A second infant used its grooming claw effectively at 7 to 8 days. By about 3 weeks it displayed the full adult pattern of self-grooming including the licking of its anogenital region for the first time.

4.3. Urine-Washing

Male and female *Galago senegalensis* were first seen to urine-wash at 20 days, in both cases displaying the full adult pattern. The sexual function of urine-washing in the male has been pointed out elsewhere (Doyle *et al.*, 1969; Doyle, 1974). Adult males invariably urine-wash when in physical contact with females and under circumstances which leave little doubt in the observer's mind that their interest is sexual. This type of urine-washing is referred to as contact urine-washing to distinguish it from urine-washing that occurs under a variety of other circumstances and for reasons not necessarily directly concerned with sexual behavior. Adult females urine-wash considerably less frequently than males and, except by coincidence, have never been observed to contact urine-wash. The distinction between the sexes in respect of this particular pattern of behavior is far greater than in respect of any other pattern of behavior observed and provides a suitable baseline for the ontogeny of sexual maturity in the adolescent of either sex.

The increase in frequency of urine-washing in general as well as contact urine-washing in particular, as a function of age, is plotted in Figs. 5 and 6, respectively, for both sexes. Figure 5 indicates that by 4 to 5 weeks both sexes are urine-washing regularly and that by 6 to 7 weeks females have reached the adult frequency and remain there. Not until 26 weeks does the male frequency exceed that of the female and by 36 weeks males are urine-washing somewhat more frequently than adult males.

Contact urine-washing appears much later, not until 15 weeks in the case of the male, and remains at the same level as the female until about 26 weeks, up till which

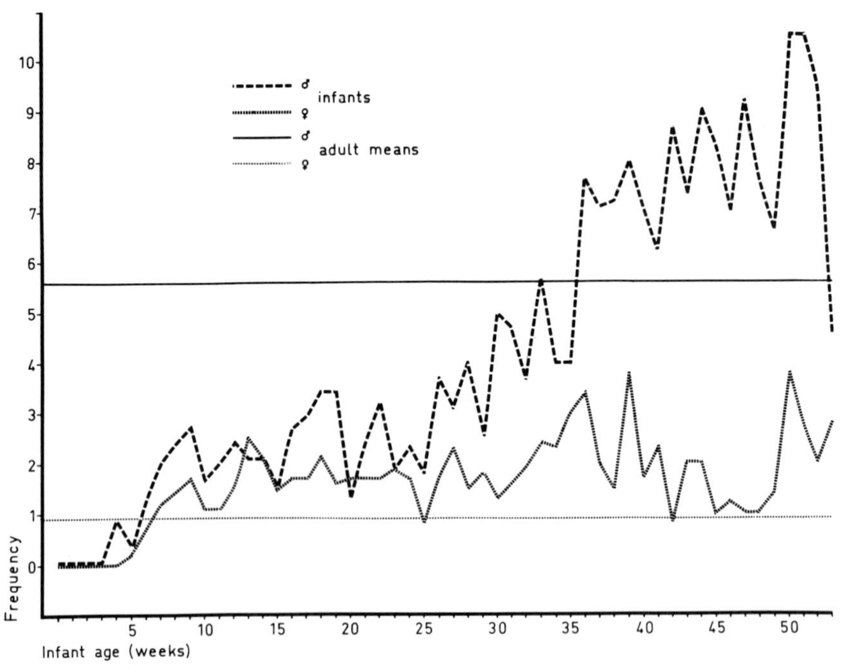

Fig. 5. Graphic illustration of the increase in the mean frequency of urine-washing as a function of age in the 2 male and 4 female primary subjects. Each point on the graphs was derived in the same manner as those for Fig. 1.

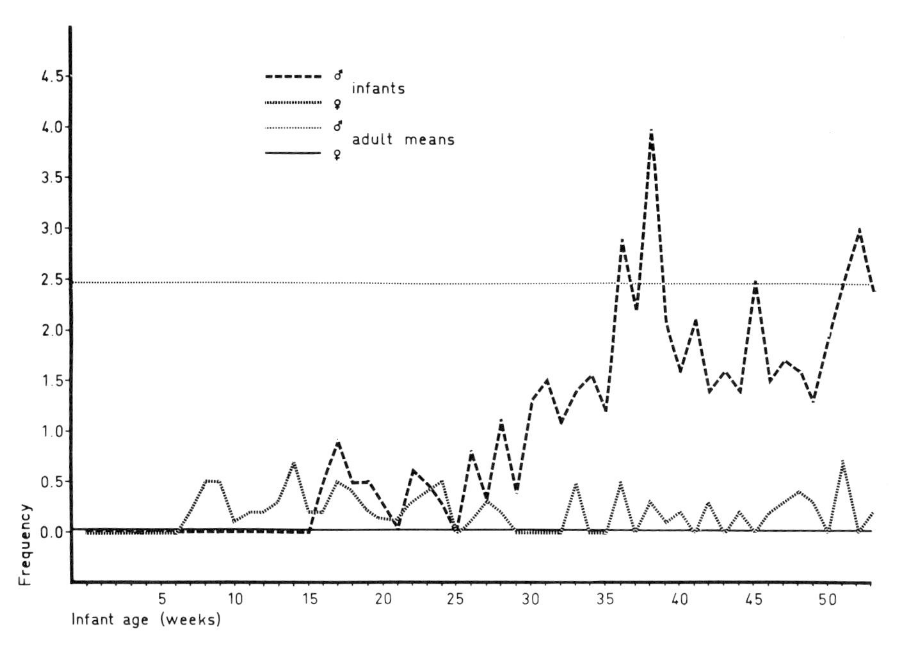

Fig. 6. Graphic illustration of the increase in the mean frequency of contact urine-washing as a function of age in the 2 male and 4 female primary subjects. Each point on the graphs was derived in the same manner as those for Fig. 1.

time contact urine-washing in either sex is probably coincidental, and after which it probably continues to be coincidental in the female. At 26 weeks the male begins to exceed the female in the frequency of contact urine-washing, and between 36 and 50 weeks males are contact urine-washing with the same frequency as adults.

Data on the ontogeny of scent-marking behavior in other species of prosimian are particularly scanty, although detailed descriptions of the behavior and its functions, including sexual functions, in the adult abound (Doyle, 1974). Woodhull and Woodhull (1969) and Brand (1977) report that urine-washing in *Galago demidovii* did not appear until 32 to 42 days of age and that it appeared first in the male (Brand, 1977). Andrew (1964) reports seeing urine-washing in *G. crassicaudatus* in the fifth week. Schilling (Chapter 11) reports that scent-marking in *Cheirogaleus major* first appeared in the sixth week.

4.4. General Activity

A quantitative measure of general activity was derived by measuring the time infants were engaged in any activity other than sleeping or resting. It thus includes most of the other behaviors already considered.

The increase in percentage time active, as a function of age, is plotted in Fig. 7 for both sexes. As has been indicated above, infant *Galago senegalensis* become active at a very early age, and Fig. 7 indicates that by 5 to 6 weeks infants have already reached the level of adult activity. Up until 53 weeks of age, at least, males are consistently and noticeably more active than adults, while female infants remain at an adult level throughout. This factor has already been related to differences in ingestive behavior and weight gain in the two developing sexes (see Section 4.1 above).

A specific measure of activity, very easy to quantify in *Galago senegalensis,* is jumping. Typically locomotion is in the form of leaps and hops, and animals very seldom either run or walk. Figure 8 illustrates the increase in the frequency of jumping in both sexes, as a function of age. These data support those from Fig. 7. By 4 weeks both sexes are jumping regularly and by 6 weeks the frequency of jumping has reached the adult level. Males very soon consistently begin to exceed the adult level while females remain at the adult level.

Male infants apparently have a need to exercise their physical abilities well beyond the requirements of simply getting from one place to another for whatever reason. Typically the adult rests unless in pursuit of food, engaged in agonistic

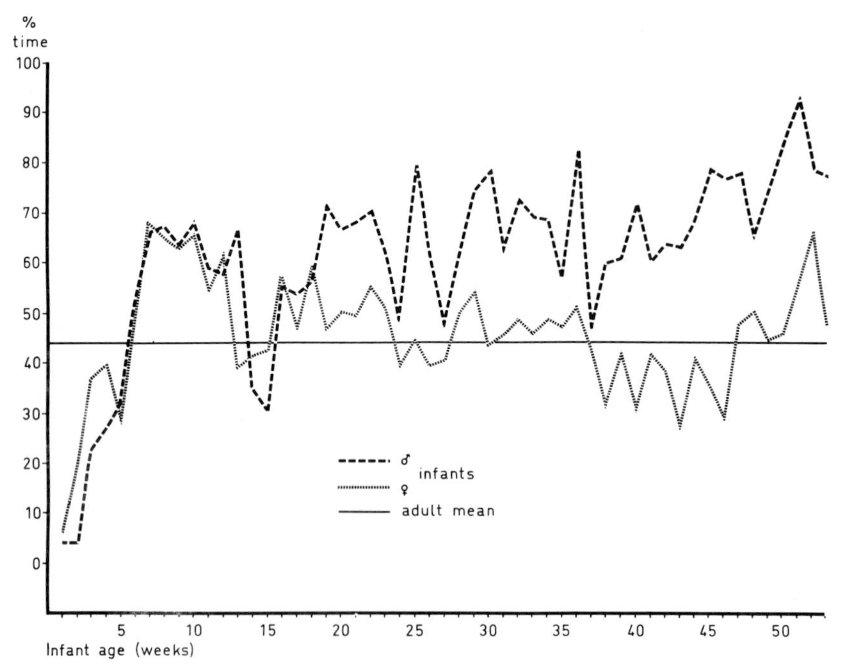

Fig. 7. Graphic illustration of the increase in the mean percentage time active as a function of age in the 2 male and 4 female primary subjects. Each point on the graphs was derived in the same manner as those for Fig. 1.

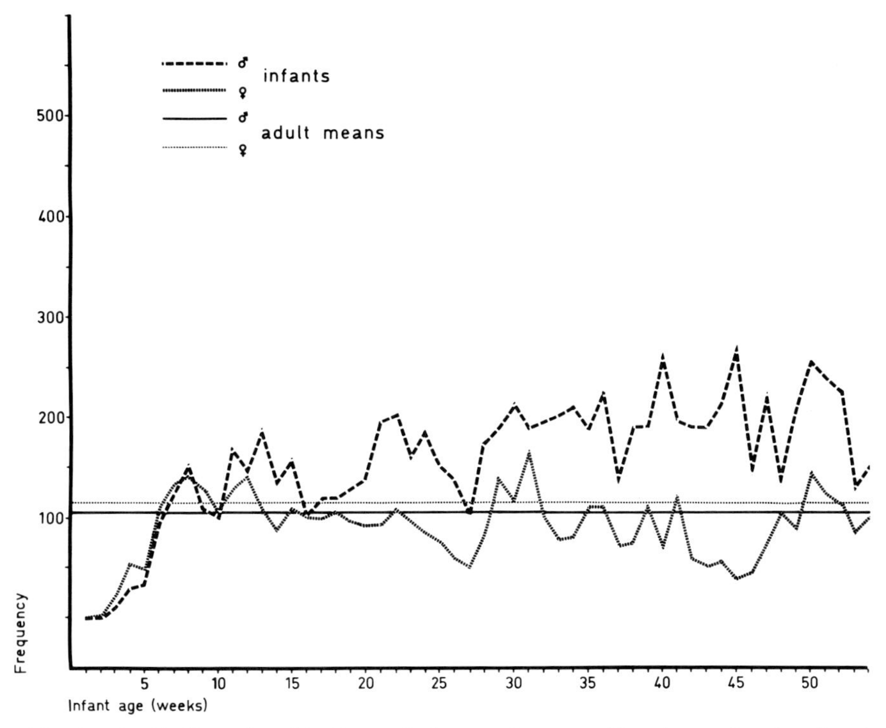

Fig. 8. Graphic illustration of the increase in the mean frequency of jumping as a function of age in the 2 male and 4 female primary subjects. Each point on the graph was derived in the same manner as those for Fig. 1.

pursuit and escape, or in a sexual situation. Movement may be said to be economical; there is always a purpose. The shortest route and the easiest and shortest jumps are always taken unless the nature of the situation dictates otherwise. This is patently not true of the adolescent and juvenile male. A great deal of individual jumping occurs that is not related to any of the activities mentioned above and it is equally clear that very often the most difficult and longest jumps are selected when there is no apparent need. It could well be that the life of the adult male is such that much practice in this form of behavior when young is a prerequisite for success in mating and even survival. Much of this activity in young animals is, of course, play of both a locomotor and social kind, the significance of which for survival in the adult world has already been discussed (see Section 3.6 above).

Although the quantitative data have been restricted to those derived from Period 2 observations, Period 1 observations (the first period of the waking day) are interesting, particularly in respect of general activity. During the first week after birth infants are active from 0 to 40% of the time during Period 2, with some animals remaining inactive during this period for the whole of the first week, but during

Period 1 activity occupies from 11 to 50% of the time. After 16 days infants are often active 100% of the time during Period 1, but during Period 2 they are active from 7 to 74% of the time. By 6 weeks all infants are active 100% of the time during Period 1 but only 45 to 80% of the time during Period 2. This difference in activity between Periods 1 and 2 is much more marked in infants than it is in adults.

5. WEIGHT GAIN

A consideration of weight gain is appropriate in a paper on behavioral development, since the development of all aspects of behavior is dependent on or highly correlated with physical growth. This was pointed out in respect of ingestive behavior and general activity in Sections 4.1 and 4.4, respectively.

Some essential data on birth weight and weight gains in both sexes for the first four weeks of life are indicated in Table II. The mean birth weights of the 56 infants reported in Table II include the weights of 42 infants reported in a previous paper (Doyle *et al.*, 1971). In that paper it was indicated that singletons tended on an average to be heavier than twins, that female twins weighed slightly more than male twins, and that in bisexual pairs the male was always slightly heavier. These differences are small, as are weight differences between the sexes, and do not appear to be significant.

Mean daily weight gains for 20 males and 23 females, from birth to 97 days, are plotted in Figs. 9 and 10, respectively. In each figure one SD about each daily mean is plotted to provide a measure of the spread of variability and the curves for females and males are plotted against the opposite sex curves for ease of comparison. Both are typically negatively accelerated weight gain curves. Differences in weight between male and female increase uniformly to 97 days, as does the variability in weight for either sex. By 97 days both sexes weigh approximately half their expected adult mean weights. The curve for females remains uniformly within −1 SD of the male curve and the curve for males remains generally within +1 SD of the female curve.

TABLE II

Vital Birth Weight Statistics for *Galago senegalensis* for Both Sexes, and Mean Time Taken to Double, Triple, and Quadruple Birth Weight

		Birth weight (gm)		Mean time in days taken to multiply birth weight		
Sex	*N*	Range	Mean	X2	X3	X4
Male	28	8.6–15.5	11.9	7	13	20
Female	28	8.5–14.8	11.7	7+	−14	21+

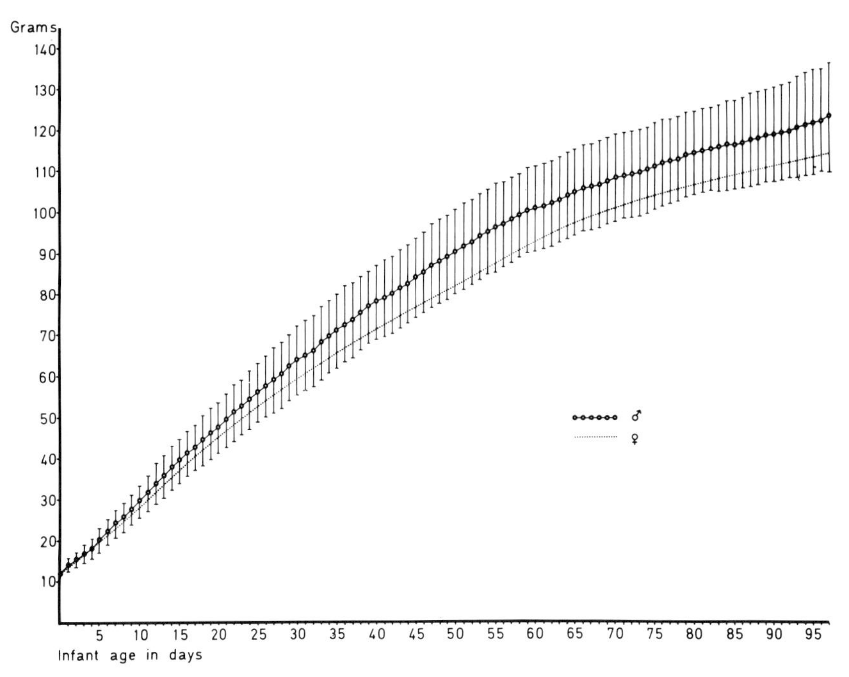

Fig. 9. Graphic illustration of mean daily weight gain ±1 SD for 20 male infants from birth to 97 days of age, plotted together with the mean daily weight gain of 23 female infants for ready comparison.

Mean weekly weight gains for 5 males and 5 females, from the first to the 107th week of life, are plotted against the mean adult weight for both sexes in Fig. 11. The way in which the mean weights are derived for the adult sample is indicated in Table III.

Two of the sample of 7 males and 5 of the sample of 10 females were wild-caught adults of unknown age, but all had been in the laboratory as adults for between 2 and 5 years and so all were between 4 and 6 years of age at the least. The sample does not include either the heaviest male or the heaviest female, which regularly weighed 350 and 280 gm, respectively, because neither of these animals was weighed regularly enough to comply with the criteria used to select the sample. Furthermore, both the adult and the infant sample excluded all weighings of females while pregnant. While on this subject it may be noted that the mean weight gain of 4 pregnant females carrying singletons was 25 gm (range = 22–31 gm), and the mean weight gain of 4 females carrying twins was 36 gm (range = 24–47 gm). These figures are based on differences in weight before and after parturition. Since the sample used for Fig. 11 was small, the weight curves are not as uniform as those for Figs. 9 and 10. For the same reason no reliable measure of variability is available. Both curves are, however, typically negatively accelerated reaching plateaux at 52 weeks and remaining there until about 72 weeks. This phenomenon has been ob-

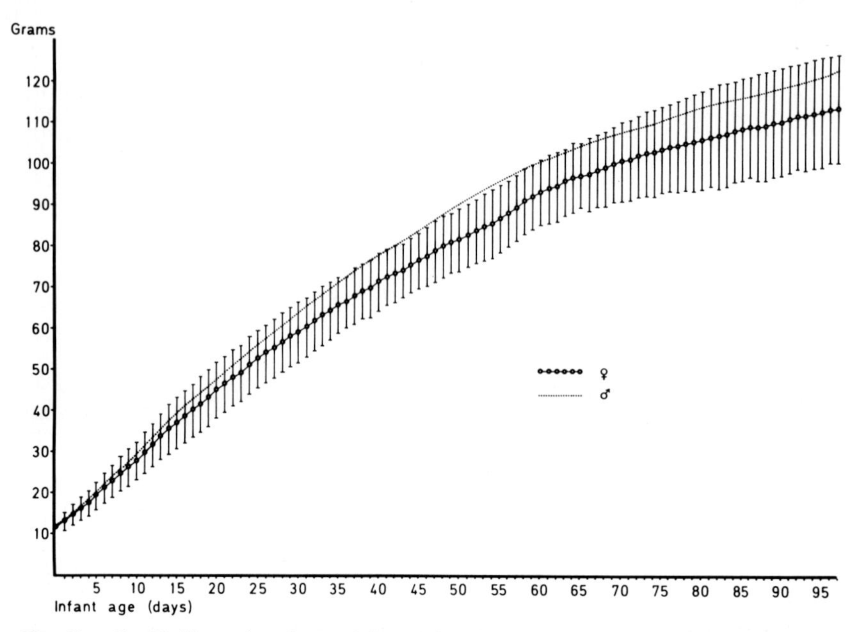

Fig. 10. Graphic illustration of mean daily weight gain ±1 SD for 23 female infants from birth to 97 days of age, plotted together with the mean daily weight gain of 20 male infants for ready comparison.

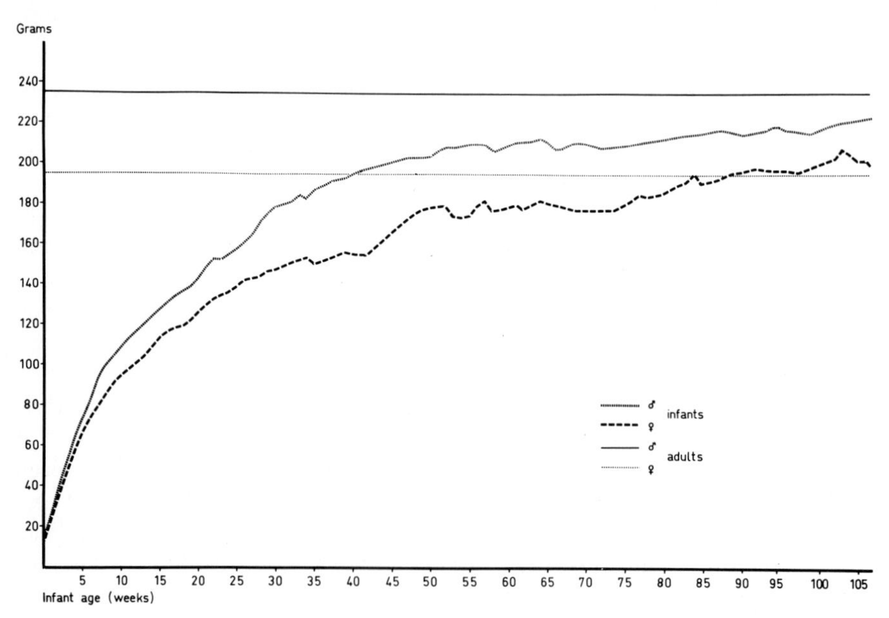

Fig. 11. Graphic illustration of mean weekly weight gain for 5 male and 5 female infants from birth to 107 weeks of age.

TABLE III

Data Showing How the Mean Adult Weight for *Galago senegalensis* for Each Sex Was Derived

Sex	*N*	No. of weighings	Weight (gm) Range	Weight (gm) Mean	Age range (years)
Male	7	47	188–301	235	2–6
Female	10	161	150–252	195	2–6

served regularly in all our animals and no explanation is offered. After 72 weeks weight gains resume slowly and regularly. Charles-Dominique (1977) reports the same phenomenon in *G. demidovii* but at a much earlier age. In this species the weight curve flattens out at 45 days and begins to increase again at 60 days. Of interest is the fact that the 5 females reached, on an average, the adult mean weight by 84 weeks of age. Two of these animals had in fact reached this weight by 75 weeks and 2 had not reached it by 107 weeks. The weight range at 107 weeks was 150–238 gm for females. Of equal interest is the fact that the 5 males, on an average, had not reached adult mean weight by 107 weeks. One male reached adult mean weight at 80 weeks and another at 88 weeks. None of the remaining three had reached it by 107 weeks. The range for males at 107 weeks is 188–258 gm. The mean birth weight for three *G.s. zanzibaricus* infants born in captivity is reported at 14.1 gm (Gucwinska and Gucwinski, 1968). Until at least about 6 weeks of age their growth curve is identical to that of *G.s. moholi,* but Gucwinska and Gucwinski (1968) report that they attain adult size at 4 months.

Charles-Dominique (1977) notes that adult body weight is reached in *G. demidovii* at 150 to 180 days, in *G. alleni* at 180 to 240 days, and in *G. elegantulus* at 240 to 300 days, all well before *G.s. moholi.* Ehrlich (1974) provides some comparative data on weight gain in nine *Galago crassicaudatus* infants and five *Nycticebus coucang* infants. Mean birth weight for the former was 46.4 gm and for the latter 46.7 gm. Approximate mean weight at 5 weeks for the nine *G. crassicaudatus* infants was 250 gm, which represents a fourfold increase, compared to a mean of 70 gm, a sixfold increase for 43 *G. senegalensis* infants over the same period. By 50 weeks the *G. crassicaudatus* infants had increased their mean birth weight by a factor of 22 and the *N. coucang* infants by a factor of 21 compared to a factor of 16 for ten *G. senegalensis* infants. Martin (1972a) reports that the *M. murinus* infant doubles its birth weight in 5 days, trebles it in 10 days, and quadruples it in 15 days.

Information on birth weight is included in Table IV (see Section 6.1). Much of it may be unreliable since reports are seldom concerned with more than a few animals. The extreme variability in weight, even at birth, in a large sample of *Galago senegalensis,* reared under uniformly the same conditions, suggests that in other prosimians too a reliable picture of weight gain from birth to maturity can be gained only from a large sample.

6. DISCUSSION AND CONCLUSIONS

6.1. Phyletic Trends in Development

Duration of development differs widely among primate species, between the very smallest like *Galago demidovii* and *Microcebus murinus,* at the one extreme, to *Pan, Pongo,* and *Gorilla* (Reynolds and Reynolds, 1965; Schaller, 1965; Napier and Napier, 1967) at the other. Several authors (Washburn and Avis, 1958; Mason, 1968; Schultz, 1956, 1969; Schusterman and Sjöberg, 1969) see in this a trend toward a prolongation of the developmental period, from prosimians to man, the more recent forms requiring a longer postnatal development period in which to acquire the more complex patterns of behavior appropriate to their more advanced phyletic status.

Ehrlich (1974) reported no such trend in comparing the data from the two species of prosimian she studied with data from three species of New and Old World monkeys from other sources. She emphasizes the need to study the nature of the developmental sequence rather than the age at which various behaviors appear. Hinde's (1971) review of the development of behavior from birth to maturity in 16 species of New and Old World monkeys and two species of ape emphasizes the considerable diversity to be found in interspecies comparisons on infant development and in the factors that influence development. He notes, for instance, that factors that may exert considerable influence on development in one species may exert little influence on another and may be absent in yet a third.

Hinde's (1971) data do reveal, however, that the apes compared to the monkeys have a noticeably prolonged developmental period just as the present review shows that the prosimians are relatively precocious in their development compared to the monkeys. Such comparisons, however, do not necessarily constitute support for a phyletic trend toward extension of the postnatal development period since an important operative variable might well be body size—the larger the species the longer the developmental period. Schultz (1956), for instance, questions whether body size is dependent upon duration of growth or upon speed of growth, and Gavan and Swindler (1966) point out similarities between the growth rates of primates and cattle, concluding that, while duration of growth was a variable, growth rate was constant.

A statistical examination of the relationship between maternal body weight of primates, on the one hand, and various indices of development, extracted from the literature (see Table IV), was made by means of a program developed by the University of Chicago (distributed and supported by SPSS, Inc.) on an IBM 370-158 OS computer at the Computer Centre of the University of the Witwatersrand. Initially the distribution for each variable was tested for normality (Filliben, 1975) and subsequent analyses were performed after natural logarithmic transformation of data. This appeared to produce a normal distribution for each subgroup separately but not for all groups of primates together.

TABLE IV

Values for Maternal Body Weight and Various Indices of Development[a]

	No.	X1 (♀ Adult weight)	Y2 (Gestation period)	Y3 (Birth weight)	Y4 (Age at sexual maturity)	Y5 (First solids eaten)	Y6 (Carried until)	Y7 (Weaned)	Y8 (Leaves mother or nest)	Y9 (Full motor dev.)
Prosimians										
A. calabarensis	1	300	134	24	274	45	105	105	11	48
L. tardigradus	2	298	167							
N. coucang	3	1014	184	45		42		70	5	
P. potto	4	1255	182	47	642	70	150	210		45
G. elegantulus	5	271	135				60			
G. alleni	6	283	140[b]				45			30
G. crassicaudatus	7	1116	131	48	365	21	70	120	21	
G. senegalensis ssp.	8	207	145	18	256					
G. s. moholi	9	201	123	12	201	19	53	61	12	25
G. s. braccatus	10		139	20	230					
G. inustus										
G. demidovii	11	51	111	7	274	31	26	53	13	32
M. murinus	12	70	61	5	256	33	18	25	21[b]	23
M. rufus	13		61				20			
M. coquereli	14	275	89	12		32	20		17	28
C. major	15	425	70	18		25	45	45[b]		
C. medius	16	124								
A. trichotis										
P. furcifer	17	419								
L. macaco	18	2046	127	69	548	30	180	135		
L. fulvus	19	2064	117	75	548	40	180	135	48	
L. mongoz	20	1799	128	53						

(continued)

TABLE IV (*continued*)

	No.	X1 (♀ Adult weight)	Y2 (Gestation period)	Y3 (Birth weight)	Y4 (Age at sexual maturity)	Y5 (First solids eaten)	Y6 (Carried until)	Y7 (Weaned)	Y8 (Leaves mother or nest)	Y9 (Full motor dev.)
L. coronatus										
L. rubriventer	21		127							
L. catta	22	2233	136	82	646	43		135	13[b]	
H. griseus	23		137	48						
H. simus										
V. variegata	24	3000	130			50		135	21[b]	50
L. dorsalis										
L. ruficaudatus	25	800	130							
L. rufescens	26	750								
L. leucopus	27	580								
L. mustelinus	28	900	130	27	548	45		120[b]		30
L. septentrionalis	29	700								
L. microdon										
A. laniger	30	650					150			
P. verreauxi	31	3480	141		913		195	180	48	90
P. diadema										
I. indri	32		137			60	225	180		
D. madagascariensis										
New World monkeys										
C. pygmaea	33	143	140	32						
Callithrix sp.	34	251	140	46	438					
Saguinus sp.	35	415	140	54						
L. rosalia	36	480	127	52						

Saimiri sp.	37	618	173	87	1314
A. trivirgatus	38	1015	130	95	
Cebus sp.	39	2144	180	240	1095
L. lagothrica	40	5750	139	450	1460
Alouatta sp.	41	5860	139	460	
Ateles sp.	42	6913	140	460	
Old World monkeys					
M. talapoin	43	783	190		
Cercopithecus sp.	44	2838	187		
M. fascicularis	45	4094	164		1205
M. radiata	46	5180	161		
E. patas	47	5591	192		
Cynopithecus sp.	48	5660	166		
C. albigena	49	6000	175		
M. maurus	50	6363	165	390	1643
M. mulatta	51	7515	163	368	1460
M. nemestrina	52	7778	176		1533
Presbytis sp.	53	8058	168	500	1460
M. irus	54	9400	166		1278
N. larvatus	55	9980	166	450	
M. sphinx	56	11500	245		1643
M. fuscata	57	12300	175	475	1460
Papio sp.	58	13000	180	420	1643
T. gelada	59	14860	180		1825
Apes					
H. lar	60	5450	210	405	2373
S. syndactylus	61	10300	233	503	
P. troglodytes	62	44950	240	1570	3103

(*continued*)

TABLE IV (*continued*)

	No.	X1 (♀ Adult weight)	Y2 (Gestation period)	Y3 (Birth weight)	Y4 (Age at sexual maturity)	Y5 (First solids eaten)	Y6 (Carried until)	Y7 (Weaned)	Y8 (Leaves mother or nest)	Y9 (Full motor dev.)
P. pygmaeus	63	58925	244	1500	2738					
G. gorilla	64	95000	270	1890	2920					
H. sapiens	65	60000	270	3480	3708					
T. glis	66	160	49	10	146					
Tarsius sp.	67	120	180	26						

[a] In many cases these values are based on reliably large samples; in some cases they are based on a single observation only and, in others, where several research workers give values based on samples, sometimes of unknown size, mid-range values have been used. The number assigned to each species identifies its position in Figs. 12–16. References consulted in compiling the table include: Anderson, 1971; Andersson, 1969; Andriantsiferana *et al.*, 1974; Asdell, 1946; Bearder, 1969, 1975; Beveridge, 1972; Blackwell, 1969; Blackwell and Menzies, 1968; Bourlière *et al.*, 1961; Brand, 1977; Buettner-Janusch, 1964; Butler and Juma, 1970; Catchpole and Fulton, 1943; Charles-Dominique, 1966, 1968a,b, 1972, 1977, personal communication; Chiarelli, 1972; Cooper, 1966; Cowgill, 1969; Crandall, 1964; de Lowther, 1940; Doyle *et al.*, 1967, 1969, 1971; Dukelow, 1974; Eaton *et al.*, 1973; Ehrlich, 1974; Fuquay *et al.*, 1975; Gucwinska and Gucwinski, 1968; Hill, 1937, 1957, 1960, 1962, 1966a,b, 1970, 1974; Hladik, Chapter 8; Holmes *et al.*, 1968; Jewell and Oates, 1969a,b; Jolly, 1966; Klopfer, 1972; Klopfer and Boskoff, this volume; Klopfer and Klopfer, 1970; Leutenegger, 1973; Manley, 1966a,b, 1967; Martin, 1972a,b,c; Morris and Jarvis, 1959; Napier and Napier, 1967; Nicholls, 1939; Niemitz, 1974; Petter, personal communication, 1965; Petter-Rousseaux, 1962, 1964; Pollock, 1975; Sacher and Staffeldt, 1974; Stephan *et al.*, 1970; Ulmer, 1963; Valerio *et al.*, 1972; Van Horn and Eaton, Chapter 3; R. N. Van Horn, personal communication; Vincent, 1969; Woodhull and Woodhull, 1969.

[b] Data not included in statistical analyses.

TABLE V

Correlations between Maternal Body Weight and Various Indices of Development in Different Groups of Primates

Index	Y1 (Maternal body weight)									
	Pearsons r					Spearmans r				
	Pros.	New World monkeys	Old World monkeys	Apes	All[a]	Pros.	New World monkeys	Old World monkeys	Apes	All
Y2 (Gestation period)										
Correlation	0.430	0.079	0.010	0.896	0.687	0.259	−0.100	0.129	0.986	0.631
S.E. of Est.	0.254	0.120	0.104	0.434	0.236	—	—	—	—	—
N	20	10	17	6	55[b]	20	10	17	6	55[b]
Significance	$P<0.05$	$P>0.05$	$P>0.05$	$P<0.01$	$P<0.001$	$P>0.05$	$P>0.05$	$P>0.05$	$P<0.001$	$P<0.001$
Y3 (Birth weight)										
Correlation	0.968	0.990	0.430	0.917	0.959	0.974	0.985	0.429	0.886	0.941
S.E. of Est.	0.226	0.158	0.119	0.367	0.484	—	—	—	—	—
N	15	10	6	6	39[b]	15	10	6	6	39[b]
Significance	$P<0.001$	$P<0.001$	$P>0.05$	$P<0.005$	$P<0.001$	$P<0.001$	$P<0.001$	$P>0.05$	$P<0.005$	$P<0.001$
Y4 (Age at sexual maturity										
Correlation	0.873	0.773	0.655	0.832	0.917	0.883	0.800	0.528	0.500	0.926
S.E. of Est.	0.248	0.426	0.997	0.783	0.348	—	—	—	—	—
N	12	4	10	4	32[b]	12	4	10	4	32[b]
Significance	$P<0.001$	$P>0.05$	$P<0.02$	$P>0.05$	$P<0.001$	$P<0.001$	$P>0.05$	$P>0.05$	$P>0.05$	$P<0.001$

(*continued*)

TABLE V (*continued*)

Index	Y1 (Maternal body weight)									
	Pearsons *r*					Spearmans *r*				
	Pros.	New World monkeys	Old World monkeys	Apes	All[a]	Pros.	New World monkeys	Old World monkeys	Apes	All
Y5 (Age when solids first eaten)										
Correlation	0.389					0.425				
S.E. of Est.	0.339					—				
N	14					14				
Significance	$P<0.05$					$P>.05$				
Y6 (Age when carrying behavior ceases)										
Correlation	0.859					0.873				
S.E. of Est.	0.445					—				
N	14					14				
Significance	$P<0.001$					$P<0.001$				

Y7 (Age when weaned)		
Correlation	0.848	0.861
S.E. of Est.	0.335	—
N	12	12
Significance	$P<0.001$	$P<0.001$
Y8 (Age when infant leaves mother first time)		
Correlation	0.530	0.587
S.E. of Est.	0.699	—
N	8	8
Significance	$P>0.05$	$P>0.05$
Y9 (Age of full motor development		
Correlation	0.757	0.760
S.E. of Est.	0.288	—
N	10	10
Significance	$P<0.005$	$P<0.005$

[a] Since distribution is not normal the significance levels may be questionable.
[b] Includes *Tupaia* and *Tarsius*.

Correlation coefficients between maternal body weight and each of the various indices of development, for the group of primates as a whole, and for prosimians where applicable, are shown in Table V. Significance levels for Spearman correlations are valid for nonnormal distributions, as well as for normal distributions, and have been shown as an insurance against the fact that any predicted normality is not borne out.

Despite small *N* values, Table V shows significantly high correlations between maternal body weight and a number of indices of development. Significant relationships were plotted by means of a CALCOMP 936 plotter and the nature of these relationships is illustrated by regression lines in Figs. 12–16.

Figure 12 shows the logarithmic relationship between maternal body weight and length of gestation in a group of 53 primates, each point representing a particular species in one of four groups of primates, and one point each for *Tupaia* and *Tarsius*. There is a fairly clear separation between the groups and a definite linear relationship between the two variables taking the primates as a whole.

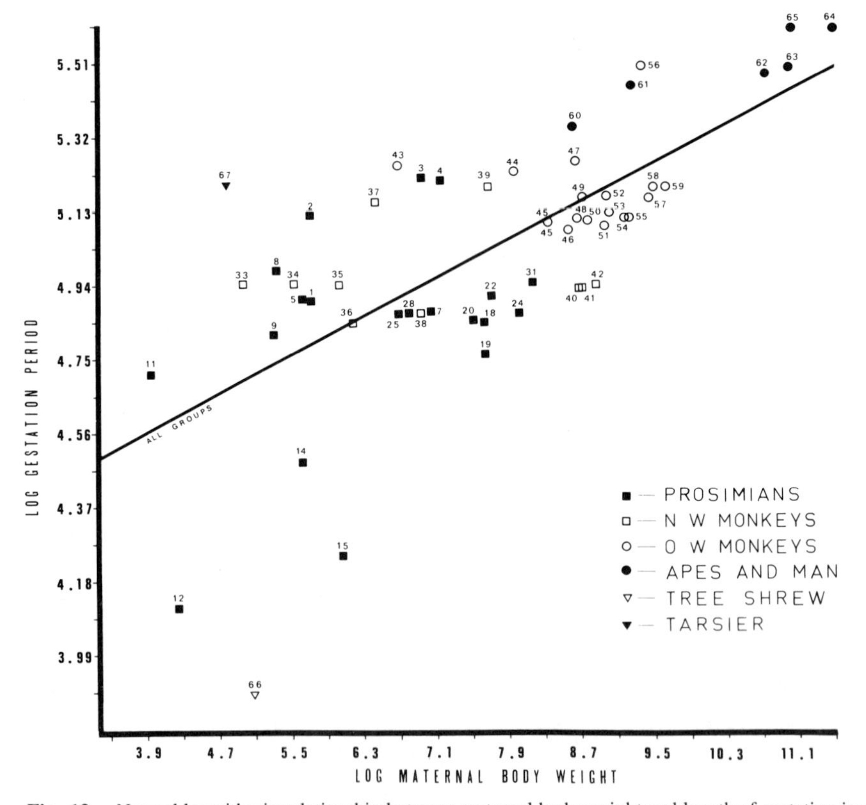

Fig. 12. Natural logarithmic relationship between maternal body weight and length of gestation in a group of primates. The single regression line is for all groups of primates taken together.

In Fig. 13 the groups are even more clearly separated from one another and the linear relationship between maternal body weight and birth weight is even more marked for the primates as a whole as well as for the subgroups.

In Fig. 14 the subgroups are also clearly separated from one another in respect of the relationship between the variables maternal body weight and age at sexual maturity. The linear relationship between the two variables for the primates as a whole, and for the prosimians in particular, is marked.

Figure 15 illustrates a clear linear relationship between maternal body weight and age at weaning in a group of 12 prosimians and Fig. 16 illustrates a similarly marked linear relationship between maternal body weight and the age at which maternal transport of the young ceases in a group of 14 prosimians.

The trend in the above figures and tables seems clear—the larger the species of primate the longer the developmental period. These data also suggest, equally clearly, that the larger the sample, the more significant would be this relationship:

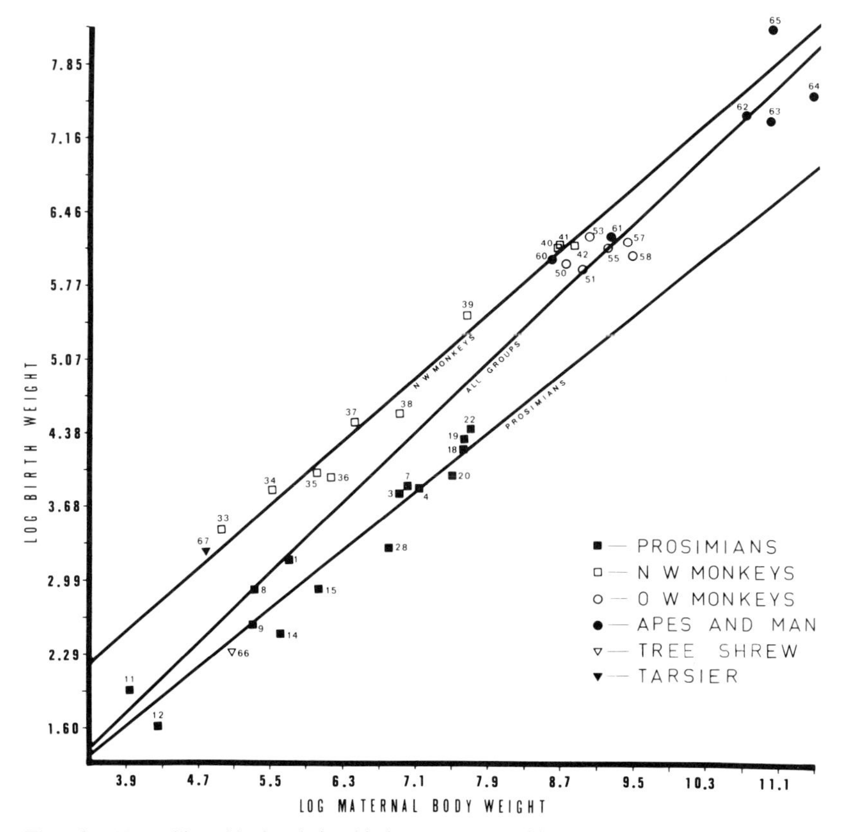

Fig. 13. Natural logarithmic relationship between maternal body weight and birth weight in a group of primates. The oblique lines are regression lines.

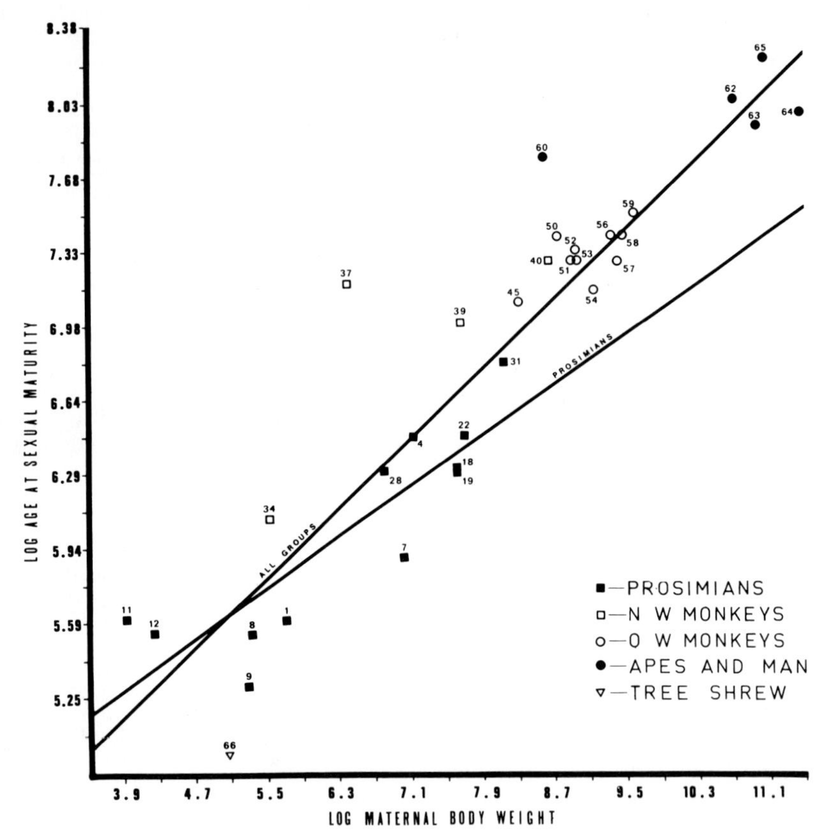

Fig. 14. Natural logarithmic relationship between maternal body weight and age at sexual maturity in a group of primates. The oblique lines are regression lines.

for those variables where the sample size is small, as in age when the infant first leaves the mother ($N = 10$), the linear trend is in the right direction.

However, the importance of the apparent phyletic trend toward a postnatal extension of the developmental period cannot be dismissed; nor, for that matter, do the above data dispute the importance of a prolonged developmental period, irrespective of growth rate, as being a prerequisite for the acquisition of those complex behaviors characteristic of more recent forms. In the Hominoidea body size appears unrelated to the length of the postnatal developmental period. All three of the Ponginae are phylogenetically of approximately the same age and the young of all three develop at approximately the same rate; yet *Gorilla,* the largest of the primates, is more than twice the size of *Pongo* and larger still than *Pan*. All three develop much more quickly than *Homo,* the most complex and phylogenetically the most recent of the primates.

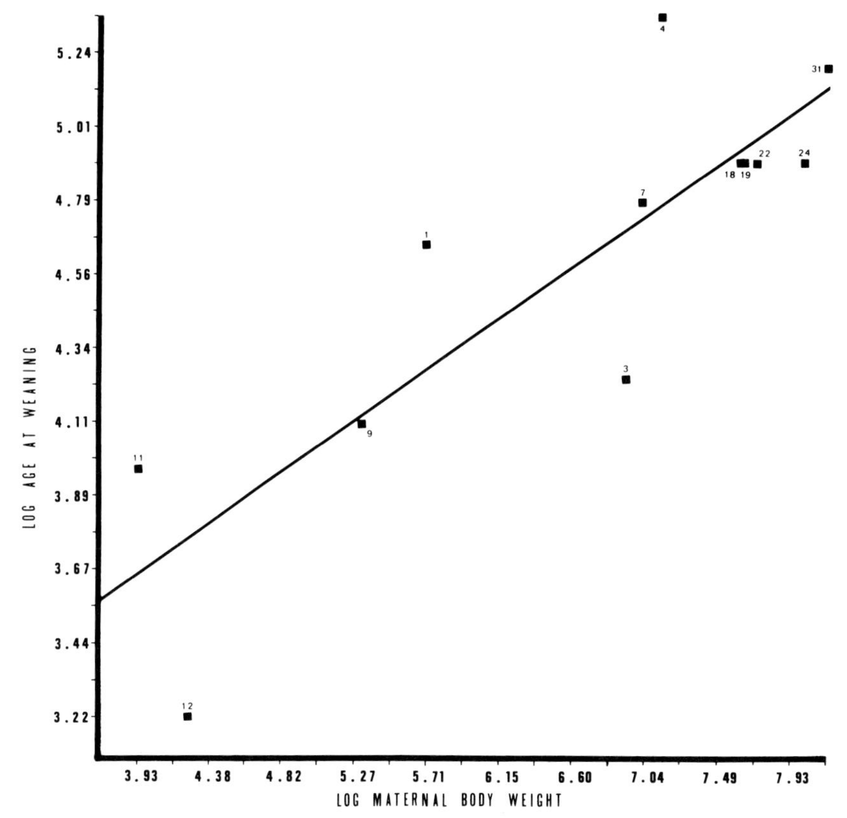

Fig. 15. Natural logarithmic relationship between maternal body weight and age at weaning in a group of prosimians. The oblique line is a regression line.

At the other extreme some tree shrew species are larger than many prosimians and even some New World monkeys and yet, in comparison, are very precocious developers reaching full adulthood in terms of both physical proportions and behavior by $2\frac{1}{2}$ to 3 months (Martin, 1968) compared, for instance, to the much smaller *Microcebus murinus* which mature in 7 to 10 months (Petter-Rousseaux, 1964; Petter, 1965).

Kim and Hutchinson (1975) compared changes in sitting height in four selected species of primate, two New World monkeys, one Old World monkey and one ape, using "instantaneous relative growth rate" as their principle measure. This gives an estimate of growth rate at an instantaneous moment in time. It makes no assumptions concerning the shape of the curve of size versus age and allows comparison of growth rates of animals irrespective of size. They concluded that differences in growth rates do exist among primates, as well as differences in duration of growth,

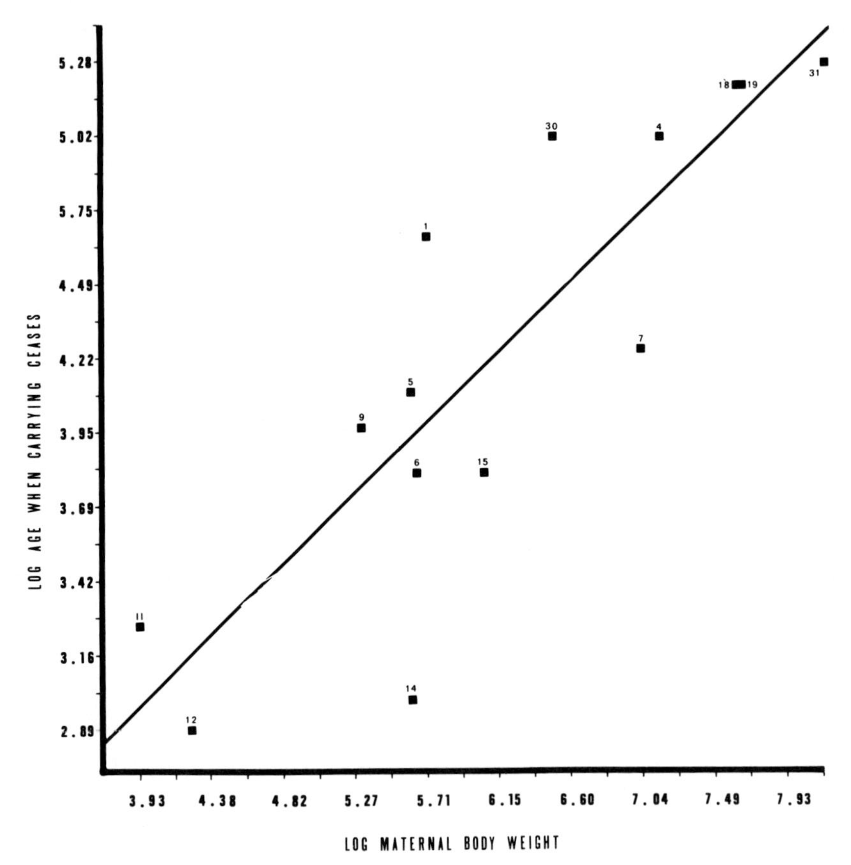

Fig. 16. Natural logarithmic relationship between maternal body weight and age when carrying ceases in a group of prosimians. The oblique line is a regression line.

even after correcting for size as Gavan and Swindler (1966) did. They noted, however, that these differences are most pronounced only in early periods of postnatal growth. They concluded that the differences in growth rate observed appeared to be a function of the phyletic status of the subjects, the closer to man phylogenetically the slower the growth rate. They concluded, however, that this hypothesis can only be tested properly when longitudinal growth data are available for more species of primate.

The need to study development from birth right through to maturity becomes quite apparent when one considers the tree shrew, for example, which up to the age of weaning at about 4 weeks develops more slowly than many prosimians, including *Microcebus murinus*. Only when a complete overview is made of as many aspects of development as possible right through the developmental period is the clear relative precocity of the tree shrew seen. This could explain why over a period of only 10 weeks, for a number of reflexlike behaviors, Ehrlich (1974) found no tendency to an extension of the postnatal developmental period, as function of phyletic status, in two species of prosimian compared with three simian species.

The emphasis in studies on development should surely be on the relationship between developing behaviors and the part they play in the overall biology of the species as the present review clearly shows with respect to a number of aspects of development. Hinde (1971), for example, focuses attention on species differences in social structure and both Hinde (1971) and Ehrlich (1974) draw attention to the relationship between the development of behavior and adaptation to a particular environment.

6.2. Summary

The achievement of adult status in *Galago senegalensis moholi* is not an all-or-nothing sudden transition from one biological state to another to be determined by a uniform set of behavioral indices. A number of indices of development have been described qualitatively and quantitatively for a sample of this species under seminatural laboratory conditions over a continuous period of a year and regular changes in weight over a 2-year period have been presented.

The initial appearance of a number of behaviors has been noted. Charted changes over a span of time show how some behaviors, representative of transitory stages in development, disappear completely over time or change imperceptibly into adult behaviors. Other behaviors change uniformly over time to reach an adult norm and yet others make their initial appearance only towards the end of adolescence.

Just as it is for any other mammal, including humans, that point in time when *Galago senegalensis* reaches the adult state can only be stated approximately. Changes in all the attributes used to arrive at a picture of the adult norm continue to occur throughout life and variations within and between animals are so great that the adult norm itself, either for any particular behavioral or physical attribute or for all together, is a somewhat arbitrary statistic.

Weight, for instance, as a physical attribute on which some attention has been focused in this paper, is probably the least important single indicator of maturity. The data show that, in terms of most if not all behavioral indices discussed, *Galago senegalensis* may be considered fully adult long before adult mean weight is reached. In terms of all the other indices taken together the male *G. senegalensis* can be said to be fully adult at just over a year and the female at just under a year.

Statistical comparison with other species of both prosimian and simian primates on the basis of information extracted from the literature suggests that, in terms of both its size and its phyletic status, it fits into an expected pattern, being neither precocious nor delayed in its development.

ACKNOWLEDGMENTS

This research was supported by the Human Sciences Research Council of the Republic of South Africa, and the University Council, University of the Witwatersrand. The author would like to thank the Computor Centre of the University of the Witwatersrand and Mr. Grady Gott, in particular, for valuable advice and assistance in the statistical analysis. Dr. Annette Ehrlich and Dr. R. D. Martin read an initial draft of this chapter and made many helpful suggestions.

REFERENCES

Anderson, M. (1971). *Discovery* **6,** 89–98.

Andersson, A. B. (1969). M.Sc. Dissertation, University of the Witwatersrand, Johannesburg, South Africa (unpublished).

Andrew, R. J. (1964). *In* "Evolutionary and Genetic Biology of Primates" (J. Buettner-Janusch, ed.), Vol. 2, pp. 227–309. Academic Press, New York.

Andriantsiferana, R., Rarijoana, Y., and Randrianaivo, A. (1974). *Mammalia* **38,** 234–243.

Asdell, S. A. (1946). "Patterns of Mammalian Reproduction." Constable, London.

Basilewsky, G. (1965). *Int. Zoo Yearb.* **5,** 132–136.

Bearder, S. K. (1969). M.Sc. Dissertation, University of the Witwatersrand, Johannesburg, South Africa (unpublished).

Bearder, S. K. (1975). Ph.D. Thesis, University of the Witwatersrand, Johannesburg, South Africa (unpublished).

Beveridge, W.I.B., ed. (1972). "Breeding Primates." Karger, Basel.

Bishop, A. (1964). *In* "Evolutionary and Genetic Biology of Primates" (J. Buettner-Janusch, ed.), Vol. 2, pp. 133–225. Academic Press, New York.

Blackwell, K. (1969). *Int. Zoo Yearb.* **9,** 74–76.

Blackwell, K., and Menzies, J. L. (1968). *Mammalia* **32,** 447–451.

Bourlière, F., Petter-Rousseaux, A., and Petter, J. J. (1961). *Int. Zoo Yearb.* **3,** 24–25.

Brand, H. M. (1977). M.A. Dissertation, University of the Witwatersrand, Johannesburg, South Africa (unpublished).

Buettner-Janusch, J. (1964). *Folia Primatol.* **2,** 93–110.

Butler, H., and Juma, M. B. (1970). *Lab. Primate Newsl.* **9,** 16.

Campbell, D., and Thompson, W. R. (1968). *Annu. Rev. Psychol.* **19,** 251–292.

Catchpole, H. R., and Fulton, J. F. (1943). *J. Mammal.* **24,** 90–93.

Charles-Dominique, P. (1966). *Biol. Gabonica* **2,** 331–345.

Charles-Dominique, P. (1968a). *Biol. Gabonica* **4,** 1.

Charles-Dominique, P. (1968b). *In* "Entretien de Chize I" (R. Canivenc, ed.), pp. 2–5. Masson, Paris.

Charles-Dominique, P. (1971). *Biol. Gabonica* **7,** 121–228.

Charles-Dominique, P. (1972). *Z. Tierpsychol., Beih.* **9,** 7–41.

Charles-Dominique, P. (1977). "Ecology and Behaviour of Nocturnal Primates." Duckworth, London.

Charles-Dominique, P., and Hladik, C. M. (1971). *Terre Vie* **25,** 3–66.

Chiarelli, A. B. (1972). "Taxonomic Atlas of Living Primates." Academic Press, New York.

Cooper, R. W. (1966). "Fourth Annual Report." Inst. Comp. Biol., Zool. Soc. San Diego, San Diego, California.

Cowgill, U. M. (1969). *Folia Primatol.* **11,** 144–150.

Crandall, L. S. (1964). "Management of Wild Animals in Captivity." Univ. of Chicago Press, Chicago, Illinois.

Davis, M. (1960). *J. Mammal.* **41,** 401–402.

de Lowther, F. (1940). *Zoologica (N.Y.)* **25,** 433–459.

Dolhinow, P. J., and Bishop, N. (1970). *In* "Minnesota Symposia on Child Psychology" (J. P. Hill, ed.), Vol. IV. pp. 141–198. Univ. of Minnesota Press, Minneapolis.
Doyle, G. A. (1974). *Behav. Nonhum. Primates* **5,** 155–353.
Doyle, G. A., and Bekker, T. (1967). *Folia Primatol.* **7,** 161–168.
Doyle, G. A., Pelletier, A., and Bekker, T. (1967). *Folia Primatol.* **7,** 169–197.
Doyle, G. A., Andersson, A., and Bearder, S. K. (1969). *Folia Primatol.* **11,** 215–238.
Doyle, G. A., Andersson, A., and Bearder, S. K. (1971). *Folia Primatol.* **14,** 15–22.
Dukelow, W. R. (1974). *Proc. Am Assoc. Zoo Veterinarians.* pp. 52–67.
Eaton, G. G., Slob, A., and Resko, J. A. (1973). *Anim. Behav.* **21,** 309–315.
Ehrlich, A. (1974). *Dev. Psychobiol.* **7,** 439–454.
Filliben, J. J. (1975). *Technametrics* **17,** 111–117.
Fuquay, F. E., Hall, A. S., and Jones, R. L. (1975). *Primate News* **13,** 2–6.
Gavan, J. A., and Swindler, D. R. (1966). *Am. J. Phys. Anthropol.* **24,** 181–190.
Gucwinska, H., and Gucwinski, A. (1968). *Int. Zoo Yearb.* **8,** 111–114.
Hill, W.C.O. (1937). *Spolia Zeylan.* **20,** 369–389.
Hill, W.C.O. (1955). "Primates: Comparative Anatomy and Taxonomy," Vol. II. Univ. of Edinburgh Press, Edinburgh.
Hill, W.C.O. (1957). "Primates: Comparative Anatomy and Taxonomy," Vol. III. Univ. of Edinburgh Press, Edinburgh.
Hill, W.C.O. (1960). "Primates: Comparative Anatomy and Taxonomy," Vol. IV, Part A. Univ. of Edinburgh Press, Edinburgh.
Hill, W.C.O. (1962). "Primates: Comparative Anatomy and Taxonomy," Vol. V, Part B. Univ. of Edinburgh Press, Edinburgh.
Hill, W.C.O. (1966a). "Primates: Comparative Anatomy and Taxonomy," Vol. VI. Univ. of Edinburgh Press, Edinburgh.
Hill, W.C.O. (1966b). *Med. Biol. Illus.* **16,** 182–186.
Hill, W.C.O. (1970). "Primates: Comparative Anatomy and Taxonomy," Vol. VIII, Part B. Univ. of Edinburgh Press, Edinburgh.
Hill, W.C.O. (1974). "Primates: Comparative Anatomy and Taxonomy," Vol. VII, Part A. Univ. of Edinburgh Press, Edinburgh.
Hinde, R. A. (1966). "Animal behaviour." McGraw-Hill, New York.
Hinde, R. A. (1971). *Behav. Nonhum. Primates* **3,** 1–68.
Hinde, R. A., and Spencer-Booth, Y. (1967). *Anim. Behav.* **15,** 169–196.
Hines, M. (1942). *Contrib. Embryol. Carnegie Inst.* **30,** 153–209.
Holmes, K. R., Haines, D. E., and Bollert, J. A. (1968). *Lab. Anim. Care* **18,** 475–477.
Jewell, P. A., and Oates, J. F. (1969a). *Zool. Afr.* **4,** 231–248.
Jewell, P. A., and Oates, J. F. (1969b). *J. Reprod. Fertil., Suppl.* **6,** 23–38.
Jolly, A. (1966). "Lemur Behavior: A Madagascar Field Study." Univ. of Chicago Press, Chicago, Illinois.
Kim, D., and Hutchinson, T. C. (1975). *Am. J. Phys. Anthropol.* **42,** 495–500.
Klopfer, P. H. (1972). *Z. Tierpsychol.* **30,** 277–296.
Klopfer, P. H., and Klopfer, M. S. (1970). *Z. Tierpsychol.* **27,** 984–996.
Kolar, K. (1965). *Zool. Gart. Leipzig* **31,** 109–117.
Le Gros Clark, W. E. (1924). *Proc. Zool. Soc. London* **94,** 217–223.
Leutenegger, W. (1973). *Folia Primatol.* **20,** 280–293.
Manley, G. H. (1966a). *Symp. Zool. Soc. London* **15,** 493–509.
Manley, G. H. (1966b). *Symp. Zool. Soc. London* **17,** 11–39.
Manley, G. H. (1967). *Int. Zoo Yearb.* **7,** 80–81.
Martin, R. D. (1968). *Z. Tierpsychol.* **25,** 409–495 and 505–532.
Martin, R. D. (1972a). *Z. Tierpsychol., Beih.* **9,** 43–89.
Martin, R. D. (1972b). *Philos. Trans. R. Soc. London, Ser. B* **264,** 295–352.

Martin, R. D. (1972c). *In* "Breeding Primates" (W.I.B. Beveridge, ed.), pp. 161–171. Karger, Basel.
Mason, W. A. (1968). *In* "Biology and Behavior: Environmental Influences" (D. C. Glass, ed.), pp. 70–154. Russell Sage Foundation and Rockefeller Univ. Press, New York.
Morris, D., and Jarvis, C., eds. (1959). "The International Zoo Yearbook," Vol. I. Zool. Soc. London, London.
Napier, J. R., and Napier, P. H. (1967). "A Handbook of Living Primates." Academic Press, New York.
Nicholls, L. (1939). *Nature (London)* **143,** 246.
Niemitz, C. (1974). *Folia Primatol.* **21,** 250–276.
Petter, J. J. (1965). *In* "Primate Behavior: Field Studies of Monkeys and Apes" (I. DeVore, ed.), pp. 292–319. Holt, New York.
Petter-Rousseaux, A. (1962). *Mammalia* **26,** 1–88.
Petter-Rousseaux, A. (1964). *In* "Evolutionary and Genetic Biology of Primates" (J. Buettner-Janusch, ed.), Vol. 2, pp. 91–132. Academic Press, New York.
Pinto, D., Doyle, G. A., and Bearder, S. K. (1974). *Folia Primatol.* **21,** 135–147.
Pollock, J. I. (1975). *In* "Lemur Biology" (I. Tattersall and R. W. Sussman, eds.), pp. 287–311. Plenum, New York.
Reynolds, V., and Reynolds, F. (1965). *In* "Primate Behavior: Field Studies of Monkeys and Apes" (I. DeVore, ed.), pp. 368–424. Holt, New York.
Rheingold, H. L., and Eckerman, C. O. (1970). *Science* **168,** 78–83.
Rosenblum, L. A. (1968). *In* "The Squirrel Monkey" (L. A. Rosenblum and R. W. Cooper, eds.), pp. 207–233. Academic Press, New York.
Sacher, G. A., and Staffeldt, E. E. (1974). *Am. Nat.* **108,** 593–615.
Sauer, E.G.F., and Sauer, E. M. (1963). *J. S.W. Afr. Sci. Soc.* **16,** 5–36.
Schaller, G. B. (1965). *In* "Primate Behavior: Field Studies of Monkeys and Apes" (I. DeVore, ed.), pp. 324–367. Holt, New York.
Schultz, A. H. (1956). *In* "Primatologia: Handbook of Primatology" (H. Hofer, A. H. Schultz, and D. Starck, eds.), Vol. I, pp. 887–964. Karger, Basel.
Schultz, A. H. (1969). "The Life of Primates." Weidenfeld & Nicolson, London.
Schusterman, R. J., and Sjöberg, A. (1969). *Proc. Int. Congr. Primatol., 2nd, 1968* Vol. I, pp. 194–203.
Stephan, H., Bauchot, R., and Andy, J. (1970). *In* "The Primate Brain" (C. R. Noback and W. Montagna, eds.), Vol. I, pp. 289–297. Appleton, New York.
Terry, R. L. (1970). *J. Psychol.* **76,** 129–136.
Ulmer, F. (1963). *Zool. Gart. Leipzig* **27,** 106–121.
Valerio, D. A., Johnson, P. T., and Thompson, G. E. (1972). *Lab. Anim. Sci.* **22,** 203–206.
Vincent, F. (1969). Thèse de Doctorat d'Etat, Université de Paris (C.N.R.S. No. A.O. 5816).
Washburn, S. L., and Avis, V. (1958). *In* "Behavior and Evolution" (A. Roe and G. G. Simpson, eds.), pp. 421–436. Yale Univ. Press, New Haven, Connecticut.
Woodhull, A. S., and Woodhull, A. M. (1969). *Lab. Primate Newsl.* **8,** 12–13.

Chapter 6

Learning and Intelligence in Prosimians

BEVERLY J. WILKERSON AND DUANE M. RUMBAUGH

1. INTRODUCTION

The contemporary products of evolution that make up the order Primates represent a wide variety of alternative adaptations. The extant primates, which constitute 11 families and 52 genera of the order (Napier and Napier, 1967; excluding the tree shrews, Tupaiidae), all possess basic characteristics that distinguish them from the rest of the animal world. Although characterized by a number of morphological similarities, such as increasing finger-thumb opposability and a generalized skeletal

ISBN 0-12-222150-8

structure, the primates are perhaps primarily distinguished by the development of their brains and the attendant increase in cognitive ability (Warren, 1974).

Neither impressive learners, nor sophisticated problem solvers, the Prosimii are the most primitive primates. In a restricted sense they can be viewed as the surviving approximations of those ancestral primate forms from whose core components the higher primates evolved. Although a careful consideration of the characteristics and capacities of these primate forms probably will not reveal the impressive cognitive capacities found in the Hominoidea, such a consideration will doubtless yield clearer understandings of important trends, the elaboration of which has produced the urban, technical *Homo sapiens sapiens*. Evolution and adaptation involve interactions of environment, structure and function: behavior patterns serve to define major parameters of these interactions.

Mason, (1968), in support of the evolutionary-comparative method, uses Le Gros Clark's view of the primate order as a series of forms which are successive approximations of the evolutionary antecedents of man. Instead of arranging primate forms in a branching phylogenetic tree, Le Gros Clark (1959) suggests that primate forms may be considered as "divergent modifications of a common ancestral type," which indicate the trends in human phylogeny. Each contemporary family is the closest approximation of the ancestral branch of which it is the terminal end. Points A–E of Fig. 1 represent the presumed transition points in human evolution. Although the contemporary primates do not represent a strict linear sequence, there are "connecting links of an approximate kind" between man and the more primitive mammals, suggesting "the general trends of evolutionary change" characteristic of the order.

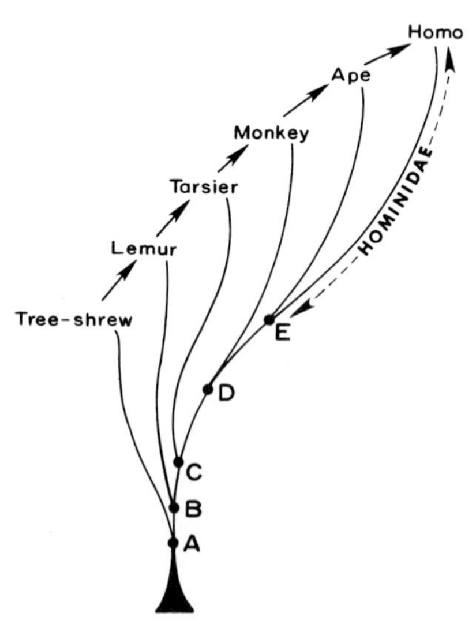

Fig. 1. Postulated sequence of hominid ancestral approximations. A–E: Hypothetical transitional stages at which different primate forms are presumed to have branched off. (From Le Gros Clark, 1959.)

An assessment of the order in terms of the nature and number of cognitive processes is vital for discerning the critical variables influencing contemporary human cognition. In addition, nonhuman primate research may shed light on the phylogenetically older behaviors from which the new behaviors of humans developed.

Mason suggests that a variety of trends and relationships exist between the primate genera, each discernible through comparative research and analysis. A restriction to intra- as opposed to interorder comparison is particularly appealing for, as outlined in the next section, methodological problems often make interorder comparisons invalid.

If the phylogenetic scheme of Le Gros Clark and the behavioral elaborations thereof offered by Mason are valid for understanding the evolution of man, a comparative study aimed at the establishment of relationships and trends among primate forms will be fruitful and enlightening.

2. PERFORMANCE CONTINGENCIES

2.1. Morphologically Determined Propensities

If the experimenter's intent is to measure cognitive competence, he must exercise great care to rule out factors that would tend to inhibit the animal's maximum ability to perform (Warren, 1974). Apparatus must not require responses that the animal is morphologically ill-equipped to make; for example, discriminanda,* manipulanda,† and reinforcers‡ should be appropriate to the species being studied.

Experimenters accustomed to larger, stronger primate forms such as macaques or to the more durable rodent species may well find prosimians difficult to work with due to their many idiosyncracies of diet, preferred postures, and generally timid or high-strung nature. An apparatus modified to accommodate one prosimian species might prove inadequate for another, even closely related form. Such difficulties make it imperative to establish the performance and morphological parameters of prosimians and to design instruments and procedures appropriate to them in order to obtain accurate behavioral measurements of capacity.

Procedures for the maintenance and testing of prosimians are difficult to summarize. As highly specialized animal forms, prosimian behavioral and morphological propensities are markedly diverse. However, general guidelines for appropriate structuring of a test situation to accommodate prosimians should take

*Discriminanda are stimuli which must be distinguished both from each other and from the surrounding environment, require different responses and have different reinforcement consequences (see English and English, 1958, for a complete rendering of psychological terms).

†Manipulanda are objects with which an appropriate predetermined response must be made—pushing, pulling, touching, licking, etc.—in the mastery of a learning task.

‡Reinforcers are response-contingent rewards.

the following considerations into account: morphological parameters, response tendencies, social behaviors, and perceptual thresholds.

Although these items are often commonsensical for primates such as macaques or anthropoid apes, with their more hominoid morphology and plasticity of behavior, prosimian subjects must be closely evaluated to determine appropriate accommodations. Apparatus which permits responses with both the snout and hand, eliminating performance restrictions resulting from limited hand dexterity and specific postural preferences, are particularly important in instrument design. Bishop (1962) has shown, for instance, that prosimians have only whole-hand control. Each genus uses a single prehensive pattern in all situations, whether it is reaching for a raisin or a branch. Their final grips, however, are determined by the shape of the object and may be even more diverse than human grips. Morphology is the chief determinant of postural preferences. The skeletal structure of torso and limbs, and in particular the hands and feet, may be specialized for a particular medium of travel—terrestrial or arboreal. Thus, locomotion is a key factor in determining horizontal or vertical postures, the use of hands in terms of opposability and dexterity, and also the orientation of the head to the vertebral column.

Doyle (1974) discusses at length the behavior of prosimians in terms of morphological aptitudes as determined from both field and laboratory observations, as well as experimentation. In addition, studies of prosimian behavior and ecology, such as those of Subramoniam (1957) and Narayan Rao (1927) with *Loris tardigradus*, provide information on feeding, social interaction, morphology, locomotor patterns, etc., which is valuable to the experimenter and animal curator. The design of the home cage and testing apparatus should reflect not only the prosimian's resting and locomotor postures, but also the associated preferences in hand usage and head placement. For example, *Lemur catta* (ring-tailed lemur) utilize a quadrupedal locomotor posture on the ground and in the trees. Bipedal postures may be adopted temporarily to increase visibility. They are good leapers. They sit on their haunches and manipulate objects with some opposibility of the hand, sometimes hopping with an object in hand for short distances. Food is retrieved or held in either hand or mouth, depending on the situation. Thus, from this kind of information it is reasonable to conclude that no special pattern or clinging structure is necessary in the test apparatus and that these lemurs will use their hands to lift or displace objects, but the selection of discriminanda should allow for responses with either hand or snout.

2.2. Attentional Skills

Attentional skills can affect performance. A task can be disadvantageous to an animal when species-relevant cues obscure task-relevant cues. Jolly (1964b) contends that there are no "ambiguous" cues. Cues are either relevant or not, their relevance depending on species-specific propensities. The relevant cues for each species, i.e., color, shape, location in the field, background or distance from the subject, should be established by a "choice of cue" battery prior to selection of discriminanda.

Although such a battery does not as yet exist, Jolly suggests that statistical analyses which set two preferences against each other will reveal species-specific preferences. For example, as cited elsewhere, (see Section 3.2 below) in discrimination problems, the reinforced variable (object), which may be an artificially induced relevant cue, is set against an experimentally irrelevant cue (position) which may be a naturally occurring preference. The result reveals the stronger preference (see Figs. 5 and 6), and may indicate how appropriate the task is to the animal's species-specific propensities. Baited object problems are another means of observing characteristic response preferences. Such evaluations may serve to indicate the species flexibility and ability to modify preferred behaviors; and conversely, the role of fixed-action patterns and other innate response sets.

2.3. Activity Patterns and Sociability

Circadian patterns, habits, and cycles are important considerations, though the general/natural behavioral repertoire by no means defines the limits of performance or intellectual capacities. Nonetheless, a knowledge of feral living patterns coupled with laboratory experimentation is essential to establish the proper testing procedures and the home cage environment conducive to maintenance and maximal test performance. The daily cycle, whether nocturnal, diurnal, or crepuscular, should determine testing schedules. In the case of nocturnal animals an artificial reversal of day/night alternation may be necessary for purposes of convenience. Duration of test sessions should be comparable with activity patterns in the awake phase, so as to obtain optimal results.

The effects of diurnal and nocturnal cycle *per se* may be broadly grouped into two areas: its correlation with intelligence and its effect upon visual acuity and color vision, which determine perceptual thresholds, i.e., salient visual cues. It is difficult to separate experimentally the effects of the former, different intellectual capacities, from effects of the latter, different behavioral/perceptual parameters. Experiments testing visual acuity typically use objects that are discriminable in one dimension, i.e., color, width of stripes, brightness, etc. Intelligence measures often use discriminanda that vary along several dimensions, i.e., stereometric objects, the correct object being chosen arbitrarily. Should more than one dimension vary together, i.e., be confounded, the test is invalid for assessing acuity. The intelligence measure, on the other hand, must consider the perceptual parameters of the animal so as to select stimuli within its perceptual and discriminable range, yet avoid the confounding of stimulus preferences and correct responses. For example, colored discriminanda should be within the animal's visual capacity, but in intelligence testing should not be the correct stimuli in a color versus white problem, if the animal is predisposed to select the colored object. If such occurs, the test is no longer one of intelligence, but one of color vision.

Correlates of increased intellectual development such as highly complex social interactions do tend to be restricted to diurnal forms. In addition, the assumed visual superiority of diurnal forms, and the subsequent enriched stimulus environment may indeed be the precursor of more complex neural functions. Ehrlich and Calvin (1967) found that *Aotus trivirgatus* (owl monkey, a New World form and the only nocturnal monkey) and *Galago crassicaudatus* (thick-tailed galago) were both color-blind as measured on a color discrimination task, and also had comparable scores on a pattern discrimination task designed to measure intelligence. Although observational information is limited, neither appear to form complex groups, although *G. crassicaudatus* (Bearder and Doyle, 1974) may engage in more diverse intragroup interaction than *A. trivirgatus* (Napier and Napier, 1967).

Since it is actually impossible to take a nocturnal form, make it diurnal, and provide it with maximum visual acuity without altering the organism drastically, experimental designs which equate performance with ability are essential for a full examination of the effects of day/night cycle on intelligence. A tool such as the Transfer Index (Rumbaugh, 1969; see Section 3.3.2 below) has not yet been used to assess the effects of day/night cycle *per se*. However, this type of method should provide some indication of the role of diurnality in learning ability.

The general activity rate is also an important consideration. For example, Ehrlich (1968), for *G. crassicaudatus* and *N. coucang*, and Pinto *et al*. (1974), for *G. senegalensis*, *G. crassicaudatus*, and *M. murinus*, have shown that activity periods in prosimians are highly diverse. Although *Nycticebus coucang* (slow loris) are so named because of their slow, laborious movements, Ehrlich (1969) determined that response rates could be improved by lessening the effort necessary to make a response. Ward and Riley (1969) determined that *Perodicticus potto* (potto) could perform quite well on an operant task* given that the response requirements were compatible with behavioral preferences. There are functional relationships between response patterns, activity, and morphological and task requirements. An understanding of these relationships often can help to modify response patterns and to improve performance.

Field observations and experimental manipulations are vital in determining optimum social maintenance conditions. Grooming, for example, may indicate relative preference for the use of hands or the mouth, and also indicate that, if animals are avid groomers in the wild, accommodation of animals in groups is more desirable than single caging. Although some forms observed as solitary can live well in groups, highly social species may suffer if deprived of social interactions. Social and especially mother–infant interactions should be of primary concern in setting up a prosimian laboratory, especially if one hopes for a successful breeding colony.

*An operant task refers to a simple learning situation, e.g., pressing a bar, in which the delivery of a reinforcer is contingent upon the production of a predetermined response.

2.4. Vision

The specific nocturnal/diurnal performance contingencies, other than the importance of testing during the animals' awake phase (see Section 2.3 above), are the result of visual perceptual parameters. DeValois and Jacobs (1971) summarize the bulk of vision research on nonhuman primates, and conclude that little can be stated conclusively about prosimian capacities. Since increased visual capacity is considered to be one of the definitive characteristics of primates (Le Gros Clark, 1959), and because almost all learning tasks involve some visual cues, an assessment of prosimian visual skills is important for appropriate experimental design and analysis.

2.4.1. Color Vision

Although nocturnal forms, such as *Galago senegalensis* (lesser bushbaby) and *Galago crassicaudatus* are assumed to be color-blind by virtue of their all-rod eye, this has not been tested. Ehrlich and Calvin (1967) found that *Galago crassicaudatus* could not discriminate between red and green. *Lemur catta* and *Lemur mongoz* (mongoose lemur), diurnal forms, could distinguish red from blue, but did not respond differentially when certain grays were substituted for the blue (Bierens de Haan and Frima, 1930). For *Lemur fulvus* (brown lemur), Glickman *et al*. (1965) found that when luminosity was controlled an established blue–red discrimination ability dropped to chance. None of these tests proves conclusively that *G. crassicaudatus* or *G. senegalensis* are completely color-blind nor that *L. catta*, *L. fulvus*, and *L. mongoz* have color vision. None of the studies encompasses more than a small segment of the color spectrum, nor were the subject pools sufficiently large to rule out the possibility of individual rather than species limitations. The most one can say is that the galagos have limited color vision, and the lemurs are more sensitive to brightness than to color.

2.4.2. Visual Capacities

Aside from color vision as understood in terms of rod and cone distribution in the eye, research in visual abilities involves a consideration of the innervation and transmitter mechanisms of the visual system as a perceptual variable, distinct from the neural/brain complex. Spectral sensitivity, critical flicker frequency, visual acuity, and depth perception are areas that reflect this research interest. Although knowledge about prosimians is limited in this regard, and methodological differences often confuse results, the following studies indicate that neither individual differences within a species nor species-specific limitations should be taken lightly.

Spectral sensitivity (luminosity function), which contributes to the presence and nature of color vision, is similar for all primates so far tested (DeValois and Jacobs, 1971), including *Galago crassicaudatus* (Silver, 1966). This does not prove that *G. crassicaudatus* have color vision, but does indicate that their visual efficiency is affected by wavelength.

Another sensitivity measure is the critical flicker frequency (CFF), the most rapidly flickering light source which can be discriminated from a continuous light source. Ordy and Samorajski (1968) and Ward and Doerflein (1971) determined comparable CFF thresholds for the nocturnal *Galago senegalensis* of 25–30 cps and 24–30 cps, respectively. This is higher than the human CFF, quoted by Ward and Doerflein (1971) as 20 cps, and indicates a smaller visual capacity for *G. senegalensis*.

Visual acuity, operationally defined as "the minimum-separable visual angle necessary for discrimination" (DeValois and Jacobs, 1971), has been evaluated for *Galago crassicaudatus* (Ordy and Samorajski, 1968). Both studies used grating patterns made of lines that varied in thickness, but which were paired such that the only discriminable difference within a pair was their orientation—horizontal or vertical. Acuity thresholds were determined by the narrowest lines that could be discriminated at greater than chance levels. Ordy and Samorajski (1968) used an 80% correct criterion.

Treff (1967) established a 4.28 acuity for *G. senegalensis* measured in minutes of the visual arc—a line width of about 1/8 inch. Ordy and Samorajski (1968) determined that *G. crassicaudatus* subjects ranged from 3.5 to 8.0—a line 1/12–1/8 inch in width. *Lemur catta* subjects ranged from 0.5 to 1.5—a line 1/128–1/24 inch wide.

Specific test situations greatly affect this acuity measure, such that comparisons are difficult to make; however, Ordy and Samorajski (1968) also tested several other primate species using the same procedures, and found that *L. catta* thresholds were comparable with two other diurnal forms, *Saimiri sciureus* (squirrel monkey) and *Callithrix jacchus* (marmoset), both New World monkeys. *Galago crassicaudatus*, on the other hand, were comparable to *Aotus trivirgatus*.

Treff (1967) tested depth perception in *G. senegalensis* and found it to be considerably poorer than in humans. The spatial arrangement and illumination in the testing situation considerably affected perception.

2.5. Response to Novelty

Prosimians have a reputation of being timid, easily distracted, or downright uncooperative, e.g., unresponsive. In view of this fact, some experimenters test animals in their home cages (Gorter, 1935). On the other hand, research by Jolly (1964a) indicates that prosimians interact with the test situation only to the extent that they are deprived of social contact during the test session. Since the effects of transport or change in environmental stimuli tend to be disruptive for prosimian performance, adequate habituation is necessary and may of necessity be quite prolonged.

After testing a large number of vertebrates, Glickman and Sroges (1966) concluded that the tendency to explore and manipulate objects in terms of total responses is more a reflection of ecological and survival pressures than intellectual endowment, as inferred from the relative development of the adult brain. His study found no correlation between brain development and positive response to novelty. The main difference he noted was a differential mode of responding and exploring objects: visual inspection and manual manipulation in the higher primates. olfactory and taste inspection and muzzle manipulation in prosimians.

Indeed, manipulatory behavior and responsiveness to novel objects may be viewed as a two-dimensional phenomenon—quantitative and qualitative. Parker (1973), approached the study of behavioral response to ''non-social, non-edible aspects of the environment'' in a comparative study of *Hylobates lar* (gibbon), *Macaca nemestrina* (pig-tail macaque), *Ateles geoffroyi* (spider monkey), *Presbytis cristatus* (silver leaf-eating monkey), and *Lemur catta* and *Lemur macaco* (black lemur). Quantitative aspects, ''the propensity to approach and contact [the] object'' were measured in terms of frequency and duration of contact. Qualitative aspects,

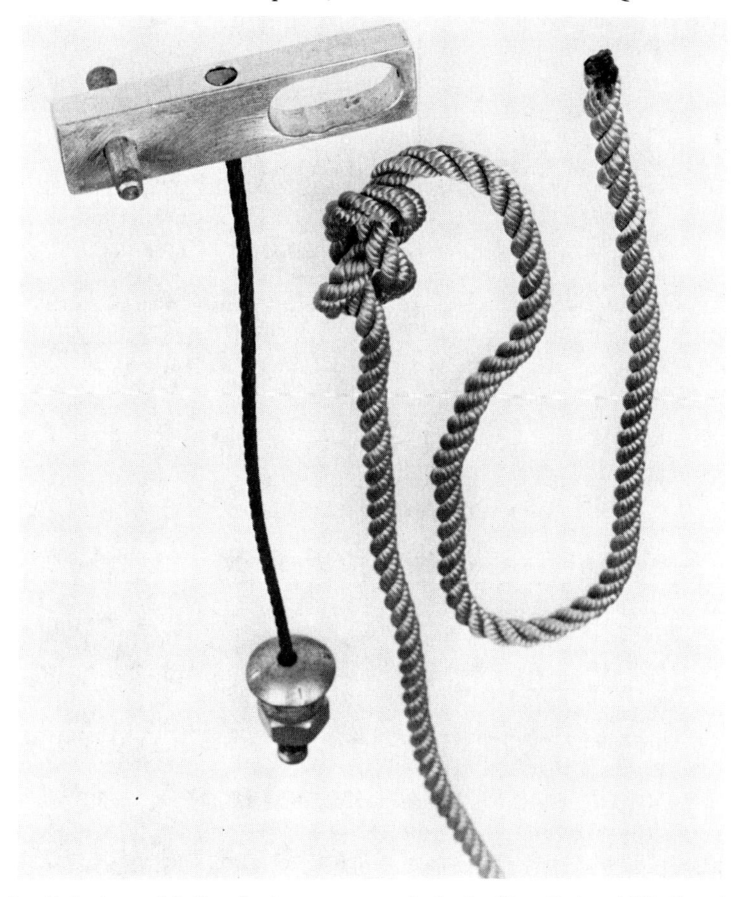

Fig. 2. Parker's metal (left) and nylon rope are manipulanda. [From Parker, 1973. Copyright (1973) by the American Psychological Association. Reprinted by permission.]

"the nature of the manipulatory behaviors performed on or with the object," were assessed according to a behavioral taxonomy of potential activities, frequency of those activities, body part, and part of object contacted, and frequency of unique combinations of body part, object, and activity. Figure 2 illustrates the two objects used in the study.

Table I ranks the species on frequency and duration of contact with the metal object and Table II ranks the species on frequency of contact with the nylon rope. Duration measures for rope contact were not taken, since the apparatus involved required that the object be electrically conductive to complete a circuit connected to an automatic timer. Only contact behaviors were recorded, thus excluding sniffing, visual orientation, etc.

A total of fourteen measures were taken. The mean rank for each species on these measures was *H. lar* (1.53), *M. nemestrina* (2.10), *L. catta* and *L. macaco* (2.78), *P. cristatus* (4.0), and *A. geoffroyi* (4.57). These mean ranks were significantly different, though Parker (1973) cautions that this finding was affected by the close correlation between performances on many of the response measures. *Lemur catta* and *L. macaco* were intermediate to the four other species. Although quantitatively more responsive than the *A. geoffroyi* and *P. cristatus* their qualitative repertoire was less extensive than the *H. lar* and *M. nemestrina*. The lemurs tended to use both hands in the same way, simultaneously to pull, touch, and hold the object, whereas the other species tended to use the hands differentially (one holding, the other picking, manipulating, etc.). All species predominantly made contact with the whole hand as a unit to push, pull, and touch the object. The range of creative performances was considerably limited. Although *H. lar* made loops with the rope, activities with the rope common to Pongidae (placing it about the body, using loops for pulling) were not observed (Parker, 1974).

TABLE I

Species of Primate Ranked in Terms of Frequency and Duration of Contact with the Metal Object[a]

Mean number of contact responses			Mean duration of contact		
Rank	**Species**	**Responses**	**Rank**	**Species**	**Seconds**
1	Gibbon	171.0	1	Gibbon	660.0
2	Lemur	85.3	2	Pig-tail	223.0
3	Pig-tail	61.5	3	Lemur	193.0
4	Spider	16.5	4	Spider	25.0
5	Silver leaf-eater	1.5	5	Silver leaf-eater	1.5

[a] Reprinted from Parker, 1973.

TABLE II

Species of Primate Ranked in Terms of Frequency of Contact with the Nylon Rope[a]

Rank	Species	Mean number of contact responses
1	Pig-tail	184.8
2	Lemur	141.3
3	Gibbon	96.0
4	Silver leaf-eater	46.8
5	Spider monkey	2.3

[a] Reprinted from Parker, 1973.

2.6. Instruments and Reinforcement

Instruments and schedules of reinforcement and testing necessarily must conform to the previously mentioned characteristics. Food, water, shock avoidance, and the opportunity to manipulate objects have been used as reinforcers (see Table III). The types of food required may be quite diverse, from manufactured, flavored pellets for *Galago crassicaudatus* (Ehrlich, 1969) to locusts for *Loris tardigradus* (slender loris; Gorter, 1935).

Prosimians also respond to a variety of schedules: Variable Interval (VI), *Nycticebus coucang* and *Galago senegalensis* (Fobes *et al*., 1973); Variable Ratio (VR), *Lemur catta, L. fulvus,* and *L. macaco* (Mitchell *et al.*, 1970, and Ward and Riley, 1969, respectively); and Fixed Ratio (FR), *N. coucang* and *G. senegalensis* (Ehrlich, 1968).*

Two reports on apparatuses show promise for prosimian work. Fobes *et al.* (1971) have devised an interesting apparatus for use with small animals including prosimians. This instrument requires little response effort and allows close approach to the discriminanda. In addition, it is easily acquired commercially or inexpensively assembled. Responses are recorded via a very weak current through the animal's body, and thus far the animals have shown no discomfort due to this method. Rumbaugh *et al*. (1972) devised both semi- and totally automated discrimination apparatuses appropriate for use with all primate species.

*Operant conditioning schedules essentially consist of two types: interval, wherein the reinforcer is delivered following the first response emitted after a predetermined time interval; and ratio, wherein a reinforcer is delivered after a predetermined number of responses. These schedules may be variable or fixed, such that the duration of the interval or the size of the ratio is varied or consistent throughout the task.

TABLE III

Evaluation of Various Reinforcers with Prosimians

Animal	Reinforcer	Source	Rating[a]
LORISIDAE			
Lorisinae			
Loris tardigradus	Raisins and mealworms	Jolly (1964a,b)	Good
	Locust	Gorter (1935)	Good
	Object manipulation[b]	Jolly (1964a,b)	Poor
Nycticebus coucang	190 mg Banana-flavored pellets	Ehrlich (1969)	Good
	Sucrose solutions (%)	Fobes (*et al.* 1973)	
	5		Poor
	10		Poor
	20		Good
	30		Good
	40		Poor
	Object manipulation	Ehrlich (1968)	Fair
		Ehrlich (1970)	Poor
		Glickman and Sroges (1966)	Poor
Perodicticus potto	Raisins and mealworms	Jolly (1964a,b)	Good
	(Noyes) 45 mg banana, sucrose, dextrose, and dog pellets	Heffner and Masterton (1970)	Good
	Water	Ward and Riley (1969)	Good
	Object manipulation	Jolly (1964a,b)	Fair to poor[b]

Galaginae			
Galago crassicaudatus	Raisins and mealworms	Jolly (1964a,b)	Good
	Fruit	Silver (1966)	Good
	190 mg Banana-flavored pellets	Ehrlich (1969)	Good
		Ehrlich and Calvin (1967)	Good
	Sucrose solutions (%)	Fobes *et al.* (1973)	
	5		Poor
	10		Fair
	20		Good
	30		Fair
	40		Poor
	Object manipulation	Jolly (1964a,b)	Poor to fair[c]
Galago senegalensis	Raisins and mealworms	Jolly (1964a,b)	Good
	(Noyes) 0.097 mg dog formula pellets	Masterton and Skeen (1972)	Good
	Avoidance of shock	Ward and Doerflein (1971)	Good
		Ward *et al.* (1970a)	Fair to good[d]
		Ward *et al.* (1970b)	Fair to good[d]
	Object manipulation	Jolly (1964a,b)	Fair
LEMURIDAE			
Lemurinae			
Lemur catta	Raisins and mealworms	Jolly (1964a,b)	Good
	"Life Savers" hard candy	Davis and Leary (1968)	Good
	Sucrose pellets	Arnold and Rumbaugh (1971)	Good
		Rumbaugh (1971)	Good
		Rumbaugh and Arnold (1971)	Good
	Avoidance of shock	Gillette *et al.*, (1973)	Good
	Object manipulation	Jolly (1964a)	Good

(*continued*)

TABLE III (*continued*)

Animal	Reinforcer	Source	Rating[a]
Lemur fulvus	Raisins and mealworms	Jolly (1964a,b)	Good
	Apple	Glickman *et al.*, (1965)	Good
	Grapes	Feldman and Klopfer (1972)	Good
	Object manipulation	Jolly (1964a,b)	Good
Lemur macaco	Sucrose pellets	Arnold and Rumbaugh (1971)	Good
		Rumbaugh (1971)	Good
		Rumbaugh and Arnold (1971)	Good
Varecia variegata	Grapes and oranges	Harlow *et al.* (1932)	Good
CHEIROGALEIDAE			
Cheirogaleinae			
Microcebus murinus	Raisins and mealworms	Jolly (1964a,b)	Good
	Food pellets	Ehrlich (1968)	Good
	Object manipulation	Jolly (1964a,b)	Poor

[a] Evaluations are direct statements in or derived from published results and discussion materials as cited for each species and reinforcer. Poor, ineffective; fair, inconsistent or moderately effective; good, successful in eliciting a consistent response.

[b] Object manipulation refers to the opportunity for the animal to manipulate an unbaited object. Manipulation frequency and quality may vary depending on the type of object (see Jolly, 1964a). Baited objects, those that net a food reward via manipulation, are generally addressed more readily and attentively. Since the reward is probably food, and not the intrinsic reward of manipulation, these latter cases are not included.

[c] Wild-born animals are much less likely to manipulate objects than captive-reared animals.

[d] These studies indicate the necessity of evaluating the reinforcer in light of the responses required to obtain it—passive versus active—and the species-specific propensities of the subject.

3. REVIEW OF LEARNING AND PROBLEM-SOLVING DATA

3.1. Shock Avoidance

Warren (1965) contends that no species differences in learning ability are reflected by performance on the continuous avoidance task described by Sidman (1953). In this design shocks are administered at regular intervals, contingent upon whether or not the subject produces a response (for example, lever pressing results in shock avoidance). On a multiple schedule, in which exteroceptive stimuli cue the subject as to when to produce a response to avoid shock, learning is rapid, not easily extinguished, and broadly generalized. Although neither schedule is illustrative of species variations in terms of learning ability, Kelleher (1965) states that "higher" forms, such as monkeys, retain the task from session to session and seldom receive a shock once the task is learned, as opposed to rats which continue to receive shocks, especially at the beginning of each session.

Since shock avoidance has proved to be a reliable task in terms of ability to learn, experimenters primarily interested in perceptual abilities have utilized it with prosimians to establish threshold levels such as critical flicker frequency in *Galago senegalensis* (Ward and Doerflein, 1971) and hearing in *Lemur catta* (Gillette *et al.*, 1973). Ward *et al.* (1970a) also utilized the visual flicker frequency stimuli to examine the effects of various light flicker frequencies on the ability to avoid shock in *G. senegalensis*. In their study, a shuttle box was used to test a "no-go" response (a passive avoidance) and a "go" response (an active avoidance response), based on 3- and 18-second flicker rates, respectively, and the reverse. "No-go" responding was rapidly learned and not differentially affected by flicker rate. However, this was not surprising since the bushbaby's characteristic response is immobility when presented with a stimulus (a light). In contrast, active avoidance was more difficult to establish, particularly at the 3-second frequency, but was facilitated at the 18-second level (see Fig. 3). Ward *et al.* (1970a) attribute this to the procedural design, in that the 3-second frequency was initially paired with the "no-go," and the "go" response with the high flicker rate.

Ward *et al.* (1970b) investigated the cross-modal transfer abilities of *Galago senegalensis*—auditory to visual and the reverse—again with a shock avoidance task.* In acquisition training, the 3- or 18-second flicker (light) or click (auditory) frequency was designated as the "go" situation, and the remaining frequency the "no-go" situation. Transfer was either direct (response to the two frequencies the same as in acquisition) or reversed (response required to the two frequencies the opposite of responses required in acquisition). The degree to which performance on transfer was facilitated was assessed by a study of total errors (see Table IV). Direct

*Cross-modal transfer is the positive transfer of learning acquired in one sensory modality to another. Positive transfer indicates that performance is enhanced as a result of prior experience in a different modality.

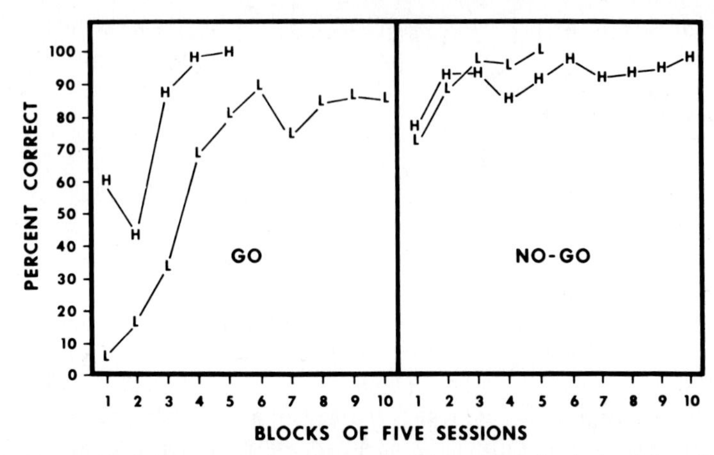

Fig. 3. Acquisition of a discrimination of two rates of intermittent visual stimulation—high (H), 18 seconds and low (L), 3 seconds—by *Galago* in a "go, no-go" task. Each point represents a mean of 100 trials, 50 per subject. (From Ward *et al.*, 1970a.)

visual transfer was the superior situation in that the error total was less than in the other condition (see Fig. 4).

Several observations were offered which may be indicative of confounding variables, not only in this study but in prosimian research in general: (1) The change of modality itself produced a disruption in performance; (2) acquisition procedures may have facilitated performance on specific stimulus values as described above; (3) bushbabies may have differential propensities for attention to a response in different modalities; and (4) more information may be gleaned from analyses of the percent of correct responses than from speed of acquisition (Warren's point, 1965).

Ward and his associates encourage research on the cross-modal transfer abilities of nonhuman primates, even with prosimians where the effect is weak, in that such abilities may be tied in with the evolution of language abilities. *Pan* (chimpanzee)

TABLE IV

Total Errors of Four *Galago senegalensis* Subjects in a Transfer Situation under Two Conditions[a]

	Subjects			
Condition	A	B	C	D
Go	51 (18 seconds)	60 (18 seconds)	156 (3 seconds)	189 (3 seconds)
No-go	28 (3 seconds)	22 (3 seconds)	35 (18 seconds)	36 (18 seconds

[a] Reprinted from Ward *et al.* (1970a).

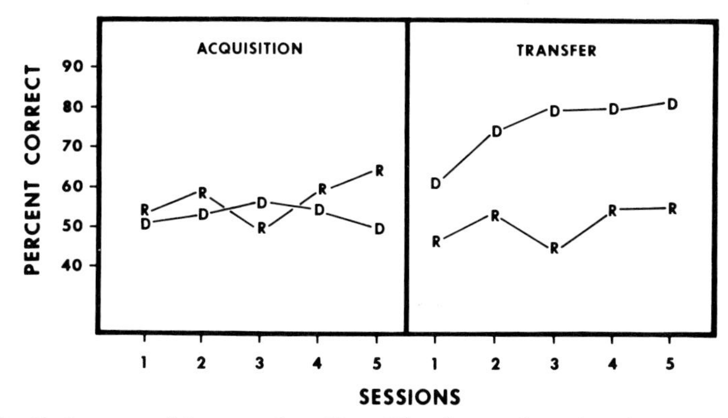

Fig. 4. Performance of direct transfer subjects (D) and reversal transfer subjects (R) in the first 100 trials of acquisition and in 100 trials of transfer over five sessions. [From Ward *et al.*, 1970b. Copyright (1970) by the American Psychological Association. Reprinted by permission.]

and *Pongo* (orangutan) can discriminate tactile stimuli, matching to a visual sample with accuracy well above chance (Davenport and Rogers, 1970). Although the relevance of this ability in language evolution cannot now be specified, possession of this meta-modal concept of stimulus equivalency does appear without a verbal (language) mediation process.

3.2. Discrimination Problems*

Several intelligence measures using a discrimination task have yielded data pertinent to the rank ordering of primate intelligence and the type of learning hypotheses employed by prosimians in problem-solving tasks. Ordy and Samorajski (1968) demonstrated problem retention and a gradual improvement of percent correct choices for *Lemur catta* and *Galago crassicaudatus* in a study of the correlation of retinal organization and visual activity. Although they report the development of a learning set,† they qualify the assertion with reference to the marked inconsistency in performance of the prosimians as individuals and as a group over time. Similar tendencies toward erratic performance have been reported by Jolly (1964b) and Glickman *et al.* (1965).

Ehrlich and Calvin (1967) tested *Galago crassicaudatus, Aotus trivirgatus,* and *Macaca mulatta* (rhesus) on color pattern and brightness discrimination. *Galago*

*Discrimination problems involve the presentation of stimuli from which the subject must select the one which has been arbitrarily and consistently designated as correct/rewarded over a number of trials.

†Learning set or "learning to learn" is the ability to acquire virtually one-trial mastery of a problem after prolonged experience with problems of a similar type. The phenomenon, as demonstrated by Harlow (1949) in discrimination learning, illustrates the ability of monkeys to go from virtually chance performance on trial one to near perfect performance on all subsequent trials.

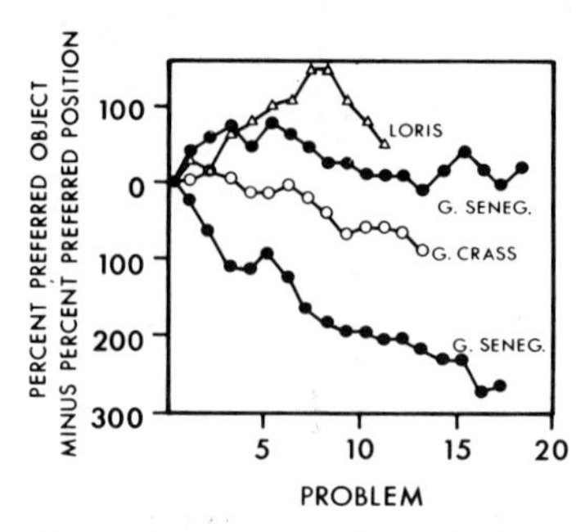

Fig. 5. Lorisiformes: Cumulative scores on discrimination problems. (From Jolly, 1964b.)

crassicaudatus and *A. trivirgatus,* being blind to the colors used, failed to learn the color task. However, all mastered the brightness and pattern discriminations (see Section 2.4 above). Although *A. trivirgatus* and *G. crassicaudatus* took many more trials to mastery than *M. mulatta, G. crassicaudatus* may be superior to *A. trivirgatus,* since they took less trials to criterion.

Jolly (1964b), on the other hand, was not optimistic about demonstrating learning skills in prosimians. The animals (*Galago senegalensis, G. crassicaudatus, Loris tardigradus*, *Perodicticus potto*, *Lemur catta*, and *L. fulvus*) received little pretraining to the task. Jolly was interested in assessing the suitability of specific "psychological" problems as reliable measures, in terms of an animal's initial success with that problem. In this regard, it was asserted that the animal's cue choice was a species-typical characteristic, part of the species variable. Subjects were, therefore tested in a "less restricted" test situation, wherein they could select the "relevant" cue as determined by ecological, adaptive propensities. On the

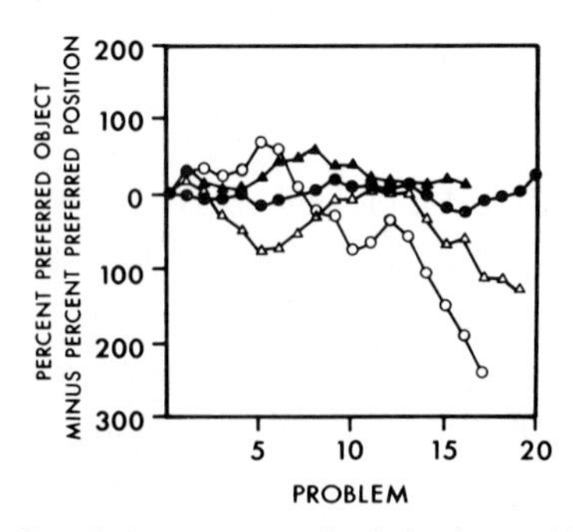

Fig. 6. Lemuriformes: Cumulative scores on discrimination problems. (From Jolly, 1964b.)

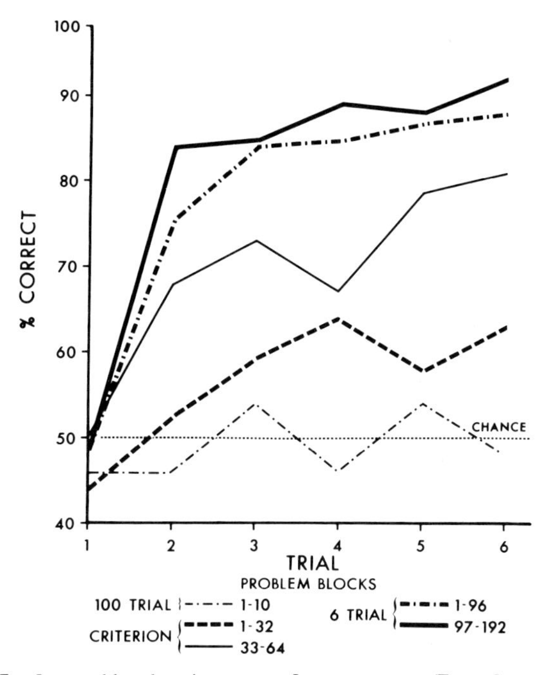

Fig. 7. Intraproblem learning curves: *Lemur macaco*. (From Cooper, 1974.)

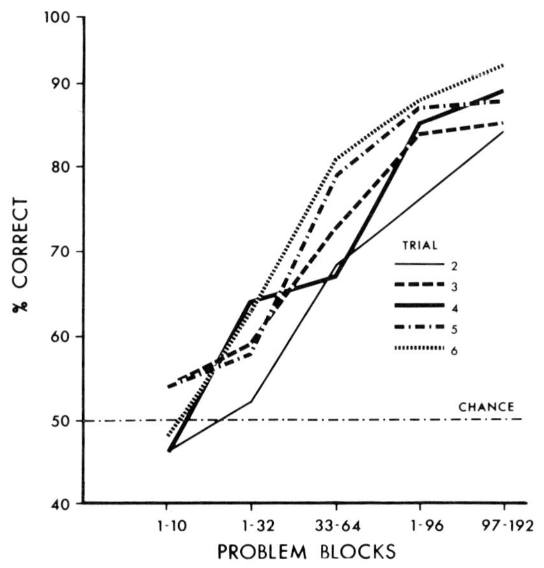

Fig. 8. Interproblem learning curves: *Lemur macaco*. (From Cooper, 1974.)

discrimination task, the prosimians demonstrated a strong position preference.* Figures 5 and 6 illustrate the score curves for each genus. The curve rises if the animal chose primarily from object cues (the reinforced cue) and falls if it chose primarily from position cues. The scoring formula used required the subtraction of the number of position-preferred choices, say, to the right side, from the object-preferred choice.

Cooper (1974) used Harlow's (1949) learning set procedure to illustrate inter- and intraproblem learning in *Lemur macaco*. Procedures allowed gradual habituation to the task, keeping frustrations to a minimum during early trials.

Figure 7 depicts intraproblem performance in terms of percent correct choice. Although performance is variable, the development of a sudden increase in percent correct between Trials 1 and 2 indicates the "one-trial learning" ability characteristic of other "higher" primate forms. Figure 8 depicts interproblem learning success as a function of the number of problems. As training progressed and the animals received more experience through exposure to more problems, performance on Trials 1 through 6 improved in a gradual, continuous manner.

This study, the first to systematically and carefully control conditions for optimal performance based on species-specific propensities, does illustrate that *L. macaco* (1) improve performance over time with practice, and more importantly, (2) demonstrate a one-trial learning typical of the more advanced primate forms.

3.3. Discrimination Reversal†

It is only recently that tasks and apparatuses have been developed that are sufficiently sensitive to qualitative differences in learning skills and equitable in terms of the diverse psychological and morphological characteristics of various primate species. Discrimination-reversal tasks and a semi-automated apparatus, as developed and implemented by Rumbaugh (1971), Rumbaugh and Arnold (1971), and Arnold and Rumbaugh (1971), are believed to answer both of these requirements for a revealing assessment of qualitatively different learning skills within the order Primates.

Rumbaugh and Arnold (1971) equated *Lemur catta* and *L. macaco* with *Cercopithecus aethiops pygerythrus* (vervet monkey) subjects on acquisition proficiency on two blocks of 100, 51-trial reversal block units. Only *C.a. pygerythrus* showed improvement on acquisition from block 1 to 2, and although both lemurs

*The repeated selection of the same spatial location, irrespective of the specific stimulus presented there, is called a position preference. In most procedures, where location of the correct stimulus is randomized, performance will thus be at chance level.

†A discrimination reversal task is a learning task in which the subject is required to discriminate between a correct/rewarded (A +) stimulus and an incorrect/not rewarded (B-) stimulus, and then to reverse their choice when the reward values of the stimuli are reversed (B + , A-).

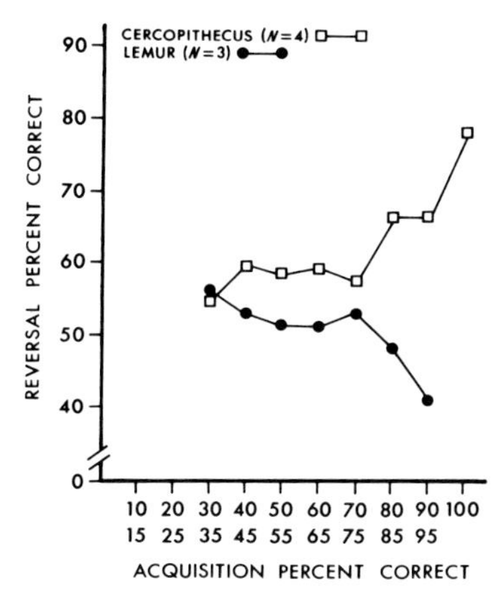

Fig. 9. Reversal percentages for the second block of 100 discrimination reversal units as a function of acquisition percentages for *Cercopithecus* and *Lemur*. (From Rumbaugh and Arnold, 1971.)

and vervets showed improvement on reversal from block 1 to 2, *C.a. pygerythrus* performance increase was significantly better than that of the lemurs.

More revealing than the overall superiority of *C.a. pygerythrus* is the relationship between acquisition and reversal performances (see Fig. 9); with improvement in acquisition performance (baseline), reversal performance improved for vervets but delayed for lemurs.

Stevens (1966) reported that *Lemur catta* subjects were unable to perform discrimination-reversal tasks due to stimulus perseveration* and a position preference. The persistent choice of the prereversal correct stimulus may be similar to the phenomenon that Arnold and Rumbaugh (1971) found in extinction trials with lemurs (see Section 3.4 below). Such a phenomenon would account for the finding that the better the prereversal performance, the worse the reversal performance. In this case the animal would be using a strategy that entailed the learning of discrete reward values associated with discrete objects, S-R associations. Cercopithecines did not develop an object preference since, on reversal trials, once the formerly correct object ceased to be rewarded they shifted to the other previously incorrect object, which was then correct and rewarded. This "win-stay, lose-shift" strategy, with respect to the object rewarded (after Levine, 1965), is assumed to indicate a more general, qualitatively superior problem-solving strategy, since animals that employ it generally master reversal problems in fewer trials and perform better on successive reversal tasks.

*Stimulus perseveration refers to a tendency to continue responding to a particular stimulus irrespective of a change in the reward value of the stimulus.

3.3.1. Tripartite Discrimination Tasks

In an attempt to clarify further the qualitative learning processes of various primates, Rumbaugh (1971) constructed a tripartite discrimination-reversal task to test for the degree of abstractness of the learning process. Three reversal conditions were given, each preceded by a uniform cue reversal trial, i.e., the positive and negative objects (A+, B−) reversed values (B+, A−). This was followed by 10 trials under the specific reversal condition. These conditions were A−, B+, C−, B+; and A−, C+, where C denotes a novel stimulus object.

The disposal of A or B on reversal would serve to facilitate reversal either by eliminating the object that had accrued habit (A+) or the object that had accrued inhibition (B-), if the prereversal learning was of an S-R type. On the other hand, a more abstract concept such as "win-stay, lose-shift" would be uninfluenced by these contingencies, rendering all tasks relatively equal and easy. This original report involved *Gorilla gorilla* (gorilla), *Hylobates lar*, and *Miopithecus talapoin* (talapoin) and indicated that *G. gorilla* performed equivalently on all conditions; whereas *H. lar* and *M. talapoin* found the conditions differentially difficult. *Hylobates lar* and *M. talapoin* performed better on problems involving C, than A−, B+ problems.

Using the same design, as outlined in Rumbaugh (1971), *Lemur catta* and *L. macaco* were tested on these reversal tasks. Analysis of variance, with arcsin transformed values of scores that expressed proportion of correct responses, revealed a significant main effect for reversal condition ($F = 213.70$, $df = 2/2, p < 0.01$) and for trial ($F = 13.58$, $df = 4/4, p < 0.05$), as well as significant block and trial ($F = 9.10$, $df = 4/4, p < 0.05$), trial and reversal condition ($F = 8.55$, $df = 8/8, p < 0.05$), and block and reversal condition ($F = 54.53$, $df = 2/2, p < 0.05$) interactions. There was no significant difference between trial one performances.

Scheffé tests (Winer, 1962, $p < 0.05$ required) revealed no significant differences between Trials 1 and 5–11, 1 and 2, and 2 and 4–7, or 3 and 8–11. Scheffé tests on trials by condition interactions for Trial 1 and 2, 2 and 3, 3 and 4–7, and 4–7 and 8–11 revealed significant differences between each for condition C−, B+. Tuckey *b* analyses showed a significant difference between A−, B+ and A−, C+; C−, B+ and A−, B+; and A−, C+ and C−, B+. The difference was largest for the first comparison and smallest for the third.

The data, therefore, indicated that *L. catta* and *L. macaco* did not find the three tasks equally difficult, as Fig. 10 reveals. It would appear that an S-R explanation of *L. catta* and *L. macaco* problem-solving processes is appropriate. One could assert that lemurs tend to approach a discrimination problem as a task of learning the reward value of specific objects.

In addition, the similarity between *M. talapoin* and lemur performance is of note. This is not inconsistent with previous research that indicates comparable performance between prosimians and New World primates (Harlow, 1932; Maslow and Harlow, 1932) and prosimians and *M. talapoin* (Rumbaugh, 1969). Perhaps this

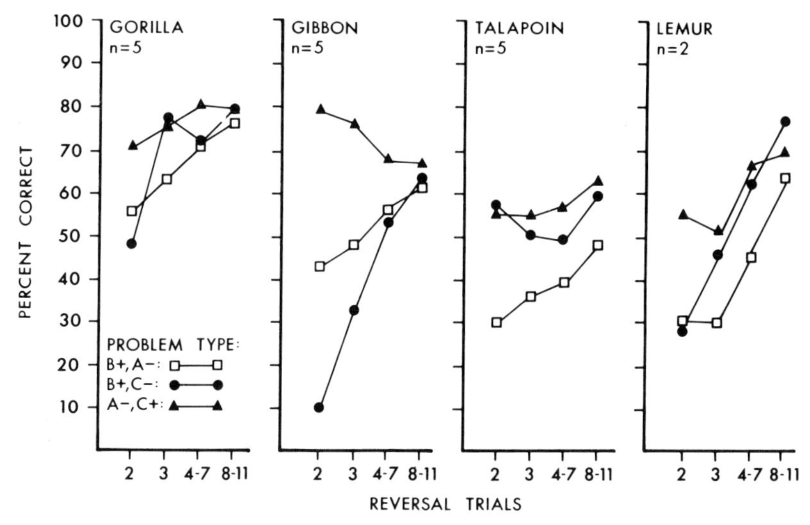

Fig. 10. Reversal performances for four primate groups on each of three types of discrimination reversal problems.

should serve as a warning for those who accept the taxonomic ordering of primate forms as a completely definitive account of a linearly increasing ability or strict ordering from lower to higher forms. More defensible relationships exist between performance on problem-solving ability and factors such as ecological adaptations (Jolly, 1964a,b; Andrew, 1962; see Section 3.7 below) and other species-specific factors such as brain complexity (Masterton and Skeen, 1972; see Section 3.5 below; Rumbaugh, 1971).

3.3.2. Transfer Index

One of the more novel and promising uses of discrimination-reversal tasks for determining differences in primate complex learning skills through comparative means is the Transfer Index (TI) developed by Rumbaugh (1969, 1970). The TI procedure involves training to a percent correct criterion on each of a series of two-choice visual discrimination problems prior to the ten test trials in which measures of prime significance for estimating cognitive capacity are taken.

TI was built upon the observation that the relationship of reversal performance (R%) to prereversal or acquisition performance (A%) was a function of the species being tested and the A% value. By equating A% values prior to cue reversal and concluding via empirical research that numbers of A (acquisition) trials to criterion *per se* are of little importance in accounting for R% differences, it was postulated that differences in reversal performance would reflect differences primarily in cognitive abilities/processes. It was assumed that performance equation would insure that prejudices inherent in the application of other standard test procedures and apparatuses to diverse groups of primate forms would be minimized.

Initial TI testing involved the training of subjects to a 67% correct criterion on acquisition. The subsequent R% correct value given as a ratio (R%/A%) yields a measure of the amount and nature, positive or negative, of the transfer of knowledge from acquisition to reversal. A 67% value for A was the result of previous research (Rumbaugh and McCormack, 1967) which indicated that R%/A% values are more widely dispersed at that point. As a measure of Harlow's learning set, TI is believed to be an aspect of complex learning capacities (Rumbaugh, 1969).

Subsequently, an 84% A value, halfway between 67% and 100% was added. If one can assume that the A% value at the end of acquisition trials is indicative of the amount of information learned about the requirements of an individual problem, then an increase in that value should infer that more information was acquired. Figures 11 and 12 illustrate the reversal performance of *Lemur catta* and *L. macaco* in relation to other primate species on a trial-by-trial basis.

Lemur catta and *L. macaco* are clearly inferior in performance to the Pongidae, but perform comparably with other monkey forms and better than *Miopithecus talapoin*. Figure 13 shows the relationship (R%/A%) on the assumed knowledge increase (A%) on postreversal performance (R%).

TI values as defined by the R%/A% ratio have been shown to be correlated with the general complex learning capacity (Rumbaugh and Gill, 1974). The ability to transfer increasing amounts of information profitably and to generalize from prob-

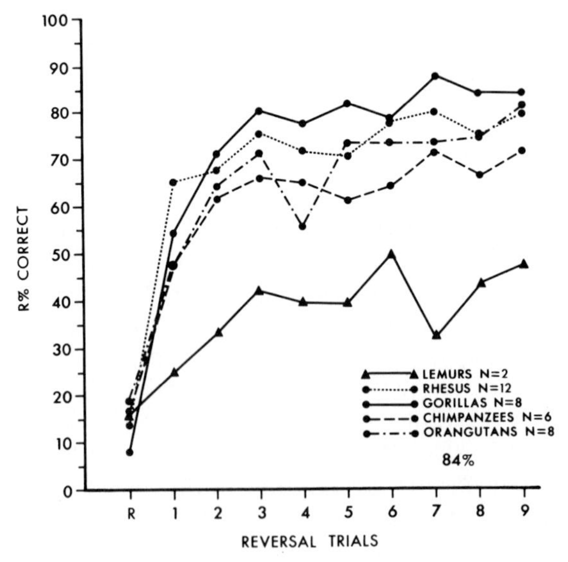

Fig. 11. Percent correct on reversal trials (R%) after 84% criterional (A) training, comparing *Lemur* with great apes and rhesus macaque.

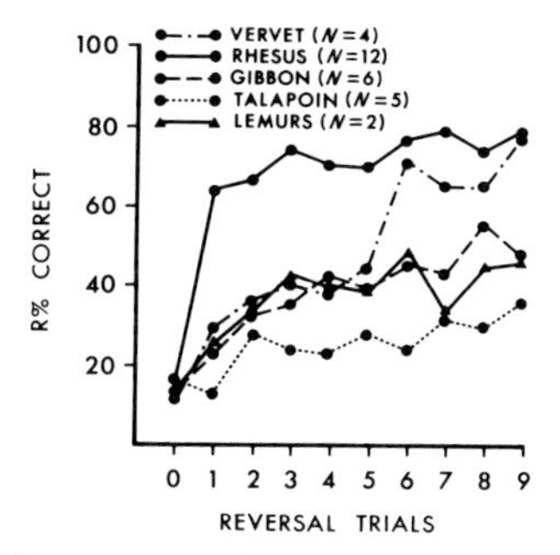

Fig. 12. Percent correct (R%) on reversal trials after 84% criterional (A) training, comparing *Lemur* with *Cercopithecus* and *Hylobates*.

lem to problem is differentially evidenced in the primate genera tested thus far. It would appear that *Gorilla gorilla* perform better on reversal when they know relatively more about the prereversal problem as reflected by the A% criteria (84% versus 67% correct). *Miopithecus talapoin,* on the other hand, do worse as the prereversal A% value is raised from 67 to 84%. *Lemur catta* and *L. macaco* are intermediate showing no change relative to the A% increase.

The full significance of the nature and amount of the information transferred with increased A% values has yet to be realized. However, if one assumes that the ability to profit from increased experience and practice is a prerequisite to the mastery of complex tasks, it follows that such an ability was no doubt necessary in the evolution of man as a quantitatively and qualitatively superior problem solver. Man's

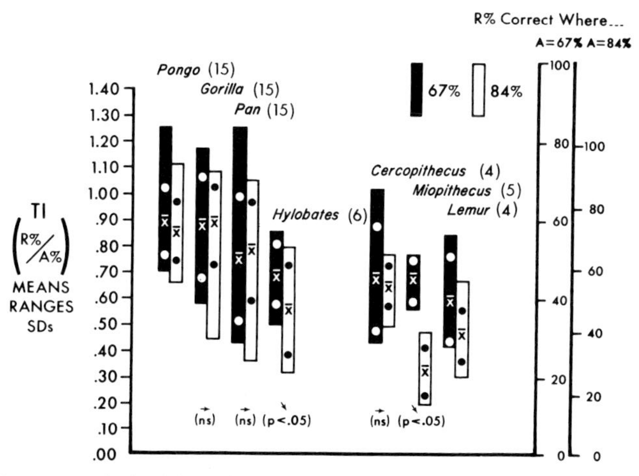

Fig. 13. TI measures obtained through use of the 67 and 84% criterional training schedules of great apes, lesser apes, monkeys, and lemurs. (From Rumbaugh, 1970.)

survival and dominance based upon mental prowess is evidence of the necessity of superior complex learning and transfer abilities. The relative position of other primate forms, particularly Old World forms, in terms of their complex learning skills in relation to brain development, indicates their relative positions along the intelligence continuum suggesting the various possible steps that led to human capacities.

3.4. Extinction*

Very little experimental research has been done on the nature of extinction. This phenomenon has alternatively been attributed to inhibition (a tendency *not* to respond), built up in relation to unrewarded energy expenditures (Hull, 1943), or to learning of a discriminative type, wherein the subject learns to distinguish between rewarded and nonrewarded responses (Guthrie, 1935; Woodworth, 1958).

Arnold and Rumbaugh (1971) conducted the only study of extinction in primates to date that included prosimians. They posited that if extinction is a learning phenomenon, it should exhibit features and reveal species differences similar to those of other choice discrimination-reversal learning tasks. One might therefore expect to obtain one-trial extinction, evidence of choice patterns based on pre-extinction correct and incorrect object value, and differential rates of response and extinction dependent upon a species variable such as neural mechanisms. Basically the study involved a comparison of *Lemur catta* and *L. macaco* with *Cercopithecus aethiops pygerythrus*. Previous discrimination-reversal training had established that the latter were superior learners to the former, and the present experiment demonstrated that the same was true of extinction abilities.

Although subjects were trained on two different procedures—"criterional," in which *S*s were equated in task mastery and "fixed reinforcement," wherein training ceased upon the acquisition of a specific number of rewards regardless of overall percent correct, neither procedure nor their order of presentation influenced extinction performance. However, the rate of acquisition to criterion and the number of trials to the requisite number of reinforcements was faster and smaller respectively for the cercopithecines. In addition, the cercopithecines displayed a "win-stay, lose-shift" strategy with respect to the object that was correct prior to the extinction procedure, while the lemurs persisted in choosing the previously correct object. Figure 14 illustrates these response patterns of lemurs and cercopithecines in relation to the correct stimuli of training days 1 and 4 and the respective extinction days for each: 2 and 3 and 5 and 6. Although rate of response on extinction trials was higher for lemurs, the pattern of response decrease was similar to the cercopithecines. Figure 15 illustrates the similarity of the patterns of response rate change on extinction trials.

*Extinction refers to the elimination of a learned response as a result of the removal of the response-contingent reward (reinforcer).

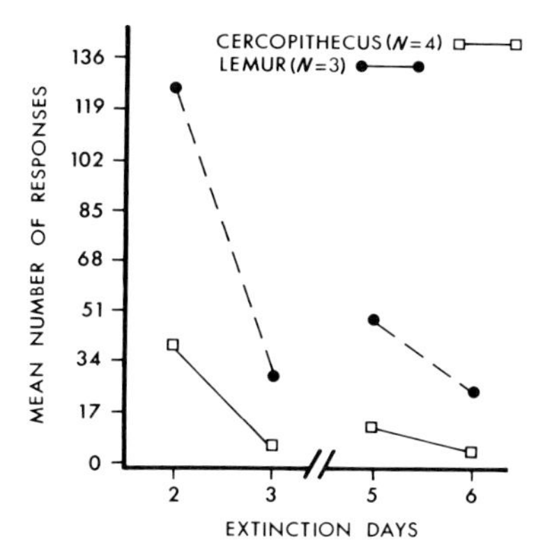

Fig. 14. Mean number of responses emitted during extinction plotted per extinction day. (From Arnold and Rumbaugh, 1971.)

The greater number of responses and the longer time required for extinction by lemurs corresponds with the longer time and greater number of trials required by them to meet "criterional" and "fixed reinforcement" schedules in training when compared with the cercopithecines. Although rate of response decrease and number of responses made to extinction may be explained by both an inhibition and a learning phenomenon, the differential pattern of response in relation to the previously correct stimuli and the learning set data of Behar (1966) on extinction in monkeys, lend support to the hypothesis that extinction is a learning phenomenon (see Fig. 16).

More research with the number of learning trials as an independent variable is warranted before any clear statement can be made concerning the nature of extinction; however, the data thus far do not rule out the concept of extinction as a

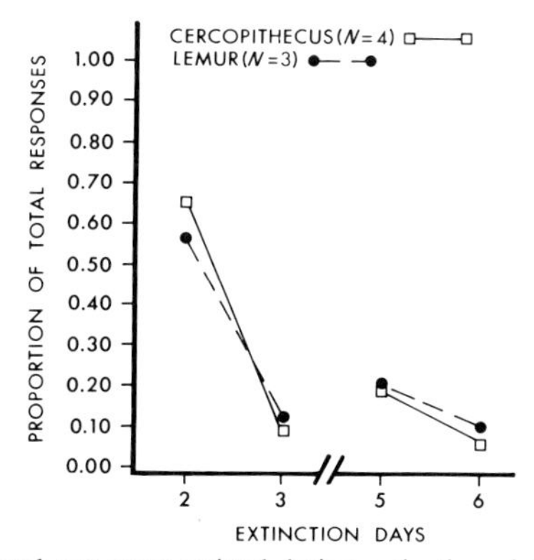

Fig. 15. Proportion of total responses emitted during extinction plotted per extinction day. (From Arnold and Rumbaugh, 1971.)

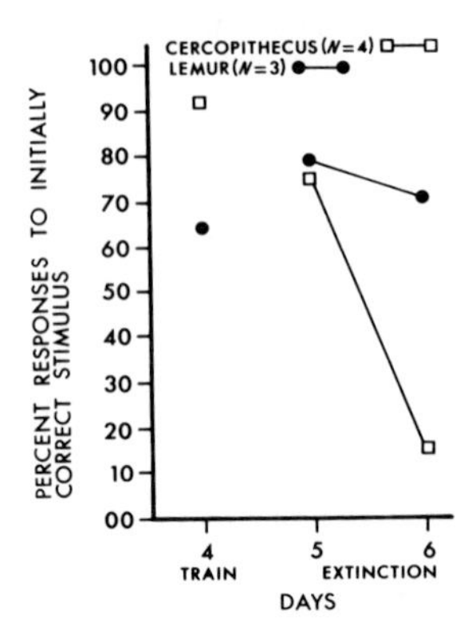

Fig. 16. Percentage of responses to the stimulus correct on all trials of training day 4 regardless of training condition and extinction days 5 and 6. (From Arnold and Rumbaugh, 1971.)

learning task that is correlated with other learning processes in their relationships to primate brain development.

3.5. Delayed Response*

The ability of an animal to delay responding and still retain the ''memory'' of the correct choice has been of interest to comparative psychologists since Hunter (1913) originally designed the task as a test of ''higher order capacities,'' indicative of symbolic conceptual abilities. More recently, Fletcher (1965) defined the delayed response problem as ''an intra-trial task in which the correct instrumental response is the overt terminal response of an orienting-response chain initiated totally, accurately, and immediately by an observing response made at the beginning of each trial'' (Fletcher, 1965, pp. 135–136). Experimentation over the years has led from the measurement of maximum delay intervals to a consideration of the subject's posture, activities, etc., during the delay period, as indicative of the overtness of orientation necessary for solution; and conversely, the degree of internalization of the stimuli over time. Both delay maxima and activity (degree of orientation) can be accurately manipulated under controlled conditions. However, attention to the qualitative nature of animal behavior during all phases of the delay trial ''yields much more data in terms of the species variable and individual differences'' (Fletcher, 1965, p. 161).

3.5.1. Traditional Delay Tasks

Harlow *et al*. (1932) and Maslow and Harlow (1932) analyzed data on the basis of maximum delay, degree of correctness at each delay, and the development of a

*In its simplest form, a delay task involves placing a reinforcer, in view of the subject, beneath or behind some opaque object but preventing the subject from retrieving it before a predetermined time interval has elapsed.

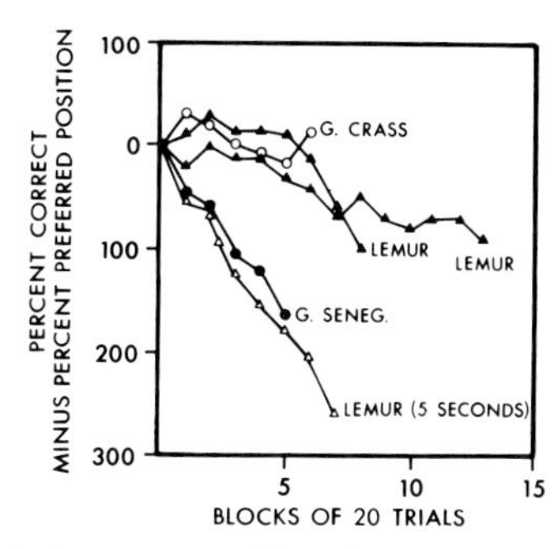

Fig. 17. Cumulative scores on delayed response tests. (From Jolly, 1964b.)

learning curve, using an array of primate forms including *Varecia variegata* (*Lemur variegatus*, ruffed lemur). Subjects were tested by a direct and indirect method, distinguished by the latter having an opaque screen lowered during the delay interval. Harlow *et al.* (1932) concluded that "the ability to solve tasks of delayed response by means of representative factors appears as a common capacity in all primate forms." Although their single lemur was not tested by the indirect method, its maximum delay on the direct method was 15 seconds after much coaxing, and was included in the generalized statement above. Low in comparison to the Old World primates tested, the lemur's performance was comparable with the New World monkeys and an immature *Mandrillus sphinx* (mandrill).

The Maslow and Harlow (1932) study indicated that *Lemur mongoz* and *L. fulvus* could delay 5 seconds. It was asserted that given more training, their performance could have reached the New World monkey's delay of 15 seconds, but only after much coaxing. Harlow (1932) was unable to test *Varecia variegata* on a complicated delayed response task involving either distraction, simultaneous problems, or position shifts. He attributed this to the subject's extreme "lethargy," but does not describe the behavior in any detail.

Jolly (1964b) noted a position preference that conflicted with the choice of the correct object following a delay. By subtracting the percent of responses to position from the percent of correct responses to the object, a graph was constructed wherein position-determined responses dipped the graph downward and correct choice-of-object responses turned it upward (see Fig. 17).

Although no animal exceeded the 5-second delay, the degree of position-determined responses increased with the number of trials for the subjects, which included *Galago sengalensis*, *G. crassicaudatus*, *Lemur fulvus*, *L. catta*, *Loris tardigradus*, and *Perodicticus potto*.

3.5.2. The "Diamond and Jane Strategy"

Masterton and Skeen (1972) utilized a combination of delayed response, alternation, and discrimination procedures to test whether prefrontal development was correlated with the ability to perform a delayed alternation task. Masterton and

Skeen (1972) based their study on the "Diamond and Jane strategy," which reserves the questions of the nature of intelligence until the behavioral dimensions associated with successively more hominid brain structures are understood sufficiently to permit the abstraction of relatively independent phylogenetic psychological trends. As illustrated in Fig. 18, such an approach utilizes a concept of the primate order similar to that proposed by Le Gros Clark (1959) and Mason (1968).

The strategy involves three broad steps. First, from the paleontological evidence animal forms must be ordered according to their location in the successive ancestry of man, as based upon their propinquity and recency of common ancestry. Extant forms are selected which, according to comparative anatomical considerations, are the best neurological approximations of the extinct forms. Second, tasks are selected which are not necessarily intellectual measures, but which evaluate the role of a particular neural structure in the determination of behavioral parameters. Third, performance on these tasks is assumed to be indicative of a change in behavior as a function of geologic time (evolution). Although the increasing "humanization" may be found to occur in the same direction and at equal rates for all structures and behaviors, the absence of such consistency is irrelevant to the strategy.

Thus, *Hemiechinus auritus* (hedgehog), *Tupaia glis* (tree shrew), and *Galago senegalensis* were selected as subjects by virtue of their similarity to man's early mammalian ancestors *vis à vis* the development of the prefrontal cortex. The delayed alternation task was selected by virtue of its assumed applicability to tests of the prefrontal system. Inferences of possible behavioral changes in the evolutionary

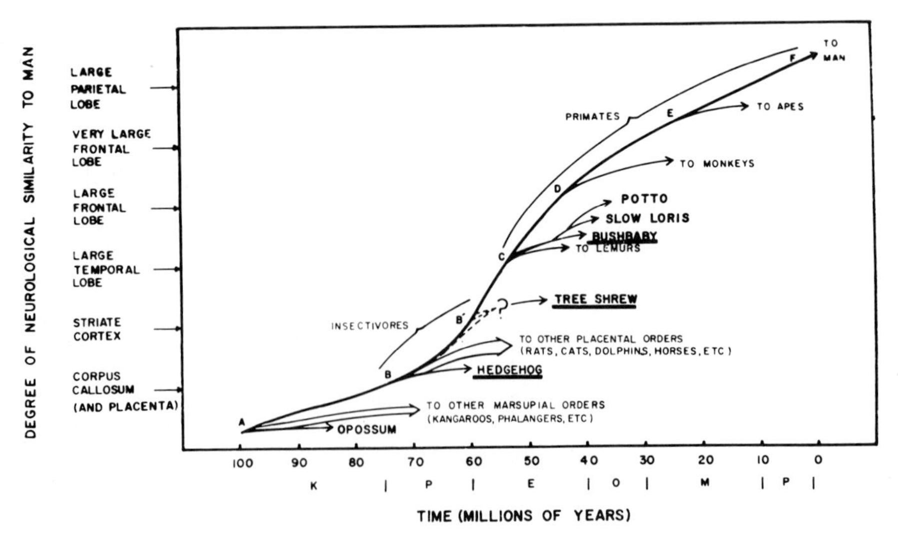

Fig. 18. Phylogenetic relationships among the experimental subjects and the extinct animals in the anthropoid line of descent. [From Masterton and Skeen, 1972. Copyright (1972) by the American Psychological Association. Reprinted by permission.]

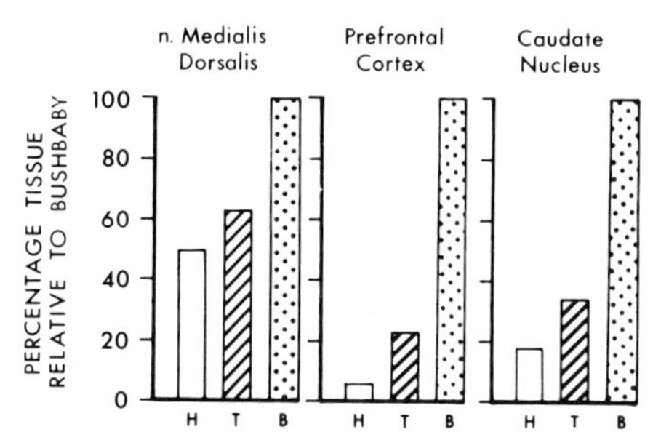

Fig. 19. Amount of tissue in the prefrontal systems of hedgehog and tree shrew, relative to bushbaby. [From Masterton and Skeen, 1972. Copyright (1972) by the American Psychological Association. Reprinted by permission.]

ancestry of man would be reserved since numerous tests with various other "grades" of development are required before such an evaluation is considered valid.

The delayed alternation task required subjects to discriminate between stimuli that indicated a no-go response and one that indicated that a response would be rewarded. The test procedures utilized various delay intervals, subsequently requiring the animal to select the opposite box of a two-box reward chamber from the one chosen on the previous trial. Intervals were randomized and delay intervals were increased until performance was at chance.

Bushbabies showed an ability to delay and alternate up through 256 seconds of delay, mastering the task within a few sessions. The results corresponded to the prefrontal development gradient for the species compared: *H. auritus*, *T. glis*, and *G. senegalensis*, in ascending order (see Fig. 19).

Two procedures were used: randomized-interval, where intervals were randomly presented; and interval-tracking, where the delay interval was increased after two

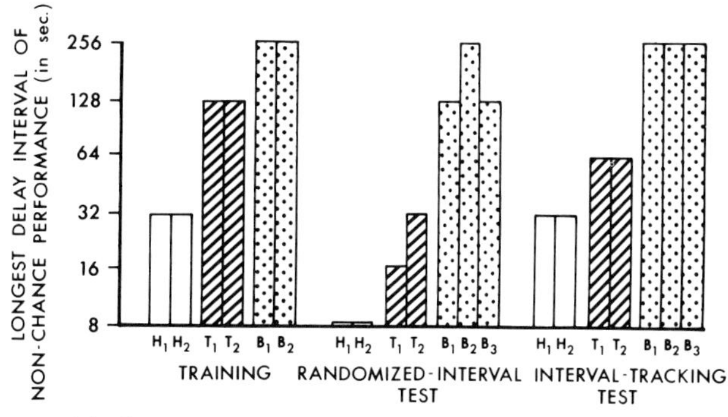

Fig. 20. Longest delay interval at which delayed-alternation performance exceeded chance ($p<.01$). [From Masterton and Skeen, 1972. Copyright (1972) by the American Psychological Association Reprinted by permission.]

successive correct responses and decreased after two incorrect, testing being terminated after 1 hour. The order of performance was the same under both conditions (see Fig. 20).

Since all subjects were successful on the task at the shorter delay intervals, the task *per se* could be performed successfully and would therefore be an accurate test of delay abilities. An evaluation of the source of errors of choice and premature response was not consistent with delayed alternation performance (see Fig. 21).

This indicates abilities to delay and not alternate and alternate and not delay, but less proficiency in the delaying with alternation than other "higher" primate forms. Masterton and Skeen (1972) concluded that these delayed-alternation results indicate a unidirectional increase in prefrontal tissues and certain learning capacities that were evolved early from the primitive placental ancestry. However, full confirma-

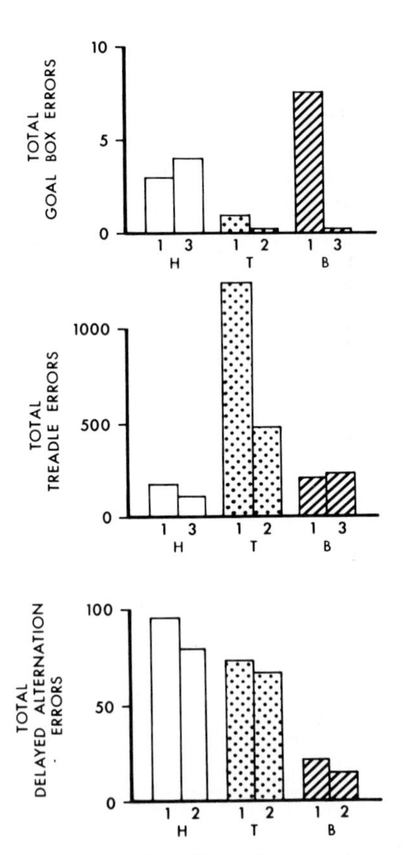

Fig. 21. Total errors in delayed responding without alternation (top and middle) compared to total errors in delayed alternation (bottom) during randomized-interval tests. [From Masterton and Skeen, 1972. Copyright (1972) by the American Psychological Association. Reprinted by permission.]

tion of delayed-alternation as a measure of increasing neural complexity was reserved until further tests of more representative types of the various "grades" of neural development rule out species-specific variables unrelated to brain evolution.

3.6. Detours and Other Problems

Köhler (1925) and Yerkes (1943) first used the terms "insightful" and "reasoning" behavior in relation to problems requiring the rapid association of unrelated experiences to solve a task in a novel way. Box stacking and the use of poles and ropes to reach inaccessible rewards were illustrative cases for *Pan*. Not all primate forms, however, have the extensive manual dexterity that enables such behaviors to develop. Other tasks have, therefore been developed to accommodate species' morphological parameters and propensities in order to test this behavior (Maier and Schneirla, 1964). Bent-wire detour problems and baited-object problems are excellent examples of these tests.

Davis and Leary (1968) conducted bent-wire detour experiments with *Lemur catta* and a number of species of Old World and New World monkeys. The procedure consisted of presentation of six bent-wire problems: two involving one-segment problems requiring lateral pushing or pulling for retrieval of reward (Life Savers); and four two-segment problems, two requiring that the subject push on the food and two requiring that the subject pull on the food to free it from the wire. Although all 21 of the Old World monkeys and one of the 15 New World monkeys achieved perfect performance, none of the four *L. catta* did. Representatives of the three groups improved performance with practice, with no sudden insightful experiences reported.

There was a significant difference between the two-segment "push" and "pull" problems, the former being more difficult than the latter. The difference was largest in lemurs and smallest in macaques. A ratio of percent correct on "push" problems to the percent correct on "pull" problems yielded values that when ordered from lowest to highest, ordered the species in terms of brain development. Davis and Leary (1968) therefore conclude that the ability for pushing a food reward away in order to attain it increases with brain development.

Loris tardigradus was the subject of an interesting study by Gorter (1935) in a simple learning and problem-box experiment. Gorter housed and tested his animals in a large mesh-wire cage with abundant perches, ropes and swings for the animal's activities. He attempted to demonstrate the learning abilities of lorises by training them to accept food through (1) the lowest of three holes in the wire cage netting, or (2) the leftmost hole of three holes in the netting. Two of the three holes were always baited on each trial; however, only a correct choice netted a reward. Although the animals mastered the procedure, performances measured in correct choices never exceeded 74%. In the second procedure, distances between the horizontally organized holes could be gradually reduced from 40 to 10 cm without a decremental effect on performance.

Gorter (1935) contends that *L. tardigradus* can only be trained to the absolute characteristics rather than relative ones. For example, when presented with one reward in the middle hole and the other in the hole left of center, the subject usually correctly chose the left one; however, when one reward was in the middle and the other to the right, performance was at chance. The latter position is one of relative leftness as opposed to the absolute leftness of the former.

When a live grasshopper was placed (1) in a box with the lid open, (2) in a box with the lid closed, or (3) underneath an upside-down box without the lid, *L. tardigradus* did not successfully open the box or manipulate it successfully in response to the food placement. Most captured and ate the live grasshoppers as they attempted to escape and would sometimes manipulate the slit between lid and box with their mouths. However, only one of the four subjects ever opened the box and then not necessarily in response to the food placement. Gorter (1935) commented that their search was ''trial and error,'' since their attempts to retrieve the reward were neither systematic nor directed. Prolonged experience of finding grasshoppers in the proximity of the box or in escaping from the box did not improve performance.

Inability to perform well in either of these experiments, although attributable to lower intelligence, could in part be accounted for in terms of motor and sensory capacities. They are lethargic movers, have limited depth perception, are easily distracted, and were considerably influenced by an alteration in the artificial lighting conditions of their cage. These limitations may well represent ecologically valid adaptations that conflicted with the laboratory requirements.

In a similar experiment, Jolly (1964a,b) drew attention to the position and stimulus perseveration tendencies of prosimians. She conducted a series of baited and unbaited object problems. The unbaited problems were analyzed for manipulation propensities, responsiveness, and duration thereof to novel objects. A hasp puzzle box, clear and opaque baffles, a glass bottle, hinged-lid box, pivoted dowell, and a string-pulling task were all baited and successes and related activities were observed.

Neither object nor problem detail were attended to by any of the prosimians (which included *Galago sengalensis*, *G. crassicaudatus*, *Loris tardigradus*, *Microcebus murinus, Perodicticus potto, Lemur fulvus,* and *L. catta*) except one *L. fulvus* and one *L. tardigradus*, after intensive training. All attended to a directional cue, thereby reaching in the direction of the bottle (beside it more often than inside), lifting upward in response to the lidded box (whether baited/attached or not). In addition, strong position habits developed in discrimination tasks within this same study. *Loris tardigradus* did not search for rewards that were placed inside an opaque box within view unless trained to do so. In fact, animals could see the food placed under one of two discriminanda, yet still persist in selecting the same object consistent with the position habit formed.

Jolly (1964a,b) notes that this directionally cued behavior is present in infant humans and gorillas. After picking up an object, both will follow through with moving the hand to the mouth even if the object is dropped on the way. Also, both are prone to look for the dropped object in the other hand much like the prosimians. Infant humans and gorillas, lacking object permanence, also fail to search for an object removed from sight, even if they see it placed there. Human infants begin to acquire the concept of object permanence when they are 8–14 months old (Piaget and Inhelder, 1956), gorillas by 2 years (Lang, 1962).

3.7. Observational and Social Learning

Jolly (1966), in a field study report of two Lemuriformes (*Lemur catta* and *Propithecus verreauxi*—the sifaka) proposed that intelligence, as measured by laboratory testing procedures, assesses only part of the intellectual processes possible in primate forms, particularly those having complex social organizations. "Monkey intelligence" or the ability to work with laboratory "gadgetry" is but one aspect of intelligence. Social behavior reflects quite another form of intelligence.

Although both of the forms Jolly (1966) studied have social organizations similar to, though much less complex than, monkeys, such as the baboons and macaques, their intellectual capacities as measured by laboratory tests are quite circumscribed: the capacity to adapt quickly and successfully to new social conditions does not correlate with the ability to learn laboratory discrimination tasks. Discrimination tasks are based in great measure, according to Jolly (1966), on "discovery" learning and are quite different from the "mimicry" capacities necessary to solve the social tasks in complex group life. The latter ability is therefore the more important one, taking precedence over the former, particularly in prosimians, unless deprived of social activities. According to Jolly, as social skills are basic for group life, social intelligence evolved long before the "cleverness" seen in higher primate forms today, including man. Thus, according to Jolly (1966), social intelligence is a more basic form of intelligence and might be fundamental to the emergence of hominid intelligence that provides for facile learning of technical problems and material.

Andrew (1962) viewed the "monkey intelligence" that Jolly discussed as dependent for development upon diurnal living and predatory threat. He asserted that intelligence is the product of an "interaction between geographic and social habitat."

A common type of social learning entails one animal learning from observing the behaviors of others. Studies of observational learning with *Pan troglodytes* (Goodall, 1965) in the field and in the laboratory (Yerkes, 1943; Darby and Riopelle, 1959), and with *Macaca fuscata* in the field (Itani, 1958), illustrate not only the ability to learn vicariously, but the method and paths of transmission of knowledge and change throughout a social group.

Only one study (Feldman and Klopfer, 1972) involves an investigation of prosimian observational learning in a laboratory task. *Lemur fulvus* subjects were paired in this experiment, two mother–infant diads and one unrelated pair (a male and female juvenile pair). One of each pair (the mother in the mother–infant diads) was designated as teacher and the other as pupil, the former getting the task first. Three experiments were conducted so as to allow the pupil to benefit, possibly, from observing the teacher solve the problems. However, there was no clear indication of direct facilitation of performance on the part of the pupils in terms of choice accuracy, even though all pupils appeared to become more task-oriented as a result of their observations.

4. DISCUSSION

Data collected on prosimian subjects in problem-solving tasks, over the last 40 odd years, indicate that their learning abilities are of a relatively low order. The reader might be inclined, therefore, to dismiss further consideration of prosimian cognition as unnecessary. Stopping here, however, implies that rank ordering prosimians along a "less than–greater than" continuum is all that can be hoped for or is useful for comparative research. This hierarchical approach presents a myriad of conceptual problems should differences in performance prove to be contradictory or less than clear-cut. Problems of this nature can be viewed as a contemporary version of those trends in animal/comparative psychology that Thorndike spoke against in his 1898 monograph. He disagreed with the concept that "animal associations [cognitive processes] are homologous with the associations [cognitive aspects] of human psychology." He further states: "If, as seems probable, the primates display a vast increase of associations and a stock of free-swimming ideas [over other animals], our view gives to the line of descent a meaning which it never could have, so long as the question was the vague one of *more* or *less* 'intelligence.' " Thorndike thus advocated a shift in emphasis from what animals can do, a quantitative assessment, to how they do it, a qualitative differentiation.

Comparative psychology, not without dissidents and pessimists within its own ranks, is currently engaged in a qualitative assessment of learning that supplements and clarifies the quantitative data that is so avidly accumulated by the participants in this behavioristic discipline.

Analytical tools already exist, such as Levine's (1965) hypothesis testing, Restle's (1958) cue theory, and Harlow's (1959) error hypothesis. To use Restle's (1958) model of learning, prosimians persist in using *type b cues* (cues relevant within each problem), never evolving to a use of *type a cues* (rules that apply across problems) in regard to discrimination reversal.

Methodological developments in controlling the effects of species-specific variables that tend to distort performance and, therefore, assessments of ability, are being implemented. The Transfer Index (Rumbaugh, 1970, 1971) and Warren's (1965) application of Bitterman's (1965) method of systematic variation are two examples of promising experimental tools. Hopefully, such efforts will eventually yield measures that will do as much for qualitative comparisons within the primates as Harlow's (1949) learning set phenomenon did for differentiation of primates as a group from other mammals (Warren, 1965).

This search for qualitative differences is an indication that comparative psychology is breaking the bonds of the strict use of the anathematic phylogenetic scale and the static consideration of absolute performance levels alone in favor of a more flexible system that views primate skills as processes differentially constituted and developed, showing varying degrees of approximation and relationship to human mental abilities. The application of models such as the "Diamond and Jane strategy" (Masterton and Skeen, 1972) may clarify these trends in intelligence as a function of evolutionary changes in neural structure and correlated behavioral propensities. These trends in comparative endeavors hopefully will prove more fruitful than the conventional analysis of the past.

A number of confusing situations may well be resolved by this qualitative approach. First, there may be no inherent need for preserving a singular ordering of ability across the evolutionary continuum. The relationships and evaluation of the efficiency of different skills do not have to correlate, since many relationships are possible. Differences in ability within and between species in various kinds of situation reflect the species variable complex that affords a variety of potentially adaptive responses to a multifeatured environment.

Second, the question of how an animal learns is more directly addressed by a consideration of learning processes rather than by simply recording the percentage of correct responses alone. Third, there is some justification for the abstraction of the particular skill in question by controlling other species-specific variables. A rationale for the comparison of specific skills can effectively justify ignoring other species-specific properties, since the isolation and understanding of such irrelevant variables is prerequisite to their control. In this way some clarification can be made of the apparent gross contradictions in various measures of learning ability.

For example, *Pan*, *Gorilla*, and *Pongo* exhibited differential amounts of performance disruption when an irrelevant foreground cue (wire mesh positioned in front of the discriminanda) was introduced into a problem on which they had already been trained to criterion (Rumbaugh *et al.*, 1973). Also, since groups of *Pan troglodytes*, differing profoundly in learning skills, did not differ with respect to the effect of the irrelevant cue, the readiness to attend to foreground seems to be clearly species-specific. In addition, the degree of disruption corresponds to the progressive arboreality of the species habitat. By inference, a better understanding of the somewhat puzzling (i.e., poor) performance of highly arboreal Hylobatidae on similar tasks with encased discriminanda may be acquired.

Fourth, qualitative differences in learning processes do not necessarily call for the making of "better/worse-than" value judgements. With reference to the particular species, each qualitative characteristic is assumed adequate for its ecological adaptation. One cannot separate completely the relative efficiency of a qualitative characteristic in the laboratory test situation from a consideration of its relationship to the ecological situation of the species. The aforementioned value judgement assumes that one must not take a completely human perspective of nonhuman forms. Just as nonhuman research data cannot be applied directly to humans, so human research data cannot be applied directly to nonhuman primates. Theoretically, qualitative characteristics can be viewed simply as alternative processes and adaptations.

Finally, qualitative considerations may further the understanding of quantitative differences by evaluating the role of species-specific propensities. Whether derived from natural observations or "choice of cue" analyses, such propensities and their effect on the animal's ability to function in experimental situations may reveal something of the nature of the learning ability.

Efforts to evaluate learning skills qualitatively may well lead to a better understanding of observations, particularly in the field, by supplying information about possible abilities. In this way a balance may be struck between the extremes of anthropomorphizing and C. Lloyd Morgan's (1894) Canon.* Prosimian research can profit considerably from an emphasis on qualitative analysis that can state more about their learning capacities by viewing them as dynamic processes, instead of simply stating their quantitative performance limits. Their distinct approaches to problem solving gain relevance through their relationship to the approaches of the more anthropoid primate forms, particularly in a consideration of the evolution of human learning. We may better understand contemporary man by understanding the natural order of which he or she is a part.

ACKNOWLEDGMENTS

The preparation of this paper was in part supported by NIH Grant RR-00165. The fine secretarial assistance of Mayvin Sinclair is gratefully acknowledged. Also, we thank Helen Wells and Frank Kiernan of the Yerkes Regional Primate Research Center of Emory University for the final preparation of the figures.

REFERENCES

Andrew, R. J. (1962). *Science* **137,** 585–589.

Arnold, R. C., and Rumbaugh, D. M. (1971). *Folia Primatol.* **14,** 161–170.

Bearder, S. K., and Doyle, G. A. (1974). *In* "Prosimian Biology" (R. D. Martin, G. A. Doyle, and A. C. Walker, eds.), pp. 109–130. Duckworth, London.

Behar, L. (1966). *Midwest Psychol. Meet.* 1966, unpublished

Bierens de Haan, J. A., and Frima, M. J. (1930). *Z. Vergl. Physiol.* **12**, 603–631.

Bishop, A. (1962). *Z. Vergl. Physiol.* **102**, 316–337.

*The now famous canon was originally stated as "In no case may we interpret an action as the outcome of the exercise of a higher psychical faculty, if it can be interpreted as the outcome of the exercise of one that stands lower in the psychological scale (Morgan, 1894, p. 53)." The delineation of such a scale and how the levels thereof are constituted has yet to be determined.

Bitterman, M. S. (1965). *Am. Psychol.* **20**, 396–410.
Cooper, H. M. (1974). *In* "Prosimian Biology" (R. D. Martin, G. A. Doyle, and A. C. Walker, eds.), pp. 293–300. Duckworth, London.
Darby, C. L., and Riopelle, A. J. (1959). *J. Comp. Physiol. Psychol.* **52**, 94–98.
Davenport, R. K., and Rogers, C. M. (1970). *Science* **168**, 279–280.
Davis, R. T., and Leary, R. W. (1968). *Percept. Mot. Skills* **27**, 1031–1034.
DeValois, R. L., and Jacobs, G. J. (1971). *Behav. Nonhum. Primates* **3**, 107–157.
Doyle, G. A. (1974). *Behav. Nonhum. Primates* **5**, 155–353.
Ehrlich, A. (1968). *Folia Primatol.* **8**, 66–71.
Ehrlich, A. (1969). *Proc. Int. Congr. Primatol., 2nd, 1968* Vol. I, pp. 119–127.
Ehrlich, A. (1970). *Behaviour* **37**, 55–63.
Ehrlich, A., and Calvin, W. (1967). *Psychon. Sci.* **9**, 509–510.
English, H. B., and English, A. C. (1965). "A Comprehensive Dictionary of Psychological and Psychoanalytical Terms." Longmans, Green, New York.
Feldman, D. W., and Klopfer, P. H. (1972). *Z. Tierpsychol.* **30**, 297–304.
Fletcher, H. J. (1965). *Behav. Nonhum. Primates* **I**, 129–165.
Fobes, J. L., Ehrlich, A., and Williams, K. (1971). *Lab. Primate Newsl.* **10**, 7–8.
Fobes, J. L., Ehrlich, A., Rodriguez-Sierra, J., and Mukavetz, J. (1973). *Anim. Learn. Behav.* **1**, 99–101.
Gillette, R. G., Brown, R., Herman, P., Vernon, S., and Vernon, J. (1973). *Am. J. Phys. Anthropol.* **38**, 365–370.
Glickman, S. E., and Sroges, R. W. (1966). *Behaviour* **26**, 151–188.
Glickman, S. E., Clayton, K., Schiff, B., Gurtz, D., and Messe, L. (1965). *J. Genet. Psychol.* **106**, 325–370.
Goodall, J. (1965). *In* "Primate Behavior: Field Studies of Monkeys and Apes" (I. DeVore, ed.), pp. 425–473. Holt, New York.
Gorter, F. J. (1935). *Arch. Neerl. Zool.* **2**, 95–111.
Guthrie, E. R. (1935). "The Psychology of Learning." Harper, New York.
Harlow, H. F. (1932). *J. Comp. Psychol.* **14**, 241–252.
Harlow, H. F. (1949). *Psychol. Rev.* **56**, 51–65.
Harlow, H. F. (1959). *In* "Psychology: A Study of a Science" (S. Koch, ed.), Vol. II, pp. 492–537. McGraw-Hill, New York.
Harlow, H. F., Uehling, H., and Maslow, A. H. (1932). *J. Comp. Psychol.* **13**, 313–344.
Heffner, H., and Masterton, B. (1970). *J. Comp. Physiol. Psychol.* **71**, 175–182.
Hull, C. L. (1943). "Principles of Behavior." Appleton, New York.
Hunter, W. S. (1913). *Behav. Monogr.* **2** (1), (Serial No. 6).
Itani, J. (1958). *Primates* **1**, 84–98.
Jolly, A. (1964a). *Anim. Behav.* **12**, 560–570.
Jolly, A. (1964b). *Anim. Behav.* **12**, 571–577.
Jolly, A. (1966). *Science* **153**, 501–506.
Kelleher, R. T. (1965). *Behav. Nonhum. Primates.* **I**, 211–245.
Köhler, W. (1925). "The Mentality of Apes." Routledge & Kegan Paul, London.
Lang, E. M. (1962). "Goma the Baby Gorilla." Victor Gollancz, Ltd., London.
Le Gros Clark, W. E. (1959). "The Antecedents of Man." Edinburgh Univ. Press, London.
Levine, M. (1965). *Behav. Nonhum. Primates* **I**, 97–127.
Maier, N. R. F., and Schneirla, T. C. (1964). "Principles of Animal Psychology." Dover, New York.
Maslow, A. H., and Harlow, H. F. (1932). *J. Comp. Psychol.* **14**, 97–107.
Mason, W. A. (1968). *Sci. Psychoanal.* **12**, 101–118.
Masterton, B., and Skeen, L. C. (1972). *J. Comp. Physiol. Psychol.* **81**, 423–433.
Mitchell, G., Gillette, R., Vernon, J., and Herman, P. (1970). *J. Acoust. Soc. Am.* **48**, 531–535.
Morgan, C. L. (1894). "Introduction to Comparative Psychology." Methuen, London.

Napier, J. R., and Napier, P. H., (1967). "A Handbook of Living Primates." Academic Press, New York.

Narayan Rao, C. R. (1927). *J. Bombay Nat. Hist. Soc.* **32**, 206–208.

Ordy, J. M., and Samorajski, T. (1968). *Vision Res.* **8**, 1205–1225.

Parker, C. E. (1973). *In* "Gibbon and Siamang: A Series of Volumes on the Lesser Apes" (D. M. Rumbaugh, ed.), Vol. 2, pp. 185–206. S. Karger, New York.

Parker, C. E. (1974). *J. Comp. Physiol. Psychol.* **87**, 930–937.

Piaget, J., and Inhelder, B. (1956). "The Child's Conception of Space." Routledge & Kegan Paul, London.

Pinto, D., Doyle, G. A., and Bearder, S. K. (1974). *Folia Primatol.* **21**, 135–147.

Restle, F. (1958). *Psychol. Rev.* **65**, 77–91.

Rumbaugh, D. M. (1969). *Proc. Int. Congr. Primatol., 2nd, 1968* Vol. I, pp. 267–273.

Rumbaugh, D. M. (1970). *In* "Primate Behavior" (L. Rosenblum, ed.), Vol. 1, pp. 1–70. Academic Press, New York.

Rumbaugh, D. M. (1971). *J. Comp. Physiol. Psychol.* **76**, 250–255.

Rumbaugh, D. M., and Arnold, R. C. (1971). *Folia Primatol.* **14**, 154–160.

Rumbaugh, D. M., and Gill, T. V. (1973). *J. Hum. Evol.* **2**, 171–179.

Rumbaugh, D. M., and Gill, T. V. (1974). *In* "The Rhesus Monkey" (G. Bourne, ed.), pp. 303–321. Academic Press, New York.

Rumbaugh, D. M., and McCormack, C. (1967). *In* "Progress in Primatology" (D. Starck, R. Schneider, and H. J. Kuhn, eds.), pp. 289–306. Fischer, Stuttgart.

Rumbaugh, D. M., Riesen, A. H., and Wright, S. C. (1972). *Folia Primatol.* **17**, 397–403.

Rumbaugh, D. M., Gill, T. V., and Wright, S. C. (1973). *J. Hum. Evol.* **2**, 181–188.

Sidman, M. (1953). *Science* **118**, 157.

Silver, P. H. (1966). *Vision Res.* **6**, 153–162.

Stevens, D. A. (1966). *Diss. Abstr.* **26**, 5567.

Subramoniam, S. (1957). *J. Bombay Nat. Hist. Soc.* **54,** 387-398.

Thorndike, E. L. (1898). *Psychol. Monogr.* **2**, No. 8.

Treff, H. A. (1967). *Z. Vergl. Physiol.* **54**, 26–57.

Ward, J. P., and Doerflein, R. S. (1971). *Psychon. Sci.* **23**, 43–45.

Ward, J. P., and Riley, R. S. (1969). *Lab. Primate Newsl.* **8**, 6–7.

Ward, J. P., Doerflein, R. S., and Riley, R. S. (1970a). *Psychon. Sci.* **18**, 265–266.

Ward, J. P., Yehle, A. L., and Doerflein, R. S. (1970b). *J. Comp. Physiol. Psychol.* **73**, 74–77.

Warren, J. M. (1965). *Behav. Nonhum. Primates* **1**, 249–281.

Warren, J. M. (1974). *J. Hum. Evol.* **3,** 445–454.

Winer, B. J. (1962). "Statistical Principles in Experimental Design." McGraw-Hill, New York.

Woodworth, R. S. (1958). "Dynamics of Behavior." Holt, New York.

Yerkes, R. M. (1943). "Chimpanzee: A Laboratory Colony." Yale Univ. Press, New Haven, Connecticut.

Chapter 7

Vocal Communication in Prosimians

J.-J. PETTER AND P. CHARLES-DOMINIQUE

1. INTRODUCTION

As is the case with most other mammals, the prosimians employ a wide range of signals in communicating with conspecifics: auditory, visual, olfactory, and tactile. In each individual case, according to the social structure exhibited, there is pre-

ISBN 0-12-222150-8

dominance of one or more of these signal modalities. However, it should not be forgotten that with every species there is integration of all of these diverse means of communication in the overall signal repertoire.

The prosimians are often compared as a group with the simians, which are generally considered to be more highly evolved in many respects. With the prosimians, olfaction, vision, and hearing are roughly of equal importance in the sensory domain, whereas with the simians there is a clear-cut dominance of vision over olfaction. With hearing, on the other hand, the role played seems to be approximately equivalent in the prosimians and the simians. Yet although the simians have been fairly well studied with respect to vocal communication, the prosimians are still poorly studied in this respect. Given the fact that the prosimians represent 30% of the primates, study of their vocal communication should yield a considerable wealth of new information for interpretation of this type of communication, which has achieved such importance in human behavior. The present chapter represents no more than an approach to this aspect of prosimian behavior, which has concerned the authors in the course of their investigations of social behavior and of the relationships between behavior and ecology. More detailed studies will eventually follow, but for the time being the data available do not permit such extensive analysis as has been possible in comparable treatments of simian primates.

According to whether they are diurnal or nocturnal in activity, the prosimians have developed quite different social structures. The diurnal prosimians, like the majority of the simians, are gregarious. Group living introduces a number of problems of social organization, such as coordination of movement, priority of certain individuals in particular situations, development of a hierarchy, emergence of leadership, and the avoidance of frequent confrontations through aggression-regulating signals. Although these diurnal forms still exhibit a significant degree of olfactory communication (Schilling, Chapter 11), their complex social behavior is largely established on the basis of visual communication (Pariente, Chapter 10) and vocalizations.

With nocturnal prosimians, which are solitary with perhaps one or two exceptions, the social structures are generally based on the relative positions of home ranges. Olfactory signals (urine marks and secretions of cutaneous glands) play a particularly important role, permitting communication (Schilling, Chapter 11); but vocal signals nevertheless contribute significantly to social and territorial relationships. Whereas with the gregarious diurnal primates many vocalizations are connected by graded intermediates, indicating subtle variations in emotional states, the vocalizations of the nocturnal species are often somewhat more stereotyped. Some vocalizations are weak in intensity and destined for communication with a conspecific at close quarters (solicitation of contact, refusal of contact, threat, etc.); they are sometimes associated with specific postures in which the tail and the ears are involved. Other calls are powerful, as is the case with alarm calls, gathering calls, and various vocalizations utilized for communication with distant conspecifics.

TABLE I

A Tentative Classification of the Major Calls of Seven African Lorisid Species Arranged to Show Possible Homologous Relationships[a]

Doyle *et al.* denomination	*Galago crassicaudatus*	*Galago senegalensis*	*Galago alleni*	*Galago demidovii*	*Galago elegantulus*	*Perodicticus potto*	*Arctocebus calabarensis*	Ch.-Dominique denomination
Agonistic situations		Sob						
	Hack							
	Spit	Spit?	Quee quee quee[b]	Spitting sound	Ki ki ki[b]	heee[c]		
		Spit chatter						Threat calls
		Spit grunt						
		Rasp						
	Scream	Fighting chatter	Hoarse growl	Hoarse growl	Hoarse growl	Groan	Hoarse growl	
		Scream	Weet call[b]	Weet call[b]	Weet call[b]	Weet call[b]	Weet call[b]	Distress calls
	Yell							
		Grunt	Groan	Groan				
Anxiety alarm situations	Sniff							Alarm calls
		Sneeze						
	Knock	Shivering stutter						
	Creak	Gerwhit cluck						
	Squawk	Whistle						
	Whistle	Yap	Kiou kiou kiou[d]	Chip α[d]	Tee-ya[d]			
	Whistle yap			β				
	Chirp	Wuff and wail		δ				
	Chatter	Caw		γ				
	Moan	Explosive cough						
	Rattle							

(continued)

TABLE I (*continued*)

Doyle *et al.* denomination	*Galago crassicaudatus*	*Galago senegalensis*	*Galago alleni*	*Galago demidovii*	*Galago elegantulus*	*Perodicticus potto*	*Arctocebus calabarensis*	Ch.-Dominique denomination
Social Cohesion	Buzz Click	Click Crackles	Click Series of clicks	Click Series of clicks	?	Click		Physical contact and contact-seeking calls
Friendly and contact situations	Squeak Cluck	Squeak Coo Soft hoot	Tsic	Tsic Gathering call Rolling call	Tsic[c] Rolling call	Tsic[b]	Tsic[b]	
	Cry	Bark	Croaking call[d]	Plaintive squeak[c]	Bird call[d]	Whistle[b]		Auditory contacts

[a] The descriptive terms for the calls of *Galago crassicaudatus* and *G. senegalensis* are those used by Doyle *et al.* (1977). Those for the remaining five species are the terms used by Charles-Dominique (1977).

[b] Short-range carriage vocalization.

[c] Medium-range carriage vocalization.

[d] Long-range carriage vocalization. The other vocalizations do not carry far in the forest (for the 5 species of Gabon).

It is, in fact, extremely difficult to establish homologies between the calls of different species purely on the basis of their physical structure (sonographic analysis). In many cases, individuals of different species respond not only by calling but also by behaving in a characteristic, uniform manner in similar situations. When different species are compared in this wider sense, the calls uttered in a given situation are often found to be very similar; yet in some cases they are utterly different. For this reason, it has proved to be preferable to classify the vocalizations in terms of the situations with which they are associated (alarms, threat, distress, etc.) rather than on the basis of physical similarities. It is inevitable that the categories selected are fairly broad and may contain several types of vocalization; however, this rough classification permits comparison of the different psychological states associated with the various calls. In some instances, more detailed analysis permits identification of similarities between calls uttered under the same circumstances. Such similarities may perhaps indicate a phylogenetic relationship between the two species considered, but one must not forget that a vocalization may evolve in close dependence on local environmental conditions (conditions of sound propagation, Chappuis, 1971) and that convergence is not impossible in this area (cf. the similarities between the calls of various birds and those of certain South American monkeys noted by Moynihan, 1967). It will, in fact, be seen that homologies can most easily be established among vocalizations of the weak type which are destined for a nearby conspecific, for which the conditions of sound propagation are generally similar in all the relevant biotopes. Table I represents an attempt to classify the calls of some nocturnal lorisids in order to illustrate possible homologous relationships.

It should be emphasized that, just like scent-marking or the adoption of specific postures, vocalizations reflect emotional states. For this very reason, conspecifics receiving the signals only respond in terms of their own emotional condition and one should not expect to find precise, immediate responses to every vocalization which is singled out for study. As Gautier-Hion (1971) has demonstrated with the talapoin, the entire species' repertoire may be performed in the space of a few seconds in particular abnormal situations. Hence, the interpretations of prosimian vocalizations given below are based on observations in which a given call was most often associated with a certain behavioral situation.

Before going on to review the calls of the various prosimian families, it is necessary to define the principal categories of vocalizations identified in terms of the associated behavior.

1.1. Contact and Contact-Seeking Calls

Contact calls include all vocalizations uttered in the course of tactile interactions. In most cases, these calls are uttered by the infant and/or the mother during their contacts with one another, but in some cases they are also heard with groups of adults.

Contact-seeking calls vary according to the circumstances, such as in mother–infant relationships, interactions at sleeping places, and sexual encounters. The calls in this category may exhibit intergradations.

1.2. Distant Communication Calls

A large number of vocalizations can be included in this category. They permit dispersed individuals (i.e., particularly in nocturnal prosimian species) or members of a moving group to communicate with one another despite their separation. In some cases, powerful calls in this category can be interpreted as territorial (spacing) signals, such as those exchanged between individuals belonging to different social groups.

1.3. Alarm Calls

These vocalizations betray a state of excitation which is generally associated with some disturbance in the surroundings (e.g., presence of a predator, occurrence of some unusual phenomenon). Alarm calls are typical for each species, and normally there is only one or perhaps two types per species. Where more than one alarm call can be exhibited by a given species, graded intermediates can occur as the animal becomes more excited and there may be weak or pronounced signals according to the circumstances.

1.4. Contact-Rejection Calls

Vocalizations in this category, which are often associated with characteristic postures (open mouth, raised hands, etc.) or with sudden movements, prevent conspecifics from approaching. In most cases, the calls consist of staccato spitting sounds which usually appear as white noise in sonographic analysis, though occasionally they take the form of a series of parabolas.

At a higher level of excitation, for example, when an animal is threatened, frightened, or involved in fighting, a second type of vocalization may emerge, interspersed with calls of the first type. This second call type is a two-phase grunt (inspiration/expiration) which is found in the same form in virtually all prosimian species. The two successive components of this latter call produce a special sound contrast which may in certain cases scare, or at least arrest, an approaching predator. When a paroxysm of excitation is reached, these two-phase grunts are associated with urination, defecation, erection of the ears, wide opening of the eyes and mouth, and spreading of the arms and hands.

1.5. Distress Calls

These characteristic vocalizations, which vary little from one prosimian family to another, are uttered when the animal is extremely frightened and especially in response to pain. With both *Galago demidovii* and *Lemur fulvus* the authors have found this call to play a part in the arrest of intraspecific fighting.

2. SYSTEMATIC SURVEY OF PROSIMIAN VOCALIZATIONS

The phylogenetic relationships between the various prosimian families are still controversial (Petter and Petter-Rousseaux, Chapter 1), and it seemed to be more appropriate at this stage to present the data on vocalizations separately for diurnal and nocturnal prosimians. Among the seven families in the Strepsirhini, the family Lemuridae is distinctive in that, in addition to a majority of diurnal or crepuscular forms, there are some species with essentially nocturnal tendencies (*Hapalemur* and *Lemur mongoz*); but the family Indriidae is the only one to exhibit a sharp dichotomy between strictly diurnal species (*Propithecus verreauxi*, *P. diadema*, *Indri indri*) and one strictly nocturnal species (*Avahi laniger*).

The nocturnal members of these two families will be discussed after the other nocturnal prosimians, since they may provide a transitional stage linking them to the two lemur groups with a majority of diurnal representatives (Lemuridae and Indriidae). In addition, comparisons between the diurnal lemurids and indriids, on the one hand, and the crepuscular/nocturnal lemurids and indriids, on the other, might perhaps allow some evaluation within each family of the respective influences exerted by specific ecological adaptation and phylogenetic continuity in the vocal repertoires of each species.

2.1. Lorisidae

The Lorisidae are divided quite distinctly into two subfamilies: the Lorisinae, with two African species and two Asiatic species, and the Galaginae, with six species all restricted to Africa. The members of these two subfamilies have followed two diametrically opposed pathways in evolution, one leading to a slow-climbing type, relying upon camouflage (Lorisinae), though Bearder (see Charles-Dominique and Bearder, Chapter 13) notes that *G. crassicaudatus* are, in this respect, intermediate between the two subfamilies. A large number of behavioral features have been affected by these divergent "strategies," especially feeding behavior and social interactions, among which vocalizations are prominent (Charles-Dominique, 1971, 1977). It is evident that an antipredator mechanism based on discretion in locomotion and on camouflage would be incompatible with powerful, frequently uttered vocalizations which are likely to betray the animal's presence. For this reason, the vocal repertoires of the Lorisinae are very poor, and consideration of this subfamily will be preceded by a discussion of the far more vocal bushbabies (Galaginae).

2.1.1. Galaginae

Among the bushbabies inhabiting relatively dry forest areas, nothing is known of the biology of *Galago inustus* (*Euoticus inustus*), but a great deal of work has been carried out with the two other species in North, East, and South Africa, *G. senegalensis* and *G. crassicaudatus*. The vocalizations of these two species have been described by Sauer and Sauer (1963), Andrew (1963, 1964), and Jolly (1966a) in fair detail, and a recent unpublished paper by Doyle *et al*. (1977) has reconsidered these various publications and incorporated a valuable sonographic and behavioral analysis.

As far as the bushbabies inhabiting rain forest areas of West Africa are concerned, there have been some previous descriptions of the vocalizations of *G. demidovii* (Vincent, 1969; Struhsaker, 1970), but most available information derives from a major field study (Charles-Dominique, 1977) of the five lorisid species in Gabon (*G. demidovii*, *G. alleni*, *G. elegantulus* (*Euoticus elegantulus*), *Perodicticus potto*, and *Arctocebus calabarensis*).

2.1.1.1. Contact Calls and Contact-Seeking Calls. ***a. Mother-infant calls.*** The neonate bushbaby will utter, as a result of any excitation, very short, high-pitched ''clicks'' in isolation or in series (Fig. 1a) (which are referred to as ''crackles'' in *Galago senegalensis*). At a higher level of excitation, the ''click'' will gradually give way, through a whole range of intermediates, to a ''squeak'' (Doyle *et al.,* 1977) which has also been described as ''tsic'' (Charles-Dominique, 1977). This latter vocalization, which is composed of a parabola with numerous harmonics, represents the highest level of excitation of the bushbaby infant. According to the situation, the mother will always respond to this call by retrieving the infant (if it is isolated) and/or by exhibiting maternal care. The mother may also elicit this call from the infant by uttering a call herself. As soon as the infant responds in this way to the mother's call, she may localize and retrieve it. The mother's call varies considerably from species to species. With *G. crassicaudatus*, it takes the form of a ''cluck,'' which is homologous with the ''coo'' and ''soft hoot'' of *G. senegalensis* and with the ''croaking call'' of *G. alleni* (Fig. 1f). *Galago demidovii* utter a faint ''rolling call'' (Fig. 1g) which is doubtless homologous with a ''tsic'' usually uttered by *G. elegantulus* (Fig. 1c) (the latter call being quite similar to that of the infant itself, but more powerful).

b. Calls exchanged between adults. It is not only mothers and infants that meet up to return to the sleeping site. Mothers and their adult daughters are united by social bonds, with the result that one finds quite large sleeping groups containing several adult females, subadults, and sometimes one or more adult males. All of these individuals are dispersed during the night and they regroup in the morning, as soon as the first signs of dawn appear, by uttering powerful gathering calls. With *G. demidovii* this call emerges by gradual transformation of the infant's ''tsic'' (Fig. 1d), which develops eventually to form a crescendo composed of 15 to 20 components. Great individual differences exist in the form of this vocalization and in captivity this permits recognition of each animal by its gathering call. The call is uttered when the animals awake, on occasions during the night, but primarily at dawn. When uttered at dawn, the call evokes vocal responses of similar form from

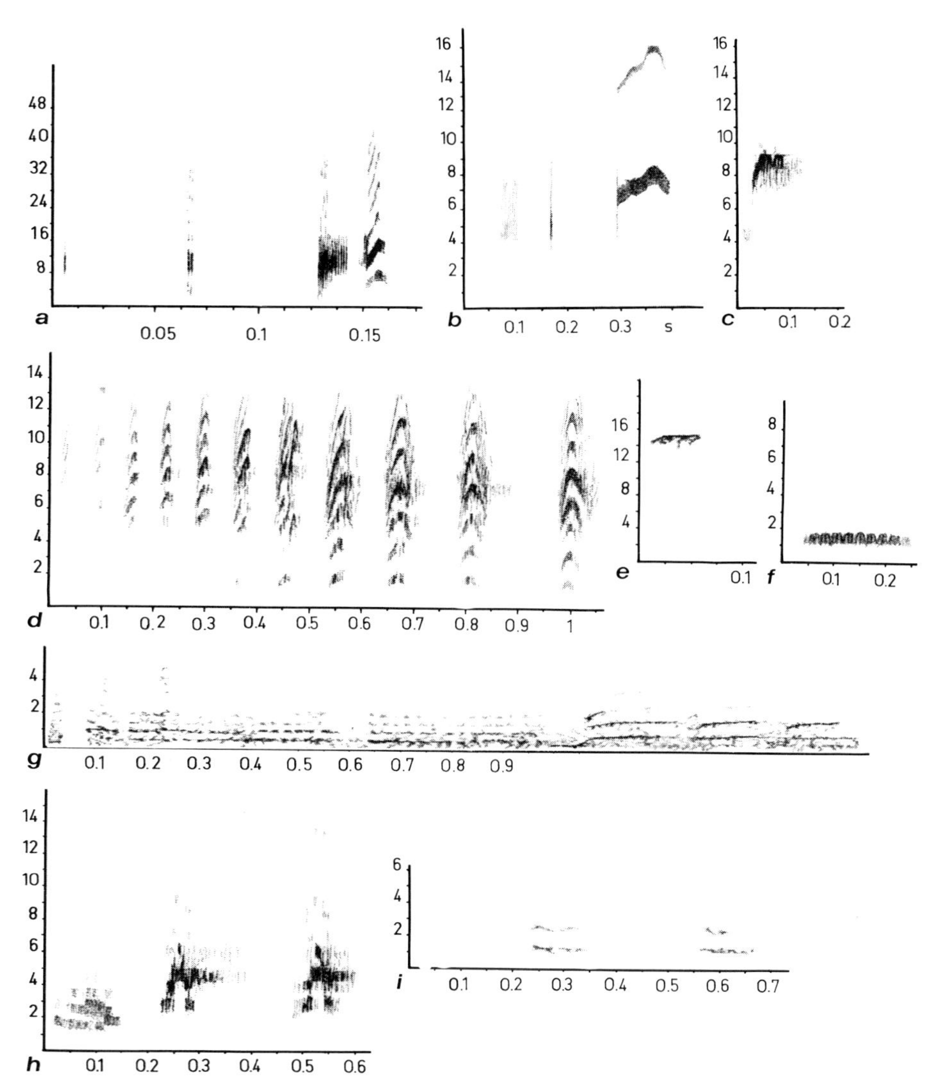

Fig. 1. Contact-seeking and distant communication calls in the Lorisidae. (a) Call of newborn *Galago demidovii* (click and tsic); (b) call of young *Perodicticus potto*; (c) call of *Galago elegantulus*; (d) gathering call of *Galago demidovii*; (e) sexual call of the ♀ *Perodicticus potto*; (f) croaking call of *Galago alleni*; (g) rolling call of *Galago demidovii* changing to gathering call; (h) bird call of *Galago elegantulus*; (i) plaintive squeak of *Galago demidovii*.

conspecifics, and the animals may come together to form a group in the space of a few minutes.

The gathering calls of *G. elegantulus* are little different from the "tsic" calls uttered by the infant. Although this call does not carry as far as the gathering call of *G. demidovii*, it is adequate since the sleeping site is always located in a given zone less than 1 hectare in area and the individuals of the group arrive in this zone at dawn before calling to one another.

Galago alleni have no morning gathering call. In this species, each individual returns to a tree hollow and the females, associated in matriarchies, often sleep separately. (There are normally a dozen or so tree hollows which can be utilized as sleeping sites in an individual's home range.) In the same way, there do not seem to be any gathering calls in *G. crassicaudatus* or *G. senegalensis*.

In contrast to the other bushbaby species, *G. crassicaudatus* are relatively gregarious during the nocturnal activity period (Bearder and Doyle, 1974). There is a certain amount of cohesion between the mother and her maturing infants, which regularly emit "buzz" calls during their movements through the forest.

2.1.1.2. Distant Communication Calls. Each bushbaby species exhibits a powerful type of vocalization which is somewhat difficult to interpret, viz., the "cry" of *Galago crassicaudatus*, the "bark" of *G. senegalensis*, the "croaking call" of *G. alleni*, the "plaintive squeak" of *G. demidovii* (Fig. 1i), and the "bird call" of *G. elegantulus* (Fig. 1h).

With *G. alleni*, which were studied by radio tracking in Gabon, it was found that the "croaking call" could be emitted as a very weak call (mother calling her infant) or, with a series of intermediates, as an extremely powerful vocalization. In the latter case, the call was uttered as a series of 5 to 15 components, separated by intervals of 1/2 to 1 second. Such series were repeated approximately every 10 to 20 seconds, sometimes for as long as an hour. This "croaking call" elicits similar vocal responses from conspecifics in the vicinity and can be artificially evoked with a tape recording. In one population, in which every individual was followed by radio tracking, the male and several females were able to localize one another with this call (the male regularly visits different females in the course of the night). The same vocalization was also used in the course of territorial encounters between neighboring females and neighboring males. "Rival" males would approach the border area between their ranges and utter the "croaking call" in alternation. This behavior probably represents reciprocal territorial demarcation. Apparently the "croaking call" can be interpreted in different ways by conspecifics as a function of the social position they occupy with respect to the vocalizing animal.

The "cry" of *G. crassicaudatus* and the "bark" of *G. senegalensis* may well be homologous with the "croaking call" of *G. alleni*. Apparently the "plaintive squeaks" of *G. demidovii*, which are often uttered in chorus, and the "bird calls" of *G. elegantulus* also play the same role as the "croaking calls" of *G. alleni* (Charles-Dominique, 1977).

2.1.1.3. Alarm Calls. Bushbabies are extremely sensitive to any disturbance in their surroundings, such as the presence of an unusual object or living organism, appearance of a predator, occurrence of a strange noise, or arrival of a strange conspecific. Their anxiety in such situations is reflected by alarm calls, which are the most commonly uttered and therefore the best known vocalizations of these species. With both *Galago alleni* (Fig. 2e) and *G. senegalensis,* a hoarse grunt may precede the alarm calls at the onset of excitation. Whereas the alarm calls of *G. elegantulus* ("tee-ya") (Fig. 2c and d) and of *G. alleni* (Fig. 2b) ("kiou kiou kiou") rapidly take on their definitive, stereotyped form, the other species all exhibit a range of intermediates, passing from weak calls ("sniff" of *G. crassicaudatus,* "sneeze" of *G. senegalensis,* "chip" of *G. demidovii*) (Fig. 2a) to very powerful vocalizations which may be uttered for periods of 15 minutes to 1 hour once the animal is fully excited. The role of these alarm calls is less clear than that of similar calls uttered by gregarious animals, where sighting of a predator by one animal alerts the entire group. In the case of the bushbabies, the animals are dispersed and individuals hearing the alarm calls of a distant conspecific cannot benefit from this information. At the most, a young animal accompanying its mother may in this way learn to recognize dangers which she perceives and indicates.

Similarly, the intermediate forms of the alarm calls are difficult to interpret. For example, *G. alleni* and *G. demidovii* exhibit fairly similar social structures. With the first species, the alarm call is very stereotyped, whereas with the second there is a whole range of intermediates. However, in both cases the alarm calls produced are individually recognizable, and they may therefore indicate to conspecifics the presence, identity, and location of the calling animal.

2.1.1.4. Contact-Rejection Calls. When an individual rejects contact with an approaching conspecific, it will utter staccato calls which will normally be sufficient to arrest the approach. The signal can be modulated by increasing its intensity and pitch, and this allows fine regulation of relationships between individuals. The calls involved are the "ki-ki-ki" of *G. elegantulus*, the "spitting call" of *G. demidovii*, the "quee-quee-quee" of *G. alleni* (Fig. 2f), the "sob, spit," "spit chatter," "spit grunt," and "rasp" of *G. senegalensis*, and the "hack" and "spit" of *G. crassicaudatus*.

If a bushbaby is attacked by a conspecific or predator, trapped in its nest, or otherwise frightened, a second type of call is superimposed on the first. This is a two-phase grunt (inspiration/expiration) described as a "hoarse growl" for *G. elegantulus* (Fig. 2g), *G. demidovii*, and *G. alleni*, and probably equivalent to the "fighting chatter" of *G. senegalensis* and the "scream" of *G. crassicaudatus*. This latter vocalization, elicited by fear, is often associated with urination, defecation, and adoption of a body posture equivalent to the "hunched-back posture" of cats (ears distended, eyes wide open, spread hands held out in front).

2.1.1.5. Distress Calls. When extremely frightened, and particularly when in pain, bushbabies utter a high-pitched, plaintive call which is fairly constant in form from species to species: the "weet call" of *G. alleni, G. demidovii* (Fig. 2j), and *G. elegantulus*, the "yell" of *G. crassicaudatus* and probably the "scream" of *G. senegalensis*. With *G. demidovii* it has been observed that intraspecific fighting can be arrested when the subordinate animal, following some injury, utters the distress call. The call is quite similar to that uttered by many small rodents when they are injured. If this call is imitated in the forest, one can attract bushbabies which will exhibit mobbing, and also small carnivores, which are doubtless attracted by the prospect of a wounded animal which would be easy to seize.

2.1.2. Lorisinae

There is very little information on the Asiatic lorisine species, and this discussion will be limited essentially to the two African species: *Perodicticus potto* and *Arctocebus calabarensis*.

2.1.2.1 Contact Calls and Contact-Seeking Calls. ***a. Mother-infant calls.*** Although the young lorisine is able to cling to his mother's fur from birth onward, the mother begins quite early on to leave her infant hidden in vegetation at the beginning of the night, finally retrieving it definitively only at dawn. At dawn, the infant emits "clicks" and "tsics" (Fig. 1b), to which the mother responds by uttering somewhat louder "tsic" calls. These vocalizations permit reciprocal localization between the mother and her offspring so that they can reunite to go to the sleeping site.

b. Calls exchanged between adults. During courtship, the male potto follows the female uttering a "tsic" call very similar to that exchanged between the mother and her infant.

2.1.2.2. Distant Communication Calls. Manley (pers. comm.) has reported a fairly powerful call for *Perodicticus potto* and *Arctocebus calabarensis* in Nigeria. In Ceylon, *Loris tardigradus* have been heard to produce high-pitched whistling calls, which may be homologous with calls heard in captivity (Charles-Dominique, 1977) from a female *P. potto edwardsi* (Fig. 1e) when in estrus. In the latter case, the whistling calls were each 1/2–1 second in duration and were emitted in a fairly variable rhythm, often at intervals of 20–60 seconds. As a general rule, the social communication patterns of the potto are relatively discreet and rely essentially on olfactory marking.

2.1.2.3. Alarm Calls. For the reasons outlined above (see Section 2.1), alarm calls appear to be completely lacking among the Lorisinae.

2.1.2.4. Contact-Rejection Calls. Among the lorisines, encounters between conspecifics are extremely rare; they are observed only during courtship and mating

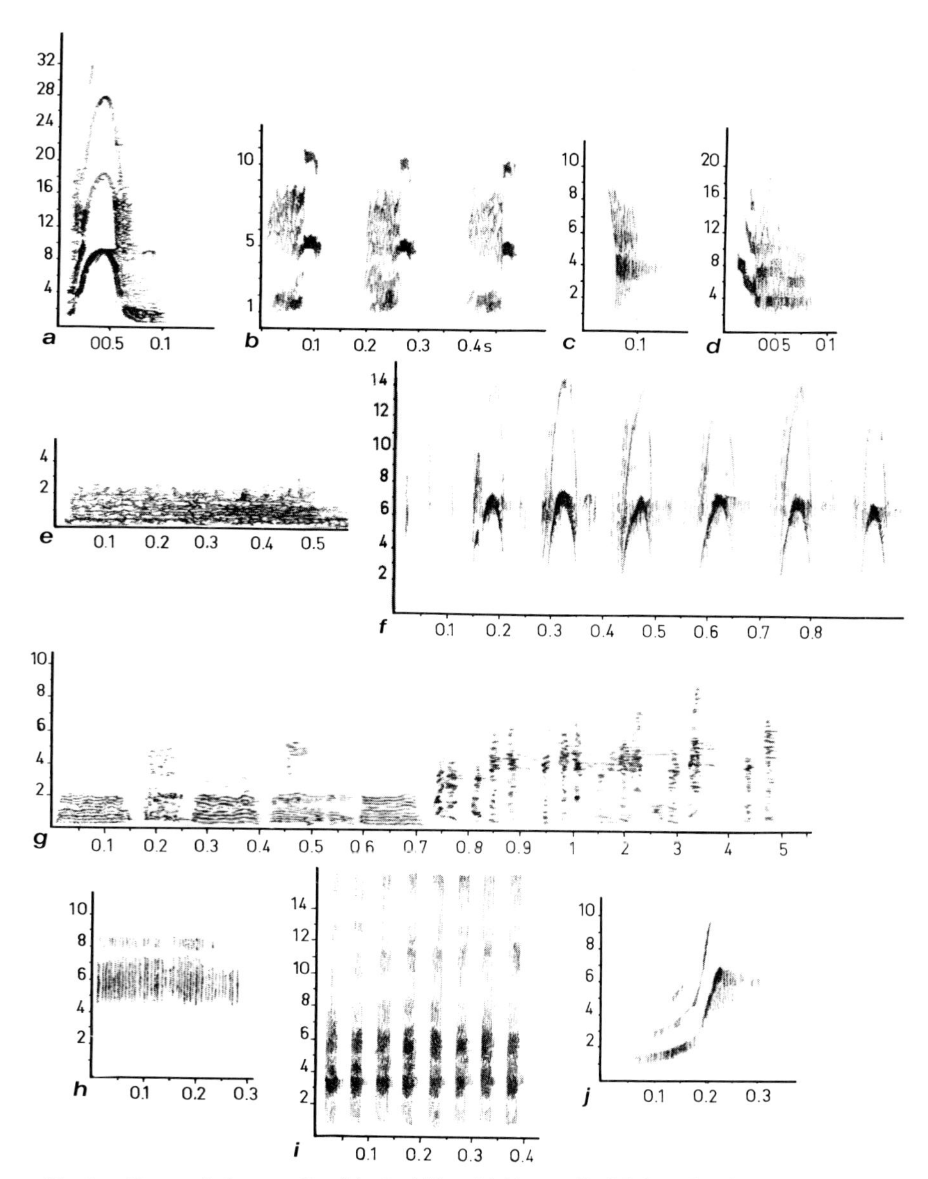

Fig. 2. Alarm and distress calls of the Lorisidae. (a) Alarm call of *Galago demidovii*; (b) alarm call of *Galago alleni*; (c) alarm call of *Galago elegantulus* (analyzed at normal speed); (d) same as (c) (analyzed at half speed); (e) grunt of *Galago alleni*; (f) contact-rejection call of *Galago alleni*; (g) contact-rejection call of *Galago elegantulus*—"ki-ki-ki-ki," preceded by double (hoarse) growl; (h) soft grunt of *Perodicticus potto*; (i) loud (staccato) grunt of *Perodicticus potto*; (j) distress call of *Galago demidovii*.

or during very occasional territorial combats. When a potto is attacked by a conspecific or a predator, it always adopts the same defensive posture, with the head retracted between the hands and the scapular shield presented to the adversary. This posture is accompanied by a grunt whose elements are transformed as excitation increases. The elements, which appear initially as a series of bursts of white noise, become intensified and more high pitched, while aggregating into groups, giving rise to staccato grunting which accentuates the sound contract (Fig. 2i). The thrusting charges of the potto, which are directed against the predator directly in front, are accompanied by this intense, high-pitched grunting, and the combination often induces the adversary to withdraw. When the potto is frightened, on the other hand, the two-phase grunt (inspiration/expiration "groan"), which is homologous with that of the bushbabies, is intercalated between the usual grunting calls (Fig. 2h).

Arctocebus calabarensis are more discrete than pottos. When attacked, they roll up into a ball and emit a "hoarse growl" (inspiration/expiration).

2.1.2.5. Distress Calls. When in pain, both *Perodicticus potto* and *Arctocebus calabarensis* utter a "weet call" which is homologous with that of the bushbabies.

2.2. Cheirogaleidae

2.2.1. Cheirogaleinae

Within the subfamily Cheirogaleinae, one can distinguish two types which to some extent parallel the distinction between bushbabies and lorises: (i) species of the genus *Microcebus*, which are rapid leaping forms; and (ii) species of the genus *Cheirogaleus*, which are running and leaping forms with slower and less agile locomotion. *Microcebus*, although less specialized, correspond to the bushbaby type, while *Cheirogaleus*, although they have not attained the high level of specialization for camouflage exhibited by the lorisines, resemble the lorises both in their discreet form of locomotion and in the absence of alarm calls.

Microcebus murinus and *M. coquereli* both exhibit social organization patterns comparable to that of the bushbabies (Martin, 1972; Petter *et al.*, 1971), with individual male territories overlapping those of several females. However, the structures of their calls are quite different from those of the bushbabies. Since *M. murinus* and *M. coquereli* themselves exhibit considerable differences in their vocalizations, they will be considered separately.

2.2.1.1. *Microcebus murinus*. Most of the calls of *Microcebus murinus* are very high-pitched, and some of them are almost inaudible to the human ear, or at least only the low-frequency components are audible. The apparatus used in the field permitted recording of only those features audible to the human ear, and supplementary work in the laboratory was therefore required. Under natural conditions, one can only hear these calls if particular attention is paid to them and many of the more high-pitched sounds are often difficult to differentiate from those emitted by insects.

Some work has been conducted to determine whether some of these calls play an echolocatory role, but the limited experiments conducted on this aspect to date (Pariente, 1974) have not revealed any such function.

Since mouse lemurs are partially insectivorous, it is also possible that some calls act to attract or set in motion certain insect prey species, but so far there have been no observations or experiments suggesting that this is the case.

With *Microcebus murinus*, one can distinguish medium frequency vocalizations including contact calls, contact-rejection calls, and distress calls, all of which can be fairly easily characterized. On the other hand, high-pitched calls which probably correspond to distant communication calls and alarm calls are much more difficult to interpret.

a. Contact calls and contact-seeking calls. *i. "Purring" of the infant.* The young animal will sometimes produce a weak "purring" vocalization when nestling against the mother. An infant hand-reared from an early age rapidly becomes tame and will often purr when held on the palm of the hand. In addition, this vocalization has often been heard with very tame subadults or adults in response to stroking. This vocalization, as has been suggested by Martin (1972), may represent a social signal in sleeping groups. No equivalent call has yet been recorded for the Galaginae.

ii. Summoning call of the infant. Under natural conditions, the young mouse lemur is deposited in a nest or in a hollow tree where it remains silent and immobile in the absence of the mother. In captivity, when an infant is isolated, cold, and/or hungry, or when it is roughly licked by the mother, it may utter a weak, high-pitched plaintive call. The mother is very responsive to this call when it is uttered forcibly, and will immediately return to retrieve her infant.

iii. Contact calls among adults. Certain calls, which may be represented as repetitive bursts of "fee-tsi" (Fig. 3e), occur in series which usually contain 2–5 components separated by intervals of 3–5 seconds. The calls themselves are separated by intervals of 0.2 second within the bursts, and they may be repeated for some time. These vocalizations are characteristic of periods of high excitation, particularly during sexual activity, and they may be interpreted as contact-seeking calls exchanged between adults.

Longer calls of more complex structure are also emitted by males at other times, especially when females are in estrus. They usually elicit a vocal response, probably from another male, over a distance of 20 m or more. These calls may in some cases be uttered with a simpler structure, incorporating only the second component of the call, and they are accordingly less intense.

iv. Mating call. When the females are in estrus, the male will emit a resonant call with a quite specific structure (Fig. 3f). The call itself can be represented as "fit tsirrrr," and there is a possibility that the female may respond with an identical call, as is the case with the homologous vocalization of *Microcebus coquereli*. While uttering this call, the male attempts to sniff at the female's anogenital region and subsequently to mount her. Initially the female may respond with high-pitched contact-rejection calls and threat grunts. While uttering this call, which is repeated at intervals of 2–3 seconds, the animal concerned shows retraction of the mouth somewhat resembling that seen in a miaowing cat.

b. Distant communication calls. As has been pointed out above, it is quite difficult to distinguish the high-pitched calls of *Microcebus murinus,* and for certain signals there are still problems of interpretation. As a preliminary measure, the calls concerned have been separated, perhaps somewhat arbitrarily, into distant communication calls and alarm calls.

Distant communication calls have been taken to be those which occur as prolonged series of very high-pitched whistles which can be heard both in the field and in captivity throughout the night. They may perhaps represent territorial signals. It has been observed in captivity that several individuals living in neighboring cages may reply to one another by uttering calls of slightly differing frequencies (Fig. 3d). When this occurs, there is slight superimposition of the calls, the beginning of the second overlapping the end of the first. When they utter this call, the animals generally appear to be very calm.

c. Alarm calls. There are various types of whistling call corresponding to varying degrees of alarm. Mouse lemurs are very attentive to all noises in their surroundings. According to their degree of excitation, there is a whole range of alarm responses which appear to climax in a call resembling the contact-rejection call. Some of these calls probably have a relatively complex function.

i. Series of short whistles of elevated frequency and regular pattern. These series of short whistles have a frequency and intensity close to those of the long whistles described above as distant communication calls. However, there is some variability in the rhythm of emission. Sometimes, the components follow one another in such rapid succession that it is very difficult to differentiate the call series from continuous whistles. Finally, in some recordings a progressive reduction in frequency has been noted. Several individuals may utter this same call type at slightly differing frequencies, yet there is no clear evidence of any kind of "dialogue" since the calls may follow one another or occur superimposed. When uttering this call, a *Microcebus murinus* is often quite active and will attentively scan its environment while now and then moving its forequarters. When this occurs, the head is moved laterally and vertically, with the mouth open and the head region trembling. If the animal is approached with a finger while calling in this way, it will usually respond in a very aggressive fashion, biting and transforming its whistles to more intense calls and grunts (i.e., a shift toward contact-rejection calls).

ii. Short reinforced whistles. Another variant of these calls is represented by short, emphasized whistles, giving the impression of two successive components (Fig. 3b). Such whistles occur in isolation and they occur at intervals of one or more seconds.

iii. Series of whistles of variable intensity and rhythm. Under certain conditions, series of whistles can exhibit a regular variation in intensity and speed of emission, which first increase, then decrease. In captivity, these vocalizations are heard when mouse lemurs are very excited, though without any evidence of aggressive tendencies (for example, when they are presented with live moths).

d. Contact-rejection calls (Fig. 3a and c). There is a complete scale of intermediates between series of brief high-pitched whistles and series of two-phase grunts comparable to the "hoarse growl" of the Lorisidae. The components become more closely packed, their intensity increases and the frequency band becomes broad. When fights occur, these calls take the form of "decrescendos" or rapid, intense components. At night, in the forest, actively moving animals utter calls of this kind if the tree trunks are tapped.

Grunts are frequently uttered when mouse lemurs are disturbed while sleeping. When this occurs, the animals exhibit all intermediate states between calm threat (with the animals attempting to hide deeper in the nest while uttering a few slow, deep grunts) and active threat (with the animal facing the adversary, uttering grunts with the mouth wide open and then lunging forward to bite). In this latter case, the

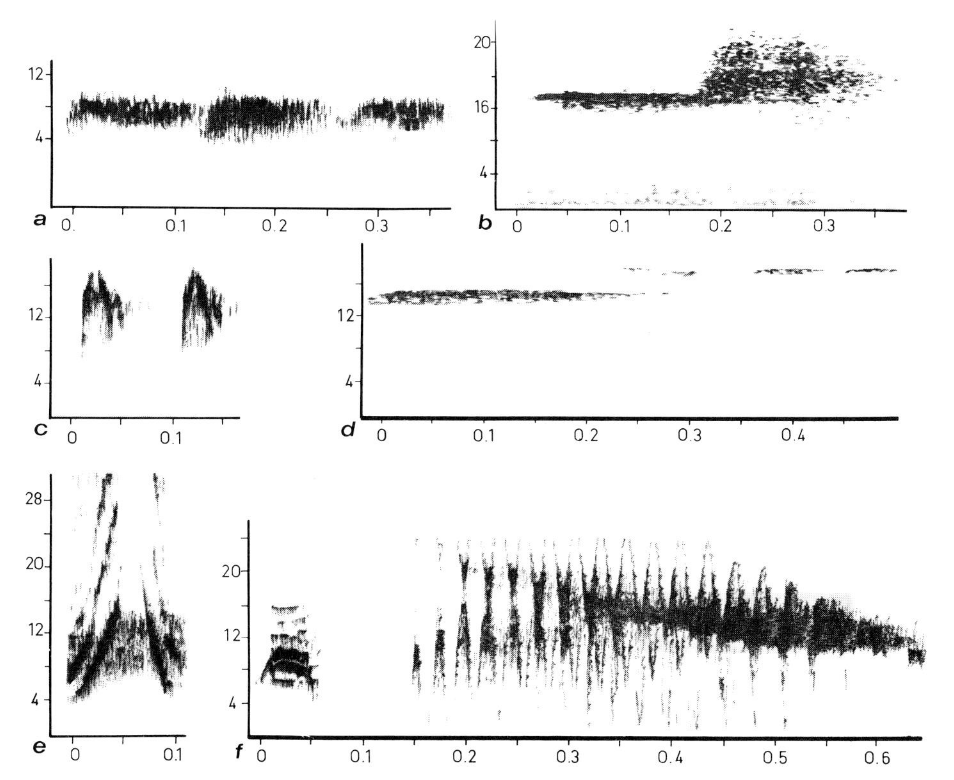

Fig. 3. Calls of *Microcebus murinus*. (a) Contact-rejection grunts; (b) accentuated brief whistle; (c) contact-rejection call; (d) long whistle followed by brief whistles emitted by another individual; (e) contact calls between adults—"Fee-tsi"; (f) sexual call.

vocal emissions are often accompanied by urination and defecation. Calls of this kind are produced by infants in the nest from the second day of life onward.

e. Distress calls. With *Microcebus murinus*, as with the other cheirogaleids and lorisids, there is a very high-pitched, intense call which can be represented as "couisc" (a transcription of one Malagasy name for this animal) and which is uttered in rare situations of distress.

2.2.1.2. *Microcebus coquereli.* Coquerel's mouse lemur is very similar to *Microcebus murinus* in its morphology, and it is distinguished primarily by its greater size and its more limited geographical distribution. Social structure seems to be fairly comparable in the two species. Recordings conducted both in captivity and in the field indicate that *M. coquereli* have a repertoire of vocalizations at least as rich as that of *M. murinus*. Although it is not possible to draw parallels between all the vocalizations of the two species, some of them are obviously directly comparable.

a. Contact calls and contact-seeking calls. *i. "Purring."* As with *M. murinus*, the infant emits a weak purring noise when licked by the mother.

ii. Mother–infant contact calls (Fig. 4e and f). A high-pitched whistle (comparable to the onset of the estrus call of this species) can be emitted under certain circumstances by the infant *M. coquereli*. However, the most common contact call is a kind of piping "pui" call of relatively low pitch. The latter type of call has often been heard under natural conditions, and in some cases it has been repeated for more than 1/2 hour without interruption. In most cases, two animals (usually a female and her offspring) were observed exchanging this call at about 0530 hours, shortly before dawn. Subsequently, the two animals would meet up to return to sleep in the same next. A comparable situation has already been described for the Lorisidae.

iii. "Mating call." This call can be regarded as homologous with the call described above for *M. murinus* (Fig. 4i), which it resembles in form and differs only in its lower pitch and more resonant quality. It is uttered by both the male and female, which may show reciprocal calling. The most characteristic part of the call is a "decrescendo" squeak, often preceded by a very high-pitched whistle. The complete call may contain as many as 19 squeak components following the whistle.

b. Distant communication calls. *i. Cohesion call.* When *M. coquereli* are on the move, a discreet "hou" of low frequency is uttered more-or-less regularly and automatically, whether or not a conspecific is in the vicinity. (In fact, this call provides one of the most reliable means of spotting *M. coquereli* in the forest.) It is a social cohesion call with a function similar to the "hong" of *Phaner* and the grunting calls of *Lemur* species.

ii. Contact calls exchanged between adults (Fig. 4d). An adult contact call, which can be represented as a drawn-out "ptiao," has been found to resemble the mother-infant contact call, though with the omission of the emphasis. Such vocalizations may occur as rapid sequences, and they operate as a system of calls and responses between adults.

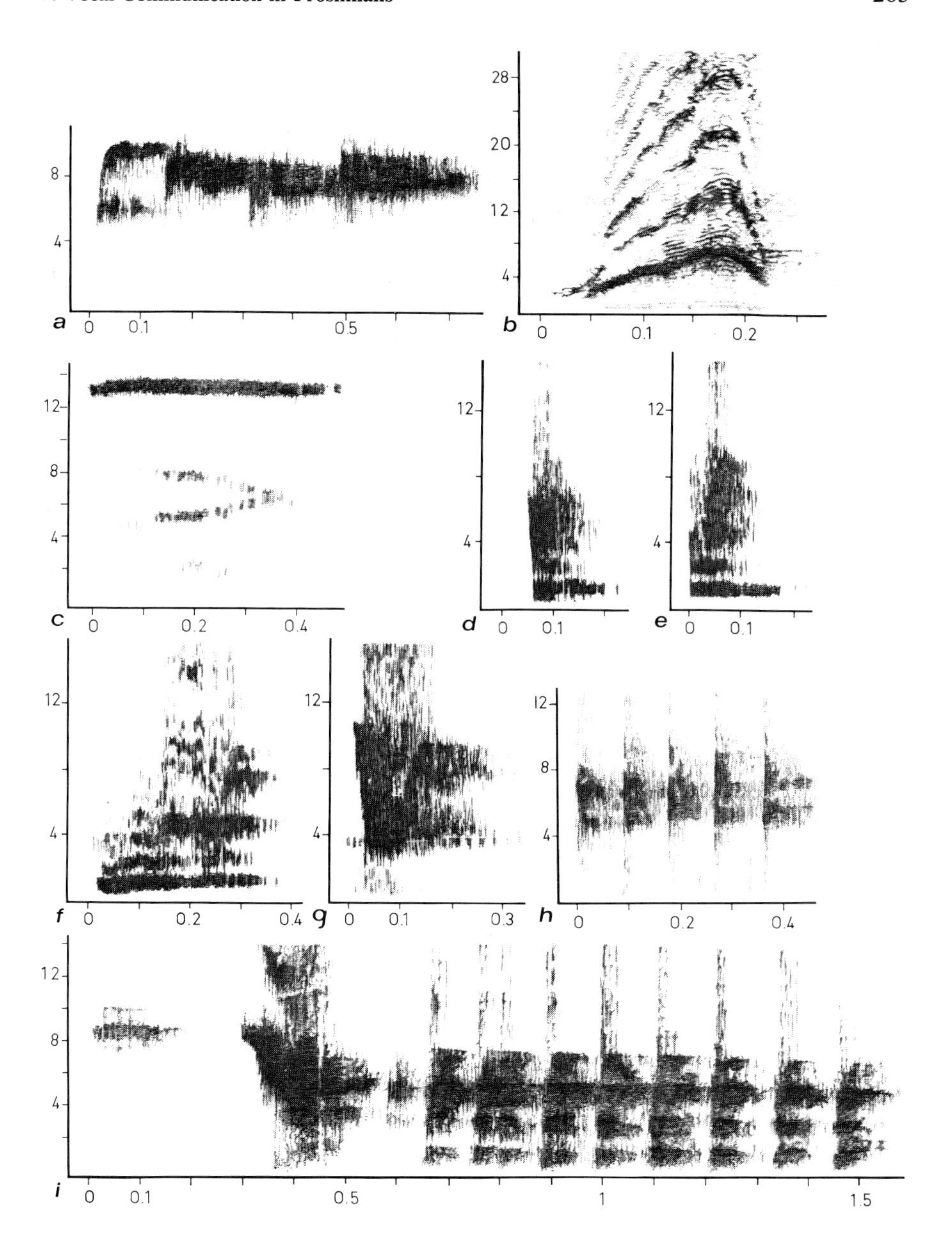

Fig. 4. Calls of *Microcebus coquereli* and *Cheirogaleus major*. (a) Contact-rejection grunts of *M. coquereli*; (b) distress call of *M. coquereli*; (c) prolonged whistle of *C. major*; (d) contact call between adult *M. coquereli* (elicits an identical response); (e) contact call of young *M. coquereli*—"Pui"; (f) distress contact-seeking (double) call of young *M. coquereli*; (g) alarm call of *M. coquereli*; (h) contact-rejection call of *M. coquereli* (during a fight); (i) sexual call of *M. coquereli*.

c. Alarm call (Fig. 4g). The alarm call, which can be given as "ptiak," is fairly similar to the homologous call of *Galago elegantulus*. It can be considered as a variant, with increased intensity, of the "ptiao" contact call exchanged between adult *M. coquereli*. It has been heard coming from an isolated male in captivity, housed next to several other *M. coquereli*. The adjacent animals were immediately put on the alert by the call and looked down toward the ground.

d. Contact-rejection call. Adult *M. coquereli* (Fig. 4a) utter aggressive calls comparable to those of *M. murinus*. During fights (Fig. 4h), these grunting calls succeed one another in a crescendo, becoming more high-pitched.

e. Distress calls. As with *M. murinus*, (Fig. 4b) the distress call is characterized by high-pitched, short calls which are repeated at high intensity, interspersed with weak grunts.

2.2.1.3. *Cheirogaleus major* and *Cheirogaleus medius*. These two species, which are relatively quiet in the forest, have a number of calls which are closely comparable.

a. Contact calls and contact-seeking calls. The only contact calls recorded concern mother-infant interactions. With its mouth closed, the infant utters weak, high-pitched, and plaintive squeaks which immediately attract the attention of the mother and often induce her to respond with an identical call. This squeaking call is also emitted frequently in the course of allogrooming between adults, sometimes developing into contact-rejection vocalizations.

b. Distant communication calls (Fig. 4c). In the course of the evening, both in the wild and in captivity, dwarf lemurs of both sexes sometimes utter drawn-out, high-pitched whistling noises which are barely audible. The animals reply to one another over some distance with this call, and it may well be a territorial signal.

c. Alarm calls. As with the lorisines, the two *Cheirogaleus* species, which are generally very discreet, do not utter any kind of alarm call. When faced with any kind of danger, their first reaction is to freeze or to take refuge in a tree hollow.

d. Contact-rejection calls. When attacked or disturbed in their nest, *Cheirogaleus* utter series of high-pitched, powerful calls which, according to the degree of excitation, may be progressively transformed into a series of grunts. These calls are defensive in nature; the animal will at the same time protect itself by retreating to a more secure position. From time to time, an animal calling in this way will abruptly lunge forward to bite; for example, if one attempts to grasp it or simply if it is disturbed by the movements of a conspecific living in the same tree hollow. These calls are very similar to the homologous vocalizations described for *Microcebus* but they are more powerful, particularly in the case of *Cheirogaleus major*.

e. Distress calls. With both *Cheirogaleus* species, the most easily aroused individuals, when grasped in the hand, will utter a succession of resonant "ouiii-ouiii-ouiii" calls, which are drawn-out and repeated without interruption.

2.2.2. *Phanerinae (Phaner furcifer)*

Before the vocalizations are described, it is worth noting a number of peculiar features of *P. furcifer* which, in contrast to all the other nocturnal lemurs, communicate almost entirely by vocalizing (Petter *et al.*, 1971, Charles-Dominique, 1978). The pattern of social organization follows the classical nocturnal lemur system: females occupy well-defined territories, which may overlap to some extent, while the males occupy separate territories which are superimposed on those of one or more females with which they are associated. An associated male and female will often move around together during the night, with the female leading and the male following. *Phaner furcifer* feed essentially on gums and the sweet secretions exuded by certain homopterans, with a small supplement of insects (Petter *et al.*, 1971), and territorial encounters take place in a small number of restricted areas where there are no gum-producing trees. In contrast to certain other lemur species, in which territorial encounters are expressed by urine-marking, encounters between adjacent *Phaner* are purely vocal. Small groups of neighboring animals regularly assemble in these small encounter zones, all calling at once, after which they will return to their normal ranging behavior. Calling choruses of this kind often include three or four individuals and sometimes as many as ten are involved. However, fights hardly ever occur, even though the animals are only separated by a few meters when calling (Charles-Dominique, 1977).

2.2.2.1. Contact Calls and Contact-Seeking Calls (Fig. 5a). Nothing is known of the behavior of the mother and her infant, which doubtless exchange vocalizations with one another. Similarly, there is no information about sexual behavior and it is possible that in this context, too, there are special vocalizations.

In Madagascar, during the month of May, females, still accompanied by their infants, were heard emitting a call of chevron form from time to time, apparently composed of a relatively acute-angled parabola, which was repeated 4 to 5 times in sequence. It is possible that this vocalization may play a part in communication between mother and infant.

2.2.2.2. Distant Communication Calls. Any kind of excitation in this species produces a vocalization which may vary in intensity and form along a smooth gradient.

a. Calls during movement. As soon as it begins to move, any adult *P. furcifer* regularly emits every 2–5 seconds a mild "hong" sound which permits the male and female to remain in contact when they are moving in association. However, the call is uttered automatically, whether or not the animal is associated with a partner.

b. Duets (Fig. 5c). As a regular occurrence, one of the partners will intensify its "hong" calls so that they become transformed into a series of "tia" calls of increasing intensity. The male and female exchange these "tia" calls in a duet, and this may elicit similar duets from neighboring territories. In the early stages of excitation, the "tia" appears as a relatively flat parabola with numerous harmonics. But as the animal's level of excitation increases, the "tia" calls become more and more intense and are repeated at ever shorter intervals. The parabola becomes more arched and the harmonics disappear (Fig. 5). When the male and female are separated from one another, and one of them utters the "tia" call, the partner will always respond. Hence, it would seem that it is possible for the call to indicate the identity of the calling animal. In the course of territorial confrontations, it is always the "tia" call which is uttered, from time to time accompanied by male calls (see below). When the excitation reaches its climax, the "tia" calls are repeated about once a second, and when such calls are analyzed sonographically they appear as bands of white noise with intensifications of certain frequencies.

c. Male call (Fig. 5b). In the course of the night, and particularly just after waking, males utter a call which is derived from the "tia," but is more low-pitched and powerful. This call is normally uttered as a single unit or as two units, and various males in the vicinity will produce the call in response. Sometimes, when a female utters "tia" calls, the associated male will respond with similar calls and then incorporate a male call which elicits other male calls from neighboring males.

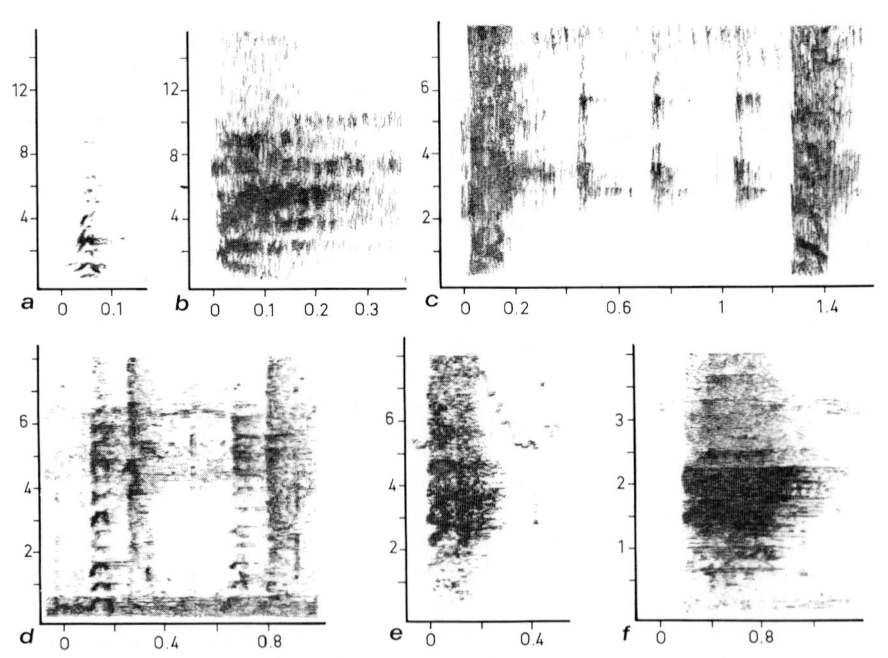

Fig. 5. Calls of *Phaner furcifer* and *Daubentonia*. (a) Contact call of *P. furcifer* (low arousal level); (b) male call of *P. furcifer*; (c) exchange of duets in *P. furcifer*; (d) low-intensity alarm call of *Daubentonia* "ron-tsi"; (e) contact-rejection call of *Daubentonia*; (f) contact call of *Daubentonia*.

Sonographic analysis shows that the male call is represented by the descending part of a parabola with several harmonics. Under experimental conditions, it has been demonstrated that this call plays a territorial role.

2.2.2.3. Alarm Calls. With this lemur species, it would seem that the alarm calls are merely an amplification of contact calls which become more powerful and more distinct.

2.2.2.4. Contact-Rejection Calls. In the course of aggressive confrontations, or when a *Phaner* is trapped in a nest, staccato calls are produced. At a higher level of excitation, these give way to two-phase grunts comparable to those of the Lorisidae and the Cheirogaleidae.

2.2.2.5. Distress Calls. In response to pain or in conjunction with some powerful emotion (e.g., when forcibly seized), *Phaner* emit a rising, plaintive call comparable to the distress calls described for the Lorisidae and the Cheirogaleidae. Conspecifics respond to this call with numerous "tia" calls.

2.3. Daubentoniidae

2.3.1. Daubentonia madagascariensis

Few observations have been conducted on the aye-aye, and the calls previously reported by various authors (Lavanden, 1933; Petter and Peyrieras, 1970a) are difficult to interpret. So far, it has been possible to identify a mother–infant contact call, threat vocalizations, anger call, distress calls, and an alarm signal (Petter and Peyrieras, 1970a), but there are probably other calls which remain to be observed.

2.3.1.1. Contact Calls and Distant Communication Calls (Fig. 5f). In the course of a systematic study of this species conducted on the east coast of Madagascar (Petter and Peyrieras, 1970a), an adult female observed continuously for part of the night was periodically heard to utter a resonant, prolonged "creeee" call, to which a subadult (also observed from time to time) responded with an identical call. This call is, therefore, interpreted as a mother–infant cohesion call, particularly since the two individuals concerned met up from time to time to make contact. Another observation, made at the time of release of captured aye-ayes on the island reserve of Nossi-Mangabe, involved a young animal 1 to 2 months old which was separated from its mother in an adjacent tree. The young animal emitted the "creeee" call every 5 seconds until the mother responded with the same call, at which point it immediately climbed down from the tree and went to rejoin its mother. It is possible that this call also plays a part in vocal communication between adult aye-ayes.

2.3.1.2. Alarm Calls (Fig. 5d). At night in the forest, when an aye-aye is approached or illuminated, it will often respond by facing toward the human observer and uttering repeated grunting sounds, which can be represented as a series of "ron-tsi" calls. This call is not particularly resonant, but it can easily be distinguished from other sounds in the forest. When the animal is only mildly alarmed, the call is quite weak, but it accelerates and becomes more powerful as excitation increases. If excitation exceeds a certain level, however, the aye-aye will then take flight. A young, tame aye-aye which was observed during its nocturnal activity responded to disturbances with expirations preceding the "ron-tsi" calls, which were sometimed emitted for more than 15 minutes. There was also some fluctuation in these calls before they came to an abrupt halt, as the animal changed its activity. In some instances, a very mild surprise, such as a whistle or the sudden appearance of its own shadow, was sufficient for the animal to produce this grunting call.

2.3.1.3. Contact-Rejection Calls (Fig. 5e). When an aye-aye is disturbed in its nest during the day, it will abruptly thrust its head out, as if with the intention to bite, and emit a powerful hissing noise. It seems likely that this abrupt, noisy appearance of the head would deter a predator, and the surprise element could be quite dangerous for an animal clinging close to the nest several meters above the ground. If the disturbance persists, the aye-aye may repeat this hissing sound two or three times, accompanying each vocalization with attacks, which can be followed by high-pitched calls. Emission of this hissing sound is fairly frequent with captive animals which are disturbed while sleeping. A tame young female which was observed by weak light at night would respond to a slight movement of the observer by abruptly stopping its exploratory or play activities and advancing slowly and deliberately toward the source of the disturbance while hissing noisily. When the female was taken unawares, a very slight noise was sufficient to make her jump. She would then abruptly whirl around to face the source of the noise, hissing loudly with the long hairs on the head and tail erect. At the age of 5 months this female aye-aye frequently played with a kitten, and the two animals would roll on the ground nibbling each other. But the aye-aye occasionally interrupted the game quite abruptly, chasing away her partner with her hair erect and hissing noisily.

2.3.1.4. Distress Calls. Animals which were disturbed on the nest in the course of capture sometimes uttered, following aggressive actions, high-pitched distress calls comparable to those of *Cheirogaleus*.

2.4. Lepilemuridae

2.4.1. Lepilemur sp.

The resonant calls emitted by *Lepilemur* vary from species to species. These differences are important, since there is little morphological difference and separation of the species is based heavily on cytogenetic distinctions. It has been possible to observe them on a number of occasions during field visits (Petter, 1962; Charles-Dominique and Hladik, 1971). Some of their vocalizations are discreet, while others are very noisy. Localized high densities of these animals are always associated with a high frequency of vocalizations. Calls are uttered as soon as the sun sets, at which time they may be very intense in association with the first movements around the home range. Thereafter, the vocalizations are very irregular in occurrence and depend upon encounters between individuals or sporadic disturbances. At the beginning and the end of the night, the light intensity is greatest and the animals can observe one another more easily, which leads to reciprocal excitation. The same applies in the course of the night when the moon rises.

Most of the calls which were noted in *Lepilemur edwardsi* in the Ankarafantsika, in *L. dorsalis* in the region of Nossi Bé, and in *L. leucopus* in the south, were found to be extremely variable in intensity and in tonal character. There is a range of vocalizations extending from weak squeaks to powerful, high-pitched calls. Considerable experience and numerous recordings in the forest were necessary for recognition of the great variety of intermediates and variants of a given call. On the other hand, the vocalizations of the other forms of *Lepilemur* in the west of Madagascar generally seemed to be less noisy and less variable. Despite numerous nights of observations, no calls have ever been heard in the *Lepilemur* species of the east coast forests (*L. mustelinus* and *L. microdon*).

In various *Lepilemur* species, mother–infant contact calls have been noted, and extremely variable call exchanges have been heard, which may be provisionally classified as distant communication calls and alarm calls. In fact, the graded calls reflect various levels of excitation; their function is comparable to that of the calls uttered by *Phaner*.

2.4.1.1. Contact Calls and Contact-Seeking Calls (Fig. 6a). From birth onward, the mother leaves her infant clinging to a branch while she goes off to feed. From time to time, she will transport the infant in her mouth from one place to another. As soon as the offspring is capable of independent movement it will follow the mother, and the two animals will remain in contact throughout the night (Petter, 1962). In *L. edwardsi*, the maternal call is a discreet "tchen," while in *L. dorsalis* and *L. leucopus* it sounds like a kiss. The infant call is a long, plaintive "on."

2.4.1.2. Distant Communication Calls and Alarm Calls. Several different vocalizations can be distinguished for the various *Lepilemur* species, in each case as a gradient reflecting different levels of excitation. It is in this category that the greatest variation between *Lepilemur* species is evident. With *L. leucopus*, which inhabit southern Madagascar, studies to date have permitted fairly complete interpretation of their vocalizations. This species utter a series of harsh "hein" calls in relatively rapid succession, usually followed by a more high-pitched "hee," which may sometimes be repeated. In some cases, however, the "hee" calls are omitted and the ''hein'' components may then become more drawn-out and widely spaced. Similarly, the ''hee'' components may be uttered in isolation or in regular sequences. When these calls are uttered, two animals are always facing each other over a distance of 3–10 m. The ''hein-hein-hein-hein'' sequences are uttered by both animals simultaneously, and each responds to the other's vocalizations. For example, one animal will begin a ''hein-hein-hein . . .'' sequence which is soon interrupted by a ''hee'' call from the neighboring animal. Subsequently, several ''hee'' calls can be exchanged by the two animals, which will then gradually become less excited while maintaining their mutual surveillance.

When the animals are excited, one call can elicit another from a considerable distance away, especially at the beginning of the night. The calling animals are usually males which face one another across territorial boundaries, and these vocalizations are apparently utilized for territory definition. More rarely, the calls may be uttered by adult females facing one another across a territorial boundary. On two occasions a male was seen directing a series of ''hein-hein'' calls at a female moving around her territory. On each occasion, the female replied with a ''hee'' before continuing on her way and the male paid no further attention.

Charles-Dominique and Hladik (1971) have interpreted the ''hein-hein-hein'' calls as aggressive in character and the ''hee'' calls as individual territorial signals. There are recognizable individual variations in these calls.

Lepilemur leucopus have also been observed uttering regular series of ''hee'' calls at higher levels of excitation.

With the other *Lepilemur* species, similar calls are uttered in these encounters. *Lepilemur edwardsi* (eastern Madagascar) utter trembling ''oooai'' (Fig. 6d) components followed by a rapid sequence of ''oui oui oui. . .'' calls. The calls of *L. dorsalis* (northeastern Madagascar) and *L. septentrionalis* (Fig. 6b) (northern Madagascar) are comparable, though there are recognizable specific characteristics. With *L. ruficaudatus* (western Madagascar), on the other hand, the homologous calls uttered during encounters are quite different: ''boako-boako'' (Fig. 6c). As with various other species, these particular calls are often preceded by grunts. However, there does not seem to be the usual second type of call to follow the ''boako-boako'' components.

2.4.1.3. Contact-Rejection Calls. In captivity, if individual *Lepilemur* are approached by a hand or a stick, they will strike out with their hands, attempt to bite, and utter a series of resonant hissing calls, often followed by a two-phase vocalization. It has been observed during interspecific fights that high-pitched calls of the same intensity accompany blows with the hands and biting. Under natural conditions, these calls have also been heard on several occasions when two individuals have been in close proximity to one another.

2.4.1.4. Distress Calls. Distress calls, very similar to those uttered by other lemur species, were heard during capture of *Lepilemur leucopus* for marking and

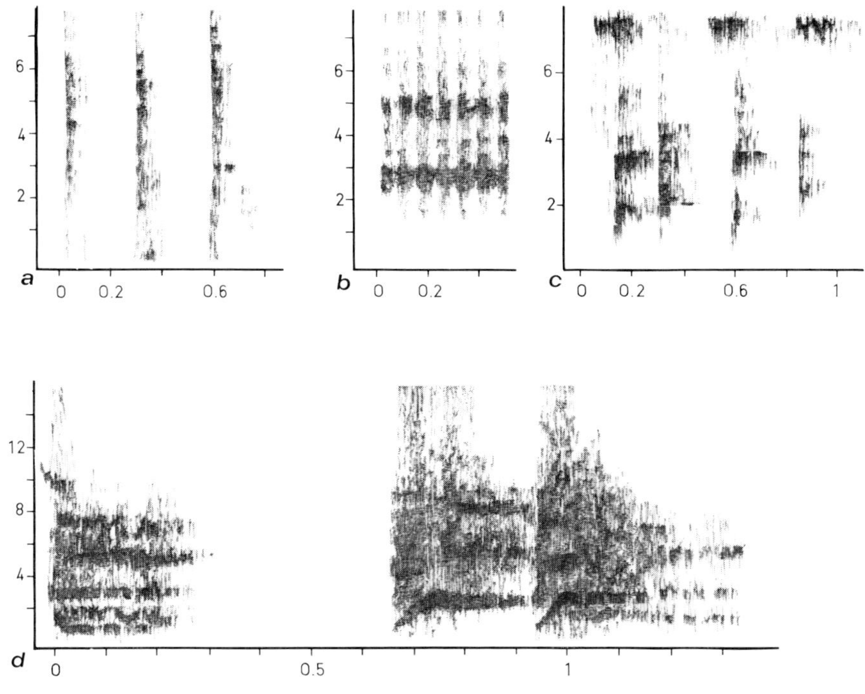

Fig. 6. Calls of *Lepilemur*. (a) Contact call of mother *L. edwardsi*; (b) contact call of *L. septentrionalis*; (c) contact call of *L. ruficaudatus*—"Boako-boako"; (d) contact call of *L. edwardsi*—"O-o-o-o-ai" followed by "oui-ou."

release (Charles-Dominique and Hladik, 1971). However, no such calls have been uttered by other *Lepilemur* species during capture.

2.5. Indriidae

2.5.1. Avahi laniger

The avahi is an extremely discreet lemur species in terms of its general behavior. It is the only nocturnal representative of a family which otherwise contains the most strictly diurnal of the lemur species. It is fairly difficult to observe this species under natural conditions. During the daytime, it sleeps rolled up in a ball in a clump of foliage, often in groups of two or three. While active, *Avahi* move around by leaping from one vertical trunk to another, making very little noise in the foliage, and they are usually difficult to spot with a headlamp. In most cases, one finds a pair of adults (probably a male and a female) and the young from the previous birth season.

2.5.1.1. Contact Calls and Contact-Seeking Calls. This group of vocalizations essentially consists of infant calls, plaintive whistlelike noises, which attract the mother's attention.

2.5.1.2. Distant Communication Calls (Fig. 7c and d). These calls are modulated, prolonged, and very high-pitched whistles, which are heard quite frequently during the night. A whistle uttered by one individual is usually followed by a similar call from another individual a short distance away. In some cases, there may even be a call from a third individual and there can be partial overlap between the vocalizations of the different animals involved.

These calls can be provisionally interpreted as territorial signals. They may perhaps be a high-frequency equivalent of the long, wailing calls of *Indri*.

2.5.1.3. Alarm Calls. When an avahi is mildly disturbed it may utter faint, discreet grunting sounds followed by a weak snorting sound which may be transformed into a cooing call (Fig. 7e). If the animal's excitation is increased (Fig. 7f), this call may progress through a whole series of intermediates to become a louder, trembling call in which the pitch rises very rapidly to terminate on a powerful high-pitched note. It is this call which is imitated in the local Malagasy names of "Ava Hy" (Fig. 7g) or "Ampongy."

When an avahi is disturbed in the course of its nocturnal activity in the forest, it becomes immobile and utters snorting and cooing sounds before fleeing from one trunk to another. Sometimes, the presence of an avahi is indicated merely by a sudden, high-pitched call which evokes a similar call from another individual in nearby foliage. This call can thus be interpreted as a powerful alarm vocalization, but it also serves as a cohesion call which permits the animals to remain in vocal contact and to locate one another even if they are separated by a distance of 50 m or more in the foliage. This particular call may in fact be equivalent to the "honk" emitted by Indri when they are dispersed following an alarm.

2.5.1.4. Contact-Rejection Calls. When an attempt is made to grasp an avahi, it will usually utter aggressive grunting sounds, while attempting to bite. If the animal's excitation is increased, once it has been grasped, the initial call becomes transformed into a resonant snorting sound which reaches quite a high pitch and which is in fact an intermediate between the usual snorting sound and the typical alarm call.

2.5.1.5. Distress Calls. No distress call has yet been recorded for this species.

2.5.2. Propithecus sp.

Propithecus species (sifakas) live in small groups. In some areas where the forest is relatively untouched, groups of three seem to be the most common, whereas in other areas larger groups of up to eight individuals have been seen (Jolly, 1966a;

Richard, 1973; Petter, 1962). These animals are most active during full daylight, and their vocalizations seem to be markedly more varied than those of *Avahi*. However, this observation may simply be a reflection of the fact that the diurnal sifakas are easier to observe and therefore better known.

2.5.2.1. Contact Calls and Contact-Seeking Calls. ***a. Infant call (Fig. 7i).*** In order to attract the attention of its mother, the infant utters a weak, high-pitched call which is rather like a fainter version of the contact-rejection call of the mouse lemur. On hearing this call, the mother will immediately lick her infant to calm it. The infant may also emit a low-pitched, grunting call and a "purring" sound similar to that of *Avahi*.

2.5.2.2. Distant Communication Calls. ***a. Cohesion calls during rapid movement (Fig. 7m and n).*** On each landing, during rapid locomotion from trunk to trunk, sifakas will often produce a short "hon" sound. In many cases, this call has been heard even after sifakas have leaped away out of sight. It seems very likely that this vocal signal is important in maintaining cohesion between group members when they are in dense vegetation.

b. Barking calls (Fig. 7l). *Propithecus verreauxi* sometimes utter a very resonant call, composed of a series of guttural, very noisy barks: "roa-roa-roa-roa." The number of barks in a series varies considerably, but the sequence is always uttered without interruption by all members of the group. These calls, which are very characteristic and can be heard from some distance away, permit fairly accurate localization of sifaka groups in the forest. They may be uttered at any time during the day, but they are most common early in the morning and at the beginning to middle of the afternoon. Their frequency varies according to the season, and according to the sensitivity of individual animals, and there may be some hormonal determination of calling frequency. When this call is produced, all the animals sit with their mouths open and their bodies erect, looking at the sky. In many cases, calls from the members of one group will evoke similar calls from neighboring groups. The main role of these calls is probably that of distant communication between groups. Indeed, they are often associated with territorial confrontations. In this respect, they resemble analogous vocalizations produced by various diurnal simian primates which live in small to medium-sized groups (e.g., the howls of gibbons and the raucous calls of the howler monkeys). The sifakas may also utter this call when a bird of prey is sighted, and in such cases they doubtless have an intimidating function. With *P. diadema*, barks seem to be far less common than with *P. verreauxi*, and it therefore seems unlikely that they could play a part in localization. Distant communication may also be subserved by some alarm calls (see Section 2.6.2.3 below).

c. "Vouiff" call. If a group of *P. diadema* is suddenly surprised in the forest, and the group members are separated from one another, resonant "vouiff" calls may be uttered for some time, at intervals of about 10 seconds, interspersed with ''shim-poon'' grunts and cooing noises. All of the members of the group join in the vocalization. These calls are probably homologous with the high-pitched calls of *Avahi*, but there does not seem to be an equivalent call in the repertoire of *P. verreauxi*. The ''vouiff'' calls of *P. diadema* seem to play a dual role as alarm calls and as cohesion calls. As with *Avahi*, it would seem that a group taken by surprise generally attempts to disperse, each animal fleeing in a different direction. (Such behavior has not been observed with *P. verreauxi*.) A cohesion call of this kind is probably necessary for the animals to be able to regroup in the dense vegetation typical of the east coast rainforest.

2.5.2.3. Alarm Calls. ***a. Mild alarm.*** When *P. verreauxi* are resting during the daytime, a mild disturbance, or the sight of a predator some distance away, may elicit a weak snorting noise. This low-pitched, drawn-out ''aaar'' call is usually uttered first by one individual and then the other members of the group may follow suit. Production of this call places all members of the group on the alert, and in some cases it may be uttered in the course of slow locomotion. If the level of excitation increases, the snorting sound becomes more pronounced and may be accompanied by rapid grunting sounds. In such cases, the vocalization will often be transformed into a kind of ''rrroouou'' call, just like the end of the cooing sequence of the domestic pigeon. This vocal signal also acts as a cohesive factor within the group. This is clearly seen when one or two individuals are (unusually) separated from the group, for example, when group progression is interrupted by a human observer on a pathway through the forest. Following some preliminary grunting calls (mild alarm), the members of the group keep a watchful eye on the observer and then attempt to join up by some roundabout route. In so doing, they may occasionally lose sight of one another, but they will periodically exchange snorting calls. In one instance, a female with her infant, after having been isolated in this way from two other adults, periodically responded to the snorting calls of the latter with ''heuh'' sounds.

In captivity, this kind of snorting-cooing call is uttered by *P.v. verreauxi*, *P.v. coquereli*, and *P.v. deckeni* on hearing the vocalizations of a bird of prey (or an imitation thereof). Thus, this call is probably the homolog of the snorting vocalization of *Avahi laniger*.

Under similar conditions, both *P.d. diadema* and *P.d. perrieri* utter a kind of weak grunting call from time to time. This may be represented as ''roo-han'' (fig. 7o), with the second part of the call uttered at a much lower pitch. When these sifakas are active, they exchange calls of this kind very frequently.

b. Moderate alarm. The most typical alarm call uttered by *P. verreauxi* in response to some marked disturbance is a series of ''omb-tsi'' sounds produced at approximately 1-second intervals without interruption. The ''omb'' component is rather like a hiccup, while the very short ''tsi'' component occurs as an inspiration. This is the call (Fig. 7k) which has given rise to the Malagasy name ''sifaka'' (pronounced: ''sheefac''). The call is actually quite variable; sometimes the ''omb'' component is dominant, but sometimes it is scarcely audible. The rhythm may also vary according to the circumstances. When a group of sifakas is disturbed, this grunting vocalization is uttered by all group members, but there is no synchronization of the individual vocalizations. If the source of disturbance persists, the animals in a group may continue uttering these calls for as long as 2 hours. The call is usually initiated by a single individual and the other members of the group will immediately follow suit even if they cannot see the source of the disturbance. Sometimes, the call is sparked off merely by a falling branch, and the vocalizations then rapidly come to an end. In the subspecies *P. verreauxi* in the south of Madagascar, this grunting call seems to be noticeably more resonant and clear-cut. The Malagasy name "sifaka" would appear to fit this subspecies better than the others.

Propithecus d. diadema utter a similar call, but it is duller and is quite well represented by the Malagasy name "simpona" (pronounced: "shimpoon," with the accent on the first syllable). The homologous call of *P.d. perrieri* may be represented as "on-ee." This particular grunting call seems to be primarily associated with moderate alarm situations where the disturbance concerned is not too sudden in its appearance. If the disturbance is more marked, the initial part of the grunt may be replaced by a resonant cooing sound, which turns the call into something closer to "rourourou-tsi." Another variant of the call occurs in the form of sequences of the first component ("omb") repeated at a more rapid rate, approximately twice a second. This latter signal, although it is less intense, is very similar to the characteristic grunting calls of *Lemur macaco*. As a rule, it is only uttered by one animal in a group when it is more excited than the others, sometimes on approaching the human observer to take a closer look. It is interesting to note that there does not appear to be an homologous call in *Avahi laniger*.

c. Pronounced alarm (barking calls). The barking sequences discussed under the heading of distant communication calls may also play an important part in alarm situations and in intimidation, as numerous observations indicate. In fact, barks are most commonly uttered in the presence of a bird of prey (e.g., *Gymogenys radiatus* or *Milvus migrans parasiticus*).

At the Zoological Gardens in Tananarive the sifakas kept in captivity are frequently heard to emit barking calls when a kite is flying overhead or simply when a small heron flies too close to their cage. It is also quite easy to evoke this call deliberately by imitating the long, whistling vocalizations of birds of prey. (Indeed, the other large lemur species in adjacent cages also respond with their own particular alarm calls in almost perfect synchrony.) On occasions, this alarm call also seems to be associated with the arrival of a human being or a dog, when the sifakas are taken by surprise or when they are already somewhat excited. In general, it seems to correspond to a relatively high level of excitation. Richard (1973) has heard this alarm call in a group of sifakas when heavy lorries passed along a road 3 km from the forest, and on one occasion while a snake (*Ichthycyphus miniatus*) was seen eating an adult mouse lemur nearby.

2.5.2.4. Contact-Rejection Calls (Fig. 7j). Sifakas live in close-knit social groups and contact-rejection between individuals can be observed when a female is not receptive during the mating season, when a mother is protecting her young infant from the curiosity of other members of the group, and during play (e.g., when a young animal annoys an adult). The contact-rejection call heard in these situations is a series of angry grunts ("greh-greh-greh") uttered with the mouth open and accompanied by an aggressive posture or even by blows delivered with the hands.

A male *P.v. coronatus* in captivity was heard emitting a more intense anger call, a rapid, resonant "chah-chah-chah" while chasing a young male which had bitten its hands as a solicitation to play.

2.5.2.5. Distress Calls. No distress call has been noticed during the capture and handling of sifakas, though the strident call emitted by the infant when one attempts to remove it from the mother's fur may be placed in this category.

2.5.3. Indri indri

The indri is the largest living representative of the family Indriidae. Although it is active during the day, it is difficult to observe and it is primarily the loud, harmonious calls which attract attention. *Indri indri* live in family groups usually composed of two or three members, but sometimes containing up to six individuals (Petter, 1962; Petter and Peyrieras, 1974; Pollock, 1975.)

2.5.3.1. Mother–Infant Communication. The infant is initially carried on the mother's belly and then later on her back. As with other indriid species, there are doubtless mother–infant contact calls, but none have been recorded as yet.

2.5.3.2. Distant Communication Calls. ***a. Group cohesion calls.*** Like the sifakas, weak grunts accompany the leaps of indri and permit the animals to remain in contact during locomotion. However, these grunts are difficult to hear, as the animals disappear rapidly into the forest when fleeing and in any case produce quite a lot of noise in leaping.

b. Modulated, plaintive calls (Fig. 7a and b). The most typical and well-known call of *Indri* is a very resonant, plaintive howl, by which their presence is indicated some distance away in the forest. This call can be divided into two sections. It is usually initiated by three or four loud barks which are emitted simultaneously by several members of a group. The individual voices then become distinct, and there is a decrease in intensity before the howls reach their full amplitude. One at a time, or sometimes in pairs, the members of the group then produce a long, modulated howl which differs in character from individual to individual. The pitch rises and then falls away again, though sometimes the pitch may remain constant for some time and then fall somewhat abruptly toward the end of the call. The frequency range covered by these howling vocalizations is 0.5–4.0 kilocycles per second.

These howls have sometimes been likened to the howling of dogs, but they are more powerful and more distinctly modulated. After some time has elapsed, often more than a minute, the howls emitted by an indri group progressively die away. Usually, after a short period of silence, another group on a nearby ridge will produce the same calls, and this is in turn followed by vocalizations from other groups further and further away in the forest. "Concerts" of this kind may last up to 20 minutes. In the east-coast rain forest zones where *Indri* occur, never a day will pass without these calls being heard somewhere in the forest. However, it is frequently difficult to localize the calling groups. In fact, the howling call does not seem to have a clear-cut origin because the animals often turn their heads in a variety of directions while calling.

A protocol made in December in the Rantabe Valley (northeast Madagascar) indicates that the calling of a particular indri population was most frequent when the sky was clear in the afternoon. This is shown in the following tabulation:

Time (hours)	Sky conditions	Shade temperature (°C)	Vocalizations
Morning	Sunny	32	Sporadic calls
1100	Clouds and light rain	31	No calls
1200	Clouds and light rain	31	No calls
1300	Heavy rainstorm	—	No calls
1400	End of storm; sky temporarily clear	24	Numerous calls from all directions, approximately every 5 minutes
1520–1540	Sky cloudy	27	No calls
1550	Sky black	27	No calls
1555	Sky clears and the sun appears	28	More calls from various directions, though less numerous than at 1400 hours

According to Pollock's observations (1975), *Indri* calls are most numerous in March. Local wood fellers often claimed that *Indri* would call as a result of horsefly bites. This simple interpretation doubtless has some foundation, in that excessive biting from these insects may raise the level of excitation of the indri and thus eventually elicit calling. In the months of October, November, and December, *Indri* were often observed with binoculars while plagued by horseflies and they were seen to make sweeping motions with their hands and feet. (Although the observers themselves were only a dozen meters away, they were never attacked by the horseflies, which only attempted to settle on black surfaces.)

The long, howling calls can be interpreted as a means of vocal advertisement of territory. There is also the possibility suggested by Pollock (1975) that the calls transmit information about group composition. The actual carriage of the calls through the forest is difficult to define precisely, since both the pitch and the intensity vary greatly. Variation in frequency of the call is doubtless an adaptation for better propagation through the forest. In this connection, it is interesting to note that other primate species living in dense forest, such as gibbons, have comparable vocalizations and that even certain Malagasy bird species, (such as *Leptosomus discolor* and some birds of prey) produce modulated calls at the tops of trees or while gliding. Indeed, the local Betsimisaraka wood fellers communicate with one another with modulated chants, often over considerable distances.

An attempt was made to evaluate the carriage of *Indri* calls using two different methods. The first, direct, method involved the cooperation of several observers located at different distances from a localized, calling indri group. The second, indirect, method was based on tape recordings of the calls, which were compared with others conducted under identical conditions where the distance of the calling animals had been accurately measured.

Under conditions where the population density seemed to be representative, all of the indri producing the howling vocalizations were located at the tops of trees on a ridge in the forest. It seems likely that optimal conditions for both production and reception of the calls prevail in such positions. However, in situations where the population density was judged to be abnormally high, *Indri* were sometimes observed calling from valleys. In one case, three groups were heard calling from three different valleys around a particular ridge.

In dense forest, the howling call, under good weather conditions, can usually be heard over a distance of more than a kilometer by an observer on the forest floor. However, when the sound is crossing a valley it probably carries much further (2 km or more) while calls actually produced in a valley probably do not carry further than 500 m.

Indri have much better developed ear pinnae than the other members of the family Indriidae. The auditory acuity of *Indri* is probably better than that of the human ear, and it is probably sufficient to permit distant reception of calls uttered by other groups, despite the wind, rain, and screening effect of the undulating landscape. According to Pollock (1975), *Indri* can hear each other's calls at least over a distance of 3 km, since an observer situated midway between two groups separated by that distance could hear one group responding to the calls of another. His calculations indicated that in areas of population density one group of *Indri* can be potentially in vocal contact with a hundred other groups.

Fig. 7. Calls of the Indriidae. (a and b) Contact call of *Indri indri*; note in (a) the modulation of intensity (variation in darkness in the sonogram); in (b) the ordinate scale has been increased to show the variation in frequency; (c and d) contact call of *Avahi laniger* (variation of modulation in intensity and frequency is shown in the same way as for *I. indri*); (e) low-intensity alarm call of *Avahi laniger*; (f) medium-intensity alarm call of *Avahi laniger*; (g) high-intensity alarm call of *Avahi laniger*—"Ava Hy"; (h) alarm (contact-restoration) call of *Indri indri* after dispersal of the group (series of barks); (i) distress call of a young *Propithecus verreauxi*; (j) contact-rejection (snorting) call of a young *Propithecus verreauxi*; (k) low-intensity alarm call of *Propithecus verreauxi coquereli*—"Shee-fac"; (l) contact calls (barks) of *P.v. coquereli*; (m and n) group cohesion call emitted during progression of *Propithecus diadema perrieri*—"ou-arr"; (o) low-intensity alarm call of *P.d. perrieri*—"rooo-han."

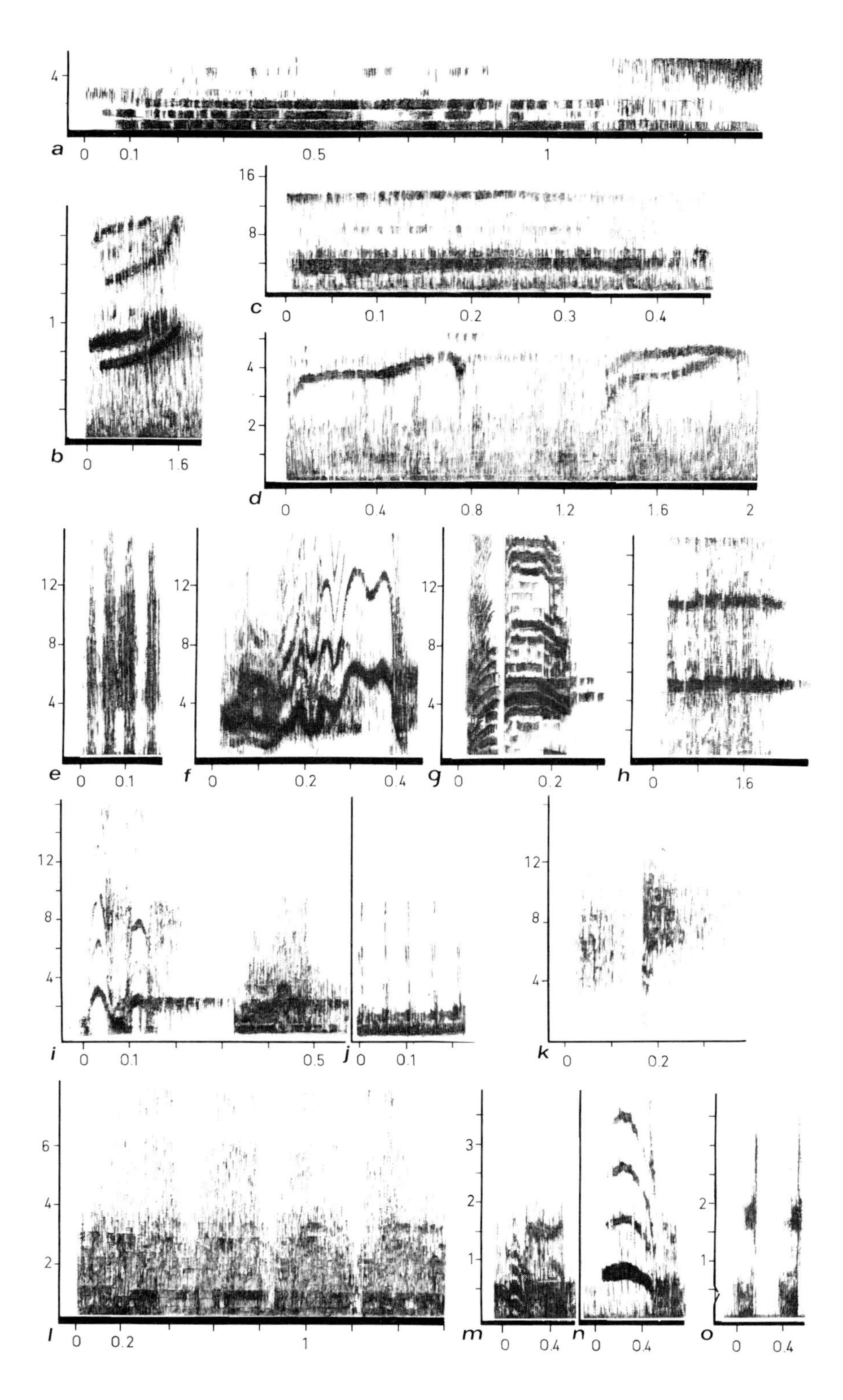
a
b
c
d
e
f
g
h
i
j
k
l
m
n
o

2.5.3.2. Alarm Calls. When an *Indri* group is disturbed in the forest, the response observed varies according to circumstances (time of day, climatic conditions, season, and accidental features such as additional disturbance by horseflies). In some cases, if the indri are resting, they merely look at the source of disturbance without reacting, but on other occasions they show signs of excitation and may utter a number of weak grunts in rapid succession. If excitation is even more pronounced, they will raise their heads and produce a kind of rapid blowing, snorting sound. If the source of disturbance persists, they will eventually take flight and disappear into the vegetation with a few powerful leaps. The barking calls at the beginning of the modulated plaintive call may correspond to an alarm call for response to aerial predators (Pollock, 1975). As with the call of *Varecia (Lemur variegatus)*, these barks may act to deter predators.

If a group of indri is suddenly surprised, for example, if an observer arrives while the group is on the ground, the animals flee rapidly and disperse, as is the case with *Propithecus diadema* and *Avahi laniger*. If an indri is greatly disturbed in this way and separated from the rest of the group, it may emit a very resonant "honk" (Fig. 7h), with the body held rigid and the head raised. As a rule, another member of the group will respond with an identical call. If no such response is forthcoming, the animal may allow itself to be approached to some extent, particularly if it is an adult, but it calls roughly once a minute while staring fixedly at the source of the disturbance. When one is close to the animal concerned, this call appears to be very powerful, but it does not seem to carry through the forest as well as the modulated howls. If a recording of this call is played to an individual isolated from its group, a response is elicited. According to Pollock (1975), this call represents a warning against terrestrial predators. It is uttered by the adult male and occasionally by the oldest subadult male in a group.

The barking calls which precede the modulated, plaintive howl are not necessarily linked to the latter call. They are sometimes uttered in isolation, in which case they occur as an alarm/warning call uttered by all group members. In this form, the barking call apparently represents a warning against aerial predators (Pollock, 1975).

2.5.3.5. Contact-Rejection Calls. Pollock (1975) observed aggressive behavior characterized by approaches and attempts to bite, accompanied by blows with the hands in *Indri* groups. Such behavior was often accompanied by vocalizations. In order of increasing intensity, Pollock distinguished grunts, kisslike sounds, and wheezes.

2.6. Lemuridae

Like *Avahi* among the Indriidae, *Hapalemur griseus* and *Lemur mongoz* provide a transition, among the Lemuridae, from nocturnal to diurnal life. In fact, the

Lemuridae are generally much less specialized in terms of their activity rhythms than are the Indriidae. Various species which are usually diurnal may remain very active in twilight conditions or even exhibit activity during the night. And it is among the Lemuridae, where the most complex patterns of social organization are found, that the maximum degree of variety in vocal communication is present.

2.6.1. *Hapalemur griseus*

This species, very discreet in their behavior, live in small groups and only occur in areas where either bamboos or reeds are abundant. Being one of the smallest representatives of the Lemuridae, they are fairly vulnerable to predation and many of their behavioral features seem to be associated with this (Petter and Peyrieras, 1970b). The vocalizations of *H. griseus*, although they are numerous and varied, are generally relatively discreet and adapted to a predominantly nocturnal existence in dense vegetation.

2.6.1.1. Mother–Infant Contact Calls and Contact-Seeking Calls (Fig. 8g, h and i). When licked by its mother or stroked, the young *H. griseus* emits a "purring" sound comparable to that of the infant *Microcebus murinus*. "Purring" of this kind has also been observed in tame adults as a response to stroking.

In contrast to species of the genus *Lemur*, the young *H. griseus* is not carried around by the mother continuously from birth onward. The mother may carry her infant in her mouth, but usually deposits it in a tangle of foliage on a branch, to which it is able to cling quite easily from birth onward. In the absence of the mother, the infant generally remains motionless and silent. However, when the mother returns the infant emits a high-pitched, trembling "oeeeee," to which the mother responds with short "hon" sounds and low-pitched grunts. This behavior is comparable to that of many nocturnal prosimian species which exhibit "baby-parking."

2.6.1.2. Distant Communication Calls. Among the multitude of vocal signals exchanged by *Hapalemur griseus*, it is difficult to define the exact functions of individual calls, especially since some may exhibit gradations from a discreet, simple form to a more complex one.

a. Cohesion calls. When a group of *H. griseus* is active, weak grunting calls of brief duration ("hon") (Fig. 8c) are commonly uttered continually by all individuals. In this way, the group members remain constantly in contact, despite the density of the vegetation and the low light intensity. In some cases, these grunts may be very variable in strength, rhythm, and pitch. They may thus grade into bleating or plaintive vocalizations. The latter type of call, which evoked similar calls in response, is uttered particularly from time to time when two animals meet up to groom one another.

b. Distant contact calls. In addition to the weak contact grunts discussed above, *H. griseus* will occasionally utter stronger "coooee" sounds, which are prolonged and occur as single calls. Such calls are probably involved in communication between animals some distance apart. When producing this call, an individual usually keeps still and gazes blankly ahead. Another individual will often respond with an identical call or with a simple grunt.

c. Mating calls (Fig. 8a and b). During periods of high sexual arousal, both male and female *Hapalemur griseus* utter a very characteristic call. It begins with a weak, high-pitched and trembling component which gives way to a rapid sequence of low-frequency, resonant sounds.

2.6.1.3. Alarm Calls (Fig. 8d, e, and f.). Signaling of mild alarm always begins with weak, rapidly uttered grunts which become transformed into a very characteristic double grunt, which can be represented as ''co-dot.'' All the individuals in a group utter this alarm call while staring intently at the intruder. The intensity of the alarm grunts varies according to the degree of the disturbance. At moderate levels of alarm, the grunts may become very resonant and interspersed with another call, ''rou-fou.'' If the intensity of excitation rises still further, powerful ''co-dot'' grunts are followed by ''co-ouiiiiiiiii'' calls which are sometimes very prolonged and are usually uttered in unison by several individuals at once. This call is quite similar to the alarm ''crouiii'' of *Lemur* species.

2.6.1.4. Contact-Rejection Calls. With *Hapalemur griseus*, aggressive excitation is often displayed by tooth-grinding, which is very audible with captive animals. Sounds of this kind, accompanied by feinted attacks, are numerous at the time of birth of the infant. At such times, the mother becomes very aggressive and will not permit anyone to approach.

When approached by a strange animal (e.g., a tenrec or a cat), *Hapalemur griseus* will initially display numerous signs of alarm and may then carry out a mock attack or a rapid biting attack accompanied by an aggressive call. This is a single, abrupt ''co-dot'' sound which is far more intense than the usual ''co-dot'' alarm call. If the animal feels itself to be more seriously threatened, this anger call may be transformed into a short, very powerful ''creee'' call. This transformation may be regarded as a progressive shift from an alarm call to an intimidation call.

Male *H. griseus* often display aggression by exhibiting marking behavior with the tail, accompanied by high-pitched grunting sounds rather like those of *Lemur catta*.

Fig. 8. Calls of *Hapalemur*. (a and b) Sexual call of the ♀ *H. griseus* [(b) end of call]; (c) contact-maintenance calls of *H. griseus*—''hon''; (d) medium-intensity alarm call of *H. griseus*—''Rou-fou''; (e) mild distress call of an adult *H. griseus* following loss of contact with the group—''croee'' followed by an alarm call ''co-dot''; (f) high-intensity alarm call of *H. griseus*—''creuuu''; (g) contact call of a young *H. griseus* on seeing mother from which it has been separated, followed by an ''arr'' emitted by the mother; (h) distress call of a young *H. griseus*; (i) answer by an adult ♂ *H. griseus* superimposed on the call of a young *H. griseus*; (j) mild-intensity alarm call of *H. griseus*; (k) contact call of *H. simus* homologous with the ''bleating'' of *H. griseus*; (l) alarm call of *H. simus*.

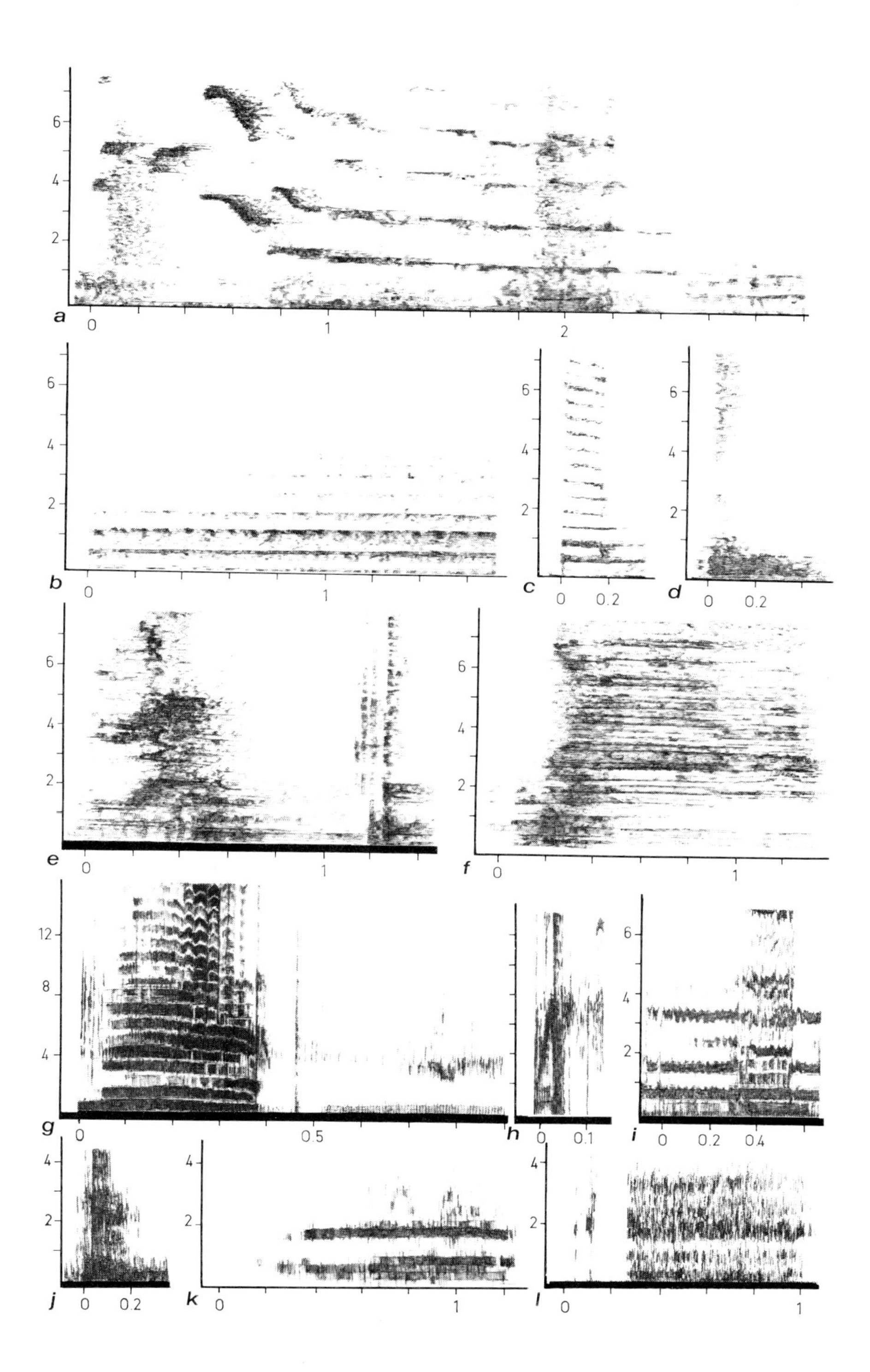
a
b
c
d
e
f
g
h
i
j
k
l
0
0.5
1
2
0.1
0.2
0.4
4
6
8
12

2.6.1.5. Distress Call. No distress calls were noted in the course of capture of *H. griseus*, but a high-pitched call uttered by the infant when separated from the mother, though still in visual contact with her, may represent such a call.

2.6.2. Hapalemur simus

Very little information on this extremely rare species is available as yet. However, it recently proved possible to observe a few specimens of this species in their natural habitat in bamboo forest and to record some of their vocalizations (Petter *et al.*, 1976). *Hapalemur simus* are considerably larger in body size than *H. griseus*, but like the latter they live in small groups of 3 to 5 individuals. On several occasions, a group of *H. simus* was seen in association with a group of *H. griseus* or *Lemur fulvus rufus*. The calls of *H. simus* are clearly distinct from those of *H. griseus*. So far, it has been possible to distinguish two types which can be interpreted as contact calls and alarm calls, respectively.

2.6.2.1. Contact Call (Fig. 8k). This call is a plaintive, yelping sound which is quite powerful, with the intensity rapidly rising and falling as the call progresses. This call seems to be emitted in situations equivalent to those associated with the bleating call of *H. griseus*, and it therefore seems to act as a group-cohesion signal.

2.6.2.2. Alarm Call (fig. 8j and l). When *H. simus* are disturbed in the forest, they utter powerful, low-pitched roars of decreasing intensity. These roars, which may be represented as ''grrraaa,'' are sometimes divided into two parts (''ouik-grrraaa'') and they may be produced in long, rapid sequences.

2.6.3. Lemur mongoz

Mongoose lemurs, which are the smallest members of the genus *Lemur*, live in family groups of 2 to 4 individuals. They occur on the Comoro Islands as well as on Madagascar, and in the latter country they are sympatric with *Lemur fulvus*. Some observations suggested that they are strictly nocturnal, and this has been confirmed by a recent study (Tattersall and Sussman, 1975). During the daytime, they spend the day rolled up into a ball in groups in a dense patch of vegetation at the top of a tree. When active, they may become mixed with troops of *L. fulvus*, which may accompany them for the early part of the night. However, the vocalizations of *L. mongoz* are quite distinct from those of *L. fulvus* (Petter, 1962; Tattersall and Sussman, 1975).

2.6.3.1. Contact Calls and Contact-Seeking Calls (Fig. 9d and f). When separated from its mother, the young mongoose lemur utters a discreet call ''oum.'' This is notable in that the contact calls of the other *Lemur* species are by no means as discreet. Adult *Lemur mongoz* respond to the infant's call with a low-pitched ''ararar'' snort.

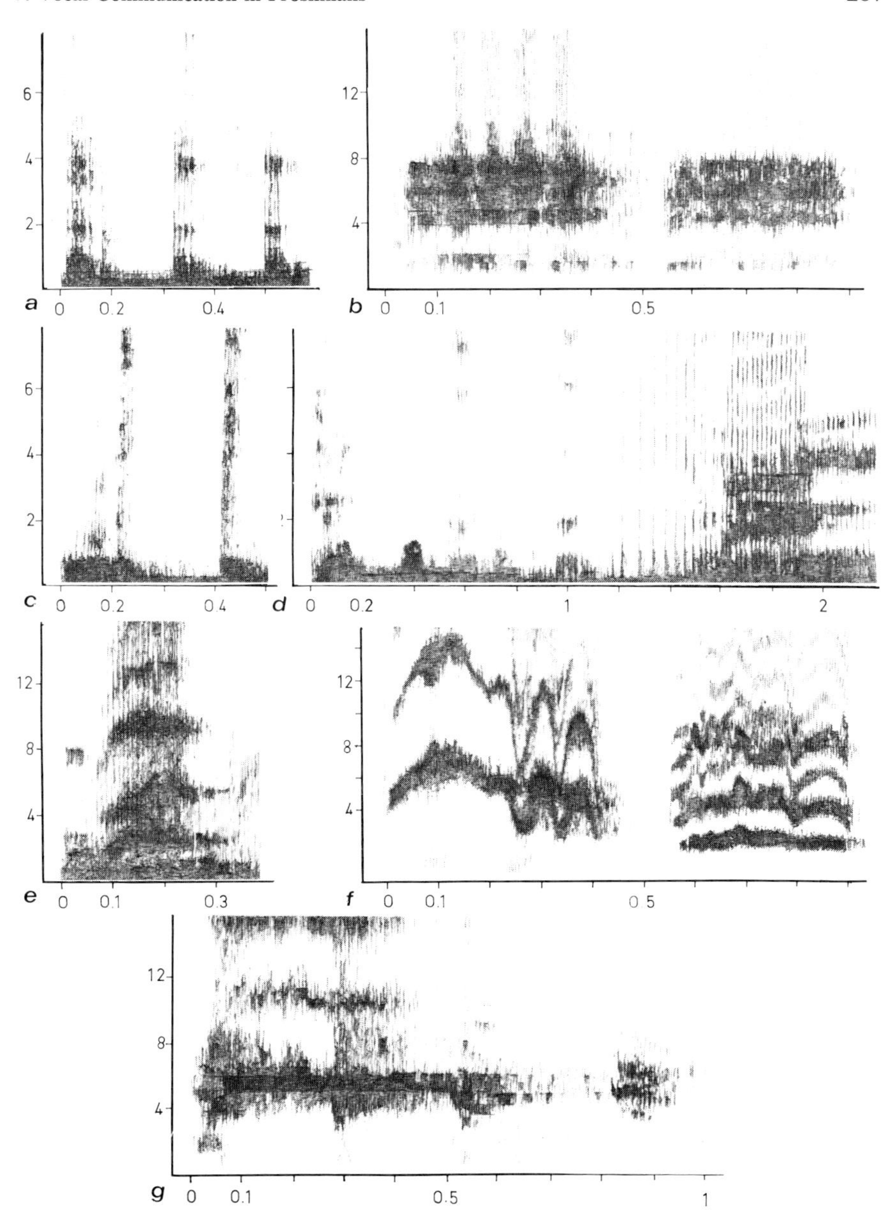

Fig. 9. Calls of *Lemur mongoz* and *L. coronatus*. (a) Medium-intensity alarm call of *L. mongoz*; (b) high-intensity alarm call of *L. mongoz*—"creeeee"; (c) medium-intensity alarm call of *L. mongoz;* grunts interrupted by sneezes: in this case a grunt is followed by two sneezes; (d) distress call of a young *L. mongoz* (at 0.1, 0.4, and 0.7 seconds) with an answer by the parents; (e) medium-intensity alarm call of *L. mongoz* (similar to a pig's grunt); (f) two successive cries of distress emitted by a young *L. mongoz* isolated from its mother; (g) high intensity alarm call of *L. coronatus*—"cuiiiiii ciot ciot."

2.6.3.2. Distant Communication Calls. While moving around actively, *Lemur mongoz* frequently utter weak grunting calls, "hon," which can be interpreted as cohesive signals.

2.6.3.3. Alarm Calls. When mildly alarmed, mongoose lemurs utter cohesion grunts more rapidly and more distinctly (Fig. 9a). At a higher level of excitation, the repeated "hon" calls are interspersed with drawn-out grunts which are fairly powerful "greeee" sounds somewhat like the grunting of a pig (Fig. 9e). The "hon" calls may also be alternated with sneezing sounds (Fig. 9c), which appear as white noise covering a wide frequency band. Following a series of alarm grunts, when excitation is extremely high, the entire group in unison may utter a fairly high-pitched, drawn-out "creeeeeeeeeee" (Fig. 9b) interspersed with the usual alarm grunts. This additional call resembles that which is produced under the same conditions by *Lemur fulvus* and *L. macaco*, but it is more high-pitched.

2.6.3.4. Contact-Rejection Calls. Although mongoose lemurs are often less aggressive than other *Lemur* species, particular animals (when very excited) may exhibit intimidatory behavior comparable to that described below for *L. fulvus*. Although males may show this behavior, it is more common in females. In fact, female *L. mongoz* appear to be markedly less tolerant of other females than is the case with *L. fulvus*. The calls produced during such intimidatory behavior are comparable to those described below for *L. fulvus*.

2.6.3.5. Distress Calls. Mongoose lemurs utter very high-pitched distress calls when they are captured or in the course of fights in captivity.

2.6.4. Lemur rubriventer

In terms of social structure, *L. rubriventer* seem to be quite similar to *Hapalemur* and *L. mongoz*. *Lemur rubriventer* do not seem to form large social groups of the kind found with the remaining *Lemur* species. Usually, one finds only small groups of 3 or 4 individuals. In fact, they are a relatively discreet species about which very little is known, either in terms of their activity rhythm or in terms of their specific behavior patterns.

The normal alarm grunts of *L. rubriventer* are very similar to those of *L. fulvus*, but they will also utter, usually following a number of typical grunts, a special two-phase grunt ("gre-ii") (Fig. 10j), somewhat resembling the grunt of *Daubentonia*; only the second part of this special grunt is resonant, and it is variable in duration. The intensity of this call is also variable. It characteristically occurs in situations of moderate alarm. If the alarm becomes more intense, the animals as a group will utter a series of high-pitched calls in chorus homologous with the "crouiiii" (Fig. 10i) call of *L. fulvus* and *L. macaco*, but shorter. These calls are repeated five or six times, thus resembling to some extent the rhythm of *L. catta* barking calls.

2.6.5. Lemur fulvus and Lemur macaco

All of the various subspecies of *L. fulvus* and *L. macaco* share essentially the same, fairly rich repertoire of vocalizations (Petter, 1962; Andrew, 1963; Klotz, 1966). All of these forms live in social groups larger than those found with the other lemurid species considered thus far. In some cases, groups of up to 20 individuals have been sighted. Particular difficulty is encountered with the social contact calls, which are very variable and hard to interpret.

2.6.5.1. Contact Calls and Contact-Seeking Calls. ***a. "Purring."*** A kind of "purring" indicating contentment is characteristically heard in association with reciprocal grooming (Fig. 10d). The intensity and duration are variable, with the calls generally occurring in series as a function of grooming activity. With an adult of either species which has been tamed, this "purring" can easily be elicited by stroking. The same type of "purring" is heard with young animals when they are licked by an adult, or if they are stroked by a human foster-parent.

b. Infant call. The infants of these two *Lemur* species utter a very typical vocal signal when clinging to the mother's belly. It consists of a series of brief grunting sounds ("cot-cot-cot") which can be interpreted as a simple contact call for communication with the mother. This call is sometimes mingled with, or followed by, weak "oum" sounds which are often repeated several times and which can be more precisely interpreted as genuine contact calls, since they are uttered, for example, when the infant is attempting to reach the mother's teat. The mother is generally very responsive to this call, and will interrupt her activity to attend to her infant.

c. "Greeting call." With *Lemur fulvus*, close contact between group members is common and there is a typical "greeting call" which is uttered when two animals from a given group approach one another and begin to groom (Fig. 10e and f). This call sounds somewhat like a rapid series of clucking noises, and when excitation reaches its peak the clucking noises may be preceded by a more intense, high-pitched component. This type of call was first described by Andrew (1963). It is accompanied by rapid licking movements, and the two animals involved typically adopt a special posture with each passing its arm over the back of the other and inclining the head to lick the other's anogenital region. Such clucking noises are also uttered when the observer approaches a tame animal, and again the call is eventually accompanied by licking. If the animal is initially some distance away, it can emit distant clucking calls, with the tongue flicking rhythmically in and out and the head pointing toward the observer.

d. Contact call. An isolated adult will emit a characteristic, rather weak call (Fig. 10h), composed of a resonant "kiou" sound following a drawn-out "onomb" or a double "omb" sound. Similar signals are also uttered continuously by a juvenile when it is separated from its group. In such cases, the call is a drawn-out "craaeee" with a plaintive sound and mildly rising frequency. Other adults may respond to this call with alarm calls or by "recognition" (greeting) calls which seem to exert a calming effect.

2.6.5.2. Distant Communication Calls. ***a. Cohesive signals.*** While moving around fairly rapidly, the members of a group continuously emit mild grunting noises ("hon") (Fig. 10c). The rhythm of the grunts is variable and it is often possible, even without seeing the animals, to estimate their speed of movement and any encounter with obstacles simply by listening to the grunts. For example, the "hon" sounds are uttered approximately every second before a leap, whereas they cease at the time of the leap and then abruptly reappear in rapid sequence just after the leap.

b. "Recognition signals." Another very common vocalization of *Lemur fulvus* and *Lemur macaco* consists of a series of loud, drawn-out, low-pitched grunts which sometimes resemble the quacking of a domestic duck, though at a much lower pitch. Both the intensity and the intonation of this call are quite variable. Generally the call is emitted by a single individual, usually a male and rarely a female; but sometimes several individuals may utter the call simultaneously. The most drawn-out version of the call is a calmly produced "meueu-meueu-meueum-meume-meume-me" emanating from resting animals. The impression was gained that the arrival of the observer close to a relatively well-known group frequently elicited this particular vocal signal. However, the call is more often produced when individual *L. fulvus* or *L. macaco* are moving around close to resting members of their own social group. By far the most frequent occurrence of the call is associated with encounters between two groups. In this situation, the grunting sequence is mixed with several other vocal signals and it can be accelerated or abbreviated according to the animals' state of excitation.

c. Territorial signals. These signals are very similar to the powerful alarm calls, but they are less high-pitched and more drawn-out. Their intensity progressively declines at the end of the call, instead of rising: "creeeeee. . ." Such vocalizations are heard late in the evening when the animals have returned to their nocturnal sleeping site, without any apparent stimulus acting as a releaser. These long calls, which are usually produced simultaneously by all members of a fairly large group, can be heard some distance away in the forest. As a rule, similar vocal signals are emitted by other groups elsewhere in the forest, producing the effect of an echo, and this in itself suggests a territorial function. The call is repeated roughly every 2 or 3 minutes for a variable length of time. It has usually been heard between 1800 and 1900 hours, but sometimes it can be heard much later than this.

2.6.5.3. Alarm Calls. When a group of *L. fulvus* or *L. macaco* is disturbed by some persistent intrusion, even if there is no noticeable danger, the animals begin by producing grunts which are louder and more staccato than those in the cohesive signal, thus giving rise to mutual excitation. When seriously alarmed, the alarm grunts pass through a whole series of intermediates which become increasingly more powerful until the whole group abruptly produces a strident "crou-crou-crou-crouou-crouiiiiiiiiiiiii" which intensifies toward the end. According to the degree of excitation, this call will be produced at intervals of 2–5 seconds. In the meantime, the animals keep up continuous grunting noises and swing their tails from side to side, a clear-cut visual signal of excitation. In fact, it is only the final part of the pronounced alarm call which is uttered in perfect unison by the members of the group. The initial part of the call seems to be important for preparation of all individuals for the final chorus in unison, and it is the most excited animal which initiates the calling and is imitated by the others. In some cases, the remaining group members do not follow the call, whereas on other occasions, when excitation is pronounced and widespread, the initiating role of the first animal is so immediately successful that it is difficult to discern.

In captivity, this call can easily be evoked, particularly in the morning and the evening during periods of maximal activity, even with animals which are used to human presence. The call results from a certain level of excitation which can be produced simply by imitating the whistling of a bird of prey for a short while. (When cages containing other *Lemur* species and sifakas are located nearby, these species will produce their own particular, homologous call at the same time.) In the forest, this vocal signal is always heard when two groups encounter one another. Several variants of this call have been noted on various occasions, but it is not easy to provide specific interpretations of them.

2.6.5.4. Contact-Rejection Calls. ***a. Threat calls.*** A special, resonant call sequence characterizes the intimidatory behavior of the dominant male when he is at the center of a group. In captivity, when an observer approaches a caged group the dominant male will move forward, lift his head back to expose his canines in a typical threat gesture and then abruptly move away again. At that point, the male will then utter a succession of plaintive, chevron calls of increasing pitch, followed by four or five grunts in a fairly regular sequence: "oe-ai-oe-ooo" (Fig. 10g).

b. Tooth grinding. Aggressive excitation can also be indicated by grinding of the teeth, which is distinctly audible with captive animals.

c. Barking calls. When faced with an apparent sudden threat these animals sometimes utter loud grunting sounds or barks. These brief, raucous calls alert the entire group and may subsequently degenerate into alarm calls. It also happens frequently that "greeting" calls are produced in response to these vocalizations by individuals in the group, and this has the effect of reducing the general level of excitation.

2.6.5.5. Distress Calls (Fig. 10a and b). When separated from its mother, the infant becomes very noisy and utters high-pitched plaintive calls which can be interpreted as distress signals. During chasing and fights involving individuals belonging to two separate groups, the individual being chased (either male or female) may emit high-pitched, drawn-out calls, particularly if cornered. However, this call can also be uttered as an aggressive signal by a very excited individual

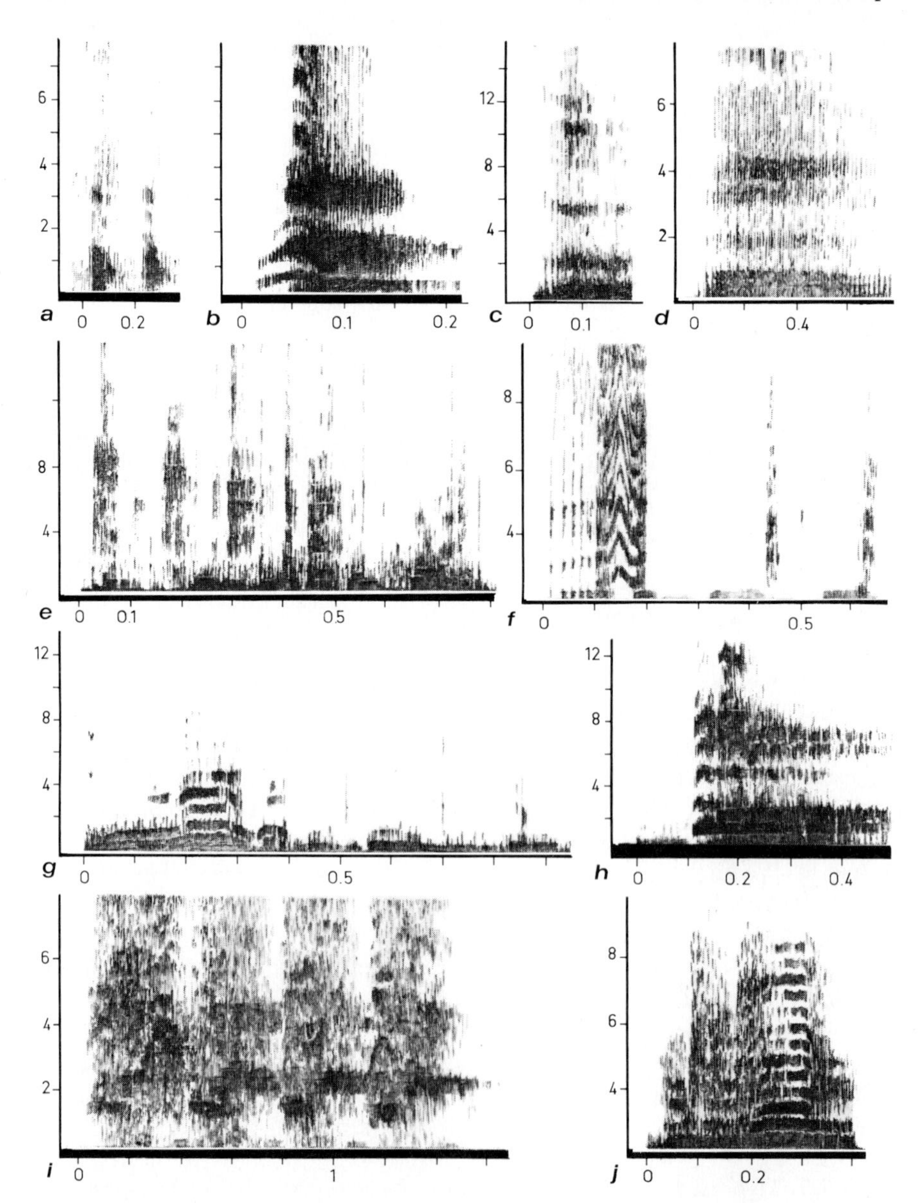

Fig. 10. Calls of *Lemur fulvus* and *L. rubriventer*. (a) Low-intensity alarm call of a young *L.f. albifrons*; (b) distress call of a young *L.f. albifrons*; (c) group-cohesion grunts of *L.f. albifrons*—"hon"; (d) purring of *L.f. albifrons* during allogrooming; (e) recognition call during meeting of two *L.f. albifrons*; (f) contact call during allogrooming of two *L.f. albifrons*; (g) contact-rejection call of a male *L.f. albifrons* after having threatened a stranger; (h) mild distress call of a *L.f. albifrons* isolated from a group; (i) high-intensity alarm call of *L. rubriventer*—"croui-croui-croui-croui"; (j) typical medium-intensity alarm call of *L. rubriventer*—"gre-ii."

when faced by a neighboring group on the other side of a cage barrier. When uttered in the course of aggressive chases, this distress call seems to produce a temporary arrest of fighting.

2.6.6. *Lemur coronatus*

Although it was formerly considered to be a subspecies of *L. mongoz*, *L. coronatus* is in fact a distinct species closer to *L. fulvus*, with which they share many behavioral characteristics. In fact, on the Montagne d'Ambre (N.E. Madagascar), *L. coronatus* and *L. fulvus* are sympatric, but the former are really better adapted for the dryer areas of wooded savannah found in the northern corner of Madagascar. In general, the calls of *L. coronatus* and *L. fulvus* are very similar. The alarm grunts are, however, generally more high-pitched in *L. coronatus*, sounding rather like series of "seo" sounds. Alarm calls uttered in unison are also more high-pitched (Fig. 9g), and often shorter, than those of *L. fulvus*. They are sometimes followed by a drawn-out grunting "ro-arrr. . . ." call which has so far not been recorded with *L. fulvus*.

2.6.7. *Lemur catta*

Lemur catta are the most gregarious of the lemurs, troops of up to 45 individuals having been recorded. Each troop occupies a very large home range (Sussman, 1972), and the vocal repertoire is large and quite distinct from that of the other *Lemur* species. The calls of *L. catta* have been studied particularly closely by Jolly (1966a), who found that there are numerous intermediates between the typical call types, reflecting the greater complexity of social relationships.

2.6.7.1. Contact Calls and Contact-Seeking Calls. As with the other *Lemur* species, the infant "purrs" when licked by its mother. The same "purring" is uttered by adults during sequences of social grooming.

When in an uncomfortable situation, the infant may emit sequences of brief, high-pitched calls, and the mother's attention is immediately aroused.

2.6.7.2. Distant Communication Calls. ***a. Cohesion grunts.*** When a *Lemur catta* troop is on the move, rapid sequences of weak grunting noises are produced (Fig. 11b). These sounds may also accompany the initiation of movement by the troop.

b. Cohesion "miaouws." In many cases, movement of a troop is preceded by reciprocal exchange of single "miaouw" calls. When a troop of *L. catta* is sufficiently disturbed for it to be dispersed, the troop members maintain vocal contact by frequently repeated "miaouw" calls. When this call is being emitted, the corners of the mouth are drawn downward and the mouth is opened, just as with the "miaouw" of a cat. The call may be accentuated to give a two-tone "miaouw" ("wails" of Jolly, 1966a).

c. Territorial signals. On certain occasions, the long, plaintive ''howls'' of *Lemur catta* (see Jolly, 1966a) have appeared to operate as intertroop signals with a territorial function, rather like the howling calls of *Indri*. The howl of *L. catta* consists of a loud, drawn-out ''miaouw-like'' call, and the accent is on the second component of the call, which is uttered with the head directed up toward the sky and the cheeks puffed out.

According to Jolly (1966a), this call can be heard at a distance of 1 km. If it is uttered in the center of the reserve at Berenty, where *Lemur catta* have been studied in particular detail, it can be heard by at least eight other troops. Apparently, only males in a troop utter this call (Jolly, 1966a).

2.6.7.3. Alarm Calls (Fig. 11d). A slight disturbance evokes the emission of weak grunts which may become more intense and eventually give rise to a more intense alarm response if the disturbance persists. Should the disturbance become more pronounced, the grunts are intensified and progressively transformed into weak ''barks'' which may be further reinforced and occur in a regular sequence. These distinct ''oua-oua'' calls are uttered by several individuals at the same time. These calls are homologous with the ''crouiiiiii'' alarm calls of the other *Lemur* species. As with the latter, they are elicited by whistles emitted by birds of prey, and in captivity they can be evoked ''sympathetically'' by the alarm calls of sifakas or variegated lemurs. When seriously alarmed, the alarm barks may be interspersed with powerful plaintive calls whose pitch drops progressively over the course of 2–3 seconds. Such calls are emitted by several individuals at the same time. Observations of animals producing these plaintive calls indicate that they tend to look down toward the ground, whereas when barks are produced they are usually looking up toward the sky. Thus, these two different call types may correspond to two different alarm situations. The plaintive call may undergo periodic reinforcement, without any interruption of the call, sounding rather like vexed cries from a baby. The call may also be reinforced with harsh grunts.

2.6.7.4. Contact-Rejection Calls. ***a. Rapid grunting.*** Rapid, staccato grunts may be emitted in the course of an encounter between two aggressive individuals. The grunts are gradually transformed into spitting noises if the level of aggression rises (e.g., if one animal is attacked by another).

b. Squeaks and snorts. A male displays anger with visual signals (waving of the tail after marking it and a specific facial expression) accompanied by high-pitched, grating squeaks interspersed with snorts (Fig. 11c). The squeaks can also occur alone, in which case they may take on a pure, plaintive tone of gradually falling pitch. Sometimes one hears only a part of the plaintive call, sounding rather like ''ou,'' in the middle of very low-pitched snorts or ''hon'' calls. Such vocalizations are generally produced by males as a manifestation of dominance. In captivity, they are frequently heard in the course of aggressive responses directed toward an

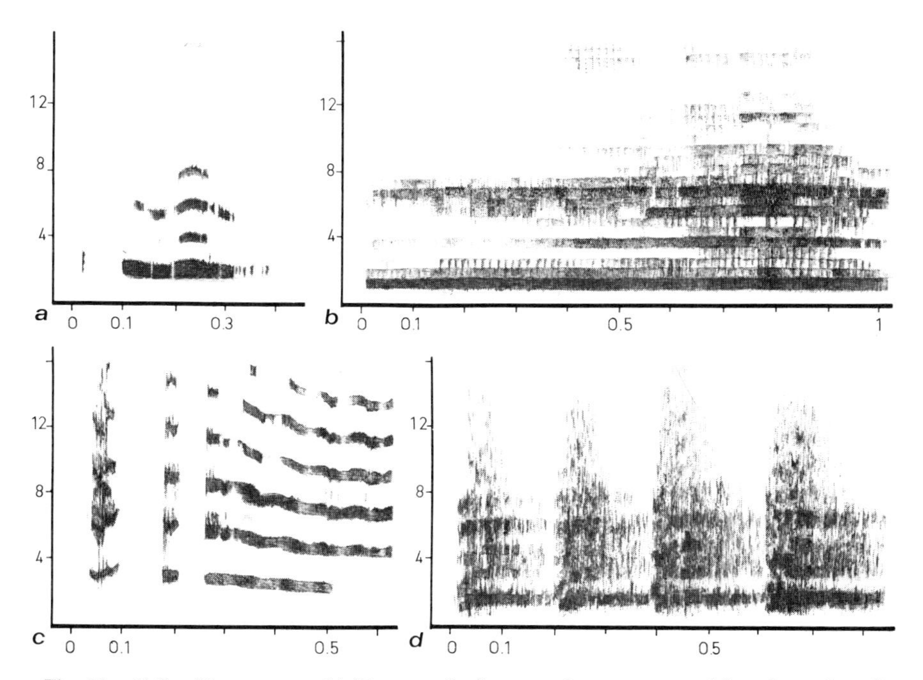

Fig. 11. Calls of *Lemur catta*. (a) Distress call of a young *L. catta* separated from its mother; (b) contact-maintenance call emitted by a group; (c) sharp call of a male preceded by snorts emitted in association with tail-waving; (d) alarm barks.

unfamiliar human being. In such cases, the calls are associated with tail-marking and attacks, and this is reminiscent of the behavior of *Hapalemur griseus*.

2.6.7.5. Distress Calls. High-pitched, extremely piercing calls can be heard in the course of fights, for example, when two males encounter one another, or when an animal is grasped in captivity. A high-pitched, modulated plaintive call emitted by the infant when it is carried by the mother or when an attempt is made to remove it from the mother's fur may also be interpreted as a distress call (Fig. 11a).

2.6.8. Varecia variegata

Variegated lemurs are shy animals which live in the dense rain forest of the east coast of Madagascar, usually confining their movements to the canopy level. They live in family groups of 2–5 individuals, and it is important to note that the female may have up to 3 infants in a litter. Very few observations have been carried out under natural conditions, but variegated lemurs can be kept and bred quite easily in captivity. The most characteristic calls of this species are the alarm calls and intergroup communication calls.

2.6.8.1. Contact Calls and Contact-Seeking Calls. At birth, the infants are left by the mother on a branch fork, in a tuft of epiphytes. In captivity, she will make use of a nest box for this purpose, and the infants only emerge when transported in the mother's mouth or when they are capable of independent locomotion.

In order to call the mother, young variegated lemurs emit a very weak plaintive call whose pitch appears to rise (Fig. 12a), a kind of "hon-eh-oum." The mother responds with a similar plaintive call and will usually return to retrieve her offspring. This call may also be exchanged between adult individuals, although such cases are rare.

2.6.8.2. Distant Communication Calls. *Varecia* produce periodic plaintive calls accompanied by clucking noises (Fig. 12c and d) with the same intensity as their alarm roars. These calls are often heard in the forest and they are also frequently emitted by animals in captivity (e.g., at the Zoological Gardens in Tananarive). Typically, 2 or 3 animals will call at the same time, producing a powerful plaintive sound upon which a number of clucking sounds are superimposed "ko-ko-ko-ko-ko-ko-ko-kie-kie-kie-kie-kie-kie-kie-kie-kieeee." The "ko" sounds at the beginning are uttered with increasing intensity and speed, while the "kie" components represent the maximum intensity part of the vocalization. Toward the end, the intensity of calling diminishes and the call ends with a fairly drawn-out "kieeee" component, which may sometimes be followed by a much weaker "kiou-kodot." In their typical form, these calls seem to serve a territorial signaling function. Production of the call by one group usually provokes similar responses from one or more neighboring groups. At Fampanambo in North-east Madagascar, variegated lemurs were heard to produce these calls at all times of the day, but particularly between 1700 and 1900 hours, when the animals seemed to be most active.

2.6.8.3. Alarm Calls. As will be seen below, the alarm call also functions as an aggressive call when it attains its maximum intensity.

When *Varecia* are encountered in the forest and are situated so high in the trees that they do not attempt to flee (or when they are resting), they remain motionless and silent, simply staring fixedly at the intruder. If one stops for a short while to observe the animals, they stare hard at the observer and utter one or two grunts ("hon") in isolation from time to time. Sometimes, such grunts are followed by a deep, drawn-out snorting "orororoon" of weak intensity (Fig. 12b). This relatively monotonous vocal signal is uttered at irregular intervals. Usually, one individual is imitated by another, and sometimes a third animal, with slight differences in pitch. It was noted that frightened animals in captivity exhibited similar responses, and this particular call was also uttered by a female when an observer approached her offspring too closely. Probably, this signal acts to alert the entire group under natural conditions. If a tape recording of the vocalizations is played close to a group of *Varecia* in the forest, they become particularly attentive, stiffening their bodies

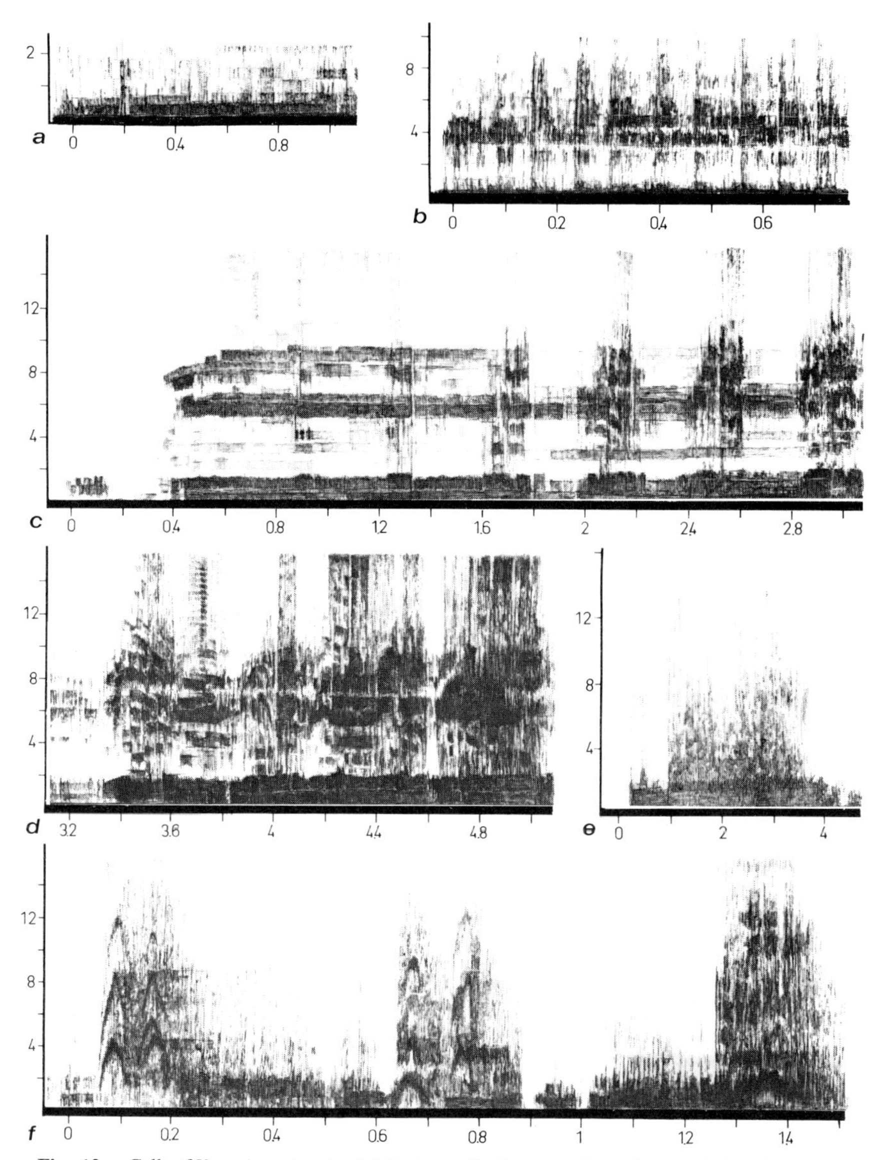

Fig. 12. Calls of *Varecia variegata*. (a) Contact call of a young *V. variegata*; (b) low-intensity alarm call (low-pitched snorting); (c) contact-maintenance call; (d) series of contact-maintenance calls with clucking amplified; (e) medium-intensity alarm call—"ar-rou"; (f) high-intensity alarm call; series of powerful roars mixed with sharp cries.

and looking all around them. When the level of excitation increases, the usual alarm grunt may sometimes be followed by more powerful "ar-rou" sounds (Fig. 12e), which represent a transition to the maximum intensity alarm call.

When the animals are at a maximal level of excitation, extremely powerful alarm calls are evoked (Fig. 12f). These can be compared to a series of very intense roars whose structure is difficult to define, since 2 or 3 animals in a group will usually be calling at once. The calls pass from a low to a high pitch, becoming progressively louder, and the emission may last for several seconds before stopping, either abruptly or with a decrescendo of weaker roars. In the forest, they can be heard over a distance of as much as 1 km. Analysis of the calls indicate a uniform low-frequency structure which is periodically reinforced before terminating with one or more high-pitched components. There is some evidence that when the animals are in groups of two or three they attempt to synchronize these calls, particularly at the higher intensity of excitation.

These calls are quite variable in terms of both intensity and duration. Sometimes, during the course of the call or toward its conclusion, one can distinguish a plaintive call or clucking sounds which correspond to another type of call. Apparently, when the animal is very excited it will mix up its vocal repertoire to some extent. The roar is uttered under the same circumstances as the barking calls of sifaka and the "crouiiii" calls produced by most *Lemur* species. Its primary role seems to be that of scaring away predators, particularly birds of prey. Although the adults, thanks to their large body size and muscular strength, are not seriously menaced by predation, the infants are exposed to considerable hazards, particularly during the first few weeks of life.

In captivity, this call is elicited from *Varecia* by an imitation whistle of a bird of prey, and the animals stare intensely at the sky while vocalizing. The response can also be elicited by a small heron flying close to the branch on which variegated lemurs are sitting.

In the forest, this call has been heard at most times of the day and night, except when the weather is very hot. In most cases, where it was possible to observe the animals, a bird of prey (usually a buzzard) was noticed in the vicinity. In one case, the disturbance arose from the felling of a tree 50 m away and on another occasion a passing plane elicited the response. It is interesting to note that, while uttering these roars, the animals exhibit all the signs of intense aggression. Their bodies are tense, the mouths open, and they stare fixedly ahead. In times of intense excitation of this kind, even a very tame animal may react by biting if approached.

Under certain circumstances, the alarm call is atypical. For example, on several occasions at Nossi Mangabe and Iaraka in northeast Madagascar, two groups located fairly close to one another uttered series of only three roars at intervals of 5–10 seconds. These calls were generally followed by exchanges of clucking calls. In such instances, the calls may be interpreted as signals of aggression exchanged between groups during a territorial confrontation.

2.6.8.4. Contact-Rejection and Distress Calls. No contact-rejection calls or distress calls have been heard in *Varecia variegata*.

3. DISCUSSION

This study has covered the vocalizations of 34 prosimian species, all observed both under natural conditions and in captivity. A great deal still remains to be done in terms of analysis of the individual calls and their ontogeny, but the field data summarized above permit preliminary consideration, for each species, of the principal calls associated with specific social and ecological situations. Viewed from this angle, comparisons between the various species permit fairly detailed interpretation of this particular mode of communication.

Whereas the simian primates include only one, little studied, species specialized for nocturnal life (*Aotus trivirgatus*), three-quarters of the extant prosimian species are nocturnal. As indicated in the introduction, some prosimian families contain both nocturnal (or crepuscular) and diurnal species. With these families, the close phylogenetic relationship between the individual species permits interpretation of the marked behavioral differences as products of divergent adaptation for nocturnal or diurnal life. For example, the prosimians have different patterns of social organization according to whether they are nocturnal or diurnal; the nocturnal forms are generally solitary in habit, while the diurnal species are all gregarious. However, as has been pointed out on several occasions (Charles-Dominique, 1972, 1975, 1977), solitary life is not incompatible with a relatively complex social system and all of the solitary nocturnal prosimian species studied to date have proved to possess patterns of social organization. In these cases, social relationships are based primarily on the exchange of indirect signals (olfactory marking), with some delay between transmission and reception of the signal. These nocturnal species also exhibit exchange of vocal signals, but the calls involved are usually more stereotyped than those produced by gregarious species. It is self-evident that a long-distance vocal signal which is deformed by the screen of vegetation through which it must pass cannot carry very complex information.

3.1. Contact and Contact-Seeking Calls

Whereas the infant is continuously carried by its mother in almost all of the gregarious lemur species, with the solitary prosimians the mother deposits her infant in a nest or in the vegetation and retrieves it in due course. There are therefore mother–infant contact calls, with the mother usually calling first and the infant responding. When the infant is left alone, it remains perfectly motionless and silent, as an efficient means of self-protection. (This is found in the Lorisidae, *Lepilemur* and even one gregarious lemur species—*Hapalemur griseus*.) At dawn, or when discomforted in some way, the infant will spontaneously emit its contact call. In addition, with certain solitary nocturnal species several adults meet up at dawn to return to the sleeping site together. In this case, the adults may make use of the same call as that utilized by the mother and infant (*Galago elegantulus*) or employ a call derived from the latter (*Galago demidovii*).

With the gregarious lemur species, the infant, which is in most cases carried on the mother's belly, from time to time emits faint calls of variable intensity. Such contact calls are emitted by the infant particularly in situations of discomfort (e.g., when hungry, when in an awkward position, when tired after the mother has been actively leaping around) and the mother usually responds by providing the necessary care. As soon as the infant is licked, it calms down. If, on the other hand, it is accidentally separated from its mother the infant usually utters loud, uninterrupted calls until the mother returns to retrieve it. These latter calls, which immediately arouse the attention of the mother and sometimes of other group members, may also be interpreted as distress signals.

With some of the nocturnal prosimian species (*Microcebus murinus*, *M. coquereli*, *Hapalemur griseus*, *Perodicticus potto*?) there is a call associated with mating, and in other species there are calls partially related to social relationships between males and females (*Galago demidovii*, *G. alleni*, *G. elegantulus*, *G. senegalensis*). These vocalizations have no clear-cut counterpart among the diurnal, gregarious lemurs, which communicate at close quarters primarily with visual and olfactory signals in this context. The situation with *Hapalemur griseus* is interesting. Although gregarious, this species is crepuscular and it does have a very characteristic mating call. However, it should be noted that in this species the social groups seem to be limited to a pair and their offspring, and that a certain degree of crepuscular or even diurnal activity may occur in addition to nocturnal behavior.

In addition to these vocalizations concerned in contact-seeking behavior, there are specific calls associated with tactile communication. In this respect, no difference between nocturnal and diurnal species has been noted; they behave like many other mammal species in which physical contact is preceded by, and facilitated by, vocalizations.

3.2. Distant Communication Calls

With most of the nocturnal prosimian species, individuals typically move around alone at night with no coordination between their movements. Hence, there is usually no vocal signal to permit cohesion between group members while moving. However, *Phaner furcifer* are to some extent gregarious in that the male and female coordinate their movements by means of a special call which is emitted for part of the night. A more pronounced degree of coordination of movements is found with *Avahi laniger*, *Hapalemur griseus*, and *Lemur mongoz*, all of which exhibit more pronounced gregarious tendencies. With these species, as with the other members of the families Lemuridae and Indriidae, cohesion calls are uttered continuously during group locomotion. In many cases, these calls exhibit varying degrees of intensity and form, which indicate to other group members a given individual's emotional state. This system of vocal communication is fairly common among gregarious forest-living animals, such as birds and monkeys, which cannot remain in continuous visual contact.

There are other types of distant communication calls which are exhibited by gregarious prosimian species when group members have abruptly dispersed in order to escape from some source of danger, or when (as with *Lemur catta*) the group is about to move off normally. The calls involved are powerful, discreet sounds such as the "honks" of *Indri* and the "miaows" of *Lemur catta*. Of course, this second type of call is not found in solitary nocturnal species and is not present even in *Phaner furcifer*. However, *Hapalemur griseus*, *Avahi laniger*, and *Lemur mongoz* (?) all emit such calls just like their diurnal relatives.

A third type of distant communication call is used for territorial signaling, both by solitary nocturnal species and by diurnal gregarious species. These calls have been interpreted as "territorial signals," since they are often associated with territorial confrontations. They may vary in frequency according to the season but, as a general rule, the higher the population density of a given species, the higher the frequency of confrontations and the more common are the associated vocalizations. With most of the gregarious diurnal species, the individuals in the social group rapidly synchronize their calls to produce choruses. Such synchronization is also found in *Lemur mongoz* and *Hapalemur griseus*, but it is less clear in *Avahi laniger* and *Phaner furcifer*. With this type of vocalization, adaptation of the structure of the call to local ecological conditions is most noticeable. Examples of such adaptation are particularly evident in the cohesion or alarm calls of *Lemur catta* (which are better adapted than simple grunts for communication between members of a large troop dispersed among foliage) and in the territorial calls of *Indri indri* and various *Lemur* species.

3.3. Alarm Calls

Like most other higher vertebrates, almost all of the prosimian species utter alarm calls when disturbed. The role of these calls is obvious in the gregarious species, where any predator sighted is thus indicated immediately to any conspecifics which have not yet noticed the danger. In solitary species, the role of such vocalizations is far less evident, especially where individuals are greatly dispersed. However, it must be remembered that many animals associate the alarm calls of other species with the presence of danger. (For example, in Gabon the alarm calls of any of the bushbaby species will alert the other two sympatric forms.) In addition, any young animal accompanying its mother for some time may learn to recognize dangers indicated by alarm calls. In nocturnal species possessing eye markings of contrasting color, the direction of gaze is clearly indicated despite the low light intensities, and this may lend a more precise social significance to vocalizations (see Pariente,

Chapter 10). In contrast to the nocturnal species, the diurnal species emit alarm calls in synchrony (e.g., the "croueeeeeeee. . ." call of *Lemur fulvus* and the roaring call of *Varecia variegata*), or at least the other members of the social group will join in the vocalization, even if they have not seen the danger (e.g., the "shee-fac" call of *Propithecus verreauxi*). These alarm calls of the diurnal species are associated with visual signals, such as nodding of the head, swinging of the tail, and clearly visible direction of gaze indicated (as with nocturnal species) by contrasting patches on the head, reminiscent of the social alarm systems of many other gregarious primates (e.g., various *Cercopithecus* species). In certain species, these alarm calls may take on an intimidatory function, as in *Varecia variegata* responding to the sight of a bird of prey. It is interesting to note that, as in the other members of the family Lemuridae, both *Lemur mongoz* and *Hapalemur griseus* synchronize their alarm calls, which are similarly associated with visual signals. In *Phaner furcifer*, the male and female will often, but not always, respond to one another when alarmed, and the same applies to *Avahi laniger*. By contrast, in the more solitary species such as the bushbabies, alarm calls from one individual will alert neighbors but will not induce them to vocalize.

3.4. Contact-Rejection Calls

The prosimians, like many other mammal species, regulate their contacts with one another by means of contact-rejection calls which counterbalance contact-seeking calls. The emotional state of the calling animal is indicated by gradations in the calls and there is, therefore, fine control of contacts so that they occur at appropriate times. Calls indicating contact rejection are often associated with characteristic postures. In general, these relatively weak calls seem to have undergone little elaboration in evolution, and they are very similar from one family to another. In some cases, these calls extend beyond their social role to take on a function of predator repulsion (e.g., the grunting call of *Microcebus murinus* and *Cheirogaleus medius*, and the threat calls of *Daubentonia* and *Perodicticus potto*).

3.5. Distress Calls

Distress calls, which are associated with pain, terror (e.g., when an animal is seized), and even agony, are also very similar from one prosimian family to another and even comparable to those of numerous other mammal species. These calls have been interpreted here as signals leading to arrest of fighting between conspecifics. When a recording of this signal is played, or when it is coarsely imitated, it attracts various mammal species (prosimians, rodents, and carnivores). Prosimians which are attracted in this way arrive rapidly, spend a few seconds observing and then move away, often while uttering alarm calls. It is also possible that this call, although of no direct value to an individual captured by a predator, may teach conspecifics to recognize predators which are really dangerous to the species. As far as the carnivores are concerned, it is likely that they arrive in search of an injured or dying prey which will be easy to capture, though it is possible that they are merely curious.

3.6. General

Table II summarizes the main differences between the vocalizations of solitary prosimians and those of gregarious prosimians. It can be seen that the few gregarious nocturnal species (*Lemur mongoz*, *Hapalemur griseus*, *Avahi laniger*, and, to a lesser extent, *Phaner furcifer*) have a mixture of vocalizations of nocturnal forms with some vocalizations more typical of diurnal forms. It is clear from the table that it is primarily the distinction between solitary and gregarious habits (rather than merely the distinction between nocturnal and diurnal life) which has exerted the main influence on vocal repertoires. The nocturnal gregarious species represent an intermediate stage between nocturnal and diurnal life. In addition, analysis of the vocal repertoires of the prosimians shows that these animals have a "classical" vocal inventory compared with that of the simian primates and even with that of numerous other mammalian orders.

In the simian primates, the repertoire is rendered more complex by intermediate forms of vocal signals, combined signals, facial expressions, and specific body

TABLE II

Principal Characteristics of Prosimian Vocalizations as a Function of Social Organization[a]

	Solitary nocturnal species	Gregarious nocturnal species[b]	Gregarious diurnal species
Baby-parking + mother-infant call	+	+ (P.f.; H.g.)	–
Mating call, or male-female contact call	+	+ (H.g.)	–
Cohesion call during movements	–	+ (P.f.; H.g.; L.m.)	+
Regrouping call, following alarm	–	+ (H.g.; L.m.)	+
Group territorial call (chorus)	–	+ (H.g.; L.m.)	+
Synchronized alarm calls	–	+ (P.f.*; H.g.; L.m.)	+

[a] Note that the few gregarious nocturnal species possess both the typical vocalizations of the solitary nocturnal species and vocalizations typical of gregarious diurnal species.

[b] Key to abbreviations: P.f., *Phaner furcifer;* H.g., *Hapalemur griseus;* L.m., *Lemur mongoz;* *, intermediate case.

postures. A similar trend, though less pronounced, is seen among the diurnal prosimian species, which have more gradations in their signals than do the nocturnal species.

ACKNOWLEDGMENTS

The authors would like to thank Mireille Charles-Dominique for her painstaking reproductions of the sonographic illustrations.

REFERENCES

Andrew, R. J. (1963). *Behaviour* **20**, 1–109.

Andrew, R. J. (1964). *In* "Evolutionary and Genetic Biology of Primates" (J. Buettner-Janusch, ed.), Vol. 2, pp. 227–309. Academic Press, New York.

Bearder, S. K., and Doyle, G. A. (1974). *In* "Prosimian Biology" (R. D. Martin, G. A. Doyle, and A. C. Walker, eds.), pp. 109–130. Duckworth, London.

Chappuis, C. (1971). *Terre Vie* **25**, 183–202.

Charles-Dominique, P. (1971). *Biol. Gabonica* **7**, 121–228.

Charles-Dominique, P. (1972). *Z. Tierpsychol. Beth.* **9**, 7–41.

Charles-Dominique, P. (1975). *In* "Phylogeny of the Primates: A Multidisciplinary Approach" (W. P. Luckett and F. S. Szalay, eds.), pp. 69–88. Plenum, New York.

Charles-Dominique, P. (1977). "Ecology and Behaviour of Nocturnal Primates." Duckworth, London.

Charles-Dominique, P. (1978). *In* "Recent Advances in Primatology, Vol. 3: Evolution" (D. J. Chivers and K. A. Joysey, eds.), pp. 139–149. Academic Press, London.

Charles-Dominique, P., and Hladik, C. M. (1971). *Terre Vie* **25**, 3–66.

Doyle, G. A. *et al.* (1977). Calls of the Lesser Bushbaby, *Galago senegalensis moholi,* and the Thick-tailed Bushbaby, *Galago crassicaudatus spp.*, in the laboratory and in the field. (unpublished).

Gautier-Hion, A. (1971). *Biol. Gabonica* **7**, 295–391.

Jolly, A. (1966a). "Lemur Behavior: A Madagascar Field Study." Chicago Univ. Press, Chicago, Illinois.

Klotz, M. (1966). *Wiss. Z. Humboldt-Univ. Berlin, Math.-Naturwiss. Reihe* **15,** 23–56.

Lavanden, L. (1933). *Terre Vie* **3**, 142–152.

Martin, R. D. (1972). *Philos. Trans. R. Soc. London, Ser. B* **264**, 295–362.

Moynihan, M. (1967). *In* "Primate Ethology" (D. Morris, ed.), pp. 236–266. Weidenfeld & Nicolson, London.

Pariente, G. F. (1974). *Mammalia* **38**, 1–5.

Petter, J. J. (1962). *Mem. Mus. Nat. Hist. Nat. Paris, Ser. A* **27,** 1–146.

Petter, J. J., and Peyrieras, A. (1970a). *Mammalia* **34**, 167–193.

Petter, J. J., and Peyrieras, A. (1970b). *Terre Vie* **24**, 356–382.

Petter, J. J., and Peyrieras, A. (1974). *In* "Prosimian Biology" (R. D. Martin, G. A. Doyle, and A. C. Walker, eds.), pp. 39–48. Duckworth, London.

Petter, J. J., Schilling, A., and Pariente, G. F. (1971). *Terre Vie* **25**, 287–327.

Petter, J. J., Albignac, R., and Rumpler, Y. (1976). "Faune de Madagascar: Lémuriens." ORSTOM - CNRS, Paris.

Pollock, J. I. (1975). *In* "Lemur Biology" (I. Tattersall and R. W. Sussman, eds.), pp. 287–311. Plenum, New York.

Richard, A. F. (1973). Ph.D. Thesis, London University (unpublished).

Sauer, E. G. F., and Sauer, E. M. (1963). *J. S.W. Afr. Sci. Soc.* **16,** 5–35.
Struhsaker, T. T. (1970). *Mammalia* **34**, 207–211.
Sussman, R. W. (1972). Ph.D. Thesis, Duke University, Durham, North Carolina (unpublished).
Tattersall, I., and Sussman, R. W. (1975). *Anthropol. Pap. Am. Mus. Nat. Hist.* **52**, 193–216.
Vincent, F. (1969). Doctoral Thesis, University of Paris (unpublished).

Chapter 8

Diet and Ecology of Prosimians

C. M. HLADIK

1. DEFINITION OF DIET IN RELATION TO FIELD AND LABORATORY RESEARCH METHODS

1.1. What Is the Diet of a Primate Species?

This question usually elicits immediate and obvious answers with reference to certain types of animals: herbivores eat only grass or leaves while carnivores look

ISBN 0-12-222150-8

for live prey. Although we intuitively refer to a "standard eating behavior" it is never as clear-cut for most prosimians and other primate species. Feeding behavior is a selective response to stimulation (Hinde, 1966), but no overall explanation for the specific food choices of closely related species has ever been proposed. Hypotheses will be presented at the conclusion of this chapter but, to formulate a clear definition of diet, we must refer to qualitative as well as quantitative data. The methods used (tests and different types of observations) yield different types of information, the value of which will be discussed.

1.2. Food Tests in Captivity

The simplest method of investigating dietary preferences consists in presenting various types of food to a caged animal and recording what it selects and what it rejects. A quantitative estimate of food preferences can be obtained for each type of food from the difference in weight between the food offered and what remains. Petter (1962) used some of these tests to supplement information on the natural diet in his first overall survey of the Malagasy lemurs.

Charles-Dominique (Charles-Dominique and Bearder, Chapter 13) demonstrated that such tests did not entirely support the results of his own field observations of the natural diet of different prosimian species. Fruits and different species of insects were given *ad libitum,* to five different lorisid species in captivity that had already been intensively studied in the wild. All five species ate large amounts of insects (mainly Orthoptera) for several months, although their natural diet includes smaller amounts of insects with differences between the species in "preferential" food choices. In three of them, *Perodicticus potto*, *Galago elegantulus* (*Euoticus elegantulus*), and *G. alleni*, insects make up only 10–25% of the natural diet and there are marked differences in choice of insects, especially between the two other species, *G. demidovii* and *Arctocebus calabarensis*, the latter feeding mainly on caterpillars and insects avoided by the other species (see Section 2.3 below).

Food preferences in natural conditions are thus determined not only by the availability of insects but also by the ability of the species concerned to locate and catch them. For example, *Arctocebus calabarensis* prefer locusts and crickets in captivity but, under natural conditions, they feed on insects avoided by the other species and which, according to our analysis, are of less nutritional value, simply because they are not able to find and catch more edible prey.

Laboratory tests can be used to determine the types of prey that continue to be rejected when nothing else is available. A recent experimental study by G. Bernardi and P. Charles-Dominique (personal communication) was conducted on the five nocturnal lorisids mentioned above, using as prey a number of butterflies and moths available in their natural habitat in the rain forest of Gabon. Most edible Lepidoptera are cryptic and are eaten by all lorisid species while many moth species, which are also edible, have ocellated lower wings (containing eyelike spots) that might frighten predators. The less edible Lepidoptera are noncryptic and may display brightly colored wings signaling their unpalatibility to potential predators. Some of these may be eaten by *Perodicticus potto* and *Arctocebus calabarensis*, but the various *Galago* species, once having tasted them, never again try to eat them. Taste, which is certainly not identical in the different prosimian species, results in *Perodicticus* and *Arctocebus* having a wider choice of the less edible prey, allowing them to utilize those that are generally available in large quantities since they are avoided by the other insectivorous species. The mechanism of food selection in *Loris* (see below) is very similar.

Food tests using a larger sample of potential prey were conducted as part of our field study on *Loris tardigradus* (Fig. 1) in Sri Lanka (Petter and Hladik, 1970). The results obtained (Table I) were in agreement with previous experiments of Still (1905) and Philips (1931). They showed that *Loris* may feed on types of invertebrates that are neglected by the other predators (birds and monkeys) living in the same area. For instance the Reduvidae, a type of Homoptera with bright bronze-green wings, which obviously act as a deterrent signal to potential predators, as well as the most common butterfly, *Euploea core* (neglected by the insectivorous birds),

Fig. 1. *Loris tardigradus* feeding on a Coleoptera Cetoniidae, which is rejected by other insectivorous species, during food tests of potential prey in the dry deciduous forest of Sri Lanka. (Photograph: C. M. Hladik.)

TABLE I

Results of Food-Choice Tests on *Loris tardigradus*[a,b]

Food choice	Prey
+	Molluscs (snails)
−	Annelidae (earthworms)
−	Myriapods (Iulidae and other types)
	Arachnids
++	Salticidae
++	Opilionidae
	Insects
+	Hemiptera (Reduvidae)
	Orthoptera
++	Forficula
++	Grasshopper and crickets
	Coleoptera
++	Cerambycidae
++	Curculionidae
+	Cetoniidae
−	Carabidae
++	Lepidoptera (imago of Sphyngidae and caterpillars)
++	Other moths
+	*Euploea core* (Danaidae)
−	Unidentified caterpillars with long hair
++	Hymenoptera (ants)
++	Diptera (flies)
	Vertebrates
−	frogs
+	geckoes

[a] After Petter and Hladik (1970).
[b] The potential prey were collected in the natural environment of *Loris*. ++, Eaten immediately; +, eaten after hesitation; −, not eaten.

were eaten, albeit reluctantly, by *L. tardigradus*. These tests demonstrate to what extent *Loris* may utilize the least edible types of prey, without proving what is actually eaten by preference in the wild. Nevertheless, the results partly explain the mechanism of specialization in feeding behavior, very similar in *Arctocebus calabarensis* and *Loris tardigradus*.

Positive results from tests on food choices have recently been obtained after a 2-year study conducted in our animal house at Brunoy (Petter-Rousseaux and Hladik, in press). Free access to food was given to five species of nocturnal prosimians which normally live sympatrically in the forest of the western coast of Madagascar—*Cheirogaleus medius*, *Microcebus murinus*, *M. coquereli*, *Phaner furcifer*, and *Lepilemur ruficaudatus*. A common set of various foods was placed in each of the cages containing one or several animals of the same species. The weights of the different foodstuffs ingested were calculated, after taking into account loss of weight by evaporation determined by means of food samples left outside the cages. The main purpose of this experiment was to collect data on the different physiological cycles, but the results concerning preferential food choices, some of which are presented in Table II, give evidence of important differences between the prosimian species, that can be related to their diets under natural conditions (see Section 2.4 below). Among these five species, *C. medius* are the most frugivorous, *M. murinus* the most insectivorous, and *L. ruficaudatus* the most folivorous.

King (1974), using the same type of food tests, found a slight difference between the diets of *Lemur macaco* and *L. mongoz* which might reflect slight differences under natural conditions.

In conclusion, food tests in captivity do not yield precise information about natural diet. At best they may help to explain some mechanisms of food selection when testing with prey and other food types occurring in the environment of a prosimian species. The tests on artificial diets, such as those presented in Table II, give evidence of differences between species, independent of social tradition, since the food samples utilized never occur in the wild. These differences are related to the way species perceive food as well as to more complex physiological adaptations (see discussion on "flexibility" in Section 4.1 below) that determine the different possible expressions of feeding behavior.

TABLE II

Relative Proportions of the Fresh Weight of Different Food Categories Ingested in One Year by Five Species of Sympatric Nocturnal Prosimian[a,b]

Food[b]	*Microcebus coquereli* (%)	*Phaner furcifer* (%)	*Cheirogaleus medius* (%)	*Microcebus murinus* (%)	*Lepilemur ruficaudatus* (%)
Leaves of lettuce and willow tree	0	0	0	0	39.9
Pulp of apples and pears	22.2	15.7	4.9	2.5	38.3
Cucumber and other fruits	1.2	1.2	1.9	2.9	1.1
Banana	42.8	40.0	76.9	48.2	0.2
Mixture of milk and flour	11.6	15.7	7.6	12.3	7.2
G3 Lemur cake (protein and fat)	21.5	26.5	8.6	31.4	12.9
Meat and insects	0.7	0.9	0.1	2.7	0.4

[a] After Petter-Rousseaux and Hladik (in press).

[b] Except for insects, the different types of food were available *ad libitum*.

1.3. Observations in the Wild

In the last 15 years more observations have been carried out on wild primates than during the entire preceding period of scientific research. Observations on wild prosimians were even more recently intensified (cf. Doyle and Martin, 1974). However, it seems that feeding behavior has been relatively neglected compared to other aspects of the socioecology of prosimians, the reason being that quantitative observations on feeding behavior are generally very difficult to carry out in the natural habitat, and indirect methods have to be used to study both diurnal and nocturnal prosimian species.

1.3.1. Studies of the Diet of Diurnal Prosimians

It is surprising that the simple method of direct visual recording has seldom been used in quantitative studies to describe the diet of diurnal species of prosimians living in the dry deciduous forest where conditions of visibility are good. A fairly accurate estimate of the natural diet can be achieved by following one target animal for an entire day and counting the number of different fruits and leaves actually eaten, after which samples of each type of food can be collected to allow for calculation of the average weight of material ingested by the animal. This method has been used on different species of primate by the author (Hladik and Hladik, 1969, 1972; Hladik, 1973, 1975) and by Iwamoto (1974a, b). Comparison of the diets of the different species cannot be made on the basis of a list of Latin names, even if the proportions of food eaten are known. A further analysis of the food samples collected (Hladik *et al.*, 1971a; Hladik, 1977b) is necessary to allow for calculation of the average composition of the diets permitting interspecific comparisons. Such results can be related to specific physiological characteristics subject to further investigation by the methods described in the preceding section.

In most recent studies of wild diurnal prosimians, time has been used as a measure of feeding behavior. The time a target animal spent feeding on different food items was directly recorded by Richard (1973), while Sussman (1972) obtained an indirect measure of this time by periodical observation of individual activity records. But feeding time is not diet and, accordingly, Sussman and Richard (1974) used their measures essentially to compare the time budgets and behavior of different species. A detailed discussion of the correlation between feeding time and diet (Hladik, 1977a) reveals that comparison of the diets of frugivorous species is essentially meaningless when based on feeding times, while for species feeding on leaves and other common foods of homogeneous structure the correlation is more accurate. The above-mentioned studies, as well as the descriptions of Jolly (1966) comparing feeding data of lemur species living on the same common resources, go a long way toward presenting an accurate picture of the diet itself.

The density of wild vegetation is the main obstacle to direct observation, and continuous recording of the feeding behavior of one target animal is particularly difficult in the rain forest. For this reason one is unlikely to obtain data on the diet of

Indri indri, for instance, that are more detailed than those of Pollock (1975, 1977) who made 3000 regular observations and attempted to complement his quantitative data by the analysis of fecal material. Since many of the diurnal prosimians of Madagascar are endangered species data cannot be collected by sacrificing animals. By contrast, in exceptional areas where conditions of visibility are good, such as the deciduous forest of the south of Madagascar, quantitative studies of diet based on direct observation of the less shy lemur species would be a highly valuable complement to the work of Jolly (1966), Sussman (1972, 1974), Richard (1973, 1974), and Budnitz and Dainis (1975). If, in addition, chemical analysis of food samples eaten over a yearly cycle were to be undertaken, the results would provide a basis for further physiological investigations.

1.3.2. Studies of the Diet of Nocturnal Prosimians

Direct observation at night requires rather exceptional conditions if the same order of accuracy is to be achieved as observations during the day. The unique conditions of visibility found in the "bush" of the south of Madagascar at night, during the dry season, constitute such conditions of obtaining direct quantitative information on the diet of *Lepilemur leucopus* (Charles-Dominique and Hladik, 1971). During this study, the feeding rates of captive *L. leucopus* were first controlled from a short distance, using plant species normally eaten in the wild. When the animals were browsing, the rate of ingestion was shown to be very constant. The information thus gained permitted accuracy in subsequent field measurements.

Indirect measures of food ingested necessitate trapping or shooting of animals. For species that are not subject to total protection, such as the prosimians of continental Africa, the collection of stomach contents gives very important results. Charles-Dominique (1966, 1971a, 1974a; Charles-Dominique and Bearder, Chapter 13) worked in a large area of primary forest located within a radius of 40 km around Makokou (Gabon), where prosimians are not hunted by local people because of the availability of larger game. The total of 174 specimens collected in three years in this area represents a very small fraction of the prosimian population (estimated at less than 1/1000). Thus this method, when correctly applied in a large area, can have no adverse effects on conservation of the species.

Analysis of the different proportions of the stomach contents is generally limited to an overall description of gross classes of foodstuff (leaves, fruits, insects, etc.). Invertebrate prey can be identified more accurately from the chitinous remains. By contrast, the pulp of many fruits, when the pit has not been swallowed, does not generally retain any obvious specific characteristics sufficient to allow for identification. Juicy parts, that might be very important in the diet, are totally omitted in this kind of description which is, nevertheless, the main source of information on the diet when direct observation is not possible.

The entire digestive tract of the sacrificed animal must be carefully examined, in addition to the stomach contents, since it may retain seeds and any other solid matter giving additional information about the food ingested. A very good example of the need for this careful approach is given by Charles-Dominique (1971a, 1974a). In the stomach contents of *Galago elegantulus* and *Perodicticus potto* there is generally no trace of the gums eaten by these animals on trees and lianas. Gums may, however, be found in the cecum (particularly in *G. elegantulus*). It seems that gums are retained in the stomach only for a few minutes while other foods (fruits and insects) stay longer. Thus, in these particular cases, the diet was calculated from the weight of the cecum content added to the weight of the stomach content.

In Madagascar, where all the nocturnal prosimian species are fully protected, quantitative information on diet is currently being compared by the analysis of feces (Hladik *et al.*, in press). During a field study carried out on the west coast (Petter, 1978), all the nocturnal prosimians were trapped for marking and prepared for radio tracking. Before release, the first feces of the animals were collected (other feces would contain the fruits used as baits). The remains identified in the feces represent only a part of what was actually ingested, even smaller than the part identified in the stomach itself. As an example, the diet of *Microcebus coquereli* includes large amounts of the sweet liquid secretion of Homoptera (see Section 2.4 below) that would have been undetected without direct observation.

Data from field work, which yield information about the "natural diet" of a given species, are not generally very accurate and must be complemented by other methods. Diet is related to environmental factors, thus local variations have to bc considered. The composition of the food ingested is the main criterion for interspecific comparison of diet.

1.4. Methods of Collecting and Processing Food Samples

Most of the foods utilized by wild primates are unknown. Food samples have, therefore, to be collected and kept in sufficiently good condition for later analysis. Different analytical tests are appropriate for different methods of preservation (Hladik, 1977b). In this respect, a few important points will be presented at the end of this technical section.

The major method of obtaining food samples located on trees is by means of a tree pruner fixed at the end of a long pole. Bamboo is the least fragile material with which to build a pole up to 15 m, made of separate parts bound together with tenons, which allows direct collection from most deciduous trees (cf. Hladik, 1977a). Collection from the higher trees of the evergreen tropical forest calls for a "climbing tree stand." A very safe model of a tree stand (Forestry Suppliers, Inc., Jackson, Mississippi) was used in field work in Gabon to climb as high as 30 m along smooth tree trunks, without any special training. The samples of leaves and fruits must be collected in large amounts (at least 200 gm wet weight) and kept in plastic bags to prevent loss of water before weighing and processing.

Drying is the most convenient process for preserving food samples in field conditions. This process must be as fast as possible compatible with maximum preservation of specimens (i.e., heat no higher than 60 ° to 80 ° C). Two types of simplified dryers are presented in Fig. 2. Basically they consist of a box of wood or metal, embodying a heat source and designed to allow sufficient air circulation compatible with the maintenance of a fairly high uniform temperature around the box. Botanical specimens, necessary for identification of food samples, can be dried simultaneously. Food samples are kept in nonglossy paper bags on which weight and any other important information is marked. After drying to constant weight, the paper bags with food samples are sealed in plastic bags.

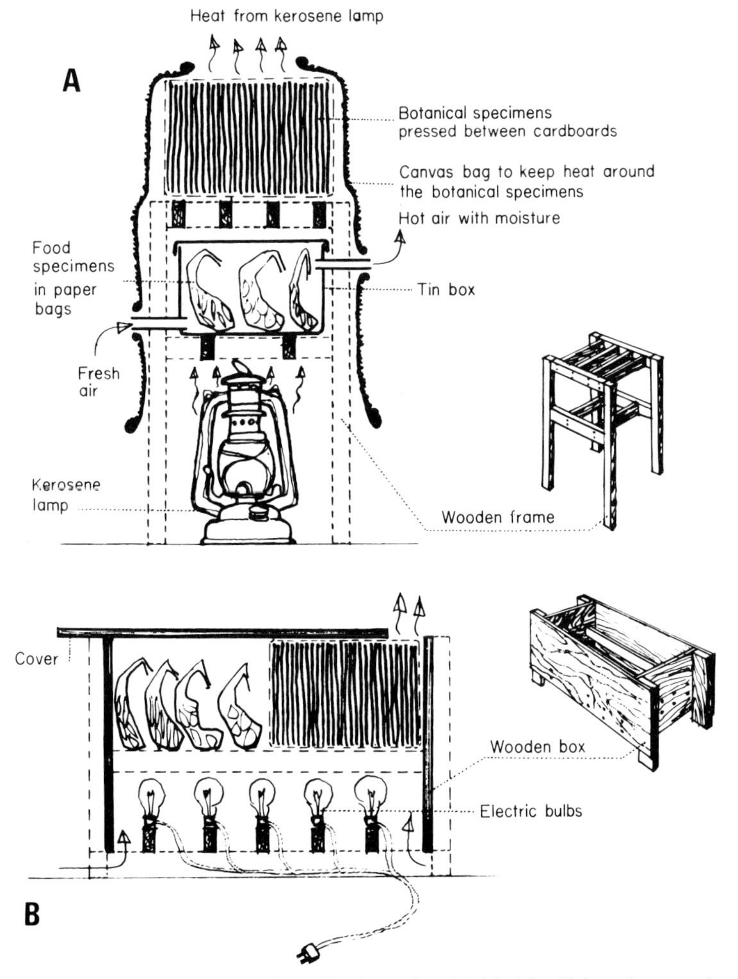

Fig. 2. Models of dryers for preparation of food samples. (A) Model utilizing a kerosene lamp to dry food specimens and botanical specimens under field conditions. (B) Simplified electric dryer used at the field station.

Many of the standard operations of analysis (titration of minerals, nitrogen, lipids, total carbohydrate, and cellulose) can be carried out on dried food samples. Further investigations require another type of field processing, such as fixation in ethyl alcohol. Here the sample is sliced in small pieces (2 mm thickness) and maintained for about 10 minutes in boiling ethyl alcohol (96 ° at alcoholometer) in a flask equipped with a condensor. The samples can be stored in plastic jars with the alcoholic solution. This process is considered one of the best for preventing enzymatic reactions: thus soluble sugars, amino acids, and lipids can be separately titrated. Deep freezing (below -30 ° C) would also allow any kind of analysis, but this method is not generally convenient in field conditions.

It is beyond the scope of this chapter to go into greater detail about food sample analysis (see Hladik *et al.,* 1971a; Hladik, 1977b): the main point is that interspecific comparison of diet requires the use of similar methods for investigating the diets of different species.

2. ECOLOGY AND SPECIALIZATION IN THE DIET OF PROSIMIANS

2.1. Dietary Specializations in Relation to Morphology

Several morphological characteristics have evolved in prosimians as a result of the selective pressures of the limited number and quantity of natural foods available in different habitats. Conversely the acquisition of certain characteristics, such as large body size, has reduced the range of possible adaptations to different diets.

2.1.1. Diet in Relation to Body Size

Variability in the composition of the most common substances eaten by primates in the wild (insects, fruits, gums, leaves) is important but restricted to limits within each category, as shown by the examples presented in Table III. The smallest species (0.1–0.2 kg) utilize insects or other small invertebrate prey as a staple food. Thus their diet, including a maximum amount of protein and fat, yields a maximum of energy (necessary for small mammalian forms, according to basic principles of metabolism). If we consider larger species (0.2–0.5 kg), the maximum amount of insect food they can obtain is approximately constant in a given habitat, since the chances of finding prey during one day (or one night for nocturnal forms) are fairly similar for different species irrespective of size, as demonstrated by Hladik and Hladik (1969) and by Charles-Dominique (1971a). These species have, therefore, to utilize other food resources, such as gums and fruits, to obtain sufficient energy in carbohydrate; the larger the species, the greater the amount of fruit necessary.

Most prosimians of large size (0.5–2.0 kg) rely on fruits as their staple food. The amount of insect food they eat is relatively small in relation to their weight, but this animal food is necessary to balance the overall proportion of protein in their diet. As

TABLE III

Examples of the Composition of Natural Foods of Primates Available in the Evergreen Forest and in a Deciduous Forest[a]

	Natural food	Percentage dry weight		
		Protein	Carbohydrate (after hydrol.)	Lipids
	INSECTS			
From Gabon (rain forest)	Insects from litter	70.2	0.5	3.5
	Lepidoptera (adult Sphingidae)	65.2	1.8	16.3
	Caterpillars (ref. An 15)	62.3	6.4	21.2
	Ants with nest (*Macromiscoïdes aculeatus*)	29.0	20.0	4.2
	FRUITS			
From Gabon (rain forest)	*Nauclea diderrichii*	4.5	47.2	5.3
	Aframomum giganteum	8.2	48.8	11.1
	Musanga cecropioides	8.8	31.5	6.3
	Pseudospondias longifolia	7.5	17.4	0.6
From Sri Lanka (dry forest)	*Drypetes sepiaria*	2.8	52.5	7.1
	Ficus benghalensis	8.4	6.8	18.1
	Syzygium cumini	6.6	4.1	2.2
	Garcinia spicata	6.2	30.6	20.3
	LEAVES			
From Gabon (rain forest)	*Baphia leptobotrys*			
	Shoots	55.0	—	—
	Young leaves	36.3	20.1	2.2
	Mature leaves	26.1	—	1.3
	Ongokea gore	19.3	24.9	1.2
	Gilbertiodendron dewevrei	10.2	13.5	0.7
From Sri Lanka (dry forest)	*Glycosmis pentaphylla*, young leaves	31.5	5.6	3.1
	Alangium salvifolium, shoots	26.2	10.3	2.6
	Walsura piscidia, young leaves	19.5	8.3	1.5
	Adina cordifolia, mature leaves	11.8	7.8	7.3

[a] After Hladik (1977a, 1978), and unpublished data.

a matter of fact, fruits yield an average of 5% protein (by dry weight) which is not sufficient to compensate for nitrogen loss. For the largest species (3–10 kg), even the protein that could be obtained from insects would not compensate for fruit protein deficiency. The diet of these large forms must include leaves or other green vegetable matter, a ubiquitous resource rich in protein. Leaf buds and young leaves are preferentially eaten because they yield more protein (25–35% by dry weight, with a maximum of 55% observed in the shoots of leguminous trees, Hladik, 1978). Thus the largest prosimian species are generally frugivorous and folivorous.

Some of these large prosimians, such as *Indri indri*, utilize leaves as a staple food (Pollock, 1977) and fruits might be regarded simply as supplementing the carbohydrate in their diet. For such folivorous animals, insects would no longer be necessary as a protein supplement since foraging for insects would yield too small an amount of protein in relation to body size and energy expenditure.

2.1.2. The Prosimian Hand and Its Ability to Manipulate Food Objects

The above statement does not apply to diurnal lemur species weighing between 2 and 4 kg. These species feed mainly on fruits and leaves (see Section 2.5 below) and it may well be asked why they do not eat insects, at least in small amounts, when this food resource is readily available. In fact, the relation between diet and body size presented above for prosimians also applies to the smaller simian primates (Hladik, 1975, 1978) and among these the forest-dwelling monkeys, weighing 2–4 kg, have a diet which includes insects as well as leaves, to supplement the low protein content of fruits. The foraging techniques of monkeys (Thorington, 1967; Hladik and Hladik, 1969, 1972; Gautier-Hion, 1971) involve a very sophisticated hand allowing complex manipulation of bark and dead leaves to find the insects and other invertebrate prey in sufficient quantity.

The parallel evolution of vision is also very important for an efficient foraging technique (see Pariente, 1976, and Chapter 10).

The prosimian hand, with its fairly fixed pattern of control (all fingers press the object toward the distal or proximal palmar pads; Bishop, 1964) does not allow such complex manipulations. Consequently prosimians have not been able to develop the foraging techniques of the simians and the feeding strategy involving leaf-eating, as a supplement to frugivorous diets, has been adopted by species of relatively small size. As suggested by Charles-Dominique (1975, 1977a), the evolution of the prosimian hand may well have stopped at an early stage in evolution because it was already highly specialized for stereotyped movements. These stereotyped movements, perfectly adapted to catch small insects moving rapidly or in flight, are used by small, primitive insectivorous forms (Fig. 3A and B) but, in larger lemur species (Fig. 3C), manipulation of food objects is limited to a hooklike gesture. In spite of a slight difference in the arrangement of the digits when they grasp a piece of food, the Lemuridae, if artificially trained to eat insects, are also able to perform a stereotyped movement resembling that of the small insectivorous species of

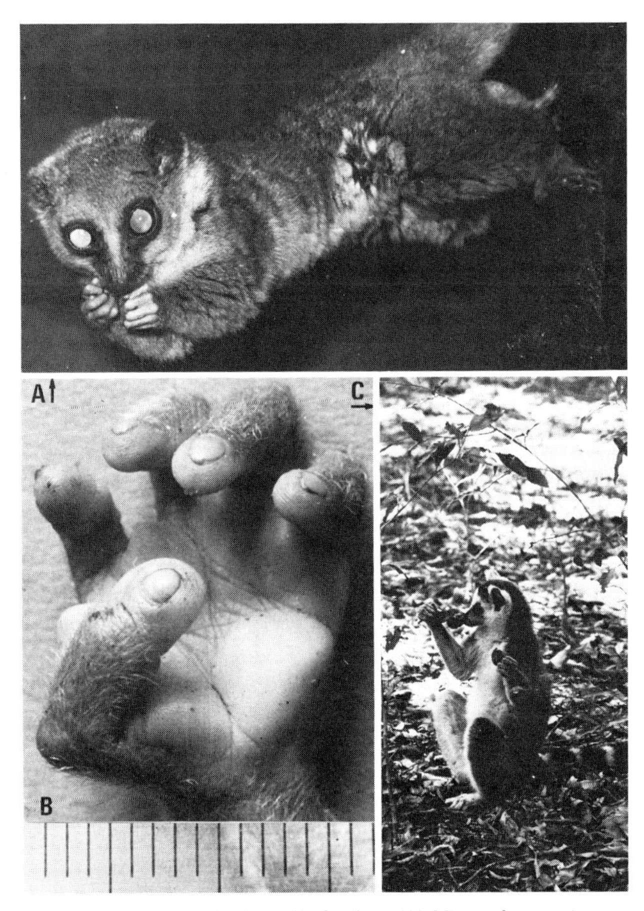

Fig. 3. Movements of prosimian hands while feeding. (A) *Microcebus murinus* catching an insect with a stereotyped movement of both hands—all fingers converge simultaneously as soon as the distal palmar pad touches the prey. Note the particular position of the body—hanging by the legs only. (Photograph: R. D. Martin and C. M. Hladik.) (B) Left hand of *Loris tardigradus*, showing the convergent position of all fingers. The scale is in millimeters. (Photograph: C. M. Hladik and J. J. Petter.) (C) *Lemur catta* handling pods of *Tamarindus indica*, with a hooklike gesture of the hand, pressing the object against the proximal palmar pad. (Photograph: A. Schilling.)

Lorisid. This type of movement has been observed in *Lemur fulvus* and *L. mongoz* (Charles-Dominique, 1975).

2.1.3. *Prosimian Teeth and the Digestive Tract*

A recent general study of primate dentition compared to that of other mammals (Kay and Hylander, 1978) illustrates the difference in function between the front teeth and the cheek teeth. Incisors, canines, and premolars are used for seizing, manipulating, and separating food while biting it; the essential function of the molar teeth, on the other hand, is mastication. Species that eat food that needs to be shredded during mastication require large shearing blades on their molars and, depending on the molar structure in different taxonomic groups, different cusps might be developed for this purpose. Consequently the rear part of the mandible of the folivorous species must be large. By contrast, the extension of the front part of the mandible is an adaptation to frugivorous/insectivorous diets.

A peculiar dental structure is found in the lorisids as well as in the Malagasy lemurs (though not restricted to them, if considering converging forms in other mammals): the two canines and four incisors of the lower jaw, are styliform and almost horizontal, forming a "tooth-scraper." This instrument is frequently called a "dental comb" (Buettner-Janusch and Andrew, 1962) because of its frequent use by prosimians in grooming. The term "tooth-scraper" (Martin, 1972b) probably better reflects its functional adaptation in the extant species, and particularly among those feeding on large amounts of gums and/or sap, such as *Galago elegantulus* and *Phaner furcifer* (see Sections 2.3 and 2.4 below). Martin (1972b), following Walker (1969), suggested that the tooth-scraper evolved in pre-Miocene times in Africa. Lorisiformes and Lemuriformes would thus have had a common ancestor for which gum eating would have been a vital strategy. The tooth-scraper is still very useful for such genera as *Cheirogaleus*, *Microcebus*, and *Galago*, allowing them to scrape inside small cracks of bark, around gum secretions, and to gather a large number of tiny droplets of gum efficiently. The large and strong tooth-scraper of *Propithecus* permits them to eat the bark and cambium of woody species (Richard, 1974).

Even more specialized is the anterior dentition of *Hapalemur* and *Lepilemur*, in which the upper incisors are greatly reduced, as in *Hapalemur*, or entirely absent, as in adult *Lepilemur*, in relation to their browsing habits. Further references and hypotheses concerning the evolution of the prosimian dental formula from the ancestral lemur/loris type ($\frac{2}{2}\,\frac{1}{1}\,\frac{3}{3}\,\frac{3}{3}$), present in the Lorisidae and Cheirogaleidae, toward the Indriidae ($\frac{2}{2}\,\frac{1}{0}\,\frac{2}{2}\,\frac{3}{3}$) and the Daubentoniidae ($\frac{1}{1}\,\frac{0}{0}\,\frac{1}{0}\,\frac{3}{3}$), are reviewed elsewhere (see Martin, 1972b; Petter and Petter-Rousseaux, Chapter 1).

One cannot expect the development of these dental characteristics to change as a function of any rapid change in the diet of the extant prosimian species. By contrast, the proportions of the different parts of the digestive tracts can rapidly follow change to an artificial diet (Hladik, 1967) and would eventually allow some flexibility (see Section 4.1 below).

A large cecum has been retained by those species feeding on gums while the folivorous species have either a very large cecum, or a large foregut, or both. By contrast, the smallest intestine sizes are found in the insectivorous species, in relation to diet concentrated in nutrients. These structural adaptations to diet and their more or less fixed patterns in the different taxonomic groups of prosimians, might be considered either as limits to the range of dietary variations in the most specialized species, or as adaptive tools enabling prosimians to compete with other species for a given ecological niche.

2.2. Specialization in Relation to Ecological Niche

The concept of ecological niche refers to specific structural adaptations as well as to the physiological and behavioral responses of the animal within its community and ecosystem. Biologically speaking the ecological niche could be presented as the ''profession'' of a given species (Odum and Odum, 1959); thus it could be defined almost entirely in terms of dietary specialization (for example, the primary consumer, or the predator, in a given ecosystem). By contrast the ''address'' of the same species would correspond merely to its particular habitat. An accurate definition of the niche concept was proposed by Hutchinson (1975), relating the multiple environmental parameters to a multidimensional hypervolume. Our present knowledge of primate ecology (Hladik and Chivers, 1978) does not allow the introduction of such a concept on a practical basis.

2.2.1. Nocturnal and Diurnal Activity Rhythms

There are a limited number of so-called ''professions'' practicable in a given habitat, because two species cannot share the same food resources, at least not to a very large extent (Gause's principle; Odum and Odum, 1959). This principle implies that during the long process of evolution, interspecific competition for food, if occurring, would result either in the total disappearance of one of the species, or to its adaptation to another type of available food. However, food resources are generally accessible round the clock and, for any particular category of diet, one can often find two different species, one feeding at night, and the other by day (Charles-Dominique, 1975, 1977a). The activity rhythm can thus act as a mechanism ensuring ecological separation. The chances of obtaining food are equal for both species in this particular case and no selection factor can favor one of them.

As a matter of fact, the ancestral prosimians would have had to compete with many diurnal bird species feeding on insects and fruits, if the two classes of animal had had the same activity rhythm. This argument was presented by Charles-Dominique (1975) to demonstrate that the early prosimians were nocturnal, of small size, and fed on insects, fruits, and gums. Later the diurnal primates evolved in Africa and Asia without competition from frugivorous/insectivorous birds, thanks to their particular foraging techniques (see Section 2.1.2 above), allowing access to invertebrate prey hidden in bark and dead leaves. In Madagascar the Lemuridae and Indriidae also evolved as diurnal forms, but essentially because they utilized particular types of food (such as seeds and hard pods), not accessible to birds and mammals lacking the necessary dexterity, as well as large amounts of leaves.

2.2.2. Use of Different Locations in a Given Habitat and Seasonal Rhythms

For different species of prosimian living in a nonhomogeneous environment, the ''address'' might be as important as the ''profession.'' Two examples will illustrate these different systems allowing several sympatric species to feed on limited food resources.

The first example concerns the evergreen forest of Gabon, near Makokou, where Charles-Dominique (1966, 1971a, 1974a, 1977a; Charles-Dominique and Bearder, Chapter 13) conducted his field work. The diets of the different species of nocturnal prosimian are presented in Fig. 4, according to the weight of stomach contents. Without going into detail (see next section) it is apparent that all the species feed on insects supplemented by a small amount of fruit, in the case of *Arctocebus*, or larger

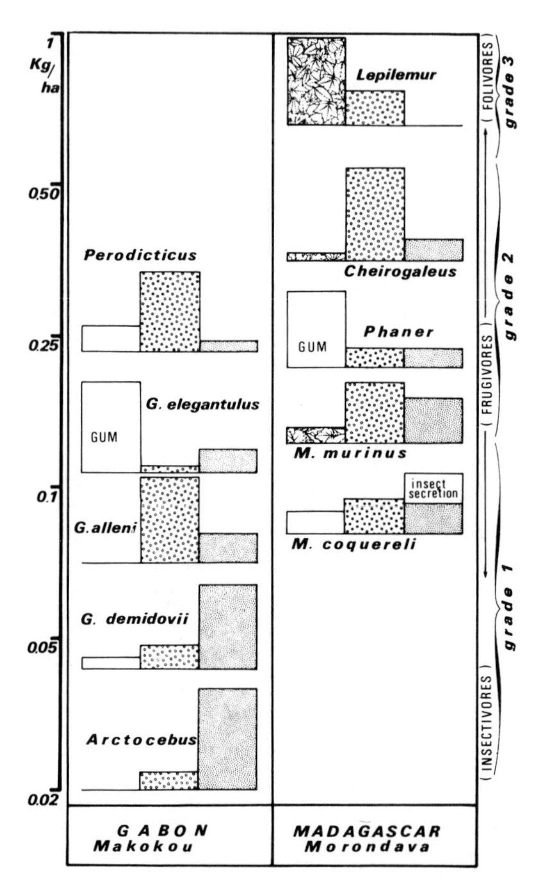

Fig. 4. Comparison of the natural diets of the five nocturnal prosimian species living in the evergreen forest of Gabon (Makokou area), with that of five species inhabiting the deciduous forest of the western coast of Madagascar (Morondava area). For each of the species the diagram represents the proportions of different food categories ingested in 1 year: leaves or gums, left rectangle; nectar, fruits, and seeds, center rectangle; insects and other prey, right rectangle. The three "grades" refer to the ecological significance of these diets, as presented in Section 2.2.3. The different species are located on a vertical scale with reference to the biomass observed in the field. The data concerning Makokou are from Charles-Dominique (1971a) and Charles-Dominique and Bearder (Chapter 13); while those from Morondava are from J.-J. Petter, P. Charles-Dominique, E. Pagès, G. Pariente, and C. M. Hladik (unpublished data) which yield a first rough estimate of the diet of the Malagasy species.

amounts, in the case of *Perodicticus* and two of the *Galago* species, or gums, in the case of *G. elegantulus*. These five species utilize similar resources to a large extent, and would compete with one another, but for the operation of other factors, an example of which is the type of support on which the animals are generally observed in the rain forest.

Arctocebus calabarensis were generally observed on small lianas and bushes in the undergrowth of the forest. *Galago demidovii* were found in small lianas and in the foliage high up in the canopy. *Galago alleni* preferred vertical supports (small tree trunks and the bases of lianas) near the forest floor. *Galago elegantulus* were more at home on the larger branches, tree trunks, and lianas from low level to canopy. *Perodicticus potto* were observed on supports ranging widely in size in the canopy.

According to these observations, summarized in Fig. 5, certain species utilize a very small part of the habitat. *Galago alleni*, for instance, stay exclusively near the forest floor. Furthermore, the primary forest is not a homogeneous habitat, but a "mosaic" of different stages of growth, only 5% of which is truly mature (see A. Hladik, 1978). *Arctocebus calabarensis* live in the very young parts of this forest, in natural clearings, where the very dense and intricate vegetation constitute their preferred habitat. The three other species share the canopy of the forest but *G. demidovii* confine themselves to the small twigs and lianas in the thickest parts of the tree tops.

Other factors separating the species have to do with their different characteristic methods of finding and catching insects (see next section). The absence of competition can be related essentially to their concentration on different insect populations and fruits located in different parts of a nonhomogeneous forest.

The second example also concerns sympatric nocturnal species feeding to a large extent on similar resources, in a fairly homogeneous habitat—the dry deciduous forest of the west coast of Madagascar, near Morondava (a preliminary report concerning this study can be found in Petter, 1978). The diets of these prosimians can also be found in Fig. 4. A fairly large amount of insects is eaten by four of the sympatric species that could lead to competition for this limited food resource.

The main characteristic of the Morondava forest is the very important seasonal variation in food production. During the long dry season, lasting about 9 months, all trees shed their leaves, creating a shortage of food for primary consumers, especially for insects. By contrast, during the luxuriant rainy season, leaves grow quickly and insects are very abundant.

One of the prosimian species, *Cheirogaleus major,* relies entirely on the excess of production during the rainy season. For the remainder of the year they hibernate inside hollow trunks (Fig. 6A and B). Thus interspecific competition is avoided by a temporal factor rather than by spatial distribution of the different species. The food eaten by *C. medius* (which results in fat storage in the tail and under the skin) is sufficiently abundant during the rainy season to preclude competition from other species.

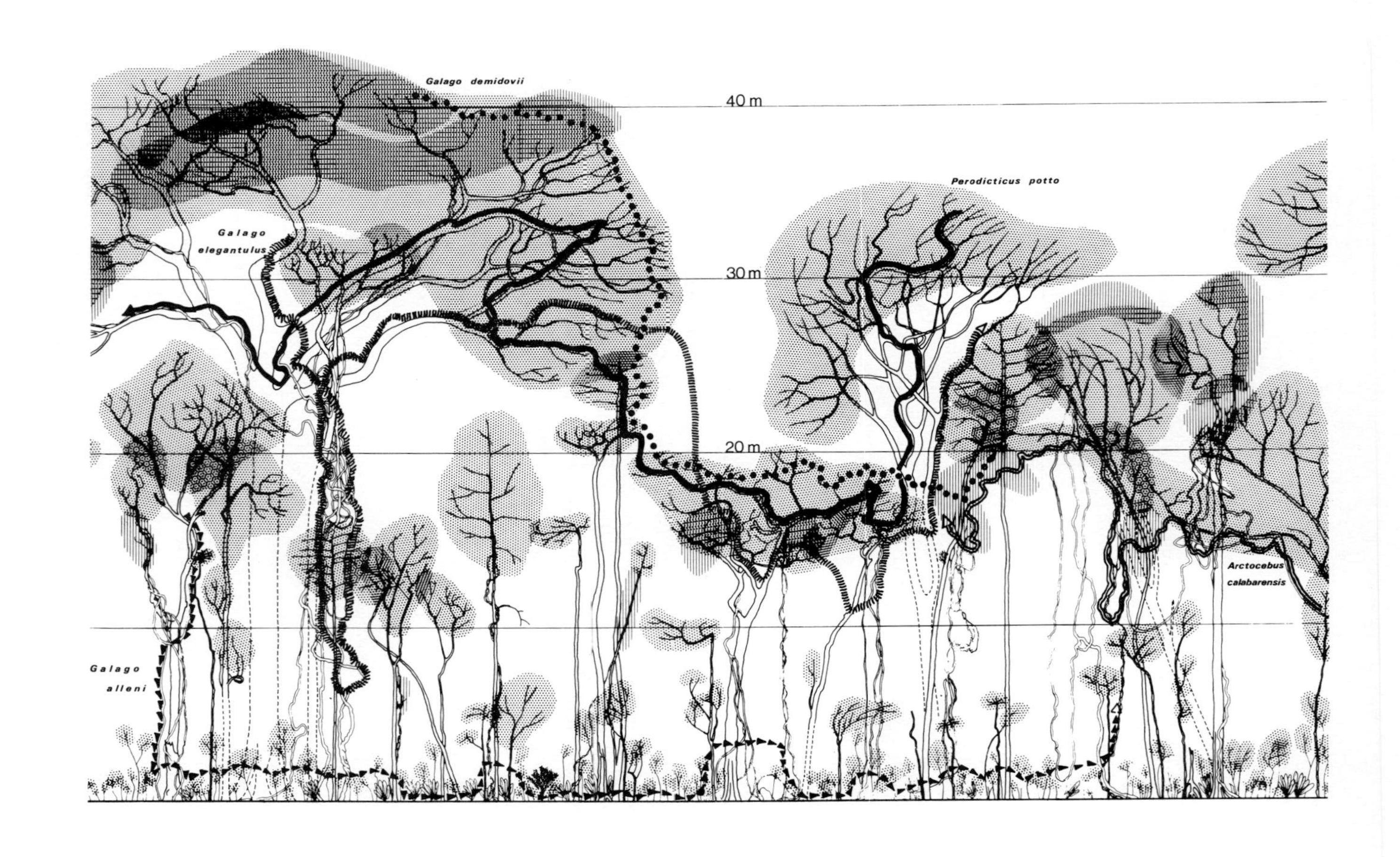
Galago demidovii
40 m
Perodicticus potto
Galago
elegantulus
30 m
20 m
Arctocebus
calabarensis
Galago
alleni

Microcebus murinus, to a lesser extent, also store fat in the tail during the rainy season, when many insects and fruits are available. Although they are less active during the dry season they do not truly hibernate.

Lepilemur ruficaudatus do not compete for food with the other prosimians since they are able to utilize tough foliage and a few fruits when available.

The other two species have developed unique strategies allowing subsistence throughout the year—utilization of the gums of some common trees by *Phaner furcifer* (Fig. 7A), and utilization of insect secretions (as well as dried secretions accumulated during the dry season) by *Microcebus coquereli* (Fig. 7B). Nevertheless, gums and insect secretions are only available in small quantities and can sustain only a limited population of nonhibernating prosimians.

2.2.3. *Food Resources in Different Ecological Niches*

According to the ecological significance of the diet, discussed in the previous section, prosimians as well as simian primates (Hladik, 1975) can be classified into three grades as follows:

Grade 1: from the typically insectivorous forms, such as *Loris* and *Arctocebus*, toward species utilizing fruits and/or gums as a supplementary source of energy, in combination with the insects available (*Microcebus*, some *Galago* species, etc.).

Grade 2: species feeding on small quantities of insects and/or other prey, young leaves, fungi, or other vegetable matter, as a protein supplement to large amounts of fruits or gums which are the major sources of energy. This intermediate grade mainly includes simian species—*Macaca*, *Cercopithecus*, *Cebus*, and some prosimians such as *Perodicticus* and *Cheirogaleus*.

Grade 3: species feeding on fruits supplemented by leaves in sufficient quantity for protein balance, such as *Lemur,* and species eating more leafy parts and/or flowers, such as *Propithecus* and *Hapalemur,* to the most specialized folivorous forms, *Indri* and *Lepilemur*.

There is no absolute distinction between these classes, as shown by the examples in Fig. 4, but this progessive ecological classification allows more accurate definitions than terms such as "frugivore" or "omnivore" (which have been used to

Fig. 5. A reconstruction of the different types of itinerary utilized by the five nocturnal species living in the primary rain forest of Makokou (Gabon). The sample of forest represents a transect of 5 m in width (some trees, in broken lines, are located out of this section, after data from A. Hladik, 1978). The dotted areas represent tree foliage and the hatched areas represent lianas. The itineraries of the different prosimian species are based on the field work of Charles-Dominique (1966, 1971a, 1977a) concerning the height and the typical types of support where the animals are characteristically observed. *Galago demidovii*, on small branches and lianas in the canopy; *G. elegantulus*, looking for gums along smooth tree trunks and liana stems; *G. alleni*, jumping on the vertical supports of the undergrowth and catching insects on the ground; *Perodicticus potto*, moving along large branches and lianas in the canopy; *Arctocebus calabarensis*, in dense vegetation, generally in the tree fall areas, where newly grown lianas and trees do not exceed 10 m in height.

describe grades 1 and 2 diets as well as any diet which includes the flesh of large prey).

Each of the three grades is characterized by a major food type:

Grade 1: Insects (and/or small prey—invertebrates as well as vertebrates).
Grade 2: Fruits (and/or seeds and gums).
Grade 3: Leaves (and/or shoots).

These major types of food are available in limited amounts according to the primary production of the habitat.

Primary production of leaves is fairly constant in the tropics at 7 tons (dry weight) per hectare per year in a rain forest (A. Hladik, 1978). A deciduous forest has a smaller production, but the order of magnitude is the same at 3–4 tons. The total production of fruits measured in the rain forest was around 0.5 tons per hectare per year. Insects collected in the litter at Makokou, representing an undetermined proportion of the total insect production (A. Hladik, unpublished data), yielded a figure of 23 kg per hectare per year. This is a classical pyramid of production, with the insects on top accounting for at least 1/200th of the leaf primary production.

At the very top of the pyramid would be the prosimians of grade 1, obtaining their energy from insects, together with other insectivorous mammals and birds sharing this resource. The biomass of such species is generally between 0.01 and 0.1 kg per hectare. The species of grade 2 have their major food type (fruits) available in larger quantity. As a result their biomass may be higher—between 0.1 and 1.0 kg per hectare. The folivorous species of grade 3 are able to reach the highest biomass—5.0 kg per hectare in the case of *Lepilemur* (Charles-Dominique and Hladik, 1971) and up to 15 kg per hectare in the case of folivorous simians.

In the examples presented in Fig. 4 the biomass of the species included in grade 1 increases in relation to the proportion of fruits in the diet (and decreases when the proportion of insects increases). There is a similar type of relationship between the food available and the biomass in the other grades. Thus, in each of these grades, the biomasses of different prosimian species follow the proportions of the major food types in their diets according to their availability in the natural environment. The intermediate levels, within each of the three ecological grades, can be indirectly shown by measuring the biomasses of a given habitat.

Other species of prosimian will be described below with reference to these ecological characteristics determined by food resources to which their specific patterns of feeding behavior have adapted.

2.3. Prosimians of Asia and Continental Africa

Whenever prosimians share the same habitat as monkeys and apes they occupy entirely different ecological niches (Bourlière, 1974). In Asia and continental Africa a limited number of nocturnal prosimians live in the same dense evergreen forest (rain forest) as well as in dryer habitats such as deciduous and semiarid shrub of the tropical and equatorial zones. Their nocturnal activity rhythm accounts for the separation of ecological niches from those of the diurnal simian primates. Other more subtle factors account for niche separation in a habitat shared by several prosimian species.

Only two Lorisinae and the three species of tarsier are present in Asia.

2.3.1. Loris tardigradus (Slender Loris: Fig. 1)

The slender loris is found in Sri Lanka and in India. Different subspecies are adapted to different habitats (cf. Hill, 1953). For instance, in Sri Lanka, *L. t. nycticeboides*, which have a very thick fur, inhabit the montane forest of the central districts of the island where low temperatures and permanent high humidity are the predominant climatic features. Three other subspecies are distributed in the various climatic zones, among which *L. t. nordicus* were studied in the field (Petter and Hladik, 1970), especially in the deciduous forest of Polonnaruwa.

This animal is relatively heavy (0.25 kg) if one considers its diet which includes only small amounts of fruit, the bulk of the diet being made up of insects and invertebrate prey (and, occasionally, geckos and small birds). This is due to the utilization of abundant prey (see Section 1.2 above) which are generally avoided by other mammals and insectivorous birds living in the same habitat. *Loris t. nordicus* utilize approximately 1 hectare of forest per individual: accordingly their biomass is about 0.2 kg per hectare, a very high figure for a predator.

Considering the distribution of males and females, one would expect the social structure of *L. tardigradus* to resemble that of *Galago* or *Perodicticus* (Charles-Dominique, 1977b), each animal being generally solitary in its individual home range or territory. They move according to irregular patterns, a strategy utilized by most insectivorous primates (Hladik, 1975), which avoids localized destruction of the invertebrate prey, and thus allows a regular supply. Further investigations are necessary to confirm and elaborate on some of these socioecological observations of *Loris*.

The habitat of *L. t. nordicus* is restricted to the dense parts of the forest, where the animals move slowly along the lianas and thin branches. This particular locomotor pattern is partly a strategy to prevent detection by predators, but it also allows them to approach large insects sufficiently closely to grab them with the stereotyped movement of the hand described in Section 2.1.2 above. Like other Lorisinae, *L. tardigradus* utilize the olfactory sense more than do simian primates (see Schilling, Chapter 11), and detection of prey, when the animal moves slowly with its nose close to the branch, depends more on olfaction than on vision. The unpleasant smell of many invertebrate species, which is generally an antipredator device, is used by *L. tardigradus* to improve feeding efficiency.

2.3.2. Nycticebus coucang (Slow Loris)

The slow loris is a larger animal, weighing about 1.2 kg, living in southeast Asia where different subspecies are adapted to different habitats. *Nycticebus coucang* might once have been sympatric with *Loris tardigradus* in the southern forest of India, where the latter species now occurs alone.

According to the field observations of Elliot and Elliot (1967) the feeding strategy of *Nycticebus coucang* is somewhat similar to that of *Loris tardigradus*—moving slowly along lianas and branches in a large home range in search of invertebrate prey, some of which have a repugnant smell and taste; but, owing to its larger size, the diet of *N. coucang* must be supplemented by large amounts of fruits. This different diet (the exact nature of which needs further field investigation) in combination with other ecological factors, such as those accounting for niche separation between *Arctocebus* and *Perodicticus* (Section 2.2.2), prevents competition in case of sympatry.

2.3.3. *Tarsius spp.*

The three species of tarsier (*T. bancanus*, *T. spectrum*, and *T. syrichta*) are distributed exclusively on different islands of the Sunda archipelago (Hill, 1955, and see distribution map in Petter and Petter-Rousseaux, Chapter 1). *Tarsius bancanus* inhabit the lower strata of secondary as well as primary rain forest in Borneo (Fogden, 1974; Niemitz, this volume). Vertical clinging and leaping (Napier and Walker, 1967) between small trunks and lianas allows very rapid movements in this particular habitat.

The diet of *T. bancanus* mainly includes invertebrates such as Orthoptera and spiders, and some vertebrates such as geckos and other small lizards. There is no evidence that fruits are included in the natural diet, since the animal was observed near fruiting trees only to catch the insects attracted to the fruit. Some tests of food preferences on *T. bancanus* (Niemitz, Chapter 14) showed that the animal was also able to catch and eat venomous snakes.

In relation to this specialized diet the biomass of *T. bancanus* is lower than 0.05 kg per hectare. The animal weighs between 0.10 and 0.12 kg, and has a home range of 3 hectares in some instances. The absence of nocturnal insectivorous competitors in the habitat of the tarsier presumably explains the persistence of such a form with an optical system of relatively poor efficiency (no *tapetum lucidum*; see Pariente, Chapter 10) which necessitates a considerable increase in the weight of the eyes and the dependent muscular system (including the neck) to allow efficient nocturnal vision.

In the rain forest of continental Africa the ecological specializations have allowed several nocturnal species of prosimian to inhabit the same area (see Charles-Dominique and Bearder Chapter 13). The dietary characteristics of the five sympatric species of Gabon will be briefly described as a complement to the other features presented in Section 2.2.2.

2.3.4. *Arctocebus calabarensis aureus (Angwantibo)*

The angwantibo is the African homolog of *Loris tardigradus*. *Arctocebus calabarensis* inhabit the densest parts of the forest. Their particularly slow locomotor habit, when moving along lianas and small branches, is an effective device against detection by large predators as well as allowing close approach to invertebrate prey. In the rain forest of Makokou the diet of *A. c. aureus* includes a fresh weight average of 85% animal prey and only 14% fruit (see Fig. 4). The particular identity of the prey, determined by Charles-Dominique (1966) from the analysis of stomach contents, showed a very high proportion of caterpillars, some of which are protected by venomous hair. Before ingesting such prey, *A. calabarensis* force it through one hand, squeezing and extending it until most of the venomous hairs are broken. The other types of prey eaten, such as beetles, crickets, and ants, are also often rejected by other animals. As with *L. tardigradus* such a diet is not necessarily determined by preference (see Section 1.2 above) but is based on the ability to detect prey predominantly by smell, as well as the ability to tolerate prey repugnant to other species (Charles-Dominique, 1971a, 1972, 1974a, 1977a).

As a result of these ecological characteristics *Arctocebus calabarensis aureus* can maintain a body weight of 0.2 kg, which is large for an animal feeding mainly on insects and quite similar to that of *Loris tardigradus*. Nevertheless, its biomass is very low at 0.005 kg per hectare (mean value) with a maximum of 0.015 kg per hectare in some areas. These low figures are deduced from the mean densities calculated by Charles-Dominique along pathways in the rain forest. If we consider, however, that *A. calabarensis* are restricted largely to natural clearings and small patches of rapid growth, that can be defined as "subhabitats" (see Section 2.2.2) in the rain forest, the biomass would be much higher in these local patches. Unfortunately there are, as yet, no reliable methods for studying separately and accurately each of the many parts of the mosaic structure of the primary rain forest. The total biomass of the prosimian species has to be considered in relation to the total food resources, in particular, the secondary production of insects and other small prey utilized by the prosimian species, for any comparison with other ecosystems.

2.3.5. Perodicticus potto (Potto)

Perodicticus rely on the same food resources as *Arctocebus* but eat a larger proportion of fruit (65%), some gums (21%), and only 10% animal prey (Charles-Dominique, 1977a). This diet accords with its larger body size (0.8–1.2 kg) and the need to supplement with fruits and gums the relatively small amount of invertebrate prey that can be found in one night's activity. Ants (especially *Crematogaster spp.*) form the bulk of the prey supplemented by large beetles, slugs, caterpillars, spiders, and even centipedes that are avoided by the other primate species. The potto is the African homolog of *Nycticebus coucang*, of similar size and shape, the same slow locomotor behavior, and a comparable frugivorous diet supplemented by insects and other small prey rejected by other species.

Perodicticus potto are found in the canopy of the primary rain forest, moving along large branches in a more open "subhabitat" than that of *Arctocebus calabarensis*. Near Makokou the mean biomass of the potto, calculated after Charles-Dominique's observations along pathways, is 0.1 kg per hectare; but in particular areas, especially in more humid patches, the maximum biomass is as high as 0.3 kg per hectare. These figures, higher than in the case of *A. calabarensis*, are obviously determined by a diet including fruits as the major source of energy (grade 2, as defined in Section 2.2.3).

2.3.6. *Galago demidovii* (Dwarf Bushbaby)

Together with the two other Galaginae of the primary rain forest of Makokou, *G. demidovii* are very different from the two Lorisinae with which they are sympatric. They are the smallest of the prosimians of continental Africa (body weight, 0.06 kg), and probably the most active, being able to leap and run very fast.

The diet of *Galago demidovii* includes an average fresh weight of 70% insects, 19% fruits, and 10% gums (Charles-Dominique, 1974a, 1977a) and a very small amount of leaves and buds. The energy required by this active small form comes from the secondary production of insects and other small prey (grade 1, as defined in Section 2.2.3) and, accordingly, the biomass is low at 0.03 kg per hectare, the animal occupying large home ranges of up to 3 hectares (Charles-Dominique, 1971b).

The "subhabitat" of *G. demidovii* is limited to the top of the canopy of the rain forest where they are usually found among small lianas and fine branches and where the potto, with its larger body weight, is not able to go. Thus *G. demidovii* have access to food resources largely inaccessible to the other sympatric Lorisidae. Nevertheless, food choice in *G. demidovii* is related largely to their strategy of rapid movement and fast running which allow them to capture large fast-moving insects such as moths and grasshoppers (cf. Charles-Dominique and Bearder, Chapter 13, Table VI). The stereotyped grabbing movement (see Section 2.1.2) of *G. demidovii* are particularly rapid and efficient, the animal being able to maintain its support by means of its legs. Furthermore, sonolocation of flying insects was deduced by Charles-Dominique (Charles-Dominique and Bearder, Chapter 13), the large mobile ears of *G. demidovii* being immediately directed toward relevant sounds. Hearing and sound detection of movements is more important than olfaction in the predatory strategy of this species.

Galago demidovii can escape rapidly from predators by leaping and running on small branches and, again, this active strategy is different from that of the Lorisinae which resort to passive avoidance by dissimulation and slow movements.

2.3.7. *Galago alleni* (Allen's Bushbaby)

Galago alleni have a feeding strategy very similar to that of *G. demidovii* but, as vertical clinging and leaping animals, they confine their activity to the undergrowth of the primary rain forest at low levels or on the forest floor (Charles-Dominique, 1974a; Charles-Dominique and Bearder, Chapter 13). The main types of prey, beetles and moths, are similar to those eaten by *G. demidovii*, but collected in a different "subhabitat," a peculiar type of niche separation which avoids competition for food resources.

The quantity of animal food found by Charles-Dominique in the stomachs of both species of *Galago* was, on average, fairly similar (about 2.0 gm). The relation between diet and body size (see Section 2.1.1) thus follows the general rule: *Galago alleni*, which are larger (0.3 kg), eat more fruits (73%) as a supplement to prey (25%). Other differences concern snails and frogs eaten only by *G. alleni*, and ripe fruits which have fallen to the ground and which are not accessible to *G. demidovii*. The biomass of *G. alleni* (0.04 kg per hectare) in the primary rain forest of Makokou is probably limited by the availability of insects. In a recent study Charles-Dominique (1977b), using radiotelemetry, demonstrated that individual home ranges are very large—about 10 hectares for females and up to 50 hectares for males.

2.3.8. Galago elegantulus (Needle-Clawed Bushbaby)

Galago elegantulus have a very specialized diet including, on average, a fresh weight of 75% gums, 5% fruits, and 20% insects (Charles-Dominique, 1974a). Gum eating is probably a primitive habit of prosimians (see Section 2.1.3 above) and, in the Makokou rain forest, *Perodicticus potto* and *G. demidovii* also feed on gums exuded by different lianas and tree species, but to a lesser extent than *G. elegantulus*. This species has peculiar morphological and behavioral adaptations, described by Charles-Dominique (1977a; Charles-Dominique and Bearder, Chapter 13), such as the clawlike nails, allowing access to gums along smooth trunks and large branches, a large "tooth-scraper," and a rigid behavioral pattern of using regular pathways to visit the different lianas and trees producing gums within the home range.

During these visits *Galago elegantulus* move in the canopy of the rain forest and may descend to visit certain lianas such as *Entada gigas*. They may catch some large insects, mainly grasshoppers, some large beetles, moths, and caterpillars. On an average, *G. elegantulus* eat as many insects as *G. alleni*, and both species have approximately the same body weight (0.3 kg); their biomass (0.05 kg per hectare for *G. elegantulus*) is also fairly similar. As already stated, the total biomass of the three Galaginae feeding on insects, which is 0.12 kg per hectare, must be taken into consideration. This is about the same as the biomass for the two Lorisinae in the primary rain forest of Makokou. Together they constitute a biomass approximating that of *Loris tardigradus* in Sri Lanka (0.2 kg per hectare). Food production has the same order of magnitude in the different forest types but, in the African rain forest, it must be distributed among a number of different specialized forms.

In the dry habitats of South and West Africa two species of *Galago* are represented, sometimes sympatrically, by different subspecies living in different climatic zones (Bearder and Doyle, 1974).

2.3.9. Galago senegalensis moholi (Lesser Bushbaby)

The lesser bushbaby is the same size as the largest *Galago* species living in the rain forest of Makokou (body weight 0.3 kg) but they have a much higher biomass, about 0.2 kg per hectare, calculated on the basis of the average home range (Charles-Dominique and Bearder, Chapter 13). If calculated on the basis of population densities the biomass may be as high as 1.0 kg per hectare. These figures approximate the total biomass of the prosimians feeding on insects in the rain forest of Makokou.

The diet of *Galago senegalensis* (Sauer and Sauer, 1963; Charles-Dominique and Bearder, Chapter 13) includes large amounts of various insects, mainly butterflies, moths, and beetles (Doyle, 1974a, b). The gum of *Acacia* trees is eaten throughout the year and may represent the main source of energy (grade 2 diet).

2.3.10. Galago crassicaudatus umbrosus (Thick-Tailed Bushbaby)

Galago crassicaudatus were studied by Bearder (Charles-Dominique and Bearder, Chapter 13) in the riparian bush of the north east Transvaal where they occur sympatrically with *Galago senegalensis*. This large animal (body weight 1.3 kg) also presents a very high biomass—about 0.3 kg per hectare (Bearder and Doyle, 1974) for one family group, but may be as high as 1.5 kg per hectare in high population density areas. The diet of *G. crassicaudatus* mainly includes fruits and gums (or sap) and only rarely large prey like reptiles and birds in some areas. Nectar and seeds are also consumed, as well as some insects. This diet, typical of grade 2 (as defined in Section 2.2.3), might tend, in some cases, toward a grade 3 diet, utilization of the primary production as the major source of energy, which would allow for the very high biomass reported.

Further field investigations will probably be necessary to specify more precisely some of these dietary parameters. In the habitats where the two bushbabies live sympatrically the mechanism of niche separation is based on the use of "sub-habitats" similar to that resulting in the sophisticated mode of sharing food resources in the rain forest of Makokou. *Galago senegalensis* are localized in orchard bush, and *G. crassicaudatus* mainly in the riparian bush (Bearder and Doyle, 1974). The latter species have also developed certain strategies reminiscent of the slow movements of *Perodicticus potto*, that would result in food preferences different from those of *G. senegalensis*. Nevertheless, both species of *Galago* rely on gums as the main source of energy, at least in winter, when fruits and insects are scarce and subtle mechanisms must exist to prevent interspecific competition.

2.4. Nocturnal Prosimians of Madagascar

The examples presented below will be limited to the five species of sympatric prosimian living in the deciduous forest of the western coast of Madagascar (cf. Section 2.2.2), and to the unique case of *Daubentonia madagascariensis*. Recent field investigations of these species illustrate the relationship between behavior and physiological adaptation. The same relationship presumably applies to many other prosimian species for which further field investigations are needed (Petter, 1978; Petter *et al.*, 1977).

2.4.1. *Cheirogaleus medius* (Fat-Tailed Dwarf Lemur)

In the deciduous forest of the west coast of Madagascar, near Morondava, *C. medius* have a very high biomass (0.5 kg per hectare) for a grade 2 species which include in their diet a large amount of insects. The high population density (about 4 animals per hectare) is a result of the seasonal availability of food resources in excess of the requirements of all the sympatric species (cf. Section 2.2.2; Petter, 1978).

The seasonal feeding strategy of *Cheirogaleus medius* (Hladik, Charles-Dominique, and Petter, in prep.) is related to their unique physiology, probably similar to that of rodent species hibernating in the temperate zones (Jameson and Mead, 1964). A very rapid increase in the body weight from 0.12 to 0.25 kg during the rainy season is due to storage of fat in the tail, which increases in volume from 20 to 54 cm^3, and under the whole skin. During the dry season *C. medius* hibernate in hollow trunks, where several animals pile together in small lodges of earth and decaying wood, a system which maintains moisture and a fairly constant temperature. They remain lethargic for 7 to 9 months, and lose about 100 gm in body weight.

The diet of *Cheirogaleus medius* (Fig. 4) includes fruits and flowers (mainly the nectar of some flowers, very abundant in certain tree species), insects (mainly beetles), and a few leaf buds and gums. Flowers and nectar are used at the beginning of the rainy season, December and January (Fig. 6A), while fruits are the staple food in February and March.

2.4.2. *Microcebus murinus* (Lesser Mouse Lemur)

Microcebus murinus also have a seasonal feeding strategy, but less varied than that of *C. medius*. The activity of *M. murinus* decreases during the dry season, with torpid periods, but no true hibernation. Body weight varies from 0.05 to 0.08 kg, and the volume of the tail varies from 5 to 20 cm^3, after accumulation of fat at the end of the rainy season. In terms of diet, feeding strategy, and ecology, *M. murinus* are very similar to *Galago demidovii*; these two species representing an archaic type, probably close to the common ancestor of Malagasy and African/Asian prosimians (Charles-Dominique and Martin, 1970). The diet of *M. murinus* consists of a large amount of insects and other small prey, mainly beetles and spiders, but also occasionally tree frogs and chameleons. Fruits and flowers (and nectar) are also eaten in large amounts, with leaf buds, gums, and insect secretions in smaller amounts. In a different environment, *M. murinus* were observed feeding on leaves of *Uapaca sp*. (Martin, 1972a, 1973). Such an animal, deriving the major source of its energy from the secondary production (grade 1), but using most of the food types available, could be the ancestor of many more specialized forms.

Microcebus murinus are generally found in the dense parts of the forest environment, running and leaping rapidly among the small branches (Fig. 6C) and lianas. They are able to grab prey with stereotyped movements of both hands and to

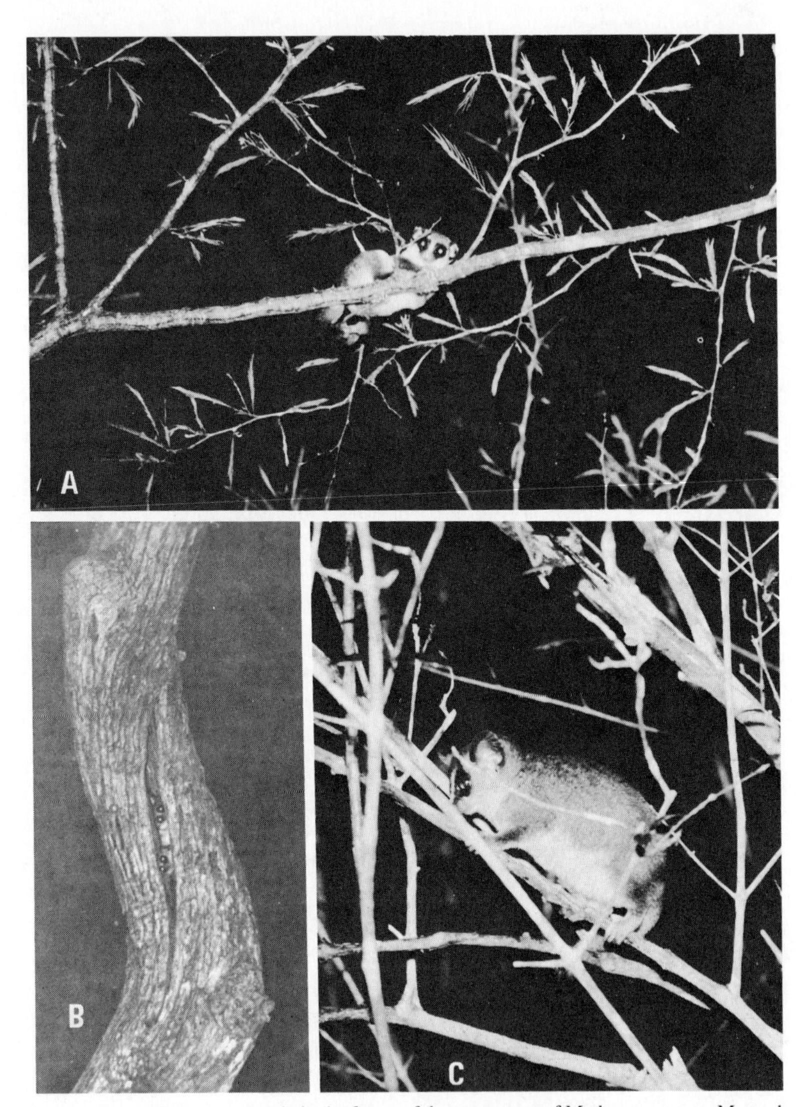

Fig. 6. Cheirogaleinae sympatric in the forest of the west coast of Madagascar, near Morondava. (A) *Cheirogaleus medius* foraging in the flowers of Mimosaceae. This animal was marked, after trapping, by shaving a part of its tail to allow individual identification. (B) At the end of the dry season *Cheirogaleus medius* were still in the hollow trunks where they had been hibernating for 7 to 9 months. (Photograph: Hladik and Charles-Dominique.) (C) *Microcebus murinus* foraging in small twigs at a low level in the forest.

maintain balance at the same time by gripping the support with their feet only (Fig. 3A). These combined rapid movements are also performed by *Galago demidovii*. Due to their greater speed *M. murinus* have access to certain types of prey, especially flying insects, unavailable to the slower *Cheirogaleus medius* in the deciduous forest of Morondava.

The population density of *M. murinus* is 4 animals per hectare in the Morondava forest, thus the biomass is around 0.2 kg per hectare. Very similar figures, 3.6 animals per hectare, were obtained in another dry habitat in the south of Madagascar (Charles-Dominique and Hladik, 1971). The biomass is very high if compared with other grade 1 species, which rarely exceed 0.1 kg per hectare. The seasonal use of excess food resources with a change in the level of activity during the season of scarcity is the physiological mechanism (although less marked than in the case of *Cheirogaleus medius*) which accounts for the wide distribution of *M. murinus* in Madagascar.

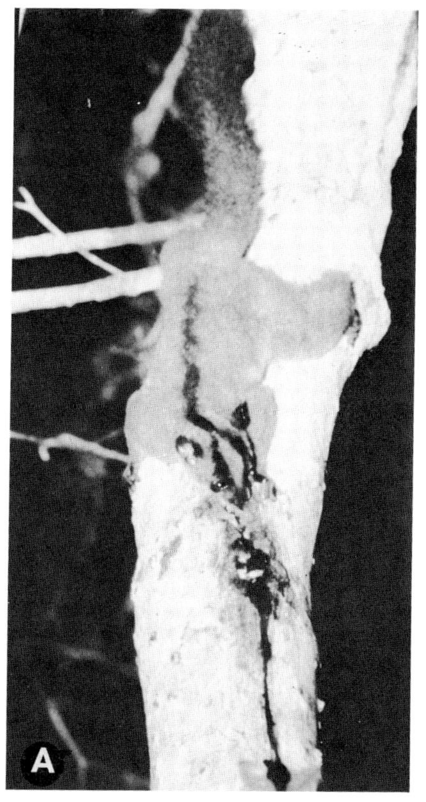

Fig. 7. Three other sympatric species of nocturnal prosimian in the deciduous forest near Morondava (Madagascar). (A) *Phaner furcifer* descending along a trunk of *Terminalia* sp., to eat the gum with the help of its tooth-scraper. (*continued*)

Fig. 7. (*continued*) (B) *Microcebus coquereli* moving along lianas to feed on the sweet secretion of Homoptera larvae: the nest is visible on the right side of the picture. (Photograph: J. J. Petter.) (C) *Lepilemur ruficaudatus* before leaping from a vertical tree trunk.

2.4.3. *Microcebus coquereli (Coquerel's Mouse Lemur)*

Microcebus coquereli, sympatric with the above two mentioned species in the forest of Morondava, have the lowest biomass (less than 0.1 kg per hectare), even if those parts of the forest used exclusively by this species are considered. For instance, Pagès (1978) found 2 males, 2 females, and 3 juveniles ranging over 17 hectares. *Microcebus coquereli* are generally found along rivers and near semipermanent ponds, where the forest is thicker and slightly higher (up to 20 m) than in the dryer parts (Petter *et al.*, 1971).

This animal, weighing 0.3 kg, feeds on insects, some fruits, and some gums, but a large quantity of its energy is derived from particular insect secretions (Fig. 4). The activity of *M. coquereli* is constant throughout the year and there is no obvious seasonal variation in body weight. During the dry season they rely mainly on the sweet secretion of the larvae of Flatidae (Homoptera), spending up to 60% of their feeding time licking the branches around these larvae, and later looking for the dried secretions left by colonies of Homoptera, which have the appearance of large pieces of sugar and are also used by local human populations. The availability of this unique food, which may be considered part of the secondary production, is probably a factor limiting the population density of *M. coquereli*. During the rainy season, when other types of food are abundant, *M. coquereli* feed from the same sources as the other sympatric Cheirogaleinae, i.e., fruits, flowers, nectar, and insects, but probably also feed on larger prey (following the results of food tests in captivity) and may also catch and eat *M. murinus*.

Microcebus coquereli spend the daytime in spherical nests made of small twigs and leaves woven together, which serve as a protection against predators especially in the case of young animals.

2.4.4. *Phaner furcifer (Fork-Marked Lemur)*

Phaner furcifer are specialized gum eaters. As for the other species mentioned above, the availability of a particular food source throughout the year allows this animal to survive in the deciduous forest without any marked seasonal variation in body weight or activity (Charles-Dominique and Petter, 1978).

Phaner furcifer have many characteristics also found in the other specialized gum eaters, like *Galago elegantulus* (see Section 2.3.8 above), a fairly similar body weight (0.3 kg), a unique nail structure allowing them to descend smooth trunks to gain access to particular gum sites (Fig. 7A), and a large and horizontal tooth-scraper. Certain behavioral patterns are also convergent characteristics related to reliance on gum as a staple food, in particular, the efficient use of regular pathways to visit gum sources at the beginning of the night, all the gum-producing trees being located in the home range (see Pariente, 1975). In the Morondava forest a common tree species, *Terminalia* sp., is the main source of gum. It might be noted in passing that, during the daytime, one bird, *Coua cristata*, feeds on the gum of these trees, but not as effectively as *P. furcifer* (Charles-Dominique, 1976). The diet of *P. furcifer* also includes small quantities of insect secretion (from Homoptera larvae and from ladybirds), some fruits, and about 10% insects necessary to compensate for the low protein content of gums. The sap of the Baobab, *Adonsonia* sp., is also eaten during the dry season. The biomass of *P. furcifer* may be as high as 0.4 kg per hectare in the Morondava forest.

The total biomass of the nocturnal prosimian species feeding partly on insects (grade 1 and grade 2) in the Morondava forest, is much higher than in the rain forest of Makokou (see Fig. 4). The higher biomass of Morondava is not due to higher primary and secondary production, but may be due to the presence of other animals in Makokou, not shown in Fig. 4, like the diurnal simian primates, which do not compete directly with the prosimians (see Section 2.2.1 above), but which include large amounts of insects in their diets (grade 2). In Morondava, the diurnal lemur species (*Propithecus verreauxi* and *Lemur fulvus*) and *Lepilemur ruficaudatus*, have grade 3 diets; thus the food resources are shared by a smaller number of species, allowing higher biomasses of the particular forms adapted to the severe local seasonal changes.

2.4.5. *Lepilemur ruficaudatus* (Sportive Lemur)

Lepilemur are able to utilize the tough foliage of different tree species as their staple food. This ubiquitous material is almost totally absent during the dry season in the forest of Morondava, and it is not yet known what type of production determines the carrying capacity of this habitat at this time. The biomass of *L. ruficaudatus* exceeds 2.0 kg per hectare; its body weight of 0.8 kg and its population density is estimated at 3 animals per hectare.

In the bush of southern Madagascar the biomass of *Lepilemur leucopus* is 2.1 kg per hectare (Charles-Dominique and Hladik, 1971) and is limited by the availability of the flowers of two species of *Alluaudia*, which constitute the main food source during the dry season, but the biomass of the same species of Sportive Lemur may exceed 5 kg per hectare in the gallery forest where leaves are available throughout the year. A unique physiological adaptation, cecotrophy, allows the different species of *Lepilemur* to digest a sufficient proportion of these foods which are particularly poor in nutrients (see Section 4.2 below). The activity of these animals is minimal and less than 10% of their energy is spent in moving to feeding sites (Hladik and Charles-Dominique, 1974) within a very small territory (see Section 3.1 below). Most of their time during the night is devoted to immobile surveillance of the borders of their territories defending small food resources.

2.4.6. *Daubentonia madagascariensis* (Aye-Aye)

Daubentonia are remarkable for their unique feeding strategy enabling them to feed on wood-boring grubs. The absence of any representative of the woodpecker family has left vacant an ecological niche for which *Daubentonia* have become specialized, enabling them to monopolize a readily available source of food (Petter and Peyrieras, 1970a).

Fig. 8. Specialized prosimian feeding techniques. (A) *Daubentonia madagascariensis* feeding on a wood-boring grub. The superficial wood of the branch is bitten off with the incisors and the grubs are crushed inside the hole with the thin third digit which is then rapidly licked. (Photograph: J. J. Petter.) (B) *Hapalemur griseus* are specialized for feeding on bamboos but can be adapted to a common grass, *Dactylis glomerata*. While feeding on this grass they handle it with stereotyped movements. (Photograph: C. M. Hladik.)

Daubentonia were observed in the rain forest of the east coast of Madagascar (Petter, 1962; Petter and Petter-Rousseaux, 1967; Petter and Peyrieras, 1970a). Their diet includes a large amount of fruits supplemented by insect larvae. The larvae are detected by smelling and hearing the insects inside decaying wood, the large ears of the aye-aye being directed toward the source. The wood is then bitten and the superficial parts are removed by means of the very strong and sharp incisors. When the larvae are found they are crushed inside the hole by means of the very thin and elongated (relative to other prosimians) third digit and extracted in the form of juice by rapid movements of this finger which is licked at each stroke (Fig. 8A). A similar technique is used to drink and eat the young pulp inside a coconut, after making a small hole with the incisors. Other types of fruit are eaten by more classical methods.

Since *Daubentonia*, weighing about 2.0 kg, include a large amount of insect larvae in their diet, the home range must be large enough to provide a sufficient amount of this resource. Petter and Peyrieras (1970a) observed one male with a female and a juvenile ranging in an area of about 5 km in length. The biomass might, therefore, be smaller than 0.01 kg per hectare, which is the order of magnitude found for other insect eaters of the rain forests.

2.5. Diurnal Prosimians of Madagascar

Where the diurnal prosimians of Madagascar fill ecological niches occupied in other parts of the world by simian primates they show a high degree of convergence but they are by no means exactly similar to simians (Charles-Dominique, 1977a). Many morphological and physiological differences, some of which are discussed in Section 2.1 above, have not permitted the evolution of certain feeding strategies characteristic of simian primates. A limited number of examples will be presented to illustrate the main types of diet of the diurnal prosimians.

2.5.1. Lemur catta (Ring-Tailed Lemur)

Lemur catta are a large frugivorous species supplementing their diet with a large amount of leaves (Fig. 3C). In natural conditions *L. catta* have only rarely been observed feeding on insects (Jolly, 1966). Young leaves and shoots are, therefore, a necessary supplement to fruit protein. In terms of number of observations their "diet" consist of 70% fruits, 25% leaves, and 5% flowers. This diet, based on primary production (grade 3), allows the very high population density observed in the gallery forest of the south of Madagascar. One group of about 20 *L. catta* was calculated to have a home range of 5.7 hectares, thus the biomass may be as high as 7 kg per hectare. This high population and the ranges of the different groups were maintained with very little change over several years (A. Jolly, personal communication).

2.5.2. *Lemur fulvus* (Brown Lemur)

Lemur fulvus were studied by Sussman (1972, 1974) in a deciduous forest of the southwest of Madagascar where they live sympatrically with *L. catta*. These two species have different feeding strategies enabling them to share the available food resources. *Lemur fulvus* live in small groups (about 10 animals) in small territories of about 1 hectare and feed in the canopy of the most common plant species (80% of the feeding observations concerned only three species of plant). Leaves form the bulk of the diet (89% of the feeding observations) but some fruits (7%) and flowers (4%) are also eaten.

In this same area *L. catta* form larger groups (an average of 18 animals), ranging over larger territories (9 hectares) and feed on some plant species that are not evenly distributed. A large part of the foraging time is spent on the ground or near the forest floor. A large amount of leaves are still eaten (44% of the feeding observations), but fruits (34%), some flowers (8%), and herbs (15%) are also eaten (cf. Section 2.5.1 above).

The very high biomass of folivorous primates in the particular habitat is surprising, reaching 25 kg per hectare (20 kg for the most folivorous species and 5 kg for the most frugivorous ones). This ecological system, enabling two sympatric species, of similar morphology and fairly similar body weight (both species weigh about 2.5 kg), to share the resources, parallels the system found in Sri Lanka, regarding two sympatric species of leaf monkey (Hladik and Hladik, 1972; Hladik, 1977a). One species, in small groups, feeds on the most common plants in small territories; the other species, in large groups, feeds on unevenly distributed resources in large territories with a total primate biomass of 27 kg per hectare. These figures are the maximum biomasses ever recorded for primates.

2.5.3. *Propithecus verreauxi* (Sifaka)

Propithecus verreauxi, another large folivorous and frugivorous species, share the food resources with *Lemur catta* in the forests of the south of Madagascar. *Propithecus verreauxi* play the same ecological role previously described for *L. fulvus*. Groups of sifakas are small (about 5 individuals), living in territories of approximately 2 hectares and sharing the most evenly distributed tree species (Richard, 1973). In terms of number of feeding observations the "diet" of *P. verreauxi* consists of 65% fruits, 25% leaves, and 10% flowers (Jolly, 1966).

2.5.4. *Hapalemur griseus* (Gray Gentle Lemur)

Hapalemur griseus are adapted to a marshy environment where bamboo grows. They live in small groups and feed mainly on bamboo shoots (Petter and Peyrieras, 1970b, 1975) and, presumably, to a lesser extent, on fruits. Their technique for handling food, and particularly the pieces of grass, is the stereotyped grabbing movement characteristic of other prosimians (Fig. 8B).

2.5.5. *Indri indri* (*Indri*)

Indri indri were recently studied in the field by Pollock (1975, and Chapter 9) in the rain forest of the eastern coast of Madagascar. *Indri*, the largest of the extant prosimians, with a body weight of over 10 kg, have a folivorous diet but a fairly low biomass of about 1.0 kg per hectare (calculated from the data of Petter and Peyrieras, 1974). This low biomass can be explained by a folivorous diet which includes only a very small quantity of mature leaf material. From Pollock's (1975, 1977) observations 50–75% of the feeding observations concerned young leaves and buds, 25% fruits (including 10–15% unripe seeds), and mature leaves featured in less than 1% of the feeding records. Some earth was also occasionally eaten (see Section 6). Mature leaves are an ubiquitous source of food in the forest, but young leaves are represented only in small amounts generally scattered on deciduous trees and locally available for short periods only (A. Hladik, 1978).

Nevertheless, if one considers the biomass of all arboreal folivores in this rain forest, it will be around 6 kg per hectare, a classical figure which is partly explained by the complexity of this environment divided into subtle ecological niches.

3. SOCIAL LIFE IN RELATION TO DIET AND ECOLOGY

3.1. Home Range, Territory, and the "Supplying Area"

The notions of home range and territory are generally clearly understood in different field studies concerning individual prosimians as well as groups (see elsewhere in the present volume). The inventory of food resources available to each

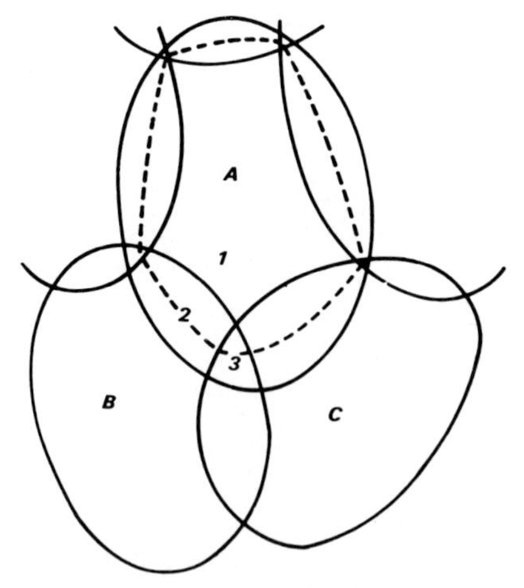

Fig. 9. The "supplying area" of group A, as determined by the limits of the home ranges of the other groups B and C, etc., is shown by the broken line.

group (or to individual solitary prosimians) calls for a clear definition of what is actually available to each of them in the overlapping parts of the home ranges.

If Fig. 9 is taken as a theoretical example of the home ranges of three groups of primates, A, B, and C, then A will have exclusive access to the food resources in the core area 1. This central area, if defended, will be considered a territory. In area 2, where the ranges of A and B overlap, an equivalent amount of food will be

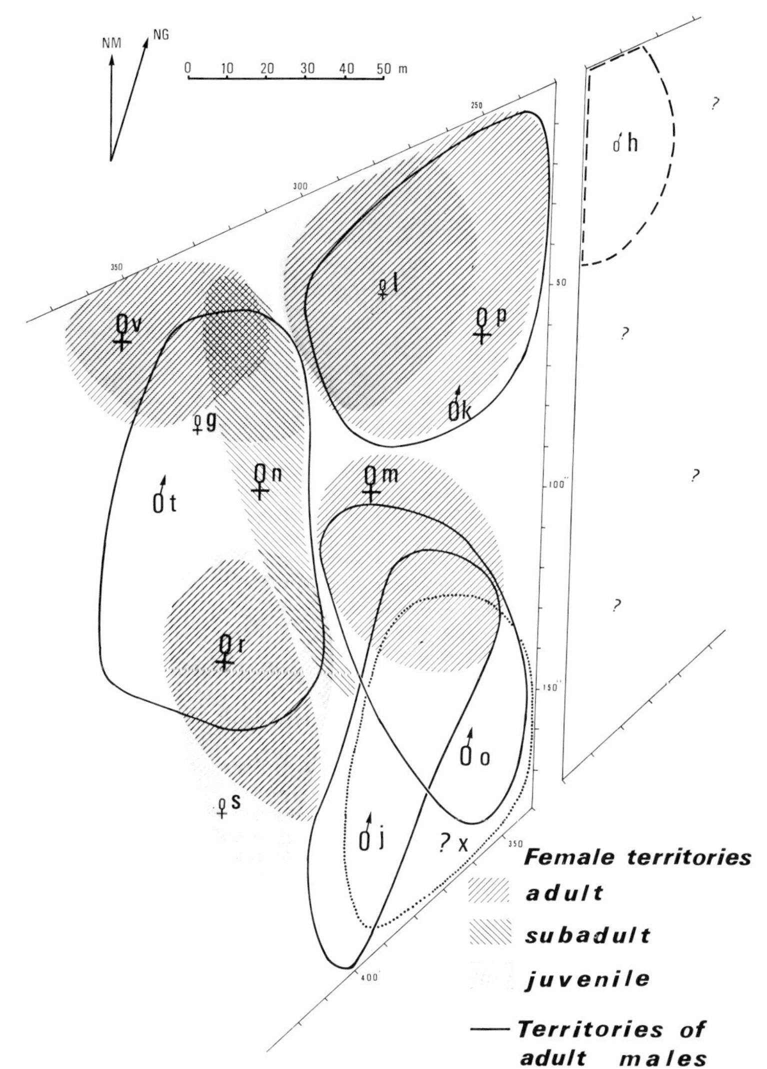

Fig. 10. Habitat utilization by *Lepilemur leucopus* at Berenty in southern Madagascar, as shown by the limits of the individual territories of females and males. (After Charles-Dominique and Hladik, 1971.)

available to each group, irrespective of the dominance relationship between the groups, as shown by their territorial behavior, since field observations have shown that groups ranging in the same area eat approximately the same quantity (Hladik and Hladik, 1972; Hladik, 1977a). In area 3, where the home ranges of all three groups overlap, one third of the food will be available to each group. The limit of the "supplying area" (dashed line in Fig. 9) is purely fictitious, but is convenient and necessary to calculate the quantity of food available to each group in an homogeneous environment.

The size of the supplying area of one group of diurnal prosimians (as well as that of the supplying area of one individual nocturnal prosimian) is determined by the diets of the different species allowing even distribution of food resources in the population, even during periods of shortage. For example, it has been demonstrated (Charles-Dominique and Hladik, 1971; Hladik and Charles-Dominique, 1974) that, in a population of *Lepilemur leucopus*, in the territories which are shown in Fig. 10, the supplying area of each individual may contain only 1.6 times the amount of food that is actually eaten during the period of maximum shortage. Such a social system, which allows a maximum population density according to the size of the individual territories, prevents any significant mortality during the period of food scarcity.

The biomasses in relation to the different grades of diets, illustrated by many examples in Section 2, have been determined as a result of the selection of social mechanisms regulating the size of the supplying area. The maximum biomasses in habitats where populations are stablized, are generally determined by the production of the major food types (Section 2.2.3 above), but the pattern of distribution of food resources also influences the distribution of the species feeding on these resources.

3.2. Social Patterning in Relation to Habitat Utilization

Any type of food source (mainly fruit trees) may be considered as evenly distributed only in terms of a particular scale. For example, approximately the same number of trees of a very common species may be found in each of two adjacent home ranges of 1 hectare, while a rare tree species might be found by chance in only one of them. Rare tree species are likely to be evenly distributed between adjacent home ranges only if they are very large, say 100 hectares or more. Because type of diet has a limiting effect on the maximum possible biomass of a given habitat, a prosimian species living in large groups will have a large supplying area that may include several uncommon tree species in sufficient number to be utilized as a food source. By contrast, another prosimian species, of the same ecological grade, but living in small groups, will have a small supplying area in which only the most common tree species can be considered as a potential food supply.

The example of the two sympatric *Lemur* species studied by Sussman (Section 2.5.2) illustrates this principle of a feeding strategy depending directly on social structure. This interdependence must have been a strong selective pressure for the evolution of social structure in groups of prosimians as well as in groups of simian primates with converging structures (Hladik, 1975, 1977a, 1978). There are, of course, more complex combinations. A very large home range with a relatively small supplying area might permit the use of scattered resources by several groups of relatively small size. Even in nocturnal prosimian species, there is a relationship between diet and social structure because of their different overlapping systems (Charles-Dominique, 1974b, c, 1978).

3.3. Food Traditions and Learning

The prosimian infant clinging to the fur of its mother starts its feeding education by taking fragments of fruits or insects from the hand or from the mouth of the mother. These fragments generally have little nutritional value. For instance, a young *Perodicticus potto* takes only the leg of an insect and chews or sucks it (P. Charles-Dominique, personal communication), but the perception of a particular taste is almost certainly of paramount importance as a basis for future feeding behavior. The acceptance of "repugnant" prey by the adult *P. potto* probably starts from this early conditioning.

The juvenile *Microcebus coquereli* approaches its mother with its head turned on its side below its mother's head and is allowed to take some parts of the prey (Fig. 11A). It thus learns the taste of a large number of different kinds of prey before it reaches adult size (E. Pagès, personal communication). The juvenile *Galago elegantulus* was directly observed by Charles-Dominique following its mother along the woody stems of *Entada gigas* to eat the gum exuded by this vine. As soon as the mother began to feed on the gum, the juvenile rushed to her side to collect some. The young *Lemur*, while still suckling, plays around the mother and occasionally collects and tastes a few leaves (Fig. 11B).

During the major part of its first year, the young prosimian may learn the various types of food available during different seasons in its habitat by observing the behavior of its conspecifics. The young *Galago demidovii*, for instance, for a period of 6–8 months, may follow its mother at a close distance, or it may follow another female associated with its mother, or it may follow the dominant male (Charles-Dominique, 1971a, 1977a, personal communication).

Behavior is very flexible during the early period of learning and feeding behavior may alter radically following an unpleasant experience. For instance, after a young *Galago alleni* had been badly irritated when trying to eat a female butterfly *Anaphe sp.*, it avoided all butterflies for several weeks. This last observation of Charles-Dominique (1977a) follows the general principle of aversive conditioning which applies to any vertebrate and can be induced by an "internal" reaction to food (*i.e.*, the illness following absorption, as discussed by Garcia *et al.*, 1974; see also Section 5.1). As a matter of fact, learning supposes more subtle possibilities of discrimination and extinction.

Fig. 11. Young prosimians during their first natural food tests. (A) *Microcebus coquereli* taking some pieces of a locust being eaten by its mother. (Photograph: E. Pagès.) (B) *Lemur catta*, still clinging to the fur of its mother in a tamarind tree at Berenty in southern Madagascar, can grab pieces of food and taste them. (Photograph: C. M. Hladik.)

Although there seems to be very little difference in visual discrimination learning among the various prosimians (and among many other vertebrates), significant differences appear to exist in the ability to adapt learned responses to changes in the visual situation. This ability seems to be the beginning of a higher degree of behavioral flexibility developed in some nocturnal forms (Cooper, 1978 and personal communication) and most diurnal forms thus far tested (Wilkerson and Rumbaugh, Chapter 6). Interspecific differences in learning and extinction have probably been selected as a result of the need for each species to adapt to the particular characteristics of its own ecological niche.

The gregarious species have inherent advantages for learning during longer periods than the more solitary forms. In groups the young may learn food choices, temporal patterns, and other important strategies from other juveniles and adults. A "group memory," which may partly substitute for individual memory, plays an important part in the development of social tradition. Habitat utilization can be very sophisticated, particularly the seasonal exploitation of a large number of uncommon food species.

4. PARTICULAR ASPECTS OF DIETARY SPECIALIZATION

4.1. Flexibility in Different Types of Diets

Many field studies of prosimian ecology have shown that several types of habitat can be utilized by a given species. Some species are more adaptable than others and, in many cases, populations of the same species are sufficiently different in a large enough area such as to constitute different subspecies, e.g., subspecies of *Loris tardigradus*, *Galago senegalensis*, and *G. crassicaudatus* (see Section 2.3).

Some other species, such as *Microcebus murinus*, occurring in the south of Madagascar and in different deciduous forests along the west coast, and *Galago demidovii*, occurring in large areas of the rain forest in Central and West Africa, do not show such obvious local differentiations. Nevertheless, there are many differences in the habitats that certain species can use, especially in terms of the available food, its rate of production, the number of edible plant species, etc. Such prosimian species can be considered as more adaptable than those localized in a restricted geographical range and a particular habitat.

There are several other possibilities of variation in the natural diet of a given species. If we consider the five sympatric nocturnal prosimian species studied by Charles-Dominique in the rain forest of Makokou (Sections 2.2.2 and 2.3.4–2.3.8), all the examples given were concerned with the primary rain forest. The five species also inhabit the secondary forest and Charles-Dominique, in the various sources cited, presents the data collected in this particular habitat. In spite of the close proximity and similar climatic conditions in these two habitats, the plant species are almost entirely different, and the entomological fauna partly different; thus the composition of the diets of each of the prosimian species varies between the two habitats.

Seasonal variations, in relation to availability of food, are also very important: in many cases the seasonal difference in the diet of a given species is greater than the difference between two species at a given time. This has been shown for simian primates (Hladik, 1977a) and, according to the feeding records and feeding times reported by Jolly (1966) and Richard (1973) for *Lemur catta* and *Propithecus verreauxi*, respectively, it also appears true for lemur species. In the Cheirogaleinae, with the exception of *Microcebus coquereli*, these marked seasonal variations in the diet (Sections 2.4.1 and 2.4.2) are correlated with physiological cycles (Perret, 1972, 1974; Petter-Rousseaux, 1974) and, in captivity, the seasonal variations in food intake still occur even when the food supply is maintained unchanged (Andriantsiferana and Rahandraha, 1973; Petter-Rousseaux and Hladik, in press).

Important changes in the diet necessarily occur when prosimians are maintained in captivity. Major changes can be made without difficulty in species generally considered most adaptable in the wild, such as those of the genera *Lemur* and *Galago*. Other species, such as *Indri*, *Propithecus*, *Lepilemur*, *Phaner*, and *Cheirogaleus*, more specialized in their natural diets, have difficulty in adapting to, and rarely reproduce, in captivity. *Lepilemur ruficaudatus, Propithecus verreauxi,* and *Cheirogaleus medius* have only recently reproduced in the laboratories of Duke University and Brunoy, and *Phaner furcifer* have never bred in captivity.

Species feeding on ubiquitous resources, such as leaves, or on food available throughout the year, such as gums, do not require much flexibility in their diets, provided the composition of the food ingested is sufficiently regular to maintain bacterial flora in the gut. Although they are not adapted to sudden changes, this does not mean that they are more vulnerable than the more adaptable species: on the contrary, in their natural habitat, they are more efficient than any competitor and can maintain a very high biomass.

Adaptability of prosimian species depends on particular aspects of their digestive tracts. The chewing movement was acquired at an early evolutionary stage, according to the shape of the teeth (Kay and Hiiemae, 1974), and there is little possible change in the way food is masticated (see Section 1.2 above). The intestinal tract, on the other hand, may undergo marked changes (Hladik, 1967) as an adaptation to dietary change, in particular, the less differentiated digestive tract (without large chambers of bacterial fermentation). The general proportions of the digestive tract (Amerasinghe *et al.*, 1971; Chivers and Hladik, in prep.) are correlated with the diet, the insectivorous and frugivorous species having a long small gut, while the extension of the cecum and the hindgut characterizes the species feeding on gums and leaves (Fig. 12). But further physiological and behavioral adaptations, such as cecotrophy, are as important as these morphological characteristics.

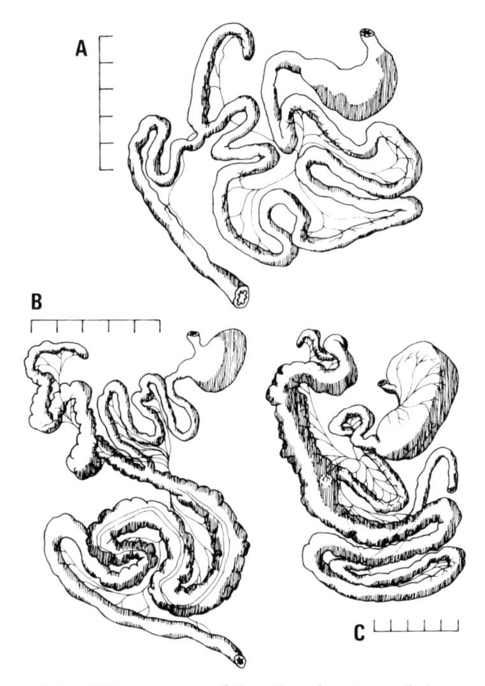

Fig. 12. Proportions of the different parts of the digestive tract of three prosimian species. These drawings were made from samples collected in the wild, partly dissected to show the different sections with special attention paid to accurate dimensions; measurements were taken after complete dissection to verify the lengths indicated in the drawings. The scale for each specimen represents 5 cm. (A) *Arctocebus calabarensis*. (B) *Galago elegantulus*. (C) *Lepilemur leucopus*.

4.2. Cecotrophy in *Lepilemur*

The particular physiological and behavioral mechanism of the digestion of leaves by *Lepilemur leucopus* has been described by Charles-Dominique and Hladik (1971) and analyzed by Hladik *et al*. (1971b). Recent observations on *Lepilemur ruficaudatus* (J. J. Petter, personal communication) reveal a similar behavior, suggesting that it is a feature common to the different species of the genus *Lepilemur*.

Lepilemur leucopus, which feed mainly on crassulescent leaves, flowers, and some fruits, reingest a part of the fecal material, just as the rabbit does (Taylor, 1940); but this material is very different from that of the rabbit. The food ingested under natural conditions by *L. leucopus*, during the period of observations, was a mixture of the flowers of two species of *Alluaudia* (cf. Hladik and Charles-Dominique, 1974) which was found to have an average composition of 15.1% protein, 2.7% lipids, 5.8% soluble sugars, 16.7% cellulose, 9.1% minerals, the rest of the dry matter being made up of fiber, mainly hemicellulose and lignin. Very little hemicellulose is hydrolyzed in the stomach and the small gut after the animal starts feeding at night. Part of the cellulose is hydrolyzed in the cecum. The hemicelluloses are slowly hydrolyzed afterward and absorbed in the colon. A portion of the food, partly decomposed in the cecum, passes quickly and is eaten by the animal, licking its anus at intervals during the daytime. This particular food contains a larger proportion of protein than the original food (45.8%), as well as a large proportion of hemicelluloses. The soluble components resulting from the bacterial fermentation are absorbed during this second cycle. This unique physiological strategy allows *L. leucopus* to make maximum use of what is ordinarily a very poor diet, in terms of the energy that can be obtained from the soluble components. In fact, it has the poorest diet known among primates with little variation in the food available in two different habitats (Fig. 13).

The extreme shortness of the small gut of *L. leucopus* would not allow sufficient absorption of the soluble components in one direct passage only. According to the histological structure of the cecum (Hladik *et al.*, 1971b) it is likely that a complementary absorption occurs in this part.

The digestive tract of *Galago elegantulus* resembles that of *Lepilemur leucopus* and has a similar function of bacterial fermentation of highly polymerized carbohydrates (gums are compounds of pentoses) (Fig. 12). One would, therefore, expect

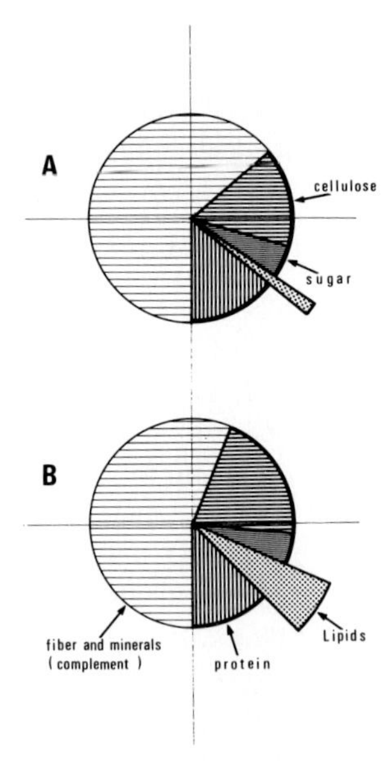

Fig. 13. Proportions of nutrients in the diet of *Lepilemur leucopus* in two different habitats in the south of Madagascar during the dry season. The larger amount of lipids is proportional to its higher calorific value. (After Hladik and Charles-Dominique, 1974.) (A) In the Didiereacae bush. (B) In the gallery forest.

similar behavior (cecotrophy) in both species. However, one *G. elegantulus*, closely observed for 36 consecutive hours in captivity in Gabon, and fed with gums eaten in the wild, showed no signs of cecotrophy. Among the primates this particular behavior appears to be strictly limited to *Lepilemur*.

4.3. Artificial Diets: the G3 Lemur Cake of Brunoy

A general problem of primate care in captivity is to maintain a minimum standard of diet (Wackernagel, 1966, 1968). This has been done successfully in several zoos, for example, in Basel, where species of leaf monkey, reputed to be unable to adapt to captive conditions, are kept in excellent health. Success is partly due to the introduction of raw grasses, improperly called "monkey cake," containing minerals and vitamins to offset any deficiency in the diet.

At the animal house in Brunoy, different types of dietary supplements have been tested. The "monkey cake" was rejected by the small species of prosimian. A composition, more refined in texture, made of cooked semolina, was finally accepted. The composition of this cake was based on the natural foods of the smallest species of Callithricidae (Hladik *et al.*, 1971a) because no data on the natural food of prosimians were available at that time.

The G3 lemur cake (third formula worked out by Mrs. Grange) includes large amounts of proteins and lipids (Tables IV and V) and is accepted, sometimes after a few days, by all the prosimian species presently at Brunoy. It is given only in small quantities, as a supplementary and not as a substitute diet, and has been found to be very efficient in maintaining a good rate of reproduction in some species considered

TABLE IV

Recipe for the G3 Lemur Cake[a]

In a 12-liter stew pan, pour successively:
- 5 liters of water (heat)
- 2 kg of semolina (slowly mixed in the water)
- 1 kg of white cheese, 40% fat
- 1 kg of white cheese without fat
- 250 gm of butter (slowly stir the hot mixture, but stop heating)
- 36 eggs (about 1800 gm)
- One tin of sweetened concentrated milk (397 gm)
- 10 gm of salt
- 100 gm of pure fructose
- 300 gm of pure glucose (continue stirring until the mixture is cool)

The mixture is transferred in 25 plastic boxes containing about 500 gm each, then add:
- 5 drops of "Vitapaulia" (Proligo-78800 Houilles, France) in each of the boxes when the mixture is cold but not yet hardened

[a] 12 kg fresh weight which can be stored in a freezer.

TABLE V

Composition of the G3 Lemur Cake

Water : 64.9% of the fresh weight		
Percentage of the dry weight:		
Carbohydrate	54.1	
Protein	22.2	
Lipids	22.1	
Minerals	1.53	(including calcium 0.38, and phosphorus 0.404)

"delicate" in captive conditions, such as *Lepilemur ruficaudatus* and *Microcebus coquereli*, for which the available natural food would have been insufficient.

5. FOOD COMPOSITION AND FEEDING BEHAVIOR OF PROSIMIANS

5.1. Regulation of Feeding Behavior

The role of the hypothalamic nuclei in the regulation of hunger and satiety has been demonstrated in many mammalian species including primates. Taste stimulation also plays an important role (Le Magnen, 1966; Aschkenasy-Lelu, 1966) in parallel with this system of regulating food intake according to the demands of the "milieu interne."

The long-term effects of food intake on the state of the "milieu interne" of the animal could originate by a process of conditioning, positive in the case of correct types of food resulting in beneficial effects, and negative in the case of noxious substances, unbalanced or inadequate diets, resulting in illness. Nevertheless, positive conditioning by taste stimulation is likely to be involved in most cases and can explain many differences in the particular feeding behavior of the different prosimian species.

5.2. Food Conditioning under Natural Conditions

The long-term beneficial effect is obviously the major criterion determining diet in the Lorisinae, since they feed on insects rejected by other prosimians because of their taste, but their preferential choice of more palatable prey, when available (see Section 1.2 above), shows that the immediate response to pleasurable taste stimulation is not absent in determining food choices.

Conditioning as a result of the immediate effect of taste stimulation might be very important in natural conditions, in increasing the efficiency of a species having a grade 2 diet, as well as those having a grade 1 diet, which includes large amounts of fruits. These species generally have a large supplying area in which scattered food resources must first be located. Vision is the predominant sense involved in food detection, as shown by Pariente (Chapter 10), the moving prey and the stationary prey providing different kinds of stimulation. (Hearing also plays a role in some of these cases; see Sections 2.3.6 and 2.4.6). Color vision, though imperfect, is very important in diurnal species for increasing the contrast of certain fruits against the green background of foliage. Whatever the case, interest in a particular kind of food may be increased by taste stimulation of such soluble substances as sugars and organic acids, which may constitute a reward in itself and induce positive conditioning.

Prosimian species which have grade 3 diets (exclusively utilizing the primary production) are probably conditioned to dietary preferences by their long-term effects. This is apparent for example, in operant conditioning tests on *Hapalemur griseus*: a reward of a piece of cucumber, even though it represents only a small return in energy, may be preferred to higher energy foods such as sweet fruits and insects (H. Cooper, personal communication). Nevertheless, among the species feeding on fruits and leaves, the role of taste can explain differences in feeding strategy, particularly between the sympatric *Lemur* species, *L. fulvus* and *L. catta* (Section 2.5.2). This example, as well as other cases of sympatry between *Lemur* and/or *Propithecus* species, needs more investigation into the composition of preferred plant species. The parallel example of sympatric simian species utilizing the same type of habitat in Sri Lanka (Hladik, 1977a) shows that interspecific differences in food preference cannot be due exclusively to the long-term effects. It can be predicted that interspecific differences in taste perception (or in taste stimulation) of the same substances will be found. The evolution of these different sensory characteristics (some of them already demonstrated in simian primates by Glaser and Hellekant, 1977) could have resulted in different feeding strategies: some species with greater taste sensitivity specializing in scattered and rich resources, while others with less sensitive tastes developed long-term preferences for the more common food resources.

6. DISCUSSION AND CONCLUSIONS

The primary factors which condition prosimian species to the various dietary grades are the beneficial effects of the nutritional constituents.

Many substances in plants, which are not nutritious, may play a role, generally a nociceptive one, in the physiological and sensory reaction of the animal which feeds on them. These substances, described as "secondary compounds" or "allelochemics" by Whittaker and Feeny (1971), may act either to repel or attract insects. The same type of substance is also found in the invertebrates. The glandular secretions of different insects may be one of these secondary compounds taken from a plant on which the insect feeds and which it then concentrates or otherwise chemically modifies in some way. For some prosimian species, feeding on these insects, the "repellent" substance actually functions as an attractant.

Secondary substances found in many plant species, such as alkaloids, tannins, saponins, glucosides, may play a similar double role in the orientation of feeding behavior of different prosimian species. In fact, very little is known of the composition of the plant species used as food by prosimian species, and most of these suggested effects of the so-called "secondary" substances are speculative.

In the data collected on simian food (C. M. Hladik, 1977a, b, 1978; A. Hladik, 1978), there are very few examples of strong concentrations of toxic or noxious substances in fruits and/or leaves which would lead to avoidance responses. In most cases the concentrations are very low and secondary substances do not play a nociceptive role. Many types of food eaten by simians are also eaten by prosimians and their effect is probably similar, except that, as has been noted, some species are actually attracted to certain foods because of taste and/or smell which repels other species. Several plant species have selected fruit attractants as a mechanism for seed dispersion and other vital functions (Levina, 1957; Hladik and Hladik, 1967).

Efficient conditioning to a particular diet, due to its long-term beneficial effects, is further reinforced by the smell and taste of these secondary compounds. Thus the presence of these substances must be very helpful in maintaining the feeding strategies of prosimian species, with grade 2 or 3 diets, which use scattered resources in a large supplying area.

Eating earth (geophagy), observed by Pollock (1975) in *Indri indri* (see Section 2.5.5), may represent a very different feeding strategy. It has been shown, for instance, that for many folivorous primate species, which eat some earth, the required concentration of the mineral nutrients is lower in the earth than in many leaves used as a food (Hladik and Guegen, 1974). A possible function of earth eating could be that earth serves to adsorb those tannins or other secondary substances that are in excess in certain types of food.

It is confidently expected that, in a few years, our knowledge of the composition of the food used by prosimians will lead to verification of the above hypotheses. Further research work on feeding behavior must be complemented by laboratory tests on conditioning to determine the taste thresholds of different prosimian species for various nutrients and secondary compounds, and to define the basis of the behavioral adaptations to various food resources in different habitats.

ACKNOWLEDGMENTS

Many ideas expressed in this chapter were developed and discussed at Brunoy while J.-J. Petter, A. Petter-Rousseaux, P. Charles-Dominique, G. Pariente, A. Schilling, A. Hladik, E. Pagès, M. Perret, and H. Cooper were doing research or returning from field trips. Special thanks are also due to the editors, G. A. Doyle and R. D. Martin, for their active and patient cooperation.

REFERENCES

Amerashinghe, F. P., Van Cuylenberg, B. W. B., and Hladik, C. M. (1971). *Ceylon J. Sci., Biol. Sci.* **9**, 75–87.

Andriantsiferana, R., and Rahandraha, T. (1973). *C. R. Hebd. Séances Acad. Sci.* **277,** 2025-2028.

Aschkenasy-Lelu, P. (1966). *J. Sci. Cent. Natl. Coord. Etud. Aliment.* **14,** 55–101.

Bearder, S. K., and Doyle, G. A. (1974). *In* "Prosimian Biology" (R. D. Martin, G. A. Doyle, and A. C. Walker, eds.), pp. 109–130. Duckworth, London.

Bishop, A. (1964). *In* "Evolutionary and Genetic Biology of Primates" (J. Buettner-Janusch, ed.), Vol. 2, pp. 133–225. Academic Press, New York.

Bourlière, F. (1974). *In* "Prosimian Biology" (R. D. Martin, G. A. Doyle, and A. C. Walker, eds.), pp. 17–22. Duckworth, London.

Budnitz, N., and Dainis, K. (1975). *In* "Lemur Biology" (I. Tattersall and R. W. Sussman, eds.), pp. 219–235. Plenum Press, New York.

Buettner-Janusch, J., and Andrew, R. J. (1962). *Am. J. Phys. Anthropol.* **20**, 127–129.

Charles-Dominique, P. (1966). *Biol. Gabonica* **2**, 347–353.

Charles-Dominique, P. (1971a). *Biol. Gabonica* **7**, 121–228.

Charles-Dominique, P. (1971b). *Recherche* **15**, 780–781.

Charles-Dominique, P. (1972). *Z. Tierpsychol.*, Beih. **9**, 7–41.

Charles-Dominique, P. (1974a). *In* "Prosimian Biology" (R. D. Martin, G. A. Doyle, and A. C. Walker, eds.), pp. 131–150. Duckworth, London.

Charles-Dominique, P. (1974b). *Mammalia* **38**, 355–379.

Charles-Dominique, P. (1974c). *In* "Primate Aggression, Territoriality and Xenophobia: A Comparative Perspective" (R. L. Holloway, ed.), pp. 31–48. Academic Press, New York.

Charles-Dominique, P. (1975). *In* "Phylogeny of the Primates: A Multidisciplinary Approach" (W. P. Luckett and F. S. Szalay, eds.), pp. 69–88. Plenum, New York.

Charles-Dominique, P. (1976). *Oiseau Rev. Fr. Ornithol.* **46**, 174–178.

Charles-Dominique, P. (1977a). "Ecology and Behaviour of Nocturnal Primates." Duckworth, London.

Charles-Dominique, P. (1977b). *Z. Tierpsychol.* **43**, 113–138.

Charles-Dominique, P. (1978). *In* "Recent Advances in Primatology," Vol. 3: Evolution (D. J. Chivers and K. A. Joysey, eds.), pp. 139–149. Academic Press, London.

Charles-Dominique, P., and Hladik, C. M. (1971). *Terre Vie* **25**, 3–66.

Charles-Dominique, P., and Martin, R. D. (1970). *Nature (London)* **227,** 257-260.

Charles-Dominique, P., and Petter, J. J. (in press). *In* "Ecology, Physiology, and behavior of five nocturnal lemurs of the west coast of Madagascar" (P. Charles-Dominique, ed.), Karger, Basel.

Cooper, H. (1978). *In* "Recent Advances in Primatology, Vol. 1: Behaviour" (D. J. Chivers and J. Herbert, eds.), pp. 941-944. Academic Press, London.

Doyle, G. A. (1974a). *Behav. Nonhum. Primates* **5,** 155-353.

Doyle, G. A. (1974b). *In* "Prosimian Biology" (R. D. Martin, G. A. Doyle, and A. C. Walker, eds.), pp. 213–231. Duckworth, London.

Doyle, G. A., and Martin, R. D. (1974). *In* "Prosimian Biology" (R. D. Martin, G. A. Doyle, and A. C. Walker, eds.), pp. 3–14. Duckworth, London.

Elliot, O., and Elliot, M. (1967). *J. Mammal.* **48**, 497–498.

Fogden, M. P. L. (1974). *In* "Prosimian Biology" (R. D. Martin, G. A. Doyle, and A. C. Walker, eds.), pp. 151–165. Duckworth, London.

Garcia, J., Hankins, W. G., and Rusiniak, K. W. (1974). *Science*, **185**, 824–831.

Gautier-Hion, A. (1971). *Terre Vie* **25**, 427–490.

Glaser, D., and Hellekant, G. (1977). *Folia primatol.* **28**, 43–51.

Hill, W. C. O. (1953). "Primates: Comparative Anatomy and Taxonomy," Vol. I. Edinburgh Univ. Press, Edinburgh.

Hill, W. C. O. (1955). "Primates: Comparative Anatomy and Taxonomy," Vol. II. Edinburgh Univ. Press, Edinburgh.

Hinde, R. A. (1966). "Animal Behaviour: A Synthesis of Ethology and Comparative Psychology." McGraw-Hill, New York.

Hladik, A. (1978). *In* "The Ecology of Arboreal Folivores" (G. G. Montgomery, ed.), in press. Random House Smithsonian Inst. Press, New York.

Hladik, A., and Hladik, C. M. (1969). *Terre Vie* **23**, 25–117.

Hladik, C. M. (1967). *Mammalia* **31**, 120–147.

Hladik, C. M. (1973). *Terre Vie* **27**, 343–413.

Hladik, C. M. (1975). *In* "Socioecology and Psychology of Primates" (R. H. Tuttle, ed.), pp. 3–35. Mouton, The Hague.

Hladik, C. M. (1977a). *In* "Primate Ecology: Studies of Feeding and Ranging Behaviour in Lemurs, Monkeys and Apes" (T. H. Clutton-Brock, ed.), pp. 323–353. Academic Press, New York.

Hladik, C. M. (1977b). *In* "Primate Ecology: Studies of Feeding and Ranging Behaviour in Lemurs, Monkeys and Apes" (T. H. Clutton-Brock, ed.), pp. 595–601. Academic Press, New York.

Hladik, C. M. (1978). *In* "The Ecology of Arboreal Folivores" (G. G. Montgomery, ed.), in press. Random House, Smithsonian Inst. Press, New York.

Hladik, C. M. and Charles-Dominique, P. (1974). *In* "Prosimian Biology" (R. D. Martin, G. A. Doyle, and A. C. Walker, eds.), pp. 23–37. Duckworth, London.

Hladik, C. M. and Chivers, D. J. (1978). *In* "Recent Advances in Primatology, Vol. 1: Behaviour" (D. J. Chivers and J. Herbert, eds.), pp. 433–444. Academic Press, London.

Hladik, C. M., and Guegen, L. (1974). *C. R. Hebd. Séances Acad. Sci.* **279,** 1393-1396.

Hladik, C. M., and Hladik, A. (1967). *Biol. Gabonica* **3**, 43–58.

Hladik, C. M., and Hladik, A. (1972). *Terre Vie* **26**, 149–215.

Hladik, C. M., Hladik, A., Bousset, J., Valdebouze, P., Viroben, G., and Delort-Laval, J. (1971a). *Folia Primatol.* **16**, 85–122.

Hladik, C. M., Charles-Dominique, P., Valdebouze, P., Delort-Laval, J., and Flanzy, J. (1971b). *C. R. Hebd. Séances Acad. Sci.* **272,** 3191–3194.

Hladik, C. M., Charles-Dominique, P., and Petter, J. J. (in press). *In* "Ecology, physiology and behaviour of five nocturnal lemurs of the west coast of Madagascar" (P. Charles-Dominique, ed.), Karger, Basel.

Hutchinson, G. E. (1957). *Cold Spring Harbor Symposia on quantitative Biology*, **22**, 415–427.

Iwamoto, T. (1974a). *Primates* **15**, 241–262.

Iwamoto, T. (1974b). *In* "Contemporary Primatology" (S. Kondo, M. Kawai, and E. Ehara, eds.), pp. 475–480. Karger, Basel.

Jameson, E. W., and Mead, R. A. (1964). *J. Mammal.* **45**, 359–365.

Jolly, A. (1966). "Lemur Behavior: A Madagascar Field Study." Univ. of Chicago Press, Chicago, Illinois.

Kay, R. F., and Hiiemae, K. M. (1974). *In* "Prosimian Biology" (R. D. Martin, G. A. Doyle, and A. C. Walker, eds.), pp. 501–530. Duckworth, London.

Kay, R. F., and Hylander, W. (1978). *In* "The Ecology of Arboreal Folivores" (G. G. Montgomery, ed.), in press. Random House (Smithsonian Inst. Press), New York.

King, G. (1974). *Jersey Wildl. Preserv. Trust: Ann. Rep.* pp. 81–96.

Le Magnen, J. (1966). *J. Sci. Cent. Natl. Coord. Etud. Aliment.* **14,** 1–54.

Levina, R. E. (1957). "Dispersal of Fruits and Seeds." Univ. of Moscow Press, Moscow (in Russian).

Martin, R. D. (1972a). *Z. Tierpsychol. Beih.* **9**, 43–89.

Martin, R. D. (1972b) *Philos. Trans. R. Soc. London, Ser. B* **264**, 295–352.

Martin, R. D. (1973). *In* "Comparative Ecology and Behavior of Primates" (R. P. Michael and J. H. Crook, eds.), pp. 1–68. Academic Press, New York.

Napier, J. R., and Walker, A. C. (1967). *Folia Primatol.* **6**, 204–219.

Odum, E. P., and Odum, H. T. (1959). "Fundamentals of Ecology." Saunders, Philadelphia, Pennsylvania.

Pagès, E. (1978). *In* "Recent Advances in Primatology, Vol. 3: Evolution" (D. J. Chivers and K. A. Joysey, eds.), pp. 171–177. Academic Press, London.
Pariente, G. (1975). *J. Physiol. (Paris)* **70**, 637–647.
Pariente, G. (1976). Doctoral Thesis, Université des Sciences et Techniques du Languedoc (unpublished). Montpellier (France).
Perret, M. (1972). *Mammalia* **36**, 482–516.
Perret, M. (1974). *In* "Prosimian Biology" (R. D. Martin, G. A. Doyle, and A. C. Walker, eds.), pp. 357–387. Duckworth, London.
Petter, J. J. (1962). *Mém. Mus. Natn. Hist. Nat., Paris Sér. A* **27,** 1–146.
Petter, J. J. (1978). *In* "Recent Advances in Primatology, Vol. 1: Behaviour (D. J. Chivers and J. Herbert, eds.), pp. 211–223. Academic Press, London.
Petter, J. J., and Hladik, C. M. (1970). *Mammalia* **34**, 394–409.
Petter, J. J., and Petter-Rousseaux, A. (1967). *In* "Social Communication among Primates" (S. A. Altmann, ed.), pp. 195–205. Univ. of Chicago Press, Chicago, Illinois.
Petter, J. J., and Peyrieras, A. (1970a). *Mammalia* **34**, 167–193.
Petter, J. J., and Peyrieras, A. (1970b). *Terre Vie* **24**, 356–382.
Petter, J. J., and Peyrieras, A. (1974). *In* "Prosimian Biology" (R. D. Martin, G. A. Doyle, and A. C. Walker, eds.), pp. 39–48. Duckworth, London.
Petter, J. J. and Peyrieras, A. (1975). *In* "Lemur Biology" (I. Tattersall and R. W. Sussman, eds.), pp. 281–286. Plenum, New York.
Petter, J. J., Schilling, A., and Pariente, G. (1971). *Terre Vie* **25**, 287–327.
Petter, J. J., Albignac, R., and Rumpler, Y. (1977). "Faune de Madagascar: Lémuriens." ORSTOM - CNRS, Paris.
Petter-Rousseaux, A. (1974). *In* "Prosimian Biology" (R. D. Martin, G. A. Doyle, and A. C. Walker, eds.), pp. 365–373. Duckworth, London.
Petter-Rousseaux, A. and Hladik, C. M. (in press). *In* "Ecology, physiology, and behavior of five nocturnal lemurs of the west coast of Madagascar." (P. Charles-Dominique, ed.), Karger, Basel.
Philips, W. W. A. (1931). *Ceylon J. Sci.* **16**, 205–208.
Pollock, J. (1975). Ph.D. Thesis, University of London (unpublished).
Pollock, J. (1977). *In* "Primate Ecology: Studies of Feeding and Ranging Behaviour in Lemurs, Monkeys and Apes" (T. H. Clutton-Brock, ed.), pp. 37–69. Academic Press, New York.
Richard, A. F. (1973). Ph.D. Thesis, University of London (unpublished).
Richard, A. F. (1974). *Folia Primatol.* **22**, 178–207.
Sauer, E. G. F., and Sauer, E. M. (1963). *J. S.W. Afr. Sci. Soc.* **16,** 5–36.
Still, J. (1905). *Spolia Zeylan.* **3**, 155.
Sussman, R. W. (1972). Ph.D. Thesis, Duke University, Durham, North Carolina (unpublished).
Sussman, R. W. (1974). *In* "Prosimian Biology" (R. D. Martin, G. A. Doyle, and A. C. Walker, eds.), pp. 75–108. Duckworth, London.
Sussman, R. W., and Richard, R. F. (1974). *In* "Primate Aggression, Territoriality and Xenophobia: A Comparative Perspective" (R. L. Holloway, ed.), pp. 49–75. Academic Press, New York.
Taylor, E. L. (1940). *Vet. Rec.* **52**, 259–262.
Thorington, R. W. (1967). *In* "Progress in Primatology" (D. Starck, R. Schneider, and H. J. Kuhn, eds.), pp. 180–184. Fischer, Stuttgart.
Wackernagel, H. (1966). *Int. Zoo Yearb.* **6**, 23–37.
Wackernagel, H. (1968). *Symp. Zool. Soc. London* **21**, 1–12.
Walker, A. C. (1969). *Uganda J.* **32**, 90–91.
Whittaker, R. H., and Feeny, P. P. (1971). *Science* **171**, 757–770.

Chapter 9

Spatial Distribution and Ranging Behavior in Lemurs

J. I. POLLOCK

1. INTRODUCTION

This chapter examines the relationship between the nature of social groupings in the lemur species of Madagascar and the movements of individuals about their environment. The spatial distribution of lemurs is examined from measures of population density obtained in a study area, the size and composition of ranging parties, and the tendency of individuals persistently to inhabit specific areas for long

ISBN 0-12-222150-8

periods of time. The term "ranging behavior" is used to discuss the amount of time spent in various parts of a ranging area (range utilization) as well as the patterns of movements about that area (ranging movements). Spatial distribution and ranging behavior are integrated in discussions of the dispersion of individuals within and between social groupings of lemurs.

The aim of this chapter, in addition to presenting an edited summary of all the available information on the spatial distribution and ranging behavior of lemurs, is to look for similarities and differences between species so that propositions of function of the behaviors discussed may be employed. For this purpose data derived from the author's own research in Madagascar on *Indri indri* and recent long-term field studies on *Lemur fulvus rufus* (*Lemur macaco rufus*), *Lemur catta,* and *Propithecus verreauxi* spp. are emphasized. Detailed published information on the ranging behavior of other species is generally lacking and this serves to emphasize those research areas which will prove fruitful in future comparative examinations of lemur behavior and ecology.

Naturalistic studies of lemur ecology are in their infancy and some of the information presented below was obtained from short-term studies of a few individuals. In addition to the natural behavioral variability to be expected in so large and diverse a group of animals inhabiting different environments (see Section 2), differences found between species or within species occupying a variety of habitats, may result from observations made (1) at different times of the year, (2) in different years, and (3) in different environments. Widespread observations of a strictly seasonal reproductive pattern in the lemurs of Madagascar (Petter-Rousseaux, 1962, 1964; Martin, 1972b) and extensive seasonal variation in individuals' ranging behavior (Richard, 1974a; Budnitz and Dainis, 1975; J. I. Pollock, personal observation) strongly suggest that a complete picture of behavioral variation within and between species can only be obtained with observation periods lasting longer than the intervals between births. For *Indri indri* (possibly an extreme example) this interval may be as long as three or more years.

Climatic variation between years can considerably influence the growth of vegetation in the drier parts of Madagascar where the single rainy season (November–March) not infrequently fails to materialize. The flushing of new leaves on the trees, normally synchronous with the onset of the rains, may thus be somewhat inhibited and changes in the ranging and feeding behavior of folivorous lemurs consequently affected.

Finally, the same species studied in different environments may show marked differences in behavior. *Propithecus verreauxi verreauxi*, living in gallery forest at Berenty in the south of Madagascar (Fig. 1), occupy tiny, defended ranging areas of less than 1 hectare (Jolly, 1966, 1972b), while groups of the same species inhabiting xerophytic *Didierea* forest 25 km to the east ranged in 9 hectare areas, which neighboring groups extensively overlapped in their own ranging activities (Richard, 1974b). Figure 1 shows the localities of previous studies made on lemur ecology

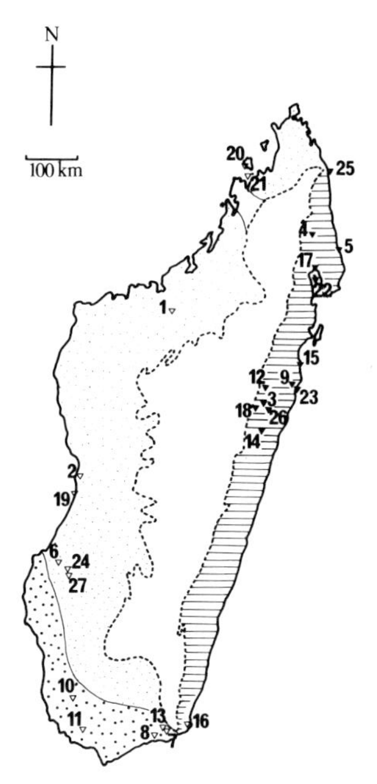

Fig. 1. Main vegetational zones in Madagascar and the location of study areas mentioned in the text. Key: Stippling, western region; fine stippling, west ("western domain"); heavy stippling, west ("southern domain"); hatching, eastern region. Dotted line marks the approximate boundary of the central plateau (uninhabited by lemurs).

Study areas: **1** Ampijoroa; **2** Analabe; **3** Analamazoatra; **4** Andapa; **5** Antalaha; **6** Antserananomby; **7** Bebarimo; **8** Berenty; **9** Betampona; **10** Ejeda; **11** Evasy; **12** Fierenana; **13** Hazafotsy; **14** Lokato; **15** Mahambo; **16** Mandena; **17** Maroansetra; **18** Moramanga; **19** Morondava; **20** Nosy Bé; **21** Nosy Komba; **22** Nosy Mangabé; **23** Tamatave; **24** Tongobato; **25** Vohémar; **26** Vohidrazana; **27** Vondrove.

and behavior in Madagascar, and Table I states the duration of the field observations and the season and type of environment in which they were conducted.

2. LEMUR HABITATS IN MADAGASCAR

2.1. Arboreal Habitats

The distribution of extant lemurs in Madagascar coincides with the presence of a suitably dense shrub and tree vegetation. The denuded central plateau (Fig. 2C), which was until recently a forested area, has become a heavily eroded savannah largely through the activities of man. Excavations in this region have produced evidence of a rich lemur fauna in relatively recent times.

TABLE I

Details of Field Studies Whose Locations Are Illustrated in Fig. 1

Species	Place[a]	Duration of study	Season or month of study	Locomotor type
Microcebus murinus	16	3 months (1968)	October–December	Vertical clinger and leaper (VCL) and quadrupedal
	16	1 month (1970)	September	
Microcebus coquereli	2	16 days	October–November	VCL and quadrupedal
	27	4 days		
Cheirogaleus major	15			VCL and quadrupedal
Phaner furcifer	2			VCL and quadrupedal
Lepilemur mustelinus	8	1½ months (1970)	September–October	VCL
	3			
	6			
Hapalemur griseus	3	Occasionally over 15 months	All	VCL
Varecia variegata	9,12	Occasionally over 15 months	All	Quadrupedal
Lemur catta	8	5½ months (1963)	Feb.–April, parts of May, July, Aug. Sept. (1963) March, April (1964)	Quadrupedal
	8	2 weeks (1970)	Sept. 21–Oct. 4	
	8	12 months	(All) April 1972–May 1973	
	8	Ongoing	1975	
	6	5 months (1970)	July–November	
Lemur macaco macaco	21	Occasionally	(1958)	Quadrupedal
	20	Occasionally	(1963)	
Lemur fulvus fulvus	6,12	Occasionally over 15 months	All	Quadrupedal
	26			
Lemur fulvus rufus	24	90 days	Dec. 1969–April, 1970	Quadrupedal
	6	72 days	July, 1970–Nov. 1970	

Diet	Activity pattern	Group size	Population density (No./km²)	References
Arachnids, insects, flowers, fruit leaves, sap	Nocturnal	Solitary	1300–2600	Martin (1972a)
				Martin (1973)
Insects, sap	Nocturnal	Solitary	250	Petter *et al.* (1971)
Fruit, nectar	Nocturnal	Solitary		Petter (1962)
	Nocturnal	Solitary	75–110	Petter and Petter-Rousseaux (1964)
Nectar, sap, insect secretions	Nocturnal	Solitary?	550–870	Petter *et al.* (1971)
Leaves, flowers	Nocturnal	Solitary	220–810 417 200	Charles-Dominique and Hladik (1971) Sussman (1972)
Fruit, bamboo	Crepuscular	3–6	47–62	J. I. Pollock, personal observation
Fruit + ?	Crepuscular	2–4	?	J. I. Pollock, personal observation
Leaves, fruit, insects	Diurnal	12–24	350	Petter (1962) Jolly (1966)
Insects, flowers			250	Jolly (1972b); Sussman (1972)
		5–20	152	Budnitz and Dainis (1975)
			156	Jolly *et al.* (1977)
Bark, herbs, cactus, leaves, fruit, insects			215	Sussman (1972, 1974)
Fruit, leaves, bark, flowers		4–15	58	Petter (1962) Jolly (1966)
Fruit, seeds	Diurnal and nocturnal	3–10	40–60	personal observation; Petter (1962)
Leaves, fruit	Diurnal and nocturnal	4–17	900	Sussman (1972)
Flowers, bark		4–17	1227	Sussman (1975)

(*continued*)

TABLE I (*continued*)

Species	Place[a]	Duration of study	Season or month of study	Locomotor type
Lemur mongoz	1	2 months	July–Aug. 1973	Quadrupedal
	1			
Lemur rubriventer	3,26	Occasionally over 15 months	All	Quadrupedal
Propithecus verreauxi verreauxi	8	As: *L. catta* (Jolly, 1966, 1972b)	Ongoing	VCL
	10,11,13	18 months	All	
Propithecus v. coquereli	1	18 months	All	VCL
Propithecus diadema diadema	3,12 26	Occasionally over 15 months	All	VCL
Avahi laniger	3,12, 26	Occasionally over 15 months	All	VCL
Indri indri	3,12, 26	15 months	All	VCL
Daubentonia madagascariensis	15			Quadrupedal

[a] Refers to numbered study areas in Fig. 1. (In general, where more than one study has been completed on the same species the *longest* period of observation was the criterion for inclusion in this table.)

All the living lemurs spend most of their time in trees, bushes, and shrubs. The most terrestrial species, *Lemur catta*, have been observed in rocky, grassland parts of the south by Buettner-Janusch (1963), but detailed information has not been published. In gallery forest vegetation at Berenty *Lemur catta* spend only 4–36% (mean 15%) of their time on the ground (Jolly, 1966; Sussman, 1974; Budnitz and Dainis, 1975), and much of this time is devoted to travelling from one part of the range to another (Sussman, 1972). All lemurs, therefore, feed on or in trees and bushes, either travelling around a ranging area primarily by moving from tree to tree or, in the case of *L. catta*, sometimes descending to the ground to travel and climbing into the trees at their destination in order to feed. It should be added that most lemurs, if not all, are capable of moving on the ground and may frequently do so to cross a treeless part of their ranging area. Lemurs invariably sleep in trees or bushes.

Diet	Activity pattern	Group size	Population density (No./km²)	References
Nectars, flowers. fruit, leaves	Nocturnal	2–4		Tattersall and Sussman (1975)
				Petter (1962)
	Diurnal	2–4		J. I. Pollock personal observation
Fruit, flowers, leaves, bark	Diurnal	4–9	110–150	Jolly (1966, 1972b); Jolly *et al.* (1977)
		4–10		Richard (1973, 1974a,b)
Fruit, flowers, leaves, bark	Diurnal	4–10		Richard (1973, 1974a,b)
Fruit + ?	Diurnal	2–5		J. I. Pollock, personal observation
?	Nocturnal	2–4		J. I. Pollock, personal observation
Leaves, fruit, flowers, seeds	Diurnal	2–5	9–16	Pollock (1975a,b, 1977)
Insects, fruit	Nocturnal	Solitary		Petter (1962); Petter and Peyrieras (1970b)

2.2. The Climate of Lemur Habitats

The island of Madagascar contains an elevated central plateau running the length of the island which separates a steep escarpment and narrow coastal strip to the east from a gently sloping western plain. The high plateau (1500–3000 m above sea level) is characterized by a moderate to heavy annual rainfall (1500–2000 mm) which falls mostly between November and March (summer). The temperatures in this region, which greatly depend on altitude, range from moderate in the summer (22°–30° C) to cool in the winter (mean < 11° C), with great diurnal variations.

In the east, moist, warm prevailing winds from the Indian Ocean deposit large quantities of rain on the coast and escarpment regions. Although these areas are continuously wet (the number of days on which rain falls each month is approximately constant), the amount of precipitation varies seasonally, peaking in the

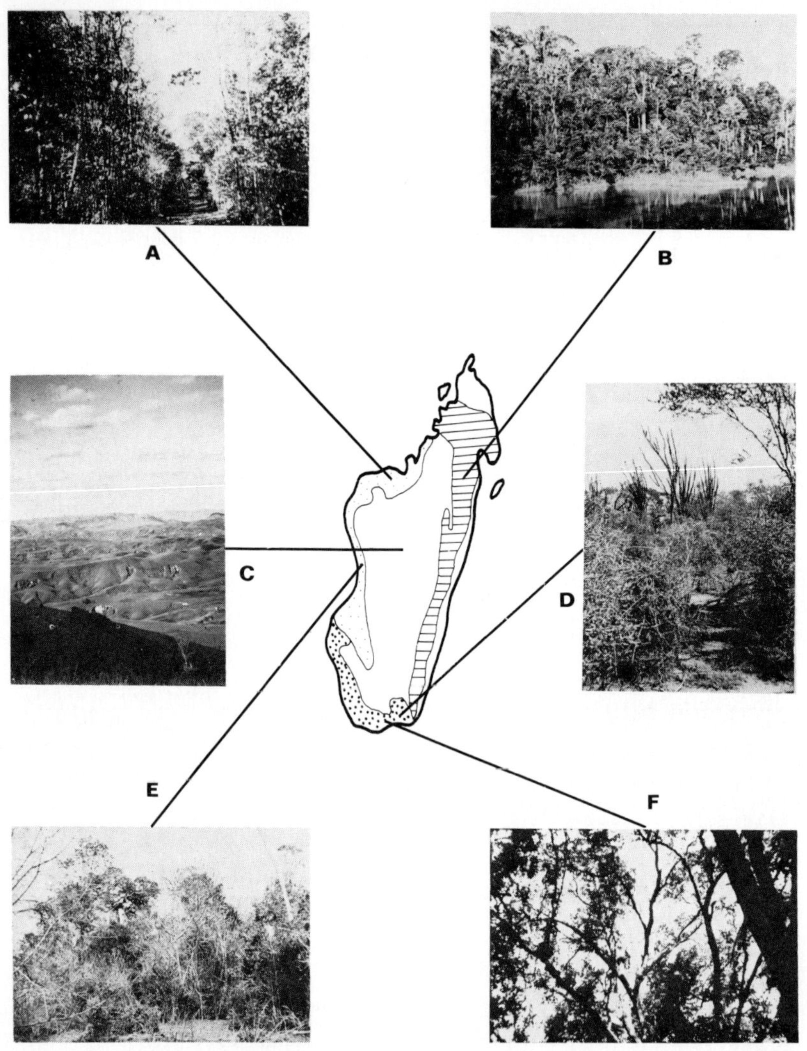

Fig. 2. Lemur habitats in Madagascar. Key: (A) Ankarafantsika; (B) evergreen rain forest; (C) central plateau region; (D) *Didierea* forest; (E) "bush and scrub" forest; (F) riverine forest at Berenty.

summer months (total annual rainfall 2000–5000 mm). Temperatures, which reduce by about 5°C for every 1000-m increase in altitude, are cold or moderate in the winter and warm or hot in the summer in the higher regions. Along the east coast, however, the climate is warm or hot and wet throughout the year.

The west of Madagascar is warm in winter but very hot in summer when the only rain falls. Annual rainfall varies from 300 mm to 1500 mm with southern parts of the west, in general, being drier than northern parts. A north–south gradient of temperature is particularly evident in the winter when the southern areas (being outside the tropics) become much colder at night than those in the north.

Seasonal variation in the climate of lemur habitats in Madagascar is thus greater in the west than in the east where continuously wet conditions buffer to some extent changes in the ambient temperature. Within both the eastern and western regions latitude (each region spans 13° of latitude) and altitude differences are responsible for most of the major climatological variations.

2.3. The Vegetation in Lemur Habitats

The relationship between the distribution of lemurs and types of vegetation in Madagascar has been recently reviewed by Martin (1972b). The five phytogeographic divisions of the island proposed by Humbert and Cours Darne (1965) adequately account for major discontinuities in the distribution of lemurs and therefore only a brief descriptive account of the botanical properties of lemur habitats is presented here. Associations between soil type, soil erosion, and supported vegetation are discussed by Perrier de la Bathie (1936) and le Bourdiec (1972).

2.3.1. Western Regions

The west of Madagascar (see Section 2.2) can be loosely divided into western and southern domains (Fig. 2). Southern domains are characterized by an indigenous, xerophytic vegetation dominated by different species of fleshy, spiny members of the families Didiereaceae (*Didierea* sp. and *Alluaudia* sp.) and Euphorbiaceae (*Euphorbia* sp.). According to Richard (1974b) over 80% of the plant species in this type of forest (at Hazafotsy) are endemic and adapted to the severe (hot, dry) conditions by possessing tree trunks and branches with large girths and by having thick, small leaves which conserve water. Tree height rarely exceeds 13 m in this region. Few lemur species inhabit these forests (Fig. 2D) and in the following text (Section 3) the only occupants mentioned are *Propithecus verreauxi verreauxi* and *Lepilemur mustelinus*.

Along the banks of rivers draining the southern domain in the west, the *Didierea* forests are replaced by a more luxuriant growth of high (< 25 m) gallery vegetation, usually consisting of a continuous canopy growth of *Tamarindus indica* (e.g., Berenty, Fig. 2F). These riparian forests are often distributed, as in Berenty, in small stretches surrounded by cultivations of sisal or endemic *Didierea* bush. Lemurs are found at relatively high densities in these areas, attracting observers who can easily study the diurnal occupants, *Lemur catta* and *P. v. verreauxi*. Observations also made on the nocturnal species, *Lepilemur mustelinus* and *Microcebus murinus*, inhabiting such environments are included in the text.

The western domain of the Malagasy west has a variety of forest types distributed according to the quality of the soil and its drainage properties. Xerophytic vegetation in this domain is formed into dense, low-stature, bush and scrub forests with no dominant plant species and the density of trees seemingly dependent on the soil's ability to retain moisture (Fig. 2E). Interspersed in the vegetation one frequently finds the familiar baobabs (*Adansonia* sp.). This very dry, woody vegetation is gradually replaced to the north by single-dominant, deciduous woodland (similar to that found at Berenty in the southern domain), which is also present throughout the western region both along rivers and in some basin areas where the ground is sufficiently moist (Sussman, 1972). Observations reported below on *Phaner furcifer* and *Microcebus coquereli* at Analabe were made in bush–scrub forests of the type described above (Fig. 2E). Information from Sussman's study of *L.f. rufus* and *L. catta* (with isolated references to other species) at Antserananomby and Tongobato refer to areas where the vegetation was mainly of the closed-canopy, riverine variety.

Petter (1962), Richard (1973, 1974a,b), and Tattersall and Sussman (1975) all worked at or near Ampijoroa in the forest of Ankarafantsika. The forest in this area is generally of the gallery type described above, varying according to local soil conditions. Thus, detailed observations on *P.v. coquereli* in this region were made by Richard in a dense, low forest (< 13 m high) composed of trees with trunks of small diameter growing on an elevated, sandy platform (Fig. 2A), while those of Petter were made on the same species and *Lemur fulvus fulvus* (*Lemur macaco fulvus*) living in more typical, but partially degraded, gallery vegetation surrounding Lac Ampijoroa. *Lemur mongoz,* studied by Tattersall and Sussman (1975), were seen to range and feed around the forestry station compound in vegetation that is largely composed of imported species (*Tectonia grandis* and *Mangifera indica*) which are absent from the autochthonous forest of the region.

Forests of the west of Madagascar are largely composed of deciduous plants, all or most of the leaves falling during the dry, winter season to be replenished just before or during the onset of the rains in October or November. Evergreen species are present, however, especially in the northern parts of the west. Flowering also normally occurs during the summer (October–February), but fruit may remain on the trees of various species for different periods of time throughout the dry season.

2.3.2. Eastern Regions

The forested parts of the eastern escarpment and thin coastal plain are characterized by a luxuriant evergreen growth of rain forest, with large numbers of woody plant species growing at low densities but forming, nevertheless, a dense and complex vegetation (Fig. 2B). Some of the observations of rain forest-living lemurs which are reported in the text were made in coastal areas (Andapa, Mahambo, Vohémar) where lowland, degraded forests are common but are continuously and increasingly threatened by cultivations of rice, manioc, maize, and bananas (*Daubentonia madagascariensis, Avahi laniger, Microcebus rufus*). All the other information presented was obtained in higher, colder, montane forests further inland (Analamazoatra, Fierenana, Vohidrazana, Antalaha), which differ chiefly in the presence of bamboo and widespread epiphytic parasitism by tree ferns, mosses, and orchids. Seasonality in the phenology of both littoral and montane regions is not marked and different species and different individual trees of the same species may flower, fruit, and flush new leaves at widely different times of the year.

The montane structure of the higher altitude eastern forests, where observations on *Indri indri, Propithecus diadema diadema, Avahi laniger, Microcebus rufus,* and *Lemur fulvus fulvus* were made, causes an extensive physiognomical variability in forest structure to exist between valley bottoms and ridge tops. Although the growth of the major common groups of species of Lauraceae (*Beilschmedia* sp., *Ravensara* sp., *Ocotea* sp., and *Cryptocarya* sp.) and Euphorbiaceae (*Euphorbia* sp.) occurs at all levels, trees growing in the wet valleys are usually buttressed, tall (< 45 m high), with slim, straight trunks and wide, horizontal canopies which shade the open forest floor. In contrast, the exposed tops of the ridges support a mass of short, twisted, epiphyte-infested trees with intermingled networks of superficial root systems.

Many forested areas in the east of Madagascar (especially in the north) are subjected to heavy morning mists or long periods of light rain in the winter months, and periods of cyclonic rain and wind during the summer.

3. POPULATION DENSITY, THE DISTRIBUTION OF INDIVIDUALS IN THE POPULATION, AND LEMUR RANGING BEHAVIOR

3.1. Introduction

The gross structures of lemur populations are discussed in this section. Information about the numbers of individuals found in an area is presented and the way in which the local distribution of individuals of a species is clustered or spread out is described in terms of the geographical apportioning of the environment into areas of varying degrees of exclusive use by individuals or groups.*

Measures of population density are stated below as estimates of the number of individuals in a square kilometer (100 hectares) of the habitat. Primates are not

*In the following text territoriality in primates is considered to exist where it has been shown that individuals or groups of individuals occupy a part of a continuous habitat that is suitable for and denied to other members of the same species for a period of time. This may or may not be the consequence of overt displays (Ellefson, 1968; Chivers, 1974) commonly cited as 'territorial' behavior. The suggested use of the word "territorial," therefore, accepts that the ecological function of territorial behavior is mostly associated with obtaining an exclusive portion of the habitat and everything it contains (i.e., including mates). This implies that the means carried out to maintain exclusive use have no other function. Thus although the dramatic, energetic displays of gibbon territoriality manifested regularly in Ellefson's study (Ellefson, 1968) may have additional functions, they are regarded here as being primarily concerned with maintaining a territory. Local ecological conditions may cause different territorial strategies of behavior to develop at different times or in different environments (Jay, 1965; Yoshiba, 1968; Jolly, 1966, 1972b; Richard, 1974b). To be useful, therefore, territoriality should be viewed with temporal and environmental as well as phylogenetic components.

uniformly distributed through the environment, however, and these figures are not meant to imply that in a typical or average square kilometer one might expect to find the stated number of individuals. The use of this standard is thus translative and not representative, as it is frequently computed from counts of a few individuals in small areas or from counts made of a dense population chosen for study because of its high animal concentration (Jolly, 1972a).

3.2. Cheirogaleidae

3.2.1. Microcebus murinus and M. rufus (Lesser Mouse Lemurs)

Two distinct forms of lesser mouse lemur are found in Madagascar, a large, gray, western species, *Microcebus murinus,* and a smaller, brown, eastern species, *M. rufus* (see Martin, 1972a).

In the east Petter and Petter-Rousseaux (1964) quantitatively estimated the abundance of mouse lemurs by noting the frequency with which they were encountered alongside forest paths. Assuming a narrow depth of vision into the forest on either side of the path, the number of counted individuals could be attributed to a known area of forest. Using this technique, Petter and Petter-Rousseaux found that at Mahambo on the east coast individual *M. rufus* lived in a widely dispersed pattern at an overall population density of 250 individuals/km^2. Six weeks after the first survey a figure of 262 animals/km^2 was obtained by the same method.

Near Mandena, at the southeastern tip of Madagascar, however, the western (gray) mouse lemur was found to inhabit forest edges and the strips of forest bordering paths and clearings at relatively high concentrations (Martin, 1972a, 1973). A small population of mouse lemurs was found to live in the 1.5 hectare Mandena study area at a density equivalent to 1300–2600 animals/km^2 but, according to Martin, discontinuous, local concentrations of these animals occurred. Nearby at Bebarimo, *Didierea* forest was found to be occupied by gray mouse lemurs at a density of 360 individuals/km^2 following counts made throughout the 2.5–hectare study area by Charles-Dominique and Hladik (1971).

Although both western and eastern forms are frequently encountered in shrubs and bushes under 5 m high, *Microcebus* in both regions can successfully occupy the fine-branch niche afforded by dense, canopy vegetation at the tops of tall trees (Martin, 1973; J. I. Pollock, personal observation). All levels of the forests are, therefore, used by mouse lemurs, the time spent at each height perhaps varying according to the vertical distribution of insects. In both west and east, mouse lemurs have been seen feeding on insects on the ground (Martin, 1973; J.I. Pollock, personal observation).

During nocturnal observations, *M. murinus* are usually seen singly in the forest. During the day, however, *M. murinus* may be found sleeping in small aggregations, usually of 2 to 9 individuals, although as many as 15 mouse lemurs have been observed in such ''dormitories'' (Martin, 1972a). Collecting animals from daytime leaf nests or tree hollows enabled Martin (1972a) to measure variation in *M. murinus* nest compositions. He showed that nesting patterns fell into two groups: solitary males or male pairs, and female groupings (ranging from 2 to 15 individuals, mean size 4.3), indicating that sexual segregation frequently occurred when the animals were not active. A subsequent study at a different time of the year suggested that during the mating season female nesting-group size decreased and that individual males had an increased tendency to nest with one or more (median 3) females (Martin, 1973).

During the course of two short field studies Martin estimated that females were 3 to 4 times more frequently observed in localized populations than males, although the sex ratio of infants captured with their mothers in nests appeared to be 1:1. Following observations made at night on the ranging movements of individuals, it appeared that local populations or population "nuclei" were structurally composed of a central core of males each occupying a ranging area that often overlapped the ranging areas of one or more females. Within sexes, ranging areas were individually separated while between sexes a small degree of overlap was sometimes observed (Martin, 1972a). Furthermore, individuals encountered on the fringes of local concentrations of mouse lemurs (population nuclei) were mostly males, which were probably of lower weight than central core males. This information has been interpreted by Martin as evidence for a central breeding aggregation of many females and a few males which probably range more or less independently within small areas; marked mouse lemurs were never seen by Martin more than 50 m away from the position of their first sighting. Surrounding this central concentration were isolated males which had been excluded from access to females. The lower weight of these "peripheral" males suggests either that they were juvenile or that, for ecological or social reasons, they were less successful in feeding.

Details on individual mouse lemurs' nocturnal movements have not been obtained and although it is certain that leaf nests, tree hollows, and birds' nests are repeatedly used for sleeping, it is not known that they house the same individuals on different occasions. However, in order that females nest in sleeping aggregations, some communal constraints on ranging behavior must occur, even if it applies only to converging movements, at dawn, of individuals which subsequently sleep together.

In *Galago demidovii* communities, powerful assembly calls emitted toward the end of the night's activity period effect the formation of sleeping parties very similar to those seen in *M. murinus* (Charles-Dominique and Martin, 1970; Charles-Dominique, 1972). A "social contact" call has been described for a hand-reared, captive *M. murinus* by Petter (1962), and Martin (1972a) suggests that high-pitched squeal vocalizations have some spacing function " . . . as it is the only call regularly heard when the animals are dispersed for feeding. . . ." There is no direct evidence, however, to show that spacing is effected or maintained by the use of these calls. No behavioral mechanism accounting for the limitations of individuals' ranging movements has been proposed, but Petter (1962) describes a persistent intolerance between adults in the wild and aggression in the laboratory between captive males introduced at different times to the same cage.

Little is known, as yet, about the distribution of food sources in the *M. murinus* habitat. A large proportion of the diet is insects and these may be distributed more randomly than plant species about a ranging area. Whether or not ranging strategies are employed for obtaining insect food is not understood, but it seems likely that nocturnal travel about an area (as opposed to local movements within bushes or trees) might depend more on social than nutritive factors.

Patterns of spacing and ranging in *M. murinus* might be expected to vary with changes in the animals' general levels of activity. Starmühlner (1960) cites Grandidier's statement that *M. murinus,* like *Cheirogaleus* sp. (see Section 3.2.3), undergo a period of dormancy during the winter months (March–September). Martin (1972a), however, observed no change in sighting frequency at different times of the year for either the western or eastern species, pointing out that by using fat stored in the tail these animals may be supplementing their energy supply during the winter rather than totally depending on it.

The environmental temperatures experienced by *M. murinus* appear to influence the general level of the animals' activity. Laboratory experiments have shown that *M. murinus* living at ambient temperatures of less than 18° C are less active than those maintained in warmer environments (Bourlière and Petter-Rousseaux, 1953; Bourlière *et al.*, 1956). Recently, seasonal changes in lethargy and the ability to maintain homeothermy have been demonstrated for captive *M. murinus* (Russell, 1975).

The implications of these results for mouse lemurs' ranging behavior in the wild are not clear. Martin (1972a) remarks that as daytime temperatures in forests inhabited by the western mouse lemur rarely fall below 18° C at any time of the year, such effects are unlikely to cause major seasonal changes in activity. However, night temperatures frequently drop below 18° C, both in the west and east of Madagascar and torpidity may act more on a day-to-day basis according to small fluctuations in the climate rather than resulting from an all-or-nothing seasonal switch.

The use of nests (which are conceivably of thermoregulatory significance) might, if other suitable sleeping spots were uncommon, impose restrictions on ranging activities. More information on individuals' visits to nests are required in order to examine this possibility. In the eastern rain forest of Analamazoatra mouse lemurs regularly slept in birds' nests (after they had been vacated by fledglings) until they decomposed (J. I. Pollock, personal observation). Although some birds' nests may remain usable sleeping spots for mouse lemurs throughout the year, their seasonal availability might be expected to influence ranging movements to some extent.

3.2.2. *Microcebus coquereli*

This large species of mouse lemur, which inhabits parts of western Madagascar, has been observed until recently on few occasions and there is very little information available on its natural history. A 750-m stretch of gallery forest at Analabe near Morondava was found to contain approximately 250 *M. coquereli*/km² (Petter *et al.*, 1971). This population density estimate was based on counts of only 14 individuals and may not be representative of concentrations of *M. coquereli* in a large area.

There is good evidence, however, that *M. coquereli* build leaf nests and do not sleep in naturally occurring holes in trees (Petter *et al.*, 1971). The effects of nest utilization on the ranging activites of *M. coquereli* and *M. murinus* may be similar.

3.2.3. ***Cheirogaleus*** **sp. (*Dwarf Lemurs*)**

Virtually no information is available on the ecology of the western dwarf lemur *Cheirogaleus medius*. A single individual of this species was seen in a 10-hectare patch of forest at Antserananomby (Sussman, 1972). The larger, eastern dwarf lemur, *Cheirogaleus major*, was encountered by both Petter and Petter-Rousseaux (1964) and the author at Analamazoatra. At Mahambo, individuals of this species were found to be evenly distributed in the forest bordering a path, the animals being sighted on average at intervals of several hundred meters, a density equivalent to 75–110 individuals/km² (Petter and Petter-Rousseaux, 1964).

Dwarf lemurs sleep during the day either in hollows in trees or, probably less often, in leaf nests (Petter, 1962). No data are available on the distribution, frequentation, or social significance of their sleeping sites.

There is good reason to believe that *Cheirogaleus* sp. undergo considerable changes in activity level during the year. It was first reported by Shaw (1879) that during June (midwinter in Madagascar) a captive dwarf lemur was asleep, cold to touch, and could only be induced to become active by vigorous rubbing and artificial warming. Shaw also observed an "unfortunate" enlargement of the tail, an enlargement which was probably less unfortunate for the animal then it was for Shaw's sentiments as, according to Petter (1962), it becomes ". . . packed with fatty tissue and remains swollen for a week or more. . ." in *C. medius*. Petter also describes periodic variation in winter dormancy lasting from two or three days (*C. major*) to one month (*C. medius* in September) and a general interspecific difference in the duration and incidence of torpor during the cold season. Short-term cycles in tail-fattening have also been seen to occur as a response to fluctuations in ambient temperatures.

It is not clear for how long each year dwarf lemurs in the wild may be partially or totally inactive. Neither species of dwarf lemur was sighted by Martin (1972a) between July 14 and September 14 (1968) in forests known to contain these animals. After the middle of October of the same year, however, sightings became frequent. Martin states that it can be safely assumed that at Mandena *C. medius* remain inactive in their nests for part of the year.

Bourlière and Petter-Rousseaux (1953) have shown that *Cheirogaleus* sp., like *M. murinus,* exhibit imperfect homeothermy. A progressive drop in body weight associated with the elimination of fatty tissue in the tail occurs during the winter period of low activity and the tail fat presumably acts, therefore, as an energy store.

As for the other cheirogaleines, one may only suppose the implications of major changes in physiology for dwarf lemur's ranging activities. It seems likely, however, that there are both (1) considerable seasonal differences in the amount of energy expended during the period of activity each night, and (2) immediate constraints on activity according to fluctuations in air temperature which may act throughout the year. These two effects may give rise to both long-term and short-term variability in the ranging activities of dwarf lemurs.

3.2.4. *Phaner furcifer* (Fork-Marked Lemur)

A population density estimate of 550–870 *Phaner furcifer*/km^2 was reported in gallery forests near Analabe and Vondrove in the west by Petter *et al*. (1971).

Rand (1935) states that *P. furcifer* were "usually found in pairs" and, according to Petter *et al*. (1971), although individuals were most commonly encountered on their own, two animals were found licking each other, two were seen moving together, and two or three fork-marked lemurs were observed feeding in the same tree. These authors also report that a spacing or warning vocalization is emitted when animals approach too closely and that violent interactions are frequent.

A "home range" of about 1 hectare was said to be occupied by some ten *P. furcifer* during a study period lasing 16 nights near Analabe in the west (Petter *et al*., 1971), individuals apparently moving independently within this area but frequently meeting in trees where gum exudates were eaten.

Although *P. furcifer* sometimes occupy leaf nests during the day they apparently parasitize those made by other cheirogaleines and do not construct their own (Petter *et al*., 1971).

3.3. Lepilemuridae (Sportive Lemur)

The sportive lemur, *Lepilemur mustelinus*, is widely distributed about the forests of both east and west of Madagascar, where it is frequently found living at high densities. In the east (Analamazoatra) 5 *L. mustelinus* were seen by the author in a 2-hectare area, a density equivalent to 417 individuals/km^2. In the *Didierea* forests near Berenty, *L. mustelinus* population density was found to be 220–350 individuals/km^2, while the adjacent riverine forests were inhabited by the same species at a density of 270–810 individuals/km^2 (Charles-Dominique and Hladik, 1971). According to Sussman (1972) 20 *L. mustelinus* occupied a 10-hectare patch of forest at Ansterananomby (equivalent to a density of 200 individuals/km^2).

Lepilemur mustelinus, like the Cheirogaleidae, are strictly nocturnal and, except in the case of females with their single young offspring, are found on their own in the forest (Petter, 1962; Charles-Dominique and Hladik, 1971; J. I. Pollock, personal observation). Unlike the smaller mouse lemurs and dwarf lemurs, *L. mus-*

telinus have a folivorous (rather than insectivorous/frugivorous) diet. *Lepilemur mustelinus* also seem to differ from the other small-bodied nocturnal lemurs by engaging in frequent social interactions in the wild. Petter (1962) describes an aggressive interaction between two adults which faced each other 8 m apart, calling alternately for 2 hours, and finally disappearing into the forest during a violent chase. This behavior was interpreted by Petter as "territorial" defense, a view supported later by Charles-Dominique and Hladik (1971) who described a similar episode and mentioned "mutual surveillance," "tree shaking," and "head shaking" displays subsequently shown to have occurred at stable, territorial borders.

During the intensive 6-week study carried out by Charles-Dominique and Hladik (1971) on *L. mustelinus* at Berenty, marked animals were followed and the position of ranging limits and encounters with conspecifics geographically plotted. Adult female *L. mustelinus*, some with their single, young offspring, were found to maintain ranging areas from which they excluded other adult females (although a single exception to this general rule is cited below). Adult males also ranged in areas to which other adult males were denied access. Male territories were found to be slightly larger (0.2–0.46 hectares) than female territories (0.15–0.32 hectares) and geographically coincided with or extensively overlapped those of the females. In general, no male's territory overlapped the territory of a female already partially covered by the ranging area of another male. An essentially complete intrasexual territorial separation was accompanied, therefore, by an intersexual regional monopoly of varying degrees.

The distances separating active, adult female *L. mustelinus* from their offspring were found to vary according to the latter's ages. Adult female/young sleeping aggregations were observed by Rand (1935) and Petter (1962), the latter in studies of *L. mustelinus* near Lac Ampijoroa during the month of January. This close physical proximity probably diminishes rapidly as "by September . . . immature and adult females had separate sleeping holes 5 to 15 m apart. . . ." (Charles Dominique and Hladik, 1971).

It is possible that social relationships between *L. mustelinus* of different generations may persist down the female line of descent for some time. One case where an old female, a (presumed) primiparous female, and an immature female had overlapping ranges and engaged in interactions of an affiliative nature suggest that these individuals were closely related to each other (Charles-Dominique and Hladik, 1971). Apart from differences in size and age no evidence was presented by these authors, however, to show that a close genetic relationship existed.

The complete movements of one adult male *L. mustelinus* were recorded for a single night by Charles-Dominique and Hladik (1971). They found that, despite the minute ranging area (approximately 40-m radius), physiological limitations on individuals' movements appeared to be operating. During the night the subject animal covered a distance of 270 m in about 180 leaps interspersed with short feeding bouts and long periods of inactivity lasting, on occasions, more than 2 hours. The total duration of the period that the animal was active (i.e., not in a sleeping position) during the night was 630 minutes, but only 74 minutes (or 11.7% of this time) were spent in feeding activities.

By comparing conservative estimates of the nutritional value of consumed food with an estimate of the energy requirements of the individual, Charles-Dominique and Hladik (1971) concluded that the generation and expenditure of energy could be very finely balanced. The products of bacterial action on undigested cellulose materials in the cecum of *L. mustelinus* are reingested (cecotrophy) and probably constitute a vital nutritional supplement (see Hladik, Chapter 8). Charles-Dominique and Hladik argue persuasively for the presence of modest physiological limitations on the activity of this species. Very little is known, however, about long-term nutritional parameters in *L. mustelinus*. It is not understood, for example, if nocturnal movements in some months depend on metabolic stores replenished at other times of the year or, more frequently, when there are fewer demands on available energy, or when a more energy-rich diet can be obtained (see Hladik, Chapter 8).

The issue of *L. mustelinus* territoriality at Berenty and some of its consequences for individual ranging behaviors were clarified by Charles-Dominique and Hladik (1971) in the course of their 6-week study period. They describe a category of behavior, "surveillance," consisting of vigilance and awareness of neighboring conspecifics' movements by visual scanning from a vantage point high in a tree. For the duration of the study period *L. mustelinus* were found to eat the leaves and flowers of the most common tree species in the forest, *Alluaudia procera* and *Alluaudia ascendens*, and during the single complete night's study one *L. mustelinus* male moved to only eight different sources of food. Much of the night (5 hours, 50 minutes) was spent by this individual in a small region near the borders of two neighboring males' different ranging areas, surveying their movements. It seems possible, therefore, that during this study territorial maintenance constituted a major activity for *L. mustelinus*, accounting for much of the animals' ranging movements. Certainly the removal of one male *L. mustelinus* from its territory resulted in its colonization by a neighboring male, which immediately established a nonaggressive relationship with the females ranging in that region and maintained the original male's territorial borders.

Territorial defence is clearly facilitated, particularly if critical restrictions in energy budgets are acting, by knowledge of the location of potential intruders (visual surveillance). This might be expected to have important consequences on ranging behavior. Furthermore, the ranging behavior of *L. mustelinus* may be dependent on the stage of the lunar cycle. Charles-Dominique and Hladik (1971) found a positive correlation between the level of ambient light and the frequency of vocalizations, suggesting that when visibility was good the location, identity, and perhaps intentions of neighbors could be better perceived. Ranging movements associated with territorial defense may be increased, for example, when the moon is full.

Finally, it has been suggested by Petter (1962) that *L. mustelinus* at Ankarafantsika and Lac Ampijoroa live in small population clusters (*noyeaux de populations*). Although Charles-Dominique and Hladik (1971) state that *L. mustelinus* inhabit the *Didierea* forests in the south homogeneously, they claim that this is due to the more or less contiguous distribution of these concentrations of animals. Such concentrations may reflect the distribution of resources in the environments or may develop solely as a result of social pressures. Whichever is the case, the influences on individuals' ranging activities must be considerable. Long-term studies in a variety of habitats are required to clarify the nature of these influences.

3.4. Lemuridae

3.4.1. *Hapalemur griseus* (Gentle Lemur)

No published information exists on the population density of *Hapalemur griseus* in the wild. In the eastern rain forest of Analamazoatra, two study areas, each covering an area of 18 hectares, contained a total of 17–22 *H. griseus*, equivalent to a density of 47–62 individuals/km^2 (J. I. Pollock, personal observation). It was not possible, during the author's own observations, to determine the extent to which ranging areas were limited or exclusive, as individuals could not be distinguished.

Hapalemur griseus are most commonly seen in small groups of 3 to 5 individuals (Rand, 1935; Petter and Peyrieras, 1970a; J. I. Pollock, personal observation). One group of at least 6 individuals was briefly encountered by the author in the high, primary rain-forest near Fierenana, an area rich in bamboo which forms an important part of *H. griseus* diet. Throughout the author's period of research in Madagascar (see Section 3.5.3) groups of *H. griseus* were encountered on 38 occasions. If the fewest certain numbers of individuals counted in each group are taken as the size of the group (minimum group size) a mean figure of 2.8 individuals per group is arrived at. As group counts were usually precise, this figure is unlikely to be far below the true mean group size.

These small groupings were interpreted by Petter and Peyrieras (1970a) as constituting *families* (one adult breeding pair and their young). The present author's own observations indicate that most if not all groups contained at least one animal clearly smaller than the rest. The age and sex of *H. griseus* in the wild are, however, very difficult to determine.

Hapalemur griseus groups are spatially tight, the animals being usually found within a 5- to 10-m radius, moving in single file between 3 and 10 m above the ground. Daily travel distances and factors influencing the ranging movements of *H. griseus* have not yet been determined.

3.4.2. *Varecia variegata* (*Lemur variegatus*, Ruffed Lemur)

Groups of 2 to 4 animals were reported by Petter (1962) and, at Fierenana, the present author observed three groups of 2 animals each. According to Rand (1935) *V. variegata* are "usually seen in pairs."

Petter (1962) has suggested, from group size information, that ruffed lemurs live in small family groups. He also suggested that infants are not carried by their mothers in the wild but are deposited in "parking spots." The tendency for captive *V. variegata* to make use of nesting boxes for this purpose is cited as evidence for this behavior. If they exist, parking spots may impose strict limitations on adults' ranging behavior after the birth season.

3.4.3. Lemur sp. (True Lemurs)

3.4.3.1. ***Lemur catta*** **("Ring-Tailed Lemur").** Ring-tailed lemurs live only in the south and southwest of Madagascar where they are commonly found at high densities. From a survey at Antserananomby Sussman (1972) estimated the population density of *Lemur catta* to be 215 individuals/km^2. In the Berenty Reserve, a somewhat isolated 100-hectare patch of forest mostly composed of *Tamarindus indica* dominated riverine vegetation, changes in the *L. catta* population have been studied over a period of 12 years. In 1963 the whole reserve was thought to contain about 350 *L. catta,* equivalent to a population density of 350 individuals/km^2 (Jolly, 1966). Seven years later Sussman censused a slightly different part of the reserve at Berenty, including part of the forest in which Jolly made her intensive study, and concluded that the population density approached no more than 250 individuals/km^2. Most recently, however, Budnitz and Dainis (1975) have produced evidence that vegetationally poor parts of this reserve (bush and scrub) are populated by *L. catta* at a lower density than the "closed canopy" forested parts (see below) on which most researchers had previously concentrated their efforts. Budnitz and Dainis censused the whole Reserve twice, once during the beginning of their study period (May–September, 1972) and again toward the end (February–May, 1973); the period in between censuses included a *L. catta* birth season. In the first census the reserve was found to contain 152 *L. catta* (= 162 individuals/km^2) and in the second 153 *L. catta*, despite the birth of 42 infants. Budnitz and Dainis (1975) discovered that this difference arose from a loss of young animals, mostly juveniles and infants, from the population, presumably due to their differential mortality. The data obtained by these authors are more accurate than those of previous researchers, as no extrapolation was utilized in the census, all the troops of *L. catta* in the reserve having been found and counted.

In 1975 a further census of the whole reserve showed that the total *L. catta* population numbered 156, indicating a remarkable stability over the last 11 years (Jolly *et al.*, 1977).

The Berenty ring-tailed lemurs are found in groups of 5–24 individuals although groups of less than 7 or more than 20 are occasionally encountered (Jolly, 1966; Budnitz and Dainis, 1975; Jolly *et al.*, 1977). At Antserananomby, Sussman (1972) found one *L. catta* group to contain 19 individuals.

Groups of *L. catta* at Berenty have been said to occupy territories on the basis of their exclusive use of large parts of a ranging area and the nature of aggressive encounters between neighboring groups in range overlap regions (Jolly, 1966). Both Jolly (1966) and Klopfer (Klopfer and Jolly, 1970) describe confrontations between neighboring groups during which scent-marking behavior, spats, and other agonistic displays were observed. More common, however, were observations of *L. catta* groups mutually avoiding each other when meeting in overlap regions.

The size of ranging areas of different *L. catta* groups varies considerably. At Berenty these areas have been measured from 5.7 hectares (Jolly, 1966) up to 23 hectares (Budnitz and Dainis, 1975). These latter authors suggest that the size of *L. catta* ranges at Berenty may depend on the local vegetation type, as one study group inhabiting a bush and scrub and open forest part of the reserve ranged over an area three times as great as another group living in closed-canopy riverine forest. The range of the one *L. catta* group studied at Antserananomby by Sussman (1972) practically filled the 10-hectare study area.

Since Berenty has been regularly visited by many researchers over the last twelve years, some aspects of temporal stability in *L. catta* ecology have been examined. Klopfer visited Berenty in 1969 and found a *L. catta* group (Group F, containing two identified marker males from Jolly's Group 1 observed in 1963) ranging in the same part of the reserve as it had six years previously (Klopfer and Jolly, 1970). Recent information on the frequency of transfers of *L. catta* males between groups must cast some doubt on the importance of this observation (see below). It shows, in any case, that the geographical position of ranging limits may be stable over time whether or not the same animals are observing them.

In 1970 Jolly herself returned to this part of the Berenty Reserve to find that although approximately the same number of *L. catta* ranged there, they were distributed in twice as many ranging parties (each of approximately one-half the original group size) which each moved about the forest, ignoring old boundaries and maintaining less than 50% exclusive use of their new ranges (Jolly, 1972b). Major sources of food and water, and siesta and sleeping sites were used by many of the *L. catta* groups on a time-sharing basis. Jolly attributed these (and other) changes either to an increasing population density or as an "adaptive social reaction to short-term food shortage" at the end of a particularly severe dry season. It is now known, however, that throughout this period the population density of *L. catta* at Berenty was probably very stable and short-term influences are, therefore, the most likely causes of these changes. This view has been recently supported by the discovery that time-plan organized resource-sharing and considerable changes in group composition occur in some years at about the time of the *L. catta* birth season, that time of the year when Jolly (1972b) made her observations of changes in ranging and other behaviors (Budnitz and Dainis, 1975; Jolly *et al.*, 1977).

The ranging movements of *L. catta* have thus been shown to vary according to season and year. Outside the disturbed periods of mating and male migration (see below), the ranging behavior of *L. catta* at Berenty has been described as repeated visits for three or four successive days to small sectors of the ranging area, which are then varied so that "after a week or ten days" most parts had been passed through (Jolly, 1966). At Antserananomby, however, 3-day ranging cycles of the kind described by Jolly were not exhibited by Sussman's main study group, which showed a marked preference for specific daily siesta and night sleeping locations over a much longer period of time (Fig. 3; Sussman, 1972).

Lemur catta do not travel about their range throughout the day. One early progression to a feeding site and a midday group movement to a siesta point are described as a typical morning ranging pattern for *L. catta* at Berenty (Jolly, 1966). Afternoon progressions were less well-defined, but generally just before sunset another group movement took the animals to their sleeping trees. Active behavior but not group progressions occurred at night. Sussman (1972, 1974) found that the *L. catta* study group at Antserananomby organized its daily movements in a similar temporal fashion as *L. catta* at Berenty. Mean daily travel distances of single *L. catta* study groups at Antserananomby (920 m) and Berenty (965 m) were similar despite the 47% larger ranging area of the group at Antserananomby (Sussman, 1972).

The ranging movements of a whole group is obviously an aspect of social behavior and *L. catta* groups, the largest social aggregations of lemurs, have been well enough studied for information on social patterns of movement to emerge. Klopfer has described two forms of group movement in *L. catta* (Klopfer and Jolly, 1970): (1) "*amoeboid* progressions where the group is dispersed as a broad front lateral to the direction of travel with no overt control of departure or arrival. . . ." and (2)

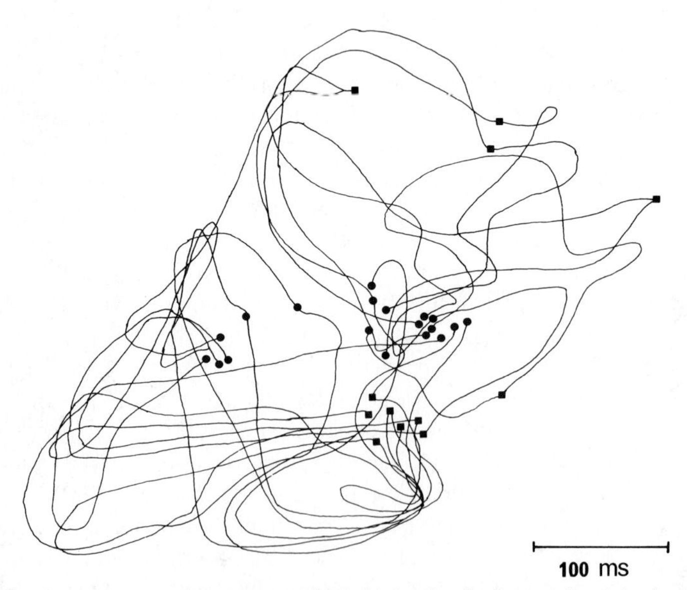

Fig. 3. *Lemur catta* ranging patterns at Antserananomby. Preferences for night sleeping sites (black circles) and siesta sites (black squares) are shown. (From Sussman, 1972.)

"*linear* progressions where an essentially single file procession is suggestive of a certain leadership or social patterning of movement." Linear progressions were regarded as common *L. catta* ranging configurations by both Jolly (1966) and Budnitz and Dainis (1975).

These authors agree that adult females, accompanied by juveniles and some adult males, generally head such progressions, with subordinate male stragglers ("drones club") strung out behind. Very often this means that the latter receive a second choice of food and water at sources where these are limited in abundance. Budnitz and Dainis (1975) emphasize, however, that the polarity of the group is much clearer during excursions into neighboring group ranges or when confronting invading groups. In these cases, an adult female always leads the group.

The *L. catta* group, spread over 20 to 30 m and several trees when traveling about its range, moves in wide circles rather than sharp angles. In general, dominant males (which are all subordinate to adult females) are more widely spaced than females although no standard interindividual distances are maintained. Jolly (1966) describes preprogression vocalizations for *L. catta* ("clicks . . . series of clicks . . . moans and wails . . . meows . . . "). The meows, Jolly claims, act as contact calls, as they are emitted when factions of the group become separated and are inevitably answered by one or more group members.

It now seems clear that at least in some years considerable seasonal changes occur in the social integrity of *L. catta* groups at Berenty. Changes in the spatial distribution of animals within groups were first noticed by Jolly (1972b), who described a more acute peripheralization of subordinate males, a greater spread of individuals within the group, and reduced behavioral synchrony between *L. catta* group members in September/October, 1970, than had been seen during other months of 1963 and 1964. Associated with these changes were an increased frequency of encounters between groups, more howl vocalizations, which acted as spacing calls both within and between groups, and a greater tendency for excursions or attempted excursions by both males and females into the ranging areas of nearby groups.

For the period September/October, 1972, Budnitz and Dainis (1975) describe similar changes in the behavior of *L. catta* at Berenty observing the process by which males (either singly or in small aggregations) changed groups. These authors suggest that in all groups of *L. catta* at Berenty there were transfers of males (but not females) toward the end of 1972. It has been proposed that about one-third of the males transfer from one group to another each year. Recent findings that some identifiable males, which changed groups in 1972, were still in their new group in 1975 does not imply that reassortment of male *L. catta* at Berenty in 1972 was necessarily exceptional (Jolly *et al.*, 1977).

3.4.3.2. *Lemur macaco* (Black Lemur).

a. Lemur macaco macaco. The distribution map of *L.m. macaco* on Nosy Komba published by Petter (1962, p. 57: but see new distribution maps: Petter and

Petter-Rousseaux, Chapter 1) shows that the 29 individuals in three groups (see below) occupied an area of about 50 hectares, equivalent to a population density of 58 individuals/km^2.

From the combined data obtained by Jolly (1966) and Petter (1962), the 15 groups of *L.m. macaco* censused on Nosy Bé and Nosy Komba had a mean group size of 9.2 (mode 9) ranging from 4 to 15 individuals per group.

Petter (1962) describes the presence of "... *groupes élémentaires et des rassemblements nocturnes*...." The latter are claimed to occupy a more or less fixed sleeping area implying that those subgroups which subsequently aggregate at night otherwise follow independent ranging patterns. These subgroups are said to maintain distinct, individual territories during the day (Petter, 1962).

Lemur macaco macaco group progressions were found by Petter (1962) to be led usually by an adult female with a male at the rear of the progression. According to Petter, high temperatures considerably reduce the ranging activities of black lemurs on Nosy Komba; it was found, for instance, that on hot days the animals rested more in the shade.

Petter (1962) noticed that changes occurred in the adult composition of some *L.m. macaco* groups between visits to Nosy Komba in May and November, 1956, and in April, 1957. Between May and November, 1956, one black lemur group lost two adult males, another lost one adult male and gained an adult female, and a third group gained two adult males. A fourth group lost one adult male. Between November, 1956, and April, 1957, the first three groups (above) remained unchanged in adult composition (assuming that changes were not continuously occurring in these groups), but one group gained at least one adult male.

It seems likely that all *Lemur* spp. throughout Madagascar reproduce at similar times of the year (Petter, 1962; Jolly, 1966; J. I. Pollock, personal observation) with births occurring just before or at the beginning of the wet season (August–October). Observations of the timing of group composition changes in *L.m. macaco* are consistent with the view that these may occur at the same stage of the reproductive cycle as they do in *Lemur catta*.

3.4.3.3 *Lemur fulvus* (Brown Lemur).

a. Lemur fulvus fulvus. Groups of about 10 of these lemurs, are described by Petter (1962) from observations made in the forest of Ankarafantsika, noting that larger aggregations of 30 or so individuals occurred in the evenings. This same subspecies is found in the wet rain forests of the east and constitutes, thereby, an exception to the general zoogeographical rule outlined in Section 2.3. In the forests of Analamazoatra, Vohidrazana, and Fierenana, *L.f. fulvus* were seen in groups of 3–10 individuals at an approximate population density of 40–60 individuals/km^3 (J. I. Pollock, personal observation).

In the forest of Ankarafantsika *L.f. fulvus* group progressions were found to be headed by adult females with their clinging infants (Petter, 1962). It was not possible, during the author's own research, to identify the sex of the leading *L.f.*

fulvus at times other than when the females' infants were physically dependent, as males resemble females so closely. At Analamazoatra it was not obviously the case that females led the groups, although leading adults seen without infants could have been females.

Encounters between brown lemur groups at Analamazoatra were repeatedly observed at night in specific areas at the edge of the forest or in *Eucalyptus* plantations adjoining the reserve. A characteristic loud call is made by *L.f. fulvus* at these times, groups confronting each other and alternately emitting powerful, rasping cries which are taken up and reinforced by several group members at a time. These calls are probably analogous (or identical) to Petter's (1962) "crou-crou-crouii . . ." description of calls made by *L.m. macaco* and the "high-pitched rasp" of *L.f. rufus* described by Sussman (1972) as a "distance-maintaining signal."

b. Lemur fulvus rufus. Sussman (1972) censused twelve groups of *L.f. rufus* at Antserananomby (excluding one 2-male consortium) and five groups at Tongobato. The overall mean group size of 9.5 (range 4–17) did not differ significantly between the two study areas. Population densities of 1227 individuals/km^2 at Antserananomby and 900 individuals/km^2 at Tongobato were computed from the survey's data (Sussman, 1972, 1975). Describing a group of 15 *L.f. rufus* at Antserananomby, which often formed independently ranging subgroups, Sussman (1972) suggested that, when groups approach a critical size, fission might occur. In this case subgroups had varying compositions on different days.

Lemur fulvus rufus groups inhabited small ranging areas at Antserananomby (mean 0.75 hectare) and Tongobato (mean 1.0 hectare) that overlapped extensively and were not rigorously defended. The concept of group, when applied to *L.f. rufus* studied by Sussman (1972, 1974, 1975) seems to depend heavily on the synchrony of movement rather than individuals' affiliations either for a ranging area or for other individuals. This view is supported by the frequent, simultaneous sightings of at least two groups of *L.f. rufus* that temporarily slept or fed in adjacent trees or even at opposite sides of a single large tree (Sussman, 1972). Overt interactions between *L.f. rufus* groups were rarely seen and did not necessarily involve one group displacing another.

Average daily travel distances for three groups of *L.f. rufus* studied by Sussman (1974) at Antserananomby varied from 125 to 150 m. Sussman suggests, however, that seasonal variation in daily travel distance and home range size may be significantly greater for *L.f. rufus* than for *L. catta*, which had a much larger ranging area and daily travel distance (Section 3.4.3.1). Compared to *L. catta*, *L.f. rufus* at Antserananomby had a less diverse diet, moved less, and rested more for the duration of the study (Sussman, 1972, 1974).

3.4.3.4. *Lemur mongoz* (Mongoose Lemur). In the deciduous forests of Ankarafantsika Petter (1962) observed two *L. mongoz* groups containing 6 and 8 individuals, respectively. More recently *L. mongoz* have been the subject of a short but intensive study at Ampijoroa in the forest of Ankarafantsika (Tattersall and Sussman, 1975). These authors report that *L. mongoz* in this area were strictly nocturnal during the 2-month study period and lived in small groups of 2 to 4 individuals (mean 2.6). These groups, it was suggested, are composed of an adult male, an adult female, and offspring (nuclear family groups).

The main *L. mongoz* study group at Ampijoroa was found to range about an area of 1.15 hectares around the administration post in the reserve where a large number of introduced tree species grow. The ranging areas of a number of *L. mongoz* overlapped extensively and no defended ranging boundaries were evident. Although encounters between groups frequently occurred, the sites of these interactions were not apparently significant or consistent. The horizontal distance traveled by the main study group each night varied from 460 to 750 m.

During the movement of a *L. mongoz* group about its ranging area no obvious leader emerged. Tattersall and Sussman (1975) observed the adult male at the front of group processions on 24 occasions, the adult female on 21 occasions, with animals frequently overtaking each other.

One subadult female (T) periodically left the group it normally ranged with and rejoined it from time to time throughout the night. During the day this female slept with the group. On one occasion an adult male was seen to leave its group, meet up with the T female and briefly copulate with her. A certain flexibility in the pattern of social ranging behavior is therefore indicated for this species.

3.4.3.5. *Lemur rubriventer* (Red-Bellied Lemur). A group of five and a group of four *Lemur rubriventer* were seen by Petter (1962) in the forest near Moramanga. This lemur is very rare in parts of the eastern rain forest. During the author's own research period in which 3500 hours were spent observing or searching for lemurs, *L. rubriventer* were seen on only three occasions: one group contained two or three individuals, one contained a total of three, and the third contained four animals, two of which were clearly smaller than the others. R. D. Martin (personal communication) also reports very small groups of *L. rubriventer* at Perinet in the forest of Analamazoatra.

3.5. Indriidae

3.5.1. Propithecus verreauxi (Sifaka)

Propithecus verreauxi are found in forests throughout the western side of Madagascar living in small groups of 2–13 individuals. Thanks to the efforts of Petter (1962), Jolly (1966, 1972b), and Richard (1973, 1974a, b) more is known about the natural history of this species than any other lemur. Longitudinal studies have been performed on the *P.v. verreauxi* population living in the Berenty Reserve (Jolly, 1966; Richard, 1973; Jolly *et al.*, 1977) and a major field study on *P.v.*

verreauxi at Hazafotsy and *P.v. coquereli* at Ampijoroa has been completed (Richard, 1973, 1974a, b; Richard and Heimbuch, 1975). Petter's (1962) observations are largely concerned with the *P.v. coquereli* subspecies in the Ankarafantsika forest.

Propithecus verreauxi verreauxi at Berenty live at a very stable population density of about 110–150 individuals/km ^{2} (Jolly *et al.*, 1977), the exact figure depending on whether or not infants are included in the calculations. Even within the small Reserve at Berenty, however, local densities of *P.v. verreauxi* vary greatly (Jolly, 1966; Jolly *et al.*, 1977). In the Antserananomby study area, where Sussman (1972) observed *Lemur catta* and *Lemur fulvus rufus*, 40–50 *P.v. verreauxi* lived, a density equivalent to 400–500 individuals/km^2.

The only long-term information on *P.v. verreauxi* populations stems from data obtained at Berenty. The total *P.v. verreauxi* population of the reserve has probably changed very little over the past twelve years (since observations began) and increases in the number of adults seem to be prevented by a high mortality of the infants, as it does for *Lemur catta* (Richard, 1973, Jolly *et al.*, 1977).

Richard (1973) carried out surveys of *P. verreauxi* group composition at Berenty, Hazafotsy, Ampijoroa, Evasy, and Ejeda in 1970 and 1971. The 43 groups censused were normally composed of 1–4 (usually 2 or 3) males, 1–4 (usually 2 or 3) females and young individuals of both sexes. This overall mean group size of 5.9 varied (probably nonsignificantly) from 5.0 to 7.0 according to the region sampled. It is probable that groups of *P. verreauxi* at Berenty did not change in their mean size or average composition between 1963 and 1975. It is worth noting that Petter reported smaller group sizes for *P.v. coquereli* in the forest of Ankarafantsika (mean 3.9, mode 4, $N = 27$) than Richard (1973) in her survey of the same forest 14–15 years later (mean 5.0, $N = 6$). Petter (1962) claimed that *P.v. coquereli* lived in family groups, citing as evidence the observation of single infants or juveniles in each group. Richard (1973), however, has shown that *P. verreauxi* groups cannot be satisfactorily considered as family groups and it is possible that Petter's (1962) results may have been influenced by high infant mortality. Alternatively, group size and composition of *P.v. coquereli* at Ankarafantsika may fluctuate over long periods of time.

Observations on the spatial distribution and ranging behavior of *P. verreauxi* must be clearly separated into those made at Berenty (a high, stable concentration of animals) and those made in the endemic *Didierea* forests nearby and on other *P. verreauxi* spp. At Berenty *P.v. verreauxi* live in small ranging areas of about 1–3 hectares, a large central part of which is actively defended against intruding groups. At these high concentrations *P. verreauxi* may travel across the exclusive part of their ranging area (called "territories" by Jolly) twice in a day (Jolly, 1966).

In contrast, two groups of *P.v. verreauxi* at Hazafotsy (southern Madagascar) and two groups of *P.v. coquereli* at Ampijoroa (northern Madagascar) were found by Richard (1974b) to range about areas measured at 9 hectares. Richard was able to ascribe the term ''territorial'' meaningfully to *P. verreauxi* in the south, but not to those in the north. The two northern study groups made exclusive use of only 24 and 29% of their respective ranging areas and 33% of the observed interactions between groups involved ritualized battle displays. In the south, however, the two study groups had exclusive use of 54 and 51% of their respective ranges and engaged in battle displays in 69% of the encounters between groups. Furthermore, areas of exclusive use (called ''monopolized zones'' by Richard) within *P. verreauxi* ranges were small, dispersed sections in each of the northern groups' ranging areas but a large contiguous block in those of the southern groups. Monopolized zones were not obviously defended in either study area.

According to Petter (1962) *P.v. coquereli* live in ''territories'' although no form of territorial defense was observed; areas of exclusive use were not determined and, according to a distribution map (Petter, 1962, p. 99), *P.v. coquereli* ranging areas in the region were not even adjacent.

The available evidence suggests that *P. verreauxi* are capable of inhabiting diverse environments at differing densities and with different social relations between groups. Petter (1962) attributes these differences to the quality of the vegetation in the habitat and recently an interesting correlation between ''territory'' size and the availability of shade in *P. verreauxi* ranging areas at Berenty has been reported (Jolly *et al.*, 1977).

The five study groups of *P.v. verreauxi* observed by Jolly (1966, 1972b) at Berenty exhibited a ranging pattern essentially similar to that of *Lemur catta* in the same area, a concentrated exploitation of a small part of the range for 3–4 days repeated in different areas so that every 7–10 days most regions had been visited for feeding. Unlike *L. catta*, however, *P.v. verreauxi* often make wide detours so as to avoid moving on the ground.

In the forest of Ankarafantsika Petter (1962) reports that *P.v. coquereli* range about in search of suitable ''sunning trees,'' sleeping trees, and food locations. One main study group moved at least 500 m daily, the individuals slowly following each other in a leisurely progression, often taking similar arboreal routes from one day to the next. Petter (1962) claimed that *P. verreauxi* ranging behavior was affected by atmospheric conditions.

Detailed information is, however, available on range utilization and travel patterns of *P. verreauxi* following the intensive study made by Richard (1973, 1974a, b). By cutting a network of paths at 50-m intervals in the home ranges of the four main study groups, accurate data on ranging movements could be obtained. Richard's study groups were found to visit most parts of their home ranges in 10–20 days, spending periods of 2–3 days when certain local areas were repeatedly visited for feeding. More time was spent in some parts of each group's home range than in others. Although there was some variation between groups in this measure, *P. verreauxi* spent on average about 60% of the time in 20% of the range and about 90% of the time in 50% of the range.

During the summer wet season Richard found that the distance traveled daily increased for all study groups although the number of different quadrats (50-m sided squares) entered varied little throughout the year. Richard (1973) concluded that *P. verreauxi* ranging behavior in the wet season was characterized by a more rapid passage over the same areas rather than increased movements to new parts of the range. These changes may have been due to the heavy selection of fruit in the diet and the widely scattered distribution of fruit sources during the wet season (selectivity in diet and ranging increased). An alternative explanation is that decreased production of food in the dry season (when many trees lose their leaves) caused *P. verreauxi* to eat a greater number of plant species and, in order to conserve energy, range less (selectivity in diet and ranging decreased).

Petter (1962) noticed that *P.v. coquereli* groups frequently appeared to have changed their composition between successive censuses and that single *P.v. coquereli* were often seen moving about on their own. Richard (1973, 1974a) also observed this phenomenon in her southern study area (*P.v. verreauxi*). During the observation period preceding the mating season (Jan. 24–Feb. 28) Richard observed a breakdown of the normally cohesive pattern of group members' ranging behavior. In addition to the relatively high frequency of intergroup encounters at this time two of the adult males under observation left their group (Group IV) a total of nine times in 28 days, once departing together. This same group was approached by "foreign" adult male *P.v. verreauxi* on five occasions, four of these being by the same individual. During the subsequent copulatory period (March 3–March 6) three different "foreign" males mated with the two females of Group IV. One of these males remained in this group with dominant status until the study ended (6 months later). A second "foreign" male mated with one of the Group IV females during an intergroup encounter, and the third male, having followed Group IV and mated with one of its females, returned to his former group.

Increases in the roaming behavior of males or pairs of males took them on long forays into neighboring groups' home ranges. Richard observed a significant increase in the frequency with which groups interacted over the precopulatory and copulatory period and found that some *P. verreauxi* were probably spending more time near the borders of neighboring groups' ranges at this time. Similar findings of male (and, to a lesser extent, female) transfers between *P.v. verreauxi* groups have been seen at Berenty (Jolly, 1966; Jolly *et al.*, 1977).

3.5.2. Propithecus diadema

The eastern species of *Propithecus* was seen by the author in groups of 2–5 individuals at Analamazoatra, Fierenana, and Vohidrazana. No accurate measure of population density was obtained, but the infrequent sightings probably meant that each study area maintained only small populations.

Three recognizable groups of *Propithecus diadema diadema* were repeatedly seen to range apparently exclusively in specific regions of the forest. Part of the ranging area of one group measured at least 20 hectares in size. No information is available on this species' ranging behavior.

3.5.3. *Indri indri (The Indris)*

Indri indri live in small groups of 2–5 individuals, an adult pair and their young (Petter, 1962; Pollock, 1975a). Measures of *I. indri* population density at Analamazoatra, Fierenana, and Vohidrazana, calculated separately from surveys of groups living in known areas, calling frequency and extrapolated estimates from study groups' ranging areas, all lay between 9 and 16 individuals/km^2 (Pollock, 1975a). Petter and Peyrieras (1974) found much lower densities of *I. indri* near Maroansetra and Lokato, although their estimates may have excluded groups which vocalized less than others. At Betampona near Tamatave, however, it is certain that *I. indri* live at much lower concentrations (J. I. Pollock, personal observation).

Groups of *I. indri* occupy defended territories which form a large central part of their ranging areas. Ranging areas of the two main study groups covered 17.7 and 18.0 hectares, respectively. A narrow band (approximately 50 m in the three measured cases) surrounding the territory is normally used for ranging and feeding by both the home and neighboring group and it is in this region that intergroup encounters occasionally occur. In 2500 hours of direct observation on three goups of *I. indri* no foreign individual or group was seen inside the territorial limit of the group being studied and on reaching their ranging limit each group would turn along the boundary or back into a central direction, whether a neighboring group was nearby or not. After each of the six occasions that intergroup encounters were observed, the two confronting groups returned slowly to the center of their respective ranges.

The author conducted a quantitative assessment of *I. indri* ranging behavior by recording the frequency with which animals moved to a different tree and how often new quadrats (50-m sided squares in a painted grid system) were entered. In this way information about movement both within and between quadrats was obtained. Data were collected from two groups (studied for 10 and 6 successive days, respectively, every 6 weeks) for 1 year. *Indri indri* were found to move between 300 and 700 m daily about their range. Although observation periods were too short to identify any cyclical organization of ranging about their area, two distinct daily ranging patterns emerged as the study progressed:

1. When certain plant species flushed into leaf, flowered, or bore fruit, *I. indri* groups visited them to feed for 1–3 hours per day. Early progressions to these trees were followed by continuous feeding for this period of time. In the early afternoon a series of short feeding bouts on a diverse array of plant species ended usually in a central sleeping area (Fig. 4A).
2. When no concentrated source of food was present, *I. indri* ranged in a less predictable fashion with small feeding progressions scattered throughout the day (Fig. 4B).

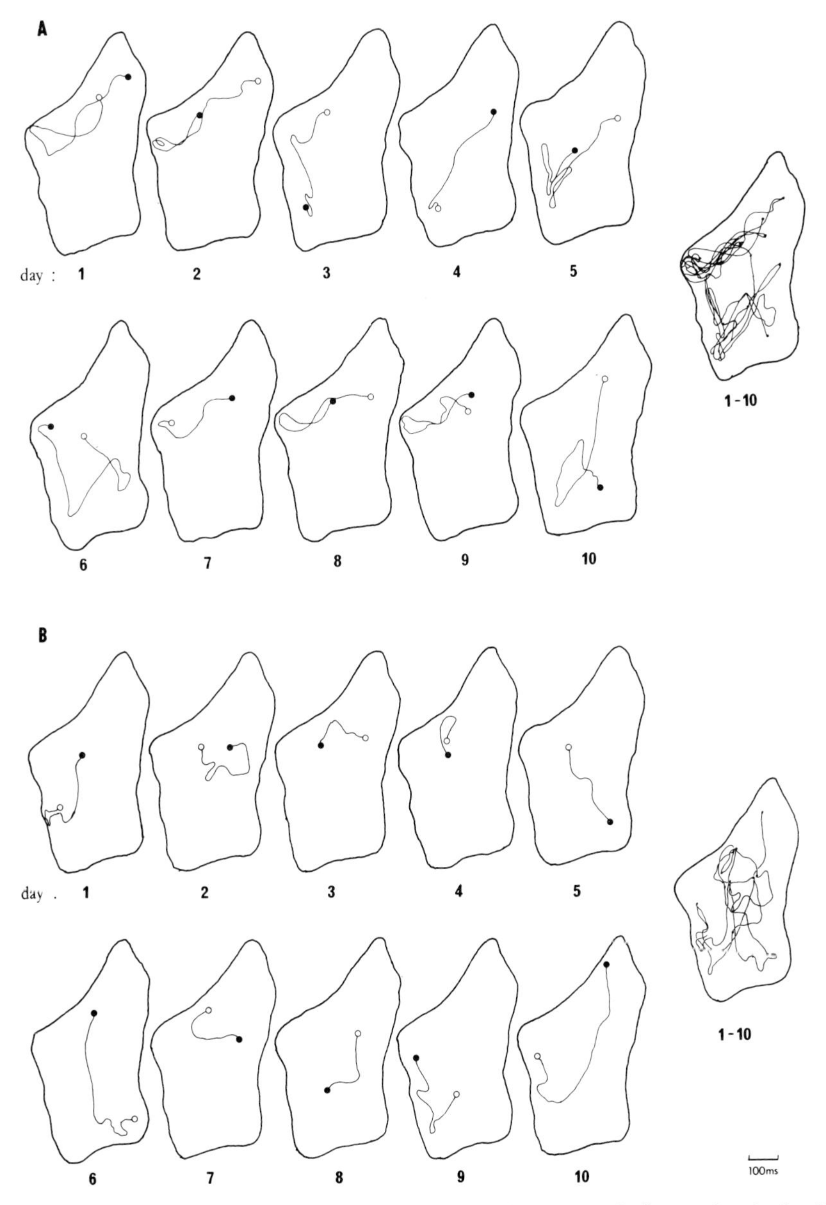

Fig. 4. Ranging patterns of *Indri indri* (Group P) at Analamazoatra. Daily travel paths for 10 successive days (and overall) are shown for two times of the year. In (A) repeated visits to a few fruiting trees resulted in predictable daily travel routes from central sleeping sites (open circles). When no food plant species was in fruit or new leaf a more randomlike pattern of movements occurred (B) to widely distributed sleeping sites (black circles).

Indri indri persist in ranging according to class (A) patterns while a particular kind of food is available. Group P, for example, was observed between June 13 and July 14, 1972, and during this time entered a 1/4–hectare patch of forest and fed on *Ravensara* fruit on the following dates: June 18–22, 26, and 27, and July 1, 2, and 5. On this last day the author could see no more fruit on the trees. Group P did not subsequently return to this area for several months; it is likely that all the available food had been consumed.

Too small a proportion of the ranging area of each *I. indri* study group was visited during each observation period (10 days for Group P, 6 days for Group V) to determine the presence of seasonal variation in parts of the range entered. The average number of different quadrats entered in 6 days by the two study groups was similar, corresponding to 41.4% of the Group P home range and 43.5% of the Group V home range. In 10 days Group P visited on average 58.1% of its ranging area. During these short observation periods, however, some quadrats were visited on more than one occasion. The mean number of times that a quadrat was entered during 6 days was 1.6 for Group P and 1.7 for Group V (Fig. 5).

The mean number of all quadrats and the mean number of different quadrats entered daily increased for both study groups during the summer wet season (November–March). This difference may, however, have been due to the overall increased activity of *I. indri* at this time of the year (Pollock, 1975b). Compared to the middle of the cold dry season, *I. indri*, on average, fed and moved about their range for 59.1% (Group P) and 51.1% (Group V) longer each day during the midsummer months. Seasonal fluctuations in daylength and mean temperatures were closely related to this variation in activity (Pollock, 1975a, b). It was also the

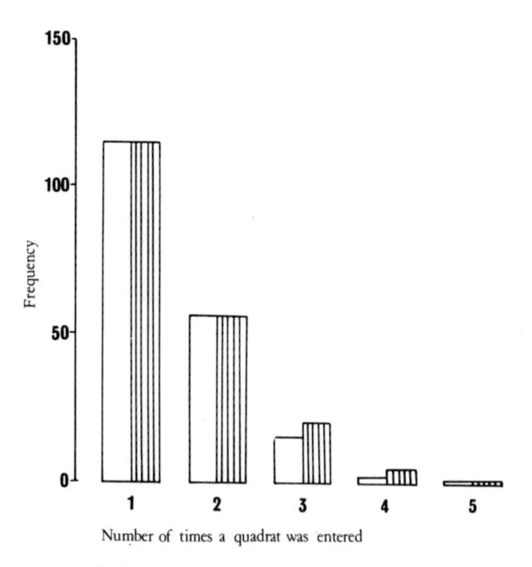

Fig. 5. The tendency for two *Indri indri* study groups (□, Group P; ▥, Group V) at Analamazoatra to revisit parts of their home ranges. The frequencies with which 50-m sided quadrats were entered by each group in 48 days (distributed throughout the year) are shown.

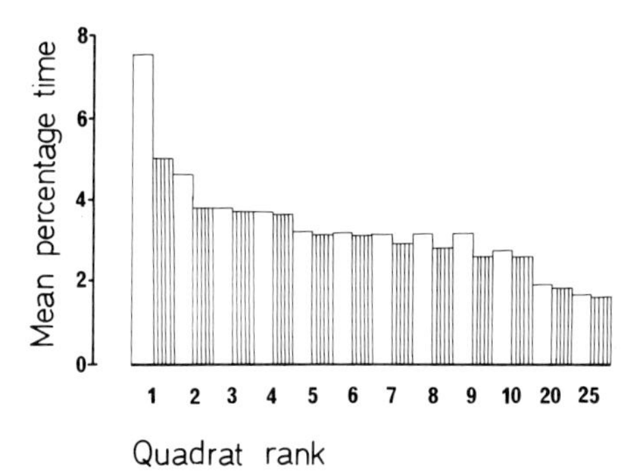

Fig. 6. Variation in the amount of time each *Indri indri* study group (□, Group P; ▥, Group V) at Analamazoatra spent in different parts of its range. The mean proportion of time spent per observation period in those 50-m sided quadrats in which most overall time was spent is shown. Quadrats are ordered in decreasing utilization or rank.

case, however, that most fruit was eaten at the time of year when *I. indri* moved the furthest (Pollock. 1977).

Although all parts of their ranging areas were visited by the two *I. indri* study groups during the research period, they both spent much more time in some parts of their range than in others (Fig. 6). Groups P and V spent 48 and 45% of their time, respectively, in 20% of their home ranges, and 78 and 76% of their time, respectively, in 50% of their home ranges. *Indri* groups were found to occupy parts of their range for relatively long periods, moving only between chosen regions and ignoring a large proportion of their ranging area for many days. Dependence on particular parts of the range could change rapidly and *Indri* might completely switch their attention to another area over a few weeks, confounding the possibility that different parts of the range were used at different times of the year (Fig. 7).

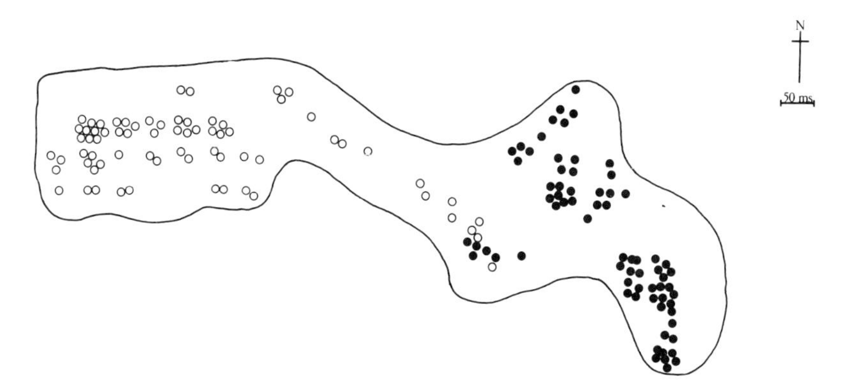

Fig. 7. Monthly variation in home range utilization by Group V. Each spot represents 30-minute situation for 6 days in May (○) and June (●) 1973.

A significant difference between the two *Indri* study groups' range utilization concerned the relationship between the frequency of quadrat visits and the time spent in quadrats during each visit. According to the animals' behavior, preferences for certain parts of the range could have been manifested in two ways: (1) by spending longer in those parts each time they were visited, or (2) by visiting them more frequently. For *Indri* this was examined by relating the amount of time spent in a quadrat each time it was visited to quadrat rank (the total amount of time spent in each quadrat). Group P was found to enter high and low ranking quadrats for equal amounts of time and thus varied their use of the range by visitation frequency. Group V, however, spent more time per visit in high ranking quadrats than in low ranking quadrats and were therefore varying range utilization, at least partly, by visit duration (Pollock, 1977).

A crude quantitative measure of directional aspects of *Indri* ranging behavior was obtained by calculating the frequency with which a group moved into an adjacent quadrat by changing its previous direction of movement while entering from the previous quadrat (angled turn). Although both *Indri* study groups made approximately the same proportion of angled turns when changing quadrats over the whole year (20% Group V, 19% Group P), it was found that for one group (Group V) changes in the frequency of angled turns between observation periods was correlated negatively with the proportion of time that group spent feeding. Group V, therefore, appeared to feed less when they made more angled turns (Spearman ranked correlation coefficient $r = 0.75$, $p < 0.05$). A further analysis of local movements (movements within quadrats) confirmed the impression gained in the field that individuals of Group V moved much more frequently than those of Group P (Fig. 8). Frequencies of tree-changing in Group V were three times as great as those in Group P and no individual differences within groups was discovered.

Attempts were then made to measure the actual distance moved by individuals of each group. One animal in Group P moved a horizontal distance of 1222 m in two days while ranging through 23 quadrats (an estimated minimum distance of 1190 m). The subject *Indri* in Group V, however, moved a horizontal distance of 1390 m in one day while ranging through nine quadrats (an estimated minimum distance of 430 m). The Group V *Indri* was therefore moving two to three times as much in order to get to the next quadrat as the Group P *Indri*, a difference equal to that obtained in observation of tree-changing frequency.

The difference between *Indri* groups in the frequency of local movements and the effect of angled turns and visit durations could have been due to different strategies of food attainment or widely different forms of food distribution. The latter explanation is considered to be unlikely as the two groups occupied neighboring regions of a fairly homogeneous forest. Different strategies of ranging for the same type of food could result from the groups' different experiences, needs (Group V had an infant and Group P did not) or knowledge. It seems more likely, however, that Group V had a poorer knowledge of their range than Group P. There was reason to believe (from information supplied by local people living in the area) that Group V had

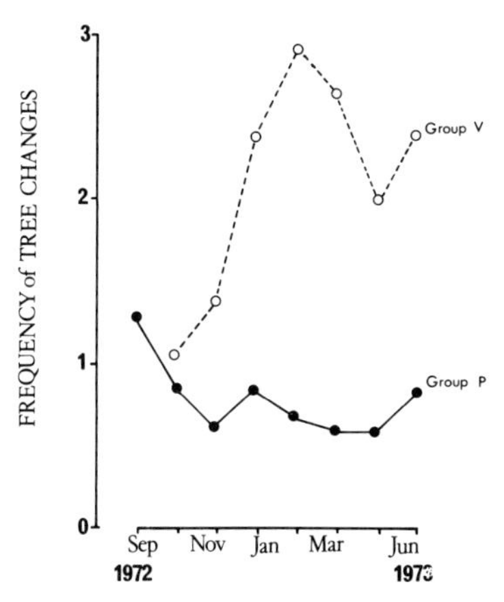

Fig. 8. Movement between trees in *Indri indri* at Analamazoatra. Local movements (movements within quadrats) of both study groups are represented as relative frequencies with which individuals changed trees per unit period of observation time. As data were recorded as a Hansen frequency measure, no absolute frequencies (vertical axis) can be computed.

recently procured their territory and may have been spending more time in the few areas they knew contained food while searching within quadrats for food by frequently visiting new trees (Pollock, 1975a).

Indri are vocally active animals, each group emitting loud morning calls on most days. Calls are contagious and groups call and answer each other over distances up to 3 km. From time to time individual voices could be recognized from vocal idiosyncracies and the callers' positions could be accurately plotted on a map of the region. In the forest of Analamazoatra at one time of the year one individual was clearly moving across the territory of three other groups, frequently giving characteristic long, low, male howls (Pollock, 1975b). It is not clear what happened to this animal as his voice was not heard in 1973. It seems, though, that not all *Indri* respect territorial limits and some individuals (possibly males in particular) may range freely about the forest.

3.5.4. Avahi laniger (Woolly Lemur)

The smaller nocturnal indriid, *Avahi laniger*, has been seen in small groups of 2–4 individuals (Rand, 1935; Petter, 1962; Martin, 1972b; J. I. Pollock, personal observation). It is, however, very difficult to be certain of group counts during the night in dense vegetation and larger groupings may exist. According to Petter (1962), *Avahi* live in family groups.

No information is available on *A. laniger* ranging behavior. It may be significant, however, that the author encountered five groups of *Avahi* in trees bordering a 150-m stretch of path during one nocturnal survey at Analamazoatra.

3.6. Daubentoniidae (Aye-Aye)

Petter (1962) suggests that this species is essentially solitary, having seen only single individuals in any scan of forest. One observation of three *Daubentonia madagascariensis* separated by only 50 m of the forest is, however, described by Petter (1962). Extensive ranging areas are used by *D. madagascariensis* according to Petter and Peyrieras (1970b): ". . . an animal may be found again 5 km from the place of the original sighting. However, in certain places we have found a male, a female and a young one all within a zone of diameter 5 km. . . ." This observation requires that animals could be individually recognized but these authors offer no criteria for such identification.

4. DISCUSSION

Those features of lemur groupings and patterns of ranging which contribute to a general classification of sociospatial types are reviewed in this section. They are presented as evidence pertinent to questioning whether lemur social organizations, spatial distributions, and ranging patterns are essentially similar, varying according to environmental contingencies, or whether the ranging behaviors of lemurs reflect fundamentally different categories of reproductive strategy peculiar to species, families, or other taxonomic grades. In the following discussion the observed ways and the possible ways in which lemurs describe the space they live in and those ecological factors influencing their ranging behaviors are reviewed. Following this is a description of social factors which are or may be influential in the composition of lemur groupings and their subsequent ranging behavior, and finally, in attempting to classify the sociospatial characteristics of lemurs, an emphasis is put on some environment-dependent features of lemur social structures and their ranging movements.

4.1. Space Definition by Lemurs

Mammals may defend or define the space they use in two ways, whether that space moves with the animals or whether it is a static geographical entity: (1) By presiding over the space (presence) they may by their activity either physically or ritually (usually by a combination of the two) prevent conspecifics from entering an area (e.g., territorial defense); Long-term definition of occupation may depend on the potential intruders' powers of memory. (2) Their occupation of an area may be advertised by means of signs (which persist during their absence) in the whole area, or, along the borders with a neighboring area, or both, with concentrations of signs placed at border regions. Long-term definition need not necessarily depend on intruders' memory (vocalizations, scent deposition, environment alteration).

Although most mammals apparently use both methods of space occupation advertisement, the degree to which (1) or (2) above is present in their behavior might be expected to vary according to the size of the space, the geometry of the space, the proportion of the space shared with conspecifics, the fading properties of the signs in the environment, season of the year, etc.

Physical defense of a ranging area in the fashion similar to that described by Ellefson (1968) for *Hylobates lar* has been observed in *Lepilemur mustelinus* (Charles-Dominique and Hladik, 1971), *Lemur fulvus fulvus* (J. I. Pollock, personal observation), *Propithecus verreauxi verreauxi* at Berenty (Jolly, 1966) and *Indri indri* (J. I. Pollock, personal observation). In these cases individuals or groups confront each other at range boundaries and mutual displays, vocalization, scent-marking, chasing, and physical combat may ensue.

Similar stimuli may also participate in evoking changes in the ranging behavior of neighboring conspecifics attempting to intrude into a home range across an unguarded border region. In these cases space occupation may be exhibited by the (absent) animal(s) whose range is being entered by visual, sonorous, or olfactory means. Visual effects may, as in the case of *Lepilemur*, involve sighting the "owner" of the range (Charles-Dominique and Hladik, 1971), or might conceivably be provided by long-term environmental alteration such as cropped plants (Oppenheimer and Lang, 1969) or broken vegetation (Schaller, 1961).

Sonorous effects, including vocal and nonvocal sounds, have been regarded as home-range delineators in many primates due to their claimed action as distance-maintaining signals (Marler, 1968; Chivers, 1969). It is, however, insufficient to claim that primates with loud regularly emitted calls are thereby defining and/or defending a territory (Jolly, 1966; Buettner-Janusch, 1973; Tembrock, 1974), unless neighboring groups are shown to react accordingly. Indeed one may question the concept of vocal demarcation especially in highly attenuating media such as the rain-forest environments inhabited by *Alouatta*, *Hylobates*, *Siamang*, and some *Cercocebus* and *Presbytis* species, as of low potential accuracy due to rapid short-term changes in these environments' sound attenuation characteristics (Allee, 1926; Harris, 1966; Allen *et al.*, 1972). Experiments and observations on *Cercocebus albigena*, for example, have shown that their loud "whoop gobble" calls clearly influence the distances maintained between the groups. These influences did not, however, depend on the location of the calling group relative to the receiving group's ranging area (Waser, 1975).

Many (if not all) lemur species can emit vocalizations capable of carrying to other groups or individuals ranging in other parts of the forest. *Lemur catta* "howls," according to Jolly (1966), carry as far as 1000 m. Loud calls are present in several lemur species, *Lemur macaco macaco* (Petter, 1962), *Lemur fulvus rufus* (Sussman, 1972), *Lemur fulvus fulvus* (J. I. Pollock, personal observation), *Varecia variegata* (J. I. Pollock, personal observation), *Propithecus verreauxi coquereli* (J. I. Pollock, personal observation) and *Indri indri* (Petter, 1962). All lemurs appear to engage in vocal interactions during encounters with conspecifics and these may be at border areas of groups' ranging limits.

The most obviously vocal lemur is undoubtedly *Indri indri*, the subject of the author's own research interests in Madagascar. *Indri* groups produce loud calls on two out of every three days on average, from all parts of their range (Fig. 9). These calls, lasting from 1 to 5 minutes, are usually emitted only once a day (although a group may "sing" up to seven times daily), generally during the morning and often in response to other groups' calls. The contagious nature of calling behavior results in a forest resounding with calls for up to 30 minutes. As the extremely sensitive auditory perception of *Indri* permits them to hear calls 2000–3000 m from their source (Pollock, 1975a), any group might be in direct communication with 50–100 others throughout the population. In order to examine the possible effects on an *Indri* group of calling by nearby groups, the ranging behavior of one group (Group P) was described in terms of the frequency with which it entered quadrats in relation to the distance of the quadrat from the source of the call. In order to maintain an accurate level of call source locations and to deal with convenient amounts of data, only calls emitted from within 1000 m of the study group were analyzed. According to the symmetry of the network of quadrats, had there been no directional influence on ranging behavior, the relative frequencies with which quadrats nearer and further the calling group were entered should have been 1:1. The observed values were found to be 0.9:1, a nonsignificant difference. The animals' immediate behavior also suggested that other group calls did not affect ranging behavior: morning calls were generally relaxed social events with little evidence of arousal or tension and few animal movements. If, however, a neighboring group's call emanated from a position inside the subject group's ranging area, or if two groups were in the border area together, a "singing battle" would ensue, lasting for between 10 and 30 minutes. Only a few border encounters were observed, but in most cases they bore no obvious relationship to a group's previous or subsequent ranging movements. Furthermore, great seasonal variation in *Indri* calling frequency was not correlated with changes in patterns of ranging behavior (Pollock, 1975b).

In addition to visual displays and vocalizations some primates leave behind olfactory clues which in the animals' absence may persistently communicate space occupation to an intruder. Such olfactory clues consist of the odors of animals and the scents they deposit (generally excreta and scent glands exudates). The long nasal fossa (which give lemurs their prognathous appearance) and relatively large olfactory bulbs suggest that, among the primates, the Malagasy lemurs rely significantly on olfaction in their perceptual world (Petter, 1962; Andrew, 1964; also see Schilling, Chapter 11). Very little information is available, however, on the relationship between marking behavior and use of space. In *Indri* the frequency of scent-marking in certain locations over the year's observations was found to be highly correlated with the amount of time spent in those locations ($r = 0.66$, $p < 0.0005$: the infrequently used quadrats were omitted from the analysis so as not to increase the

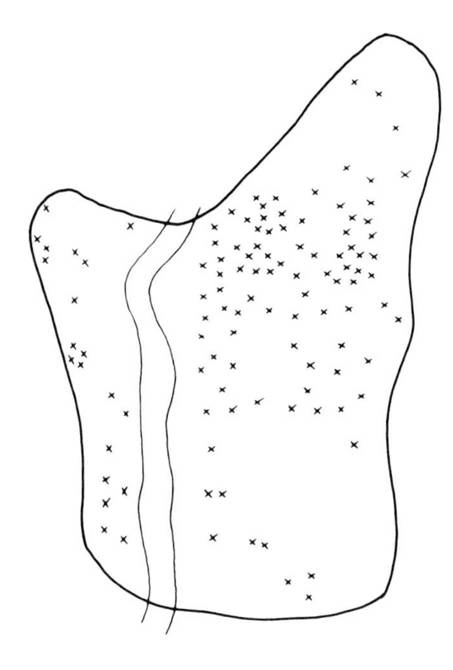

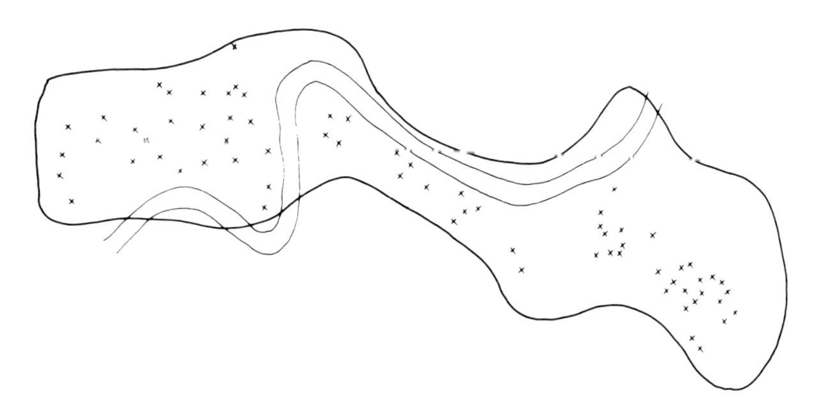

Fig. 9. The source of emittance of *Indri indri* calls (''song'') in their home ranges throughout the year. Each cross represents the site where each study group sang (duration 60–250 seconds) for Group P (77 days' observations) and Group V (48 days' observations).

significance of the correlation artificially), many infrequently used quadrats being located at peripheral or border regions in the range.

It is clear that prosimians frequently scent mark during encounters with conspecifics (Doyle, 1974) and this is also true for lemurs as far as these have been observed. This does not mean, however, that the function of scent-marking is primarily one of space definition, even if group confrontations are typically range-overlap or range-border phenomena, although it might incidentally act as such.

The extensive contextual variety of incidents appearing to evoke scent-marking implies that unitary interpretations of the function of this behavior are doomed (Ralls, 1971; Johnson, 1973). Some evidence has accumulated, for example, on the use of scent marks in orientation by cats (Rosenblatt *et al.*, 1969), rats (Gregory and Pfaff, 1971), black rhinoceros (Goddard, 1967) and, perhaps, lemurs (Schilling, 1974, and Chapter 11). These observations are in accordance with the author's experience of scent-marking in *Indri*. *Indri* were frequently seen to mark identical parts of repeatedly used arboreal routes. It may be that lemur scent marks, one of whose (possibly several) functions is associated with space utilization, will be found to imply identification rather than definition of an occupied area.

4.2. Ecological Characteristics of Lemur Ranging Behavior

Unlike the haplorhine primates, the Malagasy lemurs appear to have primarily nocturnal activity patterns (Charles-Dominique and Martin, 1970). All species, with the exception of *Indri*, are normally active or can become or remain active when night falls. Adaptations possessed by lemurs for night activity (e.g., the reflective *tapetum* at the back of the eye which seemingly improves nocturnal vision)* may make the consequences of nocturnal ranging difficult to ascertain. Night activity presumably affects aspects of visual perception more than any other and lemurs may, for this reason, have come to rely more on olfactory processes than other modes of sensation. Insectivorous, nocturnal lemurs (Cheirogaleidae, *Daubentonia madagascariensis*, for example) which move so as to catch insects may be less affected than folivorous species (*Lepilemur mustelinus* or *Avahi laniger*, for example) which range so as to locate and feed selectively on specific plants. *Indri*, which are diurnal, sometimes appeared to be able to see trees flushing into new leaves (which are bright red or orange) from far off (200–300 m) and subsequently move to them to feed. Precise color vision depends on light intensity and such clues may, therefore, be denied nocturnal folivorous forms. Again social factors affecting ranging behavior may depend on good vision, as Charles-Dominique and Hladik (1971) suggest for *Lepilemur mustelinus* (see Section 3.3 above).

*A reflective *tapetum* is present in all Malagasy lemurs except *Lemur fulvus spp*. It does not perfectly correlate with nocturnal activity habits. Thus *Indri indri* have a reflective *tapetum* but are strictly diurnal and *Lemur fulvus fulvus*, without a *tapetum*, may be very active at night (but see Pariente, Chapter 10).

During the daily or nightly activity period an uneven temporal distribution of movement by lemurs is widespread. This may, however, result from a reduction of all activities as in the case, for example, of midday siestas (*Lemur catta, Propithecus verreauxi, Lemur fulvus fulvus*). Progressions commonly occur shortly after activity commences in both nocturnal and diurnal forms. First progressions after waking may involve movements by group members or individuals to a "lavatory" area where communal urination/defecation sessions take place (*Cheirogaleus medius*, *Indri indri*, possibly *Hapalemur griseus*). Further progressions typically precede early and late feeding bouts (*Propithecus verreauxi*, *Lemur catta*, *Lemur macaco*, *Lemur fulvus*), although this arrangement may be liable to seasonal fluctuations (Richard, 1973). For *Indri*, however, the hourly proportion of their short diurnal activity period spent traveling was greater in the afternoon than in the morning with no intervening midday activity break (Fig. 10).

Seasonal variation in the distance traveled daily by lemurs has been observed in *Propithecus verreauxi* (Richard, 1973) and *Indri indri* (Pollock, 1975b). In both cases this variation is probably related in part to seasonal differences in the duration of the animals' daily activity period. For *Cheirogaleus medius, Cheirogaleus major, Propithecus verreauxi*, *Indri indri*, and, probably, other lemurs, the austral summer

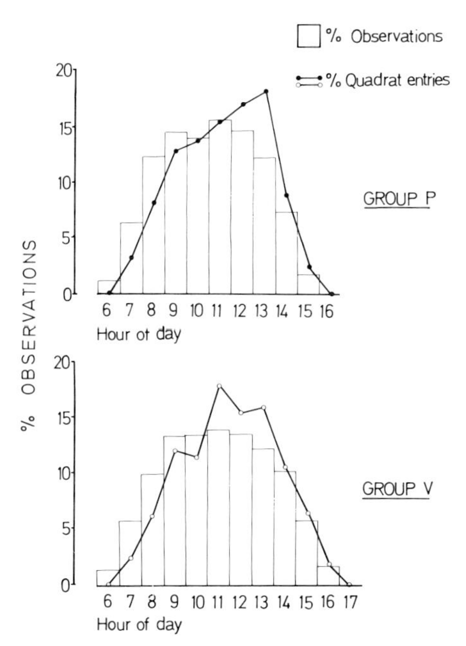

Fig. 10. Hourly distribution of movement in *Indri indri* at Analamazoatra. For each study group the proportion of the days' quadrat entries occurring in each hour of the day are compared to the hourly proportion of behavioral observations.

is a time of increased activity. This season is, for the whole of Madagascar, a period of higher temperatures, longer daylight hours, wetter conditions, and the associated changes these factors bring about in the vegetation, generally leaf, flower, and fruit production. If daylight alone were a limiting factor, one would hardly expect to see *Propithecus verreauxi* feed until well after dark in the summer and sleep from 2 to 3 hours before dusk in the winter (Richard, 1974b). Furthermore, *Indri*, which are 50–60% more active in the summer, sleep from several hours before nightfall to well after dawn throughout the year. These observations suggest that, as light levels in the tropics fall and rise rapidly each day to their asymptotic levels, other factors may be influencing the duration of these species' activity periods. Daylight length, however, is an important environmental parameter for some lemurs and may affect ranging behavior considerably. For example, nocturnal forms such as *Phaner furcifer* (Pariente, 1974) and *Lepilemur mustelinus* (Charles-Dominique and Hladik, 1971) appear to organize their activity very strictly according to levels of ambient light (see Pariente, Chapter 10). In the summer as the days grow longer so the nights get shorter and it is, therefore, conceivable that nocturnal lemurs (especially in southern Madagascar) have a reduced activity period while sympatric diurnal forms are active longer. Furthermore, changing astronomic daylength appears to stimulate reproductive behavior in some (and perhaps all) of the Malagasy lemurs (Petter-Rousseaux, 1962) and this will consequently influence patterns of ranging behavior (see Section 4.3).

Lemur activity and external temperatures participate in a significant relationship: imperfect thermoregulation in the Cheirogaleinae (Bourlière and Petter-Rousseaux, 1953; Bourlière *et al.*, 1956; Russell, 1975), "sunning behavior" (*L. catta*, *L. fulvus rufus*, *P. verreauxi*), the use of sleeping holes or nests (Cheirogaleidae, *Lepilemur mustelinus*, *Daubentonia madagascariensis*, although nests and holes could, of course, have a quite different protective function in these species), "huddling behavior" (Cheirogaleinae, *Lemur* sp., *I. indri*, *P. verreauxi*), in addition to seasonal variation of daily activity period duration, all support this contention. For this reason one might expect shorter lemur activity periods in the winter months when heat dissipation is greatest during the day. In the light of present knowledge this line of reasoning fails, however, in attempting to reconcile daylight-dependence of the nocturnal lemurs' activity patterns with shorter winter activity periods but longer winter nights.

The effect of vegetational changes on lemur ranging behavior at different times of the year has not been clearly demonstrated. Interpretations of seasonal variation in primate ranging behavior suffer from a poor conceptual background of its functions. It is not clear, therefore, whether mammals range further to find scarce and widely distributed food or adopt the alternative strategy of conserving energy supplies by moving less. Orangutans spent less time travelling and more time feeding during the fruiting season at Segama in Borneo (Mackinnon, 1974) but the presence of fruit seemed to be associated with increased daily travel distances by *P. verreauxi* and *I. indri*.

Richard's study of *P. verreauxi* spp. in the north and south of Madagascar (see Section 3.5.1) showed remarkably conservative ranging patterns in the different subspecies, *P.v. verreauxi* and *P.v. coquereli*, despite their different environments (Richard, 1973, 1974b). *Indri* occupying evergreen rain forest in the east of the island, however, lived in smaller groups occupying ranges twice as large yet moving each day half as far as *P. verreauxi*, a difference which probably reflects (at least in part) the considerable ecological contrast of eastern and western Madagascar (see Section 2). The two *Indri* study groups varied in the time spent in parts of their ranges once they were entered, in the pattern of movements about the range, in the frequency of visits to trees, the distance traveled within quadrats and the angle of direction taken between 50-m sided quadrats. This was considered to be due to differences between groups in familiarity with their ranging area. Further analysis of Richard's data shows that the northern study groups (*P.v. coquereli*) moved significantly more within 50-m sided quadrats than the southern study groups (Mann-Whitney U-test, $U = 24$, $n_1 = 12$, P_2-tailed < 0.02) although within each study area the groups did not differ significantly (*P.v. verreauxi*, south, $p > 0.155$; *P.v. coquereli*, north, $p > 0.531$). This difference may be related to feeding selectivity since, although the number of plant species and vegetation density was greater in the north, the proportion of consumed species was smaller (Richard, 1973, 1974b). This seems to suggest, therefore, that ecological differences between north and south may have affected *P. verreauxi* spp. ranging behavior while home range familiarity might have been more responsible for similar differences observed between neighboring *Indri* groups' ranging behavior. Both factors must be considered in future studies on the ecology and behavior of lemurs and other primates.

A special feature, though interestingly not confined within the order Primates only to the Lemuroidea (Goodall, 1962; Schaller, 1965; Rodman, 1973), concerns the use of nests or holes. The possible effects of nests or preferred sleeping places on the ranging behavior of *Microcebus*, *Phaner furcifer*, *Lepilemur mustelinus*, *Cheirogaleus*, *L. catta*, *V. variegata*, and *Daubentonia* remain unresolved. If such sleeping sites are hard to find, scarce, or take a long time to prepare they might (1) limit the duration of the activity period and consequently ranging behavior, or (2) be revisited on subsequent occasions. Nest construction for *Microcebus*, *Daubentonia* (and possibly *Varecia*) may be a lengthy process. One *Daubentonia* female released on the island of Nossi Mangabé following her capture on the mainland immediately began to build a nest (Petter and Peyrieras, 1970b). This individual completed the nest in about 1 hour, using about 60 branches in its construction. These authors state that *Daubentonia* usually possess from 2 to 5 nests and that old nests are often inhabited, which suggests considerable limitations on the diversity of their ranging behavior. It appears likely, in these cases, that frequentation of the same nest may result, thus affording at least partial definition (perhaps repetition) of the animals' patterns of ranging and limited abilities to visit new areas. It is quite possible that, as for many birds, nest building varies seasonally and that young are safer and warmer in such constructions.

4.3. Social Characteristics of Lemur Ranging Behavior

In this section the distinction between solitary and social ranging activities is discussed and its probable implications for the reproductive behavior of individuals and the genetic structure of the population are described.

A fine analysis of ecological correlates with lemur social organization is premature due to the primitive state of knowledge of lemur natural history and the variety of their habitats in Madagascar. For the purpose of discussing ranging behavior and spatial distribution, the social organization of lemurs is regarded in two different ways:

1. The relationships, especially reproductive relationships, between individuals within and between neighboring populations—primary or reproductive social structure.
2. The behavior synchrony and geographical integrity of certain small packets of individuals in a population—secondary social structure.

It will be shown below that neither of these two dimensions of social structure is sufficient in itself to account for all social relationships in all lemur species. Taken together, however, they provide a model in which the ecological and reproductive requirements of individuals can be contrasted and compared. This distinction does not imply that the ecology and reproductive behavior of an individual should be conceptually separated. The binary classification of social organization is introduced as a guide to where those behaviors relating to the individuals' health and fitness and those relating to the number and viability of its offspring can be in conflict.

For *Lepilemur mustelinus*, *Microcebus murinus*, and possibly other members of the Cheirogaleidae, adult individuals probably move alone about specific, largely separate, parts of the environment each night, engaging in maintenance activities such as feeding. These individually organized ranging areas (except in the case of mothers and young) may or may not be strictly exclusive within the same age/sex class but are not, generally, exclusive between them. Males may occupy ranging areas which overlap, coincide with or contain those of one or more females to whom they gain reproductive access at the appropriate season. In *Microcebus murinus* it appears, from evidence of sleeping aggregation, that males and females may remain closer (perhaps ranging together) during the mating season. It is not known how exclusive ranging areas are between males (and in some cases females) or how this propriety is maintained. It seems likely that in the case of *Lepilemur*, for example, large parts if not all of the ranging area are actively defended.

In both *Microcebus murinus* and *Lepilemur mustelinus* males in one part of the population appear to have permanent access to more than one female. *Lepilemur mustelinus* males probably have larger ranging areas than females (Charles-Dominique and Hladik, 1971). Martin (1972a) suggests that, in the case of *M. murinus*, peripheral nonreproductive males, located geographically around a central core of a few males and many females, may be involved in transfers between populations, thus ensuring an increased degree of heterozygosity within populations (Fig. 11). For these species, therefore, essentially individual ranging movements occur in a population the structure of which has great reproductive importance (primary social structure).

This form of social organization (frequently called solitary) is characteristically interpreted as a *primitive* mammalian condition (Crook and Gartlan, 1966). If this is true, it has been phylogenetically retained in several primate species including *Galago demidovii* (Charles-Dominique, 1972), *Pongo pygmaeus* (Mackinnon, 1974) and to some extent in *Pan troglodytes schweinfurthii* (Wrangham, 1975). Chimpanzees at Gombe Stream Reserve in Tanzania, however, range in numerically changing groups with males often traveling together. The composition and ranging behavior of small parties of chimpanzees depends greatly on the reproductive state of the females (Wrangham, 1975; Tutin, 1975). For both chimpanzees and orangutans, males ranged further and more widely than females and in the latter species, as in *Lepilemur mustelinus*, there is evidence that males defend their ranging areas against one another.

Unlike *Lepilemur mustelinus*, *Microcebus murinus* are found in social sleeping aggregations, suggesting that a primary structure does not satisfactorily describe all aspects of their social organization. The significance of large female dormitories is

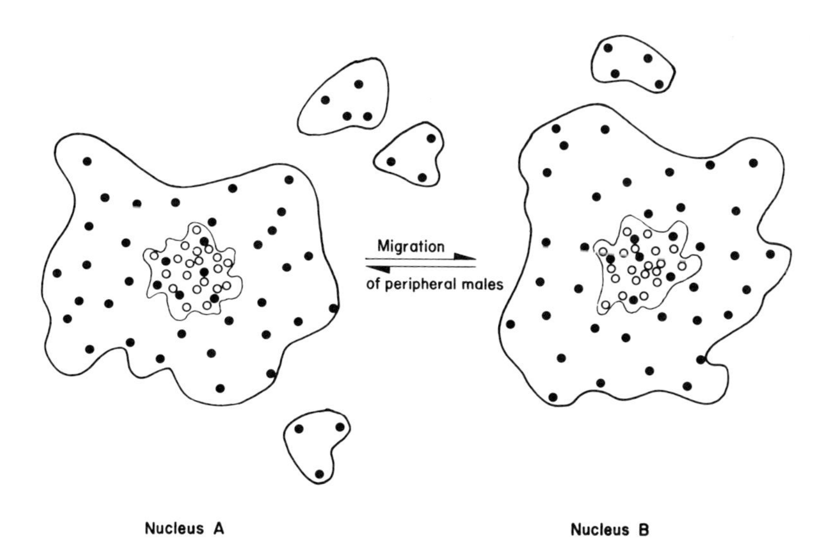

Fig. 11. Diagram of the possible population structure of *Microcebus murinus* at Mandena (after Martin, 1972a) and, perhaps, other lemur species. Local population concentrations or *noyeaux* (Nucleus A and Nucleus B) consist of a central core of a few males (●) and many females (○), surrounded by a peripheral male distribution. Migration of peripheral males between nuclei may ensure a certain degree of exogamy.

unknown (Martin, 1972a), but it suggests that some form of secondary structure exists in the form of nonsexual social relationships between individuals. Examination of the identity and genetic relationships of members of sleeping aggregations might prove fruitful in establishing the nature of such relationships.

The Lemuridae and Indriidae form ranging parties of 2–30 individuals which move together about small to large areas often of exclusive or partially exclusive use, engaging in frequent communicative acts within and between the parties. The secondary structure of some of these species appears to be the "nuclear family group" an adult pair and their dependent and/or maturing offspring (*Lemur mongoz*?, *Lemur rubriventer*, *Varecia variegata*, *Hapalemur griseus*, *Indri indri*, *Propithecus diadema* and *Avahi laniger*).* In these cases, members of the family range cohesively about an area that may be largely exclusive (e.g., *Indri indri* and possibly *Lemur mongoz*). In the Hylobatidae, Callithricidae, and many bird species, monogamous family groups of this kind seem to be correlated with ("territorial") ranging about exclusive areas and well-developed forms of paternal aid in rearing offspring (Trivers, 1972). Larger ranging aggregations found in *Propithecus verreauxi*, *Lemur catta*, *L. macaco*, and *L. fulvus* generally occupy areas which overlap extensively with those of other ranging parties. Either local ecological conditions or subspecific genetic differences account for the exclusive ranging observed in *Propithecus verreauxi verreauxi* at Berenty and Hazafotsy and the contrasting range-sharing behavior of *P.v. coquereli* at Ampijoroa. Variation in the degree to which ranging areas are exclusive is also found in some simian species (Ripley, 1967; Yoshiba, 1968; Gartlan and Brain, 1968) and may derive from similar ecological differences. *Lemur catta*, *L. macaco*, and *L. fulvus* are the most social species in numerical terms, with ranging party aggregations of up to 30 individuals.

Two further social characteristics of the secondary structure of the larger lemur ranging aggregations may be described. First, the size and composition of the ranging parties may be variable from day to day, despite the individuals' identical total ranging areas. Sleeping aggregations of separate ranging parties have been described for *L. macaco macaco* (Petter, 1962), *L.f. rufus* (Sussman, 1972) and *Propithecus verreauxi* (R. D. Martin, personal communication) and, in the case of *L.f. fulvus* and *L.f. rufus*, different individuals may associate in ranging parties of varying daily composition. In simian species similar observations have been frequently observed in *Cercopithecus aethiops* (Aldrich-Blake, 1970), *Papio ursinus* (Stoltz, 1972), *P. anubis* (Rowell, 1966), *Theropithecus gelada* (Crook, 1966), *P. hamadryas* (Kummer, 1968), *Pan troglodytes* (Van Lawick-Goodall, 1968), *Gorilla gorilla* (Schaller, 1965), and *Cercocebus galeritus* (K. Homewood, personal communication).

Second, the distribution of individuals within larger ranging parties might reflect the nature of social leadership and thus the factors which influence the direction and

*With the exception of *Indri*, the inclusion of these species is made entirely on the basis of observations of group size.

distances of travel. Group progressions consistently led by adult females have been reported for *L.m. macaco*, *L.f. fulvus*, *L. catta*, and *I. indri*, typically with peripheral and/or male animals bringing up the rear. (Female leadership is not necessarily the rule in these species. It may vary from group to group and from season to season. However, no consistent male leadership has been reported in the literature.) Although in *L. macaco* and *L. fulvus* no dominance hierarchies have been reported, adult females are invariably dominant over males and other individuals in *L. catta* and *I. indri* and may be dominant in *P. verreauxi*. In *Indri* ranging leadership by females has been observed only in groups with infants (J. I. Pollock, personal observation) and females with young offspring have also been reported to lead ranging parties in *L.m. macaco* and *L.f. fulvus* (Petter, 1962). *Lemur catta* females lead their groups in intergroup encounters and play the most active part in troop defense (Budnitz and Dainis, 1975). Compared to most simian species, sexual dimorphism in lemur size is very small and it is, perhaps, limited intrasexual selection that has enabled females to adopt a more controlling role in society (Pollock, 1975b, 1977, and see below).

What then is the primary social structure of species which range in relatively large parties? To answer this question, the degree to which ranging parties are also reproductive groupings must be established. It is now becoming clear that lemur ranging parties may have a very limited sexual integrity. Observations on *Propithecus verreauxi, Lemur catta, Lemur mongoz,* and possibly *L. macaco macaco* suggest that premating season group transfers of individuals (especially of males) are very common (see Section 3). It appears that it is essentially females and young of these species that establish a reproductive connection with a specific geographical region, while males wander more widely around the population, particularly before the mating season begins. It is likely, therefore, that females of these species may become fertilized by males with which they have not previously formed a ranging aggregation. In both *P. verreauxi* and *L. catta*, males may subsequently range in a new area with their new group or in the case of *P. verreauxi*, may return to their old group. It has been suggested by Richard (1974a) that this pattern of mating ensures outbreeding in *P. verreauxi*. This is true only if the males of such species are normally related to the females with which they range. The observed frequency of male transfer, however, is so great that it is unlikely that adult males bear a close genetic relationship with the females of their ranging party. If this is the case, one might expect young, maturing males to roam, but adult males to remain with females once these had been proved reproductively capable. This is presumably what occurs in nuclear family groups such as *I. indri* and the hylobatids, *Hylobates*, and *Symphalangus*.

According to the model proposed by Trivers (1972), males might be expected to optimize their chances of fertilizing as many females as possible. However, if this is associated with allowing females that are available (i.e., own ranging party ♀♀) to be fertilized by foreign males, male transfer, especially before the mating season, might prove valueless. Male transfer could be important for the dispersal of genes about a population, resulting in individuals' better overall reproductive chances in slightly varying local conditions. Having found satisfactory conditions, however, one might expect females to be retained rather than abandoned by males. Male transfer could ensure heterozygosity, however, if incest were a common occurrence. Most lemurs in Madagascar become mature within one, two, or three years and the intervals of male transfer between groups may approximate this duration, suggesting that by moving to a new group males ensure that they do not fertilize their mothers, sisters, or daughters.

Were the pursuit of heterozygosity a character which had been selected intensively in the evolutionary history of lemurs, rather than that of the mainland primates, the disposition of lemurs to be found in family groups (one-for-one matching of breeding adults of each sex), the lack of sexual dimorphism in size, the absence of adult male dominance in social relationships, and a high rate of transfers of animals between ranging parties might be expected. Furthermore, social characteristics such as female leadership in ranging movements, the relative geographical permanence of females, and female preferential access to sources of food might be predicted.

With the limited amount of information available, this evolutionary perspective might prove useful as a framework in which to compare different lemur species' ranging behavior and spatial distributions. Questions to which profitable answers might be sought are: How are ranging behavior and reproductive potential ecologically limited from the individual's point of view? To what comparative extent are ranging parties reproductive units in different species, and how are technical difficulties in exploitation of the environment solved by group existence?

ACKNOWLEDGMENTS

The author gratefully acknowledges receipt of funds from The Royal Society of London (Leverhulme Studentship), the Medical Research Council, Central Research Fund of the University of London, The Explorer's Club of America, The Boise Fund of Oxford, The Sigma Chi Society, and the Emslie Horniman Anthropological Scholarship Fund.

Research in the Malagasy Republic was made possible with the help and interest of the Département des Eaux et Forêts in Tananarive and in particular M. J-P. Abraham and M. Andriampianina. Ms. P. Calabi of Yale University generously helped in the collection of data on *Indri indri*. The technical and scientific aid of the French Research Institute (O.R.S.T.O.M.) at Tsimbazaza, Tananarive and the hospitality of the members of the British Council and British Embassy contributed greatly to the research project.

This review relied heavily on research performed by a few biologists. Special thanks go to Dr. J-J. Petter for the only overall account of lemur behavior and ecology, to Dr. A. Jolly for her informative articles and enthusiasm, and Dr. R. D. Martin for a continuous flow of support, advice, and information. I am particularly grateful to Drs. A. Richard and R. W. Sussman from whose largely unpublished dissertations much relevant and unique information has been extracted.

Studies of wild lemurs in Madagascar may be both the fastest growing and fastest dying area of field zoology unless steps are taken to provide a rational program for habitat conservation and animal protection. I acknowledge the concern of the Ministry of Rural Development in Tananarive and express the hope that they, with the aid of international conservation organizations, will be able to encourage the planning and execution of necessary projects for survey and census, research, and protection.

REFERENCES

Aldrich-Blake, F. P. G. (1970). *In* "Social Behaviour in Birds and Mammals" (J. H. Crook, ed.), pp. 79–102. Academic Press, New York.

Allee, W. C. (1926). *Ecology* **7**, 273–302.

Allen, L. H., Lemon, E., and Müller, L. (1972). *Ecology* **53,** 102–111.

Andrew, R. J. (1964). *In* "Evolutionary and Genetic Biology of Primates" (J. Buettner-Janusch, ed.), Vol. **2**, pp. 227–309. Academic Press. New York.

Bourlière, F., and Petter-Rousseaux, A. (1953). *C. R. Séances Soc. Biol. Ses. Fil.* **147,** 1594–1595.

Bourlière, F., Petter, J. J., and Petter-Rousseaux, A. (1956). *Mém. Inst. Sci. Madagascar* **10,** 303–304.

Budnitz, N., and Dainis, K. (1975). *In* "Lemur Biology" (I. Tattersall and R. W. Sussman, eds.), pp. 219–235. Plenum, New York.

Buettner-Janusch, J. (1963). *In* "Evolutionary and Genetic Biology of Primates" (J. Buettner-Janusch, ed.), Vol. 1, pp. 1–64. Academic Press, New York.

Buettner-Janusch, J. (1973). "Physical Anthropology: An Evolutionary Perspective." Wiley, New York.

Charles-Dominique, P. (1972). *Z. Tierpsychol., Beih.* **9**, 7–41.

Charles-Dominique, P., and Hladik, C. M. (1971). *Terre Vie* **25**, 3–66.

Charles-Dominique, P., and Martin, R. D. (1970). *Nature (London)* **227**, 257–260.

Chivers, D. J. (1969). *Folia Primatol.* **10**, 48–102.

Chivers, D. J. (1974). *In* "Contributions to Primatology" (H. Kuhn *et al.*, eds.), Vol. 4, pp. 1–131. Karger, Basel.

Crook, J. H. (1966). *Symp. Zool. Soc. London* **18**, 237–258.

Crook, J. H., and Gartlan, J. S. (1966). *Nature (London)* **210**, 1200–1203.

Doyle, G. A. (1974). *Behav. Nonhum. Primates* **5**, 155–353.

Ellefson, J. O. (1968). *In* "Primates: Studies in Adaptation and Variability" (P. Jay, ed.), pp. 180–199. Holt, New York.

Gartlan, J. S., and Brain, C. K. (1968). *In* "Primates: Studies in Adaptation and Variability" (P. Jay, ed.), pp. 253–292. Holt, New York.

Goddard, J. (1967). *East Afr. Wildl. J.* **5**, 133–150.

Goodall, J. (1962). *Ann. N.Y. Acad. Sci.* **102**, 455–467.

Gregory, E. H., and Pfaff, D. W. (1971). *Physiol. Behav.* **6**, 573–576.

Harris, C. M. (1966). *J. Acoust. Soc. Am.* **40**, 148–159

Humbert, H. and Cours Darne, G. (1965). *Trav. Sect. Sci. Tech., Inst. Fr. Pondichéry* **6,** 1–156.

Jay, P. (1965). *In* "Primate Behavior: Field Studies of Monkeys and Apes" (I. DeVore, ed.), pp. 197–249. Holt, New York.

Johnson, R. (1973). *Anim. Behav.* **21**, 521–535.

Jolly, A. (1966). "Lemur Behavior: A Madagascar Field Study." Univ. of Chicago Press, Chicago, Illinois.

Jolly, A. (1972a). "The Evolution of Primate Behavior." Macmillan, New York.

Jolly, A. (1972b). *Folia Primatol.* **17,** 335–362.

Jolly, A., Gustafson, H., Mertl, A., and Ramanantsoa, G. (1977). Bull. Acad. Malgache (in press).

Klopfer, P., and Jolly, A. (1970). *Folia Primatol.* **12,** 199–208.

Kummer, H. (1968). "Social Organisation of Hamadryas Baboons." Univ. of Chicago Press, Chicago, Illinois.

Le Bourdiec, P. (1972). *In* "Biogeography and Ecology in Madagascar" (R. Battistini and G. Richard-Vindard, eds.), pp. 201–226. Junk, The Hague.
Mackinnon, J. (1974). *Anim. Behav.* **22,** 3–74.
Marler, P. (1968). *In* "Primates: Studies in Adaptation and Variability" (P. Jay, ed.), pp. 420–438. Holt, New York.
Martin, R. D. (1972a). *Z. Tierpsychol. Beih.* **9**, 43–89.
Martin, R. D. (1972b). *Philos. Trans. R. Soc. London, Ser. B* **264**, 295–352.
Martin, R. D. (1973). *In* "Comparative Ecology and Behaviour of Primates" (R. P. Michael and J. H. Crook, eds.), pp. 2–68. Academic Press, New York.
Oppenheimer, J. R., and Lang, G. (1969). *Science* **165**, 187–188.
Pariente, G. (1974). *In* "Prosimian Biology" (R. D. Martin, G. A. Doyle, and A. C. Walker, eds.), pp. 783–998. Duckworth, London.
Perrier de la Bathie, H. (1936). "Biogéographie des Plantes de Madagascar." Soc. Ed. Geogr. Marit. Colon., Paris.
Petter, J. J. (1962). *Mém. Mus. Natl. Hist. Nat., Paris n.s.* **27,** 1–146.
Petter, J. J., and Petter-Rousseaux, A. (1964). *Terre Vie* **4**, 427–435.
Petter, J. J., and Peyrieras, A. (1970a). *Terre Vie* **24**, 356–382.
Petter, J. J., and Peyrieras, A. (1970b). *Mammalia* **34**, 167–193.
Petter, J. J., and Peyrieras, A. (1974). *In* "Prosimian Biology" (R. D. Martin, G. A. Doyle, and A. C. Walker, eds.), pp. 39–48. Duckworth, London.
Petter, J. J., Schilling, A., and Pariente, G. (1971). *Terre Vie* **3**, 287–327.
Petter-Rousseaux, A. (1962). *Mammalia* **26**, 1–88.
Petter-Rousseaux, A. (1964). *In* "Evolutionary and Genetic Biology of Primates" (J. Buettner-Janusch, ed.), Vol. **2**, pp. 92–132. Academic Press, New York.
Pollock, J. I. (1975a). *In* "Lemur Biology" (I. Tattersall and R. W. Sussman, eds.), pp. 287–311. Plenum, New York.
Pollock, J. I. (1975b). Ph.D. Thesis, University of London (unpublished).
Pollock, J. I. (1977). *In* "Primate Ecology: Studies in Feeding and Ranging Behaviour in Lemurs, Monkeys and Apes" (T. Clutton-Brock, ed.), pp. 37–69. Academic Press, New York.
Ralls, K. (1971). *Science* **171**, 443–449.
Rand, A. L. (1935). *J. Mammal.* **16**, 89–104.
Richard, A. F. (1973). Ph.D. Thesis, University of London (unpublished).
Richard, A. F. (1974a). *In* "Prosimian Biology" (R. D. Martin, G. A. Doyle, and A. C. Walker, eds.), pp. 49–74. Duckworth, London.
Richard, A. F. (1974b). *Folia Primatol.* **22**, 178–207.
Richard, A. F., and Heimbuch, R. (1975). *In* "Lemur Biology" (I. Tattersall and R. W. Sussman, eds.), pp. 313–333. Plenum, New York.
Ripley, S. (1967). *In* "Social Communication among Primates" (S. A. Altmann, ed.), pp. 237–254. Univ. of Chicago Press, Chicago, Illinois.
Rodman, P. S. (1973). *In* "Comparative Ecology and Behaviour of Primates" (R. P. Michael and J. H. Crook, eds.), pp. 171–209. Academic Press, New York.
Rosenblatt, J. S., Turkewitz, G., and Schneirla, T. C. (1969). *Trans. N.Y. Acad. Sci.* **31**, 231–250.
Rowell, T. E. (1966). *J. Zool.* **149**, 344–364.
Russell, R. J. (1975). *In* "Lemur Biology" (I. Tattersall and R. W. Sussman, eds.), pp. 193–206. Plenum, New York.
Schaller, G. B. (1961). *Zoologica (N.Y.)* **46**, 73–82.
Schaller, G. (1965). *In* "Primate Behavior: Field Studies of Monkeys and Apes" (I. DeVore ed.), pp. 324–267. Holt, New York.
Schilling, A. (1974). *In* "Prosimian Biology" (R. D. Martin, G. A. Doyle, and A. C. Walker, eds.), pp. 347–362. Duckworth, London.
Shaw, G. A. (1879). *Proc. Zool. Soc. London* **1879** 399–428.

Starmühlner, F. (1960). *Natur Volk* **90**, 194–204.
Stoltz, L. P. (1972). *Zool. Afr.* **7**, 367–378.
Sussman, R. W. (1972). Ph.D. Thesis, Duke University, Durham, North Carolina (unpublished).
Sussman, R. W. (1974). *In* "Prosimian Biology" (R. D. Martin, G. A. Doyle, and A. C. Walker, eds.), pp. 75–108. Duckworth, London.
Sussman, R. W. (1975). *In* "Lemur Biology" (I. Tattersall and R. W. Sussman, eds.), pp. 237–258. Plenum, New York.
Tattersall, I., and Sussman, R. W. (1975). *Anthropol. Pap. Am. Mus. Nat. Hist.* **52**, 193–216.
Tembrock, G. (1974). *In* "Gibbon and Siamang" (D. M. Rumbaugh, ed.), Vol. 3, pp. 176–205. Karger, Basel.
Trivers, R. L. (1972). *In* "Sexual Selection and the Descent of Man. 1871–1971" (B. Campbell, ed.), pp. 136–179. Heinemann, London.
Tutin, C. E. G. (1975). Ph.D. Thesis, University of Edinburgh (unpublished).
Van Lawick-Goodall, J. (1968). *Anim. Behav. Monogr.* **1**, 165–311.
Waser, P. (1975). *Nature (London)* **255**, 56–58.
Wrangham, R. (1975). Ph.D. Thesis, University of Cambridge (unpublished).
Yoshiba, K. (1968). *In* "Primates: Studies in Adaptation and Variability" (P. Jay, ed.), pp. 217–242. Holt, New York.

Chapter 10

The Role of Vision in Prosimian Behavior

GEORGES PARIENTE*

1. INTRODUCTION

1.1. Light Detection as a Biological Phenomenon

For all living organisms, the sun's rays provide the primary source of energy and one to which they are continuously exposed. As a result of their constant exposure to this flow of energy (since night-time represents only a reduction of the phenome-

* Deceased.

ISBN 0-12-222150-8

non and not its eclipse), living organisms have evolved to draw maximum benefit from this source. The eye is one particular organ which permits utilization of some forms of radiation. The extremely circumscribed, precise mechanism which permits light detection in a wide variety of animal species (utilization of light-sensitive pigments and optical systems) suggests that the evolution of the eye has been constrained to follow certain restricted pathways in the animal kingdom.

Some living organisms possess only one photoreceptor, and the mechanism developed in the long course of evolution may be either quite simple (as in *Euglena*) or relatively complex (the scanning system of *Copilia*). In the various arthropod species, the eye is constructed according to a precise pattern and in some cases a considerable degree of perfection has been obtained, as is illustrated by a dragonfly hunting on the wing. With the vertebrates and the cephalopod mollusks, on the other hand, a different pathway has been followed in evolution, leading to an even higher level of perfection. It is among the primates, including man himself, that we find the most interesting developments of this particular system.

It is worth emphasizing that, despite the enormous variety of adaptations which have emerged in particular ecological niches, the detection of radiations from the universe by living organisms has always followed one of the three directions cited above. Even then, the main differences reside in the degree of complexity. In each of the three systems, a maximum of information is extracted and decoded from incident radiation. Within the immense range of wavelengths which in fact bombard the animal world (from $\lambda = 10^4$ μm to $\lambda = 10^{-10}$ μm), the eye has been adapted for only a very narrow spectrum and remains blind to radiation of all other wavelengths. It has already been said that man may be regarded as occupying the summit of visual evolution in the animal kingdom. The prosimians are quite close relatives of man, and it is not surprising to find that prosimian visual systems are not greatly different from our own. Thus, a study of the details of prosimian vision must naturally cover some of the most subtle visual adaptations to be found among animals.

Let us return, however, to a brief consideration of the concept of the eye bathed in a universe of electromagnetic waves and particles, to examine its function in more detail. First of all, analysis of environmental radiation is an extremely elaborate process, and this emphasizes the fact that the eye is nothing without an associated brain. It is necessary to carry out a veritable dissection of the radiations and to identify their amplitude (= light intensity) and their frequency (= color perception). The optical construction of the eye also permits recognition of the direction from which the radiation is emanating and even the distance from the point from which the radiation was last reflected. In addition, vision is not usually based on a unidirectional analysis but on instantaneous analysis of the multitude of radiations in the visual field which are impinging simultaneously at a given point in space (construction of an image). Beyond this, one extreme case is foveal vision, which represents even more precise analysis concerned with only a restricted zone in space. Finally, and perhaps most strikingly, the analysis of each radiation is integrated with the overall product of analyses of radiations emanating from all other directions. Such integration permits the establishment of a schema of material space, and this overall product is what is known as "vision."

1.2. The Role of Vision in Primates

Following this brief survey, it is unnecessary to underline the extreme complexity of the process involved and the essential role played by the nervous system. Suffice it to say that in the prosimians one can already find the highly developed mechanisms which characterize the vision of the simian primates and man (Wolin and Massopust, 1970). The few differences which have been observed between the prosimian and simian primates concern fine analysis of frequencies (the color vision of prosimians is less elaborate or nonexistent) and selective analysis of specific zones in space (the fovea is poorly developed or lacking in the prosimian eye).

Apparently these differences between prosimians and simians are mainly attributable to differences in ecological conditions; in the prosimians, the evolution of the eye has only progressed to a certain level because the environment and/or the habits of the prosimians did not require further developments. Any attempt to explain these small differences between prosimians and simians in terms of different ancestral stocks within the order Primates would seem to be extremely tentative. One must always bear in mind the enormously complex apparatus constituted by "the eye plus the brain," which is shared by all primates, when comparing two animals which differ only in the opening through their corneas or the presence/absence of a *tapetum*. However, these relatively slight differences of detail, which are a consequence of precise ecological adaptations, provide considerable information about the ecological niches in which a given species has evolved. It is in this respect that comparative study is of greatest interest. Whether we are dealing with the primates in general or the prosimians in particular, the most obvious characteristics of all of these animals is incontestably their excellent vision. The eyes are tools which are essential for survival, and their high quality is witness to their great importance. Regardless of whether a particular primate species has evolved for nocturnal or diurnal vision, in most cases the visual system has been adapted to maintain a high level of performance. The best explanation for this no doubt lies in the great preponderance of arboreal life in the evolution of the primates. The forest is a discontinuous material environment and in the trees rapid locomotion similarly requires a discontinuous phenomenon: leaping. Such an adaptation is only possible if there is perfect knowledge of the complex environment represented by the forest. Hence, two particular functions were automatically selected and elaborated in primate evolution. First, it was necessary for primates to extract a maximum of information from visual "messages," particularly with respect to distance (mechanisms for convergence, accommodation, stereoscopy, etc.). Second, the information collected had to be stored as memories (participation of the brain). These two functions have played a large part in the evolutionary processes culminating in modern man and in the various surviving prosimian species (that is to say, in living organisms which are essentially visually oriented and possess a well-developed memory associated with a relatively complex brain).

As will be seen, this advanced degree of evolution of the visual system has been attained by all primate species (de Valois and Jacobs, 1971) and variation in an ecological parameter as important as light intensity has not been a limiting factor. The development of the other main senses (hearing and olfaction) seems to have occurred only as a secondary source of information with respect to feeding behavior or systems of social behavior (Ashley Montagu, 1943).

2. REVIEW OF ANATOMICAL AND PHYSIOLOGICAL ASPECTS

2.1. Anatomy

In order to provide a better basis for understanding behavioral features associated with vision, it is necessary to review briefly current knowledge of the prosimian visual apparatus.

In most prosimian species, the eyes occupy a frontal position, with the angle between the optical axes of the two eyes varying from 10° to 25° according to the species and, apparently, according to age. Young animals have more divergent axes than adults. Analysis of the refraction of the eye (Rochon-Duvigneaud, 1943) demonstrates that in young, healthy animals emmetropy (the capacity of the eye to form sharply focused images on the retina) is the rule, and the author's own measurements confirm this. It would seem that the angwantibo is the least well-equipped in this respect, with an apparently systematic myopia.

As Prince (1953) has pointed out, convergence of the eyes in carnivores has not been attended by orbital closure. Prince regards this retention of primitive morphology as correlated with these animals' needs for a specialized large gape and massive temporal muscles. The convergence of the eyes in strepsirhine prosimians permits these primates to have relatively large eyes as well as a powerful temporalis and coronoid process, by moving the large eyes forward and away from the masticatory apparatus. In *Tarsius* the temporalis is small and the eye, though huge, is relatively immobile, making it possible for the tissue between eye and temporal musculature to become ossified.

As a result, the prosimians have essentially binocular vision; according to Prince (1956), their total visual field covered by the individual fields of each eye amounts to about 250°. In the nocturnal species, the overlap between the visual fields of each eye is somewhat less than in the diurnal species; hence, the total visual field in nocturnal forms is somewhat larger. As a rule, this peculiarity of nocturnal prosimians is considered to be a primitive feature. However, it would seem that under low light intensity conditions an animal would benefit from the possession of a large visual field. This would permit better surveillance of the surroundings. Such an adaptation would also be useful in that the surroundings are visually explored over a much more limited radius than with diurnal species, as a result of the poor availability of light and the weak penetration of light rays into the forest at night (cf. the concept of the "sphere of vision").

It is difficult to measure the visual field in nocturnal prosimians. The cornea is very open (e.g., 125° in *Microcebus*) and it is also very prominent. In addition, the eye lens is located so far forward in the anterior chamber of the eye that the iris in myosis takes on the form of a forward-directed cone as in *Galago* (Fig. 1A). With *Microcebus murinus,* it seems likely that the visual field of each eye exceeds 210° when the pupil is maximally open. Thus, the total visual field for both eyes combined would be more than 230° (50° more than in man; Le Grand, 1956).

There is considerable variation in body size among the prosimians and it is found that the interpupillary distance varies according to body size. However, the large size of the eye in small-bodied prosimians means that this distance is relatively greater (Fig. 1B). Perception of relief is associated with this. Taking into account the relatively large distance between the pupils in small-bodied forms, one can calculate that the "sphere of vision" over a radius of 3–4 m in *Microcebus murinus* (interpupillary distance = 154 mm) is comparable, in terms of perception of relief, to that of a *Lemur* species with a radius of 10–15 m (interpupillary distance = 40 mm).

Schultz (1940) and several other authors have considered the size of the eye in relation to other anatomical characters (body weight, body size, or cranial volume). They have shown that nocturnal vision is correlated with a relative increase in the size of the eyeball. The case of *Tarsius* is generally cited as a demonstration of the importance of the visual apparatus in nocturnal forms.

This problem has already been discussed in detail (G. F. Pariente, unpublished research), and it has been shown that, in order to be efficient, a well-adapted nocturnal eye must attain a certain size. For example, in the case of *Microcebus murinus* the ratio of the weight of both eyes to the total body weight is 1:50. For *Tarsius*, the ratio is 1:35, and for *Phaner* it is 1:197. Yet all three prosimian types can see equally well at night and it is hence of little value to attempt, as so often in the past, to correlate the quality of vision simply with the relative volume of the eye. Further, there is a special explanation for the low ratio of 1:35 in *Tarsius*, in that there is no *tapetum lucidum*. (The role of this particular structure will be discussed below.) The investigations carried out by Dartnall *et al*. (1965) permit us to calculate that a tarsier equipped with a *tapetum* would be able to operate with a smaller eye, closer in size to that of a mouse lemur (G. F. Pariente, unpublished research).

The mouse lemur (*Microcebus murinus*) has arrived at the most economical solution in terms of eye weight required for an equivalent visual performance. It is worth noting in this respect that, for a small-bodied animal such as this, a low ratio of eye weight to body weight represents a saving in overall weight (and hence in musculature) and hence an increased ability to exploit an extended ecological niche (i.e., very fine branch extremities bearing many flowers and insects). An additional advantage resides in the considerably greater mobility of the eye of *Microcebus*:

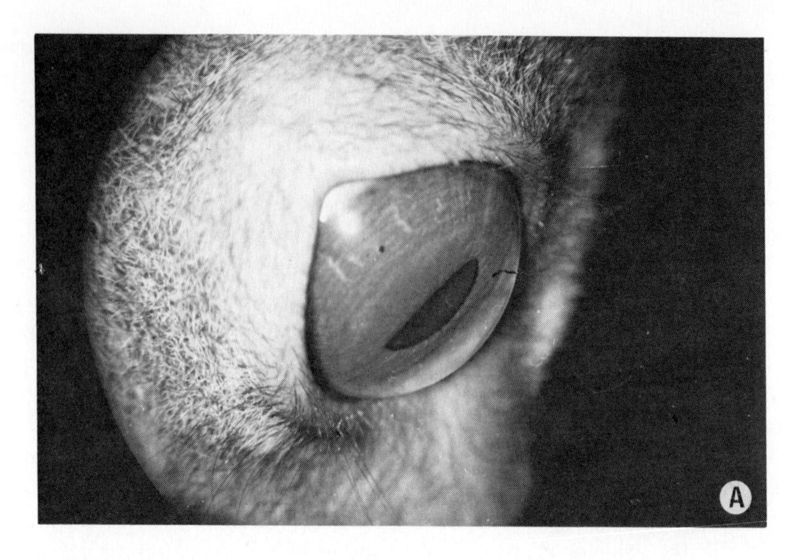

Fig. 1. (A) Right eye of *Galago elegantulus*, seen in profile. (B) Facial view of *Galago alleni*, showing the enormous development of the eyes.

such mobility eliminates the need for head movements which are absolutely necessary for visual tracking in *Tarsius*.

The problem of the mobility of the eye in the orbit is closely correlated with that of convergence, and it is therefore worthwhile to consider this aspect in more detail. Lindsay-Johnson (1900) claimed that the lemurs were incapable of convergence and could not maintain prolonged observation. The author's own investigations in the field and in captivity lead to an entirely different conclusion which applies to both nocturnal and diurnal forms. It will be seen below that with diurnal species fixation of the gaze is far more a question of behavioral repertoire than of anatomical possibilities. This is borne out by the identification (Pariente, 1975a) of a fovea in

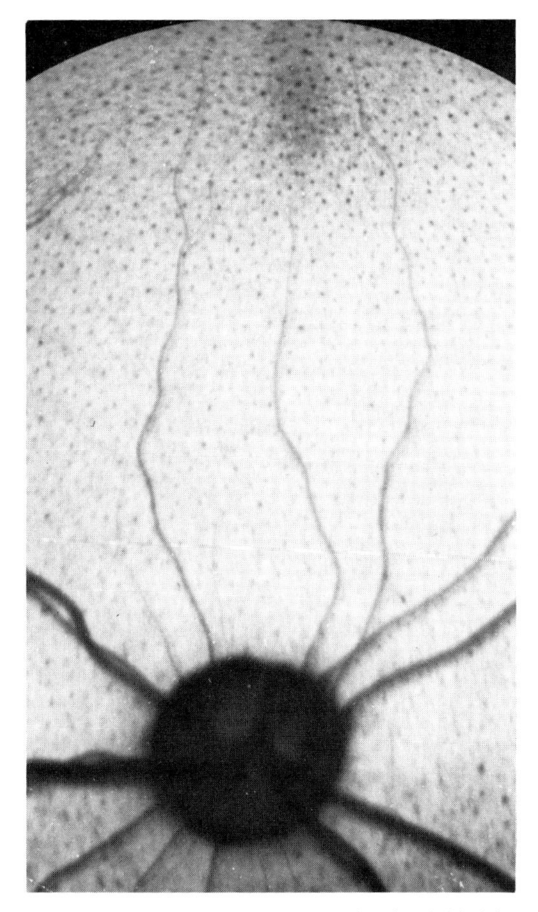

Fig. 2. Fundus photography in *Lemur catta*. The *tapetum lucidum* behind the retina is highly reflective. Note the dark patch indicating the point of emergence of the optic nerve and the existence of a foveal spot, whose position is well defined by the pattern of terminal blood vessels.

Fig. 3. The overall slowness of movement in *Perodicticus potto* (especially that of the head) is compensated for by an efficient ability to scan the environment, due both to protrusion of the eyes and an increase in the amplitude of ocular movements, as can be seen in the photograph.

Lemur catta and *Hapalemur griseus* (Fig. 2). Our observations have demonstrated that the direction of the gaze is often quite independent of the direction of the head (Fig. 3). With *Phaner,* it has been found that the eye may be rotated in the orbit over an angle of at least 30° in the horizontal plane, and with *Lemur* species the angle is greater than 60°.

Observation of the *fundus oculi* is a very important element in the comparison of mammalian eyes and in understanding their functioning. At a time when photoretinography had not yet been developed, Lindsay-Johnson (1897, 1900) carried out an exceptional piece of work in representing through water colors the appearance of the *fundus* in several prosimian species, including the aye-aye. Much later, Wolin and Massopust (1967), with the same aim in mind, undertook a comparative survey of retinographic records from numerous vertebrates, including several prosimian species. The author (Pariente, 1970) has extended this study in Madagascar by conducting retinographic analysis of several little known lemur species (Fig. 4).

From all of these investigations, it has emerged quite clearly that the prosimians, almost without exception, possess a *fundus oculi* of the nocturnal type (see also Hill, 1935). The only exceptions are provided by certain species within the genus *Lemur*

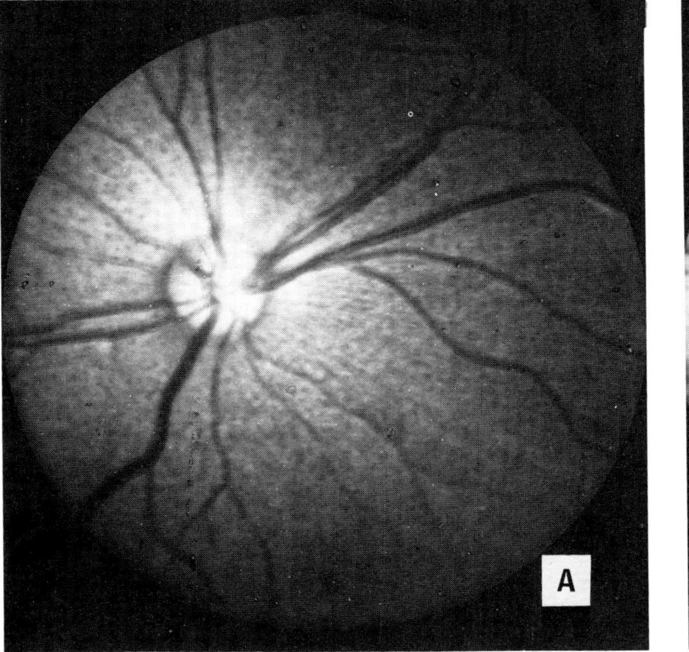

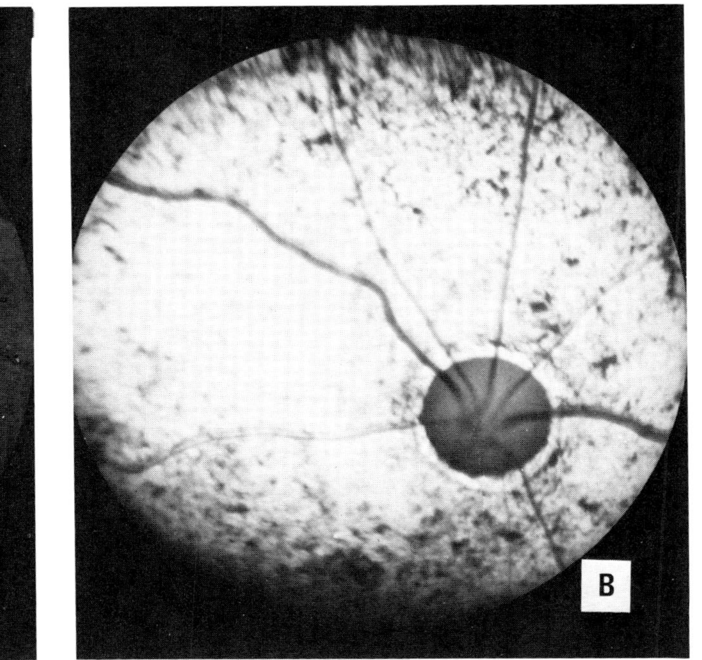

Fig. 4. General appearance of the *fundus oculi* in prosimians, showing the point of emergence of the optic nerve (*papilla*). (A) *Lemur macaco macaco*: This diurnal form has no *tapetum lucidum*. The *fundus* is dark pigmented and the *papilla* is bright. The blood vessels can be seen emerging from the *papilla*, and a number of optic nerve fibers can be seen as straight, thin, white lines. (B) *Hapalemur griseus*: In this crepuscular form the *papilla* is darker. The *tapetum lucidum* gives a bright golden reflection, but scattered spots of dark pigment mask this in places.

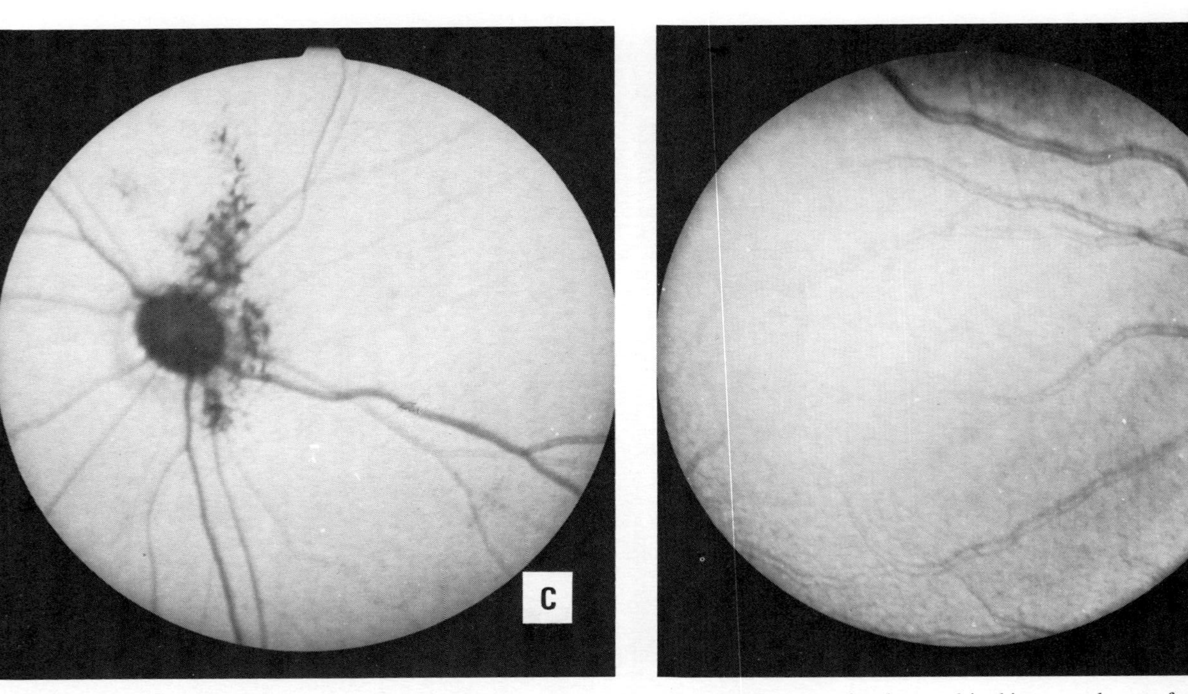

Fig. 4 *continued.* (C) *Daubentonia madagascariensis*: This nocturnal form also has a golden *tapetum lucidum,* and in this case only very few pigment spots can be seen. In some respects, this type of *fundus* is similar to that found in various *Lemur* species. (D) *Microcebus coquereli*: This typical nocturnal *fundus* shows no pigmentation and an extremely bright *tapetum lucidum*. The *papilla* is dark and there is discrete convergence of blood vessels to an *area centralis*, which is quite obvious in this particular case. African and Asian prosimians also exhibit this type of *fundus*.

and by *Tarsius*. In all the other prosimian species, a highly reflective *tapetum lucidum* is found, sometimes in association with black pigmentation which obscures the reflection (Pariente, 1975a). The pattern of vascularization is generally homogeneous in all species, and there are usually numerous ramifications anterior to the point of emergence of the retinal *papilla*. These ramifications almost always surround a clear-cut *area centralis*, marked off by very fine vessels. [Incidentally, *Tupaia* have a *fundus oculi* quite different from that of the true prosimians, and this provides a further argument for distinguishing the tree-shrews clearly from the Primates (Wolin and Massopust, 1967).]

Fig. 5. The border of the reflecting zone (*tapetum lucidum*) and the pigmented area near the *ora serrata*. (A) Clear-cut appearance in *Phaner furcifer*; the *tapetum* appears to consist of round golden spots. (B) The zone of separation is much more extensive in *Hapalemur griseus*; black interdigitations penetrate a long way into the patchy *tapetum*.

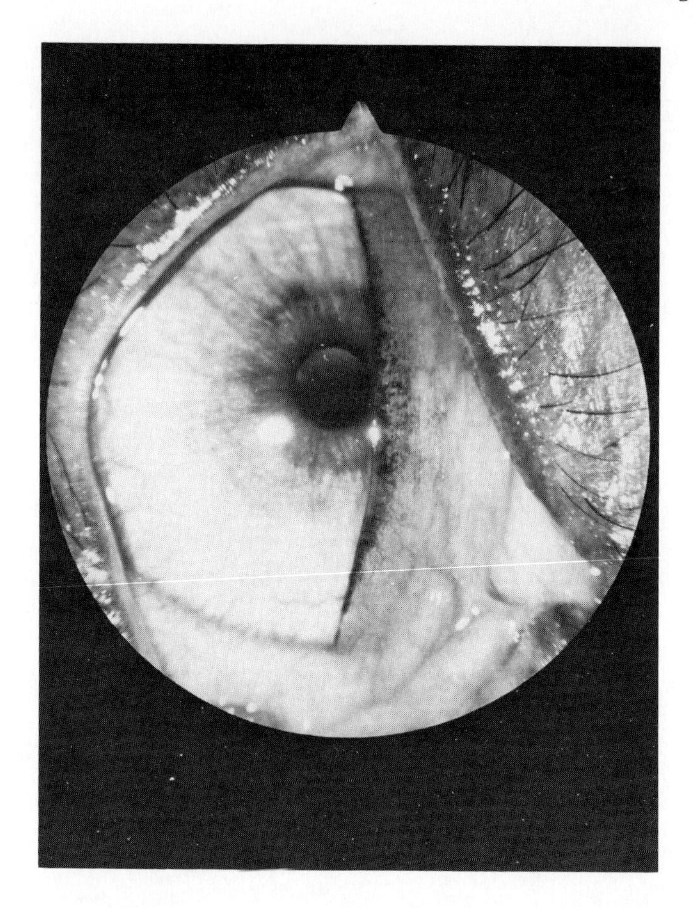

Fig. 6. The nictitating membrane (third eyelid) of *Propithecus verreauxi* is quite mobile and helps prevent bright light reaching the eye (such as in this case of a close-up photograph requiring strong illumination). The exact role of this structure is unknown. In this figure the lower and upper eyelids remain open, while the nictitating membrane, which has a heavily pigmented vertical edge, moves over the pupil. Usually the animal moves the pupil behind the nictitating membrane.

The highly reflective *tapetum lucidum* gives a golden coloration to the *fundus oculi*. When the *tapetum* is partially masked by pigmentation (as in *Lemur catta* and *Hapalemur*), there is a mottled pattern of brilliant zones intermingled with dark areas. At the *ora serrata*, the edge of the *tapetum* can exhibit one of two characteristic conditions, according to the species (Pariente, 1976; Fig. 5). A foveal depression has only been systematically observed with *Lemur catta* and *Hapalemur griseus* (i.e., these two species do not generally exhibit a macular ring reflection). Polyak (1957) has described a fovea in *Tarsius*, but this structure is extremely difficult to observe. All prosimian species possess a well-developed nictitating membrane.

2.2. Physiology

At the present time, visual processes represent one of the most extensively studied fields in the realms of neurophysiology and sensory physiology. A large number of the studies which have been conducted have involved simian primates. Among the prosimians, very few studies have been conducted and what little has been done is based on a few *Galago* species and certain Madagascar lemurs. The areas which have been most studied are those of spectral sensitivity, thresholds of nocturnal vision, visual acuity, and visual centers of the brain. Such studies have involved histologists, physiologists, and electrophysiologists.

The following publications deal in detail with histological aspects of prosimian vision: Woolard (1927), Detwiller (1924, 1939, 1940), Winckler (1934), Walls (1942), Rochon-Duvigneaud (1943), Chacko (1954), Polyak (1957), Rohen and Castenholz (1967), Ordy and Samorajski (1968), Tigges and Tigges (1969, 1970), and Nakamura and Tigges (1970).

In terms of neuroanatomy and neurophysiology, work has been mainly concentrated on African lorisid species. The major publications in this area are Kolmer (1930), Winckler (1934), Prince (1953), Tansley (1956), Polyak (1957), Luck (1958, 1963), Hassler (1966), Bauchot and Stephan (1966), Ordy and Samorajski (1968), Laemle (1968), Noback and Shriver (1969), Stephan (1969), Laemle and Noback (1970), Tigges and Tigges (1970), and Lane *et al*. (1973).

In the field of electrophysiology, the following papers can be cited: Dartnall *et al*. (1965), Graveline *et al*. (1967) on electroretinography of *Galago senegalensis*; Ordy and Samorajski (1968) on visual acuity in *Lemur catta* and *Galago crassicaudatus*; Serbanescu *et al*. (1968), Bresson *et al*. (1969), Ward and Doerflein (1971) on critical fusion frequency in *Galago senegalensis*; Alfieri *et al*. (1974) on electroretinography of *Lemur mongoz*, and J. Faidherbe and M. Goffart (personal communication).

2.3. Electrophysiology

The author (unpublished material) has also carried out electroretinography with *Microcebus murinus*, *Phaner furcifer*, and *Hapalemur griseus*, using different monochromatic lights. At present, it would seem that only diurnal or crepuscular prosimian species exhibit a mixed retinal response. With *Lemur mongoz*, for example, there is no doubt that the retina has a dual composition (Fig. 7). The existence of a *tapetum lucidum* in the majority of prosimian species (Walker, 1938), including crepuscular and diurnal forms (e.g., *Propithecus, Indri, Lemur catta*), as well as purely nocturnal forms, constitutes a double advantage with respect to nocturnal vision. First, the second passage of light through the photoreceptors represents an important gain in incident energy. Second, the *tapetum lucidum* of lemurs and lorises is composed of riboflavin, a substance which is made to fluoresce by ultraviolet radiation. Most work on the composition of the *tapetum* has been con-

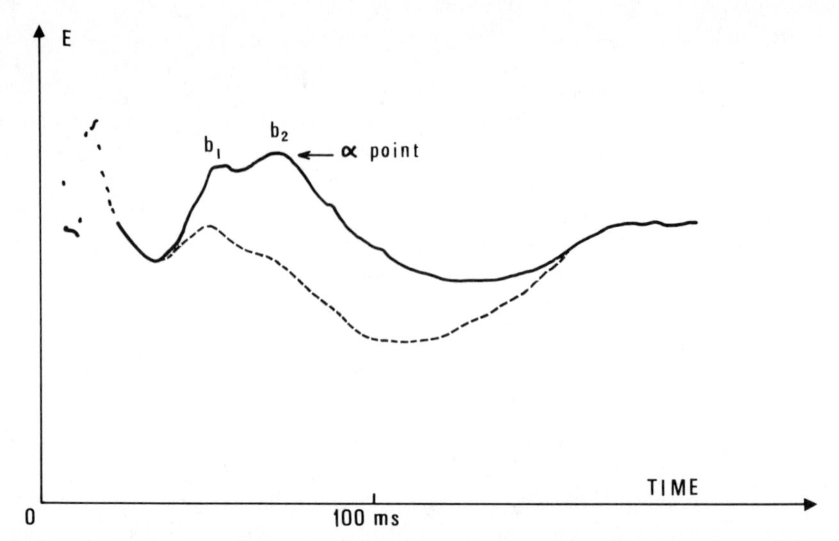

Fig. 7. Dynamic electroretinography of *Lemur mongoz* in monochromatic light (yellow = 553 nm). The point represents an equal response of cones and rods occurring in the fourth minute after a white dazzling stimulus [b_1 wave (cones) = b_2 wave (rods)]. The curve indicated by the broken line was recorded just after the white dazzling stimulus. (Modified from Alfieri *et al.*, 1974.)

ducted with *Galago crassicaudatus* (Pirie, 1959; Pedler, 1963; Dartnall *et al.*, 1965; Silver, 1966), but the author has also been able to demonstrate that the *tapetum* contains a substance exhibiting the typical absorption peaks of riboflavin in *Perodicticus potto*, *Microcebus murinus*, and *Hapalemur griseus*. The fluorescent properties of riboflavin permit the eye to utilize a portion of the ultraviolet spectrum which is otherwise invisible. In fact, ultraviolet light is transformed to light of longer wavelength which can stimulate the rods of the retina at the maximum of their sensitivity.

2.4. Behavioral Physiology

Another approach to investigation of prosimian vision is that of behavioral studies in which the animal may exhibit responses under nontraumatic conditions. Such studies, which are unfortunately very rare (Bierens de Haan and Frima, 1930; Fobes *et al.*, 1971), are most appropriate for comparison with data collected directly in the biotopes of the animals concerned (Ehrlich and Calvin, 1967). Silver (1966) carried out a remarkable study on the spectral sensitivity of *Galago crassicaudatus* and measured the absolute thresholds of vision. Treff (1967), using the same methods, established precise measures of visual acuity in *Galago senegalensis*. The author has also used these methods to study thresholds of color vision for several Madagascar lemur species (*Phaner furcifer*, *Hapalemur griseus*, *Lemur catta*). The results obtained with *Phaner* (G. F. Pariente, unpublished research) indicate that the threshold values determined under experimental conditions are somewhat higher than those which are actually found for the animal's vision in the forest.

It is to be expected that conditions in captivity are somewhat unfavorable for the achievement of maximal sensory performance, and experience shows that this expectation is fulfilled. It must be emphasized that dietary deficiencies arise easily in captivity. In addition, decisive psychological factors intervene. Finally, the nutritional motivation which is utilized in this type of experiment is by no means directly comparable with the motivation for spontaneous locomotor activity in the natural environment under conditions of almost complete darkness (Pariente, 1973).

To summarize the quantitative aspects of all of the studies conducted on prosimians to date, one can say that the absolute thresholds found in the laboratory are little different from those determined for men. For example, in the author's experiments with *Phaner*, a threshold of 1.12×10^{-12} $\mu W/cm^2$ was determined at a wavelength of 530 nm (Pariente, 1975b) while for man the threshold given for 500 nm is 2×10^{-12} $\mu W/cm^2$. Hence, given the fact that laboratory conditions are by no means optimal, one may reasonably conclude that under natural conditions nocturnal lemurs can see quite well when the light intensity is lower than that which would be just sufficient for acceptable human vision. However, at these levels (between 5 and 25 μlux under scotopic conditions in the forest, according to our direct measurements), a very small amount of light is sufficient for complete modification of the visual performance of an eye. Under such conditions it is obvious that the animals concerned are unable to see colors. In fact, the strictly nocturnal prosimian species do not have sufficiently differentiated cones for them to be included under the usual definition of these color receptors. All of the authors cited above have reached the same conclusions.

With crepuscular species (Pariente, 1975a) and diurnal species (Autrum, 1960), the situation is different; clearly defined cones are present. However, the relatively low frequency of cones in the retinas of crepuscular and diurnal prosimians indicates that color vision is restricted and is only prevalent at very high light intensities (see Mervis, 1974, for *Lemur catta*). In one *Lemur catta* studied by the author, the maximum sensitivity of the eye was found to be in the region of 500 nm (i.e., the maximal sensitivity point for the rods). Stated more simply, color vision in *Lemur catta* must be poor at the red end of the spectrum. However, to some extent as a result of the fluorescence of the *tapetum lucidum,* color vision is relatively somewhat better at the blue end of the spectrum, more so than would be indicated by the single spectral absorption curve obtained with visual pigments extracted from the retina (Silver, 1966).

2.5. Conclusion

This review can be concluded with a summary of the work carried out by Treff (1967) on *Galago senegalensis* with particular emphasis on visual acuity and depth perception, two functions which are of special importance in the life of an arboreal mammal. With *G. senegalensis*, visual acuity is inferior to that of a human being (4 feet 28 inches in *Galago*, as compared to 1 foot in man) and the perception of relief is also less developed (42 inches in *Galago*, compared to 10 inches in man). However, when one takes into account the distance between the pupils in *Galago senegalensis* and auxiliary visual behavior (head movements), these differences decline in importance. Overall, one can say that this bushbaby species has a well-developed mechanism for perceiving distances and depth (necessary as an adaptation for active leaping).

Overall, it is worth underlining the fact that all of these anatomical and physiological characters are associated with appropriate behavioral adaptations. It is evident that the broad sphere of prosimian behavior, within which vision plays an important part, can only be properly understood if one has detailed knowledge of the performance of the visual organs. Thus, it is vital to realize that prosimians are able to see the material world surrounding them with precision and clarity. The lack of (or poor development of) color vision in prosimians is in keeping with their adaptation to dark forest environments or even to a completely nocturnal existence. The great sensitivity of their eyes implies great convergence of the nervous connections from the photoreceptors on ganglion cells, and hence summation of information at the expense of analysis of detail (see also Rohen and Castenholz, 1967). Given the ecological requirements of these prosimian species, this represents more of an advantage than a disadvantage.

Observations of the various prosimian species and their behavior patterns under natural conditions will now be interpreted in the light of available information on anatomy and physiology.

3. VISION AND BEHAVIOR

3.1. Vision and Activity

3.1.1. The Relationship between Light and the Eye

Of all naturally occurring cycles, the light cycle is the most regular and possibly the most simple. A crude classification into two types of phenomena may be made: (a) Circadian variation (day/night alternation produced by the rotation of the earth on its axis); (b) seasonal variation (produced by rotation of the earth around the sun on its inclined axis). In the final analysis, both of these phenomena amount to a cyclical distribution of radiation (sources of energy) of solar or stellar origin. (High-energy extraterrestrial particles and the earth's own intrinsic radiation may be disregarded for our present purposes.) Behavioral adaptations associated with these cyclical variations may be divided, for convenience, into two general categories:

overall activity (motor activity, with its central nervous counterparts) and reproductive activity.

3.1.1.1. Overall Activity. There can be no doubt that the only sense organ involved in the determination of overall patterns of activity is the eye (Pirenne, 1961). The eye permits the animal to distinguish day from night (Tenaza *et al.*, 1969). For this reason, it was thought to be important to relate light levels with activity periods under natural conditions with certain nocturnal Madagascar lemur species (Pariente, 1974a). Even the initial observations in the field showed that there is great regularity in the timing of onset and arrest of activity in these species. This led on to the assumption that there must be either a very precise internal clock or an association of some kind with a variable in the environment (e.g., light intensity or temperature). Data obtained with *Phaner furcifer*, the species which have been most intensively studied to date, can be plotted on curves to illustrate the great regularity involved (Figs. 8 and 9). With this species (Pariente, 1975b), the onset of activity occurs when the ambient light intensity falls to approximately 1 lux. Times of arrest of activity are somewhat less regular, but there is a complicating factor in that the animals quite often spend some time close to the nest, thus making determination of the exact time of arrest of activity difficult. Nevertheless, all activity has ceased before the light intensity attains a level of approximately 1/1000 lux. This great difference between the light level initiating activity at dusk

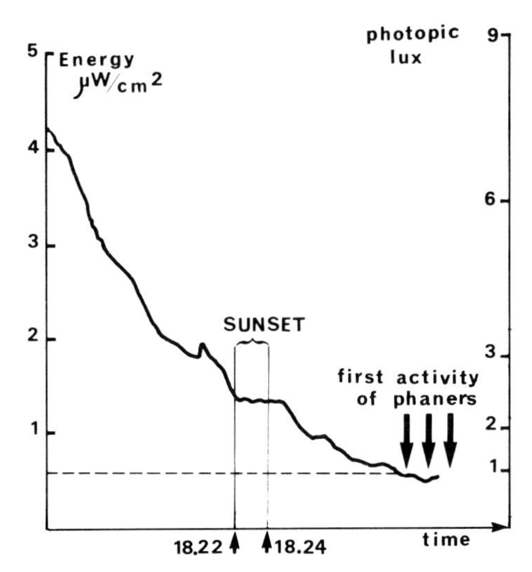

Fig. 8. Continuous recording of the gradual decline in light intensity at sunset in forest conditions toward the south of Madagascar (Analabe, 3/9/74). The light range measured is 200–650 nm. *Phaner furcifer* in this region exhibit onset of activity at about 8 minutes after sunset, when the light level reaches approximately 1 lux.

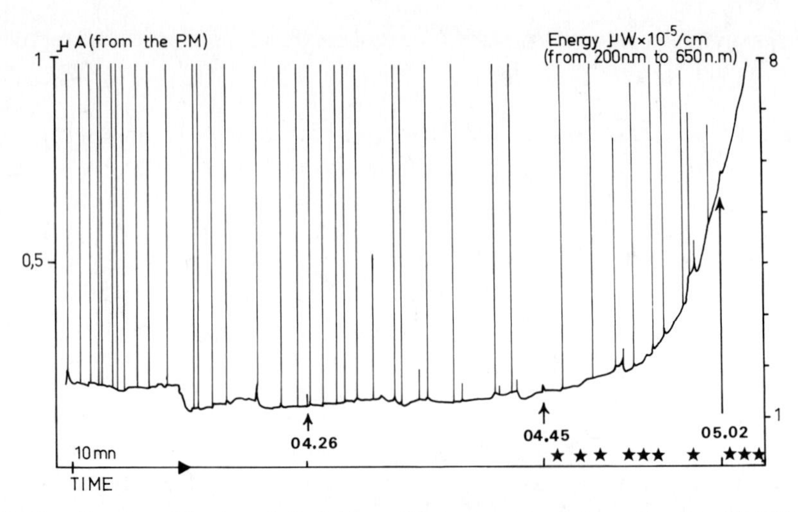

Fig. 9. Abrupt modification of the light conditions occurring at dawn (Analabe, 2/24/74). The recording was conducted during a storm, and the vertical lines represent flashes of light in the west, whose amplitude was cut off at 1 μA. Agitation of leaves during the storm and the passage of clouds are responsible for the minor variations in amplitude on the trace [starry sky; measurements conducted in the forest with a photomultiplier (without filter) directed toward the south]. The effective time of sunrise was 0600 hours and *Phaner furcifer* were observed to return to their nests at 0530 hours. The stars represent salvos of loud shrieks produced by *Phaner*.

and the level arresting activity at dawn seems to be a result of the great difference in adaptation states of the retina in the two cases. (An attempt was made to compare the light levels at dusk and at dawn required for reading a printed text, but the errors involved exceed the ratio of 1:1000 at such low light intensities.)

Since it proved difficult to conduct direct observations on *Phaner* in the field, the most frequently recorded activity noted at specific times of the night consisted of salvos of loud shrieks which almost always occur at the onset and arrest of activity. These calls, which doubtless represent contact calls with a social or territorial function, are also heard frequently during the course of the night. At dawn, however, it was particularly evident that the first signs of increase in light intensity were accompanied by numerous, repeated salvos of shrieks. This slight increase in light intensity involves a change of only a few μlux under scotopic conditions, which is completely imperceptible to the human eye and can only be detected with the aid of an elaborate photomultiplyer device (Fig. 9). It seems likely that *Phaner* can in fact perceive this change as a signal which elicits a sequence of behavior, including increase in frequency of the loud shrieks, leading to arrest of activity at a sleeping site some 30 minutes later. The sun only rises one hour after this first slight change in light intensity (Le Grand, 1942).

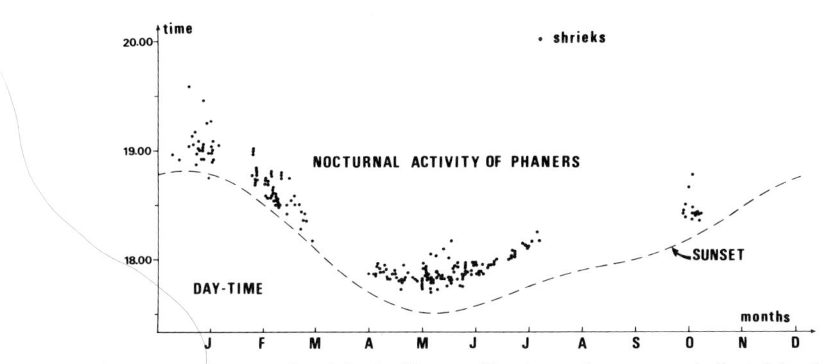

Fig. 10. Onset of nocturnal activity in *Phaner*. The times of sunset are indicated by the broken curve, while the first recorded salvos of shrieks are indicated by black dots. In most cases, the first signs of activity appeared a few minutes before the first shriek, as soon as environmental light conditions were favorable.

3.1.1.2. Reproductive Activity. The curve in Fig. 10 illustrates the sequence of events over the course of the year. The annual cycle of variation in the times of sunrise and sunset is closely followed by cyclical variation in the times of onset and arrest of activity in *Phaner*. As a result, the ratio between the sleeping period and the activity period varies according to the time of the year. The animals are subject to a slow, but very regular, variation in day length. Petter-Rousseaux (1970) has demonstrated that with *Microcebus murinus* reproductive functions are associated with the annual day length (photoperiod) cycle. At the present time, however, one cannot state whether the receptors involved are purely visual or whether more central components (e.g., the thalamic-pituitary system) are also involved (cf. Menaker, 1972). Nevertheless, it is important to emphasize at this stage the close link between visual functions (detection of solar radiation), overall activity and reproductive behavior.

With the African prosimian species which have been studied in Gabon, in an equatorial zone which is exposed only to the circadian light cycle, reproduction occurs all the year round at least in most cases (Charles-Dominique, 1968, 1971). In fact, a slight peak of births is observed during the rainy season, but births are not generally restricted to a particular time of the year. Further, a colony of *Galago demidovii* exposed to a pronounced cyclical variation in day length has so far proved to exhibit reproduction throughout the year (P. Charles-Dominique, personal communication). Thus, day length does not seem to play a regulatory role in the reproduction of these equatorial prosimian species.

It is, *a priori,* paradoxical that nocturnal animals which spend the day hidden in a retreat away from daylight (e.g., in long cavities in trees) should be able to use light intensity as a physiological marker. However, numerous observations have indicated that these animals awake well before the actual onset of nocturnal activity (perhaps as a result of hunger or some similar influence). They are therefore able to observe the ambient light intensity of the environment so that they can move off when the light level falls to a degree of "visual comfort." Under full night-time conditions, they are able to perceive sufficient light to see comfortably, to leap around and to feed. From their point of view, it is "daytime," and all of their physiological mechanisms are regulated for this sensation. (Photoreceptors, either directly visual or more deep-lying, can operate quite satisfactorily under these conditions and hence exert cyclic physiological influences.) For such nocturnal prosimians, their night-time, better referred to as their resting period, consists of the period when full sunlight is present. The notion of a night in which vision is impossible only exists for strict diurnal species. Indeed, nocturnal life is more a shift in the thresholds of sensitivity than a completely new "inversed" mode of existence. It is well known that the nocturnal prosimians can move around perfectly well in full daylight, using their vision in the usual way. Hence it can be seen that there is no reason why nocturnal prosimians should not be able to regulate their activity/rest cycle by using ambient light intensity as the releasing stimulus.

3.1.2. Vision and the Environment

All prosimian species are arboreal, and as a rule they are typically forest living. It can be stated in a general fashion that the various levels of the forest which they utilize are composed of large-diameter supports which are essentially vertical in orientation (trunks) and of finer supports (branches and twigs) without any clear-cut directional tendency (Charles-Dominique, 1971). The broad vertical supports are suitable for large-bodied, leaping prosimians, while the finer, interwoven supports are more suitable for the lighter forms (see Walker, Chapter 12, for illustrations of some locomotor patterns). The body posture maintained in the course of leaping is generally vertical for the large-bodied species and more inclined towards the horizontal for the smaller-bodied forms. The position of the head and the direction of gaze are naturally associated with the body posture adopted in locomotion. Species which move from one trunk to another generally follow a long, flat trajectory (Petter, 1962; Petter *et al.*, 1971). In such cases, estimation of distance is relatively unimportant, particularly since the body is usually maintained in a more-or-less vertical posture throughout the leap (*Propithecus*, *Indri*, etc.). The smaller-bodied prosimian species can also carry out horizontal leaps, but their body postures are then more inclined in the direction of the leap (*Lemur* spp., small *Galago* spp.; cf. Petter, 1962). In some cases, such leaps have been analyzed in considerable detail (see Charles-Dominique, 1971; Hall-Craggs, 1974). Some of the smaller-bodied prosimian species, such as *Phaner*, also perform leaps from an elevated point to a point some distance lower down. In one case, a downward leap made by a *Phaner* was measured and found to cover a distance of 4 m. In this particular case, the animal hesitated for 1 or 2 seconds before taking off, doubtless in order to improve the orientation of the leap. Since the point of departure was at a height of 12 m, it is likely that the slightest error could have resulted in a serious fall. At the end of the leap, the animal concerned landed just at the leafy tip of a very supple branch, which absorbed much of the shock of the landing. The choice of this point of arrival at such a distance, under night-time conditions, provides a perfect illustration of the complex role played by vision in integrating all of the factors defining the ballistic parameters of the leap: weight, air resistance, wind, distance, initial velocity, balance and position of the body on arrival, and elasticity of the supports involved.

Another feature which has often been associated with the structure of the forest environment is the form of the pupil (Hartridge, 1919). In actual fact, it seems quite likely that a pupil in the form of a vertical slit, compared to a circular pupil, would permit an increase in the amount of light passing through the pupil without reducing visual acuity in the horizontal plane. (The latter would be important, for example, in distinguishing vertical supports.) Thus, visual acuity in the vertical plane would be sacrificed, but this should not be of great importance in an environment composed predominantly of vertical supports. In effect, all of the nocturnal prosimians (with one exception—the tarsier) have a slitlike, pyriform pupil with a roughly vertical orientation. Nevertheless, two objections may be raised against this general interpretation: first, the tarsier has a horizontal, slitlike pupil which is difficult to interpret, except perhaps in that such a pupil must increase the brightness of peripheral images seen on the horizontal plane during the daytime. Second, at night when most of these prosimians are active the pupil is opened widely (mydriasis) so that it becomes very large and circular in outline. Thus, if the visual acuity interpretation is correct, the slitlike form of the pupil must be an adaptation of a nocturnal animal for "sequences of diurnal life." However, such a hypothesis remains to be tested.

With the diurnal and crepuscular prosimian species, the pupil is almost circular in myosis (maximum degree of closure), and this permits optimal definition in the retinal image as a whole. In mydriasis, the pupil is quite round in both the diurnal and the nocturnal forms: the surface ratios (surface ratio refers to the ratio between the area of the constricted pupil and that of the dilated pupil) are 40 in the nocturnal *Phaner*, 22 in *Propithecus*, and 21 in *Lemur catta* which are diurnal. Hence, one can see the extent to which the nocturnal species must protect themselves from the light when the appropriate circumstances occur.

This discussion may be concluded by comparing the sphere of vision of a small nocturnal species, moving around in tangled, dense vegetation where light is subject to repeated reflection, with the sphere of vision of a larger species moving among more widely spaced, large-diameter supports, where there is much less reflection of light. It is likely that, at weak light intensities, the light reflected from objects in the natural environment is rapidly reduced to subthreshold levels with respect to the sensitivity of the eye. This would tend to diminish the contrast of such objects against a dark background. On the other hand, objects which are adequately illuminated or which are close enough to be clearly seen will appear at a greater contrast with respect to the very dark background. As a result, at low light intensities, the small-bodied nocturnal species must live in a visual world composed of small spheres of vision, and even the larger species (*Lepilemur*, *Avahi*, *Galago alleni*) have spheres of vision which are scarcely larger. In any event, the dimensions of the spheres of vision of all the nocturnal species are vastly inferior to those of the crepuscular or diurnal species. The latter are adapted to profuse light conditions which permit extremely good perception of shades of gray or, in another realm, distinction of colors, regardless of the distances involved. With the diurnal species, it is the size of the retinal image which is the limiting factor; with the nocturnal prosimians, it is the very existence of that image.

Given these conditions, one can suppose that visual orientation in space (e.g., in the recognition of a territory) can only be achieved in nocturnal species if there is continual storage of visual information as memories. The latter will become increasingly more numerous as the size of the sphere of vision decreases, and any given recognition site will serve for a shorter length of time in the course of nocturnal locomotion. This necessity may well explain the considerable development, among nocturnal species with a limited sphere of vision, of alternative systems of localization such as olfactory recognition and auditory communication.

3.1.3. Visual Memory and Territory

Numerous observations of prosimians in the course of movement around their home ranges (Petter, 1962; Jolly, 1966; Charles-Dominique, 1971; Richard, 1973; Martin, 1972; Sussman, 1972) indicate that sight is a major element in recognition of specific localities in the forest. On one occasion, the author and two other observers took turns to follow on foot over a distance of 800 m a *Microcebus coquereli* which was released during the daytime after being captured the previous night. One observer took detailed notes of the pathway followed by the animal. At the end of its passage through the forest, the *Microcebus coquereli* went straight into its nest. The record of the pathway followed showed that the animal had consistently traveled directly toward the nest. Since movements were effected very rapidly as a series of leaps, there was at no time any indication of olfactory orientation on the branches, as would be expected if these had been previously "labeled" with marking substances. Unless one is to postulate a hypothetical internal compass mechanism, the only rational explanation of this behavior lies in visual recognition of the surroundings. Somewhat later, in the same sector of forest, Pagès (1978) utilized radio tracking to investigate the extent of the territory occupied by *Microcebus coquereli* and the complexity of the pathways followed. Such locomotor performance in the forest implies that these leaping prosimians possess considerable specialization of the higher nervous centers for the storage of a large body of information about the environment. Even the nonleaping prosimians, such as the lorisines, seem to memorize features of their environment and to rely heavily on sight in their movements. For example, sight seems to be important for determining pathways to be followed through the trees (''detour problems''). Charles-Dominique (1971) has studied such behavior in detail in *Perodicticus potto*. It seems highly likely that special central nervous mechanisms are involved, necessitating the utilization of visual data tŏ construct alternative ''possible pathways'' in order to make an appropriate selection. According to Charles-Dominique, olfaction does not seem to be directly involved in the movements of pottos around their home ranges. *Arctocebus calabarensis*, which have also been studied in detail by Charles-Dominique, exhibit quite different behavior, and this prosimian would seem to represent an extreme case of the limitation in size of the sphere of vision, despite the existence of a *tapetum lucidum*. (The author found that one healthy angwantibo in captivity had myopia exceeding -2 diopters.) In the case of the angwantibo, olfaction, and particularly hearing, seem to play a major part in orientation. The same seems to apply to *Loris tardigradus* in Ceylon (Petter and Hladik, 1970). The fine hearing and the slow locomotion typical of the lorisine species permits them to detect insect prey or predators with ease while remaining quite silent themselves. In connection with the limited role played by vision in the angwantibo and the slow loris, it seemed that it might be profitable to investigate in *Microcebus murinus* whether hearing might possibly replace vision under conditions of insufficient light intensity. The possibility of echolocation was naturally considered, since mouse lemurs do emit ultrasonic sounds (Pariente, 1974b). The experiments did demonstrate the part played by ultrasound in the detection of insects or of conspecifics, but in no instance was vision replaced by echolocation. Vincent (1973), in a parallel investigation of the orientation of *Galago demidovii*, under attenuated light conditions, arrived at the same conclusion.

3.2. Visual Communication

Up to this point, the mechanism of vision has been dealt with primarily as composed of the eye as receptor and light as the carrier of information. When we come to actual visual communication, however, one must bear in mind that the animal itself is the origin of images bearing a particular significance. Visual signals may be based on material form (posture and mimics), the light intensity emanating from material surfaces (patterning) and, in some cases, coloration patterns. The signals which are most easy to modulate are those relating to form, and it is hence through this medium that transitory psychological states are generally expressed. The other signal types are essentially constant features related to the evolution of each individual species. It is obvious that animals which lack color vision (nocturnal animals) can only make use of signals which concern form and light intensity. Only diurnal and crepuscular species with some degree of color vision can make use of all three channels for visual communication.

3.2.1. Invariable Signals

3.2.1.1. Body Size. Within any given species, the general shape of the body provides some information about body size and hence an indication of age. This basic channel of communication seems to be quite significant among the mammals generally; the young often benefit from a special status and perception of the body size of other individuals seems to play a fundamental role which is sometimes supplemented by other perceptual components (e.g., vocalizations, odors, postures). Behavior relating to the perception of body size is very clear-cut and is particularly easy to observe with prosimians living in structured social groups.

3.2.1.2. The Individual's Appearance. Perception of an individual's appearance among the prosimians is essentially a matter of pelage patterns. Two categories can be distinguished, according to whether a species is nocturnal or diurnal. With the nocturnal species, the most obvious markings are based on white and black hairs or banded hairs which, mixed together in the fur, give different gradations of gray. For animals whose retina consists exclusively of rods, this is scarcely surprising. However, it can be seen from Tables I and II that rufous hair is also frequently encountered among nocturnal species, and that often the fur is entirely rufous or brown in color. No other clear-cut color exists with the nocturnal species.

As a general rule, the ventral pelage is much paler than the dorsal, and in some cases it is actually white. The result is that an animal viewed from below against the clearer background of the sky is more difficult to observe. If one takes account of the fact that gray is produced by juxtaposition of white and black hairs, or by black-and-white annulation of the hairs themselves, it can be deduced that there are at least some white zones of the pelage which will reflect all incident radiation without selection. This gives rise to a certain degree of stimulation of the rods of the retina, whose maximal sensitivity is located in the region of 510 nm. In the case of nocturnal species lacking color vision, white hairs (or white areas of hairs) may represent the best means of blending in with a generally dark background which is rich in reflected green light (wavelength between 500 and 550 nm), which is little absorbed in the forest. To the nocturnal eye, the prosimians must appear as gray against a generally gray background. On the other hand, species with a dominantly rufous coloration reflect a selected portion of white light. To simplify matters, it can be taken that light of long wavelength is reflected while light of short wavelength is absorbed. Since nocturnal mammals perceive light of long wavelength only with difficulty, this probably means that, in a poorly illuminated forest, rufous animals are even more difficult to spot. If these prosimians were actually black in color, they would be more easily noticed because of the contrast effect against the faintly illuminated background. The same reasoning may be applied both in terms of visibility to conspecifics and with respect to sighting by nocturnal predators with the same type of vision. This factor of sighting by predators is particularly important for the vulnerable small-bodied species.

TABLE I

Color and General Appearance in the Leaping Forms of Prosimians in Relation to Their Distribution and Activity Dimension[a]

Species	Geographical area	Dominant color	Activity	Appearance
Lemur spp.	M	White/gray/rufous/black	DD+D+N	C
Varecia variegata	M	White/—/rufous/black	DD	C
Lepilemur spp.	M	—/gray/rufous/—	N	g
Hapalemur spp.	M	—/gray/rufous/—	D	g
Cheirogaleus spp.	M	—/gray/rufous/—	N	g
Microcebus murinus	M	—/gray/rufous/—	N	g
Microcebus coquereli	M	—/—/rufous/—	N	g
Phaner furcifer	M	—/gray/—/—	N	g
Allocebus trichotis	M	—/—/rufous/—	N	g
Avahi laniger	M	White/gray/—/—	N	g
Propithecus spp.	M	White/—/rufous/black	DD	C
Indri indri	M	White/—/—/black	DD	C
Galago spp.	Af	White/gray/rufous/—	N	g
Tarsius spp.	As	—/gray/rufous/—	N	g

[a] Key to abbreviations: M, Madagascar; Af, Africa; As, Asia; N, nocturnal; D, crepuscular; DD, diurnal; g, graded; C, contrasted.

TABLE II

Color and General Appearance in the Nonleaping Forms of Prosimian in Relation to Their Distribution and Activity Dimension[a]

Species	Geographical area	Dominant color	Activity	Appearance
Loris tardigradus	As	—/Gray/rufous/—	N	g
Nycticebus coucang	As	White/gray/—/black	N	g
Arctocebus calabarensis	Af	White/—/rufous/—	N	g
Perodicticus potto	Af	—/—/Rufous/—	N	g

[a] Key to abbreviations: As for Table I.

In conclusion, one can establish the hypothesis that in a relatively well-illuminated environment, under conditions of nocturnal vision, gray is more visible than various shades of brown (Fig. 11). Certain prosimian genera in Madagascar (e.g., *Microcebus* and *Lepilemur*) are plainly divided into species whose color (gray or rufous) depends upon the geographical region they inhabit. The rufous forms are generally common to the denser, darker, and hotter rain forest of the east coast of Madagascar. In the west of Madagascar, the forests are typically more open, dry, and colder and the nocturnal lemur species are predominantly gray in color. This agrees well with the hypothesis in that under such conditions one would expect selection for the least conspicuous color—rufous in the east and gray in the west (Gloger's Rule). However, we shall return to take another look at this situation below in considering the requirements operating when the animal actually needs to be recognized at night (visual message).

With the diurnal and crepuscular species, the problem of pelage coloration is clearly more complex. The species concerned belong to the genera *Lemur*, *Propithecus, Indri, Varecia (Lemur variegatus)* and *Hapalemur*. The high light levels available under natural conditions have permitted the development of color vision and hence the utilization of all kinds of refinements of pelage patterns. However, the interpretation of the role of such color patterns presents considerable difficulties. According to the observations conducted by C. M. Hladik (personal communication), it would seem that the black patches against a white background which occur in the pelage of *Indri* provide an excellent means of camouflage in the forest, but it is

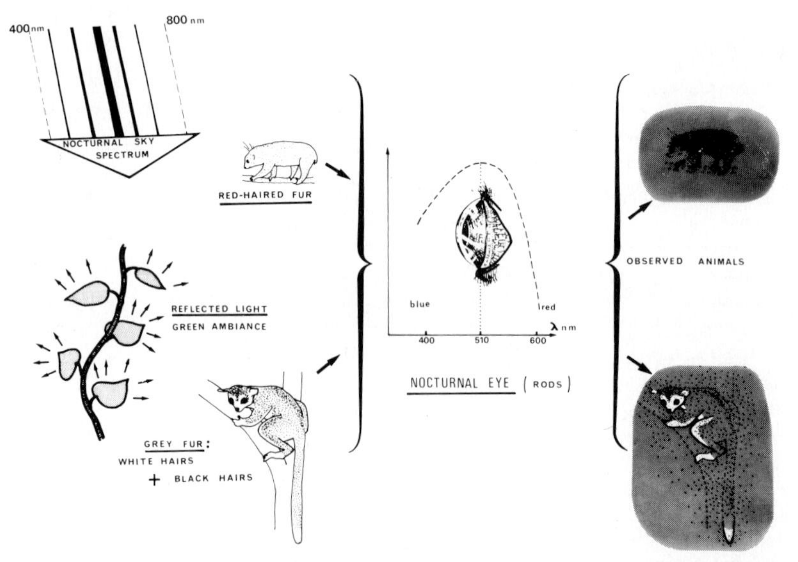

Fig. 11. Appearance of two nocturnal prosimian species with differing types of fur, as seen with a nocturnal eye on a dark night.

difficult to see under present circumstances which predators might be avoided in this way. Nevertheless, it is perfectly true that the sun, filtering irregularly through the foliage, produces on tree trunks and on the ground a pattern of dark and pale patches comparable to that seen on the pelage of *Indri*. With the *Propithecus* species and subspecies, according to Petter and Peyrieras (1972), there is a clearcut gradient in melanization passing from the south to the north of Madagascar. With the sifakas, which are actually separated into subspecies on the basis of pelage coloration, it is seen that the ventral area of the body is more subject to pigmentation than the dorsal area. In the west coast areas of Madagascar, the dorsal pelage is almost always pale in color and plays a direct part in protection against incident solar radiation. Numerous observations, however, indicate that *Propithecus* species generally seek out solar radiation for warmth and adopt a specific stretched-out body posture with the belly directed up toward the sky. Under these conditions, ventral pigmentation constitutes a definite advantage for the absorption of warmth. The pale coloration of the dorsal surface, by contrast, provides protection during the extensive periods of activity in locomotion and feeding. In contrast to man, where pigmentation of the skin provides protection against the penetration of solar radiation, mammals with a thick pelage containing an insulating layer of air must be protected by a pale, reflecting coloration. Those areas which are not provided with hair, such as the snout, the hands, and the feet, are always heavily pigmented and hence directly protected.

There are thus two simultaneously acting physical mechanisms which permit adaptation to solar radiation, either through the screening effect of cutaneous pigments (primarily acting against ultraviolet light) or through the utilization of the absorbing properties of dark bodies (a black pelage forms a very mat and therefore efficient surface in which incident energy is transformed into heat which is retained by the air imprisoned between the hairs).

On the east coast of Madagascar, in dense and humid forest conditions, the *Propithecus* subspecies have even darker patterns of fur coloration. In some species, the entire pelage is black in color. Two mountain forest forms in the southern and northern regions of the east coast rain forest, *Propithecus diadema holomelas* and *P.d. perrieri*, are completely black (Petter and Peyrieras, 1972). Only in the northeastern rain forest zone, where the climate is permanently warm, are the *Propithecus diadema* forms relatively pale in coloration.

Thus, one feature of importance which is common to all of these diurnal lemurs is rigorous adaptation to local climatic conditions. There is a resulting contrast effect in the coloration of the pelage. With *Propithecus* it would seem that other mechanisms have also been operating with respect to selection of patches of coloration in the head region. In Tables I and II, the general appearance of the pelage in the various prosimian species has been considered and compared to the way of life. It can be seen that a contrasted pelage coloration pattern is always found among the diurnal species, whereas the nocturnal forms generally have more discreet coloration patterns, though the ventral surface is typically paler in order to blend in with the sky. The only exception is provided by the *Hapalemur* species, but their mode of life and origin remain obscure. (It should be noted that, like *Lemur catta*, the gentle lemurs have a fovea associated with a very well-developed *tapetum lucidum* which is partially masked by pigment.) J. J. Petter (personal communication) has also observed with *Hapalemur griseus* and *H. simus* habitual "sunning behavior" which lasts for considerable periods of time.

3.2.2. Variable Signals

The emission of variable visual signals depends upon the psychological state of the animal concerned, either in terms of sporadic fluctuations in emotional conditions or in association with natural cycles such as those involved in reproduction. Variable signals can be arbitrarily divided into two categories: (a) the individual's overall body posture, and (b) the position of the head and the facial expression.

3.2.2.1. Body Posture. Without entering here into the problems inherent in behavioral interpretation, it can be simply accepted as fact that the prosimians utilize a whole range of body postures which are of great significance as visual signals for conspecifics (Jolly, 1966). Of course, such postures are best developed among those species which live in social groups and are diurnal in habit. In the overall pattern of relationships among the individuals of a given social group, sight is of major importance, as with the various simian primate species. This predominance of vision in diurnal, social species has even led to the transfer to the visual sphere of behavior patterns which originally involved other sensory modalities. In this way, according to Schilling (Chapter 11), certain patterns of olfactory marking behavior have probably come to lose their original significance (deposition of a scent) such that only the visual part (the "gesture") is preserved. Certain lemur species exhibit rubbing motions with the forearm or the forehead although no secretory glands are actually present in these body areas. Since other species (e.g., *Hapalemur griseus* and *Lemur catta*) exhibit the same gestures in association with functional marking glands, the interpretation that an olfactory signal may have become a visual one in some species would seem to be sound.

One might also interpret as a derivative of original olfactory marking behavior the actions of *Lemur catta* in scoring the bark of trees in the home range with a spur associated with the brachial marking gland (Schilling, 1974). In this case, there is obviously a clear-cut visual signal superimposed upon the olfactory signal. However, if such visual signals (scored bark) are compared with purely olfactory signals, they seem to be far less rich in terms of the information transmitted. Scoring of tree trunks can only operate as a distant signalling mechanism which in itself must remain anonymous. Since *Lemur catta* have a fovea, and hence a quite elaborate visual system, distant perception of score marks on trunks is doubtless quite possible. Other behavior patterns among diurnal prosimian species also seem to involve a mixture of olfactory and visual signals, as for example in manipulations involved in scent impregnation and raising and waving of the tail. In addition, the annulated pattern of the tail in *Lemur catta* (alternating black and white rings giving a maximum of contrast) is correlated with the terrestrial habits of this species, where locomotion on the ground is typically accompanied by raising of the tail in a question-mark posture. Sussman (1972) found that 65% of *Lemur catta* travel occurs along the ground.

With *Propithecus*, the typical backward movement of the head would seem to be a visual signal. However, one must consider the fact that this gesture may in fact act to release (or at least stimulate production of) substances from the marking glands situated at the ventral base of the neck in males.

One might even inquire whether marking with feces exhibited by *Cheirogaleus* species, which is such a visually conspicuous form of scent marking, has not acquired an additional signalling component. Turning to a somewhat similar, but far more specialized, species on the African mainland, Charles-Dominique (1971) has demonstrated that, in contrast to claims made by many previous authors, *Perodicticus potto* primarily rely upon vision for their movements through the forest, despite the fact that they also exhibit a great deal of olfactory marking behavior. In this species, olfactory marking certainly has some function other than that of orientation in locomotion along the branches.

There are also postures which have a purely visual significance. When alarmed, prosimians generally adopt a characteristic posture in addition to uttering special vocalizations. When in a state of alert, an animal is ready to flee, with its gaze fixed on the potential source of danger and the head and neck tensed in the direction of gaze (Fig. 12). It seems likely that, in both nocturnal and diurnal species, this posture attracts the attention of other individuals, thus permitting them to localize the source of danger.

Rhythmic waving of the tail in prosimians, generally in a plane perpendicular to the potential source of danger, is also an easily identifiable visual signal of alarm. It is even possible that the sight of all the individuals in a group suddenly waving their tails in unison may discourage a predator, though this explanation would not apply to nocturnal species. It should also be noted that piloerection of the tail, which is associated with various behavior patterns, is very important as a visual signal in *Lemur* and *Hapalemur* species.

Thus, it would seem that the tail, which may play an active role in balancing the body during leaping, also plays an important part in visual communication among prosimians. J. J. Petter (personal communication), for example, states that in *Lemur fulvus* the position of the tail provides a good indication of the psychological state of any particular individual. Further, P. Charles-Dominique (personal communication) notes that in *Galago demidovii* the position of the tail varies according to the behavioral situation (corkscrew posture, vertical posture, horizontal posture, and question-mark posture).

Pagès (1978), who has studied *Microcebus coquereli* both in the field in Madagascar and in captivity, has described several postures which are specifically associated with the motivational state of an individual with respect to another. These

Fig. 12. Alarm posture adopted by a *Propithecus* when confronted with some potential danger. The neck is stretched, the head is inclined to one side, the gaze is fixed, and the eyes are kept wide open. The animal is ready to flee.

postures include food begging, play invitation, and mating invitation. Pages postulates that, in the course of encounters between individuals of opposite sex in both *Phaner furcifer* and *Microcebus coquereli*, the postures adopted by the female (suspended from the hindlimbs with the forelimbs outstretched, or presenting the genitalia) include a visual component which is important in determining the male's behavior.

Lepilemur (the sportive lemur) derives its common name from the characteristic posture which is adopted when it is attacked. Another example is provided by various species of *Lemur*, *Microcebus*, and *Galago*, which all exhibit a grooming invitation gesture consisting of the raising of one arm and presentation of the flank. Finally, one can cite as another possible visual signal the large skinfolds which are present around the anal area of males of *Lemur* species. M. Grange (personal communication) has observed rhythmic contractions of these skinfolds, which occur when the males are in a state of excitation. In such cases, they adopt a presentation posture, with the tail raised and the rear end of the body elevated. This may act as a visual signal to other members of the social group.

The above selection represents only a small part of the body postures which can be observed among the prosimians. Numerous other body postures provide visual signals. In fact, any individual provides a complex visual impression through the overall attitude of its body. Nevertheless, it is convenient to discuss separately the various body postures and facial expressions which involve the head in visual signalling.

3.2.2.2. The Head and Facial Expressions. According to Lindsay-Johnson (1897, 1900), in prosimians the face cannot play a role in the expression of psychological states and therefore does not contribute to visual communication. If this were in fact the case, it would amount to a clear-cut distinction between prosimian and simian primates. This is clearly not true of diurnal prosimians (Andrew, 1963, 1964) though it may be at least partly true of nocturnal species.

Among the nocturnal prosimian species it is difficult to distinguish any obvious variety of facial expressions. The most easily observed facial expression is without doubt that which is exhibited by animals which are frightened and on the defensive. The mouth is opened widely (amounting to presentation of the teeth, perhaps with some intention to bite), the ears are folded and retracted, and the eyes are opened wide. It is not known whether this situation is accompanied by reflex dilation of the pupil, but G. Vuillon (personal communication) has observed in the course of studies on *Lemur fulvus albifrons* that there are variations in pupil size apparently correlated with modifications in the state of alert (as indicated by electroencephalography) during somnolence. P. Charles-Dominique (personal communication) has observed *Galago demidovii* blinking the eyelids, which may lead to almost complete closure of the eyes in submissive animals or those faced with imminent attack, and A. Schilling (personal communication) has observed the same phenomenon in *Microcebus murinus* and *medius*. Further, movements of the ears, which are very frequent and complex in *Galago* and *Microcebus*, may be considered as a particularly conspicuous visual signal betraying the degree of arousal of an individual animal.

The final example which should be mentioned here, since it is common to many prosimian species, is that of swaying, a pattern which is also found with the only nocturnal simian primate, *Aotus trivirgatus* (Moynhihan, 1964). This behavior pattern is particularly well developed in *Lepilemur*; as soon as an animal's attention is aroused, horizontal to-and-fro movements of the head occur (Fig. 13). Such behavior is usually exhibited for only a very brief period of time, in the form of sequences of three or four head movements alternating with short pauses without any relaxation of the fixed gaze. *Lepilemur* also exhibit vertical swaying, often intercalated with sequences of horizontal swaying. Among the diurnal species, e.g., *Propithecus*, and *Hapalemur*, which are crepuscular, these vertical movements seem to be the most common form of swaying. In fact, there are marked movements of the body as well during swaying, since such movements could not be produced by movement of the neck alone. Vertical swaying is also common among nocturnal prosimian species and has been observed in *Phaner furcifer*, *Microcebus murinus* and *Galago demidovii*, in addition to *Lepilemur*. Of course, it is immediately brought to mind that this head-swaying behavior permits better spatial localization of any object which arouses the animal's interest. There would be the equivalent of improvement in the perception of relief as a result of these head movements.

Among the numerous mechanisms which are involved in the refinement of perception of relief, two deserve special attention here. The first is based on monocular

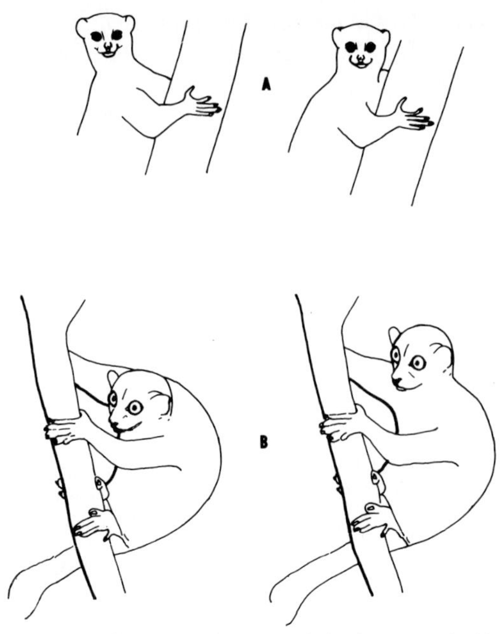

Fig. 13. Head swaying in *Lepilemur*, a very common behavior pattern in this genus. (A) Horizontal swaying; (B) vertical swaying.

parallax data emerging from the head movement, and the second involves stereoscopic vision utilizing the disparity between the retinal images originating from the two eyes (Le Grand, 1956). To take the latter possibility first, it should be immediately emphasized that such a mechanism is necessarily different (in terms of the behavior on which our attention is focused) from that involved in normal stereoscopic vision, which operates by means of instantaneous comparison of two slightly different images from the eyes (i.e., retinal disparity).

It should be noted that, according to Barlow *et al*. (1967), certain cells of area 18 in the visual cortex do not respond if homologous stimuli emanating from the two retinas are either identical or markedly different. On the other hand, these particular cells are stimulated if the homologous stimulus patterns originate from closely similar zones of the left and right retinas. These are the cells which probably intervene directly in stereoscopic vision. Now, with swaying it is obvious that the additional parameter of time is introduced, since in the course of the movement there is a whole series of paired perceptual images emanating from the left and right retinas. If it is possible for the brain to store successive images for subsequent comparison of data produced at the extreme positions achieved by the head, then swaying would evidently yield a considerable advantage in terms of stereoscopic perception. Treff (1967) attempted, without success, to carry out an experiment on

Galago senegalensis in this connection, by preventing the formation of a clear image on one of the two eyes and measuring the resulting acuity (= swaying + monocular vision). This experiment could not produce conclusive results, since, according to the schema already mentioned, the cells of area 18 involved in stereoscopic vision always require information derived from homologous regions of the retinas of the two eyes. Nevertheless, provided that some kind of mechanism exists for memory storage or a built-in delay in the transfer of information to the visual centers, stereoscopic vision involving a time-lag element is possible if both eyes are functional.

The other significant mechanism for the perception of relief is based, as stated above, on monocular parallax of movement (appreciation of parallax due to head movement), which applies to monocular vision and involves the relatively different movements of images of objects located at different distances from the eye, arising when the head is moved. In man, a strong impression of depth is produced by this mechanism, and it seems highly likely that it also plays an important role in the swaying behavior of prosimians. The author has often observed swaying behavior preceding relatively difficult leaps with various prosimian species. Although this has occasionally been observed with diurnal species, it is particularly prominent among the nocturnal forms and it is almost certainly involved in the refinement of depth perception.

It should be mentioned here that the efficiency of stereoscopic vision, based on the disparity between the two retinal images, increases with the degree of detail in the retinal images and hence with the prevailing light intensity (up to a certain level). In the case of nocturnal animals, very low light intensities necessarily limit image formation to a crude pattern with little contrast or detail and therefore unsuitable for good perception of relief. One can thus easily understand why a supplementary mechanism should be involved (monocular parallax movement) in order to permit actively leaping prosimians to assess distances with sufficient accuracy. This doubtless accounts for the frequent and obvious performance of swaying by nocturnal prosimian species.

Another behavior pattern apparently related to swaying is easily observed with young prosimians. It consists of rotation of the head around its longitudinal axis, with the animal maintaining its gaze fixed in a particular direction. It is not unusual to see the head rotate through a total arc of 100 ° between the two extreme positions. As a crude approximation, it can be said that the animal's snout does not change position during the movement. No explanation appears to have been advanced for this behavior, which is widespread among prosimians, particularly during early life. However, it seems likely that this is a behavioral manifestation of a purely visual learning process leading to appreciation of space and the localization of objects in space (visual imprinting). The well-known investigations of Hubel and Wiesel (1965, 1968) on cats have demonstrated that animals reared in an environment composed of vertical lines were incapable, as adults, of ''seeing'' horizontal lines. Between the retina and the cortex of the young animal, units are formed consisting of groups of visual nerve cells which eventually (via a series of connections) stimulate one particular cell. These units respond selectively to moving light stimuli, which are perfectly characterized in terms of their direction, orientation, etc. However, such units are not innate; they require visual experience for their formation.

Rotation of the head, which produces a concentric sweep of images across the retina (whose importance increases with the length of the object), could be interpreted as a form of hyper-self-stimulation accentuating the fixation of visual schemas (particularly for elongated objects). As has been discussed previously, any forest is in fact characterized by a collection of objects which are more-or-less vertical, according to their size. Thus, rotational movements of the head may perhaps be a behavioral adaptation finely suited to the forest environment, which provides the primary source of visual stimulation for most prosimian species.

With diurnal prosimians, the facial musculature is better developed than in the nocturnal forms, and there is therefore greater scope for facial expressions. Such development of the facial musculature is even more important in the simians and has achieved its greatest degree of development in man with additional fine control of lip movements, eyebrows, and other features.

Among the diurnal prosimians, the mouth, the direction of gaze, and the mobile accessory structures of the eye region provide the main elements of facial expression (Lightoller, 1934; Andrew, 1963, 1964; Jolly, 1966). For example, when these primates are threatened the mouth is opened widely, exposing the teeth as a threat gesture. Among species of the genus *Lemur*, there are numerous other facial expressions. One obvious example is that of a male attempting to intimidate an adversary (conspecific or predator) by raising the muzzle and displaying, without opening the mouth, the prominent canines (which are always found to protrude in adults), as shown in Fig. 14. If the situation develops to a more serious encounter, the skin covering the dorsal surface of the snout is seen to pucker (Fig. 15). In such cases, the displaying animal rarely looks directly at its adversary, as is particularly obvious when the display is directed toward a human being. This behavior gives the impression of a diversionary tactic, in that it permits the displaying lemur to launch a ferocious attack on its adversary if the latter similarly shifts its gaze or relaxes its vigilance. Such gaze avoidance provides one of the best practical indications of the mobility of the eyeball possessed by diurnal lemurs.

As a general rule, a dominant *Lemur* male rarely tolerates a direct gaze from a human being, and frequently a yawning action is observed. In addition, in *Lemur catta* and several other species, the ears are clearly seen to be retracted in any potentially dangerous situation (Fig. 16).

The sifakas (*Propithecus spp.*) seem even more prone to exhibit differing facial expressions, and in actual fact such expressions can be recognized even more easily than with *Lemur* species. The photographs in Fig. 17 illustrate some of the different facial expressions which can be seen in a short space of time with a single sifaka. Great mobility of the cheek region has also been observed in captivity with *Indri*

Fig. 14. Aggressive stare of *Lemur macaco albifrons*. The eyes are kept wide open, the snout is raised, and the lower lip is pulled downward. The animal is displaying its canines and is ready to bite.

indri. The cheeks also serve to form a kind of resonance horn when *Indri* utter their characteristic wailing vocalizations. Finally, in contrast to the *Lemur* species, the indriids seem to be much less affected by the direct gaze of a human being.

Varecia variegata also adopt a very characteristic posture, with the head tilted backward and the eyes looking up at the sky, when uttering the characteristic alarm call in a situation of danger (i.e., in response to the presence or calls of diurnal birds of prey).

One frequent facial expression found among some diurnal lemurs is partial closure of the eyelids when an animal is sleepy and sufficiently confident to remain immobile. Under such conditions, one can often see the animal looking around its surroundings, with the eyes half-closed. However, if any kind of alarming stimulus intervenes, the eyelids are immediately opened wide.

This brief review of various signals involving the head region in prosimians can be included by reference to the problem of color and variegation patterns in the pelage, which occur in many different forms on the facial region among the prosimians (Figs. 18 and 19). Of course these characters represent invariable visual signals in themselves, but the close relationship between such markings and the expression of the eyes requires some discussion of the phenomenon at this point.

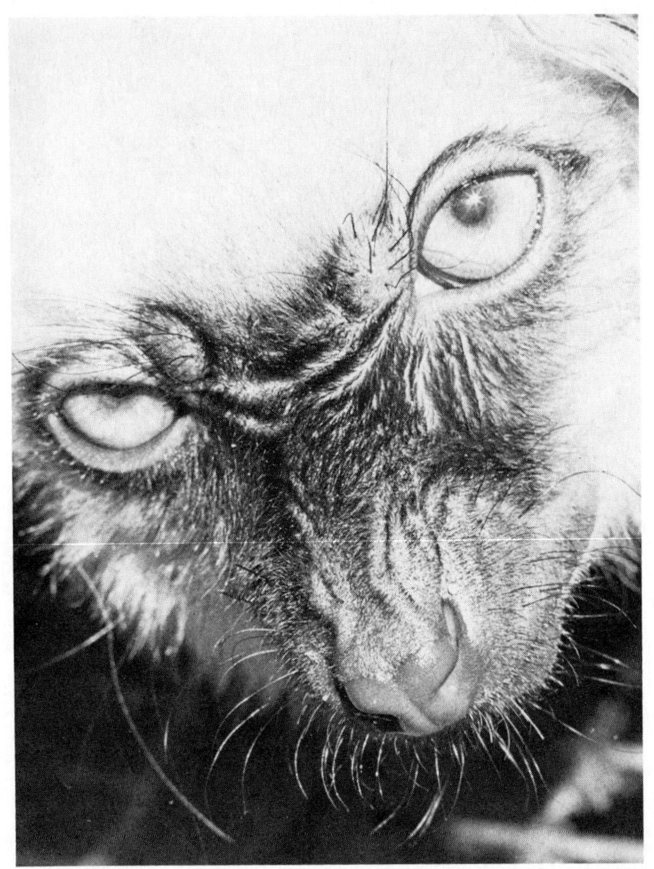

Fig. 15. Puckering of the lips and snout region as an aggressive display in the male *Lemur macaco albifrons*. While displaying in this way, the animal avoids staring directly at its adversary. This illustration provides some indication of the variability in facial expression permitted by the underlying muscles.

Fig. 16. Modification of the position of the ears in *Lemur catta* during aggressive behavior. This frequently occurring pattern is as much a protective reflex as a visual signal. (A) Normal ear position; (B) reflex modification of ear position.

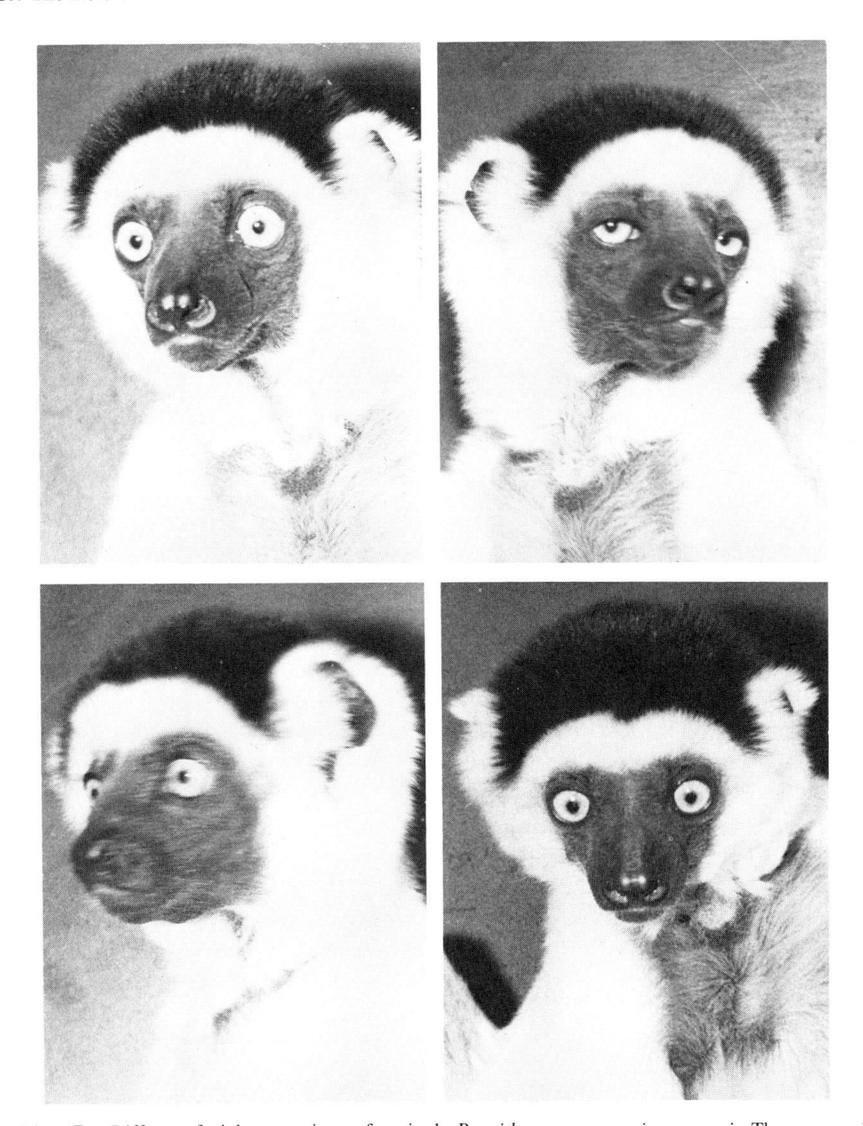

Fig. 17. Different facial expressions of a single *Propithecus verreauxi verreauxi*. The eyes and eyelids are mainly responsible for the variation in facial expression.

Fig. 18. Nocturnal prosimians in which the eyes are accentuated by a black surround. Since the pupils are large and black at night, the eyes are confused with the black surrounds and appear as two large black patches. (A) *Cheirogaleus major*; (B) *Loris tardigradus*; (C) *Daubentonia madagascariensis*; (D) *Phaner furcifer*; (E) *Nycticebus coucang*.

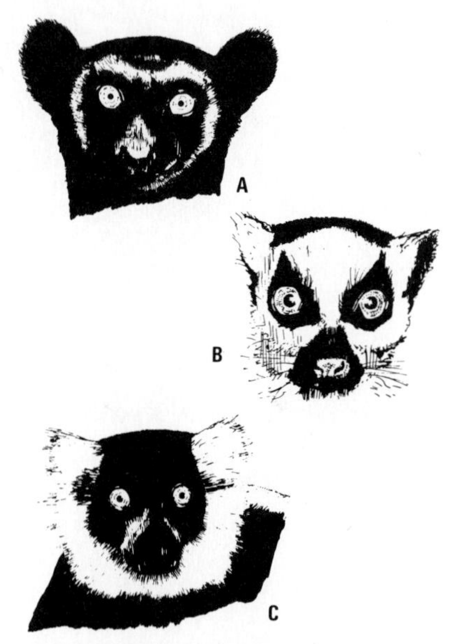

Fig. 19. Diurnal prosimians, showing the eye markings. In the daytime, the eye appears as a clear spot surrounded by a dark area. (A) *Indri indri*; (B) *Lemur catta*; (C) *Varecia variegata*.

Overall, it can be said that the pigmentation of the hair covering the facial region, both in diurnal and in nocturnal species, always bears a relationship to the eyes themselves. There does not seem to be any generally applicable principle, excepting the fact that there is a general accentuation of the eye, such that at the limit the face virtually represents two contrasted spots in space (Coss, 1972). Among the nocturnal genera, such as *Loris*, *Cheirogaleus*, *Daubentonia*, *Nycticebus*, and *Phaner* (Fig. 18), the eyes are each encircled by a separate black patch. It must also be remembered that at night all of the nocturnal prosimian species have a very expanded, black pupil; the reflection from the *tapetum* is only produced when an observer shines a light source at an animal's eyes. Hence, the black patches surrounding the eyes effectively increase the apparent size of the eyes with respect to the head, or the animal's entire body (Fig. 18). It is possible that this may generally act as an extra deterrent to predators. Other prosimian species, in contrast, have evolved instead toward increased discretion; such markings are absent and the pelage is generally brown (e.g., *Perodicticus potto*).

Avahi laniger, the only nocturnal indriid, have a gray-brown pelage which is in keeping with the nocturnal habits of this species. However, the eyes are strikingly outlined in white and there is an additional emphasis in terms of a rufous border on

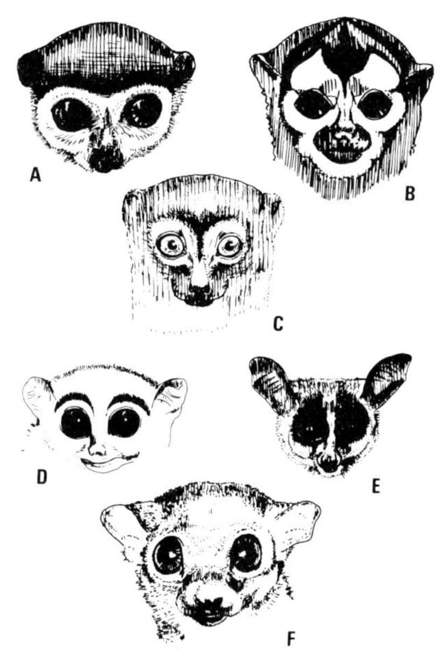

Fig. 20. In *Avahi laniger* (A) the contrast between eye and facial pelage is accentuated by the white surround. *Aotus* (B), the only nocturnal simian, has the same characteristics: both species may be derived from the same diurnal ancestor. *Hapalemur* (C), which are crepuscular, have the same characteristics but less marked. *Tarsius* (D) and the various *Galago* (E) and *Lepilemur* (F) species have not developed the same contrast between eye and face.

the forehead (Fig. 20). It is possible that at night, since the eyes themselves are black, the overall effect produced is that of a nocturnal bird of prey. (*Aotus trivirgatus,* the only nocturnal representative among the simian primates, also has a large white patch surrounding the eyes.)

As a conclusion to this discussion of markings of the nocturnal prosimians, it is worth noting the strange behavior of *Loris* recorded by Still (1905). Thanks to the two large black circles surrounding the eyes, the slender loris, in a dangerous situation, to some extent resembles a cobra in the striking position. The general posture of the head, the typical slow locomotion, and the whistling sound emitted in face of danger all contribute to this resemblance.

Among the diurnal prosimians (Fig. 19), the color of the iris can also be taken into account, since the eye is predominantly in myosis during the daytime. With the sifaka and the indri, the iris is yellow in color and a maximum of contrast is typically achieved by means of a black circle around the eye. However, with the sifakas there is considerable variation in pattern from one subspecies to another, according to the color of the forehead, the ears, and the cheeks. A golden yellow iris associated with a black circle around the eye is also found in *Lemur catta* and *Varecia variegata*. On the other hand the crepuscular gentle lemurs (*Hapalemur*) always have a gray surround to the eyes, while the rest of the face tends to have a rufous tint. The iris has a brown-red color, as with the nocturnal lemur species (Petter and Peyrieras, 1970).

The greatest variability in this feature is found among the *Lemur* species, since some forms (e.g., *Lemur macaco macaco*) have a completely black face, whereas others exhibit various combinations of black, gray, white, rufous, etc.

As a general rule, the eyes are paler in color than the face in the various *Lemur* species and subspecies. In some forms, there is also sexual dimorphism which is particularly evident in different facial markings in males and females. There can be no doubt that this dimorphism is primarily a visual phenomenon. As a result, sexual recognition can be achieved at a distance, and this doubtless contributes to the maintenance of a complex system of social relationships between the members of the group in certain diurnal species. Such visual signals, which are easily perceived at a distance, remove the necessity for actual physical contact (e.g., of an aggressive kind) when they are associated with facial expressions (indicating intimidation, submission, etc.) involved in the maintenance of social organization. In line with this, it is noteworthy that the most conspicuous visual signals are primarily exhibited by males and that their definitive form (maximum length of tufts of colored hairs) only emerges at puberty.

3.3. Vision and Feeding Behavior

The survival of an individual animal, and ultimately of the species itself, is entirely dependent upon the availability of food in the environment. When food is available, the animal must be equipped to perceive it, and this represents the primary role of the sense organs. It is therefore to be expected that, in the course of evolution, there would have been selection for sensory systems finely adapted for the identification of available food.

Much research has been done on the roles of olfaction and taste in detecting and selecting food (Hladik, Chapter 8) and it seems highly likely that vision in the primates has come to play a part in this respect. However, the wide range of functions of the visual apparatus which have emerged in the course of primate evolution have to some extent "liberated" the eyes from a predominantly dietary role. Nevertheless, it is interesting to examine the extent to which food gathering may have influenced the evolution of the eyes, and for this reason it is useful to consider two main dietary categories available to the prosimians in the forest (immobile plant food and mobile animal food), in order to see whether there is any correlation with visual developments. As far as the eye is concerned, improvement in food detection necessitates a system which maximizes the contrast between the food source and the rest of the environment. The distinction between immobile and mobile food (Table III) is hence of considerable importance.

TABLE III

Major Food Preferences in Prosimians in Relation to the Activity Dimension [b]

Species	Activity	Stationary food (plant matter)	Mobile food (mainly arthropods)
Lemur spp.	D or DD or N	+++	
Varecia variegata	D	+++	
Lepilemur spp.	N	+++	
Hapalemur spp.	D	+++	
Cheirogaleus spp.	N	+	+
Microcebus spp.	N	+	++
Phaner furcifer	N	++	+
Allocebus trichotis	N	?	?
Avahi laniger	N	+++	
Propithecus spp.	DD	+++	
Indri indri	DD	+++	
Daubentonia madagascariensis	N	+++	
Galago spp.	N	+	++
Galago elegantulus	N	++	+
Tarsius spp.	N		+++
Loris tardigradus	N	++	+
Nycticebus coucang	N		
Arctocebus calabarensis	N	++	+
Perodicticus potto	N	++	+

[a] Key for abbreviations: As for Table I.

3.3.1. Diet Composed of Stationary Items

This category involves the vast majority of prosimians, since plant components such as leaves, fruits, buds, gums, together with any stationary arthropods resting on them, contribute in some way to most prosimian diets. (The only exceptions are certain nocturnal species.) The main distinction to be expected here is in the different means utilized by diurnal or nocturnal forms in the localization of essentially identical food sources.

During the daytime, maximal contrast in the location of food is achieved through color vision. Even with imperfect color vision, valuable information is added to that obtained from perception of shape and smell. In certain diurnal lemur species, color vision has definitely been demonstrated, but only a few species have been studied and there is considerable variation in the results obtained. To date, no behavioral tests or electrophysiological investigations have been conducted with *Varecia variegata* or any of the indriid species. This is unfortunate, because it would be of particular interest to examine all the diurnal lemur species with a view to finding relationships between color vision, which doubtless vary in degree of expression from species to species, and behavioral/physiological aspects such as the diet (Hladik and Pariente, in prep.; P. Marshak, personal communication).

The nocturnal lemur species which have the same diet as diurnal forms must in some way compensate for their lack of color vision (Silver, 1966; G. F. Pariente, unpublished research) with greater development of olfactory mechanisms. In certain cases, however, it is possible that the forms of the plant items consumed by the herbivorous nocturnal prosimians are sufficiently distinctive to permit recognition of suitable food at a distance. In fact, Charles-Dominique and Hladik (1971), in the course of detailed observation of the feeding behavior of *Lepilemur leucopus* in the south of Madagascar, noted that this species had virtually no difficulty in finding and eating its staple food in the natural environment. But this is not true of those species which eat gums (e.g., *Phaner furcifer*, *Galago senegalensis*, *Galago elegantulus*, *Perodicticus potto*), nor of those which lick nectar or the exudations of certain insects (e.g., *Phaner furcifer*; Petter *et al*., 1971).

There is, of course, a fundamental relationship between feeding behavior, the dispersion of food, and its abundance. For example, in the case of a food source which is scarce and dispersed, locomotion over considerable distances is required and hence visual powers are of prime importance. This applies to *Phaner furcifer*, which doubtless utilize smell as well in the localization of food, but greatly simplify the task of food finding by means of visual memory of points of food availability and the associated establishment of ''circuits'' which are visited regularly every night. It seems likely that, without this aptitude, *Phaner* would not be able to rely on their present diet and that they would be unable to find sufficient food by random exploration each night. Once again, the importance of the brain in a complex environment such as a forest is clearly emphasized.

The final point relating to immobile food sources concerns immobile insects which are encountered at random. It is notable that only the nocturnal prosimian species prey on insects in this way, while the diurnal species are strictly herbivorous. According to A. Schilling (personal communication), it would seem that the detection of immobile insects is largely dependent upon olfaction. For instance, a mouse lemur can detect the odor of a cricket without difficulty even from some distance away. Numerous insect species utilize repulsive odors as a defense mechanism. Yet several prosimian species (e.g., *Loris tardigradus* and *Arctocebus calabarensis*) do not seem to be affected by these odors, and they may even act as attractants in these cases. It is also noteworthy that slender lorises, and perhaps all of the prosimians which move slowly and explore branches systematically, can readily recognize sleeping birds or their eggs and will take them unhesitatingly (Petter and Hladik, 1970).

3.3.2. Diet Composed of Moving Animals

It can be seen from Table III that this kind of diet is only typical of a small minority of prosimian species. The various species of *Galago*, *Microcebus*, *Cheirogaleus*, and (possibly) *Phaner* supplement a basic diet of immobile food items with actively moving small animal prey. Only *Tarsius* feed exclusively on animal prey, and in this case vision plays an extremely important role. In this connection, it should be remembered that the tarsiers are the only nocturnal prosimians which possess a *fovea* and pronounced centralization of the retinal photoreceptors (Rohen and Castenholz, 1967).

It is also important to note that in certain nocturnal prosimian species which are able to hunt mobile animal prey, there has been considerable development of the apparatus of hearing, particularly for the detection of high frequency sound (Pariente, 1974b). In fact, the reduction of the sphere of vision imposed by very low light intensities renders nocturnal vision insufficient for this type of diet, and without detection of the ultrasonic noises produced by insects in flight it would be impossible for such prey to be captured reliably. In such cases, it would seem that vision operates primarily in the regulation of the final actions of prey capture when a flying insect is close at hand. Hearing, on the other hand, alerts the hunting prosimian and guides it toward its prey, modulating the behavior patterns which precede capture. It is not unlikely that *Tarsius* also rely on hearing in this way.

In conclusion, it would seem that the diurnal species have followed an evolutionary pathway leading to a certain degree of color vision which is sufficient for them to effectively locate their food in the forests of Madagascar. On the other hand, the nocturnal species exploit the plants of the forest by means of quite different methods of reconnaissance, doubtless in association with their smaller body size (Hladik and Hladik, 1969; Charles-Dominique, 1971). With the few species which engage in active hunting of live prey, such behavior has only been permitted by development of hearing, since the restricted sphere of vision available to them does not allow exclusive reliance on visual information.

4. VISION AND EVOLUTION: CONCLUSIONS

Although the simian primates are relatively uniform in terms of their adaptations for vision, the prosimians exhibit a variety of anatomical developments which have been selected as a function of various distinctive ecological niches (Collins, 1921; Cartmill, 1975). This is particularly true of the lemurs of Madagascar. Everywhere else, the prosimians have been confined to nocturnal life (Fig. 21), and it is doubtless no chance phenomenon that these nocturnal species have been able to survive to the present day despite the major evolutionary developments which have occurred in other, sympatric mammal groups. This could be taken as evidence that the nocturnal prosimian eye was already sufficiently developed in some species several million years ago to permit the occupation of ecological niches which might *a priori* appear to be unsuitable (i.e., active leaping in trees at night). Olfaction and hearing, which were most probably predominant initially, doubtless provided a point of departure for the evolution of the eye, culminating in animals capable of high calibre visual performance (de Valois and Jacobs, 1968; Spatz, 1970; Harting *et al.*, 1973).

One may ask whether diurnal prosimians existed in areas other than Madagascar at some previous time, but eventually become extinct. This seems to be a distinct possibility (e.g., among the Adapidae of the Eocene), though we of course have no information about the type of eye which might have been present. By analogy with living prosimians, it would appear from study of Early Tertiary prosimian skulls that

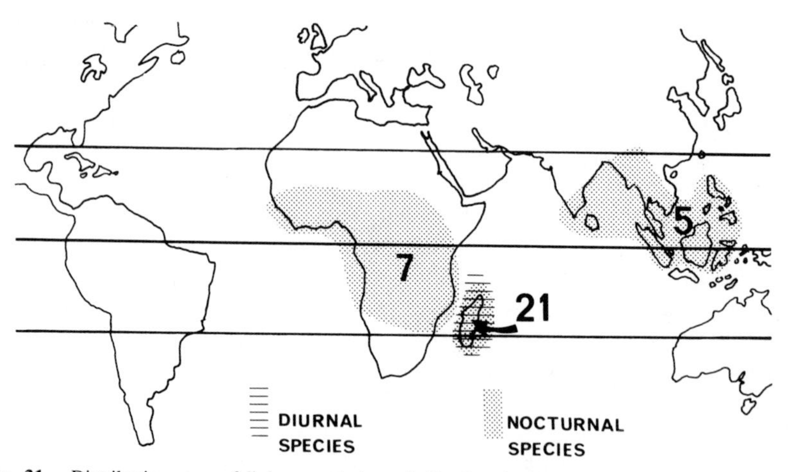

Fig. 21. Distribution map of living prosimians, indicating the occurrence of nocturnal or diurnal behavior. Diurnal species are only found among the Madagascar lemurs (11 out of 21 species). All other species (10 in Madagascar, 7 in Africa, 5 in Asia) are strictly nocturnal.

there were both diurnal and nocturnal forms outside Madagascar (Mahé, 1972). In fact, the only indication that we have of diurnal or nocturnal adaptation is based on the relative size of the orbit, and it has been shown above that the absolute size of the eye is of great importance in terms of the sphere of vision within which any prosimian moves and of physiological refinements in visual processes.

Overall, it seems to be justifiable to regard the eyes of living prosimians, even in the most "primitive" species, as fulfilling all visual requirements. This raises the problem of the evolutionary potential of a given organ, such as the eye, rather than providing clear-cut answers about the pathways followed in evolution. Nevertheless, it can be stated that the diurnal Madagascar lemurs could easily be derived from nocturnal ancestors, or from ancestors which simply had particular features of advantage in nocturnal vision. These diurnal species still exhibit marked advantages for nocturnal vision. On the other hand, the modern nocturnal prosimians are also quite capable of leaping around in the forest in full daylight. If one goes on to consider a hypothetical ancestor common to all primates, the same interpretation can be applied, except that the simians have evolved further for diurnal life (Glickman *et al.*, 1965; see also Fig. 21).

Since the nocturnal prosimians can in fact see quite well during the daytime, one gains the impression that the eye in some way retains a degree of plasticity which would permit fairly rapid transition from nocturnal to diurnal life (or vice versa). For this reason, instead of attempting to describe a hypothetical ancestor as either nocturnal or diurnal, it would seem to be more logical to think of early ancestors which needed to see well both by day and by night. Such a requirement would have led to the evolution of a dual-purpose eye at the very outset, and such an eye would have subsequently become limited in one direction or another because of specific specializations. The general environment and specific phenomena such as predation, diet and adaptations for social life would have been the principal factors influencing evolution of new features (Bearder and Doyle, 1974).

If this dual-purpose hypothesis is well founded, it should perhaps be possible to find traces of this ambivalence in the visual systems of primitive species lacking obvious specializations. This would accord well with the concept of a small-bodied ancestral form, perhaps arboreal in habit, with well-developed vision and an appreciable amount of activity in the daytime, at the origin of the order Primates. This ancestral form would already have been able to decode both the intensity and frequency of light involved in retinal images.

The complexity of the visual processes evident in living primates indicates that a long period of evolution would have been necessary prior to the appearance of the most ancient prosimians known from the fossil record (Walker, 1967). By this latter stage, the various means of exploitation of light stimuli would already have been crudely defined and specializations for ecological niches with increasingly restricted limits would have provided a broad basis for the range of visual adaptations seen among living primates. From this stage onward, it is very likely that highly refined visual mechanisms, such as dynamic binocular vision, precision in accommodation and fixation, maximal definition of the image, adaptation to light contrasts, etc., would have slowly made their appearance (Jannerod, 1974). All of these specializations depend, as we have seen, primarily on the development of the brain (Hecker and Grunwald, 1926; Radinsky, 1970).

Other developments may also have occurred in the evolution of the eye, for example, permitting better adaptation to nocturnal life. Relative increase in the surface of the cornea, elongation of the photoreceptors, elaboration of a *tapetum lucidum* with complex functions, and (above all) improvements in the nervous connections in the retina itself would all have emerged in parallel with refinement of central nervous centers.

Hence, according to this view, the modern prosimians are extremely well adapted to their environments in visual terms, whether they are nocturnal or diurnal in habit, and the necessary adaptations have a long evolutionary history. Naturally, with the exception of nocturnal vision, it is still true that if the prosimians are compared to the simian primates, their visual apparatus is generally somewhat inferior. Yet the most difficult problem of interpretation lies not with the prosimians, but with the simians. The reasons for their visual superiority remain obscure. It remains for us to establish the particular ecological niches, and the associated selection pressures, which are responsible for the parallel, but more extensive, visual evolution of the simians leading to the advanced condition now found in the monkeys, apes, and man.

ACKNOWLEDGMENTS

The editors would like to thank Dr. Lee R. Wolin, Laboratory of Neuropsychology, Ohio Department of Mental Health and Mental Retardation, and Dr. Matt Cartmill, Department of Anatomy, Duke University Medical School, for their assistance in resolving some difficult technical queries referred to them after the untimely death of the author.

REFERENCES

Alfieri, R., Pariente, G. F., and Sole, P. (1974). *Proc. Int. Soc. Clin. E.R.G. Symp., 12th, 1970.*

Andrew, R. J. (1963). *Science* **142**, 1034–1041.

Andrew, R. J. (1964). *In* "Evolutionary and Genetic Biology of Primates" (J. Buettner-Janusch, ed.), Vol. 2, pp. 227–309. Academic Press, New York.

Ashley Montagu, M. F. (1943). *Nature (London)* **152**, 573–574.

Autrum, M. (1960). *Fortschr. Zool.* **12**, 176–205.

Barlow, H. B., Blackmore, C., and Pettigrew, J. D. (1967). *J. Physiol. (Paris)* **193**, 327–342.

Bauchot, R., and Stephan, H. (1966). *Mammalia* **30**, 160–196.

Bearder, S., and Doyle, G. A. (1974). *J. Hum. Evol.* **3**, 37–50.

Bierens de Haan, J. A., and Frima, M. J. (1930). *Z. Vergl. Physiol.* **12**, 603–631.

Bresson, Y., Bellosi, A., and Godet, R. (1969). *C.R. Séances Soc. Biol. Ses. Fil.* **163,** 270.

Cartmill, M. (1975). *Science* **187**, 456.
Chacko, L. W. (1954). *J. Anat. Soc. India* **3**, 11–23.
Charles-Dominique, P. (1968). *In* "Entretien de Chizé. I." (R. Canivenc, ed.), pp. 2–5. Masson, Paris.
Charles-Dominique, P. (1971). *Biol. Gabonica* **7**, 121–228.
Charles-Dominique, P., and Hladik, C. M. (1971). *Terre Vie* **25**, 3–66.
Collins, E. T. (1921). *Trans. Ophthalmol. Soc. V. K.* **41**, 10–90.
Coss, R. (1972). Ph.D. Thesis, University of Reading (unpublished).
Dartnall, H. J. A., Arden, G. B., Ikeda, H., Luck, C. P., Rosenberg, M. E., Pedler, C. M. H., and Tansley, K. (1965). *Vision Res.* **5**, 399–424.
Detwiller, S. R. (1924). *J. Comp. Neurol.* **37**, 481–489.
Detwiller, S. R. (1939). *Anat. Rec.* **74**, 129–145.
Detwiller, S. R. (1940). *J. Opt. Soc. Am.* **30**, 42–50.
de Valois, R. L., and Jacobs, G. H. (1968). *Science* **162**, 533–540.
de Valois, R. L., and Jacobs, G. H. (1971). *Behav. Nonhum. Primates* **3**, 107–158.
Ehrlich, A., and Clavin, W. H. (1967). *Psychon. Sci.* **9**, 509–510.
Fobes, J. L., Ehrlich, A., and Williams, K. (1971). *Lab. Primate Newsl.* **10**, 7–9.
Glickman, S., Clayton, K., Schiff, B., Guritz, D., and Messe, L. (1965). *J. Genet. Psychol.* **106**, 325–335.
Goffart, M., Missotten, L., Faldherbe, J., and Watillon, M. (1976). *Arch. Int. Phys. Biochm.* **84**, 493–516.
Graveline, J., Quere, M. A., and Bert, J. (1967). *Ophthalmologica* **154**, 143–150.
Hall-Craggs, E. C. B. (1974). *In* "Prosimian Biology" (R. D. Martin, G. A. Doyle, and A. C. Walker, eds.), pp. 829–845. Duckworth, London.
Harting, J. K., Glendenning, K. K., Diamond, I. T., and Hall, W. C. (1973). *Am. J. Phys. Anthropol.* **38**, 383–392.
Hartridge, H. (1919). *J. Physiol. (Paris)* **53**, 6–8.
Hassler, R. (1966). *In* "Evolution of the Forebrain" (R. Hassler and H. Stephan, eds.), pp. 419–434. Thieme, Stuttgart.
Hecker, P., and Grunwald, E. (1926). *C. R. Séances Soc. Biol. Ses. Fil.* **94,** 1358–1361.
Hill, W. C. O. (1935). *Nature (London)* **135**, 584.
Hladik, A., and Hladik, C. M. (1969). *Terre Vie* **1**, 25–117.
Hubel, D. H., and Wiesel, T. N. (1965). *J. Neurophysiol.* **28**, 229–289.
Hubel, D. H., and Wiesel, T. N. (1968). *J. Physiol. (Paris)* **195**, 215–243.
Jannerod, M. (1974). *Recherche* **41**, 24–32.
Jolly, A. (1966). "Lemur Behavior: A Madagascar Field Study." Univ. of Chicago Press, Chicago, Illinois.
Kolmer, W. (1930). *Z. Anat. Entwicklungs-Gesch.* **93,** 699–722.
Laemle, L. K. (1968). *Anat. Rec.* **160**, 380–381.
Laemle, L. K., and Noback, C. K. (1970). *J. Comp. Neurol.* **138**, 49–61.
Lane, R. H., Allman, J. M., Kaas, J. H., and Miezin, F. M. (1973). *Brain Res.* **60**, 335–349.
Le Grand, Y. (1942). *C. R. Hebd. Seances Acad. Sci.* **214**, 180–182.
Le Grand, Y. (1956). "Optique physiologique," Vol. 3, pp. 1–390. Masson, Paris.
Lightoller, G. S. (1934). *Proc. Zool. Soc. London* **2**, 259–309.
Lindsay-Johnson, G. (1897). *Proc. Zool. Soc. London,* 183–188.
Lindsay-Johnson, G. (1900). *Philos. Trans. R. Soc. London, Ser. B* **194**, 1–82.
Luck, C. P. (1958). *Nature (London)* **181**, 719.
Luck, C. P. (1963). *Biochem. J.* **89,** 78. (abstract).
Mahé, J. (1972). Doctoral Thesis, University of Paris (unpublished).
Martin, R. D. (1972). *Philos. Trans. R. Soc. London, Ser. B* **264**, 295–352.
Menaker, M. (1972). *Sci. Am.* **226**, 22–30.

Mervis, R. F. (1974). *Anim. Learn. Behav.* **2**, 238–240.
Moynihan, M. (1964). *Smithson. Misc. Collect.* **146**, 1–84.
Nakamura, S., and Tigges, J. (1970). *Z. Zellforsch. Mikrosk. Anat.* **106**, 550–555.
Noback, C. R., and Shriver, J. E. (1969). *Ann. N.Y. Acad. Sci.* **167**, 118–128.
Ordy, J. M., and Samorajski, T. (1968). *Vision Res.* **8**, 1205–1225.
Pagès, E. (1978). *In* "Recent Advances in Primatology, Vol. 3: Evolution" (D. J. Chivers and K. A. Joysey, eds.), pp. 171–177. Academic Press, London.
Pariente, G. F. (1970). *C. R. Hebd. Seances Acad. Sci.* **270**, 1404–1407.
Pariente, G. F. (1973). *Optometry* **5**, 7–21.
Pariente, G. F. (1974a). *In* "Prosimian Biology" (R. D. Martin, G. A. Doyle, and A. C. Walker, eds.), pp. 183–198. Duckworth, London.
Pariente, G. F. (1974b). *Mammalia* **38**, 1–6.
Pariente, G. F. (1975a). *Mammalia* **39**, 487–497.
Pariente, G. F. (1975b). *J. Physiol. (Paris)* **70**, 637–647.
Pariente, G. F. (1976). *Vision Res.* **16**, 387–391.
Pedler, C. (1963). *Exp. Eye Res.* **2**, 189–195.
Petter, J. J. (1962). *Mem. Mus. Nat. Hist. Natl., Paris, Ser. A* **27,** 1–146.
Petter, J. J., and Hladik, C. M. (1970). *Mammalia* **34**, 394–409.
Petter, J. J., and Peyrieras, A. (1970). *Terre Vie* **3**, 356–382.
Petter, J. J., and Peyrieras, A. (1972). *J. Hum. Evol.* **1**, 379–388.
Petter, J. J., Schilling, A., and Pariente, G. F. (1971). *Terre Vie* **3**, 287–327.
Petter-Rousseaux, A. (1970). *Ann. Biol. Anim., Biochim., Biophys.* **10**, 203–208.
Pirenne, M. (1961). *Endeavour* **20**, 197–209.
Pirie, A. (1959). *Nature (London)* **183**, 985–986.
Polyak, S. (1957). "The Vertebrate Visual System." Univ. of Chicago Press, Chicago, Illinois.
Prince, J. H. (1953). *Br. J. Physiol. Optics* (Oct.), 143–153.
Prince, J. H. (1956). "Comparative Anatomy of the Eye." Thomas, Springfield, Illinois.
Radinsky, L. B. (1970). *Adv. in Primatol.* **1,** (C. R. Noback and W. Montagna, eds.), pp. 209–224.
Randolph, M. (1971). *J. Comp. Physiol.* 74–115.
Richard, A. (1973). Ph.D. Thesis, Univ. of London (unpublished).
Rochon-Duvigneaud, A. (1943). "Les yeux et la vision des Vertébrés." Masson, Paris.
Rohen, J. W., and Castenholz, E. A. (1967). *Folia Primatol.* **5**, 92–147.
Schilling, A. (1974). *In* "Prosimian Biology" (R. D. Martin, G. A. Doyle, and A. C. Walker, eds.), pp. 347–362. Duckworth, London.
Schultz, A. H. (1940). *Am. J. Phys. Anthropol.* **26**, 389–408.
Serbanescu, T., Godet, R., Orsini, J. C., and Naquet, R. (1968). *J. Physiol. (Paris)* **60**, 391–398.
Silver, P. H. (1966). *Vision Res.* **6**, 153–162.
Spatz, W. B. (1970). *Acta Anat.* **75**, 489–520.
Stephan, H. (1969). *Proc. Int. Congr. Primatol. 2nd, 1968* **3**, pp. 34–42.
Still, J. (1905). *Spolia Zeylan.* **3**, 155.
Sussman, R. W. (1972). Ph.D. Thesis, Duke University, Durham, North Carolina (unpublished).
Tansley, K. (1956). *Br. J. Ophthalmol.* **40**, 178–182.
Tenaza, R., Ross, B., Tanticharoenyos, P., and Berkson, G. (1969). *Anim. Behav.* **17**, 664–670.
Tigges, J., and Tigges, M. (1969). *J. Comp. Neurol.* **137**, 59–70.
Tigges, J., and Tigges, M. (1970). *J. Comp. Neurol.* **138**, 87–102.
Treff, H. A. (1967). *Z. Vergl. Physiol.* **54**, 26–57.
Vincent, F. (1973). *Ann. Fac. Sci. Cameroun* **13**, 163–167.
Walker, E. P. (1938). *Smithson. Inst., Annu. Rep.* 349–360.
Walker, A. C. (1967). *In* "Pleistocene extinction" (P. S. Martin and H. E. Wright, eds.), Vol. 6, pp. 425–432. Yale Univ. Press, New Haven, Connecticut.
Walls, G. L. (1942). "The Vertebrate Eye and its Adapting Radiation." Cranbrook Inst. Sci., Bloomfield Hills, Michigan.

Ward, J. P., and Doerflein, R. S. (1971). *Psychon. Sci.* **23**, 43–45.
Winckler, G. (1934). *Arch. Anat., Histol. Embryol.* **14**, 303–386.
Wolin, L. R., and Massopust, L. C. (1967). *J. Anat.* **101**, 693–699.
Wolin, L. R., and Massopust, L. C. (1970). *Adv. in Primatol.* **1**, (C. R. Noback and W. Montagna, eds.), pp. 1–27.
Woolard, A. M. (1927). *Proc. Zool. Soc. London* **1**, 1–17.

Chapter 11

Olfactory Communication in Prosimians

A. SCHILLING

1. INTRODUCTION

To understand fully the mechanisms of animal communication is quite impossible in the sense that the observer, unable to enter into the system himself, can only relate the emission of what he believes to be a signal to what he believes to be the effect produced in the supposed receiver. To overcome the inherent complexity of such a task requires an objective inventory of the means of signaling (origin, nature, and characteristics), decoding the system (conditions of utilization, specificity, and mode of action), and separating the message itself from its redundancies and from associated repercussions in the central nervous system during its integration. In the sphere of olfactory communication this area of investigation, initially concerned with the invertebrates, has recently been directed toward the mammals, including

ISBN 0-12-222150-8

the primates (Mykytowytcz, 1970; Müller-Schwarze, 1971; Michael *et al.*, 1971; Signoret, 1970; Johnston, 1975), but has only just begun in the prosimians.

With several possible exceptions like *Phaner, Lepilemur*, and *Indri*, prosimians devote a large part of their time to "olfactory" behavior, depositing scent and investigating the scent of others, and the functional anatomical structures involved are particularly well developed. It is of note that the anatomy of the nasal apparatus constitutes a major criterion for separating prosimians (with the exception of *Tarsius*) from the simian primates: Strepsirhini (complex rhinarium) are macrosomatic; Haplorhini (simple rhinarium) are microsmatic.

In addition, even though the use of scent as a means of communication is not confined solely to the prosimian primates (Hill, 1938, 1953), nevertheless this group possesses a diversity of means of olfactory communication of an extraordinary behavioral and anatomical richness.

This discussion will concentrate largely on olfactory communication within a social context, i.e., the use of odoriforous substances designed to elicit responses from one or more conspecifics or, in a larger sense, to obtain some social advantage for the emitter. The role of olfactory marking in orientation, mentioned by Sauer and Sauer (1963) in *Galago senegalensis* and by Seitz (1969) in *Nycticebus coucang*, will be briefly discussed. Unpublished experiments by A. Schilling with *Microcebus murinus* suggest that, even for complicated routes, olfaction appears to play only a minor role. Bearder (1969) and Charles-Dominique (1971, 1974a, 1977) present evidence strongly suggesting that in the wild *G. senegalensis* and *Perodicticus potto*, respectively, rely very little on olfactory cues in orientation (see also Pariente, Chapter 10). The importance of olfaction in locating food will not be discussed.

2. FUNCTIONAL AND ANATOMICAL BASIS OF OLFACTORY COMMUNICATION

2.1. Olfactory Structures in Prosimians

The olfactory apparatus in prosimians consists of a sensory mucosa similar to that of other vertebrates: the olfactory receptors send their axons directly to the olfactory bulbs, passing through the cribriform plate of the ethmoid bones by several *foramina* in the Lorisiformes and Lemuriformes, and by only a single foramen in *Tarsius*.

2.1.1. Peripheral Structures

If olfactory acuity is related to the surface of the olfactory mucosa (Negus, 1958; Cave, 1967), and if this depends on the greater or lesser extension of the arch and the walls of the nasal fossa, as well as on the development of bony processes characteristic of the nasal fossa (the turbinals), then the following conclusions pertain.

TABLE I

The Two Main Morphological Types of Prosimian Nasal Fossa

Nasal fossa	Lemuriformes and Lorisiformes	Tarsiiformes
General type	Insectivore	Platyrrine
Nasal cavity	Long (about ½ of head length) and more or less low	Short (about ⅓ of head length) and high
Lamina transversa	Present	Absent
Posterior olfactory recess	Present	Absent
Ethmoturbinals		
Ento	4 or 5	2
Ecto	2	0
Nasoturbinal	Well-developed	More or less reduced to agger nasi
Maxilloturbinal	Well-developed	Reduced to a narrow fold
Olfactory mucosa extension	Lateral wall on most of its posterior half	0
	Most of the posterior part of the septum	Spot in front of ethmoturbinal II
	All ethmoturbinals	Ethmoturbinals and nasoturbinals

2.1.1.1. Special Position of the Tarsiiformes. The anatomy of the nasal fossa separate the Lorisiformes and Lemuriformes so distinctly from the Tarsiiformes that Cave (1967) refers to those of the former as "nonprimate nasal fossa," and those of the latter as "primate nasal fossa." There is less difference between the nasal formations of the tenrec (an insectivore) and *Arctocebus*, or between *Tarsius* and *Leontideus* (a New World monkey), than there is between that of *Tarsius* and *Arctocebus*. Moreover, Lorisiformes and Lemuriformes have generally conserved a nasal structure quite similar to that of other mammals referred to as macrosmatic, the insectivores in particular, while *Tarsius* have acquired, either by reduction or loss of the interorbital region of the fossa, an architecture curiously resembling that of certain New World monkeys, e.g., *Callithrix, Saimiri* (Cave, 1967). The differences between these two morphological types are summarized in Table I. For Cave (1967) the presence of an olfactory posterior recess, due to the presence of the sphenoidal lamina transversa, is the distinguishing characteristic of the "nonprimate nasal fossa." The "primate" type, on the other hand, is characterized primarily by the existence of an interorbital septum.

2.1.1.2. Lorisiformes and Lemuriformes. The classification of the typical prosimian nasal fossa as an insectivore type is evidently only a generalization. Loo

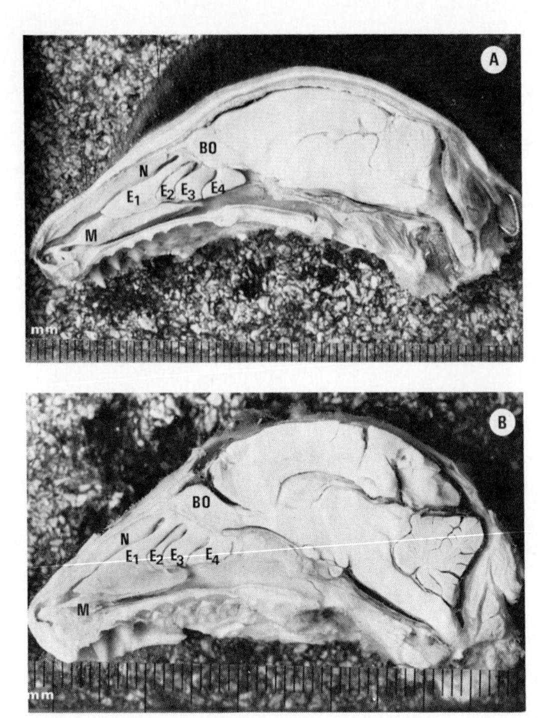

Fig. 1. (A) Parasagittal section of the head of a lemuriform, *Cheirogaleus major*. (B) Parasagittal section of the head of a lorisiform, *Galago elegantulus*. E_1–E_4, ethmoturbinals; N, nasoturbinal; M, maxilloturbinals; BO, olfactory bulb.

(1973), for example, demonstrated a relative reduction of the nasal fossa of *Nycticebus coucang* compared to that of *Tupaia glis*. On the other hand, the nasal structures of the Lorisiformes and Lemuriformes are far from identical. In the only available comparative study Kollman and Papin (1925) clearly separate the two groups. The Lorisiformes are distinguished from the Lemuriformes by greater development of the ethmoturbinals and, particularly, the first ethmoturbinal (Fig. 1). In addition, while the Lorisiformes are homogeneous in this respect, the Lemuriformes are not and special status must be given to the Cheirogaleidae which have not yet been fully studied.

2.1.1.3. Structure, Physiology, and Behavior. In the absence of sufficient evidence on comparative olfactory physiology in the prosimians, caution must be exercised in correlating structure with function or pattern of life. If eventually the increased surface area of the olfactory mucosa can be shown to be linked to increased olfactory acuity, it would remain only to compare the surface area of one species with that of another, and to demonstrate the behavioral implications of the remaining physiological differences that follow from these variations. Charles-Dominique (1977) presents some evidence to show that the sense of smell is extremely well developed in *Arctocebus calabarensis* and *Perodicticus potto*, the latter being able, for instance, to locate a cricket hidden at a distance of 1 m. This olfactory ability is at least as well developed in some Lemuriformes. For example, *Cheirogaleus major* are capable of detecting, among five glass containers, the one in which a cricket has been placed, provided that the insect has been inside for at least 45 seconds. It seems that in spite of increased development of the visual sense (see Pariente, Chapter 10) the diurnal prosimians possess olfactory structures which have not degenerated (Kollman and Papin, 1925). Finally, although it is known (Woollard, 1925; Cave, 1967) that the olfactory apparatus of *Tarsius* is reduced (see Table I) little information is available about their behavior (but see Niemitz, Chapter 14). It should be mentioned, however, that among the platyrrhines, with which *Tarsius* share a similar morphology, numerous behaviors are linked to olfaction (Nolte, 1958; Epple, 1970, 1973).

2.1.2. Central Structures

Stephan and Andy (1964) studied the olfactory bulb in three species of lorisid, *Loris tardigradus*, *Perodicticus potto*, and *Galago demidovii*, and showed that its structure seems to be intermediate between that of the insectivores and platyrrhines. The authors indicate, however, that the olfactory bulb in *G. demidovii* is relatively more developed than that of the two lorisines. This finding is somewhat paradoxical in that available behavioral evidence (Charles-Dominique, 1977) suggests that the lorisines are more dependent on the olfactory sense than are the galagines; for example, in prey-catching.

The secondary olfactory projections have been studied in *Galago crassicaudatus* by Ferrer (1969). The efferent fibers of the olfactory bulb target or relay in the anterior olfactory nuclei, the olfactory tubercle, the prepyriform, pyriform, and periamygdaloid cortex, as well as in the corticomedian zone of the amygdaloid complex. These projections are thus distributed exactly as in other mammals ranging from marsupials to the macaque (Ferrer, 1969). It seems currently accepted (Le Magnen, 1960; Adey, 1970; MacLeod, 1971) that, in addition, the input from the olfactory bulb by various pathways touches the anterior hypothalamus, other nuclei of the amygdaloid complex, the prefrontal neocortex, the septum (which is an important relay to the thalamus), the hypothalamus, certain parts of the brain stem, and the hippocampus. Conversely, a series of centrifugal fibers terminates in the olfactory bulb (MacLeod, 1971). From the behavioral point of view these neurophysiological findings are interesting in that they show that the rhinencephalic functions involve more than olfactory activity and that other cerebral centers, particularly autonomic centers, enter into close relationship with the olfactory center (see Fig. 17).

A brief summary of the interrelationships between the olfactory centers and nonolfactory centers (cerebral, medullary, and visceral) follows. Olfactory–

alimentary interrelationships have been discussed by Le Magnen (1970); olfactory–sexual interrelationships have been discussed by Rosedale (1945), Faure (1956), Signoret and Mauléon (1962), Heimer and Larsson (1967), Pfaff and Pfaffman (1969), Whitten and Bronson (1970), Scott and Pfaff (1970), Pfaffman (1972), and Michael *et al*. (1972); and the interrelationship between the olfactory sphere and emotional states have been discussed by Haug (1970), Karli (1971), Ropartz (1967b), Douglas *et al*. (1969), Walker (1969), and Haug and Ropartz (1973). In all these interrelationships Le Magnen (1960) emphasizes the importance of overlapping sensory cues as well as the emotional nature of responses to olfactory stimuli. The variety of terms employed by physiologists to describe the emotional nature of the responses in which the rhinencephalon plays a role (appetitiveness, fear, aggression, etc.) illustrates the complexity of olfactory communication and the resulting problems of interpretation; the entire psychophysiological state of the subject at any one time testifies to the complex relationship between the internal and external environments.

2.2. Odoriforous Substances

To confine olfactory communication to "scent marking" and the origin of olfactory signals to "scent marks" is an oversimplification resulting in the exclusion of a large part of the olfactory repertoire which does not occur as a precise, active, or stereotyped behavior. The sexual chase of *Propithecus*, for example, involves an entire sequence of olfactory behaviors, receptive (sniff-approach) and emittive ("endorsing," urine-marking of the neck, Richard, 1973). However, from the point of view of the message itself very little information is available on the precise nature of the exchange which occurs between two individuals.

The male touches the anus of the female with his nose, but also sniffs her anogenital region or urine-marks, before marking in turn. This apparently simple olfactory episode poses a number of distinct problems. First, from the point of view of information content, the female scent deposited is complex, not only in its origin—urine, secretions of the accessory glands, vaginal secretions (demonstrated in the macaque by Michael and Kerverne, 1970), secretions of the genital-labial glands, as well as of the perineal or anal glands; but also in its nature—urinary pheromone, chemical modification of the vaginal milieu, attractive secretion, as in the case of *Perodicticus potto* (Charles-Dominique, 1966); and, finally, in its mode of action—whether the female signal has the same significance when the act of deposition is directly perceived by the male, or whether it is subsequently perceived as a scent mark. If it is, as it probably is, of the "signalling" type (Bronson, 1971), how is the behavioral response elicited, etc.? Second, from the point of view of its expression, is the olfactory response of the male to the female stimulus represented by a quantitative increase or a qualitative difference in the secretion, or a general increase in marking behavior, or by a modification in the frequency and kind of marking behavior itself?

TABLE II

Odorant or Chemical Material Believed to Play a Role in Prosimian Communication[a]

Prosimian	Saliva + − other buccal secretions	Secretions: Head, neck, and chest glands	Secretions: Forelimb glands	Secretions: Volar glands	Secretions: Ano-genital glands	Excretions: Urine	Excretions: Feces
Cheirogaleidae							
Microcebus murinus	+ −	−	−	?	+ (10–11)	+	+
Microcebus coquereli	+ (2)	?	−	?	+ (10)	+	+
Cheirogaleus medius	?	−	−	?	+ (10)	+	+
Cheirogaleus major	?	−	−	+ −	+ (10)	+	+
Phaner furcifer	+ −	+ (4)	−	?	?	0	0
Lemuridae							
Lemur catta	0	+ (1)	+ (6–8)	+ −	+ (10)	0	0
Lemur fulvus	+ −	+ (1)	+ −	+ (9)	+ (11)	+ −	+ −
Lemur macaco							
Varecia variegata	?	+ (4)	?	?	+ −	0	?
Hapalemur griseus	+ −	−	+ (6–8)	?	+ (10)	+	0
Hapalemur simus	?	+ (4)	+ (7)	?	?	?	?
Lepilemur sp.	?	?	−	?	+ −	+	+ −
Indriidae							
Avahi laniger	?	+ (3)	−	?	?	?	?
Propithecus verreauxi	?	+ (4)	−	?	+ (10)	+	+
Indri indri	?	+ −	?	?	+ −	+ −	?
Daubentonia madagascariensis	?	−	−	+ −	?	+	?

(*continued*)

TABLE II *(continued)*

Prosimian	Saliva + − other buccal secretions	Secretions: Head, neck, and chest glands	Secretions: Forelimb glands	Secretions: Volar glands	Secretions: Ano-genital glands	Excretions: Urine	Excretions: Feces
Galaginae							
Galago demidovii	+ −	−	−	+ −	+ (10)	+	+
Galago senegalensis	+	?	−	+ −	+ (10)	+	+ −
Galago alleni	+	−	−	+ −	+ (10)	+	0
Galago crassicaudatus	+ −	+ (5)	−	+ −	+ (10)	+	0
Galago elegantulus	?	−	−	?	+ (10)	+	0
Lorisinae							
Perodicticus potto	0	−	−	+ −	+ (10)	+	0
Nycticebus coucang	+ −	−	0 (2)	+ −	+	+	+
Arctocebus calabarensis	?	−	−	?	+ (10)	+	+
Loris tardigradus	?	−	0 (2)	+ −	?	+	?
Tarsius	+	−	−	+ −	+ −	+	?

[a] +, Actively dispersed material (n° in parentheses; see Fig. 3); + −, material thought to be dispersed actively but sporadically; 0, material not actively dispersed; −, no known material; ?, no information.

All this illustrates that olfactory communication extends far beyond the simple notion of scent-marking, which may play an important role without provoking a response, i.e., by simply providing information. Information may even pass in the absence of observable emitting behavior, in which case it may be completely unnoticed by the observer. This situation may be simplified as follows for both olfactory emission and reception: (1) active or even "directed" olfactory communication occurring in a scent-sniffing dyadic situation; and (2) nondirected olfactory communication occurring spontaneously as in normal respiration, incidental deposition (feces or urine, sweat, and various glandular products), and aerial diffusion.

2.2.1. Origin and Nature of Odoriforous Material

2.2.1.1. Scent Marks. Eisenberg and Kleiman (1972) stress, on good grounds, the deep-rooted phylogenetic nature of mammalian marking behavior (which is too often restricted in the literature to glandular secretions) in the physiology of excretory behavior—defecation, urination, salivation, etc. This is also true of the prosimians which, in addition, possess some unique glandular structures, including those of the sweat glands, as well as excretory functions, which play an important role in scent deposits, as in *Propithecus*, for instance (Jolly, 1966b; Richard, 1973).

The sources of odor subserving scent-marking may be divided into five categories:

a. Urine. All of the Lorisiformes extensively mark with urine. For the Lemuriformes the situation is more complex (see Table II) since, among the Cheirogaleidae, Lemuridae, and Indriidae, some species practice urine-marking regularly, while others do so rarely (*Phaner*, *Lemur*, *Indri*). The active dispersion of urine has also been reported in *Tarsius* (Wharton, 1950) and *Daubentonia* (Petter and Peyrieras, 1970b).

b. Feces. The active deposit of feces is rare (Table II) except in the genus *Cheirogaleus* in which this appears to be an important behavior, at least in terms of the amount of time devoted to it (Ilse, 1955; Petter, 1965; Schilling, 1978). Indeed in *Cheirogaleus medius* and *C. major*, the recessed anatomical position of the anus, in relation to the first caudal vertebra (Hill, 1953) and its protuberant formation (a sort of cutaneous swelling, distinctly protruding in relation to the ventral plan of the tail), forms an efficient marking apparatus perfectly adapted to "dragging" defecation (see Fig. 6). In the following species defecation appears to be only sporadically involved in scent-marking behavior: *Microcebus murinus* (Andrew, 1964), *M. coquereli* (E. Pagès, personal communication), *Galago demidovii* (Vincent, 1969), *Propithecus verreauxi* (Jolly, 1966b), and *Nycticebus coucang* (Tenaza *et al.*, 1969). In other species the same behavior, such as in *L.f. albifrons* (Petter, 1965), or the localized deposit of feces by *Hapalemur*, *Lepilemur* (Charles-Dominique and Hladik, 1971), and *Phaner* (M. Grange, personal communication), is even less open to exact interpretation. In this regard P. Charles-Dominique (personal communication) makes two points: first, the possibility of the behavior in question being distorted by artificial conditions of captivity, which seems to be the case in *Phaner furcifer*; second, the fact that the mass of deposits found on certain branches in the territory of *Lepilemur* could be explained by their habit of remaining stationary at particular ''surveillance'' points for long periods of time (Charles-Dominique and Hladik, 1971). Finally, as will be seen below, it seems important to take into consideration what may be defined otherwise as ''incidental deposition.'' In the case of feces the stamping or smearing by the anus during marking, which is essentially glandular in nature, may lead the observer to regard this behavior as ''marking'' when, in fact, it is merely stereotyped defecation or, at least, marking in which the feces *per se* do not necessarily play any role in communication.

Even more so than for urine the nature of odoriforous substances in the feces, which can serve as messages, is as yet unknown. If such odor signals exist, they may possibly be described as odors peculiar to the individual resulting from the digestive process itself, or by the mixture of feces with particular glandular secretions. Hill (1955) notes that the feces of *Tarsius*, normally odorless, release a strong smell during the reproductive period. Among the following species no clearly differentiated gland has been found in the internal region of the anal canal: *Lemur catta*, *Galago crassicaudatus*, *Microcebus murinus*, *Loris tardigradus*, *Nycticebus coucang*, *Perodicticus potto* (Schaffer, 1924; Ortman, 1958, 1960), *L. mongoz*, *L. catta*, *Hapalemur* (Hill, 1958), and *Cheirogaleus major* (Y. Rumpler, personal communication). Possible exceptions, noted by the same authors, are *G. crassicaudatus*, which have a glandular concentration in the area of the external sphincter; *L. catta*, which have a layer of remarkably intricate glands in the muscular tissue of the anal sphincter, and in which the evacuating canals open into a cul-de-sac directed toward the rectum; and *M. murinus*, in which a residual proctodeal gland is found between the sphincter and the rectum.

c. Saliva and secretions of the labial glands. There is no evidence that saliva is involved in olfactory communication or that it is part of the olfactory or vomeronasal complex (Schilling, 1970). In addition, except for *Galago senegalensis* (Yasuda *et al.*, 1961) and the lemurs (Montagna, 1962), there is little information concerning the possible role of the labial gland in olfactory communication, though Bolwig (1960) suspects that it has such a role in some prosimians. Several lines of evidence suggest that this problem merits further investigation. First, dispersion of saliva has been observed in several prosimians, in *G. senegalensis* (Doyle, 1974a), *G. alleni* (M. Grange, personal communication), *Microcebus coquereli* (Pagès, in press) and in *Tarsius bancanus* (Niemitz, 1974). Of interest too are the hypertrophied salivary glands in *Indri* (Pollock, 1975). Second, a discoloration has been shown to exist at the internal angle of the labial commissure in *M. coquereli* (E. Pagès, personal communication). Finally, the wiping of the labial commissures or rubbing of the muzzle is relatively frequent in prosimians (see Table II).

d. Glandular secretions of the anogenital region. The anogenital region is a source of complex odors due to a rich external cutaneous glandular apparatus (glands on the margin of the anus, the perineum, and the external genitalia), to which may be added all the secretions of the accessory glands of the reproductive apparatus and those of the internal anal canal mentioned above. Discussion will be limited to the glands of the external genitalia and of the periphery of the anus (Table II).

i. The glands of the genitalia. These glands, generally apocrine glands, are located on the surface of the scrotum in males and around the vulva in females (Fig. 2A). According to Charles-Dominique (1977) these formations, which may be more or less well developed, are uniform in the Lorisidae and Lemuridae of both sexes. They are especially evident in males, in *Perodicticus potto* (Hill, 1953; von Fiedler, 1959; Montagna and Ellis, 1959), *Galago demidovii* (Machida and Giacometti, 1967), *G. crassicaudatus* (von Fiedler, 1959; Montagna *et al.*, 1961a), *Loris tardigradus* (Hill, 1958; von Fiedler, 1959; Montagna and Ellis, 1960), *Galago elegantulus* (*Euoticus elegantulus*) (Hill, 1958), *Arctocebus calabarensis* (Hill, 1953, 1958; Montagna *et al.*, 1966; Machida and Giacometti, 1967), *Lemur catta*, *L. macaco*, *L. mongoz* (Montagna *et al.*, 1961b; Montagna, 1962; Montagna and Yun, 1962b), with a particular subpenile sebaceous area in *L. catta* (Evans and Goy, 1968) and *Microcebus murinus* (von Fiedler, 1959). In addition the present author

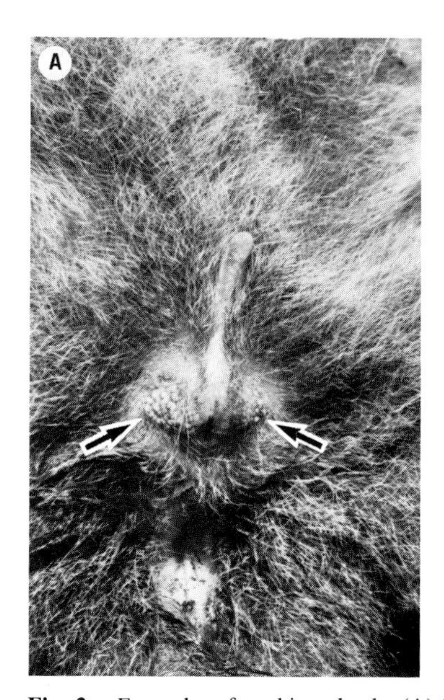

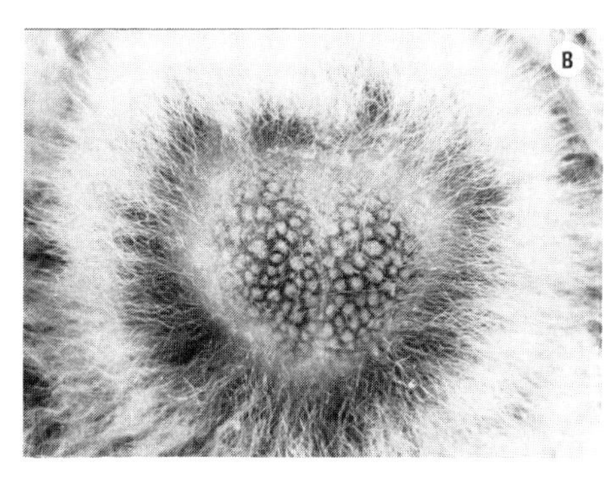

Fig. 2. Examples of marking glands. (A) Glands of the external genitalia in a *Galago alleni* female (Photograph: A. Devez). (B) Glands of the external genitalia in a *Cheirogaleus major* male.

(*continued*)

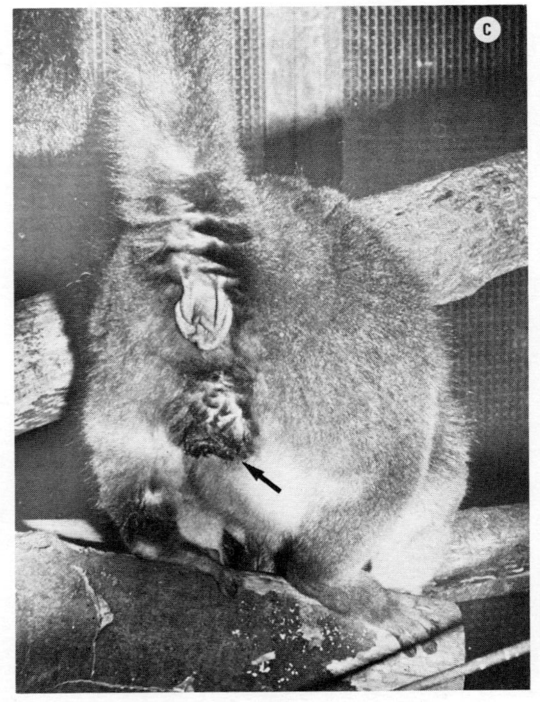

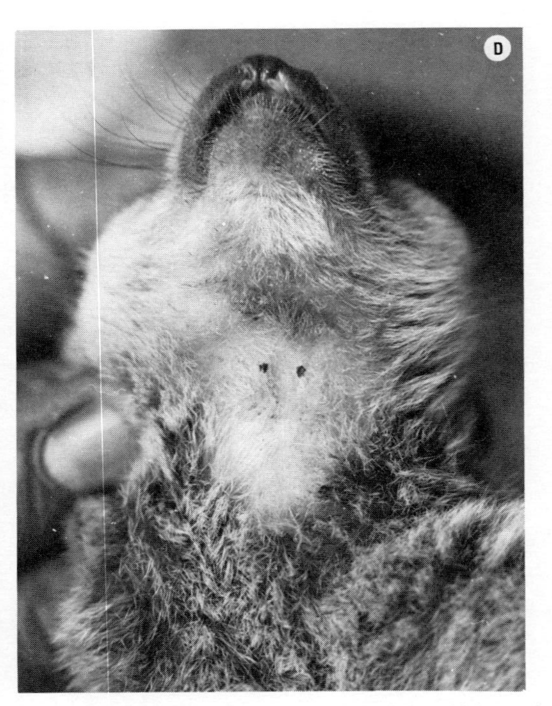

Fig. 2. (*continued*) (C) Circumanal glands in a *Lemur fulvus albifrons* male. Note the curious glandular skinfolds which may be more or less dilated and pulsating, and also the ''brush'' formed by the hairs impregnated by glandular secretions from the posterior region of the scrotum. (D) Neck glands in a *Phaner furcifer* male (Photograph: G. Pariente).

has found them in *Cheirogaleus major* (Fig. 2B) and in *M. coquereli*. Hill (1958) does not mention their presence in the Indriidae, *Daubentonia*, or *Tarsius*. Petter (1962, 1965) argues convincingly in favor of the existence of such glandular areas in certain *Propithecus* and *Lepilemur* species. Female *Perodicticus potto*, in addition to genital-labial glands forming a pseudoscrotum, possess "two preclitoridian glands" (Montagna and Yun, 1962b). These exceptional formations secrete a thick white substance which releases a strong and disagreeable odor (Charles-Dominique, 1966), probably the result of bacterial decomposition of the secreted product.

ii. The glands of the periphery of the anus. These are clearly characteristic of the genus *Lemur*. Schaffer (1940) distinguishes between circumanal glands in *L. fulvus* and *L. catta* and infracaudal glands in *L.f. albifrons*. Montagna *et al*. (1961b) and Montagna and Yun (1962b), however, indicate that this zone varies in aspect according to species; in *L. catta* and *L. mongoz* it is covered with fur while in *L. fulvus* (Fig. 2C) and *L. macaco* it is naked and has a glandular appearance. *Lemur macaco* and *L. fulvus* have been studied by Rumpler and Oddou (1970), who found a difference in the number of sweat glands in the two species and the absence of glandular formations in females of the former species. These authors note that in two cases of hybridization of the two species concerned, the hybrids inherit "the characteristics of marking of the parent that possesses the greater amount" with respect to both histological and behavioral criteria.

e. Other glandular secretions. In addition to the above-mentioned formations, prosimians possess specialized glands (see Fig. 3) called marking glands, because of their obvious relation to scent-marking behavior; as well as nonspecialized glands which seem to be used, in certain cases, for scent-marking.

i. Specialized glands. These are found principally on the neck and on the upper limbs, and are always less well developed, or atrophied, if not entirely absent, in females. Neck glands (see Table II) have been found in *Avahi laniger* (Bourlière, 1956) and in *Phaner furcifer* (Rumpler and Andriamiandra, 1971) in both sexes (see Fig. 20), and in the male *Varecia variegata* (*Lemur variegatus*; Rumpler and Andriamiandra, 1971), as well as in *Propithecus verreauxi* and *P. diadema* (Petter, 1962), while in *Hapalemur simus*, for which no histological material is available, they seem to resemble those of *P. furcifer* (J. J. Petter, personal communication). They may also exist in *Microcebus coquereli* (E. Pagès, personal communication). These glands occur singly and medially, except in *Avahi laniger*, where they are found under the angles of the jaws. A comparative histological study of these glands has been published (Rumpler and Andriamiandra, 1971).

These glandular formations are generally absent in the Lorisiformes although the behavior of rubbing the bare zone of the chest on branches, etc., leads to frequent references in the literature to "chest glands" in some species of galagine. In *Galago alleni* (P. Charles-Dominique, personal communication) and in *G. senegalensis* (Yasuda *et al*., 1961) histological examination has confirmed the absence of glands in this bare zone. Although Montagna and Yun (1962b) reported no sebaceous glands on the chest of *G. crassicaudatus*, Clark (1975) describes well-developed apocrine glands on the chest of the male *G.c. argentatus* which secrete a distinctly yellow substance deposited during chest-rubbing. Subspecific anatomical variations could be invoked to explain these apparently contradictory findings.

Glands of the upper limbs (see Table II) are better known (Gray, 1863; Beddard, 1884, 1891; Bland-Sutton, 1887; Pocock, 1918; Affolter, 1937). Studies originally focused attention on the true marking apparatus that *Lemur catta* and *Hapalemur griseus* (see Fig. 3) possess on the upper limbs. This is fully developed only in males and includes a brachial gland located near the shoulder and, at the wrist level, an antebrachial complex itself consisting of an hairless oval zone, rich in tactile elements, covering an antebrachial gland and bordered by a specialized horny formation. In *L. catta* this consists of a spur; in *Hapalemur griseus* it is more of a "brush" (Petter, 1962). Less well-developed glands of the same kind have also been found in *L. fulvus rufus* and in *L. macaco macaco* (Rumpler and Oddou, 1970). Within the Lemuridae some interesting variations are indicated in Table III.

Analogous topographical structures are found in numerous mammals (Affolter, 1937) and may be interpreted as ancestral characteristics of the prosimians. From the phylogenetic point of view, it would be interesting to re-examine this question once information becomes available on the anatomy of other species of this group, like *Hapalemur simus* and *Varecia variegata*. According to J. J. Petter (personal communication) glands should be present in the region of the elbow in *H. simus* but lacking in *V. variegata*. The work of Montagna *et al.* (1961b) and Montagna and Yun (1962b) has demonstrated the histochemical properties of these glandular structures. The work of Affolter (1937) on *H. griseus* and Rumpler and Andriamiandra (1968) on *Lemur catta* and of Rumpler and Oddou (1970) on *L. fulvus* has drawn attention to the fact that these glands contribute to sexual dimorphism, since Andriamiandra and Rumpler (1968) have demonstrated the role of testosterone in their development. Seasonal variations in the size of the brachial glands in *Lemur catta* (Petter, 1962) and in *Hapalemur griseus* (Affolter, 1937), as is found in other marking glands, may be explained by such an endocrinological mechanism. Moreover, Resko and Evans (cited in Evans and Goy, 1968) have found a seasonal variation in the plasma testosterone level in *L. catta*. Hill (1956) reported glands of the brachial organ in *Loris tardigradus* and in *Nycticebus coucang* which were later described by Ellis and Montagna (1963), but no marking behavior has been associated with these formations.

ii. Nonspecialized glands. It is also possible that prosimians use for marking the secretions of nonspecialized cutaneous glands which are scattered but less characteristic. Schaffer (1940) drew attention to the possible role of the glands of the palmar and plantar pads. According to Schaffer all mammals which possess these well-developed pads, which include all the prosimians he studied, could use these pads for scent-marking. In any event glandular activity seems to be present on "the volar side of palm and feet" (Hill, 1953) in *Lemur catta* (Montagna, 1962), *Microcebus murinus* and *Cheirogaleus* sp. (Schaffer, 1940), *L. macaco* (Schaffer, 1940; Rumpler and Oddou, 1970; Montagna *et al.*, 1961b; Montagna, 1962), *L.*

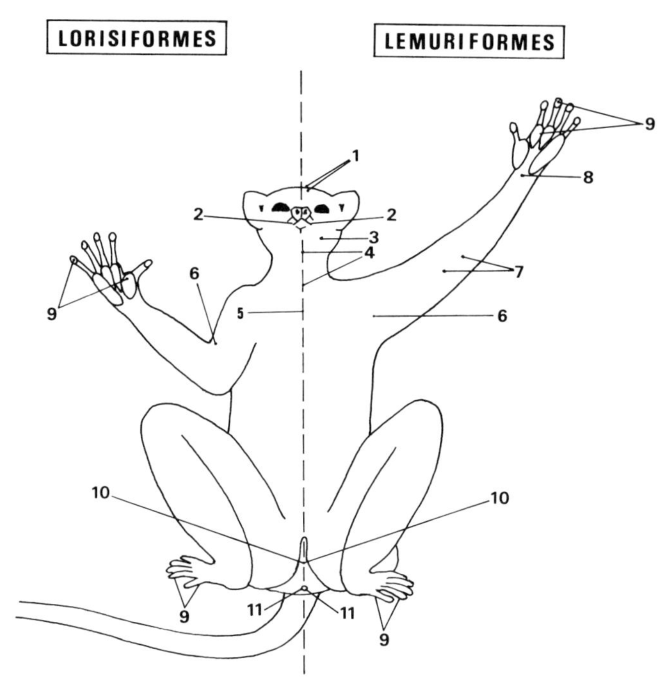

Fig. 3. Topography of the known cutaneous glands in prosimians. 1, Nonspecialized glands of the crown and forehead; 2, glands of the lips or commissures; 3, submaxillary glands; 4, neck glands; 5, chest glands; 6, brachial glands; 7, glands of the elbow region; 8, antebrachial glands; 9, nonspecialized glands of the palmar and plantar regions; 10, glands of the external genitalia; 11, glands of the anal region.

TABLE III

Variations in Specialized Glands in Three Species of Lemuridae

		Antebrachial complex		
Species	**Brachial gland**	**Antebrachial gland**	**Oval zone**	**Horny formation**
Lemur catta	+ (♂)	+ (♂ ♀)	+ (♂ ♀)	Spur (♂)
Hapalemur griseus	+ (♂)	+ (♂ ♀)	+ (♂ ♀)	Brush (♂)
Lemur fulvus	—	+ (♂)	—	—

fulvus (Rumpler and Oddou, 1970), *L. mongoz* (Montagna, 1962), *Hapalemur* and *Daubentonia* (Schaffer, 1940), *Galago senegalensis* and *G. crassicaudatus* (Schaffer, 1940; Montagna and Yun, 1962a), *Perodicticus potto* (Montagna and Ellis, 1959), *Nycticebus coucang* (Montagna *et al.*, 1961a), and *Loris tardigradus* (Montagna and Ellis, 1960). Ellis and Montagna (1963) have considered this question concerning the Lorisidae, and have shown important histochemical and histological variations which have significant phylogenetic implications.

In all prosimians the surface of the scrotum or of the genital labia, the perineal zone, and the periphery of the anus form another region of great glandular activity. These appear in a variety of forms, from macroscopically distinguishable scrotal and genital-labial glands, mentioned above for *Perodicticus potto*, and anal glands in the genus *Lemur*, to a simple microscropic concentration of sebaceous or sweat glands in a normally appearing epidermis. That this could be true of the anal region is suggested by Ortmann (1960), who confirmed the results of Schaffer (1940) in *Microcebus murinus*, *Daubentonia*, *Galago crassicaudatus*, *Loris tardigradus*, and *Nycticebus coucang*. It could also be true of the genital region in *G. senegalensis* (Yasuda *et al.*, 1961), or even, as is probably the case in the majority of species, at least at the microscopic level in both regions.

It seems that for several reasons, caution should be exercised before excluding the possibility of the existence in any prosimian of glands in the anogenital region. First, the glandular activity may be a seasonal phenomenon. This is particularly true of *Microcebus murinus* where the average diameter of the scrotum may change from 8 to 20 mm during the active period. Second, the presence of fur does not necessarily signify that the underlying skin is not glandular. In *Lemur*, for example, the presence of glands has been demonstrated not only in *L. fulvus* in which the periphery of the anus is naked, but also in *L. catta* in which it is furred. Conversely the scrotal glands are present not only in *L. catta*, whose scrotum is naked, but also in *L. fulvus* where the scrotum is furred. This is important since J. J. Petter (personal communication) notes that, in *L. catta*, either behind or, in other species, on the scrotum itself (see Fig. 2C) the hair becomes impregnated with glandular secretions, and sometimes with urine, forming an efficient marking "brush." In *L. catta* Petter (1962) describes a particular type of marking using these hairs. In *L.f. albifrons*, where the hair covering and secretions are particularly abundant, it is possible to recognize traces of "brush strokes" on the backs of marked conspecifics before they dry. Other nonspecialized glandular fields may also be involved in marking, particularly those of the face (Montagna and Yun, 1962b) in *Lemur catta* and of the scalp (Rumpler and Oddou, 1970) and forehead in *L. fulvus* and *L. macaco*.

2.2.1.2. Diffused and Undirected Odors. The research in this field is described below.

a. Diversification and nature of odors. The diversity in origin and nature of odoriforous substances actively dispersed in olfactory communication in the prosimians also includes exchanges in which olfactory communication occurs by means other than scent-marking. The substances involved may be the same as those in scent-marking (glandular excretions, in particular), or they may be of a different origin. Glands which may be referred to as odoriforous, although no active dispersal behavior has, as yet, been related to them, have been described in *Loris tardigradus* by Hill (1956) and Montagna and Ellis (1960), and in *Nycticebus coucang* by Montagna *et al.* (1961a). These glands, which are located in the region of the elbow, contain a structure identical to the axillary glands of man, the histochemical character and innervation of which are remarkable (Montagna and Ellis, 1960; Montagna *et al.*, 1961a), and are capable of releasing a strong, "disagreeable" odor (Hill, 1956; Petter and Hladik, 1970) in the case of *Loris*. However, these cases may be the exception and it is not yet known whether body odor, which may be extremely important in certain prosimians (for example, *Microcebus murinus* may sometimes be located in the field by their odor), originates from specialized glands or whether it is due to a single secretion, or to a mixture of various odors. In the case of *Perodicticus potto* certain authors (Manley, 1974; Charles-Dominique, 1977) pose the question of the possible role of these glands in defense against predators.

b. Pheromones. Research has not yet been initiated in prosimians on what are conventionally referred to as pheromones. It has been shown, however, mainly in rodents, that odoriforous signals are capable of acting on the receiver, in the absence of any special behavior on the part of the emitter, simply by being dispersed in the air while the animal is engaged in normal locomotor activity (Ropartz, 1967a). Odors dispersed in this manner have been shown to be involved in interindividual discrimination (Bowers and Alexander, 1967) in conveying individual information about emotional states such as fear and stress (Müller-Velten, 1966; Valenta and Rigby, 1968; Carr *et al.*, 1970; Caterelli *et al.*, 1974), as well as group information concerning aggression (Ropartz, 1968; Archer, 1968, 1969) and dominance (Krames *et al.*, 1969; Jones and Nowell, 1973), and in conveying information about reproductive states such as estrous synchronization (Whitten, 1956), and blocking of pregnancy ("Bruce effect"; Bruce, 1959) merely by the odor of strange males (Chipman and Fox, 1966). It would be of primary importance to ascertain whether the same aspects of physiology and behavior are regulated by such pheromones in the prosimians. According to most reports on prosimians olfactory interest in the receptive state of females increases in males when females come into estrus. This occurs indirectly by sniffing scent marks left by females and directly by sniffing the anogenital region of females (see Fig. 12). It may be presumed that the olfactory message involved is of the type demonstrated by Michael and Keverne (1968, 1970) in macaques, since odoriforous vaginal secretions have also been reported in prosimians (Sauer and Sauer, 1963). The activity of the male *Microcebus*, for instance, changes noticeably prior to the female coming into estrus even if the two sexes are in separate cages (A. Petter-Rousseaux, personal communication). Although the evidence is not yet conclusive (A. Schilling, personal observation) gas chromatography carried out on body odors of females, both in estrus and anestrus, seems to show chromatograms which differ according to the two states. Synchronization of estrus has been shown to occur in females of the same troop as well as in troops in the same area in *Lemur catta* (Jolly, 1966b) and also in *Microcebus murinus* captured in the same area (Martin, 1973). A peak of estrogens in the urine of *M. murinus* has been shown to occur at the point of estrus (A. Petter-Rousseaux, personal communication). Finally, it has been confirmed that social stress is responsible for a number of difficulties encountered in the breeding of certain nocturnal species (*Cheirogaleus* and *Microcebus murinus*) resulting in such things as renal disorders, low level of reproduction, etc. (Perret, 1975). It is tempting to relate social stress and its supposed causes (overpopulation, etc.) to a pheromone of the primer or priming type (Wilson and Bossert, 1963; Bronson, 1971). It should also be pointed out that Ropartz (1968) demonstrated the direct effect of an olfactory signal on a physiological target (the adrenals) in the mouse. In addition, von Holst (1972, 1974) has shown the influence of social stress on renal and adrenal pathology in the tree shrew (*Tupaia belangeri*), and that this stress may be provoked by the scent marks of conspecifics even in their absence.

2.3. Use of Odoriforous Substances by Prosimians

2.3.1. Different Types of Scent Diffusion

The diffusion of odoriforous substances by prosimians poses several questions. How is the scent diffused? In what form is it diffused? To whom is it directed? In order to answer these questions certain distinctions must be made.

First, active deposition of scent involves a highly specific behavior pattern as opposed to nondirected dispersion or diffusion of scent marks into the air resulting from spontaneous or incidental deposition. Second, active deposition may be directed on to the substrate or on to a (novel) object in the environment, on the one hand, or it may be directed at impregnating the animal itself or be for the benefit of a conspecific that it is facing, on the other hand (see Tables IV and V for categories of scent diffusion). Third, collection of scent, either one's own or that of a conspecific, may be either active or passive.

2.3.1.1. Nondirected Emission. Dispersion of scent may be effected by the spontaneous or incidental behavior of the animal, i.e., unaccompanied by any characteristic pattern of scent-marking behavior and either randomly on to the substrate or by aerial diffusion.

a. Incidental diffusion. It may be that this type of diffusion is as important as it is unknown in olfactory communication in prosimians. It is probably involved in reproduction as well in individual recognition, and dominance, etc. Indeed it is evident that certain olfactory messages are normally perceived during naso–nasal and naso–genital contact, by sniffing of the source of the odor itself (e.g., glandular areas), or sniffing from a short distance. Information is probably transmitted during direct contact between conspecifics. In such a case communication need not involve characteristic scent-marking behavior and it thus becomes difficult to relate a particular behavioral response to any indentifiable releasing stimulus.

Also entering into this category are automatic deposits of scent in the environment occurring incidentally during normal progression or on conspecifics during social contacts. There is some support for the hypothesis of automatic marking of branches by secretions of the sebaceous glands of the palmar and plantar surfaces. For instance, Ropartz (1966) has shown that the foot pads of the mouse are responsible for an individual odor, and Walker (1969) has suggested that the ''basal resistance level'' would be influenced by a pheromone having the same origin. An odorant dispersion of the incidental type may also exist in the Cheirogaleinae in which feces are used for scent-marking (particularly in *Cheirogaleus*) by walking over former deposits. The idea suggested by Harrington (1975), with respect to females of *Lemur fulvus fulvus*, of an automatic dispersion of olfactory stimuli on a particular area of branches where the animals have remained for a certain time, may be generalized to other prosimians.

The possibility of an olfactory exchange between conspecifics during allogrooming, play, and other social contacts, as well as during group resting or sleeping, seems evident (see Table V). The frequency and intensity of close physical contact have often been reported in lemurs (Petter, 1962; Jolly, 1966b) which literally pile on top of one another during sleep or rest. Another means of contact called ''locomotives'' has been described in *Propithecus* (Jolly, 1966b) and *Indri* (Pollock, 1975). Contact groups and contact in the nest during sleep (see Table V), where the animals are tangled together, are typical of certain Galaginae (Sauer and Sauer, 1963; Bearder, 1969; Charles-Dominique, 1972, 1977) and Cheirogaleinae (Petter, 1965; Martin, 1972a). During social contact and particularly during mother-infant contact (Klopfer and Gamble, 1966) a mutual scent impregnation may occur naturally.

b. Aerial diffusion. If the scent deposited is not directed onto a material substrate, even when characteristic marking behavior is involved, then this cannot be termed scent-marking behavior in the strict sense of the term. This mode of diffusion is employed by *Loris* and *Nycticebus* for brachial secretions. Other typical examples of aerial scent diffusion are due to movements of the tail, for example, waving of the tail after being marked by the antebrachial complex in *Lemur catta*, and tail-lashing during urine-marking in *Propithecus verreauxi* (Jolly, 1966b). *Microcebus coquereli* periodically release distinctive odors which can be detected in the forest by the human observer (Pagès, in press) (see Fig. 14).

2.3.1.2. Active Deposit of Scent on an Inert Substrate. Marking of the environment is the most spectacular and best known form of olfactory communication in the prosimians, both because of its generalized usage and its frequently stereotyped nature (see Figs. 4 and 5). Table IV summarizes behaviors known or suspected to play a part in the dispersion of odorant or chemical substances in the environment for each species.

TABLE IV

Behaviors Believed to Play a Role in the Dispersion of Olfactory or Chemical Material in the Environment

Prosimian	1[a,g]	2[b,g]	3[c,g]	4[d,g]	5[e,g]	6[f,g]
Cheirogaleidae						
Microcebus murinus	UP, UW	UD, US, G	FP	M, C, A	H, FS, S	T, L, FT
Microcebus coquereli	S	UD, G		M, H, A	N	T, L
Cheirogaleus medius	FD		UD			
Cheirogaleus major	FD	UD, G		M		T
Phaner furcifer	N			M		T
Lemuridae						
Lemur catta	W	G		A	P	T
Lemur fulvus						
Lemur macaco	A, G	P	UP	M, H, W	FP	T,L
Lemur mongoz						
Varecia variegata	N			M, AG		
Hapalemur griseus	W	G	UP	M	P	T
Hapalemur simus					N, W	
Lepilemur sp.			UP	M	G	
Indriidae						
Avahi laniger					AG, N	
Propithecus verreauxi	UP	N, G	FP		A, C	
Indri indri		G, UP		C		

Daubentonia madagascariensis	UD			AG		T
Galaginae						
Galago demidovii	UW, UP	G	FP	M	A	HT
Galago senegalensis	UW, UP	S, US, G	FP	M, H, P	A, FS	FT
Galago alleni	UW	S, UP, G		M, C	A	
Galago crassicaudatus	C	UW, UP, G		M, H	A	FT, HT
Galago elegantulus	UP	G		M	A	
Lorisinae						
Perodicticus potto	UD	UP, G		M		
Nycticebus coucang	UD	UP, FD		M, H, AG		
Arctocebus calabarensis	UP, UP	C				
Loris tardigradus	UW	UD, UP				
Tarsius sp.	S	UP		M, C	G	

[a] 1, Distinctive dispersal behaviors involving a clearly identified excretion or secretion—most typical examples.
[b] 2, Distinctive dispersal behaviors involving a clearly identified excretion or secretion—other examples.
[c] 3, Scent deposit observed but not related to a distinctive dispersal behavior.
[d] 4, Distinctive dispersal behavior but not related to a known or identified deposit.
[e] 5, Doubtful scent deposits and dispersal behaviors.
[f] 6, Frequently associated behaviors which are not olfactory.
[g] Key to abbreviations: Buccal region—S, wiping of the muzzle and labial commissures with salivary deposit; M, wiping of the muzzle, chin, and cheeks without salivary deposit; L, licking; T, biting, chewing, and scraping with the teeth. Head and neck—H, rubbing of the neck; C, rubbing of the ventral surface of the body. Limbs—W, wrist twisting; P, palming; FT, forelimb tapping; HT, hindlimb tapping. Anogenital region—G, rubbing of the external genitalia; A, rubbing of the anus; AG, anogenital rubbing. Urine—UP, punctuated urinating; UD, urination with dragging; UW, urine-washing; US, urine sole-wiping. Feces—FP, punctuated defecation with smearing; FD, defecation with dragging; FS, feces sole-wiping.

a. Head and neck. The wiping of the muzzle, lips, and cheeks has been observed in numerous prosimians, either accompanied by a salivary deposit, as in *Galago senegalensis* (Andersson, 1969) and in *G. alleni* (M. Grange, personal communication), or unaccompanied by a salivary deposit as in *Microcebus murinus* (A. Schilling, personal observation), *G. demidovii*, *G. elegantulus*, *Perodicticus potto* (P. Charles-Dominique, personal communication). Other species may also exhibit this behavior which clearly deserves the term scent-marking, at least in the case of *M. coquereli*. In this species Pagès (in press) distinguishes between a wiping of the commissures as far as the ears with abundant salivation, and a behavior resembling sharpening of the canines with the mouth wide open.

In addition, some buccal behaviors, which are not olfactory, have been reported in several species during other types of scent-marking; for example, biting or chewing of branches or other objects as in *Galago senegalensis* (Doyle, 1974a), *G. crassicaudatus* (Bearder and Doyle, 1974b), *Microcebus murinus* (A. Schilling, personal observation), *Lemur fulvus* (Petter, 1962), and *L. catta* (Schilling, 1974). Petter and Peyrieras (1970a) have reported a curious behavior in *Daubentonia* of placing the incisors against a vertical support and then holding on and pulling with the hands.

In the prosimians in general, but more particularly in *Lemur*, various forms of licking occur which apparently are unrelated to grooming but which may be integrated into a behavioral sequence that includes sniffing, biting, yawning (Bailey, personal communication), and *Flehmen*.* Licking may be directed at surfaces or objects in the environment which have been previously marked (see Fig. 9A), or at the pelage of conspecifics (see Fig. 16C), or it may be simply self-licking of the rhinarium (see Fig. 12C). Interpretation of this behavior is difficult because it may concern taste as well as olfaction, and it may involve the vomeronasal receptors at the same time (Poduschka and Firbas, 1968; Schilling, 1970; Bailey, 1978 and personal communication).

Marking with the neck glands is accomplished by rubbing the glandular region against a trunk or branch, sometimes during progression, as in *Phaner furcifer* (Petter *et al.*, 1971), *Varecia variegata*, and *Propithecus verreauxi* (Petter, 1965; Jolly, 1966b). This behavior has also been observed in *Microcebus coquereli* (E. Pagès, personal communication), *Hapalemur simus* (J. J. Petter, personal communication), and *Indri indri* (Pollock, 1975) in both sexes, although no glandular structure has been discovered. In *Avahi laniger*, on the other hand, this behavior has never been observed although suitable glandular structures do exist. Chest-rubbing, described by Jolly (1966a) and Bearder (1975) in *Galago crassicaudatus*, by Andersson (1969) and Doyle (1974a) in *G. senegalensis*, and by Pinto *et al.* (1974) in

*"Flehmen" is a widely distributed behavior pattern first described by Schneider (1930) as a characteristic facial expression in some species of ungulates and carnivores after scenting various olfactory stimuli, particularly the urine of females. The sequences include, opening of the mouth, exposing the teeth by retracting the upper lip, opening and closing the nostrils, and jerking back the head (Heymer, 1977).

Microcebus murinus, as well as chin-rubbing in *Lemur fulvus fulvus* (K. Boskoff, personal communication), can only be interpreted as scent-marking behavior if a glandular secretion is deposited such as that described for *G. crassicaudatus argentatus* by Clark (1975).

Rubbing of the top of the head or forehead in *Galago senegalensis* (Bearder and Doyle, 1974a), *Microcebus murinus* (A. Schilling, personal observation), *Lemur fulvus* (Petter, 1965; Harrington, 1974; Chandler, 1975) (see Fig. 4A), *L. macaco* (Petter, 1962), and *L. mongoz* (Bolwig, 1960) is probably of the same type. Other interpretations are possible (see below).

b. Limbs. The marking behavior of *Lemur catta*, involving the antebrachial complex, has been described in detail (Petter, 1962; Jolly, 1966b; Andrew, 1964; Schilling, 1974). It bears a remarkable resemblance to the antebrachial marking of *Hapalemur griseus* (Petter, 1962; Andrew and Klopman, 1974; see Fig. 7A), but assumes different forms in *H. simus* (J. J. Petter, personal communication). In *L. macaco* (Petter, 1962), *L. fulvus* (Andrew, 1964), and, perhaps, *L. mongoz* (Bolwig, 1960), the characteristic behavior is palmar-marking, which involves rotation of the wrist after holding the support in the palm, but which may be limited to a movement of the fingers (Rumpler and Oddou, 1970), although the anterior part of the wrist may also be involved (Petter, 1962; see Fig. 4B). The rubbing of the palms described by Vincent (1969) in *Galago demidovii*, and of the feet, described by Andrew and Klopman (1974) in *G. crassicaudatus*, may not necessarily be related to the deposit of scent. It might be more correct to interpret the former as a drumming of the hands and the latter as a trampling of the feet (Bearder and Doyle, 1974a): in other words, as behaviors which are auditory and visual in nature.

c. Anogenital region. Rubbing the anogenital region on some object in the environment (branches, etc.) has been observed in most prosimians, but there is seldom any certainty as to whether it is the anus or the genitalia which are involved, particularly in those species studied only in the field, like *Propithecus verreauxi*, or in species in which marking is effected by lowering the hindquarters onto the substrate (''rump-dragging''; Eisenberg and Kleiman, 1972) characteristic of the Lorisinae and the Lemuridae (see Fig. 4C). The most curious behavior of this kind is probably marking that is accompanied by a complete about-face in certain lemurs. In *Lemur catta* this is more in the nature of genital-marking (Petter, 1962; Jolly, 1966b) and, in *L. fulvus*, it is more in the nature of anal-marking (Petter, 1962; Andrew, 1964; see Fig. 4D).

d. Urine. Urine-marking is typical of all prosimians except perhaps the Lemuridae and *Phaner furcifer*, which are distinguished by other behavioral and anatomical traits as well. In prosimians, the behaviors involved in depositing urine are of a greater variety and degree of specialization than other behaviors concerned with olfactory communication. The Lemuridae and Lepilemur simply urinate on branches (Petter, 1962; Charles-Dominique and Hladik, 1971) but it is not known if this is a purely excretory function or whether scent deposition is involved. In captivity *Lemur fulvus albifrons* occasionally leave traces or drops of urine after

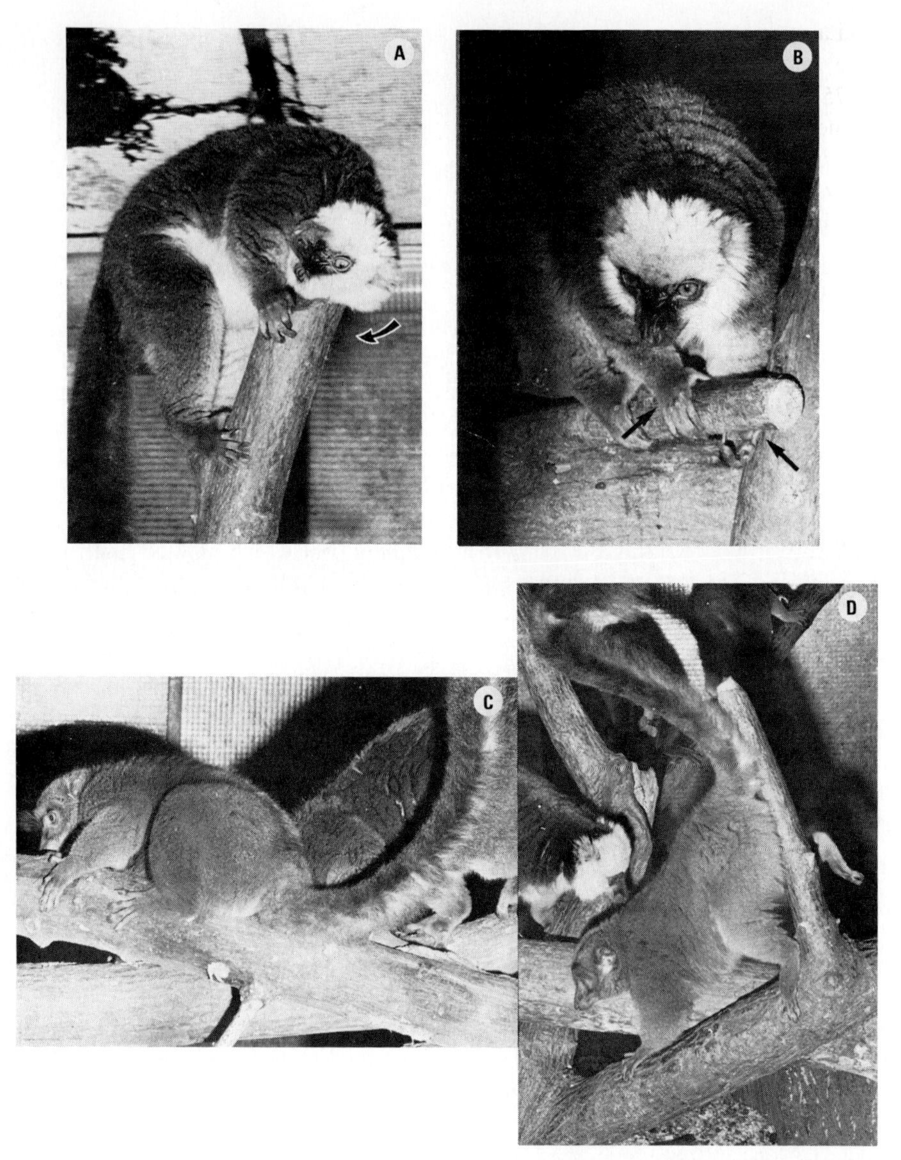

Fig. 4. Examples of environmental marking in *Lemur fulvus albifrons*. (A) Head-rubbing by a male. (B) Rubbing of the hands by the same male. The photograph shows not only palmar-rubbing (essentially the fingers of the right forelimb) but also the rarer rubbing of the forearm (left forelimb). (C) Generalized form of anogenital marking in a female. (D) Specialized form of anogenital marking in the same female. Note the support by only one of the legs.

genital-marking. However, this urinary discharge may be due simply to the excited state of the animal (A. Schilling, personal observation). According to Chandler (1975) urine-marking in *L. fulvus* is preceded by anogenital-marking.

The Lorisidae in particular have developed highly stereotyped patterns of urine-marking which include rhythmic micturation, urine-washing, and, in *Galago elegantulus*, the directing of a jet of urine with accuracy at a distance of 1 m (Charles-Dominique, 1977). These behaviors may be differentiated into direct and indirect marking.

i. Direct urine-marking. In direct marking the urine is directly deposited on the substrate by the penis or clitoris which is usually lowered on to the support. Several categories of direct urine-marking may be distinguished. Seitz (1969) distinguishes two categories in *Nycticebus coucang*. The first is "punctuated marking" (*Punktmarkierspuren*) confined to successive lengths of less than 15 cm and more frequent in males. The second is trail-marking (*Schleispuren*), distributed over a larger area, relatively less frequent, and more common in females. Other authors seem to attach more importance to the marking behavior than to the marks themselves. Thus "rhythmic micturation," described for the first time by Ilse (1955) in *Loris tardigradus*, is characteristic of a number of Lorisidae. It permits a significant economy of urine since small quantities are emitted while the animal is able to advance over a relatively long distance. Doyle (1974b) makes a distinction between this stereotyped behavior and less elaborate forms of urine-marking. All these categories may merit theoretical consideration only, owing to the polymorphism inherent in direct marking, even if stereotyped. Table IV distinguishes between "punctuated" marking, for example, urine-marking characterized by sparse or successive drops, and trail-marking or dragging characterized by a continuous trail.

A *Microcebus,* for example, may deposit a succession of drops of urine or it may deposit a continuous trail. A trail of urine may be laid in two ways. The Lorisidae and certain Malagasy prosimians, *Microcebus coquereli* and *Daubentonia* (see Fig. 5C), urinate while touching the support with the clitoris or penis. Other prosimians, like *M. murinus*, punctuate the substrate with drops of urine but spread it over a distance with "perineal wiping," as Eisenberg and Kleiman (1972) noted for the Viverridae. The important difference between the two behaviors is that in the second case (see Fig. 5A) there may be simultaneous marking by the glands of the anogenital region which may result in a complex mixture of scents—urine, secretions of the genital glands, and feces (especially in *Cheirogaleus*).

ii. Indirect urine-marking. Indirect marking involves an active behavior in which the genital organs do not deposit the urine directly on the support but rather with the aid of another part of the body, usually the palms or the soles. Urine-washing is the best known means of indirect marking, the phylogenetic importance of which has been stressed by Hill (1938), Eibl-Eibesfeldt (1953), Charles-Dominique and Martin (1970) and Andrew and Klopman (1974). Urine-washing is characteristic of the Galaginae, *Galago elegantulus* being the only exception. It is rarer in the Lorisinae and the Malagasy lemurs. In the Lorisinae it has been con-

Fig. 5. Examples of urine-marking in some Malagasy prosimians. (A) *Microcebus murinus*: generalized form or direct urine-marking; the entire anogenital region is pressed against the branch. (B) *Microcebus murinus*: specialized form or indirect urine-marking, urine-washing; note the similarity between this behavior and that of *Galago* (Photograph: F. Ponge). (C) Urine-marking and the resulting trail in *Daubentonia madagascariensis*; the length of the trail-mark is striking (Photograph: J. J. Petter).

firmed only in *Loris* (Andrew and Klopman, 1974). It was not observed in *Nycticebus* (Tenaza *et al*., 1969) nor in *Arctocebus* (Manley, 1974). In fact Manley (personal communication to Andrew and Klopman, 1974) is convinced that urine-washing has been lost in all the lorisines with the exception of *Loris*. In the Malagasy prosimians it has been observed in *Microcebus murinus* (Petter, 1962). Videotape recordings (A. Schilling, personal observation) verify that the urine-washing action in *M. murinus* is essentially the same as in *Loris* (Ilse, 1955), *G. crassicaudatus* (Eibl-Eibesfeldt, 1953), *G. senegalensis* (Sauer and Sauer, 1963), *G. demidovii* (Vincent, 1969), and *G. alleni* (Charles-Dominique, 1977). *Microcebus murinus* (see Fig. 5B) support themselves homolaterally, place the hand under the urogenital region, urinate into the shallow of the palm, and then rub the

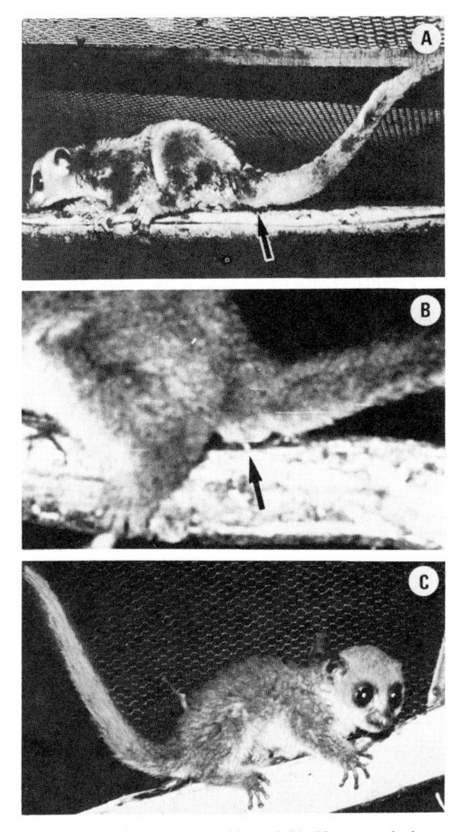

Fig. 6. Fecal-marking in *Cheirogaleus major*. (A and C) Characteristic posture of fecal-marking; the anus is applied against the support while the tail is held rigid and elevated. The complete pattern of fecal-marking is developed by 6 weeks of age whether the young have been raised with adults (A) or without adults (C). (B) The peculiar position of the anus (at the first caudal vertebra) and its capacity for dilation results in an efficient marking organ.

palmar and plantar surfaces together several times, before returning them to the support. This behavior which is practiced by both sexes, is extremely rapid, often taking only a few seconds, and is usually repeated on the other side. The deposition of urine is thus accomplished by the palms and the soles, either incidentally during locomotion or by an active rubbing of the soles of the feet on the support. This "sole-wiping" on the substrate has also been described in *G. senegalensis* (Doyle, 1974a, b), but seems more frequent in *Microcebus*, either during urine-washing or, presumably, on previous deposits of urine. Urine-washing is the most stereotyped of all forms of urine-marking, the smaller species, *G. demidovii* (Vincent, 1969) and *M. murinus* (Pinto *et al.*, 1974), performing it more rapidly than the larger species.

e. Feces. The active deposit of feces, which is different from normal defecation, is frequent only in *Cheirogaleus*. *Cheirogaleus medius* and *C. major* mark in "trails." The morphology of the anal region in these species enables the anus to make contact with the substrate without the need to lower the hindquarters noticeably, and permits defecation during locomotion, sometimes relatively rapidly (see Fig. 6). The marks form elongated cylinders which in *C. major* may measure as much as 75 cm, though they may be flattened by secondary trampling of the feet, and are often mixed with urine. This behavior has also been observed without feces being deposited, in which case there may be a glandular secretion involved, or it may be simply a stereotyped wiping of the anus.

2.3.1.3. Active Scent Deposition on a Conspecific (See Table V). Marking of a conspecific is a behavior which is more complex, rarer, more subject to variation, and often linked to social or other behaviors, making it difficult to distinguish it from the sort of incidental marking mentioned above. Although it is often reciprocal, this behavior is apparently more frequent in males. It has not been described in the Indriidae, nor in *Lepilemur*, and only rarely in the Cheirogaleidae, except for *Phaner furcifer* in which the male marks the female with the neck glands (Charles-Dominique and Petter, in press). On the other hand, it seems to play an important social role in the Lemuridae (especially in *Lemur fulvus*, *L. macaco*, and *L. mongoz*) and in the Lorisidae. Anogenital marking of a conspecific may be direct or indirect, depending on whether or not the structure responsible for the scent is in direct physical contact with the partner. Manley (1974) provides examples of indirect and direct marking of conspecifics in *Perodicticus* and *Arctocebus*, respectively. During allogrooming between the sexes *Perodicticus* scratch the scrotum or pseudoscrotum with the fingers, which become charged with glandular secretion. The hand is then applied to the fur of the partner and the process is repeated with the opposite hand. Manley referred to this as "genital-scratching-grooming." In *Arctocebus* "passing-over" consists of the male "moving forward over the entire length of the female's back" and thus directly transferring scrotal secretions during movement. Conspecific marking is also well developed in certain lemurs, *Lemur mongoz* (Bolwig, 1960), *L. macaco macaco* (Petter, 1962; Jolly, 1966b), *L. fulvus collaris* (Jolly, 1966b), *L.f. rufus*, *L.f. fulvus*, and *L.f. albifrons* (Petter, 1962) in

TABLE V

Known Self-Marking and Examples of Scent Exchange between Social Conspecifics[a]

Species	Active scent deposition		Collection of scent from a social conspecific (outside the reproductive period)		
				Incidental	
	Self-marking	Allomarking	Active	During sleep or rest	By allo-grooming during activity
Microcebus murinus	♂ ♀ UW		♂ H, M/gum	CG.; N	+
Microcebus coquereli	♂ ♀ S	+ − Licking		N	+
Cheirogaleus medius *Cheirogaleus major*				ε (N)	+
Phaner furcifer		♂N / ♀		N	++
Lemur catta	♂ W /bg (anointing) ♂ W / W ♂ W + − bg / tail			CG. ; Cl	+++
Lemur fulvus	♂ ♀P/genitals or tail ♂ H/gum or fm ♂ scratching/scrotum (hand or foot)	♂ ♀ G / ♂ ♀ ♂ A / ♂ ♀ ♂ G (passing over)/♂ ♀	♂ H / genitals ♀ ♂ H / gum or fm ♂P / gum	CG. ; Cl	+++
Varecia variegata				CG. ; Cl	++
Hapalemur griseus	♂ W / bg (anointing) ♂ / W or C ♂ W + − bg/ tail			CG. ; Cl	++

(*continued*)

TABLE V (*continued*)

Species	Active scent deposition		Collection of scent from a social conspecific (outside the reproductive period)		
				Incidental	
	Self-marking	Allomarking	Active	During sleep or rest	By allo-grooming during activity
Lepilemur sp.				ε (N)	ε
Avahi laniger				CG. ; N	+
Propithecus verreauxi			♂ N + − C / Um ♀	ε (Cl)	+
Indri indri	♂ G / N marks			ε (Cl)	+
Daubentonia madagascariensis				ε	ε
Galago demidovii	♂ ♀ UW	♂ gum / ♀	♂ H / gum ♀	CG. ; N	++

Galago senegalensis	♂ ♀ UW	♂ gm / ♀ ♂ A / ♀	♂ C / ♀ gum (chest-rubbing)	CG. ; N	++
Galago alleni	♂ ♀ UW	♂ gum / ♀		ε	++
Galago crassicaudatus		♂ C / ♀ (chest-rubbing)	♂ H. M / ♀ gum ♂ C / ♀ (chest-rubbing)	N	++
Galago elegantulus				CG.	++
Perodicticus potto	♂ ♀ / genitals (genital-scratching)	♂ ♀ UP / ♂ ♀ ♂ ♀ G / ♂ ♀ (genital-scratching-grooming)		ε	ε
Nycticebus coucang				ε	ε
Arctocebus calabarensis		♂ G / ♀ (passing-over)		ε	ε
Loris tardigradus	♂ ♀ UW			ε	ε
Tarsius sp.	♂ ♀ S	♂ ♀ gum / ♂ ♀		ε	ε

[a] Key to abbreviations: CG, contact group; N, contacts in nest, tree holes, etc.; CL, contacts during activity, e.g., clustering, "locomotives," ε, little or no contact; +++, very frequent; ++, frequent; +, occasional; Um, urine marks; Fm, fecal marks; gum, genital and/or urine marks; bg, brachial glands: /, on, against. For other abbreviations see Table IV.

the form of a direct marking by the anogenital region in males (see Fig. 16A) and females. Passing-over has also been seen in *L.f. albifrons* in captivity (A. Schilling, personal observation). Although this behavior has only been observed in males, it is not necessarily directed at the opposite sex. According to Petter (1962; also A. Schilling, personal observation) the males of certain species indirectly allomark their social partners of both sexes by means of the head (forehead or crown) which has previously been rubbed in the marker's own urine or feces.

Urine-marking of conspecifics is also sometimes direct, for example, the marking of females by males in *Galago senegalensis* (Doyle, 1974a, b), or the marking of males by females in *Tarsius syrichta* (Wharton, 1950), or it may be indirect, for example, urine-washing during allogrooming in *G. senegalensis* (Andersson, 1969), *G. demidovii*, and *G. alleni* (Charles-Dominique, 1977).

2.3.1.4. Active Self-Impregnation (See Table V). Diffusion or deposition of odors represents only one aspect of prosimians' utilization of odoriforous material. An important aspect of olfactory communication is self-impregnation involving the animal's own odors or those of its conspecifics. This tendency to self-impregnate various parts of the body is fairly generalized in mammals (Andrew and Klopman, 1974; Eisenberg and Kleiman, 1972) and has the obvious advantage of increasing scent diffusion.

a. Self-marking. Actively impregnating the body with one's own odors is frequent in prosimians, urine-washing being the most widespread example. Included in this category is tail-marking with the antebrachial complex by *Lemur catta* (Pocock, 1918; Evans and Goy, 1968) and by *Hapalemur griseus* (Affolter, 1937; Andrew, 1964), and palmar-marking of the tail in *L. macaco* (Petter, 1965) and in *L. fulvus* males and females (see Fig. 8A). Also prominent in *L. catta* and *H. griseus* (Affolter, 1937) is the action of flexing the forearm in such a way as to bring the brachial and antebrachial glands into direct contact, which results in a mixture of the two scretions eventually deposited during marking. Figure 7B shows that this movement also results in an impregnation of secretions of the chest. Finally, by bringing the two forearms into contact *H. griseus* (von Fiedler, 1959) and *L. catta* (J. J. Petter, personal communication) are able to mix secretions of the two forearms.

Salivary impregnation using the hands (similar to the toilet behavior of cats with their paws) has been observed in *Microcebus coquereli* (Pagès, in press), *Galago senegalensis* (Sauer and Sauer, 1963), *G. alleni* (M. Grange, personal communication), and *Tarsius* (Niemitz, 1974). Poduschka and Firbas (1968) describe a similar "self-anointing" behavior with saliva in the hedgehog, which they conclude is olfactory in nature and not a grooming behavior. In *G. senegalensis* rubbing of the foot in feces may not be olfactory marking (Doyle, 1974b) while in *Lemur fulvus*, rubbing of the head in feces or urine and, possibly, in anogenital marks probably is marking behavior (Petter, 1962). This species, in which the range of marking behavior is particularly rich, may impregnate the tail with glandular secretions by pressing it against the genitalia during palmar rubbing (A. Schilling, personal ob-

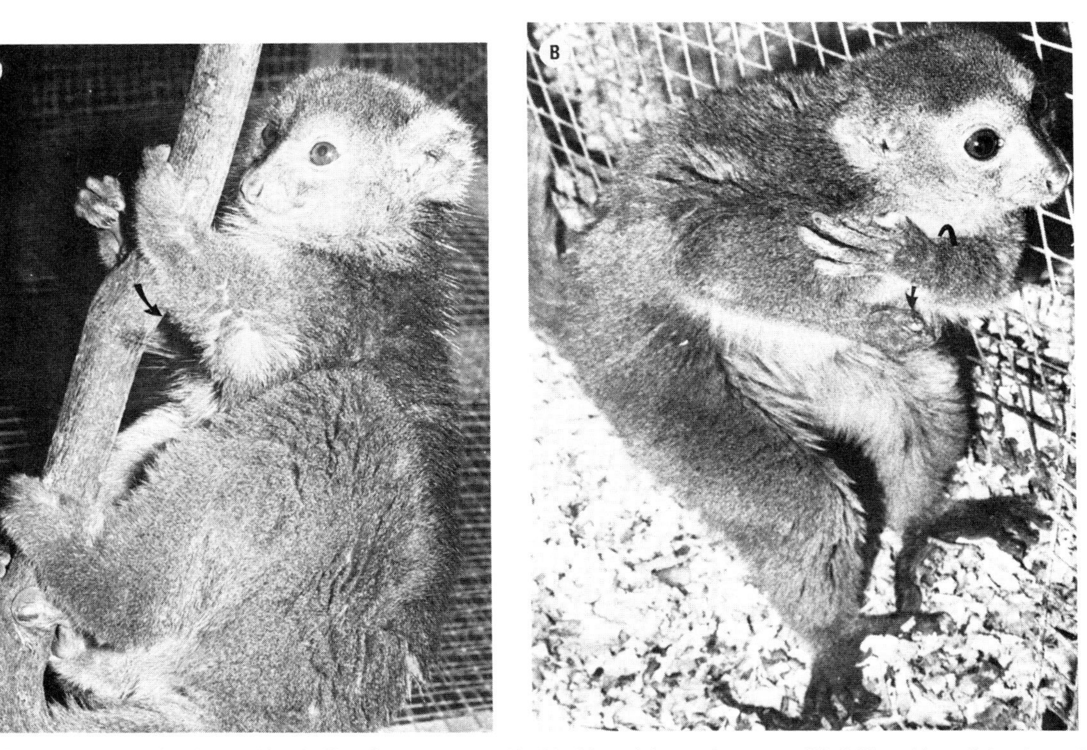

Fig. 7. Two forms of wrist-marking in *Hapalemur griseus*. (A) Marking of the environment. (B) Self-marking of the chest. After folding the forearm and thus mixing the secretions of the brachial and antebrachial glands (''anointing''), the wrists are alternatively twisted and at the same time rubbed against the chest. Occasionally the tail is passed between the legs and pressed by the wrists against the chest; the movement of the wrists and rapid retraction of the tail results in an odorant deposit on both sides of the tail.

servation; see Fig. 8B). In *L.f. albifrons* the male has been observed rubbing the internal face of the heel on the scrotum (J. J. Petter, personal communication). This may be accompanied by handling behavior (see Fig. 8C) reminiscent of the self-impregnation pattern reported by Manley (1974) for *Perodicticus*. Epps (1974) also noticed an indirect urine self-marking during grooming in *Perodicticus* which results in a mixture of odors between partners.

b. Collecting scent from conspecifics. Active self-impregnation (certain incidental forms of which have been mentioned above) may also involve odoriforous substances foreign to the animal, generally those of conspecifics and, more particularly, those of social partners. Such odors may be gathered either directly from the partner or indirectly from previous marks. The ''chest-rubbing'' of certain Galaginae, *Galago crassicaudatus* (Jolly, 1966a), *G. senegalensis* (Doyle, 1974b) and *G. alleni* (P. Charles-Dominique, personal communication), could be interpreted as a collecting of odor, since it involves the chest, the sides of the chin, the insides of the arms, and often the whole underside of the body (Bearder, 1975), although Bearder (1975) and Andersson (1969) interpret this behavior, in *G. crassicaudatus* and *G. senegalensis*, respectively, as the depositing rather than the collecting of scent. Muzzle-wiping by male *G. demidovii* (Charles-Dominique, 1977) and *Microcebus murinus* (A. Schilling, personal observation) on the anogenital mark of a female may well be odor-collecting. The interpretation of these

Fig. 8. Self-marking in *Lemur fulvus albifrons*. (A) Tail-marking in a female, palmar form. (B) Tail-marking in the same female, genital palmar form. (C) Genital self-marking of the internal side of the foot in a male, in this case accompanied by manipulation of the foot.

Fig. 8. *(continued)*

behaviors is problematical because it is difficult to determine whether these impregnations are, in fact, made in order to collect foreign odors, or whether they are simply a secondary consequence of a poorly understood behavior (casual marking, visual display, emotional release, etc.). For example, the peculiar behavior of *Lemur fulvus* males rubbing the head in their own feces or urine (Andrew, 1964; Petter, 1962, 1965; Harrington, 1975; see Fig. 4A) is open to several possible interpretations:

1. A direct marking not preceded by active impregnation if, in fact, the glands of the scalp (Rumpler and Oddou, 1970) secrete actively. The deposit may be addressed to the environment (branch, familiar object) or to a social partner. In the latter case the marker actively disperses the glandular secretions of the head on the marked subject—direct allomarking.
2. An indirect marking preceded by self-marking. The animal first impregnates its head with its excrement, and then marks either the environment or a social partner. Again, in the latter case, the marker indirectly disperses its urine or feces on to a conspecific—indirect allomarking.
3. An indirect marking preceded by active impregnation of odoriforous substances of a conspecific. The marker rubs its head in the excrement or anogenital marks of a conspecific, and then marks the environment or a conspecific.
4. An active impregnation with the odors of a conspecific not preceded by self-marking (see 1 above). The collection of odor may occur directly, the most characteristic example being rubbing of the head on the female genitalia or, indirectly, rubbing of the head on the excrement or anogenital marks of a conspecific.
5. A direct or indirect overmarking of a previous mark by a conspecific either on an environmental object or the substrate, or on a conspecific.
6. A visual display not involving olfactory communication.
7. A combination of two or more of the above.

2.3.2. Characteristics of Scent Diffusion by Prosimians

The importance of olfactory communication in the prosimians is due not only to their particular macrosmatic status, but also the variety of behaviors involved.

2.3.2.1. Polymorphic and Rigid Behaviors. The similarities of the behavioral sequences, between urine-washing of *Loris* and *Microcebus*, between brachial-marking in *Lemur catta* and *Hapalemur griseus*, between rubbing the neck gland in *Phaner furcifer* and *Propithecus verreauxi*, between passing-over in *Arctocebus calabarensis* and *L. fulvus*, or between muzzle-wiping in *Microcebus coquereli* and *Galago alleni*, pose some interesting systematic and phylogenetic problems. Conversely, a significant polymorphism exists in the collection of foreign odors, anal-marking, and direct urine-marking (e.g., position of the genital organs in relation to the substrate, type of urination, movements of the hindquarters, etc.).

2.3.2.2. Primitive and Specialized Characteristics. Scent diffusion in the prosimians is characterized not only by the persistence of some behaviors which have changed little, if at all, over time, but also by the extreme specialization of others. Eisenberg and Kleiman (1972) and Andrew and Klopman (1974) point out that "perineal- and anal-dragging," or the rubbing of certain regions of the ventral surface of the body, as well as of the head (cheeks, top of the head) exist in many mammals, including ruminants, rodents, carnivores, insectivores, and tupaiids. On the other hand, urine-washing, rhythmic micturation, brachial- and palmar-marking in the Lemuridae seem to be specializations which, if not unique to the prosimians, since urine-washing is found in some New World Monkeys (Hill, 1938; Nolte, 1958; Andrew, 1964), are at least part of the adaptation to arboreality. In the case of brachial-marking, the adaptation is functional as well as anatomical. The "brachial complex" is structured so that a mere movement of the wrist applied to a branch results in the secretion of glandular substances and their impregnation in the bark by means of a spur in *Lemur catta* (Schilling, 1974) or by means of a brush in *Hapalemur griseus* (Petter, 1962). The end result is what may be called a "vaccination" of the bark since the glandular deposit is accompanied by scarring by the horny formation.

Urine-washing in *Galago* and trail-marking in *Potto* represent the best functional adaptations to a three-dimensional space such as the forest biotope (Charles-Dominique, 1977). Indeed, the multiplication and spreading of marks increases the possibility of conspecifics meeting. In addition, from a physiological point of view, rhythmic micturation and urine-washing, in particular, represent an economy in the use of urine as well as a remarkable adaptation in the direction of economy of movement. The evolved adaptations also include scent deposits of anogenital origin (see Table VI). In the Malagasy lemurs Petter (1965) distinguishes two types of locomotion; one type which is generalized and quadrupedal, when the axis of the body is horizontal (e.g., the *Cheirogaleidae* and *Lemur*); and a second type which is specialized for leaping, also described by Napier and Walker (1967), where the body axis is vertical (e.g., the Indriidae and *Lepilemur*). Anogenital marking occurs virtually automatically by vertical rubbing in animals in the latter category. On the other hand, in the quadrupeds the marking of vertical supports has given rise in *Lemur* to a behavioral "readaptation," so that two forms of the same behavior are present. First, there is a generalized form practiced on more-or-less vertical supports and accompanied by the lowering of the hindquarters toward the support, the principal hold being by means of the hindlimbs. This corresponds to "gland-dragging" described by Eisenberg and Kleiman (1972) in the Viverridae. Anogenital marking in *Lemur fulvus* is also an example (see Fig. 4C). The important characteristic of this generalized form is that marking occurs without the animal changing its direction and often without an interruption in progression. Second, there is a specialized form which may be practiced on all types of support but most frequently on a vertical support, a branch, or a conspecific. The subject interrupts its progression,

TABLE VI

Behaviors Related to Anogenital Marking in Prosimians

Degree of specialization	Types of scent deposit	Relative importance of the secretory or excretory component	Relative importance of the locomotor component	Urine	Genital and anal gland secretions	Feces
Nonspecialized marking	Deposition by squatting	Marking behavior dominated by the secretory or excretory component	Locomotor component absent	Urination *Lepilemur* sp. ? *Hapalemur griseus* *Propithecus verreauxi* *Indri indri* *Perodicticus potto* *Arctocebus calabarensis* *Tarsius* sp.	Genital rubbing Lemuridae *Avahi laniger* ? *Propithecus verreauxi* *Indri indri* *Tarsius* sp.	Defecation with more or less smearing *Microcebus murinus* *Galago demidovii* *G. senegalensis*
Specialized marking (more or less) stereotyped	Deposition by squatting and/or dragging	Both components are functionally integrated		Urination with rhythmic micturation *Microcebus murinus* *Microcebus coquereli* *Cheirogaleus major*	Genital-rubbing with dragging *Microcebus murinus* *M. coquereli* *Cheirogaleus medius*	Defecation *Cheirogaleus medius* *Cheirogaleus major* *Nycticebus coucang*

				Daubentonia madagascariensis Lorisidae	*C. major* Lemuridae Galaginae *Perodicticus potto* *Arctocebus calabarensis*	
Highly specialized and/or stereotyped marking	Deposition without squatting or dragging	Marking component dominates behavior	Locomotor component is present or absent	Urine-washing with or without sole-wiping *Microcebus murinus* *Galago demidovii* *Galago senegalensis* *Galago alleni* *Galago crassicaudatus* *Loris tardigradus* Deposition on a partner during urine-washing *Perodicticus potto* Directed urination *Galago elegantulus* Tail-lashing (during urination) *Propithecus verreauxi* Head-rubbing (on urine-mark) *Lemur fulvus*, *L. mongoz*	Marking with about face *Lemur catta* *L. fulvus* *L. macaco* *L. mongoz* Genital-scratching *Perodicticus potto* *Lemur fulvus* Scrotal-marking of the tail and of the internal side of the foot *Lemur fulvus*	Sole-wiping *Galago senegalensis* *Microcebus murinus* Head-rubbing (on fecal marks) *Lemur fulvus*

usually does an about-face and then reverses in order to bring the hindquarters (anus or anogenital region) into contact with an object which has usually been identified beforehand by sniffing. In the case of a vertical support the hindquarters are usually above the level of the head, the principal hold being with the hands. An example of such anogenital marking in *Lemur fulvus* is shown in Fig. 4D. Often one or both of the legs are raised above the horizontal level, either gripping the vertical support or even remaining suspended in midair. In the latter instance the animal has only three points of support, the hands and the anogenital region. The important aspect of this form of marking is that it interrupts progression and usually involves a change in the initial direction of the animal which must turn about again before continuing along its original path. It can thus be seen that, as for urine-marking, the adaptation of anogenital-marking to the arboreal habitat may be expressed, even in the same species, by the appearance of a behavior more clearly specialized than the primitive or generalized form. Using the scheme of Eisenberg and Kleiman (1972) for carnivores (Viverridae), but without any evolutionary implications, Table VI lists the different types of scent deposits involving the anogenital region of prosimians according to the degree of behavioral specialization.

2.3.3. Perception of Odors

2.3.3.1. The Special Nature of Olfactory Perception in Prosimians. The scent signal, subject to the control of the neuroendocrine–exocrine system (Le Magnen, 1970), is as complex as behavior under the control of the association and neocortical areas (see Pariente, Chapter 10 on the importance of memory), which are well developed in prosimians (Stephan and Andy, 1964). The signal may have general semantic value (relating to sex, group membership, species, etc.) but will always be strongly influenced or distorted by individual components. The characteristics of the phyletic position of social communication in the prosimians with respect to lesser evolved mammals is the enlargement of the sensory range utilized in communication. It is not the superiority of one particular sensory category which is responsible for the richness of prosimian communication, but rather the diversified, flexible, and convergent manner in which the different senses are used, especially at the level of emission: physical contact, vocalization, postures, mimicry, etc. The resulting advantage is the multiplicity and especially the plasticity of the means of expression. The odoriferous signal in prosimians, which may be interchanged with, reinforced by, or mixed with other behaviors, has lost much of its rigidity of meaning compared to other mammals. Signoret and Du Mesnil Du Buissen (1962), for example, have shown that the immobilization response of the sow in estrus is largely dependent on the appropriate olfactory signal.

2.3.3.2. Modalities of Perception. Two types of perception correspond to the two principal modes of scent diffusion, i.e., directed or nondirected perception of odors.

a. Automatic or nondirected perception. This occurs when, under favorable environmental and physiological conditions (attention, motivation, etc.), perception of the signal does not depend on the perceiver having to change its behavior in any way, such as deliberately sniffing a particular source. Nondirected perception, however, does not signify passive perception. Observations of *Lemur catta*, *Cheirogaleus* (A. Schilling, personal observation), *Arctocebus*, and *Perodicticus* (Charles-Dominique, 1977), progressing along a branch, for example, in unfamiliar territory, provide convincing evidence that the olfactory sense of prosimians is constantly on the alert and that information received in this way is not "accidental" (as it may be, for example, in man) but is continually being sampled.

b. Directed perception. It appears that prosimians detect odors only at a relatively short distance (Charles-Dominique, 1977) but, when the odoriforous source is determined or even suspected, directed perception is physiologically very different from nondirected perception. Directed perception involves sniffing in short, rapid, successive bouts (the smaller the species the more rapid the bout), as close as possible to the source and designed to saturate the olfactory mucosa at the necessary pressure and with a sufficient quantity of odor molecules (Negus, 1958). This mode of perception is used in indirect social communication for sampling marks left by a conspecific, for example, *Propithecus verreauxi* (Richard, 1974a), *Lemur fulvus* (Harrington, 1974), *L. catta* (Schilling, 1974), or marks left by the subject itself, for example, *L. catta* (Schilling, 1974). It is also used in direct social communication on occasions when physical contact characteristically occurs:

1. Naso-nasal contact—*Lemur macaco, L. catta, Propithecus verreauxi* (Jolly, 1966b), *Galago senegalensis* (Andersson, 1969; Bearder and Doyle, 1974a).
2. Nose-muzzle or nose-cheek contact—all the Lorisidae (Charles-Dominique, 1977).
3. Naso-genital contact—*Lemur macaco, L. catta, Propithecus verreauxi* (Jolly, 1966b), *Galago senegalensis* (Doyle, 1974a), *G. demidovii* (Charles-Dominique, 1977), *Microcebus coquereli* (E. Pagès, personal communication).
4. Naso-anal contact—*Propithecus verreauxi* (Richard, 1974a), *Galago senegalensis* (Sauer and Sauer, 1963; Andersson, 1969).
5. Naso-anogenital contact—*Lemur fulvus* (Harrington, 1974), *Microcebus murinus* (Petter-Rousseaux, 1964), *Cheirogaleus major* (A. Schilling, personal observation), *Galago senegalensis* (Bearder and Doyle, 1974a).

2.3.4. Some Consequences for Olfactory Communication

From the point of view of the sensory signal itself, the means available for olfactory communication in prosimians fall into two categories. First, direct signals which involve an immediate communication or, at least, in which the delay does not exceed the time necessary for propagation of the stimulus; for example, signals involved in contact between animals, signals not directed on to the substrate, etc. Second, indirect signals which involve a communication delayed in time and/or mediated in space; for example, environmental marking, marking of conspecifics, etc. (Marler, 1965). As for other mammals, olfactory communication, as well as other chemical types of communication of both the gustatory and vomero–nasal type, is the only means of transferring information between conspecifics concerning the neuroendocrinological system. Le Magnen (1970) describes it as "a very typical loop of positive and negative feed-back mechanisms." Another advantage of communication by chemical signals is its perfect adaptation to nocturnal life (Le Magnen, 1970) which is the case for the majority of prosimians. The advantage over auditory communication is particularly apparent when there is a risk of predators. Charles-Dominique (1977) stresses the sparseness of acoustic signals compared to the richness of olfactory signals in the Lorisinae, in which defense strategies are based upon their unobtrusive habits.

The uniqueness of olfactory communication compared to other types of sensory communication lies in the temporal dimension introduced by the chemical message. Compared to visual and acoustic signals, olfactory signals persist, which means that there is less need to repeat the signal for it to remain effective, and it permits the establishment of a system of deferred communication which is extremely useful for solitary species (Charles-Dominique, 1968). Evans and Goy (1968) believe that the brachial and genital secretions of *Lemur catta*, for instance, remain effective for several days.

3. THE RANGE OF OLFACTORY COMMUNICATION

If a sensory message may be defined as signal + meaning = message, then one is forced to recognize that knowledge of olfactory communication is extremely limited, in the majority of situations regarded as social, to describing the type of emission, or reception, of certain olfactory signals without necessarily grasping the full meaning.

Since there is even less information on the content than there is on the origin and nature of the olfactory signal, one's approach is reduced to relating olfactory occurrences (perception or production of odors) to a physiological or behavioral situation contiguous in space and time. Objectively, however, there is a big difference between simply describing the contiguity of two behaviors and deducing one from the other. As an example, it seems that in certain gregarious lemurs a significant correlation exists between the frequency of certain scent-markings and the social hierarchy. This may also be true of some nongregarious species, like *Galago demidovii* (Vincent, 1969) and *G. senegalensis* (Andersson, 1969), in which urine-washing is more frequent in dominant males. However, until further evidence is available that a certain olfactory substance is a releasing stimulus, scent-marking can only be said to play a general role in the social hierarchy; and, in particular, it cannot be assumed that an animal marks more often because it is or becomes more dominant.

Although Andersson (1969) showed that, in the male *Galago senegalensis*, both chest-rubbing and urine-washing decrease as dominance is lost, and increase in the female if she becomes dominant, it has, nevertheless, not been proved that it is the olfactory stimulus that is involved. The crucial factor could well be the visual stimulus of the scent-marking movement itself. Even if it can be shown that it is the olfactory stimulus itself that is efficient, it would still be necessary to know whether the function of the stimulus is qualitative, i.e., whether it has a special meaning, or whether it is quantitative, i.e., whether it is the frequency (or strength) of the signal, which is the critical factor in a given social context. In the case of social dominance, where the message may be carried by specific odors, the problem remains as to whether the signal (for example, the content of the glandular secretion) is solely responsible for the information concerning the social status of the individual animal, whether it is dependent on the prevailing social context, or even on the physiological state of the eventual recipient of the message. Until the necessary analyses are made, one can only describe the behavioral component, studied in the laboratory or in the field, where behavioral studies suggest olfactory involvement.

3.1. Nonsocial Factors in Scent Deposition

It is important to stress that many factors foreign to the social context may be responsible for the same behaviors that prosimians use for both receiving and emitting olfactory signals. At this elementary level, the olfactory behavior, especially in the form of active marking, does not have the value of a signal which has significance for a conspecific, but may be interpreted as a response to the environment, or to the internal environment of the subject itself.

3.1.1. Factors Related to the Environment

3.1.1.1. Climatic or Environmental Factors. According to Welker (1973) the frequency of urine-washing in *Galago crassicaudatus* increases when (by manipulating the temperature or humidity) the relative humidity of the ambient air or the substrate, on which the animal finds itself, is lowered. Light intensity may also be an external factor affecting the frequency of scent-marking. Seitz (1969) has demonstrated that, in *Nycticebus coucang* males, the frequency of scent-marking clearly increases when the light intensity changes from 2.0 to 0.6 lux. The influence of humidity and light intensity on scent-marking in a pair of *Cheirogaleus major* has been examined by measuring the dry weight of scent marks left in an experimental situation and comparing this with locomotor activity (number of passes). Table VII shows the total results for 6 days during which two identical experiments were carried out under two conditions of humidity and light intensity. It is evident that the weight of marks increases considerably in darkness and under increased humidity conditions. Also the increase in weight is relatively greater than that of locomotor activity and, consequently, scent-marking and locomotor activity are not propor-

TABLE VII

Total Locomotor Activity and Weight of Scent Marks over Six Days for Each of Two Identical Experiments under Two Different Conditions of Light and Humidity

	Humidity		Light intensity	
	65%	95%	Red light[a]	Darkness
Total locomotor activity	363	1972	164	2090
Total weight of scent marks (gm)	0.30	3.72	0.28	6.85
Marking index Im_1	83	188	17	33

[a] Red light, 1 lux at 1 m distance from the light source. Marking index, Im_1 = ratio of weight of marks to locomotor activity derived by the formula $Im_1 = M/A$.

tional. This relationship is expressed as the ''marking index'' Im_1. An increase in the marking index Im_1 means that, for equal locomotor activity, the weight of the scent marks in darkness or under conditions of high humidity, increases by comparison with their weight under red light and normal humidity.

Climatic factors have, in addition, an indirect effect on olfactory behavior by modifying the emotional disposition of prosimians. This is clear for the gregarious forms in which rain, wind, and especially thunderstorms raise the general level of excitation. *Lemur*, in particular, increase their frequency of sniffing and marking under these conditions. In the wild as well as in captivity, *Lemur* and other prosimians do not scent-mark randomly. There is a correlation between scent-marking behavior and the nature of the substrate, for example, the shape or diameter of a branch, its orientation, and height above the ground. This has been demonstrated for wrist-marking in *L. catta* (Schilling, 1974) and for urine-washing in four species of nocturnal prosimians in the laboratory (Doyle, 1975).

3.1.1.2. Familiarization of the Environment. Numerous authors have considered Hediger's (1950) hypothesis of making the environment familiar by scent-marking. The role of scent as an agent of ''security'' is only one interpretation of the preference of prosimians, particularly solitary species, for an environment impregnated with their own odors or, conversely, of the caution shown in an unfamiliar environment, either odorless or in which the odor has been disturbed.

For a period of 6 months a *Microcebus rufus* (*Microcebus murinus rufus*), which was captured blind and had lived happily for several years in the laboratory, was kept in an isolated cage consisting of four identical polyurethane nest tubes in the form of a cross and opening into a central cage. A photoelectric cell system recorded the time spent in each of the tubes. The animal showed a clear preference for one of the tubes. After careful washing of the preferred tube, exploratory behavior increased but, after several days, the animal finally returned to sleep in the initially preferred tube. When the spatial location of the tubes was changed in a systematic manner without modifying the odor, the animal followed the chosen tube. In animals with normal vision similar results were obtained only when the tubes were washed. This sort of experiment suggests that even if olfactory cues do not suffice entirely to explain certain behaviors, they nevertheless play an important role. Using a system of wooden supports, Seitz (1969) reported that *Nycticebus coucang* preferentially followed the supports previously marked. A similar experiment with *Cheirogaleus major* did not produce the same results (A. Schilling, personal observation).

The usual olfactory response of prosimians to the introduction of foreign objects or odors, or to modifying existing marks, or to placing animals in an unfamiliar environment, is an increase in the frequency of sniffing in *Nycticebus coucang* (Seitz, 1969) and *Galago crassicaudatus* (Ehrlich, 1970), and by an increase in marking in *Loris tardigradus* (Ilse, 1955), *N. coucang* (Seitz, 1969), *Lemur fulvus* and *L. catta* (Andrew, 1964), *Microcebus coquereli* (E. Pagès, personal communication), *Galago senegalensis* (Sauer and Sauer, 1963; Andersson, 1969; Bearder, 1969), *G. crassicaudatus* (Andrew, 1964), and *G. demidovii* (Charles-Dominique, 1974a). Andrew and Klopman (1974) have also shown that *G. crassicaudatus*, *G. senegalensis*, and *G. demidovii* may urine-wash after sighting a strange object at a distance (see below).

Figure 9 illustrates the reaction of a group of *Lemur fulvus albifrons* to the introduction into their cage of a branch marked by conspecifics of another group (Fig. 9A) and of a tissue impregnated by the scent marks of a civet (Fig. 9B). In this connection certain odors, for example, onions, tobacco, cigarette smoke, etc., invariably release tail-marking in some *L.f. albifrons* of both sexes (J. J. Petter, personal communication). This may be a specific response to a particular olfactory stimulus, or it may result from a general state of excitation indirectly caused by the stimulus.

In a *Cheirogaleus major* pair the author recorded locomotor activity and weight of fecal marks on supports symmetrically placed on either side of a partition separating an experimental cage into two parts, both before and after opening of the partition. The results showed that access to an unfamiliar environment leads to a change in the relationship between locomotor activity and the weight of fecal marks (Table VIII).

Although most locomotor activity occurs in the original space (I), most of the marks are found in the new space (II). The increase in the "marking index" Im_1 suggests that *Cheirogaleus*, like other prosimian species, have a tendency to scent-mark more in a new environment, or to "address" their scent marks to an unfamiliar environment. This familiarization of the environment by scent-marking naturally also involves the familiar home range in which the smell of the individu-

Fig. 9. Reactions of a group of *Lemur fulvus albifrons* to the introduction into their cage of (A) a branch marked by a group of conspecifics; note the licking behavior of the animal on the right and the marking with simultaneous sniffing of the animal on the left; and (B) a cloth impregnated with the glandular secretion of a civet.

al's scent is regularly maintained (Hediger, 1950). In the field Bearder (1969) observed that *Galago senegalensis* urine-wash more frequently in the sleeping tree and immediately adjacent area. In captivity scent-marking, involving indirect anogenital urine or fecal deposition in *Nycticebus coucang* (Seitz, 1969), *G. demidovii* (P. Charles-Dominique, personal communication), *Microcebus murinus, Cheirogaleus medius,* and *C. major* (A. Schilling, personal observation), is often concentrated on the top and sometimes the walls of nest boxes. Urine-washing in *G. senegalensis* (Andersson, 1969) and wrist-marking in *Hapalemur griseus* (A. Schilling, personal observation) often occurs on returning to the nest.

TABLE VIII

Changes in the Relationship between Locomotor Activity and Fecal Marks as a Function of Exposure to an Unfamiliar Environment[a]

	Partition closed		Partition open	
	I	II	I	II
Locomotor activity (%)	100	—	63.2	36.8
Weight of fecal marks (%)	100	—	38.3	61.7
Marking index Im_1	20.4	—	28.2	196

[a] The formula Im_1 expresses the relationship between the two.

3.1.2. Physiological and Psychophysiological Factors

3.1.2.1. Ontogeny of Marking Behavior. Precise information is lacking on the behavioral development of prosimians (see Doyle, Chapter 5) but scent-marking, which is related to sexual maturation, may be distinguished from that which appears before or during the juvenile period. The former concerns scent-marking which involves the genital glands in males and females in both the Lorisidae and the Malagasy prosimians, as well as scent-marking which involves glands considered as secondary sexual characteristics (see above). These include anal-marking in certain lemurs and marking with the antebrachial complex in *Lemur catta* and *Hapalemur griseus*. According to Chandler (1975) it also includes rubbing of the head and palms in *Lemur fulvus* since these behaviors appear just after the first matings. The latter include primarily scent-marking by means of urine and feces and, probably other little understood behaviors; for example, muzzle-wiping observed in a *Microcebus murinus* at 3 months of age (A. Petter-Rousseaux, personal communication) or chest-rubbing observed after the twelfth week in *Galago crassicaudatus* (A. B. Clark, personal communication).

Urine-marking appears very early in all Lorisidae. Marking by dragging has been observed in *Perodicticus potto* at 2 to 3 months of age (Charles-Dominique, 1974b) and in *Nycticebus coucang* (Seitz, 1969) at the same age. Urine-washing has been observed as early as 3 weeks of age in *Galago senegalensis*, though not until about 36 weeks, at least in the case of the male, does the frequency of urine-washing correspond to that of the adult, (Doyle, Chapter 5). Petter and Peyrieras, (1970a; see Fig. 5C) report scent-marking in a *Daubentonia* female before puberty. The author personally studied the ontogeny of fecal-marking in two *Cheirogaleus major major* litters of which one, brought to the laboratory at the age of 4 weeks without its parents, required personal care, while the other was born in the laboratory. The first indications of marking were observed several days after the animals began accepting their first fruits and insects, at which time they had already explored all of their cage. A diet based entirely on milk results in feces which are not of a solid enough consistency to be effectively deposited on branches. The interesting fact is that marking appeared in both litters during the sixth week which suggests that an innate maturational factor is involved (see Fig. 6C). The behavior rapidly becomes identical to that of the adults, with marks reaching more than 10 cm in length at the end of the sixth week. In addition, the behavioral sequence of raising the tail, lowering the hindquarters, and creeping may occur in the absence of fecal deposition.

Other prosimian species do not appear to be as precocious in acquiring marking behavior. In *Lemur*, for example, the social group seems to be an important factor in the acquisition of certain olfactory–visual behaviors (J. J. Petter, personal communication). Rubbing of the anus, the hands (see Fig. 16B), and the head on branches, or self-marking of the tail, seem to be, in part, learned by imitating adults. Chandler (1975) studied the development of anogenital rubbing of the head and palms in *L. fulvus*, raised either in social groups or in male–female pairs. All of these behaviors, except anogenital rubbing, appeared first in the male raised in the social group. It is difficult, however, to determine the age at which these behaviors become significant and play a role in olfactory communication, because they seem initially to be limited to the gestural component; for example, rubbing in the case of anal-marking in *Lemur* without any scent being deposited. In other words, the practice of certain marking behaviors may be acquired perhaps by social learning before marking becomes fully functional.

3.1.2.2. Elementary Physiological Variations in Olfactory Behavior.

a. State of hunger. Generally, the state of hunger directly influences olfactory behavior by lowering the perceptual threshold for certain odors. In the case of fecal or urinary marks nutrition has a double role, acting on the quantity or frequency, and also on their distribution during the periods of activity. Quantitatively, it appears that animals will mark more when they eat and drink more; this may vary on a day-to-day basis. In some nocturnal Malagasy prosimians there is also a seasonal variation in the amount of food and water ingested which, among other things, causes seasonal variations in levels of activity (research in progress). In a pair of *Cheirogaleus medius,* in which activity was recorded over a year (Schilling, in press), the curves for food consumption and fecal marks are roughly positively correlated (see Fig. 11A).

b. Circadian cycle. It has also been observed in certain nocturnal species—*Galago senegalensis*, *G. crassicaudatus*, *G. demidovii*, and *Microcebus murinus* (Pinto *et al.*, 1974; G. A. Doyle, personal communication)—that urine-washing is more frequent in the period following waking. The same is true of fecal-marking in *Cheirogaleus major*. This fact is related, first, to the high level of general activity of the animals during this period, in *Galago demidovii* (Charles-Dominique, 1972), *G. senegalensis* and *G. crassicaudatus* (Pinto *et al.*, 1974), *Nycticebus coucang* (Seitz, 1969), *Microcebus murinus* (Pinto *et al.*, 1974; Pagès and Petter-Rousseaux, in press), *Phaner furcifer* (Charles-Dominique and Petter, in press), and *M. coquereli* (Pagès, in press). Second, it is related to the physiological state of pressure on bowel and bladder as a result of accumulation of waste matter during the resting period which, in turn, stimulates the autonomic centers controlling micturation and defecation. Apparently several types of scent-marking behavior, not only excretory (Andrew and Klopman, 1974) but also glandular, are explained in some cases by pressure due to accumulation of marking substances or even the simple irritation of the organ or cutaneous region concerned. These physiological factors should not be overlooked in behavioral interpretations of marking behavior.

c. ***Seasonal cycle of activity.*** One of the characteristics that distinguishes the Lemuriformes from the Lorisiformes is the existence in the former of a seasonal cycle of activity related to the photoperiod (Petter-Rousseaux, 1970). This cycle is particularly evident in *Cheirogaleus*, which hibernate for several months, and in *Microcebus murinus* (Petter, 1962). It would be of interest to determine whether frequency of scent-marking also varies seasonally. In *M. murinus* the mean hourly frequencies of three types of scent-marking in a group of animals observed in either an active or relatively inactive locomotor period have been compared. Figure 10 shows that the frequency of scent-marking increases during the active period and the proportion of different types of scent-marking, particularly anogenital marking, varies from one period to the next.

Although *Cheirogaleus medius* neither hibernated nor reproduced under the laboratory conditions of observation, the recording of locomotor activity, food consumption, and scent-marking (Schilling, in press), led to the following conclusions:

1. The curve for marking roughly follows that of food consumption (see Fig. 11A). The ratio of the weight of scent marks to food consumption, termed the "marking index" Im_2 (M/N), is not constant but contains certain peaks particularly at the time when the species would be coming out of hibernation (corresponding to early August in Beroboka, Madagascar), as well as at the beginning and, especially, at the end of the wet season (end of November and end of March; see Fig. 11B).
2. The curves of marking and of locomotor activity, the ratio that we define as "marking index Im_1" (M/A), are differentiated outside of the period between April and July. There is a curious but clear delay between the period when scent-marking reaches its highest frequency (February in Madagascar) and the period of greatest locomotor activity (December–January; see Fig. 11A).
3. Comparisons between the two "marking indexes" (see Fig. 11B) reveal that in October–November, these change inversely to one another, Im_1 decreasing and Im_2 increasing. This change in relationship corresponds to the mating period. In summary, the marking activity of *Cheirogaleus medius* shows an isolated annual

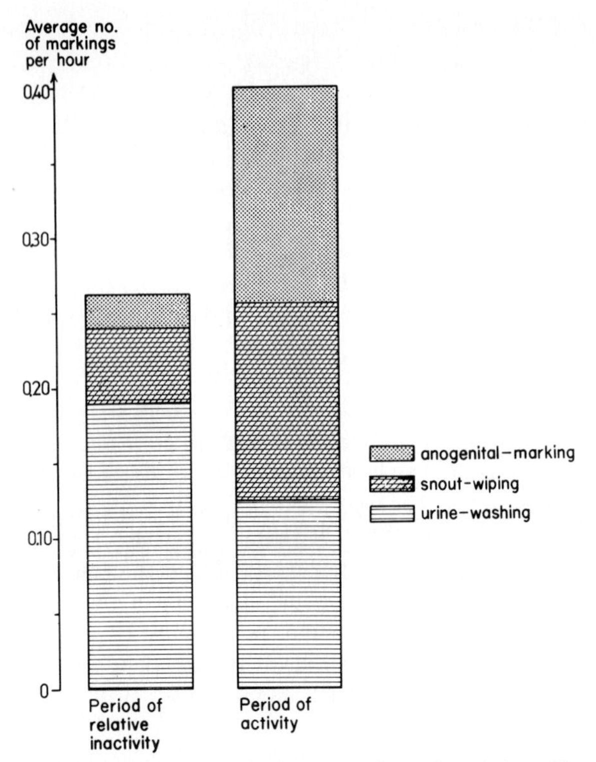

Fig. 10. Observed mean frequencies, in a group of five *Microcebus murinus*, illustrating seasonal variations in different types of environmental marking under laboratory conditions.

peak characterized by the high indexes Im_1 and Im_2 which implies that, during this period (February), the frequency of marking behaviors changes independently of both locomotor activity and the quantity of food ingested. Similar changes in the frequency of marking behavior have either not been demonstrated at all in other Malagasy species, for example, in *Lemur catta* (Evans and Goy, 1968) and *Microcebus coquereli* (E. Pagès, personal communication), or have been demonstrated only in relation to the period of reproduction (see below).

d. Individual and sexual variations. Apart from marking behaviors considered to be secondary sexual characteristics or related generally to sex (see above), the scent-marking behavioral repertoire of a given species is not practiced with the same frequency, to the same degree, or in the same manner between the sexes or between individuals. Individual variations, although recognized, have not yet been studied systematically except in relation to questions of dominance and the social hierarchy in which some animals have been shown to mark more often than others (Harrington, 1974; Chandler, 1975).

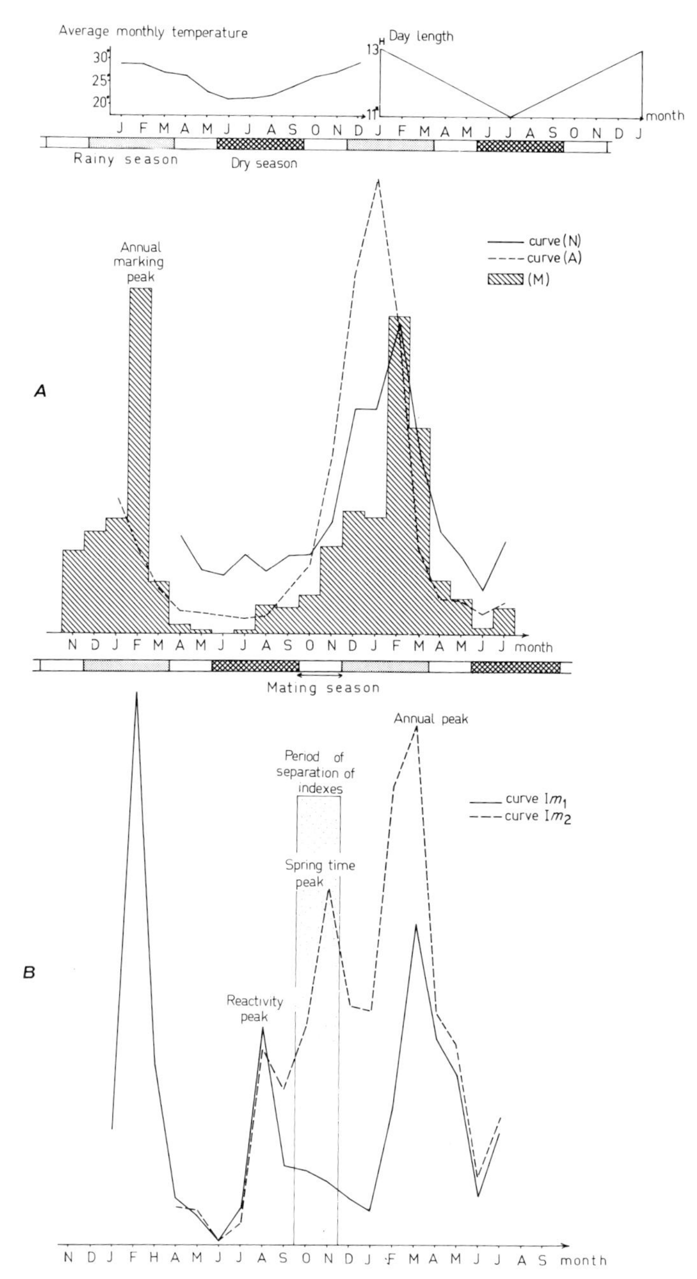

Fig. 11. Monthly variations in marking behavior in a pair of wild-caught *Cheirogaleus medius* under simulated climatic laboratory conditions. (A) Weight of fecal marks collected at a fixed location (M), locomotor activity level (A), and weight of food (N). (B) Comparison of marking indices: $I m_1 = M/A$; $I m_2 = M/N$.

Some authors have, however, shown sex difference in scent-marking behavior. In *Galago senegalensis* urine-washing is more frequent in males than in females (Doyle *et al.*, 1967; Andersson, 1969; Doyle, 1975) under laboratory conditions. The same applies to *G. demidovii* (Doyle, 1975). In *G. crassicaudatus* and *Microcebus murinus*, under similar conditions, females urine-wash more (Doyle, 1975) although in neither of these cases do these differences appear significant (G. A. Doyle, personal communication). In *Propithecus verreauxi* (Jolly, 1966b) and *Indri indri* (Pollock, 1975), urine-marking is practiced more often by the males and anogenital marking is practiced more often by the females. In those species in which both sexes mark with the glandular region of the neck, it is more frequent in males than in females, for example, *Phaner furcifer* (Charles-Dominique and Petter, in press) and *Indri* (Pollock, 1975).

The varied scent-marking repertoire in numerous species of *Lemur* is not practiced in the same manner by males and females (Petter, 1962; Rumpler and Oddou, 1970). Affolter (1937) and Jolly (1966b) report the same finding for *L. catta* and Harrington (1975) for *L. fulvus rufus*. In *L.f. albifrons* and *L. mongoz* anogenital marking of conspecifics, as well as rubbing the head, rubbing the palms and "passing-over," are essentially male behaviors. Both sexes of the same species, but more often the males, practice muzzle-wiping and tail-marking. Anogenital marking of objects, at least in *L.f. albifrons* in captivity, (see Fig. 10) is practiced with complete about-face more often by males, while females practice the nonspecialized form more often (A. Schilling, personal observation).

e. Level of excitation. Andrew (1964), Ehrlich (1970), and Andrew and Klopman (1974) have noted that several species of *Galago* often urine-wash on seeing an unfamiliar object from a distance. This behavioral reaction, which manifests itself in other ways in Malagasy lemurs when isolated from conspecifics, for example, brachial-marking in *Hapalemur griseus* (J. J. Petter, personal communication), urine-marking in *Daubentonia* (Petter and Peyrieras, 1970a), and anogenital-marking in *Lemur fulvus albifrons* (A. Schilling, personal observation), suggests that a mere increase in the level of excitation (attention, curiosity, fear, etc.) is also a major factor responsible for releasing scent-marking behavior in individuals. According to this hypothesis, scent-marking may be interpreted as a means by which prosimians may reduce the level of emotional tension, outside of social conflict situations. In other words it may be regarded as "displacement" behavior, several examples of which have been presented for *L. catta* (Schilling, 1974).

In summary, an important part of olfactory communication in prosimians takes place outside of the social context and should be considered as an individual olfactory relationship with the environment. This relationship, which has been demonstrated in all species studied, seems to play a vital role enabling these macromatic animals to locate and be located in their environment. Understood from this viewpoint, scent-marking behavior may be conceived of as an elementary and perhaps primitive language, giving the prosimian a range of olfactory expressions to affirm its identity, and to express itself and any changes in its mood. Captive Lemuridae, which often exhibit pathological, stereotypic behaviors, characteristic of captive conditions, often indulge in fits of scent-marking which illustrate their inability to adapt completely to captive conditions.

3.2. The Olfactory Dialogue

Among all the eco- and ethophysiological distinctions that can be made between the prosimians, their separation into diurnal or nocturnal forms, and solitary or gregarious forms, is fundamental, since the different neurosensory adaptations involved which characterize these forms necessarily imply certain modes of social communication (Charles-Dominique, 1975).

Sensory adaptation in prosimians is not characterized by the extreme development of one of the sensory systems at the expense of the others, as is the case in certain other mammalian forms, but rather by the convergent adaptation of the entire sensory system to a particular mode of life. Thus, in the diurnal forms, while the visual sense is well developed (see Pariente, Chapter 10), the olfactory sense remains undiminished in importance. Conversely, although in the nocturnal forms olfaction is of major importance, it functions in harmony with the other equally well-adapted sensory modalities, like audition and vision.

Concerning social communication in general, the olfactory sensory system is the only sensory modality able to deal with discontinuity in time between stimulus emission and reception. This is the major and probably a primitive advantage of scent-marking. The scent mark in olfactory communication is what writing is to language; the message may be conveyed to the receiver long after its emission. Particularly in those solitary species which range over large territories, like *Perodicticus potto*, *Arctocebus calabarensis*, *Nycticebus coucang*, and *Daubentonia madagascariensis*, this constitutes a unique means of communication. It is possible, therefore, that olfaction is the most fundamental of the sensory modalities in the more nocturnal and solitary species. It is equally possible that in gregarious forms organized in troops, such as in certain *Lemur* species, or in small family groups, such as the Indriidae and certain Lemuridae, scent-marking is also practiced as an indirect means of communication, not only in intergroup relations but also in interindividual contacts within the group, for example, in the acquisition and maintenance of the social hierarchy. However, olfactory communication is capable of assuming so many varied forms (volatile odors, pheromones, nondeposited odors, incidental traces, etc.) that scent-marking behavior may be regarded as a category or, rather, a specialization, of olfactory communication.

Olfactory communication may be said to involve two successive levels of social relations, an interindividual level and a more "organized" level in which the message is addressed to, or has meaning for, a number of individuals.

3.2.1. Interindividual Level

3.2.1.1. Mother–Young Olfactory Relations. The frequency of reciprocal sniffing after birth, for instance, in *Tarsius* (Niemitz, 1974), suggests that olfaction plays an important role at this time. This role is certainly facilitated by the constant and close physical contact between the mother and her young, particularly grooming and licking, as well as in sleeping and oral transport of the young (Doyle, 1974b).

In a brief laboratory experiment on recognition of the young *Galago crassicaudatus* by the mother, Klopfer (1970) concluded that the dominant sensory modality involved was not visual, but rather auditory and olfactory, without specifying the relative importance of each. Some informal laboratory observations concerning this problem may be helpful. The first (M. Perret, personal communication) concerns precocial olfactory interactions between mother and young. A bilateral olfactory bulb oblation did not prevent a female *Microcebus* from giving birth normally, but her maternal behavior was disturbed. The day after giving birth, the infant, which had fallen from the nest, remained undetected by the mother, which would not subsequently accept it. The second observation concerns recognition of the young and maternal behavior after the mother–young olfactory relationship had been established. Two *Cheirogaleus major* infants at 14 days of age (observed by A. Schilling at Maroantsetra, Madagascar, 1970) were placed in a container which opened into two bamboo tubes 30 cm long. These tubes, which served as nest boxes, were impregnated either with the odor of the infants themselves or with that of a young male 13 weeks of age from a different area. The mother, searching for her young, investigated both tubes (the young remained immobile and silent), carefully sniffed each entry, and finally recovered her young from the correct tube without having entered the other. This did not necessarily constitute proof that olfactory stimuli predominate in maternal discrimination, since it proved impossible subsequently to obtain positive evidence from the same subject. Nevertheless, the reaction appears significant not only from the point of view of the discrimination itself but also from the point of view of the relatively short distance involved. It seems that in prosimians, recognition of the young may be less dependent upon olfactory stimuli than it is in other nonprimate mammals, e.g., sheep (Klopfer and Gamble, 1966; Bouissou, 1968) and rodents (Noirot, 1969; Gandelman *et al.*, 1972). Thus, newborn *Microcebus*, after being manipulated by human hands, are accepted by the mother or even by another lactating female. The possible role of the modification of odors emitted by prosimian young in such things as determining ages of separation from the mother, in olfactory imprinting as a basis of species recognition, and in sexual preferences, are important but, as yet, only speculative questions. In *Perodicticus* and *Arctocebus* Manley (1974) suggests the possibility of an olfactory alarm signal emitted by the mother which immediately causes the young to stop all activity and cling.

3.2.1.2. Olfaction and Reproductive Behavior. The relationship between reproductive physiology and olfaction in prosimians has long since been emphasized (Hill, 1953; Petter, 1962). However, experimental investigations have only recently been attempted. First, as has been shown, behaviors which are presumably olfactory (sniffing and marking) usually occur together with other behaviors (licking, chewing, salivation, swallowing, even ingestion; see Figs. 9A and 16C). They are thus part of a "general investigation by the muzzle" in which other sensory organs (Jacobson's organ, gustatory and tactile senses) may be involved. The sensory modality involved in any particular suspected stimulus, as well as the precise origin of the stimulus, remains to be verified experimentally. In the case of olfactory investigation of the female genitalia by the male, which most authors label "anogenital," it is not known whether the primary concern is the glands of the internal or external genitalia, the anus, vaginal secretions, or urine, etc. Not only have none of these odoriforous substances, suspected to be involved in prosimian communication, been isolated or synthesized, as has been done in the simian primates (Michael *et al.*, 1972), but also proof of their involvement in physiological or behavioral responses has not yet been established. It would be more prudent, therefore, to speak of chemical rather than olfactory stimuli, and signals rather than pheromones, even in hypothesizing a pheromonal action of these signals which, as Wilson and Bossert (1963) have shown, may act at the two different levels listed below:

1. An endocrine level, in which the response to the message is primarily physiological (primer effect) even if a secondary behavioral effect follows. As an example, a response of this type is suggested by Jolly (1966b) as one of the possible explanations for the synchronization of estrus in *Lemur catta*. Explanations at the physiological level may provide the answer to a number of naturalistic observations on olfactory communication, e.g., the fact that the presence of a male advances estrus in *Arctocebus calabarensis* (Charles-Dominique, 1968), and that gonadal involution occurs in subordinate *Microcebus* males (M. Perret, 1977).
2. A behavioral level, in which the response to the message is an immediate modification in the behavior of the receiver (releaser effect). Many of the reactions of males to female chemical signals enter into this category (see Fig. 12).

Olfactory stimuli may be involved at both these levels but, at the present time, the relationship between the appearance or modification (the frequency, for example) of a certain number of behaviors presumed to be olfactory, and the mating period, can be established but without being able to define the nature of this relationship and its precise role. Olfactory communication is itself deeply rooted in a complicated background, since the mechanisms of reproduction in these seasonally polyestrous primates are still poorly understood. Factors such as the photoperiod in *Microcebus murinus* (Petter-Rousseaux, 1970), ecological conditions in *Propithecus verreauxi* (Sussman and Richard, 1974; Richard, 1974b), diet and social stress in *Tupaia* (von

Holst, 1974) are also influential; olfaction probably plays an important auxiliary role in each of them. Until now this problem has been investigated only at the behavioral level and answers have been sought to clear several points.

a. The relationship between observed "olfactory" behavior and the reproductive cycle. Evans and Goy (1968) distinguish behavioral changes that occur throughout the entire period of reproduction (long-term cycle), called the precopulatory period (Richard, 1974a), or the period of courtship (Charles-Dominique, 1974b, 1977), from those associated with the femal estrous cycle itself (short-term cycle).

i. The period of reproduction. Many observers have noticed an increase in sniffing and scent-marking during the reproductive season in both the Lemuriformes (Petter, 1962, 1965) and the Lorisiformes (Sauer and Sauer, 1963). Changes in olfactory behavior have been studied quantitatively in *Lemur catta* (Evans and Goy, 1968), *L. fulvus fulvus* (Harrington, 1975), and *Propithecus verreauxi* (Richard, 1974a). In the wild, the duration of the precopulatory period is from 4 to 6 weeks in *L. catta* (Jolly, 1966b) and *L.f. fulvus* (Harrington, 1975), about 6 weeks in *P.v. verreauxi* (Richard, 1974a) and, more importantly, corresponds to changes in the female genitalia. These quantitative studies show that: (1) an increase in observed olfactory behavior during the precopulatory period concerns both sexes but mainly males of *Lemur fulvus fulvus* (Harrington, 1975) and only the males of both *L. catta* (Evans and Goy, 1968), and *P.v. verreauxi* (Richard, 1974a): (2) in all three species studied, sniffing of the female genitalia increases. (3) Changes in scent-marking frequency appear to be complex, and cannot be ascribed simply to

Fig. 12. Courtship in *Lemur fulvus albifrons*. (A) Genital marking by the male accompanied by direct sniffing of the female genitalia. (B) Indirect sniffing of the female genital odors. The sexually excited male (note the erect penis) sniffs the mark just made by the female before "over-marking." (C) Licking the rhinarium accompanied by anogenital marking in the male. This behavior resembles one of the sequences of "Flehmen" described by Bailey (in press) in *Lemur catta*.

Fig. 12. *(continued)*

TABLE IX

Frequency of Nonsexual Behavior in Relation to the Female Cycle (Evans and Goy, 1968)[a]

Item	Vaginal condition[b] Diestrus (105)	Estrus (45)	$P^{c,d}$
Investigation of females	1.4 ± 0.1	6.1 ± 0.7	0.001
Tail display	2.3 ± 0.9	8.5 ± 1.9	0.01
Spur mark	8.3 ± 2.0	4.2 ± 1.1	0.05
Genital mark by females	4.8 ± 1.3	3.1 ± 0.4	N.s.

[a] Average score ± S.E.M. per 10 minutes.
[b] Number of tests in parentheses.
[c] "t" for correlated means.
[d] N = 14 pairs.

generalized increase. For example, the mean frequency of male anogenital scent-marking increases in *L.f. fulvus*, but decreases in *L. catta*.

Evans and Goy (1968) have also demonstrated complex changes in brachial-marking and tail-waving behavior during the reproductive period in *L. catta* (see Fig. 13). The frequency of brachial-marking (which they refer to as "self-expression") steadily decreases (as does aggression between males) while that of tail-waving, an effective means for dispersing male odors (Jolly, 1966b), reaches a maximum; Evans and Goy (1968) believe that these behavior changes "contribute to the activation of neuroendocrine mechanisms controlling ovulation." In *P.v. verreauxi* males Richard (1974a) describes two activities seen only during the reproductive period. These ac-

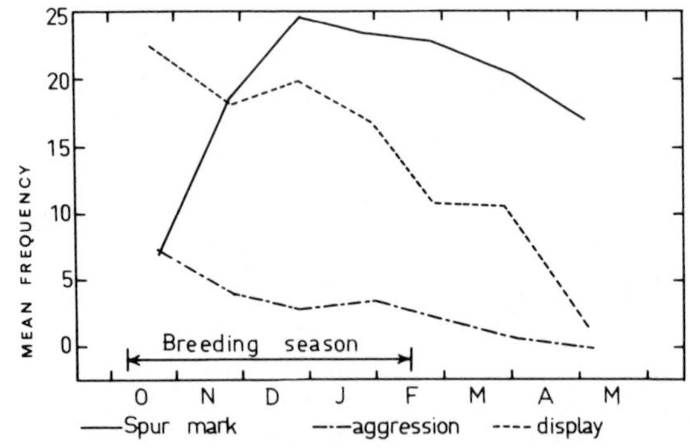

Fig. 13. Frequencies of three categories of behavior among males tested during and after the breeding season. Moving (monthly) averages per 10-minute test (N = 6 pairs). (From Evans and Goy, 1968.)

tivities consist of normal olfactory behaviors which, instead of occurring separately, are arranged in characteristic sequences: "endorsing" or neck-marking, followed by penile contact, often accompanied by urination, and followed in turn by rubbing of the perineal region on a female mark; "sniff-approach and mark" or approach of a female from behind, followed by naso–anal contact, followed by neck-marking and genital-marking of the environment. In *Lemur fulvus albifrons*, scent-marking behaviors also appear to be practiced more frequently in sequence than separately during courtship (see Fig. 12). Harrington's (1975) quantitative observations of male *L.f. fulvus* show a clear increase in scent-marking associated with female odors, sniffed either directly, or indirectly by sniffing areas which have recently been marked or simply where the female has been sitting. Although this has been quantified by Harrington (1975) only in *L.f. fulvus*, it appears that allomarking of females by males increases during the reproductive period in all *Lemur* species (Petter, 1962, 1965). This usually concerns anogenital marking, but in captive *L.f. albifrons*, allomarking by rubbing of the head on the sides or even the anogenital region of the female, in which case there is also a collection of odors by the male, has also been observed (A. Schilling, personal observation). These behaviors, as well as rubbing of the palms or muzzle, "passing-over," licking, etc., in *L. macaco macaco* (Petter, 1962), *L.f. albifrons* (A. Schilling, personal observation), which show a significant increase during the reproductive period, indicate that social communication is mediated as much through olfactory channels as, for example, visual and, at least from the point of view of its expression, should be considered as simultaneously multisensory. In the Lorisidae of Gabon, Charles-Dominique (1974b, 1977) distinguishes the period of courtship from that of mating, and indicates that odoriforous deposits of the glands situated in the genital area occur most frequently during male–female encounters. During the courtship period urine-marking increases, and a reciprocal exchange between partners occurs either by urine-washing during allogrooming (see above) or when the male rubs the cheeks on urine marks left by the female. In *Perodicticus potto* the possible role of secretions of the preclitoridian glands as a sexual attractant has been suggested (Charles-Dominique, 1966, 1975). Mutual odor exchange appears to be part of courtship in the Lorisidae: for example, urine-marking of the groomed individual in *P. potto* (Epps, 1974), which is still poorly defined since it is related to "genital-scratching-grooming" which itself is not associated with sexual behavior (Manley, 1974), and "passing-over" in *Arctocebus* males.

ii. Estrus. The period of estrus is relatively short in the prosimians (Doyle, 1974b) but is accompanied by significant behavioral changes, particularly in the nocturnal species. Systematic sniffing of the female genitalia by the males has become the basic olfactory behavior in all species studied. However, it should be noted that, according to Harrington (1971), adult male *Lemur fulvus* are incapable of distinguishing an estrous female from an anestrous female by odor alone. The time of estrus is characterized by a marked increase in the frequency of scent-marking, the deposition and collection of odoriforous material involving both the environment and conspecifics. In *Galago senegalensis*, the frequency of urine-washing remains unchanged in the females during estrus but increases by 30% in males from the beginning of estrus (Doyle *et al*., 1967). These authors have reported that inspection of the female genitalia remains at a fairly stable level throughout the year, at 1 to 9 times per 50-minute observation period, but that it increases noticeably at the time of estrus. Urinary allomarking and chest-rubbing on the female are also more frequent at the time of mating in *G. senegalensis* (Andersson, 1969). A peak in scent-marking and in the collection of odors during the time of estrus occurs in *G. demidovii* (Vincent, 1969) and in *G. crassicaudatus* (Jolly, 1966a). This is equally true for the Lorisinae in which, according to Seitz (1969), the urogenital marks of the females on the environment and on social partners reaches a peak during estrus in *Nycticebus coucang* and *Loris tardigradus*. Similar observations have been made in the Cheirogaleinae—frequent urine-washing (Petter, 1962), anal-marking (Andrew, 1964), and strictly genital-marking (A. Schilling, personal observation) in *Microcebus*. Certain behaviors such as rubbing or wiping of the labial commissures (A. Petter-Rousseaux, personal communication; A. Schilling, personal observation) licking, chewing and biting of branches, as well as rubbing of the forehead (A. Schilling, personal observation) may be observed at the time of mating. A male *Cheirogaleus major* has also been observed sniffing at urogenital and fecal marks left by a female, immediately prior to mating (A. Schilling, personal observation). Scent-marking of the environment also increases markedly in *Hapalemur griseus* (brachial-marking in the male, genital-marking in the female) when the female is in estrus (Petter and Peyrieras, 1970b). Moreover, presenting behavior has been observed in *G. demidovii* males (Charles-Dominique, 1977) and *Lemur* males (M. Grange, personal communication) and is even more striking in *Lemur* females (Andrew, 1964). This behavior, first described by Bishop (cited by Andrew, 1964) in the female *L. fulvus* ("back-turned posture"), as a sequence proceding anal-marking with a complete about-face of a conspecific (see Section 1), should be distinguished, according to Andrew (1964), from the scent-marking itself; but, in any case, it increases in the males and appears in females at the time of estrus. Andrew (1964) also notes that anal-marking of a male by a female is normally only observed before and during copulation, while the males practice this behavior during the entire courtship period. Evans and Goy (1968) have shown that the behavior of *L. catta* males changes from the time of the precopulatory period, when the female is sexually attractive but unreceptive, to the short copulatory period, when the female becomes receptive (see Table IX). Richard (1974a), who notes that certain aspects of reproductive behavior are similar in *Propithecus verreauxi verreauxi* and *L. catta*, states that in both species the scent-marking behavior of females does not change with the approach of estrus, while in the males a peak occurs in several types of behavior including olfactory behavior. This peak lasts 4 days in *P. v. verreauxi*, and involves the two particular behaviors already described above. In summary, the olfactory behavior of social prosimians appears to contribute to the efficiency of reproduction, which is characterized by a marked synchronization of estrus in the females and tends to concentrate mating within a short period at the optimal time of the annual ecological cycle.

b. Odoriforous signals of sexual origin which prosimians are able to discriminate. By comparing the time an experienced animal spends sniffing artificial urine marks with the time spent smelling those left by a male conspecific or of a female conspecific, or its own urine, Seitz (1969) concluded that *Nycticebus* can discriminate between the sexes of conspecifics by their scent. Using a similar method, but unfortunately without specifying the origin of the scent in question, Harrington (1974) also concludes that *Lemur fulvus* are capable of recognizing the sex of lemurs of the same subspecies, *L.f. fulvus*, and also other closely related subspecies, *L.f. rufus* and *L.f. albifrons*. The suggestion of olfactory identification of the receptive state of the female is based on an odoriforous vaginal discharge characteristic of many species at the time of estrus, for example, *Galago senegalensis* (Lowther, 1940), *G. crassicaudatus* (Jolly, 1966b), *Nycticebus coucang* and *Loris tardigradus* (Seitz, 1969), *Lemur catta* (Evans and Goy, 1968), *Tarsius* (Hill, 1955), to name but a few. However, the behavioral changes in males often observed in the field in *G. senegalensis* (Sauer and Sauer, 1963), in *Perodicticus potto* (Charles-Dominique, 1974b), as well as in captivity in *L. catta* (Evans and Goy, 1968) and *Nycticebus coucang* (Seitz, 1969), cannot be accepted as definitive evidence of a particular neurosensory mechanism. Even though in prosimians, as well as in other primates including man (Russel, 1976), discrimination of a certain number of male and female olfactory signals is currently being demonstrated, the problem remains to analyze their precise role in reproductive behavior and in the sexual physiology itself. In courtship, for example, it is not known what are the specific responses provoked in the females when certain glandular secretions are actively dispersed by the males; nor is it known what is the specific effect of these olfactory stimuli on the synchronization of estrus, for instance.

c. Hypotheses suggested to explain the role of olfactory communication in reproductive behavior. Since sexual dimorphism is absent in many prosimians it is not unlikely that olfactory stimuli play a major role in sexual recognition (Doyle, 1974b). Evans and Goy (1968) note that, in *Lemur catta*, a consequence of exchange of odors between the sexes would be the facilitation of mutual arousal throughout early estrus. Certain behavioral or even physiological changes related to estrus are presumed to be released as a result of direct and/or indirect olfactory signals (Jolly, 1966b; Evans and Goy, 1968; Martin, 1972a). Courtship behavior may be in part dependent upon olfactory communication (Sauer and Sauer, 1963; Seitz, 1969; Charles-Dominique, 1977) although in *Galago* (and this is not entirely contradictory) "any strange female is courted, whether immature, suckling, gestating or in any sexual state" (Charles-Dominique, 1972). The problem is complex because other nonolfactory indices of the receptive state of the female may be involved and, in this respect, solitary species may differ from gregarious species. In the former, Charles-Dominique (1971, 1974b, 1975, 1977) stresses the important lorisine characteristic of "deferred communication" by urogenital marking, especially in reproduction. The Lorisinae, utilizing scent marks left in the environment as an indication of the sexual state of the female, practice a system of communication adapted to their physiology (slow movements) and to their unique ethological characteristics, e.g., relatively large territories and strictly solitary habits. As is often the case in solitary prosimians, where the social structure is based on male territories overlapping with those of a number of females which they periodically visit (Charles-Dominique and Hladik, 1971; Martin, 1972a, b; Charles-Dominique, 1972, 1974a, 1977) the collection of odors and reciprocal allomarking may play an important role in the formation and cohesion of the sexual pair (Charles-Dominique, 1966). Charles-Dominique (1974b, 1977) has shown that the long courtship period of 1–2 months in *Perodicticus potto* is necessary for the formation of social bonds, without which acceptance of the male and olfactory control over the sexual state of the female during the visiting behavior which follows could not be accomplished; in other words, mating would not occur. The diurnal gregarious prosimians, which have no such need for deferred communication, have a more highly developed system of visual communication (see Pariente, Chapter 10), and one might, therefore, assume that an elaborate system of olfactory communication would not be necessary. However, the contrary appears to be true, at least for the Lemuridae. *Lemur* have a system of olfactory communication which is not only highly diversified, but, perhaps, even more complex and at least as frequently practiced as those of the nocturnal forms. Some suggestions are given below to explain this apparent evolutionary paradox but, like many other problems of olfactory communication, it will not be resolved until the actual content of the supposed "message" is understood.

Two conclusions may be tentatively drawn regarding reproductive behavior and communication in prosimians: (1) The synchronization of sexual activity in the more solitary nocturnal forms is due to a system of signals different from that of the more gregarious diurnal species (P. Charles-Dominique, personal communication). In the former, deferred olfactory signals and auditory signals act over long distances; in the latter, direct olfactory signals and visual signals act over short distances. (2) Sexual activity and social relations are inseparable and can only be studied together since one influences the other. This is true both in solitary prosimians (Charles-Dominique, 1975, 1977) and in gregarious species for which Evans and Goy (1968) note, concerning *Lemur catta,* "both long- and short-term fluctuations in gonadal activity were associated with changes in the frequency of expressions of several non-sexual patterns. . . ." For example, in *Propithecus* (Richard, 1974a,b) a reorganization of the social hierarchy occurs after the period of reproduction, demonstrating the effect of sexual factors on social relations; however, the females mate only with dominant males, demonstrating the impact effect of social relations on sexual behavior. Consequently, olfactory behavior involved in the sexual dialogue has *de facto* social implications, just as olfactory behaviors observed in different social situations, e.g., dominance relationships, territoriality, etc., have repercussions in the sexual context.

3.2.1.3. Olfaction and Autrui *

a. Interindividual discrimination. In general, the presence of scent marks causes active sniffing in conspecifics (Andersson, 1969). Using the method referred to in a previous section, Seitz (1969) concluded that *Nycticebus* males are capable of discriminating their own urine from that of a conspecific of the same sex. The criterion of time spent sniffing, applied within an habituation-sensitization schedule, which is much more reliable, provides evidence that interindividual olfactory discrimination applies equally well to the Lemuriformes, for example, *Lemur fulvus* (Harrington, 1974) and *L. catta* (Mertl, 1975). In addition, this discrimination includes a refractory period (Mertl, 1975) and is not dependent on the quantity of the olfactory stimulus. In neither of the above experiments is there any clear indication of the anatomical origin of the scent concerned. Clark (1975) reported interindividual olfactory discrimination in *Galago crassicaudatus* using scent from the sebaceous glands of the chest, having shown that the same subjects could not make similar discriminations on the basis of urine alone. However, a study of simultaneous olfactory discrimination of urine being undertaken by the present author in *Microcebus coquereli*, using a conditioning technique, indicates an extremely well-developed olfactory ability. Using the urine of different males as stimuli, the experimental subject (also a male) is required to respond, from a distance of about 40 cm, to a current of air which passes over urine samples of two different animals. Reinforcement follows the correct response. Preliminary results show that not only are *Microcebus coquereli* capable of distinguishing between fresh urine samples of two other males, but they can still differentiate between the two after a period of several months when the urine is totally dry. This discrimination persists even after the water content of the dried urine samples is restored with distilled water. This study has also been concerned with discrimination of urine samples from the same animal but taken at different times. Results indicate that animals can discriminate between samples deposited within as little as twelve hours of one another. Related research (in progress) shows that stimuli to which experimental animals have learned to respond positively are recognized even when mixed in the proportion of 1:24 with other stimuli.

However, no results are yet available concerning the exact chemical nature of the olfactory stimuli concerned in all the above experiments.

b. Olfactory self-expression. Whether solitary or gregarious, nocturnal or diurnal, prosimians have developed a repertoire of olfactory signals permitting communication of what may be called emotional states. These signals appear to be as complex as the visual and accoustic communication system used by the simian primates for the same purpose (Evans and Goy, 1968). However, with the exception of certain reproductive behavior patterns, it is not yet possible to demonstrate a precise relationship between different scent-marking patterns and the emotional

*There is no satisfactory English equivalent for this word which is, at best, translated as "everything which is not the self."

state (fear, satisfaction, curiosity, aggression, self-assertion, etc.) of the animal emitting them (Andrew, 1964; Jolly, 1966b; Schilling, 1974; Harrington, 1974; Doyle, 1974b). This is largely due to the fact that any one of those specific states may be expressed by several types of marking behavior, either singly or in association; and, conversely, the same marking behavior may signify completely different emotional states. For example, *Lemur catta* and *L. fulvus* investigating a strange environment anal-mark as well as brachial-mark. Conversely, tail-marking in *Hapalemur griseus* may express sexual excitation as well as aggression (A. Schilling, personal observation), or fear (Andrew, 1964), or nervous excitation (J. J. Petter, personal communication). There is a measure of agreement in the literature regarding the communication of states of "satisfaction" as opposed to "nonsatisfaction," either by leaving olfactory marks which are significant of themselves, or by combining these with the gestural part of the scent-marking behavior, in which case the message is of a visual–olfactory nature. Petter (1962) and Andrew (1964) provide examples in *Lemur catta* and *L. fulvus*, respectively, of anogenital, palmar- or brachial-marking during demonstrations of "affection" toward a conspecific or even toward the observer. Under similar circumstances, toward an observer, urine-washing has been reported in *Galago senegalensis* (Andrew and Klopman, 1974) and *Galago alleni* (M. Grange, personal communication), salivary dispersion in *Microcebus coquereli* (Pagès, in press), fecal-smearing in *Cheirogaleus major* (A. Schilling, personal observation), brachial-marking in *Hapalemur griseus* (Petter and Peyrieras, 1970b), and trail-laying with urine in *Daubentonia madagascariensis* (Petter and Peyrieras, 1970a). These interpretations are, at least, somewhat subjective since it is difficult to draw parallels between olfactory communication involving a conspecific and similar behavior in animals raised by hand and, therefore, conditioned to strange odors. Nevertheless, these behaviors are usually directed toward people with whom the animal has a special relationship. In nature, reciprocal grooming and greeting between conspecifics are activities generating satisfaction (comfort activities according to Bearder, 1975) which may be and are frequently accompanied by scent-marking (Petter, 1962; Andrew, 1964; Jolly, 1966b; Andersson, 1969; Doyle, 1974b). This may also be true for certain feeding behaviors. In *Galago demidovii*, for example (M. Perret, personal communication) urine-washing follows the ingestion of highly preferred foods like certain insects. Conversely, and still more frequently, situations which generate discomfort lead to scent-marking; for example, fear (Andrew, 1964), irritability, or nervous tension (Petter, 1965; Petter and Peyrieras, 1970b). The expression of an emotional state may be better interpreted by examining the relationship of the individual to its environment from the point of view of the degree of conflict or frustration inherent in a particular situation. In the field, although manifestations of scent-marking outside of any conflict situation have been observed in *Varecia variegata*, *Lemur catta*, *L. fulvus*, and *Propithecus verreauxi*, the present author has noted in these species, first, that in general, any conflict situation involving an individual and either its social or nonsocial environment may result in scent-marking; second and, more particularly, any conflict situation in an individual's own behavior may result in scent-marking; third, a certain number of the former and the majority of the latter instances may be interpreted as displacement activites. These displacement activites have been described in *Lemur catta* (Schilling, 1974), but more recent observations suggest that they occur in all the prosimians, even the solitary species. For example, in *Galago demidovii* males observed in Gabon and in captivity (M. Perret, personal communication), several instances of urine-washing follow what may be called "errors"; for example, in the choice of insects captured, or after grooming a conspecific which was not the normal grooming partner. It is evident that certain scent-marking behaviors are correlated with the degree of tension in prosimians. Preliminary observations of the context in which scent-marking behavior in *Lemur fulvus albifrons* occurs indicate that, at least in dominant males, the specialized form of anogenital marking of the environment occurs in situations which do not involve tension, while marking or rubbing of the muzzle, the forehead, the head, the palms, with or without biting of branches, is associated with an increase of excitation in the subject, whether the elements of this sequence occur separately or in succession (M. Grange, personal communication). In an analysis of conflictual motivation of "genital–scratching–grooming" in *Perodicticus potto*, Manley (1974) has established, at least in one case and for the first time in a prosimian, a relationship between a scent deposit, in this case of the scrotal or pseudoscrotal glands, and the state of fear of the subject, a phenomenon called *Angstgeruch* by Müller-Velten (1966), who described it in the mouse. This may also be true of *Arctocebus calabarensis* (Manley, 1974) and, in antipredator situations, in other contexts, it may characterize scent-marking in *Nycticebus coucang* and *Loris tardigradus* as well.

Jolly (1966b) proposes the concept of "self-advertisement" as the only concept which characterizes the full range of situations in which an individual scent-marks to establish or assert itself within its social environment. In the gregarious forms "self-advertisement" largely involves the direct olfactory–visual channels of communication, and in the solitary forms it largely involves indirect olfactory communication. According to this concept, information is transmitted which is of both a qualitative nature, relating to sex and psychophysiological state, for example, and of a quantitative nature, relating to age, degree of anxiety, fear, or threat, and may or may not have the function of a message. Manley (1974) suggests that *Angstgeruch* in *Perodicticus* and *Arctocebus* functions as an alarm signal between mother and infant.

3.2.2. The Socialization of Olfactory Communication

All prosimians engage in close social interactions which are complex. This has been demonstrated not only in the gregarious species (Petter, 1962; Jolly, 1966b; Sussman and Richard, 1974; Harrington, 1975) but also in the solitary forms, *Galago senegalensis* and *G. crassicaudatus* (Bearder, 1969, 1975), *G. demidovii*, *G. alleni, G. elegantulus, Perodicticus potto, Arctocebus calabarensis* (Charles-

Dominique, 1971, 1977), *Microcebus murinus* (Martin, 1972a, 1973), *M. coquereli* (Pagès, in press), *Phaner furcifer* (Charles-Dominique and Petter, in press), *Lepilemur mustelinus* (Charles-Dominique and Hladik, 1971) and in *Tarsius bancanus* (Fogden, 1974). These interactions concern, on the one hand, relations between solitary or gregarious individuals and their conspecifics, or even individuals of other species and, on the other hand, relations within the social groups in the gregarious species. The means of communication are multiple, among which olfactory signals play an important but, as yet, poorly understood role.

3.2.2.1. Olfactory Behavior and Territoriality.

a. Interspecific relations. Scent-marking in prosimians has much greater intra- than interspecific importance (Jolly, 1966b). Jolly reports the lack of interest shown by a *Lemur fulvus collaris* in the olfactory behavior of a troop of *L. catta* in which it was integrated. Harrington (1974), in experiments with *L.f. fulvus*, was not able to show that this subspecies could discriminate its own scent samples from those of *L.f. rufus* or *L.f. albifrons*, although it responded to *L. mongoz* (Harrington, 1971). However, olfactory discrimination between subspecies has been reported in *Galago crassicaudatus* (Clark, 1975). It seems to be the general rule that animals are indifferent to the scents of other species. Thus, Seitz (1969) has stated that, in his experiments with *Nycticebus coucang* in a "model territory" (wooden struts), scent-marking occurred less frequently near a *Loris* cage than near another *Nycticebus* cage. Also in captivity, and using as a measure the weight and length of fecal marks left by *Cheirogaleus major* per unit of length of the experimental wooden struts, A. Schilling (unpublished data) has shown that, first, scent-marking always occurred at the periphery of the core area most frequently used by the animals, and that this distribution pattern appeared simultaneously with the acquisition of scent-marking behavior itself, i.e., in the sixth week of life; and second, that marking increased on the struts adjacent to the areas occupied by a *Microcebus murinus* group, suggesting that *C. major* are not indifferent to the scent of another species.

b. Intraspecific relations. In a recent article Charles-Dominique (1975) concluded that the social system of gregarious primates is derived from a primitive type characterized by the nocturnal prosimians and based on defense of individual territories. Solitary prosimians defend the borders of their territory either directly by fighting, or indirectly by the means of advertising signals adapted to the size of the territory and the ecology of the habitat. Olfactory signals naturally play an important role in these mechanisms of defense, especially when the territory is relatively large compared to the size of the animal, as in *Galago demidovii* and *Microcebus murinus*, and in which the scent marks serve essentially as deferred communication.

In the gregarious prosimians (Petter, 1962; Jolly, 1966b; Sussman, 1974; Richard, 1974b; Sussman and Richard, 1974; Harrington, 1975; Pollock, 1975), the composition (*Indri indri*), distribution (*Propithecus verreauxi*), and the activity (*Lemur catta*) of the group depend on the ecological conditions. The social systems are more varied than in the solitary forms not only in the composition of the groups (family, troop) but also in their more (*Lemur catta*) or less (*L. fulvus*) hierarchical structure, as well as in their several types of intraspecific relations. For example, agonistic encounters between troops or families exist, but have more to do with the defense of the territory in the Indriidae than they do in the Lemuridae (Jolly, 1966b; Sussman and Richard, 1974). These interactions are largely limited to the reproductive season in *P. verreauxi* and *L. catta* and are primarily directed toward the maintenance of intragroup cohesion in *L. fulvus fulvus* and *P. verreauxi*. The problem is complicated by the fact that in the diurnal species, both visual and auditory signals tend to increase in importance without diminishing the importance of olfactory signals. Olfactory or olfactory–visual behaviors have been reported in territorial defense in some nocturnal species like *Galago demidovii* and *Perodicticus potto* (Charles-Dominique, 1974a) and *Nycticebus coucang* (Seitz, 1969), as well as in *Propithecus verreauxi* (Jolly, 1966b; Richard, 1973) and in *Indri indri* (Pollock, 1975); in interactions at the limit of the home range in *Lemur catta* (Jolly, 1966b; Sussman and Richard, 1974), in *L. fulvus fulvus* (Harrington, 1975), and in *P. verreauxi* (Richard, 1973), and in encounters between animals in which the home ranges overlap, as in *L.f. rufus* (Sussman and Richard, 1974). The complete range of olfactory or olfactory–visual signals from the discrete and rapid urine-washing of *Microcebus*, to the spectacular stink fights described by Jolly (1966b) in *L. catta*, may be emitted, although no particular scent-marking pattern can be related to a specific agonistic situation (Jolly, 1966b).

Observations concerning the ethology of *Microcebus coquereli* (Pagès, in press), which is quite different from other Cheirogaleinae, have shown the existence of curious "releases of odor." The locations of these olfactory releases, which recall those described by Hill (1956) in *Loris*, are mapped on the home range of one of the animals outside the period of reproduction (see Fig. 14). These releases of odor take place within the home range, but outside of the core area. The important question is whether or not these signals facilitate avoidance of conspecifics in agonistic situations. Avoidance behavior, after the introduction of an individual into the territory of a conspecific, has been observed in the wild in several prosimians in Gabon, and the role of olfactory stimuli in this behavior has been demonstrated in captivity by Charles-Dominique (1974a) in male *Galago demidovii*, and by Epps (1974) in *Perodicticus potto*. According to the former, avoidance is less evident in the females, but where encounters do take place, physical aggression is rarer than in males. It thus appears that deferred olfactory communication plays an active role in the defense of the home range in these solitary species. This possibly explains the increase in scent-marking observed at the borders of territories in *Nycticebus coucang* (Seitz, 1969), *G. demidovii* (Charles-Dominique, 1974a), and *Microcebus murinus* (A. Petter-Rousseaux, personal communication) in captivity. Using a technique of radiotelemetry, this has recently been verified in the field by Charles-Dominique (personal communication) for *G. alleni*, in which urine-marking occurs four times more frequently on the periphery of the territory than in the center. Finally, the general increase in scent-marking behavior in *Cheirogaleus major*,

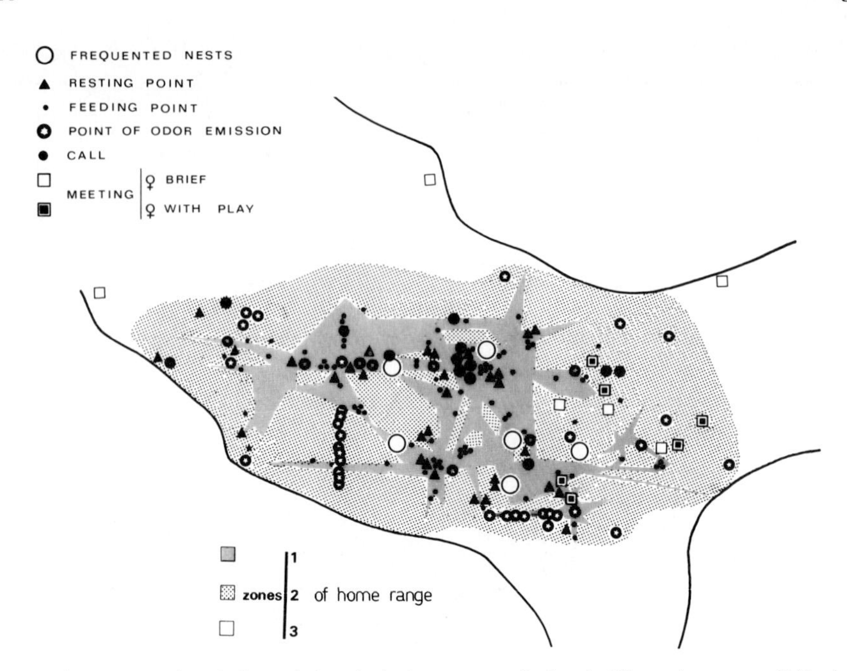

Fig. 14. Topography of odor emissions in the home range of a female *Microcebus coquereli* (Pagès, 1978). Zone 1 represents 10% of the home range and is frequented 50% of the time; Zone 2, 20% of the home range, 30% of the time; Zone 3, 70% of the home range, 20% of the time.

following introduction of a conspecific near to its territory (personal observation), in *G. demidovii* after penetration into the territory by a conspecific (M. Perret, unpublished observation), or in *G. senegalensis*, after cleaning of the animal's cage (Doyle, 1975) may also have territorial significance. It should be emphasized that caution must be used in interpreting observations made in captivity, in reduced areas bearing no relation in size to the natural territories of the animals and sometimes involving animals displaying pathological behavior as a result of captive conditions (P. Charles-Dominique, personal communication). In addition, such observations made on captive animals are complicated most of the time by forced interindividual interactions not of a social nature, by an increase in certain elementary behaviors, and especially as a result of the presence or intervention of the human observer.

Two types of mechanism have been advanced to explain the role of scent-marking in territoriality (Martin, 1968). First, scent-marking may serve to "reassure the individual animal within its own range." This hypothesis is supported by the fact that territorial species are usually successful in agonistic encounters within their own territories, for example, in *Galago senegalensis* (Bearder and Doyle, 1974a). Second, scent-marking may serve to deter or demoralize conspecific rivals from other territories. The diurnal gregarious prosimians do not need to rely solely or primarily on olfactory signals to inform them of the presence of conspecifics from neighboring territories in their home ranges nor do they need to advertise their own presence by scent-marking in order to ensure avoidance of intruding conspecifics. It is, therefore, quite understandable that scent-marking of the borders of territories has not been observed in *Lemur* (Schilling, 1974) or in *Propithecus* (Jolly, 1966b), while avoidance behavior has been reported in *P. verreauxi* after sniffing of scent marks in some instances, but not in others (Petter, 1962; Jolly, 1966b). It should also be remembered that a scent mark itself does not deter a conspecific, but is only a signal to be interpreted by the receiver. In the diurnal prosimians the gestural component of scent-marking has assumed sufficient importance to be effective as a visual signal (Jolly, 1966b). Communicative exchanges between groups of gregarious prosimians may have lost strict dependence on olfactory signals in favor of olfactory–visual signals, rather more stereotyped (P. Charles-Dominique, personal communication) than ritualized (Jolly, 1966b; Doyle, 1974b). Good examples are provided by endorsing in *Propithecus* (Richard, 1974a) and *Indri* (Pollock, 1975), and stink fights in *L. catta* (Jolly, 1966b; Schilling, 1974). In *L. catta* (Schilling, 1974) signaling by brachial marking is both olfactory and visual in the strictest sense; visual because the scarring of bark by means of the spur on the wrist leaves a mark visible at several meters; and olfactory because glandular secretions are encrusted in the scar. This double signalling theoretically permits a communication of a kind perfectly in harmony with the concept of deferred communication (inherited perhaps from a nocturnal ancestor). However, the respective roles of the olfactory and visual signals in this context have yet to be demonstrated.

It may be concluded from recent field studies that social relations, involving different types of contact in solitary and gregarious forms, change over the seasonal cycle under the influence of factors principally related to the diet and reproduction and that, consequently, the signficance of certain social signals, both olfactory and olfactory–visual, in particular, are not unequivocal but depend rather on the entire social context. Figure 15 illustrates diagrammatically one set of results of a series of 30 presentations of two *Galago demidovii* (M. Perret, unpublished observation) males to each other from adjoining experimental territories (by removing the partition separating the two territories) under two different experimental conditions: (a) following a period when no female had been present in the territory of either, and (b) following a period when a female had been present in the territory of one of them.

The left-hand column of Fig. 15 illustrates that, under condition (a), male A not only marked more but marked his conspecific's territory exclusively, whereas male B confined his marking entirely to his own territory. This strongly suggests that male A was the more dominant of the two. Under condition (b), however, and irrespective of their relative dominance under the previous condition, both males marked their own territories more when the female had been living in their own territory, and in both cases marking of their conspecific's territory increased when the female had been living in the conspecific's territory. The important finding is that both males displayed an increased tendency to mark the territory in which the

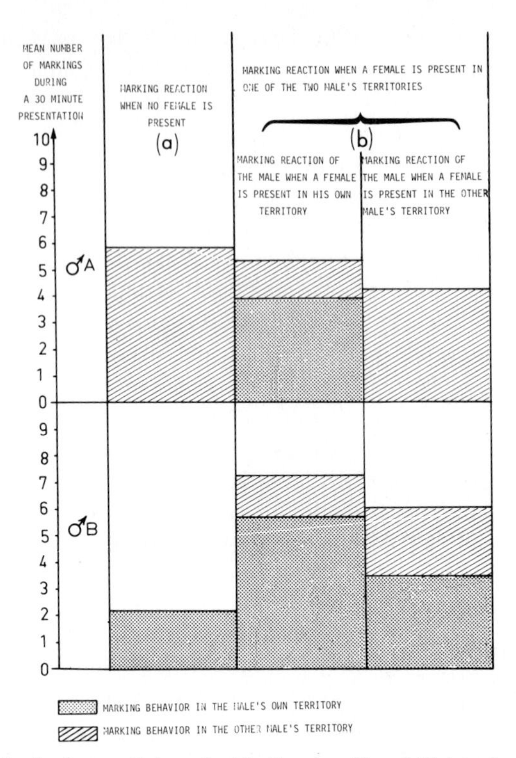

Fig. 15. Marking behavior in two *Galago demidovii* males, (A and B) introduced to one another from adjoining cages under two experimental conditions: (a) when no female had been present in the cage of either prior to presentation, and (b) when a female had been present in the cage of one of the males prior to presentation.

female had been present before the presentation irrespective of their relative dominance status. This is particularly clear in the case of B, which would not even venture into male A's territory under condition (a), when no female had been involved in either territory. The results show that different motivations, either separately or together, may subserve the same behavior, in this case scent-marking, and that these motivations may vary according to the social context. Olfactory signals in the solitary species, and olfactory–visual signals in the gregarious forms result in the substitution of stereotyped behaviors, e.g., stink fights, for actual physical engagements. These signals thus actively control intraspecific agonistic encounters (Charles-Dominique, 1974a; Sussman and Richard, 1974). Epple (1974) notes that in the New World monkeys also the control of aggression is one of the most important functions of chemical signals.

3.2.2.2. Olfactory Behavior and Social Relations in Gregarious Prosimians. If chemical signals in general, and scent-marking in particular, serve only to reinforce other signals of communication, it might be assumed that their importance would have diminished in the diurnal prosimians. This is apparently not the case, and chemical signals have assumed certain essential functions in intragroup communication. These functions may be postulated from ecophysiological and behavioral observations. Jolly (1966b) concludes her study of the gregarious prosimians by stating that these are "socially intelligent and socially dependent." This social interdependence of gregarious prosimians is an important fact that signifies the absolute necessity for an efficient system of contact signals. Bearing in mind that visual contact between the members of a group is limited both by definition and by the nature of the forest biotope, that tactile contact depends on social situations limited in both time (to resting periods) and space (to physical contact), that contact calls appear less rich in the nocturnal species (e.g., *Lemur mongoz* and *Avahi laniger*), and that during the night certain Lemuridae, like *Varecia variegata* (Petter, 1962), *Lemur catta* (Jolly, 1966b), *Hapalemur griseus* (Petter and Peyrieras, 1970b), *L. fulvus fulvus* (Harrington, 1975), and *L.f. albifrons* (A. Schilling, personal observation; see also Petter and Charles-Dominique, Chapter 7) may be relatively active. One of the functions of olfactory communication, since the stimulus persists, may be in maintaining relatively permanent contact between individuals and their social group. Intraspecific fights in Lemuriformes (Petter, 1962; Jolly, 1966b) and in the Lorisiformes (Andersson, 1969; Jewell and Oates, 1969; Walker, 1970; Charles-Dominique, 1974a) do not favor population balance. The olfactory–visual communication system of gregarious prosimians may have an important function in the control of aggression between conspecifics of the same social group either by avoiding confrontations or by stereotyped olfactory–visual

Fig. 16. Examples of olfactory components in the complex social behavior of *Lemur fulvus albifrons*. (A) Direct allomarking (specialized form of anogenital marking) of a subadult male by the dominant male of the group. (*continued*)

Fig. 16. (*continued*) (B) Palmar-rubbing by the dominant adult male imitated by a young subadult male. (C) Sniffing followed by licking of the genitalia of the dominant male by the same subadult male.

displays which substitute for agonistic contact. With the development of gregariousness, the role of scent-marking, initially confined to olfactory communication between solitary individuals, now operates to control relations between individuals within the social group by conveying information on the state of aggression, degree of tolerance, etc. In short, interindividual olfactory communication (outside of reproductive behavior) has evolved from being restricted to intermittent use over time and from a distance, to its permanent and immediate use in close contact (see Fig. 16).

a. Role of olfactory behavior in social cohesion and dependency. The permanence of olfactory contact between members of different types of social groups is well established in the gregarious prosimians (Petter, 1962; Petter *et al.*, 1977). Contact is assured not only by olfactory activities themselves such as sniffing, allomarking, and collection of odors, but also during all direct contacts between animals—allogrooming, play, contact resting, etc. There is no apparent need to invoke a concept of a group odor, as has been demonstrated in the mouse (Ropartz, 1968), to explain certain behaviors of gregarious prosimians. Olfactory interractions, which are infinitely varied and continuous between individuals of the same social group may, instead, lead to olfactory conditioning within the group. This conditioning may occur relatively early in the development of the young (see Fig. 16b) and would subsequently be continually reinforced. The social dependence of gregarious prosimians stressed by Jolly (1966b) is certainly not uniquely olfactory nor exclusive, since the author describes the case of a troop of *Lemur catta* into which a *L. fulvus collaris* had been successfully integrated. However, it is possible that among the many factors responsible for this interdependence, olfactory behavior plays an important role. Indeed, as has often been noted in the field (Petter, 1962) and in captivity (M. Grange, personal communication), when a lemur leaves the troop, or is separated from it, it is copiously marked by conspecifics and itself engages in a great deal of olfactory activity when it returns to the troop. Conversely, any attempt to disturb the social balance, i.e., to disrupt social cohesion, may elicit olfactory behavior in *Lemur* and *Propithecus* (Petter, 1962; Jolly, 1966b) and in *L. fulvus albifrons* (A. Schilling, personal observation) indicating the important integrative function of olfactory communication.

b. Role of olfactory behavior in the hierarchy of prosimian groups. In some lemurs, *Lemur fulvus fulvus* (Harrington, 1975) and *L.f. rufus* (Sussman and Richard, 1974), no social hierarchy has been observed; thus the role of olfactory and olfactory–visual communication signals is probably limited in these species. On the other hand, in *Lemur catta* (Jolly, 1966b; Sussman and Richard, 1974) and *Propithecus* (Jolly, 1966b; Richard 1973; Sussman and Richard, 1974) olfactory and olfactory–visual signals play a role in determining and maintaining the establishment of a hierarchial structure.

It would, however, be an oversimplification to relate frequency of scent-marking to this hierarchy. Indeed, added to the variation in individual reactions to the social situation, are all the factors releasing behavior at the elementary level discussed above. Dominance may be accompanied by a high frequency of scent-marking behavior, but scent-marking is not necessarily correlated with dominance status. It is thus not surprising that observations in the field (Petter, 1962; Schilling, 1974) and in captivity, show that animals which scent-mark most frequently may be either dominant or subordinate animals. Nevertheless, preliminary observations on captive *Lemur fulvus albifrons* suggest that behaviors like sniffing and licking (see Fig. 16C) and certain specific forms of scent-marking, like anogenital marking of the environment or anogenital-allomarking (see Fig. 16A) and rubbing of the head, may be correlated with social dominance. In *Lemur fulvus albifrons* (M. Paillette, personal communication) anal presentation to a dominant animal has been reported. In agonistic situations, involving members of a group occupying an adjacent cage, the subject (dominant in his own group, but subservient to the oldest male and female in the neighboring group) turns, lifts its tail toward an aggressive animal in the neighboring group, presents its hindquarters and, remaining still, turns its head in the aggressive animal's direction. In the entire sequence the subject may reverse to the point of pressing its anus against the cage separation and waiting until the dominant animal in the adjacent cage sniffs it. Whether the whole sequence runs through to completion or not it immediately reduces aggression in the dominant animal.

In Section 2 it was pointed out that the anal region in the Lemuridae in general and the area surrounding the anus in certain species, particularly in *Lemur fulvus* (see Fig. 2c), is richly endowed with various glands, and it would be particularly interesting to discover to what specific olfactory signal the dominant animal in this case reacts. It should be noted that in the dog, although the function is completely different, the anal glands contain secretions serving as chemical signals (Donovan, 1969). Within social groups of certain prosimians, olfactory communication apparently plays a role in the control of interindividual aggression, thereby contributing to social cohesion.

4. DISCUSSION AND CONCLUSIONS

4.1. The Complexity of Olfactory Communication

This review of olfactory signals and scent-marking behavior of prosimians in relation to their social and nonsocial environment leads to the conclusion that this relationship is as rich as it is complex. The problem is particularly complex since research has only recently begun in these primates and little evidence is available as to the exact nature of the information transmitted by the olfactory signal, which is all too often reduced to speculating about the relationship between a presumed olfactory activity and a given physiological or behavioral situation. The danger of false interpretations is due to numerous factors.

1. Communication by olfactory means is unique in terms of its spatio-temporal characteristics. Consequently it is difficult to determine if a signal is or is not followed by a response, and if the behavioral response studied is actually provided by an olfactory signal; in other words, whether the response is elicited by a signal at all.

2. It should be stressed that scent deposits can only play a communicative role to a receptive conspecific. Thus, for example, feces or urine deposited in response to bowel or bladder pressure or even a scent mark left by an animal on the border of its territory, will only have significance to the extent that a conspecific encounters it, focuses its attention on it, and derives some meaning from it. In summary, the two elements of olfactory communication, the deposit of scent and its detection by a conspecific, are rarely linked in time and space.

3. The particular scent serving as a signal if often difficult to specify precisely on the basis of observation alone, especially when a mixture of scents from several sources is involved. This often results in unintentional misinterpretation, for example, when anogenital-marking or sniffing is mistaken for genital-marking or sniffing, or anal-marking or sniffing.

4. In a number of behavioral observations, the exact role of olfaction is not clear. For example, in frequent naso-genital contact, taste, the vomero–nasal organ, and the tactile sense may be involved in addition to olfaction.

5. Particularly in the diurnal prosimians, certain olfactory behaviors appear to have taken on an increasing visual significance which is probably related to the profound changes in social structure that have taken place with the evolution of the diurnal forms. Communication between gregarious animals required instantaneous and simultaneous reactions which are better satisfied by visual and auditory signals than by olfactory signals. Between typically olfactory behaviors like urine-trail-marking, in many species, anointing behavior in *Lemur catta* and *Hapalemur griseus*, for instance, and behavior in which the visual component is obviously paramount, like posturing from a distance, movements of the tail and of the head in *Lemur* and *Propithecus*, and rubbing of the head and of the hands in the Lemuridae, there exists an entire series of equivocal behaviors in which the sensory modalities may be operating simultaneously, such as scent-marking of the environment in gregarious forms, allomarking in *Lemur*, tail-waving in *L. catta*, automarking of the tail in several Lemuridae, anogenital presentation, rubbing of the entire body in *Galago crassicaudatus*, or "passing-over" in *L. fulvus albifrons* and *Arctocebus*, to name but a few examples. Certain typically visual behaviors could be explained as either having been derived from an olfactory component, lost during the course of evolution, or by their change in emphasis from olfactory to gestural activities, analogous to olfactory or olfactory–visual behaviors, like brachial marking in *L. fulvus*, tapping of the hind legs in *G. crassicaudatus*, and chest-rubbing in *G. senegalensis* and *G. crassicaudatus*. Perhaps also the interindividual distance component inherent in the change to a more visual mode of communication may lead to

the loss of active allomarking in certain gregarious species like *L. catta*, *Propithecus*, and *Indri*.

4.2. The Importance of Olfactory Communication in Prosimians

Considering our present state of knowledge, the complexity of olfactory signals, their intermixing with other systems of communication (see Fig. 17) does not allow a complete understanding of the significance of the actual message communicated, except in certain limited cases, for instance, in the determination of estrus, the expression of certain psychophysiological states, the agonistic content of certain scent marks, and the probable relationship between certain types of allomarking in the social hierarchy. Consequently, it should be recognized that olfactory communication in the prosimians is, at this time, relatively little understood. On the other hand, a number of guidelines resulting from observations made in the field and in captivity may facilitate understanding the importance and general characteristics of olfactory communication in this unique group of primates.

1. With the exception of the tarsier, all prosimians have a macrosmatic olfactory apparatus. They practice, without exception, one or several forms of olfactory marking using substances which are extremely varied. Scent-marking involving the anogenital–urinary complex has been reported in all species.
2. Receptive behaviors (sniffing, various types of nasal contact, olfactory exchanges during grooming and in contact groups, etc.) are equally varied and are of an astonishingly widespread nature. Probably signals other than olfactory signals also enter into play but, in general, chemical signals assume a vital role in communication within both the social and nonsocial environment. Harrington (1974) stresses the importance of oral grooming in the prosimians, partially replaced by manual grooming in the simians. The numerous methods of dispersing saliva, rubbing of the lips and muzzle, may perhaps play a role in this area.
3. Apart from social considerations a large part of olfactory behavior may express nothing more than the psychophysiological state of individuals at an elementary level. Certain types of scent-marking, in particular, are part of the basic individual range of expression which differs from one individual to the next. Other types of scent-marking are more in the nature of displacement activites.
4. The role of olfactory communication appears to be quite different in the nocturnal species which are, in general, solitary and in which purely olfactory signals of a deferred nature predominate. The diurnal species, on the other hand, are in general gregarious and direct olfactory–visual signals predominate. It would be interesting to study this aspect of olfactory communication in species which are intermediate between the two types, i.e., which are gregarious to the extent that they live in family groups, and which are more or less nocturnal, like *Hapalemur griseus*, *Avahi laniger* and *Lemur mongoz*.

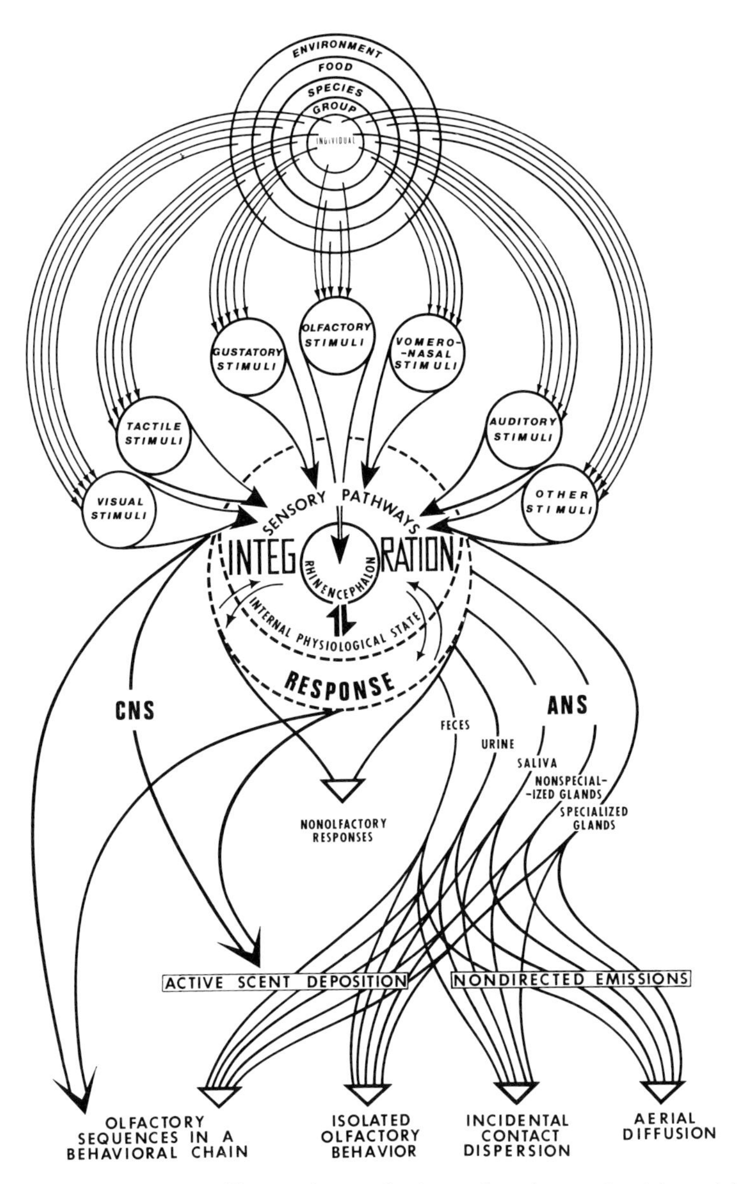

Fig. 17. Schematic diagram to illustrate the complex integration of external and internal factors subserving olfactory communication. CNS, central nervous system; ANS, autonomic nervous system.

5. Prosimians have an extremely well-developed social sense in which olfaction plays a significant role. It may be concluded that the primates in general have a communication system in part dependent on olfaction. From the evolutionary point of view, this dependence is even more interesting since certain prosimians have acquired a mode of life in which olfactory communication has lost its primary role in certain forms of communication. This may be explained by the persistence of the macrosmatic equipment of diurnal forms but also by the relative inadequacy of other means of communication, the olfactory signal permitting extremely strong interindividual relationships because of its intimate and durable character, which assures cohesion of the social group.

ACKNOWLEDGMENTS

The author would like to thank H. Cooper, who helped with the translation of the chapter into English, G. A. Doyle, who prepared the final version and checked the references, and Mrs. C. Muñoz who prepared the drawings.

REFERENCES

Adey, W. R. (1970). *Taste Smell Vertebr., Ciba Found. Symp., 1969* pp. 357–378.

Affolter, M. (1937). *Bull. Acad. Malgache* **20**, 77–100.

Andersson, A. B. (1969). M.Sc. Dissertation, University of the Witwatersrand (unpublished).

Andrew, R. J. (1964). *In* "Evolutionary and Genetic Biology of Primates" (J. Buettner-Janusch, ed.), Vol. 2, pp. 227–309. Academic Press, New York.

Andrew, R. J., and Klopman, R. B. (1974). *In* "Prosimian Biology" (R. D. Martin, G. A. Doyle, and A. C. Walker, eds.), pp. 303–312. Duckworth, London.

Andriamiandra, A., and Rumpler, Y. (1968). *C.R. Seances Soc. Biol. Ses. Fil.* **162**, 1651–1658.

Archer, J. (1968). *J. Mammal.* **49**, 572–575.

Archer, J. (1969). *J. Mammal.* **50**, 839–841.

Bailey, K. (1978). *Behaviour* **65**, 309–319.

Bearder, S. K. (1969). M.Sc. Dissertation, University of the Witwatersrand (unpublished).

Bearder, S. K. (1975). Ph.D. Thesis, University of the Witwatersrand (unpublished).

Bearder, S. K., and Doyle, G. A. (1974a). *J. Hum. Evol.* **3**, 37–50.

Bearder, S. K., and Doyle, G. A. (1974b). *In* "Prosimian Biology" (R. D. Martin, G. A. Doyle, and A. C. Walker, eds.), pp. 109–130. Duckworth, London.

Beddard, F. E. (1884). *Proc. Zool. Soc. London* pp. 391–399.

Beddard, F. E. (1891). *Proc. Zool. Soc. London* p. 449.

Bland-Sutton, J. (1887). *Proc. Zool. Soc. London* pp. 369–372.

Bolwig, N. (1960). *Mém. Inst. Sci. Madagascar, Sér. A* **14,** 205–217.

Bouissou, M. (1968). *Rev. Comportement Anim.* **2**, 77–83.

Bourlière, F. (1956). *Mém. Inst. Sci. Madagascar, Sér. A* **10,** 299–302.

Bowers, J. M., and Alexander, B. K. (1967). *Science* **158**, 1208–1210.

Bronson, F. H. (1971). *Biol. Reprod.* **4**, 344–357.

Bruce, H. M. (1959). *Nature (London)* **184**, 105.

Carr, W. J., Marorano, R. D., and Krames, L. (1970). *J. Comp. Physiol. Psychol.* **71**, 223–228.

Catarelli, M., Vernet-Maury, E., and Chanel, J. (1974). *C.R. Hebd. Séances Acad. Sci.* **278,** 2653–2656.

Cave, A. J. E. (1967). *Am. J. Phys. Anthropol.* **26**, 277–288.
Chandler, C. F., Jr. (1975). *Primates* **16**, 35–47.
Charles-Dominique, P. (1966). *Biol. Gabonica* **2**, 355–359.
Charles-Dominique, P. (1968). *Biol. Gabonica* **4**, 3.
Charles-Dominique, P. (1971). *Recherche* **2**, 780–781.
Charles-Dominique, P. (1972). *Z. Tierpsychol., Beih.* **9**, 7–41.
Charles-Dominique, P. (1974a). *In* "Primate Aggression, Territoriality and Xenophobia: A Comparative Perspective" (R. L. Holloway, ed.). pp. 31–48. Academic Press, New York.
Charles-Dominique, P. (1974b). *Mammalia* **38**, 355–379.
Charles-Dominique, P. (1975). *In* "Phylogeny of the Primates: A Multidisciplinary Approach" (W. P. Luckett and F. S. Szalay, eds.), pp. 69–88. Plenum, New York.
Charles-Dominique, P. (1977). "Ecology and Behaviour of Nocturnal Primates." Duckworth, London.
Charles-Dominique, P., and Hladik, C. M. (1971). *Terre Vie* **25**, 33–66.
Charles-Dominique, P., and Martin, R. D. (1970). *Nature (London)* **227**, 257–260.
Charles-Dominique, P. and Petter, J. J. (in press). *In* "Ecology, Physiology and Behavior of Five Nocturnal Lemurs of the West Coast of Madagascar" (P. Charles-Dominique, ed.), Karger, Basel.
Chipman, R. K., and Fox, K. A. (1966). *J. Reprod. Fertil.* **12,** 233–236.
Clark, A. B. (1975). Ph.D. Dissertation, University of Chicago, Chicago, Illinois (unpublished).
Donovan, C. A. (1969). *J. Am. Vet. Med. Assoc.* **155**, 1995–1996.
Douglas, R. J., Isaacson, R. L., and Moss, R. L. (1969). *Physiol. Behav.* **4**, 379–381.
Doyle, G. A. (1974a). *In* "Prosimian Biology" (R. D. Martin, G. A. Doyle, and A. C. Walker, eds.), pp. 213–231. Duckworth, London.
Doyle, G. A. (1974b). *Behav. Nonhum. Primates* **5**, 155–353.
Doyle, G. A. (1975). *In* "Contemporary Primatology" (S. Kondo, M. Kawai, and A. Ehara, eds.), pp. 232–237. Karger, Basel.
Doyle, G. A., Pelletier, A., and Bekker, T. (1967). *Folia Primatol.* **7**, 169–197.
Ehrlich, A. (1970). *Behaviour* **37**, 54–63.
Eibl-Eibesfeldt, J. (1953). *Säugetierkdl Mitt.* **1,** 171–173.
Eisenberg, J. F., and Kleiman, D. G. (1972). *Ann. Rev. Ecol. Syst.* **3,** 1–32.
Ellis, R. A., and Montagna, W. (1963). *In* "Evolutionary and Genetic Biology of Primates" (J. Buettner-Janusch, ed.), Vol. 1, pp. 197–228. Academic Press, New York.
Epple, G. (1970). *Folia Primatol.* **13**, 48–62.
Epple, G. (1974). *Ann. N.Y. Acad. Sci.* **237**, 261–278.
Epps, J. (1974). *In* "Prosimian Biology" (R. D. Martin, G. A. Doyle, and A. C. Walker, eds.), pp. 233–244. Duckworth, London.
Evans, G. S., and Goy, R. W. (1968). *J. Zool.* **156**, 181–197.
Faure, J. (1956). *C. R. Séances Soc. Biol. Ses. Fil.* **150,** 212.
Ferrer, N. G. (1969). *J. Comp. Neurol.* **136**, 337–348.
Fogden, M. P. L. (1974). *In* "Prosimian Biology" (R. D. Martin, G. A. Doyle, and A. C. Walker, eds.), pp. 151–165. Duckworth, London.
Gandelman, R., Zarrow, M. X., and Denenberg, V. H. (1972). *J. Reprod. Fertil.* **28,** 453–456.
Gray, J. G. (1863). *Proc. Zool. Soc. London* pp. 129–152.
Harrington, J. E. (1971). Ph.D. Thesis, Duke University, Durham, North Carolina (unpublished).
Harrington, J. E. (1974). *In* "Prosimian Biology" (R. D. Martin, G. A. Doyle, and A. C. Walker, eds.), pp. 331–346. Duckworth, London.
Harrington, J. E. (1975). *In* "Lemur Biology" (I. Tattersall and R. W. Sussman, eds.), pp. 239–279. Plenum, New York.
Haug, M. (1970). *C.R. Hebd. Séances Acad. Sci.* **271,** 1567–1570.
Haug, M., and Ropartz, P. M. (1973). *C.R Ninety-Sixth Congr. Natl. Soc. Savantes, Sect. Sci.* **96**, No. 3, 579–587.
Hediger, H. (1950). "Wild Animals in Captivity." Butterworth, London.

Heimer, L., and Larsson, K. (1967). *Physiol. Behav.* **2**, 207–209.
Heymer, A. (1977). "Ethological Dictionary." Parey, Berlin.
Hill, W. C. O. (1938). *Ceylon J. Sci.* **21**, 65.
Hill, W. C. O. (1953). "Primates: Comparative Anatomy and Taxonomy. I. Strepsirhini." Edinburgh Univ. Press, Edinburgh.
Hill, W. C. O. (1955). "Primates: Comparative Anatomy and Taxonomy. II. Haplorhini: Tarsioidea." Edinburgh Univ. Press, Edinburgh.
Hill, W. C. O. (1956). *Proc. Zool. Soc. London* **127**, 580.
Hill, W. C. O. (1958). *In* "Primatologia: Handbook of Primatology" (H. O. Hofer, A. H. Schultz, and D. Starck, eds.), Vol. III, Part I, pp. 139–207 and 630–704. Karger, Basel.
Ilse, D. R. (1955). *Br. J. Anim. Behav.* **3**, 118–120.
Jewell, P. A., and Oates, J. F. (1969). *Zool. Afr.* **4**, 231–248.
Johnston, R. P. (1975). *Z. Tierpsychol.* **37**, 75–98.
Jolly, A. (1966a). Unpublished observations, cited in Doyle (1974b).
Jolly, A. (1966b). "Lemur Behavior: A Madagascar Field Study". Univ. of Chicago Press, Chicago, Illinois.
Jones, R. B., and Nowell, N. W. (1973). *Physiol. Behav.* **11**, 35–38.
Karli, P. (1971). *C. R. Séances Soc. Biol. Ses. Fil.* **165,** 492–499.
Klopfer, P. H. (1970). *Folia Primatol.* **13**, 137–143.
Klopfer, P. H., and Gamble, J. (1966). *Z. Tierpsychol.* **23**, 588–592.
Kollman, M., and Papin, L. (1925). *Arch. Morphol. Gen. Exp.* **22**, 1–61.
Krames, L., Carr, W. J., and Bergman, B. (1969). *Psychon. Sci.* **16,** 11–12.
Le Magnen, J. (1960). *In* "Les grands activités du Rhinencéphale," (T. Alajouanine, ed.), Vol. II, pp. 67–94. Masson, Paris.
Le Magnen, J. (1970). *Adv. Chemoreception* **1**, 393–404.
Loo, S. K. (1973). *Folia Primatol.* **20**, 410–422.
Lowther, F. de L. (1940). *Zoologica (N.Y.)* **25**, 435–462.
Machida, H., and Giacometti, L. (1967). *Folia Primatol.* **6**, 48–69.
MacLeod, P. (1971). *Handb. Sens. Physiol.* **4**, Part 1, 182–204.
Manley, G. H. (1974). *In* "Prosimian Biology" (R. D. Martin, G. A. Doyle, and A. C. Walker, eds.), pp. 313–329. Duckworth, London.
Marler, P. (1965). *In* "Primate Behavior: Field Studics of Monkeys and Apes" (I. DeVore, ed.), pp. 544–584. Holt, New York.
Martin, R. D. (1968). *Z. Tierpsychol.* **25**, 409–495.
Martin, R. D. (1972a). *Z. Tierpsychol., Beih.* **9**, 43–89.
Martin, R. D. (1972b). *Philos. Trans. R. Soc. London, Ser. B* **264**, 295–352.
Martin, R. D. (1973). *In* "Comparative Ecology and Behaviour of Primates" (R. P. Michael and J. H. Crook, eds.), pp. 1–66. Academic Press, New York.
Mertl, A. S. (1975). *Behav. Biol.* **14**, 505–509.
Michael, R. P., and Keverne, E. B. (1968). *Nature (London)* **218**, 746–749.
Michael, R. P., and Keverne, E. B. (1970). *Nature (London)* **225**, 84–85.
Michael, R. P., Keverne, E. B., and Bonsall, R. W. (1971). *Science* **172**, 944–966.
Michael, R. P., Zumpe, D., Keverne, E. B., and Bonsall, R. W. (1972). *Recent Prog. Horm. Res.* **28**, 665–706.
Montagna, W. (1962). *Ann. N.Y. Acad. Sci.* **102**, 190–209.
Montagna, W., and Ellis, R. A. (1959). *Am. J. Phys. Anthropol.* **17**, 137–162.
Montagna, W., and Ellis, R. A. (1960). *Am. J. Phys. Anthropol.* **18**, 19–44.
Montagna, W., and Yun, J. S. (1962a). *Am. J. Phys. Anthropol.* **20**, 149–166.
Montagna, W., and Yun, J. S. (1962b). *Am. J. Phys. Anthropol.* **20**, 441–449.
Montagna, W., Yasuda, K., and Ellis, R. A. (1961a). *Am. J. Phys. Anthropol.* **19**, 1–22.
Montagna, W., Yasuda, K., and Ellis, R. A. (1961b). *Am. J. Phys. Anthropol.* **19**, 113–130.

Montagna, W., Machida, H., and Perkins, E. M. (1966). *Am. J. Phys. Anthropol.* **25**, 277–290.
Müller-Schwarze, D. (1971). *Anim. Behav.* **19**, 141–152.
Müller-Velten, H. J. (1966). *Z. Vergl. Physiol.* **52**, 401–429.
Mykytowytcz, R. (1970). *Adv. Chemoreception* **1**, 327–360.
Napier, J. R., and Walker, A. C. (1967). *Folia Primatol.* **6**, 204–219.
Negus, V. (1958). "The Comparative Anatomy and Physiology of the Nose and Paranasal Sinuses." Livingstone, Edinburgh.
Niemitz, C. (1974). *Folia Primatol.* **21**, 250–276.
Noirot, E. (1969). *Anim. Behav.* **17**, 542–546.
Nolte, A. (1958). *Behaviour* **12**, 183–207.
Ortmann, R. (1958). *In* "Primatologia: Handbook of Primatology" (H. O. Hofer, A. H. Schultz, and D. Starck, eds.), Vol. III, Part I, pp. 355–382. Karger, Basel.
Ortmann, R. (1960). *Handb. Zool.* **8**, No. 26, 1–68.
Pagès, E. (in press). *In* "Ecology, Physiology and Behavior of Five Nocturnal Lemurs of the West Coast of Madagascar" (P. Charles-Dominique, ed.), Karger, Basel.
Pagès, E., and Petter-Rousseaux. (in press). *In* "Ecology, Physiology and Behavior of Five Nocturnal Lemurs of the West Coast of Madagascar" (P. Charles-Dominique, ed.), Karger, Basel.
Perret, M. (1975). *Mammalia* **39**, 119–132.
Perret, M. (1977). *Z. Tierpsychol.* **43**, 159–179.
Petter, J. J. (1962). *Mém. Mus. Natl. Hist. Nat., Paris, Sér. A* **27,** 1–146.
Petter, J. J. (1965). *In* "Primate Behavior: Field Studies of Monkeys and Apes" (I. DeVore, ed.), pp. 292–319. Holt, New York.
Petter, J. J., and Hladik, C. M. (1970). *Mammalia* **34**, 394–409.
Petter, J. J., and Peyrieras, A. (1970a). *Mammalia* **35**, 167–193.
Petter, J. J., and Peyrieras, A. (1970b). *Terre Vie* **24**, 356–382.
Petter, J. J., Schilling, A., and Pariente, G. (1971). *Terre Vie* **25**, 287–327.
Petter, J. J., Albignac, R., and Rumpler, Y. (1977). "Faune de Madagascar: Lémuriens." ORSTOM - CNRS, Paris.
Petter-Rousseaux, A. (1964). *In* "Evolutionary and Genetic Biology of Primates" (J. Buettner-Janusch, ed.), Vol. 2, pp. 92–131. Academic Press, New York.
Petter-Rousseaux, A. (1970). *Ann. Biol. Anim. Bioch. Biophys.* **10**, 203–208.
Pfaff, D., and Pfaffman, C. (1969). *Olfaction Taste, Proc. Int. Symp., 3rd, 1968* Vol. III, pp. 258–267.
Pfaffman, C. (1972). *Electroencephalogr. Clin. Neurophysiol., Suppl.* **31**, 185–203.
Pinto, D., Doyle, G. A., and Bearder, S. K. (1974). *Folia Primatol.* **21**, 135–147.
Pocock, R. J. (1918). *Proc. Zool. Soc. London* pp. 19–53.
Poduschka, W., and Firbas, W. (1968). *Z. Säugetierkd.* **33,** 160–172.
Pollock, J. I. (1975). *In* "Lemur Biology" (I. Tattersall and R. W. Sussman, eds.), pp. 287–311, Plenum, New York.
Richard, A. F. (1973). Ph.D. Thesis, University of London (unpublished).
Richard, A. F. (1974a). *In* "Prosimian Biology" (R. D. Martin, G. A. Doyle, and A. C. Walker, eds.), pp. 49–74. Duckworth, London.
Richard, A. F. (1974b). *Folia Primatol.* **22**, 178–207.
Ropartz, P. M. (1966). *C.R. Hebd. Séances Acad. Sci.* **263,** 525–528.
Ropartz, P. M. (1967a). *C.R. Hebd. Séances Acad. Sci.* **264,** 1479–1481.
Ropartz, P. M. (1967b). *Rev. Comportement Anim.* **1**, 97–102.
Ropartz, P. M. (1968). *Rev. Comportement Anim.* **2**, 35–77.
Rosedale, R. S. (1945). *Arch. Otolaryngol.* **42**, 235.
Rumpler, Y., and Andriamiandra, A. (1968). *C.R. Séances Soc. Biol. Ses. Fil.* **162,** 1430–1433.
Rumpler, Y., and Andriamiandra, A. (1971). *C.R. Séances Soc. Biol. Ses. Fil.* **165,** 436–440.
Rumpler, Y., and Oddou, J. H. (1970). *C.R. Seances Soc. Biol. Ses. Fil.* **164**, 2686–2690.
Russel, M. J. (1976). *Nature (London)* **260**, 520–522.

Sauer, E. G. F., and Sauer, E. M. (1963). *J. S.W. Afr. Sci. Soc.* **16**, 5–35.
Schaffer, J. (1924). *Z. Wiss. Zool.* 79–96.
Schaffer, J. (1940). "Die Hautdrüsenorgane der Saugetiere." Urban & Schwarzenberg, Berlin.
Schilling, A. (1970). *Mém. Mus. Natl. Hist. Nat., Paris, Sér. A* **61,** 203–280.
Schilling, A. (1974). *In* "Prosimian Biology" (R. D. Martin, G. A. Doyle, and A. C. Walker, eds.), pp. 347–362. Duckworth, London.
Schilling, A. (in press). *In* "Ecology, Physiology and Behavior of Five Nocturnal Lemurs of the West Coast of Madagascar" (P. Charles-Dominique, ed.), Karger, Basel.
Schneider, K. M. (1930). *Zool. Gart. Leipzig* **3**, 183–198.
Scott, J. W., and Pfaff, D. W. (1970). *Physiol. Behav.* **5**, 407–411.
Seitz, E. (1969). *Z. Tierpsychol.* **26,** 73–103.
Signoret, J. P. (1970). *J. Reprod. Fertil. Supple.* **11**, 105–117.
Signoret, J. P., and Mauléon, R. (1962). *Ann. Biol. Anim., Biochim., Biophys.* **2**, 167–174.
Signoret, J. P., and Du Mesnil Du Buisson, F. (1962). *Proc. Int. Congr. Anim. Reprod., 4th, 1961* Vol. 4, pp. 171–175.
Stephan, H., and Andy, O. J. (1964). *Ann. Zool. (Agra, India)* **4**, 59–74.
Sussman, R. W. (1974). *In* "Prosimian Biology" (R. D. Martin, G. A. Doyle, and A. C. Walker, eds.), pp. 75–108. Duckworth, London.
Sussman, R. W., and Richard, A. F. (1974). *In* "Primate Aggression, Territoriality and Xenophobia: A. Comparative Perspective" (R. L. Holloway, ed.), pp. 49–76. Academic Press, New York.
Tenaza, R., Ross, B. A., Tanticharoenyos, P., and Berkson, G. (1969). *Anim. Behav.* **17**, 664–669.
Valenta, G., and Rigby, M. K. (1968). *Science* **161**, 599–601.
Vincent, F. (1969). Doctoral Thesis, University of Paris (unpublished).
von Fiedler, W. (1959). *Acta Anat.* **37**, 80–105.
von Holst, D. (1972). *J. Comp. Physiol.* **78**, 236–273.
von Holst, D. (1974). *In* "Prosimian Biology" (R. D. Martin, G. A. Doyle, and A. C. Walker, eds.), pp. 389–411. Duckworth, London.
Walker, A. C. (1969). *East Afr. Wildl. J.* **7**, 1–5.
Walker, A. C. (1970). *Primates* **11**, 135–144.
Welker, C. (1973). *Folia Primatol.* **20**, 429–452.
Wharton, C. M. (1950). *J. Mammal.* **31**, 260–269.
Whitten, W. K. (1956). *J. Endocrinol.* **13**, 399–404.
Whitten, W. K., and Bronson, F. H. (1970). *Adv. Chemoreception* **1**, 309–325.
Wilson, E. O., and Bossert, W. H. (1963). *Recent Prog. Horm. Res.* **19**, 673–716.
Woollard, H. H. (1925). *Proc. Zool. Soc. London* pp. 1071–1184.
Yasuda, K., Oaki, T., and Montagna, W. (1961). *Am. J. Phys. Anthropol.* **19**, 23–34.

Chapter 12

Prosimian Locomotor Behavior

ALAN WALKER

1. INTRODUCTION

It would be customary, in a review of this kind, to summarize all observations made on the locomotion of wild and captive prosimian primates and to follow that with a classificatory synthesis. There is now too much information to continue to use any simple, overall locomotor classification of the prosimians and too little information to propose any more sophisticated alternatives. With this in mind this chapter will concentrate on the problem of what sort of information might be needed to further our understanding and will add, as a kind of appendix, a summary of the observations that have been made to date on the locomotion of living prosimians.

To illustrate the problems that we face, the results of observations on the locomotion of *Galago demidovii* made in Uganda in 1968 and 1969 are presented. It is not proposed that the results of this study, in isolation, have any intrinsic worth or even tell us much about the basic locomotion of the species. The results do emphasize,

ISBN 0-12-222150-8

however, some problems encountered when dealing with locomotor studies and might be valuable in formulating concepts and developing methodologies.

Before describing this particular case history, however, a declaration of viewpoint is needed so that others who look from different directions can be reassured. If evidence gathered from their vantage point is excluded, it is not that it is undervalued, but that confusion and unnecessary controversies have been brought about in the past by different approaches leading to different syntheses. When dealing with the locomotion of living animals the viewpoint of the writer is, first and foremost, that it is the locomotion that should be studied. Considerations of functional anatomy, paleontology, and suchlike should not be introduced from the outset. Confusion has often arisen in locomotor studies when morphology and behavior have been considered together. If, after the locomotion of a species has been documented, the morphology appears not to be compatible, careful analysis may eventually show that the anatomical correlates were indeed present or that the behavior was misunderstood. The history of functional anatomy shows, after all, only a succession of revisions which have been brought about, by better understanding of the factors involved, of previous functional interpretations. In pointing this out there is no intent to belittle this discipline, but to show that one of the fascinations of the subject is that today's answer is tomorrow's unacceptable hypothesis. Before functional interpretations can be made, the first prerequisite, that of characterizing the animals' locomotion, has to be fulfilled. Any morphological inferences that seem not to be correlated with behavior are often based upon insufficient behavioral data.

It might be argued, as Stern and Oxnard (1973) have done recently, that all locomotor classifications have now outlived their usefulness, either as means of communication or for creating manageable numbers of groups. While any classification should be subject to revision as more evidence is accumulated, the level at which the classification is attempted has, however, a direct bearing on whether or not it may prove to be acceptable for a particular purpose. If enough care were taken, it would be possible to differentiate between the locomotion of any two individuals in a population, just as one can recognize the differences in locomotion between any two acquaintances. Similarly, an expected amount of variability will be found in all species. Major locomotor differences might be found between different populations of a species which occupy different habitats, but this sort of variability is often overlooked, even to the extent of dealing with the locomotion of genera that are not even monospecific. In part this is due to the fact that the overall locomotion of members of a species is, within limits, consistent. This is not to underestimate any background variability and what might be learned from its study, but we are still at the stage of trying to characterize the species' locomotion and, at our present level of understanding, it does appear that many species have similar locomotor patterns and that some species that are clearly in different genera have similar overall patterns.

At the species level, then, one could typologically classify a locomotion as, say, "*Galago demidovii*" locomotion. This really does not help in understanding the locomotion of this or any other species. Furthermore, one is constantly struck when reading through the literature by the frequency with which comparisons are made with other species or genera when a locomotion is being described. At the root of most dissatisfaction with present locomotor classifications is the fact that, on occasions, most primates will exhibit a great variety of locomotor activities. Most classifiers have, of course, been aware of this variety. What one worker regards as a fringe behavior in a species may be taken by someone else to be an important part of the locomotor repertoire. One way out of this dilemma might be to determine the frequency of individual locomotor activities. Then, for instance, it might be seen that species such as *Perodicticus potto* have a frequency of leaping that is for all practical purposes zero, while another might have a 50% frequency of leaping. Quantification of frequencies of activity should be more amenable to analysis than descriptive accounts and might lead to meaningful generalizations about locomotion within the primates.

2. THE *GALAGO DEMIDOVII* CASE

In 1968 a study was carried out with this outlook in mind. The subjects were two specimens of *Galago demidovii* that had already been in captivity for some months. The animals were transferred to a new outdoor cage in a new locality during the daytime. Locomotor counts were made at irregular intervals until the project was interrupted. The cage measured 3 × 2 × 1.2 m and consisted of a frame covered with fine wire netting. The floor was covered with leaf litter and the cage filled with a variety of branches of a fairly thin diameter that approximated as nearly as possible to the tangled vegetation that was present in the nearby forest where the animals were trapped. Measured dowels or rods were not used since a more or less "natural" branch system was thought to be more useful in this case. Counts were made at night with illumination provided by a low light source behind the cage and by moonlight. The movements of this species are extremely rapid, with up to 1400 separate locomotor movements an hour, and so the observations were recorded on a tape recorder and later transcribed to checksheets. Preliminary tests with animals in their old cage showed that it was impossible to record single movements on the tape recorder, but that changes of movements and displacements were possible to record. Similarly, it was possible to record whether the support used was a horizontal, vertical, or diagonal one. These three categories were a little arbitrary in this semi-naturalistic setting, but they were used consistently. In all, 11 hours of counts were completed at hours from sunset to midnight, during which nearly 7000 locomotor events were recorded in the form of changes of type of displacement.

The following questions can be asked even before the results are examined.

1. Is locomotor behavior under captive conditions really typical? The answer to this question is possibly yes, depending upon conditions, the species under investigation, and the nature of the study. The only sure way to answer this question would be a study that compares results from wild and captive populations of the same species. In the case of *G. demidovii*, the amount of time required to gather the information in the wild would be enormous. Also, even under natural conditions, the observer is never sure of his or her effect on the animals under observation. In this experiment, the animals were given plenty of space and confinement ought not to have greatly altered their behavior. The equivalent area in relation to body size is about the size of a football pitch for humans or a volume $30 \times 30 \times 12$ m for a medium-sized monkey like a macaque. The cage was furnished with branches that matched, as closely as was possible, one of the main habitats in which the animals had been seen in the wild.

2. Can the locomotion of a few specimens be taken as typical of the species? The answer, once again, is that this will not be known until a test of many individuals from the same species has been made. Here, in this study, just two individuals were used, but the qualitative impression was that they exhibited "normal *G. demidovii*" locomotion.

3. How does captivity change the locomotor pattern? In zoos throughout the world it is common to see cage behavior of all types and locomotor abnormalities are common. In this study the pattern was seen to change with time and, by experimental accident, some conclusions could be reached about this change and what it means in terms of locomotion. Settling in to the new cage appeared to be rapid. Although it is not possible to say whether the animals' behavior was more typical when settled or disturbed, it is just as well to be aware that the problem exists.

4. Are there any diurnal variations in locomotion that would affect the results? The study confirmed the findings of Haddow and Ellice (1964) that a pronounced biphasic activity pattern is typical of *Galago* species. The results show, however, that only the frequency of locomotor displacements varied with time of activity and not the type of displacements. It is conceivable that in some species the types of locomotion might vary with the time of observation and this should always be borne in mind.

3. THE RESULTS OF THE *GALAGO DEMIDOVII* EXPERIMENT

At the end of the observation period the count data were the numbers of the various locomotor displacements counted during each period. The time at which each count was taken allowed a check on activity from dusk to midnight and the successive counts gave a possibility for checking effects that settling in to the new environment might have on the locomotion.

The total number of separate locomotor displacements recorded was 6628 over the 11-hour period. This amounts to a mean of 602.5 changes in locomotor mode in an hour, or about 10/minute. The maximum count for one-half hour was 698, or about 23/minute, and the minimum count was 116, or about 4/minute. This great variation in count number is partly in response, however, to normal activity rhythms. Biphasic activity patterns, with high activity at dusk and dawn have been recorded for other nocturnal prosimians. This has been demonstrated using aktographs with captive galagos (Haddow and Ellice, 1964). Charles-Dominique (1972) extended this method by placing a baited aktograph in his study area in Gabon, with the same results for *G. demidovii*. In the 1968 study, activity levels clearly fell between dusk and midnight. Counts of separate activities were also correlated with time of count. One activity, hanging by the feet, was much more commonly recorded at dusk and this is taken to be a function of stretching behavior upon leaving the nest. Figure 1 shows the steady decline in activity from dusk to midnight.

In order to compare frequencies of activity at different times, the separate activity counts were transformed, therefore, into percentages of each total count (see Table I). This has a disadvantage in that correlations between the various percentages are partly masked by the percentage making procedure.

Changes with successive counts might be accounted for by acclimatization to the new environment. The animals were moved to a new cage on the day of the first count. The amount of ground activity increased significantly ($p < 0.01$) with successive counts, but neither amount of activity along horizontals nor leaping

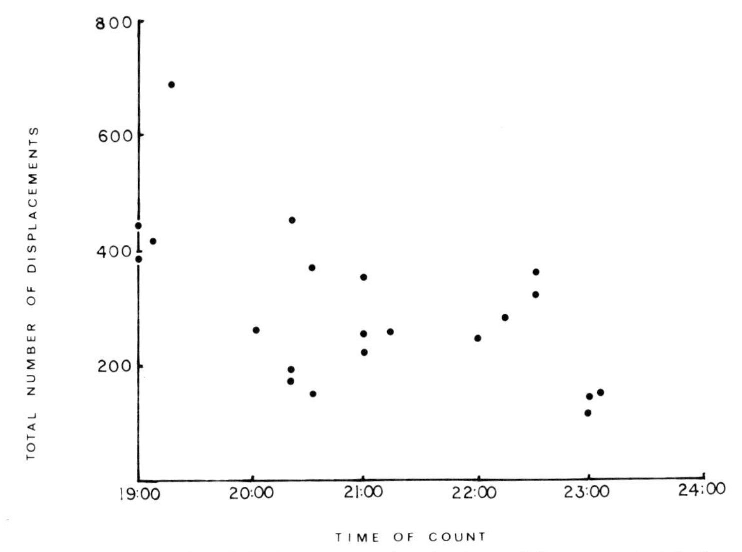

Fig. 1. Plot of number of total displacements against the times of the counts. Activity is seen to fall from its highest at dusk to its lowest toward midnight.

TABLE I

Mean Percentages and Variation of the Percentages of the Counts

Locomotor displacement[a]	$\bar{x}$	S.D.	S.E.$\bar{x}$	V
VV	14.21	7.43	1.58	52.28
AV	12.72	2.86	0.61	22.48
VH	8.74	2.66	0.57	30.43
HV	9.21	2.26	0.48	24.54
HH	13.14	5.51	1.18	41.93
AH	5.24	2.71	0.58	51.72
AD	12.82	4.64	0.99	36.19
GH	3.41	2.11	0.45	61.88
GR	3.69	2.73	0.53	73.98
GW	2.36	2.16	0.46	91.53
GG	9.46	5.45	1.16	57.45
Wire	13.14	3.17	0.67	24.12
Feet	1.82	1.00	0.21	54.94

[a] Key to abbreviations: VV, leaping from vertical to vertical; AV, climbing up or down vertical; VH, leaping from vertical to horizontal; HV, leaping from horizontal to vertical; HH, leaping from horizontal to horizontal; AH, running or walking along horizontal; AD, running or walking up or down diagonal; GH, hopping on ground; GR, running on ground; GW, walking on ground; GG, (GH + GR + GW); Wire, climbing on wire; Feet, hanging by feet.

altered significantly. However, the frequency of vertical-to-vertical leaps was negatively correlated with the amount of ground activity ($p < 0.05$). The first two recorded counts were peculiar in two respects. First, the percentage of vertical-to-vertical activity was high and, second, that of all ground activity was low, although the overall leaping figures were not significantly larger. This could be taken either as an initial response to the new cage situation after which activity dropped to ''acclimatized'' or more normal levels, or that the typical activity pattern was shown in the first two days and that activity later dropped to a level more typical of caged behavior. There is support for both interpretations. Ground activity, as we have seen, increased significantly with successive counts and this might support the second interpretation, but an interesting occurrence helped point out that vertical-to-vertical leaping might be influenced by other factors that lend support to the first interpretation. Figure 2 shows plots of the index %VV/%HH against count number and the index %VV/%GG against count number. Both plots show the first two counts as being irregular, but one other count (No. 12) was also peculiar, and in the same manner as Nos. 1 and 2. On checking the notes it was found that on the morning after count No. 12 a green mamba was discovered in the tree over the cage, and it is likely that this arboreal predator was in the tree during the time this count

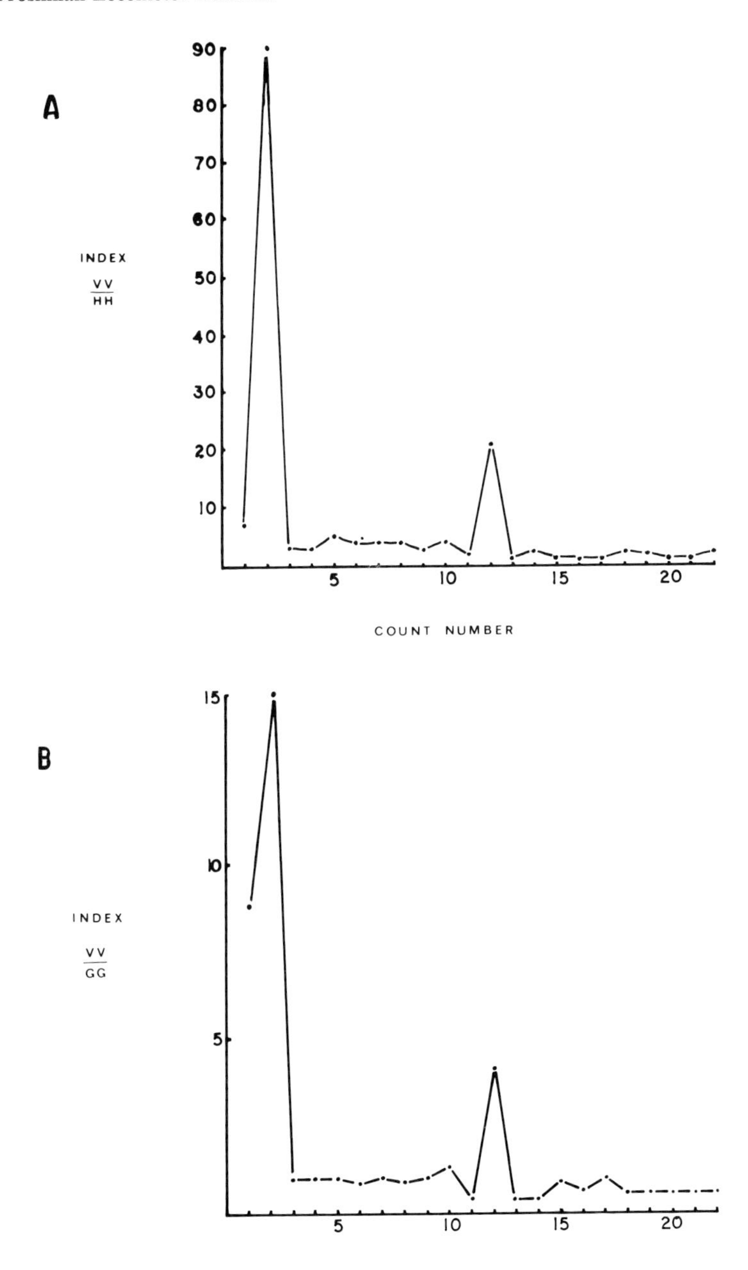

Fig. 2. (A) Plot of the index VV/HH against count to show the unusual nature of counts 1, 2, and 12 (legend as for Table I). (B) Plot of the index VV/GG against count to show the unusual nature of counts 1, 2, and 12 (legend as for Table I).

was made. It is possible that the increased vertical-to-vertical activity and decreased ground and horizontal activity was related in the first two counts to unfamiliarity with the cage setting and the observer and in count No. 12 to the snake. Whatever the real reason, this does illustrate that the cage environment should be controlled carefully in order to avoid complications such as these. An attempt was made to understand the relationship between these variables (VV, HH, and GG) by using a 3-variable linear regression analysis and it was found that using logarithmic transformations, the formula

$$\log GG = 0.25 - 0.5 \log VV + \log HH$$

accounts for 60% of all ground movements. The individual simple correlations are significant at the 1% level, whereas without logarithmic transformations they are only significant at the 5% level. This shows that vertical activity decreases disproportionately as the ground and horizontal activities increase.

The overall picture, given by the percentage means, is presented in Fig. 3. Arboreal-to-ground activity is in the ratio of about 9:1. In arboreal activity use of vertical over horizontal supports is in the ratio of nearly 9:7 and, with wire and ground activity included, in the ratio of nearly 4:3. There is a significant ($p < 0.01$) negative correlation between all leaping activity and activity along or up and down supports, but hopping on the ground is directly correlated ($p < 0.05$) with walking and running on the ground. Figures for the use of diagonals indicate that they are used as verticals rather than horizontals; activity on diagonals is directly correlated ($p < 0.01$) with activity along horizontal supports and negatively correlated ($p <$ 0.05) with activity up and down vertical supports. Up and down vertical activity is itself negatively correlated at the same level of significance with horizontal activity.

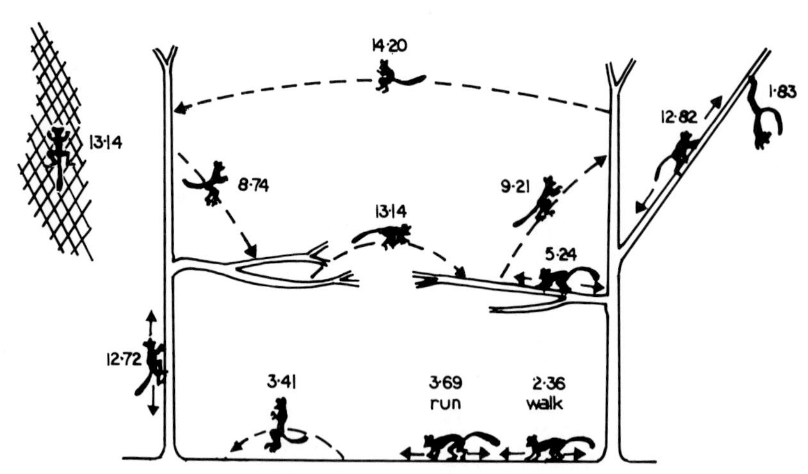

Fig. 3. Diagram to show the mean percentages of locomotor displacements over 11 hours of discontinuous observations on *G. demidovii*.

Leaping between verticals is negatively correlated ($p < 0.01$) with leaping between horizontals. This accounts for much of the variability seen in these activities. There is a strong ($p < 0.01$) correlation between horizontal-to-vertical leaps and vertical-to-horizontal leaps, and the percentage means are virtually identical for these two sets of activities. This might, in fact, indicate not only that what goes up must come down, but also that these bushbabies did it by the same routes. Activity on the wire is not correlated with either VV or AV and is thus not accounting for bias toward increase in counts on vertical supports.

Figure 4 gives indications of the variability of the separate counts and the high variability of VV, HH, and AH is seen. More importantly, the generally low variability of most of the count percentages might be taken as an indication that the method of analysis is capable of generating some confidence.

If the picture gained here is shown by further counts of the locomotion of *G. demidovii* specimens in other setting, then this species' locomotion could perhaps be characterized in the following way:

> Locomotor pattern with about 10% ground activity, consisting of hopping, running, and walking in about equal proportions. Arboreal activity includes leaping, running, walking, and climbing, with leaping about 60% of this. Leaping activity increases as walking, running, and climbing decrease. Leaps from horizontal to vertical and vertical to horizontal occur in about equal numbers. Leaping between verticals increases with decreases in leaps between horizontals. Predator avoidance is probably seen in a marked reduction in ground activity and a very marked increase in vertical to vertical leaping. Activity is high, typically up to 600 changes of locomotor mode an hour, but ranges from 1400 to 230, depending upon the activity cycles.

To this can be added the normal qualitative description of locomotion. This type of locomotor characterization might be more useful than the qualitative ones used to

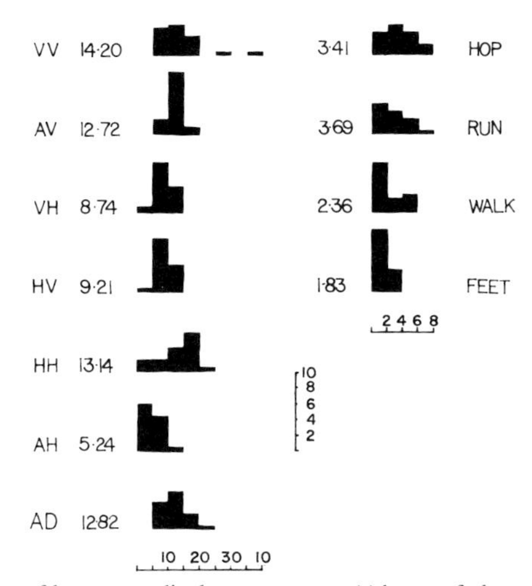

Fig. 4. Mean percentages of locomotor displacements over 11 hours of observation on *G. demidovii* with histograms showing variability recorded in the percentages of each type of displacement.

date and, taken in conjunction with the usual descriptions, the behavioral basis for locomotor studies should be much more reliable than in the past.

One aspect of locomotion that this might be useful in resolving is the amount of "fringe" activity in the total pattern. For instance, the overall frequency of hanging by the feet was found in this study to be less than 2% and the frequency of running underneath a support was only 0.3%. The major locomotor modes, and presumably these are the ones for which the animals are adapted, are clearly shown in the frequency figures.

4. DESCRIPTIVE ACCOUNTS OF THE LOCOMOTION OF PROSIMIANS

The following is a summary of what is known of the locomotion of species of prosimians in the wild and in captivity. No classification or overview is given here, but if readers are struck as they read these notes by differences and similarities of locomotion as described, they are welcome to attempt their own classification. Such a classification will be as useful as the data will allow and will probably be not at all useful if detailed morphological analysis is the reason behind wanting a locomotor classification. The observations are made, for the most part, by different observers with different outlooks and biases: as such they are fairly general and rather crude; any classification based upon these data will also be fairly general and rather crude.

4.1. Cheirogalcidae

4.1.1. Cheirogaleus (Petter, 1962; Rand, 1935; Shaw, 1879; Walker, 1974)

Cheirogaleus major run on all fours, sitting up to eat when holding food in both hands. The body is held close to the branch upon which they are running. Petter (1962) says that *C. major* move in the same manner as *Microcebus murinus*, but that they appear less agile since they are larger and heavier. This species will jump from branch to branch occasionally, but prefer to return to main trunks in order to embark on another radial path. When moving quickly the animal's gait approaches a gallop, with both fore and then both hind limbs making contact. The tail is always held straight out behind the moving animal. This species can be very sluggish in captivity.

Cheirogaleus medius are smaller and more agile than *C. major*, but always progress with rapid quadrupedal steps. Like the larger species, *C. medius* also show a tendency to creep along branches with the body following the undulations of the support. The tail is held straight out behind the body, especially when it is filled with fat. *Cheirogaleus medius* jump more frequently than *C. major* even when the tail is fat-filled. Rand (1935) observed one "running about like a squirrel rather than like its relatives of the genus *Lemur*" in the lowest three to five meters of the forest.

4.1.2. Microcebus (Martin, 1972; Petter, 1962; Petter et al., 1971; Shaw, 1879; Walker, 1974)

Microcebus murinus run about quadrupedally in a small branch milieu in a jerky fashion, rapid bouts of running interspersed with pauses. When running the body is horizontal and aligned with the support. Small gaps between branches are crossed either by reaching for the next branch and holding it with the hands before swinging the hindlimbs across, or by short leaps in which the hands as well as the feet are involved in the landing. The tail is moved from side to side as a balance in walking along fine branches. When feeding they have been seen to suspend themselves by the hind feet in order to grasp some lower food object. They also exhibit the cantilever posture in which the body can reach sideways supported only by the hind feet. On the ground progression is usually quadrupedal, but they can also hop in a froglike way, landing on all fours. When resting this species adopts a quadrupedal stance with the limbs lightly flexed.

Microcebus coquereli are much larger than *M. murinus* and do not show such rapid displacements. They move quadrupedally on horizontal branches, but have a preference for the lowest layers of the forest where they progress by running up and down large trunks and by short leaps between smaller vertical uprights.

4.1.3. Allocebus trichotis (Petter and Petter-Rousseaux, 1967)

Hardly anything is known about this rare species in the wild. They seem to progress in a quadrupedal fashion rather like a large *Microcebus murinus*.

4.1.4. Phaner furcifer (Petter et al., 1971, 1975)

Phaner have an extremely rapid locomotion. It consists of rapid quadrupedal movements interspersed with leaps between the tips of branch systems. They are also capable of climbing up and down and leaping between vertical stems and vines as well as moving on large vertical trunks, which they do by means of their keeled nails. Leaps of 4–5 m have been recorded. Feeding postures are varied and include hanging on vertical trunks by all fours and hanging by the feet alone.

4.2. Daubentoniidae

4.2.1. Daubentonia (Lauvedon, 1933; Petter, 1962; Petter and Petter-Rousseaux, 1967; Petter and Peyrieras, 1971; Walker, 1974)

This peculiar species has a locomotion that is quite distinctive and further research should provide a clear characterization of the overall pattern. The usual method of progression is by slow-climbing, quadrupedal locomotion both along horizontals and also up and down verticals. Leaping between horizontals and verticals is common. Perhaps the most peculiar feature that is commonly seen is that of walking slowly upside down on large horizontal branches. The head is held close to the branch when searching for food and forearm postures are unique since the animal uses its attenuated middle finger for the extraction of larvae from beneath bark. When drinking, this species uses the middle finger to dip into the liquid and draws it rapidly through the mouth. Resting postures are varied, ranging from being rolled in a ball to sitting quadrupedally on flexed limbs. Quadrupedalism is the rule on the ground, but because of the long digits the wrists are held in a dorsiflexed posture and the weight is taken on the thenar and hypothenar eminences.

4.3. Galaginae

4.3.1. Galago (Bartlett, 1863; Bearder and Doyle, 1974; Bishop, 1964; Charles-Dominique, 1971, 1972; Doyle, 1974; Haddow and Ellice, 1964; Hall-Craggs, 1964, 1965; Jewell and Oates, 1969; Jouffroy et al., 1974; Kingdom, 1971; Vincent, 1969; Walker, 1974)

Observations on different species of *Galago*, made during the past decade, are leading to a more complete understanding of the locomotion of this genus. It is already clear that among species of this genus there is substantial variation in overall pattern of locomotion that has been brought about, in whole or in part, by the

Fig. 5. (A) *Galago crassicaudatus*: bipedal hopping. One-second sequence taken from 16-mm cine film of a captive specimen hopping on the ground. At take off the hindlimbs are fully extended and the forelimbs are slightly lowered and the hindlimbs flexed. The hindlimbs are extended to meet the ground on landing, but the knees remain slightly flexed. The tail is trailed inertly throughout. (B) *Indri indri*: leaping from vertical to vertical. Sequences taken from 16-mm cine film of 3 leaps from vertical to vertical stems. The resting posture is one with the hindlimbs fully flexed and the foot highly dorsiflexed and holding the support. The trunk is loosely clasped by the forelimbs. At take off the hindlimbs are fully extended and the forelimbs raised to a position at the sides of the head. The hindlimbs are flexed in midleap and extended again to meet the next support. The feet strike the support first and the hindlimbs flex to absorb the impact before being extended again for the next leap. The forelimbs may only swing the body around the support and sometimes may hardly touch the support. (Film taken in the wild at Perinet, Madagascar.) (C) *Propithecus verreauxi*: leaping from vertical to vertical. Sequence taken from 16-mm film of wild specimen leaping between *Alluaudia* stems. The stems are in fact vertical and only appear to be diagonally placed because of the camera angle. The hindlimbs are fully extended and the forelimbs raised on each side of the head at take off. The hips and knees are flexed in midleap but are extended again to meet the support. The twist in the tail is probably due to a rotation of the whole body on take off. (D) *Propithecus verreauxi*: bipedal hopping. Cartoon sequence of 1 1/2 seconds taken from 8-mm cine film of wild specimen in the Berenty reserve near Ft. Dauphin, Madagascar. This species moves by a series of long bipedal hops on open ground. The forelimbs are held under the chin and the tail is trailed inertly.

success of the various species in taking over new and varied niches in Africa, whatever the ancestral pattern and niche might have been.

Galago alleni move in the lowest few meters of primary forest and their main locomotor activity is leaping from one vertical support to the next. The supports are mostly of small diameter. On the ground they use quadrupedal locomotion as well as bipedal hopping. When disturbed on the ground, *G. alleni* seek vertical supports to leap to.

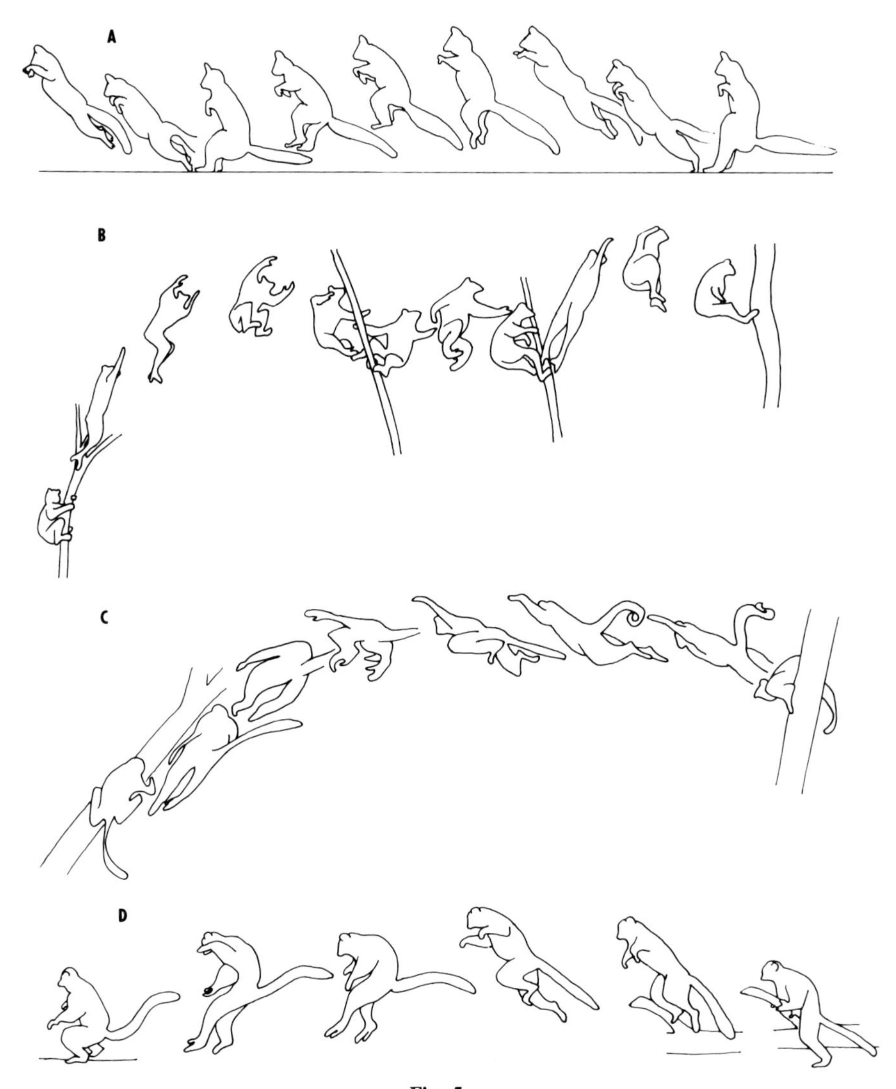

Fig. 5

Galago crassicaudatus, the largest species, move in a predominantly quadrupedal fashion, showing preferences for horizontal and diagonal supports. Nevertheless, leaping is still an important part of the locomotor repertoire. Individuals will leap from branch to branch, but since they use clearly defined routes the frequency of leaps and distance covered are extremely variable and depend upon location. Slow, cautious quadrupedal locomotion is sometimes seen in which the weight is evenly transferred from one limb to another. On the ground this species uses three basic types of locomotion; walking quadrupedally with the hindquarters and tail held high, galloping with both fore- and then hindlimbs making contact, and hopping bipedally (see Fig. 5A, and Charles-Dominique and Bearder, Chapter 13).

Galago demidovii restrict their movements to a fine branch habitat where they progress by running, walking, and climbing quadrupedally and leaping between supports. This species is extremely active and is seen to move in fits and starts, bouts of running interspersed with quiet periods and leaps of up to 2 m. On the ground slow progression is normally quadrupedal, but bipedal hopping is resorted to as soon as extra speed is needed. Feeding, sleeping, and grooming postures are extremely varied and range from hanging by the feet to cantilevering.

Galago inustus (Euoticus inustus), a little known East African species, have not been investigated fully. Preliminary observations suggest that they locomote rather like *G. senegalensis*.

Galago senegalensis are perhaps the best known species of the genus. They are renowned for their spectacular leaping ability. They prefer to hop even on horizontal branches and leaping is clearly the dominant locomotor activity. Quadrupedal progression is only used at very slow speeds. On the ground they move by hopping bipedally with the forelimbs making no contact with the ground.

Galago elegantulus (*Euoticus elegantulus*) move mainly in the higher layers of the forest and scarcely ever descend to the ground. Movement is usually along large branches, trunks, and lianas and they use their keeled nails in gripping large trunks in search for gums. They can climb vertical trunks both head upward or head downward. Horizontal movements along branches are made quadrupedally and leaps from one branch to another are common.

4.4. Indriidae

4.4.1. Avahi (Petter, 1962; Rand, 1935; Walker, 1974)

Members of this species have resting postures that involve the support of the body on a vertical trunk with the tail coiled like a watch spring. Locomotion consists primarily of hopping and leaping, whether it is up vertical trunks or from one trunk to another. Descents are made by moving slowly downwards quadrupedally or by a succession of downward leaps from trunk to trunk. Bipedal hopping is the rule for ground locomotion.

4.4.2. Indri (Petter, 1962; Petter and Peyrieras, 1974; Pollock, 1975; Rand, 1935; Walker, 1974)

Resting postures in this species are primarily those in which the body is held against a vertical trunk, but peculiar ''crouching'' postures are sometimes seen on horizontal or diagonal branches, where the branch is held by all four extremities and the long hindlimbs are flexed around the forelimbs. Sometimes sitting postures are used in which the animal is supported by a forked trunk. The main mode of progression is by powerful leaps between large vertical stems and trunks (see Fig. 5B). When vertical trunks are not available diagonals will be used. During a leap the body is held vertically and the feet are brought forward to reach the support first. Quadrupedal locomotion has been observed when animals cross long gaps in the forest by progressing slowly along thin or flexible boughs. Feeding postures vary from sitting on vertical trunks to hanging suspended by all four extremities. Bipedal hopping is the usual method of moving on the ground.

4.4.3. Propithecus (Jolly, 1966; Petter, 1962; Rand, 1935; Walker, 1974)

Of the two species of the genus only *P. verreauxi* have been extensively studied. The locomotion of *P. diadema* has received only passing reference, but does not seem to differ greatly from that of its congener. In *P. verreauxi* the locomotor pattern is dominated by the species' famous leaping abilities, but although rapid movement seems to be brought about mainly by leaping or hopping, this species exhibits a great variety of feeding and resting postures. Habitats that provide a great variety of branch types seemingly provoke a greater variety of postures. Progression through the trees is by a series of leaps from one support to another (see Fig. 5C), interspersed by short quadrupedal bouts and hopping along large horizontal branches. Resting postures are varied, ranging from the body supported on vertical trunks to the body sprawled over a horizontal branch. Feeding postures are also as varied, ranging from resting on a vertical trunk with one hand free to feed to hanging under a branch by all four limbs. On the ground, short displacements are made by short hops or by slow quadrupedal movements, but the preferred method is by a series of kangaroo-like bipedal hops (see Fig. 5D).

4.5. Lemuridae

4.5.1. Hapalemur (Lamberton, 1956; Petter, 1962; Petter and Peyrieras, 1970, 1975; Rand, 1935; Walker, 1974)

Little is known about the locomotion of the larger species, *H. simus*, either in captivity or in the wild. It is quite possible that any recorded observations on wild locomotor behavior in this species are based on a mistaken identity. *Hapalemur griseus* are found in a variety of habitats and exhibit a variety of postures. Resting postures range from clinging to vertical supports to resting quadrupedally with the limbs lightly flexed. They are capable of rapid quadrupedal locomotion in which the tail is held high and are capable of making considerable leaps, both from horizontal to horizontal and from vertical to vertical. Ground locomotion is quadrupedal and rapid.

The Alaotran subspecies, which live in reed beds that provide only vertical supports, progress by short leaps from one tuft to another and climb quite slowly in and around the stems of the reeds. It is interesting that they are more clumsy on the ground and, although they can run quadrupedally at slow speeds, will break out into a clumsy movement where the forelimbs are used for support as well as the hindlimbs (see Fig. 6A).

4.5.2. Lemur (Harrington, 1975; Jolly, 1966; Petter, 1962; Rand, 1935; Shaw, 1879; Sussman, 1974, 1975; Tattersall and Sussman, 1975; Walker, 1974)

Some species of the genus *Lemur* are better known than others and some, like *L. rubriventer*, are hardly known at all. As far as the locomotion is concerned it is clear that, on the whole, locomotion of *L. catta* is least typical of the genus. All are active, quadrupedal forms that run and walk on horizontal and diagonal branches and that are capable of leaping to and from verticals and to and from horizontals and verticals. On the whole, however, vertical supports are much less frequently used for progression through the trees than horizontals or diagonals. The frequency of leaping varies between species and even subspecies. *Lemur fulvus rufus*, for instance, seem to jump between trees only when there is no alternative route, whereas *L. fulvus fulvus* have been observed leaping from vertical to vertical trunks. *Lemur mongoz* have been observed to jump almost always despite continuous substrates of branches. Resting postures are varied and range from sitting upright to lying sprawled on a horizontal branch. On the ground progression is by quadrupedal walking, running, or galloping.

Lemur catta spend more time on the ground than any of the other species, but far from being less adept in the trees they are most accomplished leapers (see Fig. 6D). Although quadrupedal locomotion is favored as a rule, use of vertical supports is extensive. Leaps of some size are made between horizontals and verticals. Locomotion on the ground is normally by quadrupedal walking, running, or galloping (see Fig. 6E), but short bouts of bipedal running have been observed.

4.5.3. Varecia (Lemur variegatus) (Petter, 1962; Rand, 1935; Walker, 1974)

Varecia variegata are larger than any of the genus *Lemur* and the locomotion is more labored and cautious. Quadrupedal walking and running on larger branches is the normal locomotion and in a fine-branch situation these animals seem to prefer to secure handholds on the branches they intend to reach before they release their grip on those they are leaving. Leaping involves a preliminary stance in which the head

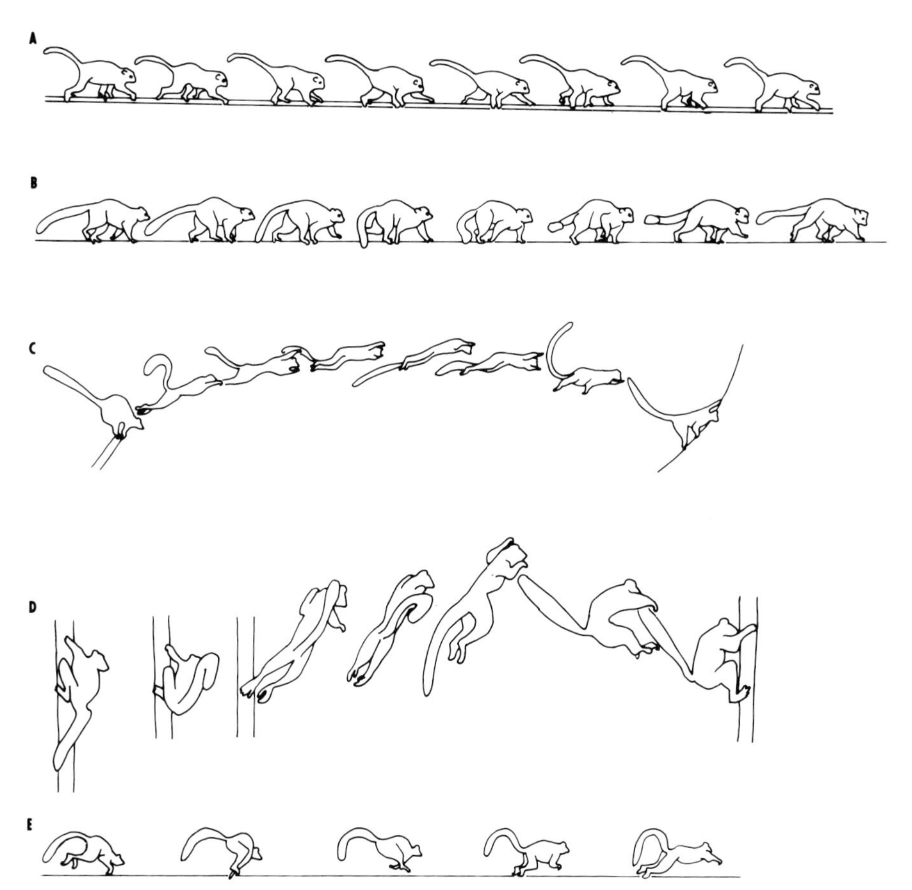

Fig. 6. (A) *Hapalemur griseus*: slow quadrupedal walking along a branch. Two-second sequence taken from 8-mm film of a captive specimen from Lac Alaotra, Madagascar. (B) *Lemur fulvus*: quadrupedal walking on an horizontal surface. Three-second sequence taken from 8-mm cine film of a captive specimen in Tananarive. The support formula is right hind-, left fore-, left hind-, and right forelimb. The side-to-side tail movement, so typical of *Lemur* quadrupedalism, is seen. The foot is inverted when reaching forward and everted at push-off. (C) *Lemur macaco*: arboreal leap. Cartoon sequence of about 1 second of a wild specimen. The head is lowered and the tail partly raised before take off. At take-off, both hind and forelimbs are extended and the tail raised in a "question mark" position. The hindlimbs become flexed and the tail is trailed behind during the leap. At the end of the leap the hindlimbs are extended and the forelimbs are brought forward to reach the support. (D) *Lemur catta*: leaping between vertical stems. Sequence of a leap of 2 m that lasted just over 1 second. The resting posture is one with the hindlimbs strongly flexed and the forelimbs preventing the trunk falling away from the support. The tail is raised before takeoff. The hindlimbs are fully extended and the forelimbs raised at takeoff. In midleap the tail is depressed and the hindlimbs gently flexed. The hips are flexed and knees extended to reach the landing support. The feet hit the support first and the forelimbs are then brought to grasp the support and the ankles and knees flex to attenuate the forces of impact. (E) *Lemur catta*: full speed gallop on the ground. Cartoon covering a little over one full movement cycle taken from 8-mm cine film of a wild animal at Berenty, Madagascar. Both fore- and hindlimbs work nearly together in this sequence. The tail is held in an arc during the gallop and the extreme flexion and extension of the spine can be seen.

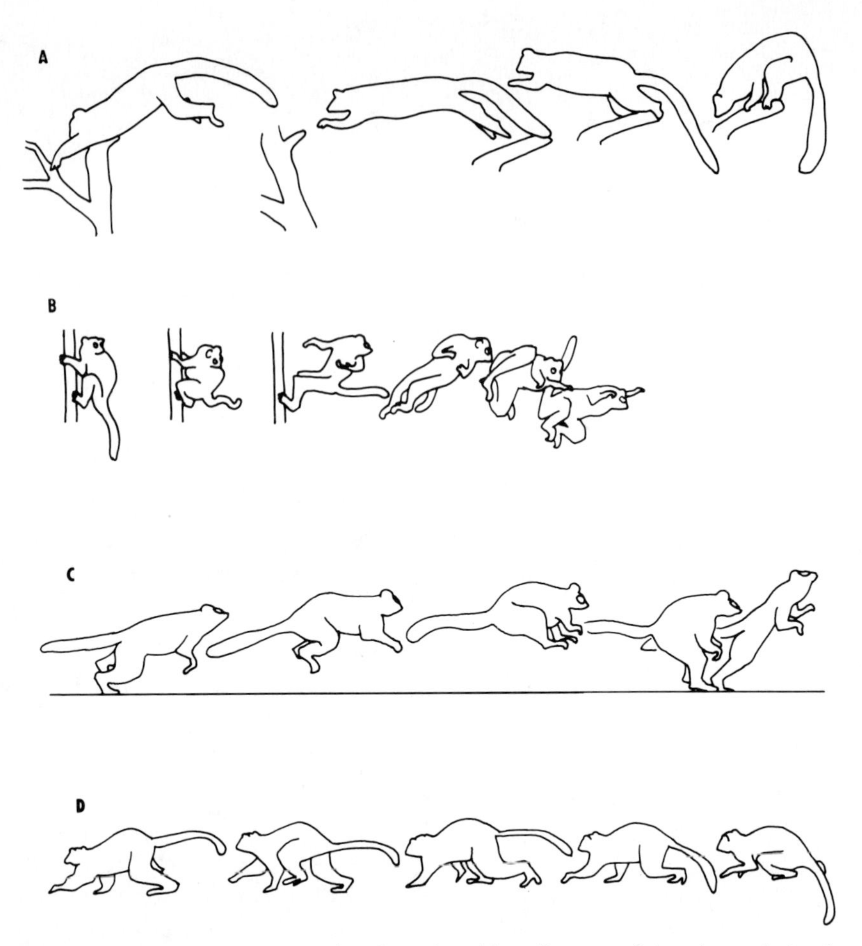

Fig. 7. (A) *Varecia variegata*: complete short arboreal leap. Sequence of about 1 second taken from 8-mm film of a captive individual in Tananarive. The head is lowered below the level of the support at takeoff. The hindlimbs are fully extended at takeoff and the forelimbs make contact with the landing support before the hindlimbs. (B) *Lepilemur mustelinus*: start of a leap from a vertical support. Cartoon sequence of just over 1-second duration taken from 8-mm film of a captive specimen in Tananarive. The forelimbs release their hold on the support before takeoff and are brought to a position on each side of the head during the movement. Shortly after takeoff the hindlimbs are flexed and the right hindlimb is brought quickly across the body, the knee lifted high, before both legs begin their extension prior to meeting the landing support. (C) *Lepilemur mustelinus*: bipedal hopping. Cartoon sequence of about 1 second of a captive animal in Tananarive. At takeoff the hindlimbs are fully extended and the forelimbs raised. The forelimbs are gradually lowered during the leap until they are close to the knees at touchdown. In midleap the lumbar spine is sharply flexed and the hands brought forward to meet the feet. The tail is trailed inertly throughout. (D) *Lepilemur mustelinus*: slow quadrupedal walking. Sequence taken from 8-mm film of a captive individual in Tananarive. Total time of sequence about 2 seconds, during which just over one complete movement cycle is carried out. The knee joint only seems to be extending to about 90 ° during walking. The foot is brought forward to reach the hand of the same side. There is marked flexion of the lumbar spine during this activity.

and shoulders are lowered and the tail raised before the main propulsive effort is made (see Fig. 7A). Resting and feeding postures are varied, but nearly always with the body held horizontally.

4.6. Lepilemuridae

4.6.1. Lepilemur (Charles-Dominique and Hladik, 1971; Petter, 1962; Rand, 1935; Walker, 1974)

Rapid, leaping locomotion from one vertical stem to another is the rule in this species (see Fig. 7B). When placed on an horizontal surface individuals will move slowly in a quadrupedal walk (see Fig. 7D), but revert to leaping as soon as they are pressed. On the ground kangaroo-like bipedal hopping is the normal method of moving (see Fig. 7C). Resting postures are almost invariably on a vertical support. The individuals sit with the hindlimbs lightly flexed and the forelimbs only preventing the trunk from falling backward. The tail hangs vertically downward. Leaps are made by powerful extension of the hindlimbs and the tail is trailed inertly during the leap. The hind feet are brought upward and meet the landing support at the end of the leap before the forelimbs. Stern and Oxnard (1973), on the basis of photographs in Charles-Dominique and Hladik (1971), suggested that *Lepilemur* flex the hindlimbs just prior to landing. This is based on a misconception and is obviously unsound biomechanically. The photographs referred to are not sequential ones of a single leap, but single ones of several leaps.

4.7. Lorisinae

4.7.1. Arctocebus (Charles-Dominique, 1971, 1972; Jewell and Oates, 1969; Walker, 1974)

Arctocebus calabarensis move by a slow-climbing, stealthy locomotion in which at least three of the extremities are involved in gripping the support at any one time. Because of their small size arboreal activity is restricted to supports of small diameter and they show no clear preference for any particular orientation of support. A position above the support is always preferred in locomotion, but when resting individuals may hang beneath a branch, especially when young. Feeding postures are varied, depending on the size and nature of the food item. This species has been observed to progress on the ground.

4.7.2. Loris (Petter and Hladik, 1970; Hill, 1953; Subramoniam, 1957; Walker, 1974)

Stealthy, slow-climbing locomotion is the rule in *L. tardigradus*. Quadrupedal progression with at least three limbs holding the support and with the body held almost invariably above the branch is normal. Feeding and resting postures are extremely varied, ranging from the individuals being rolled in a ball to hanging under the support.

4.7.3. Nycticebus (Hill, 1953; Tenaza et al., 1969; Walker, 1974)

Slow-climbing, quadrupedal locomotion is also the rule for *N. coucang* when in the trees. No leaping is observed. Usually travel in the trees is on the uppermost side of a support, but movements underneath have been reported. During normal climb-

Fig. 8. (A) *Perodicticus potto*: climbing. Sequence taken from film of a semi-captive specimen in Kampala. Three-point climbing is only relaxed during the swing from the vertical to the horizontal branch. The position of the hand acts as a guide for the foot of the same side. The long, forward reach of each stride is seen in the first stride taken on the horizontal. (B) *Perodicticus potto*: slow walking on the ground. Sequence taken from 8 mm film of a captive individual in Kampala. Outlines are taken at half-second intervals. The foot is positioned immediately behind the hand of the same side. Both hand and foot are positioned with the axis through the first and fourth digits set perpendicular to the line of movement. The cross is a stationary marker. (C) *Tarsius* sp.: leaping. Sequence taken from 16-mm cine film of a captive individual. The hindlimbs, including the feet, are fully extended at takeoff. The forelimbs are brought to a position under the chin and the hindlimbs flexed in midleap. At the end of the leap the hindlimbs are fully extended to meet the support, the forelimbs reaching it later. The tail is trailed inertly throughout.

ing a strong grasp with at least three extremities is maintained and the hind feet and fore feet of the same side are brought together before the grip is transferred from one to another. On the ground only quadrupedalism has been reported.

4.7.4. *Perodicticus (Charles-Dominique, 1971, 1974; Hill, 1953; Walker, 1969, 1974)*

Slow-climbing, quadrupedal locomotion involving the use of at least three extremities at any one time is the normal mode of progression. *Perodicticus potto* have never been observed to leap, although individuals will roll themselves into a ball and drop from a branch to escape a predator. The hind foot is brought to touch the fore foot on the same side during climbing so that the weight is transferred evenly from one limb to another (see Fig. 8A). On the ground, slow cautious quadrupedal walking is used (see Fig. 8B). Feeding postures range from a four-legged crouch to hanging by the feet. Resting postures are varied, but the most common is that where the animal sits on a support with a firm grasp with all four limbs and the back arched.

4.8. Tarsiidae

4.8.1. *Tarsius (Fogden, 1974; Le Gros Clark, 1924; Hill et al., 1952; Montagna and Machida, 1966; Niemitz, 1974; Thomas, 1896; Walker, 1974)*

The most common locomotor activity is that of active leaps between thin vertical supports (see Fig. 8C). Resting postures are commonly on vertical supports and the tail may or may not be involved in supporting the body. The tail is trailed inertly during a leap. On the ground progression is by a series of hops in *T. syrichta*, but hopping is restricted to the young and alarmed adults in *T. bancanus*.

5. SUMMARY AND CONCLUSIONS

A series of descriptive notes and bibliography of original observations have been given that describe locomotor behavior in prosimian primates. Locomotor studies using this kind of qualitative information have been useful in the past and will continue to be useful adjuncts to other behavioral studies. As far as the study of prosimian locomotion goes, it is clear that detailed, quantitative studies are now needed before any real improvement can be made in our understanding of locomotor adaptations. It seems equally clear that such studies cannot be made in isolation and must be made with due consideration given to the particular habitat and dietary habits of the species under consideration at specific localities. It might prove, for instance, that any conclusions drawn from a quantitative study of the locomotion of any wild population of *Galago demidovii* will be substantially different from those presented here. It might also prove to be that two such studies made in different areas with minor differences in habitat might also come to different conclusions. This is perhaps to be expected, but it is what the variation in behavior from place to place and from population to population can tell us that is important, not only for ecological studies, such as those on niche separation, but also for studies on morphological adaptations.

A superficial survey of the foot anatomy of the genus *Galago*, for instance, shows relatively minor differences between species, once size has been accounted for, but the locomotion of various species varies considerably and even appears, at first inspection, to be fairly stereotyped for each individual species. Does this mean that the same basic foot anatomy will sustain the many various postures and be an equally functional part of the hindlimb in all activities, whether quadrupedal, leaping, cantilevering, and suchlike? Or does this mean that the habitat imposes a certain locomotor style on a species that in another environment and without competitive pressures will behave differently? Closely related species, such as those of genus *Galago*, provide us with natural experiments with which to test hypotheses of morphological adaptation. Without detailed accounts of the locomotion that can be accomplished in different ecological situations, however, functional anatomical studies will be based upon insecure behavioral foundations.

REFERENCES

Bartlett, A. D. (1863). *Proc. Zool. Soc. London* pp. 231–233.

Bearder, S. K., and Doyle, G. A. (1974). *In* "Prosimian Biology" (R. D. Martin, G. A. Doyle, and A. C. Walker, eds.), pp. 109–130. Duckworth, London.

Bishop, A. (1964). *In* "Evolutionary and Genetic Biology of Primates" (J. Buettner-Janusch, ed.), Vol. 2, pp. 133–225. Academic Press, New York.

Charles-Dominique, P. (1971). *Biol. Gabonica* **7**, 121–228.

Charles-Dominique, P. (1972). *Z. Tierpsychol., Beih.* **9**, 7–41.

Charles-Dominique, P. (1974). *In* "Prosimian Biology" (R. D. Martin, G. A. Doyle, and A. C. Walker, eds.), pp. 131–150. Duckworth, London.

Charles-Dominique, P., and Hladik, C. M. (1971). *Terre Vie* **25**, 3–66.

Doyle, G. A. (1974). *In* "Prosimian Biology" (R. D. Martin, G. A. Doyle, and A. C. Walker, eds.), pp. 213–231. Duckworth, London.

Fogden, M. (1974). *In* "Prosimian Biology" (R. D. Martin, G. A. Doyle, and A. C. Walker, eds.), pp. 151–165. Duckworth, London.

Haddow, A. J., and Ellice, J. M. (1964). *Trans. R. Soc. Trop. Med. Hyg.* **58**, 521–538.

Hall-Craggs, E. C. B. (1964). *Med. Biol. Illus.* **14**, 170–174.

Hall-Craggs, E. C. B. (1965). *J. Zool.* **147**, 20–29.

Harrington, J. E. (1975). *In* "Lemur Biology" (I. Tattersall and R. W. Sussman, eds.), pp. 259–279. Plenum, New York.

Hill, W. C. O. (1953). "Primates: Comparative Anatomy and Taxonomy, vol. 1. Strepsirhini." Edinburgh Univ. Press, Edinburgh.

Hill, W. C. O., Porter, A., and Southwick, M. D. (1952). *Proc. Zool. Soc. London* **122**, 79–119.

Jewell, P. A., and Oates, J. F. (1969). *Zool. Afr.* **4**, 231–248.

Jolly, A. (1966). "Lemur Behavior: A Madagascar Field Study." Univ. of Chicago Press, Chicago, Illinois.

Jouffroy, F. K., Gasc, J. P., Decombas, M., and Oblin, M. (1974). *In* "Prosimian Biology" (R. D. Martin, G. A. Doyle, and A. C. Walker, eds.), pp. 817–828. Duckworth, London.

Kingdon, J. S. (1971). "East African Mammals," Vol. 1. Academic Press, New York.

Lamberton, C. (1956). *Bull. Acad. Malgache* **34**, 51–65.

Lauvedon, L. (1933). *Terre Vie* **3**, 142–152.

Le Gros Clark, W. E. (1924). *Proc. Zool. Soc. London* pp. 217–233.

Martin, R. D. (1972). *Z. Tierpsychol., Beih.* **9**, 43–89.

Montagna, W., and Machida, H. (1966). *Am. J. Phys. Anthropol.* **25**, 71–84.

Niemitz, C. (1974). *Folia Primatol.* **21**, 250–276.

Petter, J. J. (1962). *Mém. Mus. Natl. Hist. Nat., Paris* **27**, 1–146.

Petter, J. J., and Hladik, C. M. (1970). *Mammalia* **34**, 394–409.

Petter, J. J., and Petter-Rousseaux, A. (1967). *Mammalia* **31**, 574–582.

Petter, J. J., and Peyrieras, A. (1970). *Mammalia* **34**, 167–193.

Petter, J. J., and Peyrieras, A. (1971). *Terre Vie* **24**, 356–382.

Petter, J. J., and Peyrieras, A. (1974). *In* "Prosimian Biology" (R. D. Martin, G. A. Doyle, and A. C. Walker, eds.), pp. 39–48. Duckworth, London.

Petter, J. J., and Peyrieras, A. (1975). *In* "Lemur Biology" (I. Tattersall and R. W. Sussman, eds.), pp. 281–286. Plenum, New York.

Petter, J. J., Schilling, A., and Pariente, G. (1971). *Terre Vie* **3**, 287–327.

Petter, J. J., Schilling, A., and Pariente, G. (1975). *In* "Lemur Biology" (I. Tattersall and R. W. Sussman, eds.), pp. 209–218. Plenum, New York.

Pollock, J. I. (1975). *In* "Lemur Biology" (I. Tattersall and R. W. Sussman, eds.), pp. 287–311. Plenum, New York.

Rand, A. L. (1935). *J. Mammal.* **16,** 89–104.

Shaw, G. A. (1879). *Proc. Zool. Soc. London* pp. 132–136.

Stern, J. T., and Oxnard, C. E. (1973). *Primatologia* **4**, 1–93.

Subramoniam, S. (1957). *J. Bombay Nat. Hist. Soc.* **54**, 387–398.

Sussman, R. W. (1974). *In* "Prosimian Biology" (R. D. Martin, G. A. Doyle, and A. C. Walker, eds.), pp. 75–108. Duckworth, London.

Sussman, R. W. (1975). *In* "Lemur Biology" (I. Tattersall and R. W. Sussman, eds.), pp. 237–258. Plenum, New York.

Tattersall, I., and Sussman, R. W. (1975). *Anthropol. Pap. Am. Mus. Nat. Hist.* **52**, 195–216.

Tenaza, R., Ross, B. A., Tanticharoenyos, P., and Berkson, G. (1969). *Anim. Behav.* **17**, 664–669.

Thomas, O. (1896). *Trans. Zool. Soc. London* **54**, 387–398.

Vincent, F. (1969). Doctoral Thesis, University of Paris (unpublished).

Walker, A. (1969). *East Afr. Wildl. J.* **7**, 1–5.

Walker, A. (1974). *In* "Primate Locomotion" (F. A. Jenkins, ed.), pp. 349–381. Academic Press, New York.

Chapter 13

Field Studies of Lorisid Behavior: Methodological Aspects

P. CHARLES-DOMINIQUE AND S. K. BEARDER

1. INTRODUCTION

Apart from the tarsiers, the lorisids are the only prosimians which now occur in the same geographical areas as simian primates. Nevertheless, they possess a number of apparently primitive characters, many of which are shared with the cheirogaleids of Madagascar. In fact, because of considerable similarities between lorisids and cheirogaleids, their phylogenetic status is still a matter of some

ISBN 0-12-222150-8

discussion (Charles-Dominique and Martin, 1970; Martin, 1972a, b; Szalay and Katz, 1973; Szalay, 1975; Cartmill, 1975; Goodman, 1975; Tattersall and Schwartz, 1974). Perhaps because of their coexistence with diurnal simians, the extant lorisids are all nocturnal. They currently occur in Africa, Asia, and southeast Asia and are represented by ten species which fall into two distinct categories: (1) active leaping forms (galagines; 6 species all confined to Africa) and (2) slow-climbing forms (lorisines; 2 species in Africa, 1 species in Asia, 1 species in southeast Asia).

It is somewhat surprising that the active leaping category (which is theoretically better suited for dispersal) is not represented in Asia. There is possibly an ecological explanation for this in that the tarsiers, similarly active leaping forms, may have occupied a larger area of distribution at some earlier date (Walker, 1974). In Africa, the lorisines (*Perodicticus potto* and *Arctocebus calabarensis*) are confined to dense forest areas, whereas the galagines occur not only in such forests but also in semi-arid areas (wooded savannah and gallery forests). The locomotor system of the galagines, in which leaping plays an important role, permits them to exploit a wider range of habitats, in particular those with discontinuous tree cover. The two galagine species (*Galago senegalensis* and *Galago crassicaudatus*), inhabiting semi-arid forest areas to the north, east, and south of the equatorial rain forest bloc, are regarded as offshoots of forest-living species, secondarily specialized for dryer climatic conditions, as a result of dispersal or simply following retreat of the rain forest (Walker, 1974). Three other galagine species, *Galago demidovii, Galago alleni,* and *Galago elegantulus* (*Euoticus elegantulus*) occur in the dense equatorial rain forest, while *Galago inustus* (*Euoticus inustus*), whose taxonomic status has been subject to successive modifications, seem to be limited to a zone covering part of Zaire and Uganda (Vincent, 1972).

It is therefore convenient to divide the lorisids into those inhabiting dense equatorial rain forest and those occurring in the dryer surrounding areas.

The rain forest zone of equatorial West Africa is at present separated into two blocs by a strip of savannah which passes through to reach the coast between Ghana and Dahomy. This recent subdivision, which dates from some time in the Pleistocene (Moreau, 1964), has isolated the North Guinean bloc from the South Guinean bloc, which is larger and richer in terms of its primate fauna. Genetic isolation resulting from this subdivision has led to the evolution of clear subspeciation of numerous vertebrate species. Both lorisid species, which occur in the two forest blocs and even extend somewhat beyond the limits of the rain forest (viz. *G. demidovii* and *P. potto*), follow this rule.

Distribution maps indicate to the north and the east of the equatorial forest an overlap between these two species and *G. senegalensis* (see Petter and Petter-Rousseaux, Chapter 1). In fact, far more detailed distribution maps would be needed to establish genuine overlap and so far there has been no supporting evidence for sympatric occurrence (i.e., occupation of the same forest ecosystems) by *G. senegalensis* and *G. demidovii*. The South Guinean bloc of the rain forest covers a total of more than 500,000 square miles, including part or all of Nigeria, Cameroon, the Central African Empire, Equatorial Guinea, Gabon, the Congo, and Zaire. It is occupied by *G. demidovii, G. alleni, G. elegantulus, P. potto,* and *A. calabarensis*; all five species are sympatric in several ecosystems.

The two galagine species occurring in arid of semi-arid areas have a vast range of distribution to the north, east, and south of the rain forest. *Galago senegalensis* occupy the entire semicircular zone passing from Senegal to Angola and through Kenya and Mozambique to South Africa, whereas *G. crassicaudatus* only occupy suitable areas of Eastern and Southern Africa. The two species occur sympatrically in certain regions (Bearder and Doyle, 1974). There are numerous subspecies and the study of genetic problems relating to biogeographical features has only just begun (de Boer, 1972, 1973a b). *Galago senegalensis*, the species most specialized for the harsh conditions prevailing in semi-arid areas, adapt best to conditions in captivity. Living in areas where the trees are short and scattered, they are relatively easy to capture compared to the other lorisid species, and for this reason they are the most commonly encountered in zoos and laboratories. A great number of studies on the physiology, genetics, serology, anatomy, and behavior of prosimians have thus been based on this one species.

Considerable confusion has arisen in the attempts to classify the various species and races within the Galaginae and Lorisinae. Hill's (1953) divisions are based on anatomical and morphological variations in relation to geographical range and provide a useful framework for reference until other lines of evidence permit revision of the group as a whole. Various recent studies of the scant fossil record, anatomy, morphology, and comparative investigations of chromosome and serum proteins have been aimed at identifying the true phylogenetic relationships within the two subfamilies and at illustrating probable evolutionary trends (see Petter and Petter-Rousseaux, Chapter 1). At present, these studies pose almost as many questions as they answer, and the same may be said of comparative studies of ecology and behavior.

Although anatomical features and certain physiological aspects of the lorisids have been studied over a considerable period of time, their behavior under natural conditions has been relatively neglected. Of course, their nocturnal habits render them somewhat more difficult to observe than the diurnal primates. However, behavior is just as important as a subject for study. As with anatomical and physiological features, behavioral characters represent adaptive responses of living organisms, progressively evolving as they survive within a competitive situation. Indeed, it seems that behavior is more flexible and more rapidly adapting, permitting each organism to respond more profitably to prevailing environmental conditions. This applies particularly to feeding and to locomotion in search of food (in competition with other species); to protection against predators, parasites, and climatic variability; and to reproduction, involving social relationships with conspecifics and contributing to population equilibrium as a function of available resources.

Accordingly, detailed study of ecological conditions relating to a given species permits more effective interpretation of its behavior. This principle has been widely adopted for the study of simian primates, largely as a result of Carpenter's pioneering field studies of primate behavior (Carpenter, 1934). However, the first investigations of prosimians under natural conditions were initiated much later, in Madagascar (Petter, 1962a,b) and were largely confined to diurnal species. The lorisids, which are all nocturnal, were described only in brief field notes, sometimes supplemented by observations conducted in captivity (Rahm, 1960; Beaudenon, 1949; Sauer and Sauer, 1963; Petter and Hladik, 1970). It is only very recently that various African lorisid species have been studied in detail under natural conditions (Jewell and Oates, 1969; Charles-Dominique, 1971a,b, 1972, 1974, 1977; Bearder, 1969; Bearder and Doyle, 1974).

This chapter combines the results of two long-term field studies conducted continuously in Gabon since 1965 (Charles-Dominique) and in South Africa since 1968 (Bearder). The differences between the two habitats sometimes necessitated different techniques of study adapted to specific local conditions; some investigations proved to be more productive in one environment than in the other. For this reason, the results are presented in two separate sections. However, the two authors have been able to discuss their results at length on two occasions, once in the course of a joint field visit in South Africa. This has permitted analysis of the common ground that exists between the two studies.

These field studies were designed to cover as broadly as possible the biology of the species concerned, and it is therefore likely that this review of lorisid behavior and ecology will overlap to some extent with the specific behavioral topics covered in other chapters. However, it should be emphasized that this chapter deals only with those aspects of lorisid behavior most directly related to ecological conditions, which are consequently the most difficult to approach through studies under laboratory conditions. Particular attention is given to the techniques which have been utilized to study these nocturnal, solitary primates under natural conditions.

2. THE LORISIDS OF GABON

2.1. Habitat Utilization

2.1.1. The Equatorial Rain Forest and Its Fauna

In contrast to the extant Malagasy lemur species, whose ranges of distribution within the island are very localized, the lorisids exhibit very wide geographical distribution. The five species covered in this section, on the basis of 45 months of field work since 1965, occupy the enormous African bloc of equatorial rain forest, extending over more than 500,000 square miles. Two galagines (*Galago alleni* and *Galago elegantulus*) and one lorisine (*Arctocebus calabarensis*) are restricted to the

Fig. 1. *Galago demidovii* (average weight: 60 gm) (Photograph: C. M. Hladik).

Fig. 2. *Galago alleni* (average weight: 260 gm) (Photograph: A. R. Devez).

Fig. 3. *Galago elegantulus elegantulus* (average weight: 300 gm) (Photograph: A. R. Devez).

South Guinean region of the rain forest; but one galagine (*Galago demidovii*) and one lorisine (*Perodicticus potto*) occur far beyond the limits of this region to the north and the south. For example, Demidoff's bushbaby occurs from Senegal in the north to Zaire in the south and from the Atlantic coast in the west to the Indian Ocean coast of Tanzania in the east (Vincent, 1969). In the Gabon study area,* all five of these lorisid species are thus found sympatrically [*G. demidovii, G. alleni, G.e. elegantulus, P. potto edwardsi,* and *A. calabarensis aureus* (Figs. 1–5)].

Far from being a ''homogeneous'' environment, the equatorial rain forest represents a mosaic of many different biotopes. Tree-falling, which is the main natural ''destructive'' agent, opens up large clearings in which part of the canopy (masses of foliage intermingled with lianes) collapses to the ground. Large trunks are brought to the ground in this way, principally during tornadoes. Hence, regeneration of trees takes place in such tree-fall areas, preceded by rapid plant growth which particularly favors young lianes. As regeneration of the forest occurs in each area of this kind, the physiognomy of the vegetation changes. In this way, a variety

*Laboratory of Primatology and Ecology of Equatorial Forests (Centre National de Recherche Scientifique), established since 1962 in the Ogoué-Ivindo region of northeast Gabon (Makokou).

Fig. 4. *Perodicticus potto edwardsi* (average weight: 1100 gm) (Photograph: A. R. Devez).

of biotopes is presented in adjacent forest zones, according to chance occurrence of tree-falling and to the nature of the terrain, which has a dense network of rivers and streams.

This general ecological complexity is matched by a highly diversified fauna, whose evident ecological adaptations doubtless bear witness to the long history of this forest environment. Taking only species which are sympatric (that is, occur in the same geographical region), 126 mammal species and more than 300 bird species have been identified in the Makokou area of Gabon. The mammals include 17 primates (the 5 lorisids and 12 simians), 13 carnivores, 15 artiodactyls, 30 chiropterans, 33 rodents, 13 insectivores, 3 pangolins, 1 aardvark, and 1 hyrax. It is at present impossible to make a complete inventory of the lower vertebrates and invertebrates.

Fig. 5. *Arctocebus calabarensis aureus* (average weight: 200 gm) (Photograph: A. R. Devez).

As far as the flora is concerned, it has been estimated that there are 8,000 to 10,000 phanerogamic plant species in the rain forest of Eastern Gabon. Within this ecosystem, the five lorisid species play a quite negligible part, occupying very restricted ecological niches for which they are narrowly specialized. Against the background of their persistent primitive characters, they have developed novel specializations which have permitted them to survive within a highly competitive fauna.

2.1.2 Distribution of the Lorisids in the Forest

2.1.2.1. Methods. By walking through the rain forest at night with a headlamp, one can locate and observe all of the lorisid species in their natural habitat. Like many other nocturnal animals, these prosimians have a reflecting *tapetum* in each eye and they can easily be spotted when they look toward the headlamp. As a rule, the orange-red reflections from their eyes can be sighted from a distance of up to 100 m. However, leaves form a multitude of screens which reduce the likelihood of sighting as the distance increases. Hence, the ease with which the animals are discovered varies with the vegetation type (foliage density, height of the canopy, etc.). The nature of the equatorial rain forest, which is an unusual environment for anyone used to more open surroundings, leads to considerable difficulties in evaluation of distribution and abundance of the animals. It is therefore necessary to measure systematically a maximum of parameters in order to determine objective values for these aspects.

a. Width of the forest transect covered by the lamp. For every lorisid sighted, the distance of the animal from the pathway was noted. In this way, it was possible to calculate for each forest type or layer a "visibility profile" (Fig. 6). In order to express the data obtained in terms of mean values per square kilometer, it is sufficient to take the strip of forest within which no decrement in observations was observed and to count only those animals sighted within this strip. At Makokou, the following values were accordingly estimated for the width of surrounding forest strips effectively sampled and are shown in the tabulation below:

Undergrowth of primary forest (m)	30
Canopy of primary forest (m)	30
Flooded primary forest (m)	20
Secondary forest and tree-fall zones (m)	20
River banks surveyed from a canoe (m)	10

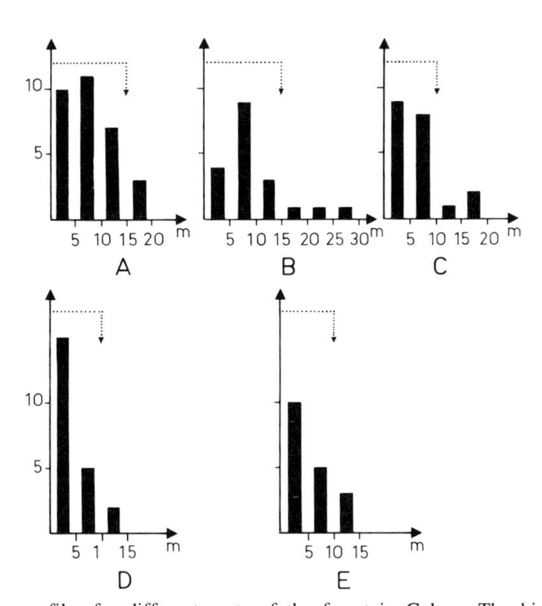

Fig. 6. Visibility profiles for different parts of the forest in Gabon. The histograms show the numbers of prosimians spotted with a headlamp (ordinate) at different distances from the path followed by the observer (abscissa). A, primary forest canopy; B, primary forest undergrowth; C, flooded forest; D, primary forest canopy viewed from a river; E, secondary forest.

b. Length of the transect covered. Evaluation of distances in the equatorial rain forest is subject to enormous variations, which may amount to as much as a fivefold difference. In general, there is a tendency to overestimate distance. The simplest solution to this problem is to measure pathways with a surveyor's tape, or with a calibrated string. At Makokou, any pathways which were regularly utilized for observations were marked at 20-m intervals with plastic labels; this permitted immediate identification of location. In the course of simple surveys of certain forest areas, which had not been calibrated in this way, it proved sufficient to note the times of all observations and the times of all stops and starts along the pathway. By taking the average speed of walking measured under identical forest conditions, it was possible to calculate the distances concerned. Similar results were obtained by using a pedometer.

c. The choice of pathways for observation. At the beginning of the study, the attempt was made to explore as many regions of the forest as possible by following, with the aid of a Gabonese guide, the narrow pathways ramifying through the forest. (Only the local people are able to use these pathways without losing their way.) In this way, during the first four years of the study (1965–1969), a total of 570 km of pathway was covered at night, 208 km in primary forest, 40 km in flooded primary forest, 80 km in a canoe exploring riverbanks, 200 km in secondary forest, and 50 km along open roads lined by secondary shrub growth in primary forest. (In fact, certain stretches were covered several times, and these figures refer to a total of 148 km of different pathways.)

At a later stage, in the light of the results from these initial surveys, specific study areas were marked out by cutting transects in series, spaced at distances of 40 m or 100 m, and forming quadrats. Each transect was numbered and marked with labels so that a detailed map could be drawn up on the basis of compass bearings and direct measurements of pathways. The study areas set out in this way provided the essential basis for all observations on ranging behavior and social relationships.

d. Trapping. Commercially available traps were rarely suitable for capturing lorisids. Special traps were therefore designed for the Gabon field study, taking into account the following requirements: large size, large entrance on the upper surface, triggering by a small weight applied to a branch (see Fig. 7).

The trap is propped open with a rod (A) attached to the trapdoor and resting on a second rod (B) which is unattached and kept in place only by the pressure applied by the tip of rod (A). The tension required to close the trap can be provided by a spring, but it was found to be more convenient to use a nearby branch of a tree with suitable curvature and suppleness, attached to the trapdoor with a thread. The tension of the branch can easily be regulated to correspond to the weight of the animal to be caught. Any trap which is too well finished will generally arouse the attention of any lorisid, and in general the animals will more readily enter traps in which the wooden struts have not been planed smooth or painted and in which the rods A and B are crudely made from branches. Each trap is usually placed at a height of 1–1.5 m, wherever possible in a tangled patch of vegetation containing plenty of lianes. The

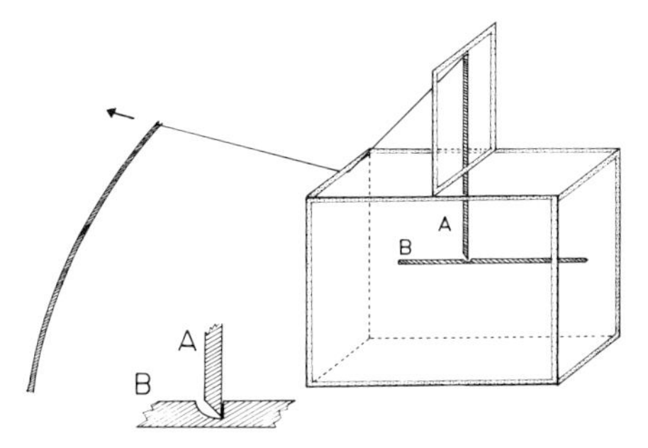

Fig. 7. Diagram of the trap used for trapping the prosimian species in Gabon.

floor of the trap is covered with litter taken from the forest floor and the cage wire walls are camouflaged with foliage. Before any trap was established for capture in Gabon, the site was baited for 6–15 days previously with bananas left in a basket made from lianes. The trap was then attached alongside the basket, and the banana bait was subsequently transferred first to the top of the trap and afterward to the interior of the trap itself, but with the releasing mechanism blocked. As soon as the lorisids began to eat regularly at the bottom of the trap, the release mechanism was set. Once a trap was set, it was inspected during the night so that no animal was kept prisoner for too long. Once any rain forest lorisid has been captured in this way, then marked and released, it will become cautious and some time will pass before it will enter a trap again. However, the same animal will go to visit another trap in its home range, or even visit the original trap if it is moved to another site, quite close by. It would therefore seem that the traumatic effect of capture is associated more with the place of capture than with the trap itself. By using simultaneously a battery of twenty traps moved from place to place according to a pre-established program, it is possible to obtain regular data on a local population of lorisids. A trap placed with fewer precautions may capture certain individuals, but they subsequently prove to be almost impossible to recapture. Further, in Gabon, it was only possible to trap reliably *G. demidovii*, *G. alleni*, and *P. potto*. (*A. calabarensis* and *G. elegantulus*, which only rarely eat fruits, could not be trapped by this method.) Between 1968 and 1973, 442 captures were made with 114 individuals belonging to these three species.* When an animal enters a trap of this type, it usually descends cautiously down the cage wire. On leaving, however, it usually moves hurriedly and it is often at this point that the closing mechanism is operated when force is applied on the rod (B). After it has been captured several times, the animal will learn to avoid touching the rod (B). At this point, several other closely packed rods pressing down on the

*The same technique has been used in Madagascar for capturing three lemur species (*Microcebus murinus*, *M. coquereli*, and *Cheirogaleus medius*) and in Gabon for trapping a fruit-eating viverrid carnivore, *Nandinia binotata* (the palm civet). It has also proved effective for trapping several squirrel and rat species.

rod (B) can be introduced. Finally, when the animal has even learned to pick its way between all of these rods, the banana can be attached to the rod (B) with a short thread, and when all else fails even the most cautious animals will not be able to resist an insect attached to the rod.

*e. **Marking.*** After routine examinations (weight, sexual condition, dentition), the captured animals were marked with triangular incisions cut into the ear pinnae. The use of various combinations of such incisions permit marking of 100 animals. This system is very useful when given populations are followed from year to year, but it does not permit identification at a distance. In order to achieve such long-range identification, several different techniques were used: shaving of various segments of the tail (allowing 20–25 possible combinations), discoloration of the pelage with hair dyes, attachment of plastic collars, and (from 1973 onward) radio-tracking. This latter technique, which has now been greatly refined, is immensely valuable and it is possible to follow several animals on a continuous basis (15–20 distinct frequencies can be utilized simultaneously).

For radiotracking, the transmitter, soldered to a mercury battery, is attached to a narrow band of flexible plastic and coated with a polymer which protects it from humidity, shocks, and any bites. The plastic band (plastic-covered nylon material) is riveted around the animal's neck or waist, as required, and the aerial is threaded along the collar. With a total collar load of 25 gm, an animal can be followed for 4 weeks until the battery is exhausted. A directional receiver aerial permits extremely precise localization. For example, it is possible to find a tree hollow in which a *G. alleni* is sleeping. Indeed, if the tree is one containing several hollows, it is possible to identify which hollow contains the animal. At night, animals carrying a transmitter are rarely approached more closely than 25 m, in order not to disturb them, and transects laid out to form a grid of quadrats allow following of the animals without necessitating movement through the vegetation.

*f. **Recording of activities.*** Activity recording can be achieved under field conditions if it is associated with regular observations. Patterns of visits to an artificial food source baited with bananas are easy to record automatically with an event recorder (Charles-Dominique, 1971a). Electrical contacts are attached to supple branches leading to the bait, the calibre of the branches being selected according to the body weight of the animal to be recorded. Large-bodied species (*P. potto* and *Nandinia binotata*, the palm civet) can be excluded from the food source by a fence with spaces only permitting the passage of smaller-bodied species. Differences in behavior (for example, between rodents and bushbabies) also permit elimination of a particular group of mammals, in that different access pathways can be provided. Finally, each individual animal which becomes accustomed to eating at the artificial food source will rapidly adopt a standardized pattern of behavior, and this will permit precise location of contacts at suitable spots after a few preliminary observations.

This method of activity recording was used in particular to record the motor activity of pottos. In the vicinity of the former biological station at Makokou, the secondary forest had been subdivided into sectors by roads and various clearings. The potto, which is an exclusively climbing species, is obliged to follow continuous pathways, and individuals occupying this secondary forest area were obliged to make detours so as to reach certain passage points with which they were familiar (branches linking forest on the two sides of a road, etc.). In order to exploit this characteristic, these access branches were cut away and replaced by lianes stretched between the two crossing points. Such bridges were rapidly utilized by the pottos, and they were accordingly equipped with electrical contacts linked to a 15-channel event recorder. The contact was attached beneath the liane (Fig. 8) with only two wooden rods projecting above. In order to pass, a potto was forced to squeeze between the two rods, thus displacing them slightly and closing the electrical contact. In order to provide more reliable information, two or three contacts were placed in series in order to indicate passage in one direction and two or three others (connected to a different channel of the event recorder) indicated passage in the other direction. As with the traps discussed above, rudimentary contacts made with wooden supports gave the best results. Elaborate apparatus constructed with metal or plastic is always avoided by the pottos.

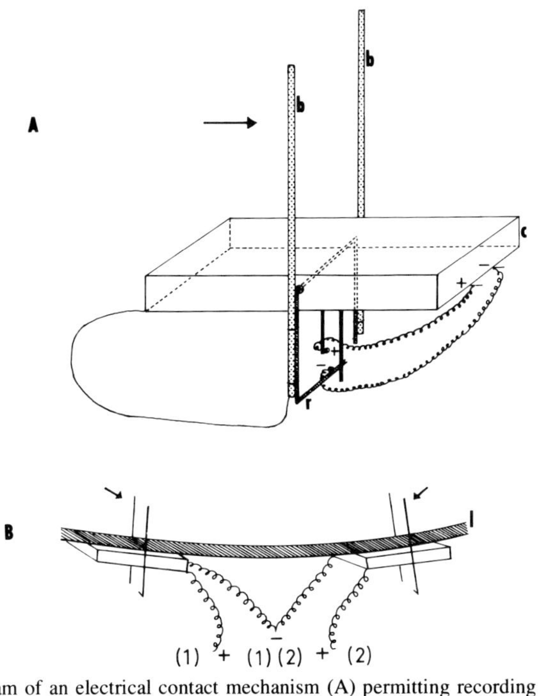

Fig. 8. Diagram of an electrical contact mechanism (A) permitting recording of passage of pottos along a liane. b, Wooden rods, between which the potto must pass; c, wooden block, to fit beneath the liane; r, spring (made from piano wire). Electrical contacts are installed on the liane (l) as shown in B (arrows indicate the direction of passage of the potto which will produce an electrical contact).

In many instances, the electrical contact was connected to a bell in the laboratory to indicate any passage of an animal across a given liane. By hurrying to the spot as soon as a bell rang, it was possible to identify the potto which had just crossed.

These different techniques, in various combinations and supported by direct observation, permitted preliminary analysis of the natural social behavior of *G. demidovii*, *G. alleni*, and *P. potto* (Charles-Dominique, 1971a, 1972, 1977).

2.1.2.2. Population Densities. Taking into account the sighting figures obtained in different biotopes, it was possible to calculate population densities per square kilometer for the five lorisid species (see Charles-Dominique, 1971a, b). However, it is necessary to consider the values calculated in greater detail because of the behavioral differences between the species and the heights at which each one characteristically moves around. On the basis of the provisional sighting values, data obtained from trapping, data from systematic marking (in particular, from radiotracking studies), and information from detailed study of social behavior, the following overall, approximate estimates have been made for population densities in primary forest (shown in the tabulation below):

Galago demidovii	50–80 animals/km ^{2}
Galago alleni	15–20 animals/km ^{2}
Galago elegantulus	15–20 animals/km ^{2}
Perodicticus potto	8–10 animals/km ^{2}
Arctocebus calabarensis	2 (or 7) animals/km ^{2}

However, it should be noted that *P. potto* are more abundant in humid forest areas (12–28 animals/km ^{2}), while the reverse applies to *G. elegantulus* (3 animals/km ^{2}). As for *A. calabarensis* in primary forest, their distribution coincides exactly with that of tree-fall zones in which the vegetation is particularly rich in small-calibre lianes. These disrupted zones, where the primary forest is particularly affected by tornadoes, depend primarily on topographic location. In secondary forest, on the other hand, which is an environment recently produced by human activity, equivalent biotopes are available and *Arctocebus* have become quite well adapted to live there. Both in tree-fall zones of primary forest and in secondary forest in general, the population density of this species is in the neighborhood of 7 animals/km ^{2} , whereas in primary forest taken as a whole it is only 2 animals/km ^{2} .

As a result of the heterogeneity of the forest, different sectors are visited to different extents by the five lorisid species. This emerges quite clearly from repeated counts in a given forest sector. The case of *Arctocebus* is particularly obvious in this respect, but such heterogeneity of occurrence is also apparent to a lesser extent for the other species. Figure 9 represents the distribution of *G. elegantulus* in the mapped study area of Ipassa, matched to that of a liane species (*Entada gigas*), a source of gums which are a preferred food for this bushbaby species. It is easy to see the clear correlation which exists between the two distribution patterns.

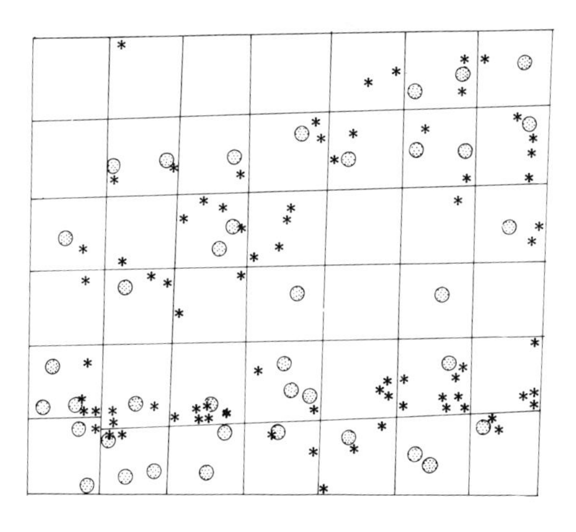

Fig. 9. Distribution map of the liane species *Entada gigas* (○) (source of gums) and sightings of *Galago elegantulus* (*) in a quadrat network (pathways,—) laid out on the Plateau d'Ipassa near the Laboratoire de Primatologie.

2.1.2.3. Spatial Distribution and Selection of Supports. In order to define more precisely the situation occupied by each species in the immense heterogeneous network of the forest, the following criteria were utilized for noting sightings of animals in the forest:

1. Height with respect to the ground*
2. Diameter of the support
3. Orientation of the support (horizontal, oblique, vertical)
4. Nature of the support (ground, narrow trunk, liane base, large trunk, large branch, foliage, mixture of tree foliage and lianes, lianes).

Only the position where the animal was first sighted was noted, since the itinerary subsequently followed might have been affected by the presence of the observer. The results have been summarized in Tables I–IV. However, as with the calculations of population densities, further analysis is necessary because of behavioral differences between the five species. Since observations were conducted from the ground, the likelihood of sighting an individual of any one species decreased with the characteristic height of movement through the trees. In the same way, it is less likely that one will spot members of the discreet lorisine species than individual

*In most cases, such heights were estimated, but from time to time direct measures were made (with a tape or a clinimeter) in order to check the accuracy of estimates.

TABLE I

Percentage Sightings of the Five Lorisid Species on Different Types of Support in Primary Forest (Makokou)

Genus	Ground	Thin trunk	Liane base	Thick trunk	Large branch	Foliage	Foliage liane	Liane
Galago demidovii (N = 242) (%)	0.4	0	0.8	0	12.8	21.5	35.5	29
Galago alleni (N = 162) (%)	7.4	33.9	36.4	0	5.6	9.3	2.5	5
Galago elegantulus (N = 95) (%)	0	0	0	3.15	49.5	5.3	2.1	40
Perodicticus potto (N = 55) (%)	0	0	0	3.6	41.8	14.5	7.3	32.7
Arctocebus calabarensis (N = 57) (%)	8.7	1.8	0	0	0	31.6	17.5	40.4

TABLE II

Distribution of the Five Lorisid Species at Different Levels above the Ground in Primary Forest (Makokou)

	Height above the ground (m)									
Genus	0–5	5–10	10–15	15–20	20–25	25–30	30–35	35–40	40–45	45–50
Galago demidovii (N = 227) (%)	11.5	20.7	24.2	22.9	9.3	9.7	1.3	0.4	0	0
Galago alleni (N = 151) (%)	73.5	21.9	3.3	0.7	0.7	0	0	0	0	0
Galago elegantulus (N = 106) (%)	4.7	22.6	18	18	16	11.3	3.8	2.8	1	1.9
Perodicticus potto (N = 44) (%)	9	25	25	16	9	9	4.5	2.2	0	0
Arctocebus calabarensis (N = 66) (%)	75.8	16.7	7.6	0	0	0	0	0	0	0

TABLE III

Percentages, According to Diameter, of Different Calibre Supports Utilized Most Frequently by the Five Lorisid Species in Primary Forest (Makokou)

	Support of diameter (cm)									
Genus	**0–1**	**1–5**	**5–10**	**10–15**	**15–20**	**20–25**	**25–30**	**30–35**	**35–40**	**40–45**
Galago demidovii (N = 123) (%)	58.5	30.9	4	2.4	2.4	0	0.8	0.8	0	0
Galago alleni (N = 78) (%)	6.4	60.3	21.8	9	1.3	1.3	0	0	0	0
Galago elegantulus (N = 59) (%)	0	39	20.3	15.2	8.5	3.4	1.7	3.4	5	3.4
Perodicticus potto (N = 45) (%)	6.7	31.1	28.9	17.6	0	4.4	6.7	0	2.2	2.2
Arctocebus calabarensis (N = 22) (%)	45.5	45.5	9	0	0	0	0	0	0	0

TABLE IV

Orientations of the Supports Utilized by the Five Lorisid Species in Primary Forest at Makokou[a]

	Support orientation		
	Horizontal	Oblique	Vertical
Galago demidovii (N = 65) (%)	18.5	40	41.5
Galago alleni (N = 126) (%)	7.1	19.8	73
Galago elegantulus (N = 56) (%)	17.9	48.2	34
Perodicticus potto (N = 34) (%)	32.4	35.3	32.4
Arctocebus calabarensis (N = 20) (%)	25	30	45

[a] Horizontal, 0°–15°; oblique, 15°–75°; vertical, 75°–90° with respect to the horizontal plane.

galagines, which frequently betray their presence by powerful alarm calls and by their greater mobility. The following outlines of the ecological distributions of each of the five species within the forest have thus been based on general observations of behavior together with any information available from radiotracking (*P. potto* and *G. alleni*).

a. Galago demidovii. This small-bodied bushbaby lives in very dense vegetation where the other prosimian species would only be able to move around with difficulty. The supports utilized are generally of small calibre (diameter less than 1 cm), and they are extremely variable in orientation. The preferred biotopes are zones of fine branches and foliage invaded by lianes (liane curtains), in both primary and secondary forest. These liane curtains hang down several meters below the branches to which they are attached. The twigs of these lianes and their branches are interwoven, forming a dense network above which foliage regenerates. From year to year, these formations increase in size, eventually forming a mass 1–2 m thick and 3–5 m in height. Such curtains are abundant low down in recently formed secondary forest, along the banks of rivers and in the tree-fall zones. In primary forest, these liane formations normally occur high up in the canopy (Fig. 10).

In secondary forest, *G. demidovii* move around in bushy vegetation a few meters from the ground, sometimes descending almost to ground level. This species is frequently encountered alongside paths and roads in hedgelike vegetation (*Aphromum* sp. and small lianes), and this has led some authors to suppose that Demidoff's bushbaby lives exclusively in low-lying vegetation (Cansdale, 1960; Jones, 1969). However, Vincent (1969) has reported that this species lives high up

Fig. 10. Liane curtain, a plant formation particularly utilized by *Galago demidovii* (Photograph: A. R. Devez).

in primary forest in the Congo. In primary forest, where they are quite abundant, *G. demidovii* move around at heights of 5–40 m in dense foliage and in liane curtains. However, it is possible for the bushbabies to be missed unless one is equipped with a powerful lamp and good knowledge of the vocalizations.

b. Galago alleni. As with the angwantibo, this species inhabits the undergrowth of the primary forest, but its distribution is independent of the tree-fall zones. It was sighted in most cases at heights of 1–2 m on vertical supports 1–15 cm in diameter, represented by small-calibre tree trunks and the bases of large lianes.

During surveys conducted through the forest at night, *G. alleni* were only rarely seen on the ground (7% of cases), since the slightest alarm caused any individual to leap to a thin trunk, ready to flee. Eight individuals were followed by radiotracking, and this permitted verification of the fact that this bushbaby species searches for insects and fallen fruit by moving slowly along the ground, sometimes making small leaps of 20–40 cm. From time to time, the bushbaby will leap to a thin trunk to cling at a height of less than 1 m above the ground, survey its surroundings, and then return to the ground. Rapid locomotion is effected by leaping from one vertical support to another, a few meters from the ground, and it is usually under such conditions that this species is observed in the forest. The nocturnal period of activity is interspersed with brief resting bouts of 15 minutes to 2 hours, which are spent in dense foliage at a height of 10–20 m. At such times, it is practically impossible to sight an Allen's bushbaby, and only radiotracking permits localization of these resting sites.

In Biafra, Jewell and Oates (1969) also observed *G. alleni* on the ground, sometimes venturing into small patches of grassland. In secondary forest, in Gabon, this bushbaby species was hardly ever sighted. In forest areas degraded by human activity, *G. alleni* are only found in the vicinity of small marshy patches (areas subject to flooding are not suitable for cultivation and are therefore not cleared).

c. Galago elegantulus. This relatively large-bodied bushbaby species lives in the forest canopy at heights of 5–35 m, sometimes going up as high as 50 m in the gigantic trees which dominate the forest. The most frequently utilized supports are large branches (49%). The nails bear small claws at their tips, and this permits *G. elegantulus* to move around on large diameter smooth branches and sometimes even on big trunks. The search for gums constrains the needle-clawed bushbaby to explore certain large-diameter lianes and trees, particularly the trunk and large branches which are inaccessible to the other prosimian species. In order to move around in the canopy, *G. elegantulus* may make use of lianes, but in many cases they run and leap on large-diameter branches somewhat like cercopithecine monkeys. When leaping from one tree to another, a needle-clawed bushbaby always lands in foliage, which serves to break a fall of 6 m or even more.

In secondary forest areas, *G. elegantulus* cxploit the highest trees which dominate the bushy undergrowth vegetation (i.e., *Albizia gummifera* and *Pentacletra eetvel-deana*), and they will only rarely descend to the ground. On a very few occasions, a needle-clawed bushbaby was spotted crossing a road through secondary vegetation.

The numerical data given for the supports utilized by *G. elegantulus* are a fairly good reflection of the real situation, since this relatively timid species often betrays its presence through its vocalizations.

d. Perodicticus potto. In primary forest, pottos are generally sighted at between 5 and 30 m above the ground. Large-calibre branches and lianes are the favorite supports for movement through the trees. The potto is an exclusively climbing species; pottos are obliged to descend along large-diameter branches to their dividing points and to move up through branches of gradually decreasing diameter in order to reach the foliage of a neighboring tree. Having effected its passage from one tree to another in this way, the potto then passes through the fine branches and back to the forks of large branches. The diameters of supports most frequently utilized lie between 1 and 15 cm. However, if a short-cut can be made along a liane, as is often the case, the potto will make use of it.

Fig. 11. *Arctocebus calabarensis* in a "tree-fall zone" in primary forest (Photograph: G. Dubost).

By following two pottos equipped with radio transmitters in primary forest, it was observed that they spent most of their time in the upper canopy level at a height of 20–40 m. In the course of their movements through the trees, they will occasionally descend to quite low levels and can then be observed directly. However, they are invisible from the ground for most of the night. Table II therefore underestimates the height of movement of the potto in the canopy, as was to be expected from initial impressions of the discretion of this species, which is specially adapted for camouflage.

In secondary forest, the potto is observed lower down because the trees themselves are of lower stature, but its activity is still largely confined to the canopy. On occasions, a potto will come down close to the ground to reach certain fruits when they are ripe, but it has some difficulty moving around in low-lying bushy vegetation. This slow-climbing lorisid species is primarily frugivorous and it seeks out fruits in the vegetation where the sun can penetrate to permit flowering and fruiting of trees and lianes. Thus, the potto is understandably most active in the upper layer of the canopy.

e. Arctocebus calabarensis. The angwantibo has never been observed at a height of more than 15 m in primary forest, and in general this species is found at heights of 0–5 m in the forest undergrowth. This situation would seem to be typical for the species, since Jewell and Oates (1969) report the same behavior for the subspecies *A. calabarensis calabarensis* in Biafra. The angwantibo in Gabon feeds almost exclusively on caterpillars, exploring the foliage of bushes in the undergrowth and utilizing intervening small-calibre lianes to move from one bush to another (Fig. 11). The diameter of the supports utilized (less than 5 cm) is in keeping with the small body size of this species. Fine lianes are more abundant in former clearings where high light penetration has favored their growth following the collapse of a large tree. It has already been pointed out above that in primary forest the angwantibo is closely associated with zones in which tree-falling is a common occurrence (tree-fall zones exposed to tornadoes because of their particular topography). Angwantibos were always encountered in such plant formations, and in almost every case they were able to flee from the small trees by utilizing the fine lianes invading them. *Arctocebus* occasionally descend to the ground, and individuals were occasionally spotted on the forest floor. In fact, 5 animals were examined after they had been killed in traps set on the ground in primary forest on trails habitually used by duikers and other terrestrial mammals.

2.1.3. Diet

In contrast to the situation with diurnal animals, it is virtually impossible to study the feeding behavior of the lorisids by direct observation. At the most, one might surprise an animal eating its prey (which usually cannot be identified from a distance) or moving around in a tree which can be seen to be carrying fruits or yielding gum (it will be seen later on how much can be said about methods of food selection). The proportions of various components in the diet, and sometimes even their qualitative description, can only be determined from analysis of stomach contents, which requires sacrifice of a certain number of animals.

During the first two years of the field study, approximately 100 individuals of the five lorisid species were shot with a rifle at night in the forest. The collection program was spread over all months of the year and arranged to cover different hours of the night, so that a broad picture of the diets could be obtained. The animals concerned were taken from a forest area some distance from Makokou, so that the populations in the main study area were left intact for studies of social behavior. Subsequently, local villagers from time to time brought in lorisids which had been killed in their traps and this brought the total of animals available for dissection to 193: 41 *P. potto*, 18 *A. calabarensis*, 55 *G. demidovii*, 17 *G. alleni*, 62 *G. elegantulus*. The overall sample was thus obtained over a period of several years from a forest region of more than 40 km radius. Despite the numbers of animals collected in this way, no difference was observed between untouched forest areas and those where animals had been collected, when the area of collection was surveyed two to three years later.

Tables V and VI, which summarize the results of stomach content analysis, reveal considerable differences between the five sympatric lorisid species. Naturally, all observations conducted under natural conditions have been incorporated into the interpretation of these results.

TABLE V

Relative Compositions of the Diet of the Five Lorisid Species in Gabon, Determined from Analysis of Stomach Contents[a]

Genus	Diet composition			
	Animal prey (%)	Fruits (%)	Gums (%)	Other
Galago demidovii	70	19	10	Some leaves and buds
Galago alleni	25	73	—	—
Galago elegantulus	20	5	75	—
Periodicticus potto	10	65	21	Some leaves and fungi
Arctocebus calabarensis	85	14	—	Some wood fibers

[a] The percentage figures are calculated from the fresh weights of the different components, which were separated out with forceps and weighed.

TABLE VI

Composition, in Order of Importance, of the Animal Prey Eaten by the Five Lorisid Species[a]

Galago demidovii	*Galago alleni*	*Galago elegantulus*	*Perodicticus potto*	*Arctocebus calabarensis*
Small beetles (45%)	Medium-sized beetles (25%)	Orthopterans (40%)	Ants (65%)	Caterpillars (65%)
Moths (38%)	Snails (15%)	Medium-sized beetles (25%)	Large beetles (10%)	Beetles (20%)
Caterpillars (10%)	Moths (15%)	Caterpillars (20%)	Snails (10%)	Orthopterans
Hemipterans	Frogs (8%)	Moths (12%)	Caterpillars (10%)	A few dipterans
Orthopterans	Ants (8%)	A few ants	Orthopterans	A few ants
Centipedes	Spiders (8%)	A few bugs (homopterans)	Millipedes	—
A few bugs (homopterans)	Orthopterans	Birds[b]	Spiders	—
A few pupae	Termites	—	Termites	—
—	Myriapods	—	Birds[b]	—
—	Pupae	—	Bats[b]	—
—	A few caterpillars	—	—	—

[a] The percentages are calculated from the fresh weights of the various prey organisms found in the stomach contents.

[b] In the case of birds and bats, one observation of predation was made directly under natural conditions, but no relevant remains were ever found in stomach contents.

1. *Galago demidovii*. This small-bodied species (average body weight: 60 gm), which habitually moves around in dense foliage invaded by lianes, feeds primarily on small insects (beetles, moths, etc.), fruits, and a small quantity of gums.

2. *Galago alleni*. This terrestrially feeding bushbaby (average body weight: 250–300 gm) has a diet consisting mainly of fallen fruit and various small animal prey.

3. *Galago elegantulus*. The needle-clawed bushbaby, which has an average body weight of 300 gm, hunts for animal prey (particularly orthopteran insects) in the forest canopy, but the main component of its diet (75%) consists of gums.

4. *Perodicticus potto*. The potto (average body weight: 1 kg) lives almost exclusively in the canopy and feeds mainly on fruits. Gums may be included in the diet (except during the long dry season), and insects (ants and relatively unpalatable forms) amount to only 10% of the diet.

5. *Arctocebus calabarensis*. The angwatibo is much more lightly built than the potto (average body weight only 200–250 gm) although its overall linear dimensions are quite comparable. This species moves around in the forest undergrowth, where caterpillars (often species with irritant properties) provide most of its animal food. Occasionally, fruits will also be eaten.

It is not possible to discuss here the detailed ecological aspects which have been examined in previous publications (Charles-Dominique, 1971a, b, 1974, 1977). It should be noted, however, that the smallest lorisid species feed mainly on insects and that the larger species, which utilize approximately the same hunting techniques as their smaller relatives in the same subfamily, capture approximately the same absolute quantities of insects. Accordingly, the larger species complement their diets either with fruits (*G. alleni* and *P. potto*) or with gums (*G. elegantulus*).* There are considerable behavioral differences because of distinct dietary adaptations in each species, but at this point only the major dietary distinctions between the galagines and the lorisines will be considered (but see Hladik, Chapter 8).

These two subfamilies are distinguished essentially by their characteristic modes of locomotion, corresponding to two different strategies which have been considered in detail elsewhere (Charles-Dominique, 1971a,b, 1975). Whereas the bushbabies move around rapidly but relatively noisily, by leaping and running, and can escape by rapid flight from any predator which is spotted, the lorisines, (angwantibo, potto, and lorises), have adopted a slow, uniform climbing pattern, permitting them to move around discreetly without arousing the attention of predators. In other words, the galagines utilize active defense (fleeing), while the lorisines utilize preventive defense (cryptic locomotion). These two different evolutionary pathways have naturally had repercussions on the anatomy of the members of the two subfamilies, and the effects on their social structures and hunting techniques have been even more marked.

*In captivity, if fruit and insects are provided *ad libitum*, all five species will eat the insects by preference and consume only small quantities of fruit.

2.1.3.1. Hunting Techniques of the Galagines. Most of the animal prey captured by the various bushbaby species are equipped for rapid escape (moths and orthopterans). In fact, capture frequently occurs at the point where an insect is taking off. The bushbaby leaps forward at this instant and seizes its prey with both hands, while (in the case of the two arboreally hunting species) maintaining its hold on the branch with its hindlimbs. *Galago demidovii* will recoil immediately after capture to their original position on the branch, whereas *G. elegantulus* remain suspended beneath the branch and will only return to their initial position after transferring the prey to the mouth. Such behavior has not been observed with the terrestrially hunting *G. alleni*, however. Bushbabies may also seize an insect resting on the same branch or on a neighboring one. In this type of situation, a bushbaby will approach slowly with its body bunched up and abruptly project itself forward as before, with its hindlimbs maintaining a hold.

Apparently, hearing plays a major role in the localization of prey. A cricket which has been rendered immobile by cooling will generally pass unnoticed if placed among foliage in a Demidoff's bushbaby's cage. However, once the cricket revives and begins to move around, the bushbaby will spot it. In fact, all of the galagines have large mobile ears which are always oriented toward the prey. In another experiment conducted with *G. demidovii* and *G. elegantulus*, a locust was passed, with its wings fluttering, behind a hardboard screen. This invariably excited the bushbabies, and they followed the movements of the concealed insect with great precision, maintaining the orientation of their ears in the correct direction throughout.

These techniques of localization and capture of prey have doubtless evolved in association with the distinctive locomotor adaptations of the galagines, since they disturb insects in the course of running and leaping along branches.

2.1.3.2. Hunting Techniques of the Lorisines. Most of the animal prey taken by the two African lorisine species fall into the category of irritant, unpalatable or noxious arthropods, which are refused by galagines if presented in captivity. The angwantibo eats large numbers of irritant caterpillars (particularly those belonging to the family Thomethopeidae), while the potto* may devour in one night more than 17 gm of ants of the genus *Crematogaster*, which secrete formic acid, or enormous millipedes of the genus *Spirostreptus,* despite their ability to secrete iodine.

The prey is always approached slowly before capture. Medium-sized prey are grasped with one hand (caterpillars, small beetles, etc.), whereas both hands are used to seize large prey. Ants are licked up directly from the branch as the animal slowly moves along the column. When an angwantibo seizes a caterpillar, it is first

*Occasionally a potto may eat small mammals or birds, which are seized with a rapid lunge of the hand (Walker, 1969; Charles-Dominique, 1971a, b). Walker has interpreted the slow, stealthy locomotion of the potto as an adaptation for predation on birds, but no remnants of bird prey were found in the 41 *P. potto* stomachs examined during the field study in Gabon.

passed to the mouth so that the head can be bitten off and then the body is vigorously massaged with both hands, which removes most of the hairs. After the pretreated caterpillar has been eaten, the angwantibo may rub its hands and its muzzle on the branch. In order to grasp a moth, on the other hand, the angwantibo will rear up on its hindlimbs and adjust its position so that its arms are in the same plane as the insect. The arms are then moved toward the moth relatively slowly to grasp the insect at the bases of its wings.

However, the lorisines do not eat noxious prey by preference. If given the choice, in captivity, they will ignore ants, caterpillars, and irritant moths to take more palatable prey (orthopterans and sphingid moths, etc.), which are usually eaten by the galagines under natural conditions.

In contrast to the situation with the galagines, in which hearing seems to play a predominant part, the lorisines apparently rely heavily on olfaction to guide them in prey detection. A cricket which has been immobilized by cooling will still be found if hidden in cage litter at a distance of 50 cm–1 m from a potto or an angwantibo. An immobile moth placed behind a metal screen with fine perforations will immediately evoke discovery and attempts at capture from an angwantibo. On the other hand, a moth placed behind a transparent plastic screen will no more elicit a response from an angwantibo than will a silhouette drawn on a sheet of paper. Yet if such a silhouette is rubbed with an insect beforehand, it will at once be noticed by an angwantibo and elicit predatory responses. It would therefore seem that the lorisines have become specialized for tolerance of irritant or even toxic products which are secreted as a defensive mechanism by many arthropod species. Such naturally protected prey are easily detected by smell, and in many cases, when threatened, they will only utilize their natural weapons rather than attempting to flee. Hence with a minimum of movement through the forest, the lorisines can capture a sufficient quantity of insects. Again, this particular adaptation (toward consumption of relatively unpalatable prey) has doubtless evolved hand-in-hand with the characteristic locomotor pattern exhibited by members of the subfamily.

2.2. The Social Life of the Lorisids

The situation of nocturnal primates is fundamentally different from that of the diurnal species, which are almost all gregarious in habits. A group of diurnal monkeys or lemurs may be located by day and then followed for observation. With such a group, it is possible to make direct observations of social behavior, which is to some extent based on communication through vocalizations and various postures, and often to recognize at a distance the sex, age class, and even identity of certain individuals. With the lorisids, on the other hand, one is dealing with animals which are all nocturnal and solitary in habits, communicating at a distance by means of olfaction as well as vocalizations. Social responses are discreet and often temporally separated from the original signalling action. Thus, direct observation cannot provide a basis for understanding the mechanisms involved in lorisid social life. In this respect, it should be noted immediately that, despite the fact that they are solitary in habits, the lorisids quite definitely are social animals. As a result of inadvertent anthropomorphism, the term "social" has often been confused with the term "gregarious". Diurnal primates are gregarious while the nocturnal lorisids are solitary, but all primates are social, although their means of communication are based on different systems.

With a social group of diurnal primates, a study of direct interactions gives an image of the social situation of a given individual and of the nature of its bonds which unite it to its fellows. With nocturnal primates, on the other hand, it is knowledge of the size of the home range and of its position with respect to those of conspecifics which permits interpretation of the mechanisms underlying social structure. In order to obtain such knowledge, it is mandatory to capture, mark, and release each individual so that its movements through its home range can be followed.

2.2.1. *Home Ranges*

The home ranges of all of the lorisid species are relatively large compared to those of their nocturnal counterparts in Madagascar. If there are only a few animal species in a given ecosystem, each one will be able to exploit a broader spectrum of dietary components available over a small area. Conversely, the greater the number of animal species present in a given ecosystem, the more the dietary spectrum is restricted and the greater is the area over which food must be sought. It is the latter situation which is encountered in Gabon, where the number of sympatric mammal species reaches one of the highest levels known.

2.2.1.1. Female Home Ranges. For females, the average range sizes were determined as shown in the following tabulation:

G. demidovii	0.8 hectares (0.6–1.4; $N = 9$)
G. alleni	10.0 hectares (3.9–16.6; $N = 8$)
P. potto	7.5 hectares (6–9; $N = 5$)

For any given species, the range size for females is thus relatively constant. The home ranges seem to be quite stable, since many of the females were followed from year to year (for up to 5 years in certain cases). From time to time range readjustment may occur, for example, when one of the females has disappeared.

Young females always remain sedentary. They maintain an association with their mothers (and with other females in the group, in the case of bushbabies) and occupy those parts of the forest which are most suited to their ecological requirements. It is only with the bushbabies, however, that one can find several adult females and their offspring associated in matriarchies. Within such groupings of maternal origin (i.e., consisting of mothers together with their adult daughters), social contacts are frequent and home ranges overlap extensively, sometimes completely. On the other hand, there are no contacts between females belonging to neighboring groups, and even if their home ranges overlap, they are never found together in the overlap area. With the potto, the females do not live in such associations. The relationships between individual females are comparable to those between female bushbabies belonging to different groups with slight range overlap. Only on one occasion was it possible to follow a young female potto at the time of transition to independence. At 6 to 8 months of age, the daughter slept progressively less frequently with her mother, and when she reached 10 months of age her mother installed herself in a new range 200 m away from her previous home range, now occupied by the daughter.

2.2.1.2. Male Home Ranges. The situation of the males is easier to interpret, since relationships between them are primarily of a territorial nature. Their home ranges are subject to far greater variations than those of the females, both in terms of size and in terms of stability (shown in the tabulation below):

G. demidovii	0.5–2.7 hectares ($N = 6$)
G. alleni	17–50 hectares ($N = 3$)
P. potto	9–40 hectares ($N = 5$)

In fact, the position of the males depends primarily on that of the females. Each male seems to seek a relationship with as many females as possible. The largest

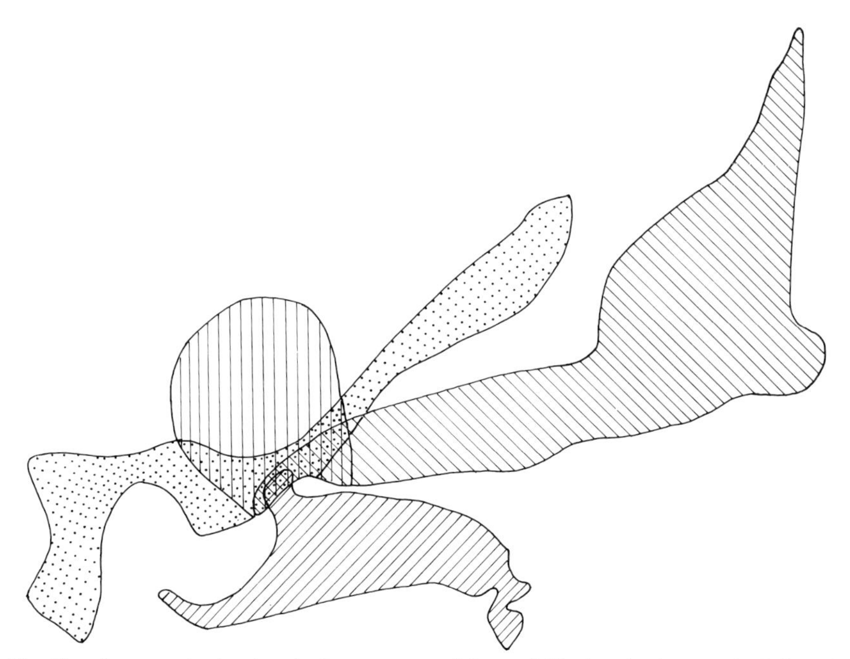

Fig. 12. An example showing the home ranges of four neighboring dominant male *Galago demidovii*. Note the central overlap zone. (After Charles-Dominique, 1972.)

males cover enormous home ranges overlapping those of their females (up to 6 in number), whereas other males are excluded and occupy marginal areas (peripheral males) in a less sedentary fashion. In some cases, however, a small-bodied male Demidoff's bushbaby occupying a restricted home range may be tolerated within the range of a large male.

The males' home ranges are never isolated from one another, since there is always a small communal zone (Fig. 12). It will be seen below that this overlap zone, which is never visited simultaneously by the different males concerned, permits mutual surveillance (probably by means of urine-marking).

In contrast to the situation found with the females, the young males, when approaching puberty, leave the maternal home range and become nomadic, thus ensuring exogamy.

2.2.2. *Territorial Relationships*

2.2.2.1. Females. It is quite difficult to demonstrate any aggressive relationships between neighboring females. In fact, the reproductive rate of all Gabonese lorisids is very low (approximately one infant per female per year) and it would seem that, given the natural mortality levels, new females can fit into the environment without active competition. Females of three species (2 *G. demidovii*, 3 *G. elegantulus*, and 4 *P. potto*) were experimentally introduced into the home ranges of other females. In most cases, the intruder avoided the occupied range and established herself in a free area; but on two occasions a resident female *G. elegantulus* was seen to chase the intruder. Faced with the attacks of the range occupant, the introduced female dropped from the canopy to the ground to flee and was chased some distance. Such conditions, produced experimentally in these cases, are probably rare under normal circumstances, and young females are doubtless integrated progressively into their social environment.

The only relationships between neighboring females which were observed without some kind of experimental intervention were quite discreet. With *G. alleni*, for example, neighboring females may exchange vocalizations for up to an hour from opposite sides of an overlap zone, at a distance of 20–30 m from one another. With the potto, females exhibit mutual avoidance in the communal zone, where urine marks probably provide for exchange of information.

2.2.2.2. Males. Aggression is much more evident with adult males, as attested by the percentages of scars recorded for individuals captured in the forest. No scars were found on 152 immature lorisids examined (except one juvenile angwantibo) whereas in 146 adults the following scar percentages were found:

A. calabarensis	5/16 ♂♂; 2/13 ♀♀
G. elegantulus	5/27 ♂♂; 0/16 ♀♀
G. alleni	6/10 ♂♂; 7/17 ♀♀
G. demidovii	8/40 ♂♂; 4/25 ♀♀

Males were found to exhibit scars 2 to 4 times as frequently as females, according to the species. Although fighting can easily be provoked in captivity, it would seem to be relatively rare under natural conditions and probably occurs mainly during periods of readjustment, when a male home range falls vacant. As in the case of the females, foreign males introduced experimentally into the home ranges of resident males were found to avoid the occupied zones. However, on one occasion, following such an experimental introduction, a fight was observed between a resident male potto and an introduced male (the intruding male escaped after the fight).

With the potto, it was possible to observe fairly clearly the behavior of neighboring males in the overlap zone. Two such males will exhibit mutual avoidance, only entering the communal area one at a time.

All of the lorisid species exhibit quite frequent urine-marking dispersed over their ranges (see Schilling, Chapter 11). In fact, such marking probably occurs in all areas of their home ranges, since the lorisids urinate regularly (either onto their hands and feet or directly on branches as urine trails). In captivity, such behavior usually occurs 4 to 8 times an hour, and any excitation will only increase its frequency. It seems likely that this scent-spreading (which probably varies in frequency in different parts of the home range) alerts any conspecific to the identity of the range occupier (see Schilling, Chapter 11).

It is impossible to observe marking behavior, still less its frequency, under natural conditions. In captivity, however, it has been observed that barriers in the middle of the cage are marked more frequently than those elsewhere (Seitz, 1969). For animals communicating with one another by means of urine-marking, it is absolutely necessary that their home ranges should overlap. Such indirect communication can only take place in the communal area, where each individual goes to verify the presence of its neighbor and to indicate its own presence on a delayed action basis. The bushbabies communicate with one another as much by calls as by urine-marking, but the lorisines base their communication almost entirely on scent-marking. Thus, this system of relationships between neighbors is quite distinctive and entirely different from that of other primates which communicate primarily by means of direct auditory and visual signals (*Lepilemur*, *Phaner*, diurnal lemurs, monkeys, and apes).

In the wild it is impossible to observe the relationships between established males and nomadic males, but it is likely that there are occasional confrontations between these two categories of males. In the Gabon study area, one newly matured male *G. demidovii* was found carrying traces of a fairly serious recent fight (ears torn, one eye collapsed, and the tail fractured in two places).

2.2.3. Social Relationships

2.2.3.1. Male–Female Relationships. Relationships between males and females are not exclusively limited to the mating periods. In fact, they are initiated by a form of courtship which is independent of the sexual condition of the female (she may even be prepubertal). Courtship begins with chasing of the female, while the latter initially flees from any contact with the male. After several days, or even weeks, the two animals will lick one another actively and thereafter their contacts become more discreet. Nevertheless, they remain linked by a specific social bond and the male will continue to visit his female (or females) from time to time throughout the year. Each female is thus under the permanent control of a given male and their mutual relationship is intensified still further just prior to and during estrus.

With the bushbabies, the male will pay a visit to his female(s) during the night. He sniffs at the female's perineal region and then, following reciprocal grooming, the two partners separate again. Such contacts are relatively short in duration and it is rare to find a male with a female in the forest. In one night, a given male *G. demidovii* was found successively in the home ranges of two females separated by 200 m. With *G. alleni*, radiotracking studies demonstrated that visits made to a given female occur at intervals of 1–4 days. On the other hand, a few days before a female enters estrus, the male will remain in her home range every night. Once estrus has terminated, the male will then resume his visits to any other females which have been neglected in the meantime.

In this respect, the potto exhibits an adaptation correlated with its peculiar mode of locomotion (slow, discreet movements, which are incompatible with daily pursuits of a female by the male). Following courtship similar to that of the bushbabies, the male and the female have no further contact. However, every 2 to 5 days the male continues to make visits to urine marks left by the female rather than to the female herself. The male, in so doing, leaves behind his own marking traces. By this means, the two animals remain in contact in an indirect fashion (deferred social communication). Before estrus, the visits made by the male are intensified and the latter will remain close to the female for the few days preceding mating. When a male potto controls only one female in this way, their social relationships are somewhat more direct: the male sometimes follows his female at a distance of 10 m or so for 1 or 2 hours each night. However, under such conditions no allogrooming has ever been observed (outside the period of courtship).

2.2.3.2. Relationships between Adult Females. Among the bushbabies, social relationships between adult females only exist within matriarchal groupings. During the night, the females of each group are dispersed and there is practically no contact between them. They do not regroup until dawn in order to return to sleep with their offspring, and at this time gathering calls are utilized. After regrouping, the individual bushbabies exhibit mutual licking, and the close contact which they have while sleeping presumably contributes to mixture of their respective odors, thus reinforcing social bonds. After reawakening, dispersion is almost immediate. Hence, bushbabies pass successively through a gregarious phase and a solitary phase in the course of 24 hours.

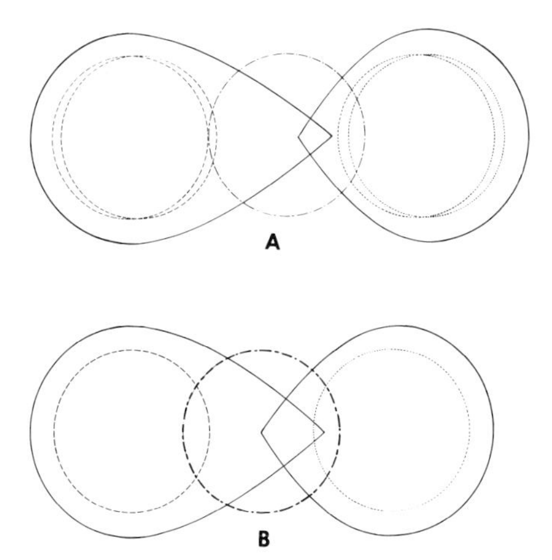

Fig. 13. Schematic representation of the home range systems of the lorisid species studied in Gabon. A, Galagines (note the formation of matriarchies); B, Lorisines (note the restricted zones of overlap between males and between females). Dotted lines, home range limits for females; continuous lines, home range limits for males.

Female pottos of common parentage do not form groups in the same way as the bushbabies (Fig. 13). If several individuals share the same home range, they are forced to cover greater distances in order to collect less densely available food. Thus, an animal which moves slowly and little like the potto will tend to occupy a small, exclusive home range. In the same way, regrouping of several individuals at the end of the night requires additional movement and mutual location by means of powerful vocalizations. Such regrouping is incompatible with the way of life of the potto, whose discreet locomotion provides its main form of protection. As with the dietary specializations of the lorisines, the particularly solitary way of life of the potto, with reduction to a minimum of social contacts, would seem to be a consequence of the locomotor pattern typical of this subfamily.

3. THE GALAGINES OF SOUTH AFRICA

3.1. Introduction

Perhaps the most familiar of the African lorisids are the two widely distributed species, the lesser bushbaby, *Galago senegalensis*, and the thick-tailed bushbaby, *Galago crassicaudatus* (Figs. 14 and 15). The smaller than squirrel-sized *G. senegalensis* are also known as night apes or ''Nagapies'' (Afrikaans). They attain a maximum weight of 300 gm with a head and body length of approximately 170 mm

Fig. 14. *Galago senegalensis moholi* (average weight: 200 gm) (Photograph: T. Bekker).

Fig. 15. *Galago crassicaudatus umbrosus* (average weight: 1300 gm) (Photograph: S. K. Bearder).

and a tail length of 230 mm. The cat-sized *G. crassicaudatus* are the largest of the Galaginae with a maximum weight of 1800 gm and measuring approximately 300 mm from the nose to the base of the tail, with a tail length of 400 mm. They are well known for their loud childlike cries, from which comes the common name of bushbaby.

The present section is based on field studies of *G. senegalensis moholi* conducted in the Northern Transvaal (Bearder, 1969) and *G. crassicaudatus umbrosus* in the North-Eastern Transvaal (Bearder, 1975). In the first case, regular field excursions totalling 6 months were made during the course of a year (1968–1969), while a full-time investigation of *G. crassicaudatus* covered a period of 15 months (1970–1971). In order to gain an unbiased impression of habitat preferences, additional data were collected for each species away from the main study areas. Observations were made on *G.c. garnetti* and *G.c. lönnbergi* in Zululand and Eastern Rhodesia, respectively, and on *G.s. moholi* in the North-Eastern Transvaal and Rhodesia, where they live sympatrically with the larger *G. crassicaudatus*.

3.2. Study Methods

3.2.1. Galago senegalensis

This species can be found during the day by searching for nests and sleeping sites. After some practice it becomes possible to recognize suitable sleeping trees, generally those which are dense and thorny. Bushbabies regularly return to particular sleeping sites and, once these are known, it is usually possible to find an animal or group by looking carefully at each one.

The effective nocturnal observation of an arboreal species in short and open vegetation is considerably less difficult than that described for rain forest lorisid species in the preceding section. In fact, the physiognomy of the study area can be chosen to best suit the needs of the observer. At night, *G. senegalensis* were observed by using red light from a simple headband torch covered with a red celluloid filter and powered by a 6-volt motor cycle accumulator battery. A 6-volt hand lantern provided a strong beam of white light when necessary. Under ideal conditions, the animals' eyes could be seen at a distance of 100 m due to the extremely reflective *tapetum*. The use of red light for watching nocturnal animals has been described by Southern (1955). Unlike white light, it does not cause bushbabies to contract their pupils, blink, or retreat rapidly; but the principal advantage is that it does not reflect strongly from nearby vegetation and thus impair the night vision of the observer. In this way the animals can be seen in relation to the surrounding vegetation and their movements followed with ease while the observer, by making maximum use of available light, is less likely to become disoriented.

When first observed, *G. senegalensis* usually become habituated to an observer after 30 minutes and subsequently they continue their activities undisturbed. It is then usually possible to follow them at a distance of 3–7 m, although they will occasionally approach to within 1 m. It is necessary to maintain such close proximity to the subject in order to avoid losing it, and behavior can be observed in some detail for several hours at a time and sometimes throughout the night. Events and sequences of behavior shown by a single individual were recorded on a notepad. The behavior of conspecifics in relation to that of the subject was noted, but in less detail. Inevitably there were intervals during which the animals were obscured from view and quantitative data derived from these observations can, at best, give only a general picture of the interrelationships between activities. Alternatively, a bushbaby may resort to extremely quiet and rapid movement which brings an observation session to an abrupt end. Range movements can be effectively plotted by placing small white marker cards on the trees at night and by pacing out and mapping the distance and direction of travel on the following day.

In order to avoid misinterpretation of the animals' activity patterns and movements, at least some observations were made at all times of the night. It proved impractical and unrewarding, however, to work in the middle of the night when the animals were least active. The majority of observations were divided almost equally between the 4 hours after sunset and the 4 hours before sunrise. Individual bushbabies were often followed away from their sleeping places and later followed back to the same sites at dawn. Unfortunately, no trapping and marking program was undertaken for this species and, although natural markings could be used to good advantage, this prevented a detailed interpretation of social dynamics. This situation is at present being remedied through the use of radiotracking.

3.2.2. Galago crassicaudatus

A prerequisite for making comparable observations on the large *G. crassicaudatus* was to find a habitat which was relatively open and where the trees were sufficiently low. Tracts of riverine vegetation confine this species to a limited area, which can be traversed by means of a network of pathways, whereas forest habitats are best tackled during the winter months when the vegetation dies back and is more easily penetrable. Using the same techniques as described for *G. senegalensis*, it proved possible to follow and record the behavior of *G. crassicaudatus* in a similar way from a standard observation distance of 5–10 m. When first observed, these slower moving prosimians remained wary for at least 1 hour and they would not approach to closer than 2 m to the observer.

Most data collected were derived from a small number of well-known individuals from an isolated population. This somewhat artificial situation, induced by the nature of the vegetation, greatly facilitated the interpretation of many aspects of behavior which would not have been possible had the same effort been distributed over a greater number of animals in an unrestricted part of their habitat.

Galago crassicaudatus were trapped using wire mesh automatic traps with honey as bait (Fig. 16). The treadle is placed sufficiently well back to allow the animal to enter completely before the door is released and the angle of the strut can be altered

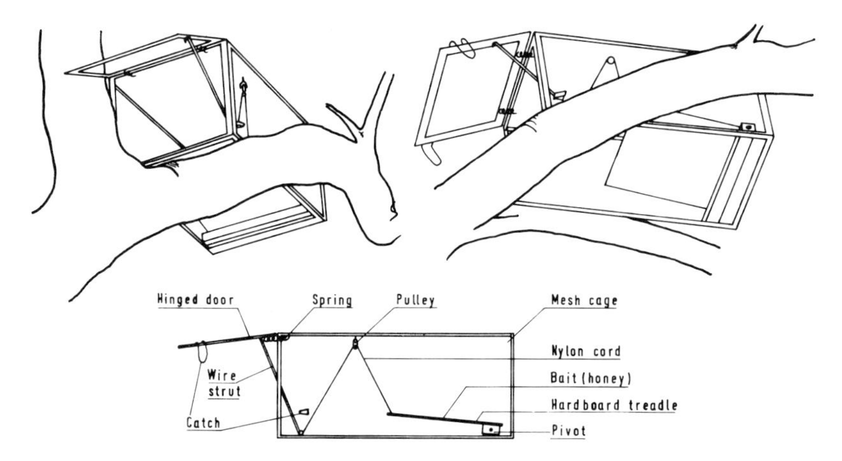

Fig. 16. An automatic trap used for the capture of *G. crassicaudatus*.

to adjust the sensitivity of the mechanism. It is thereby possible to capture two animals at one time. Individuals were handled during the daytime without anesthetization and released on the following night where they had been captured. Marking was achieved by cutting hair from segments of the bushy tail and these marks remained visible for several months. The animals showed no apparent aversion to being trapped. They did not become timid of the observer at night and would re-enter the trap on subsequent occasions.

3.3. Habitat Utilization

The behavioral adaptability of the South African galagines is illustrated by their presence in a variety of habitats where they are subject to marked seasonal variations in availability of food and cover. There is, however, a clear ecological distinction between the two species. Studies of allopatric populations reveal that each species is best adapted for life in a different type of habitat and estimates of population densities show their differential success in a number of areas (Bearder and Doyle, 1974).

3.3.1. Galago senegalensis

In general, the smaller *G. senegalensis* are able to survive and spread through tracts of dry and relatively open deciduous vegetation which are unsuitable for their larger relatives; this is reflected by the greater extent of their geographical distribution. They reach maximum numbers in semi-arid *Acacia* thornveld savanna, where *Acacia karroo, A. nilotica,* and *A. tortilis* predominate and where they have no access to surface water for most of the year.

Lesser bushbabies are remarkably active and agile animals and move considerable distances during the course of a night. Records of the frequency and duration of particular behaviors for a sample of 215 hours showed, on average, that they moved through 48 trees during each hour, while a maximum of 171 trees was traversed in that time. Direct measurements revealed that this represents an average hourly path length of 200 m, or approximately 2 km each night. The maximum distance covered in a single hour was 500 m. These records apply to adult animals, but no distinction was made between the sexes.

The heights at which these bushbabies were commonly seen and their use of particular supports appeared to be a function of the height and nature of the vegetation, rather than an adaptation to a particular zone. This is not surprising when it is considered that there are few nocturnal arboreal competitors. Rodents were encountered on rare occasions, but otherwise *G. crassicaudatus* were the only mammals to occupy a similar niche. Most activity occurred between ground level and a height of 6 m, which is the approximate height of the tree canopy, but in larger trees bushbabies would ascend to 12 m or more. The versatility of their movements ensures that they are equally at home on broad trunks or fine branches, but no precise determination of the orientations and sizes of supports used by this species was carried out.

Three fundamental methods of locomotion can be distinguished in *G. senegalensis*: jumping from branch to branch, saltation along the ground, and climbing within a tree. Rapid leaping is characteristic of this species and jumps are usually executed in a particular way. Some common locomotory postures are illustrated in Fig. 17. Horizontal leaps of almost 5 m can be achieved with the arms raised sharply on take-off. Otherwise, all but the shortest jumps are made with the arms held against

Fig. 17. Drawings taken from photographs of *G. senegalensis* to show some typical locomotor postures (arrows indicate possible sequences).

the chest. This is unusual amongst the Galaginae and is related to the fact that *G. senegalensis* raise the tail rapidly and bring the hindlimbs forward to take the initial impact of landing (Hall-Craggs, 1965). The leaping abilities of other bushbabies have been described by Charles-Dominique (1971b). The striking feature of *G. demidovii*, *G. alleni*, and *G. senegalensis* is that they are all adept at jumping vertically from one horizontal support to another by rotating their bodies in midair. This ability is lacking in *G. elegantulus* and *G. crassicaudatus*.

In moving through open woodland, *G. senegalensis* are frequently forced to descend to the ground where they may cover distances of up to 60 m at a time. Alternatively, they descend in order to forage for ground-dwelling insects. During the sample period of 215 hours, lesser bushbabies descended, on average, 7 times during each hour with a maximum of 30 times in an hour. While short distances are crossed with little hesitation, it is evident from their extreme wariness and frequent trial runs that bushbabies are vulnerable to predation when crossing large open spaces and they will use the smallest stump as a vantage point to break the journey.

3.3.2. Galago crassicaudatus

Thick-tailed bushbabies are found at highest population densities, to the exclusion of *G. senegalensis*, in humid subtropical evergreen forests where trees bearing fleshy fruits are abundant (*Ficus* spp., *Diospyros mespiliformis*, *Syzigium cordatum*, etc.). They are not confined to forest zones, however, being common in dense riparian vegetation, from which they extend into timber plantations (*Pinus* spp., *Acacia cyanophylla*, *Eucalyptus globulus*) or subtropical orchards and even into open woodland (*Acacia abyssinica*). In those areas, where they are sympatric with *G. senegalensis*, there is some degree of niche separation according to the density of the vegetation in different parts of the shared habitat.

Galago crassicaudatus are slower and less agile than the former species and the distance traveled in a night is correspondingly shorter. Minor deviations were not measured during this study but point-to-point distances for a sample of 406 observation hours varied between 0 and 400 m per hour, with an average of 84 m per hour. This indicates a travel distance of approximately 1 km each night; but since the hours of observation were least common in the middle part of the night, when the animals were least active, it is likely that this is an upper estimate of the normal distance of travel.

Throughout the literature the galagines have been characterized by their active leaping mode of progression, aided by the elongated hindlimbs. Morphological criteria have been used to classify the locomotion of these species as specialized vertical-clingers-and-leapers, as opposed to a generalized quadrupedal gait or the slow-moving quadrupedal locomotion of the lorises (Napier and Walker, 1967). Yet the behavior of *G. crassicaudatus* belies its anatomical structure and provides an interesting exception to the general rule.

Galago crassicaudatus usually progress by walking or running along the top of horizontal branches, or along the ground, in a quadrupedal fashion. This is not unlike the locomotion of *Perodicticus potto*. It is accompanied by exaggerated side-to-side movements of the body which become less noticeable as the speed increases. Jumping is performed when the occasion demands by means of rabbit-like hops (along a horizontal surface); short leaps (between supports, landing hands first); long leaps or drops of up to 5 m (landing in foliage which breaks the fall); and rarely saltation (when on the ground). These animals will not generally jump between trees if they are able to climb, but they jump in preference to descending on to the ground. The common method of moving between trees is by progressive deployment of weight across the ends of adjoining branches. A thick-tailed bushbaby usually climbs to the end of a branch and grabs the next with one or both hands before stepping carefully across. Perfect balance is maintained with the help of the long tail and the weight is transferred slowly and evenly to the new branch before the other is released.

Previous records of locomotion in *G. crassicaudatus* indicate that movement on the ground is achieved by means of saltation (Hill, 1953). On the contrary, observations during the present study show that saltation is rare. Bushbabies were observed on the ground on 75 occasions during which saltation was seen 8 times (11%). Four of these were following the release of bushbabies after they had been trapped and the only instance when more than two or three hops were made at one time occurred during a bout of intraspecific chasing. The obvious reason why saltation has been emphasized previously is that it takes place when the animals are disturbed.

A typical sequence of movements is illustrated in Fig. 18. *Galago crassicaudatus* often climb beneath a branch or use spiraling movements from a vertical trunk onto a horizontal limb. They may hang from one or both legs to reach a lower branch before dropping down. When ascending or descending narrow supports they climb head first, often stretching from one to the next. Broad trunks may be negotiated caterpillar fashion, using a looping action. Many of these postures are more typical of the Lorisinae than other members of the Galaginae and have been described for *P. potto, Loris tardigradus,* and *Nycticebus coucang* (Rahaman and Parthasarathy, 1970; Ehrlich, 1970; Charles-Dominique, 1971b).

As in the case of *G. senegalensis*, but unlike the lorisids of Gabon, *G. crassicaudatus* show no obvious preference for a particular height and, despite the tendency to use horizontal supports, they move according to the demands set by each vegetation type. Where the vegetation is more or less unbroken, thick-tailed bushbabies rarely descend onto the ground where it seems that they are most vulnerable. Nevertheless those animals which are forced to descend in order to move between trees do so with considerably less hesitation than their forest counterparts. In the open woodland habitat of Eastern Rhodesia *G.c. lönnbergi* often came down to the ground in order to cross spaces of up to 100 m or to forage: 66% of all sightings on the ground were made in Rhodesia despite the fact that only 10% of the total observations were carried out in this area. The chances of seeing a bushbaby on the ground were 18 times greater than in the Northeastern Transvaal, while in the dense forest of Zululand *G.c. garnetti* were seen on the ground only once following

Fig. 18. Drawings taken from photographs of *G. crassicaudatus* to show typical sequences of movement through the trees and on the ground.

a fight. However, it is evidently not the spacing or density of the vegetation which is the most important factor controlling the distribution of this species.

3.4. Antipredator Behavior

It can be seen that the distinction which exists between the characteristic modes of locomotion of the galagines and lorisines in the tropical forest zone also applies to some extent when comparing *G. senegalensis* and *G. crassicaudatus*, though in this case the distinction applies within the same subfamily. The active defense typical of the galagines is put to good effect by the smaller species, while the larger makes use of preventive defense similar to that of the lorisines, but not to the exclusion of the ability to flee rapidly. Indeed, many of the behavioral and ecological adaptations of bushbabies appear to be a result of the selective advantage of avoiding predators, but their survival value can only be evaluated by detailed consideration of the impact of potentially harmful species.

Both *G. crassicaudatus* and *G. senegalensis* show little interest in other species apart from their prey or potential predators. Their initial response to the observer is typical of their behavior toward a threatening stimulus. They may retreat to a vantage point from which they stare, either fixedly (*G. crassicaudatus*) or while moving the head rapidly from side to side (*G. senegalensis*). Alternatively, they approach hesitantly, making a number of calls characteristic of anxiety. The critical distance of approach is greater in the less agile species and varies according to the situation. Active disturbance will cause the subject to ascend the tree or to move away. If it is then followed it will move in circles and change direction repeatedly until, almost invariably, it is lost to view. However, unlike many diurnal primates which may take weeks to allow close approach by the observer, bushbabies become habituated to a quiet observer within one or two hours and will subsequently continue their activities to the extent of going on to the ground nearby. It is evident that they soon learn which animals to fear and to ignore those which cause no harm.

3.4.1. *Galago crassicaudatus*

In some parts of their range *G. crassicaudatus* are liable to predation from a number of large carnivores when they descend to the ground, including leopards (Turnbull-Kemp, 1967) and domestic dogs (Astley-Maberly, 1967). Raptorial birds (eagles) are a source of danger during the day, should they sleep in exposed positions, or return late to the sleeping place. Snakes (mambas, cobras, and pythons) are potentially harmful during the day and cause an intense alarm reaction at night, but otherwise this species has few nocturnal arboreal enemies.

Galago crassicaudatus are sufficiently large when adult to be unperturbed by owls or small carnivores such as genets (*Genetta tigrina*) and they have behavioral mechanisms which appear to protect the infants when they are most vulnerable (infant carriage, nest-changing). They usually sleep completely hidden in dense tangles of vegetation from which they are reluctant to move. At night they are cautious before descending onto the ground and can escape quite rapidly if danger threatens; but when moving quietly they often pass undetected. The structure of the group (see Section 3.7 below) probably ensures that the offspring learn from the mother which species to avoid. A number of mobbing vocalizations are given by one or several individuals, which orientate toward the source of the threat. These seem to warn conspecifics of the presence, position, and even the seriousness of a potential danger (Doyle *et al.*, 1977).

3.4.2. *Galago senegalensis*

As a direct result of their smaller size, *G. senegalensis* may be preyed upon by a greater number of arboreal and aerial predators, and their reactions to small carnivores and owls differ accordingly. These predator species induce avoidance and an intense alarm reaction with mobbing behavior and vocalizations which may last for upward of 20 minutes at a time. Lesser bushbabies are often hidden from sight when viewed from below during the day and make frequent use of dense thorn trees for sleeping purposes, though they may sometimes rely on the natural camouflage of the pelage to avoid detection. If disturbed, they soon become active and they will move rapidly away through the trees if the threat continues. The remarkable agility of these small prosimians and the behavior of the mother toward her young (carriage in the mouth; nest-changing) must further drastically reduce the effects of predators.

When considering the differences in the preferred habitats of the two bushbaby species, it is evident that arboreal predators are more abundant and better concealed in forest areas than in open woodland. It is therefore likely that individuals of the smaller species are least liable to predation in open areas where they are able to spot danger at a distance and react appropriately. The larger and slower *G. crassicaudatus*, on the other hand, have greater requirements for cover while sleeping during the day. Their relatively quiet movements in dense vegetation may not only enable them to avoid detection, but also ensure that they are not taken by surprise. Nevertheless, *G. crassicaudatus* will move noisily and they have retained the ability to escape danger through flight which is their usual response to a predator, although no actual attack by a predator has been recorded (A. Jolly, unpublished paper, 1966). Evidently there is a compromise between defense strategies in this species which is perhaps basic to the compromises already observed in its locomotion and also to be found in feeding habits and social structure.

3.5. Diet

The technique of following particular bushbabies at close quarters for long periods made it possible to determine by direct sight, or from sounds, when the subject was feeding. In the case of *G. senegalensis* a record was kept of the frequency and duration of feeding bouts during each hour of observation, but it often proved difficult to determine what the animal was eating. Conversely, during the study of *G. crassicaudatus*, which seldom could be seen continuously, suspect information on the frequency of feeding bouts was abandoned in favor of recording of food types taken during those bouts which were observed. In this case, it was frequently quite clear what the animal was consuming.

Galago senegalensis and *G. crassicaudatus* in captivity are known to eat a variety of animal and vegetable foods including arthropods, fruits, vegetables, gum, and small vertebrates (Lowther, 1939; Haddow and Ellice, 1964; Doyle and Bekker, 1967). A similar omnivorous tendency is reported for other galagos and lorises (Charles-Dominique, 1971b) and for a number of nocturnal lemurs (Petter, 1962a,b). Yet each species shows one or two predominant trends toward the consumption of particular foodstuffs under natural conditions (see Hladik, Chapter 8).

3.5.1. Galago senegalensis

Wild *G. senegalensis* in the Northern Transvaal fed only on arthropods and *Acacia* gum, both of which are available to a varying extent throughout the year. It would appear that this diet provides for all water requirements, since lesser bushbabies were not seen to drink or even to lick condensation from the surface of leaves.

During the driest half of the year (May–October) *G. senegalensis* subsist on gum supplemented by occasional arthropods. Gum is licked or eaten in soft lumps for periods lasting up to 15 minutes at a time. Hourly records, made toward the end of the dry season (September–October; $N = 76$), showed that an average time of 12 minutes was spent feeding during each hour (maximum = 24 minutes), with an average frequency of 21 feeding bouts (maximum = 80) per hour. At this time, periods of up to 15 minutes per hour were spent foraging on the ground.

After the first spring rains, *G. senegalensis* showed a marked decrease in the amount of time they spent feeding and they seldom searched for food on the ground. They continued to lick gum but for much shorter periods. Hourly records for midsummer (February–March; $N = 139$) gave average figures of 1 minute and 6 bouts for the duration and frequency of feeding with a maximum of 20 minutes and 43 bouts during an hour. These figures reflect an increased reliance on a smaller number of more nutritous food items, which take less time to consume; namely, a transition from gum-licking to insect predation as the availability of insects increases.

A similar diet has been recorded for *G. senegalensis* in East Africa (Haddow and Ellice, 1964) and in South-West Africa (Shortridge, 1934; Sauer and Sauer, 1963). Lesser bushbabies show no interest in birds or their eggs and occasional reports that they feed on eggs are most probably derived from their use of old birds' nests or tree holes for sleeping purposes.

3.5.2. Galago crassicaudatus

All observations of feeding in the main study area have been grouped together in order to estimate the relative use of different foodstuffs during the year. Each record consisted of feeding on a single item in one place at one time. In this way, a bout of licking at a source of gum is taken as being equivalent to eating a single fruit or insect and recorded separately from a renewed bout of licking, at the same time or a different place, or the consumption of a further fruit or insect.

The varied use of different foods by *G. crassicaudatus* is illustrated in Fig. 19. Fruits and gum represent the staple diet of this species, but seeds, nectar, and insects are also consumed. Gum is eaten throughout the year, but it is particularly important during the dry winter when soft fruits are lacking. The overall figures for the observed bouts of intake of foodstuffs are: gum 62%; fruit 21%; flower secretions 8%; seeds 4%; and insects plus a few unidentified items 5%. Moisture is occasionally licked from leaves, but otherwise thick-tailed bushbabies have not been seen to drink.

Galago crassicaudatus have been reported to be mainly predatory in their feeding habits, living on birds, eggs, small mammals, and reptiles, together with fruit (Roberts, 1951; Hill, 1953; Walker, 1954; Sanderson, 1957). During the field study, there were many independent and reliable reports of thick-tailed bushbabies killing chickens, doves, and other birds, eating only the brain, or merely licking the blood. On the other hand, there was good evidence to suggest that meat eating was

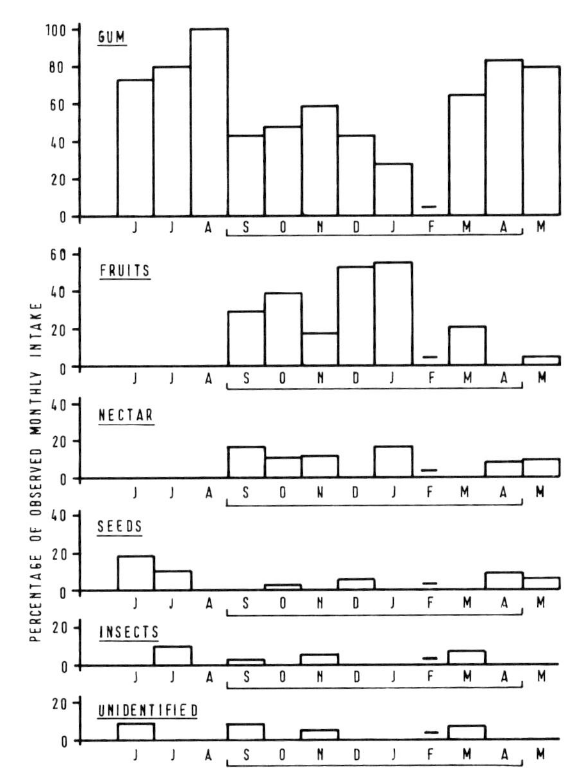

Fig. 19. Histogram showing observed monthly intakes of different foodstuffs by *G. crassicaudatus*, based on 281 bouts of feeding observed during one year. Continuous lines represent months in which average temperatures exceeded 18°C (no observations were conducted in February).

extremely rare or completely absent in the study populations. *Galago crassicaudatus* were not observed to eat vertebrates of any kind although the predatory habits of genets (*Genetta tigrina*) and marsh mongooses (*Atilax paludinosus*) were witnessed, despite the fact that these carnivore species were seen much less frequently. It is hence unlikely that regular predation by *G. crassicaudatus* would have been missed.

In order to test the predatory potential of *G. crassicaudatus*, a pair of doves was introduced to four wild-caught bushbabies in a cage of approximately 4 m^3 capacity. The doves immediately took over the next box previously used by the bushbabies and subsequently raised two clutches of chicks without harm. The complete lack of interest in birds as a source of food was confirmed in the wild, where *G. crassicaudatus* ignored roosting birds at night, even after the latter had been induced to sing by the playback of tape recordings.

The dentition of *G. crassicaudatus* and their ability to make quiet and stealthy movements indicate that they would be quite capable of catching and killing birds as large as chickens. Likewise, the habit of returning to a regular source of food coincides with reports of heavy losses from chicken houses and dove cotes over a period of several nights. The most plausible explanation for these contrasting impressions concerning predation is that meat eating is not a universal characteristic of the species, but a local habit which may be shown by certain populations.

3.6. Foraging Techniques

Both species adopt similar postures while feeding, using their hands and mouths in a variety of ways to suit the type of food which is being tackled. They will each move to almost any part of a tree or onto the ground in order to reach a desired food item.

3.6.1. Galago crassicaudatus

Acacia gum exudes from splits and wounds in the bark and from the holes made by insect borers. *Galago crassicaudatus* obtain the freshly exuded gum by licking or by first chewing away the bark with the teeth on one side of the mouth and then licking the gum which is exposed. They will also chew older gum which has the consistency of toffee, but they ignore that which has hardened completely. A remarkable number of gum-licking postures may be assumed, enabling these animals to reach gum which is inaccessible to potential competitors (Fig. 20). When feeding on the gum of *Albizia versicolor*, bushbabies were seen to bite off pieces of bark with vigorous movements of the head. The bark was then held in one hand, licked, and dropped to the ground in a similar action to that used when feeding on fruit.

Small fruits are picked directly with the mouth or first guided to the mouth with one hand and then carried away to be eaten in comfort (Fig. 21). Large fruits can be eaten *in situ*. Peel is chewed away and discarded and the soft insides licked or scraped out with the ventral anterior teeth. Unripe fruits may be dropped and ripe fruits knocked off in the course of movement, but no interest is shown in the fruit once it has fallen.

When searching for nectar and other plant secretions, *G. crassicaudatus* carefully investigate each likely source and lick those which are productive. Dry fruits, seeds, and insects are generally picked up with one hand in a slow grabbing motion and transferred to the mouth. Stationary insects are taken when the opportunity presents itself, but not usually in preference to other foods. For example, no interest was shown in the moths attracted to fruit which had been damaged while feeding. In Rhodesia, however, *G. crassicaudatus* regularly captured insects on the lawn of a floodlit house.

There is evidence that the sense of smell is the most important aid to the localization of food sources. Many trees in fruit or flower and even those bearing gum have

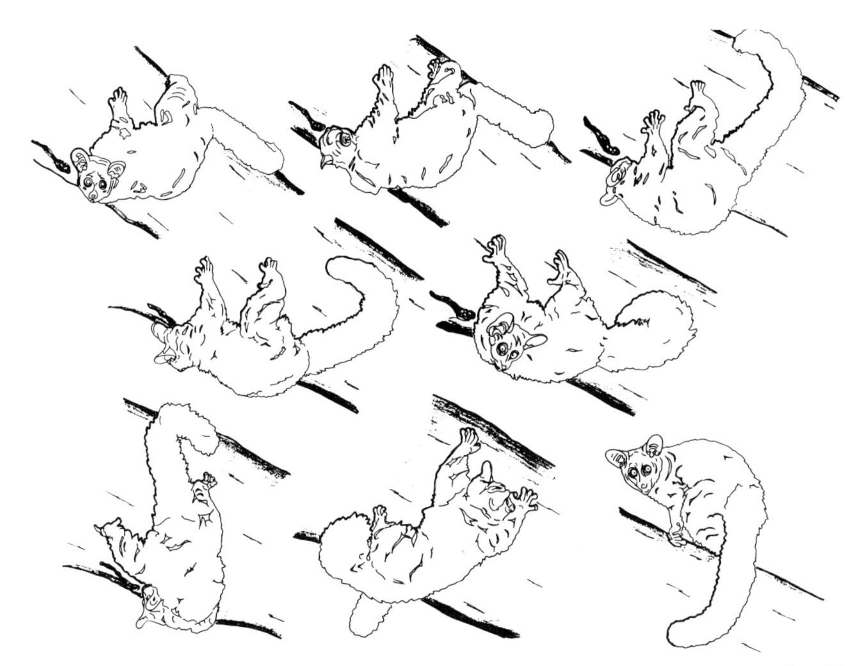

Fig. 20. Drawings taken from photographs of *G. crassicaudatus* to show a sequence of positions adopted while licking gum beneath an overhanging trunk. Note that the limbs are held wide apart, with the body close to the tree. The tail appears to be important for maintaining balance.

a distinctive odor and bushbabies were seen to go out of their way to reach such trees to the extent of going on to the ground, even though they had not been seen to visit them at any other time during the year. Honey was smeared experimentally at various points in one area and renewed at regular intervals. The bushbabies initially discovered such "licks" during the normal course of movement, but they then moved directly to others nearby. On subsequent nights, however, they visited these honey spots without hesitation, using them in much the same way as gum licks. Hence, the location of food sources is apparently remembered.

A characteristic of *G. crassicaudatus* is the stop-go nature of their movements, which can be related to use of, and movement between, well-known food sources at fixed points within the range. Thick-tailed bushbabies will sometimes remain in the vicinity of food trees for several hours at a time. They return to them frequently during the night and on consecutive nights for as long as they remain productive. Trees such as figs (*Ficus* spp.) and Jackalsbessie (*Diospyros mespiliformis*) which bear fruit ripening over a long period, or those which regularly produce gum, are of particular importance. In fact, the availability of fruit appears to be a significant factor which limits the distribution of this bushbaby species, which is absent from regions where fruits are not abundant for at least 6 months of the year.

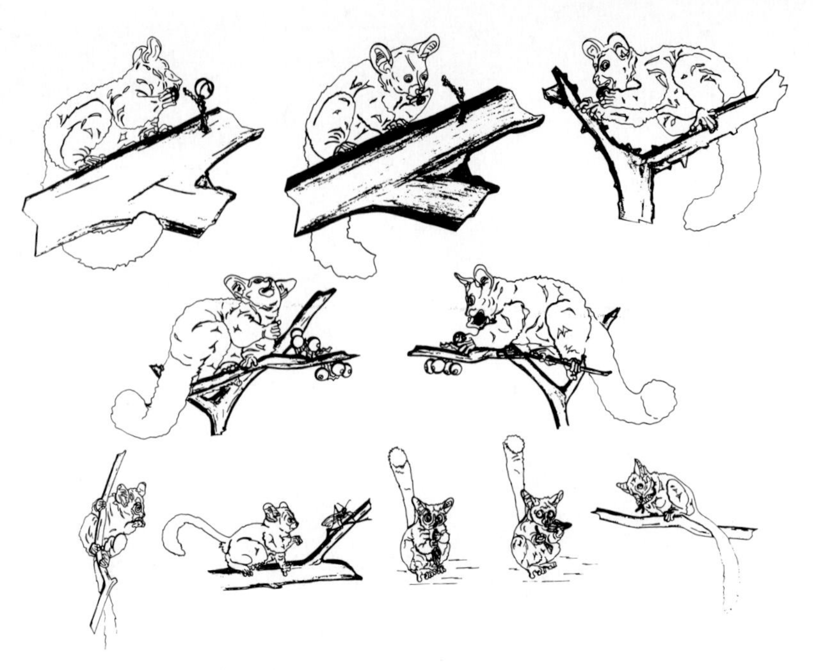

Fig. 21. Drawings from photographs, showing the use of the hands in *G. crassicaudatus* while feeding on wild figs (1–5), and in *G. senegalensis* while feeding on a locust (remaining pictures).

3.6.2. Galago senegalensis

The use of gum by *G. senegalensis* is extremely similar to that described for *G. crassicaudatus*, and the gum-licking postures are almost identical (Fig. 22). Insects are always caught in the hands with a stereotyped prey-catching strike which *G. crassicaudatus* lack (Bishop, 1964). Flying insects can be grabbed while gripping a branch with the feet, irrespective of whether the prey is above, below, or at right angles to the support. Once caught the prey is often carried in the mouth to a convenient fork before being eaten (Fig. 21). When preying on ground-living insects, *G. senegalensis* usually sit on a low branch staring at the ground. They jump down intermittently to grab a food item and immediately return to the tree to eat it; but, as noted earlier, the lesser bushbaby may spend up to 15 minutes searching for food on the ground. Haddow and Ellice (1964) note that ground-living insects were found in two-thirds of the lesser bushbaby stomachs which they examined and that they made up a considerable part of the total bulk of food.

When looking for food within a tree a bushbaby flits from one branch to another, feeding briefly here and there and moving quickly from one tree to the next. Alternatively, it moves directly to a certain tree which is used repeatedly as a source of food. *Acacia karroo* thickets are particularly attractive to lesser bushbabies, which

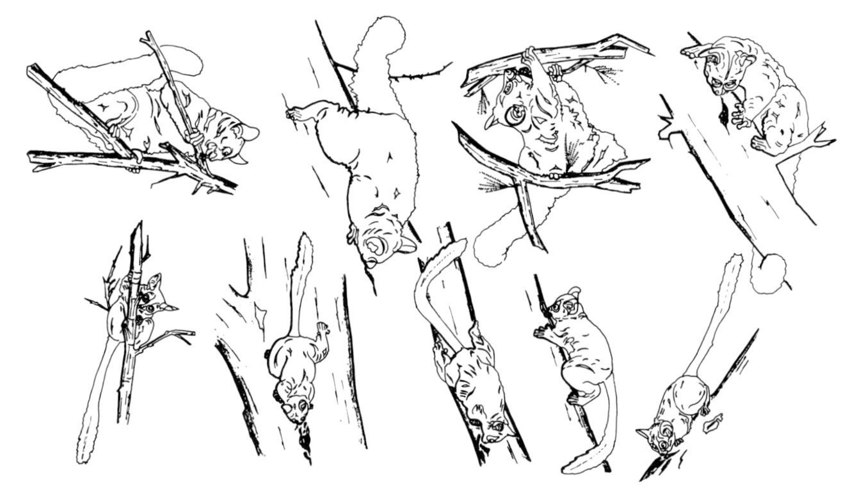

Fig. 22. Drawings from photographs showing the similarities in gum-licking postures of *G. crassicaudatus* (above) and *G. senegalensis* (below).

feed both on gum and on the numerous insects which occur there during the summer. At this time, a succession of flowering trees also attract bushbabies to feed on the insects, to the extent that they may spend up to 2 hours in a single tree. As in the case of *G. crassicaudatus*, lesser bushbabies are quickly attracted to artificial food sources, aided no doubt by the sense of smell. By contrast, observations in captivity suggest that vision and hearing are equally important aids to finding food (Lowther, 1939).

3.7. Social Structure

The relatively open conditions encountered during the study of the South African galagines facilitated regular observations of a few familiar individuals, which, in the absence of a large-scale marking program, provided the basis for an interpretation of their social structure. *Galago crassicaudatus* and *G. senegalensis* conform to the typical nocturnal prosimian pattern of a solitary or dispersed social system which is described in detail elsewhere (see, for example, Charles-Dominique and Hladik, 1971; Charles-Dominique, 1972; Martin, 1972b). The social constraints on each species are to some extent a reflection of other adaptations set by the demands of the environment. Despite extensive similarities, the South African species show a clear divergence in social structure which may be interpreted against the background of differences in ecological characteristics. Such similarities and differences are illustrated here through a general consideration of social behavior and in particular the development and care of the young.

3.7.1. Social Behavior

A. Jolly (unpublished paper, 1966) points out that in a captive group of *G. crassicaudatus* the social roles of individuals are less differentiated than those of diurnal lemurs and that social interactions are therefore relatively uncomplicated. In the wild, individuals regularly meet one another and move together at least for short periods, when they may benefit from many of the advantages of a gregarious way of life. Social organization involves a wide variety of behaviors which are much alike in the two bushbaby species. Despite their extremely sensitive night vision and excellent powers of accommodation (Dartnall *et al.*, 1965; Tansley, 1965; Silver, 1966), bushbabies make relatively little use of the visual channel of communication, although visual signals undoubtedly play a role (Andersson, 1969). The most important regulators of social cohesion and spacing involve physical contact, olfaction, and vocalizations, each of which is outlined below.

Groups of both species engage in regular amicable physical contact at the sleeping place (huddling) or during the active period (allogrooming, play). They are able to clean almost all parts of the body by autogrooming, yet allogrooming is performed frequently, suggesting that it serves important social, as well as biological, functions. In *G. crassicaudatus*, where close association between the mother and her offspring is greater and persists longer than in *G. senegalensis*, social grooming is more frequent and protracted. It can be considered together with social play as a mechanism to enhance intragroup harmony and it may aid friendly contact between compatible strangers. Social contact may also be agonistic in nature (lunging, cuffing, and fighting), but antagonism appears to be mediated by mutual recognition of the sex and status of a rival, largely by means of olfaction. Tension during social contact is therefore mild and long-range spacing mechanisms ensure that fighting between strangers of the same status is rare (Bearder and Doyle, 1974).

Six marking or rubbing actions have been described for the two species, which may be of significance in indirect olfactory communication. Five of these quite elaborate behavior patterns are common to both species (urine-washing, rhythmic micturition, head-and-mouth-rubbing, chest-rubbing, anogenital-rubbing). Foot-rubbing appears to be a peculiarity of *G. crassicaudatus* which is not found in related prosimians. It is generally assumed that these actions are of importance in a sexual or territorial context; their possible functions are discussed elsewhere (Andrew and Klopman, 1974; Doyle, 1974; Bearder and Doyle, 1974).

The vocal repertoires of *G. crassicaudatus* and *G. senegalensis* are alike in many respects and the few cases in which the call of one species is not matched by an equivalent call in the other can be explained by differences in the mother/offspring relationship and the nature of social contacts at night (Doyle *et al.*, 1977). Each species has discrete calls concerned with vocal advertisement, mother/offspring cohesion, and the avoidance of agonistic contact. A mixture of discrete and graded calls indicate anxiety or alarm on the part of the caller and they are frequently associated with a mobbing response (Petter and Charles-Dominique, Chapter 1).

The anxiety/alarm calls of *G. crassicaudatus* have the potential of giving a continuously varying array of mood indicators, while those of *G. senegalensis* are better adapted for mixed unit messages. The risk of ambiguity in the interpretation of a continuous series is probably offset by the fact that *G. crassicaudatus* are often close enough to their fellows to see either the caller or the stimulus which has evoked the call.

3.7.2. Social Relationships of the Developing Young

3.7.2.1. *Galago senegalensis.* Females usually give birth to twins, which are initially confined to a leafy nest, tree hollow, or similar retreat. After 10 to 14 days the mother returns to her usual pattern of leaving the sleeping place alone each evening to feed, but soon returns to her infants and carries them one at a time in her mouth (but never on her fur) to particular trees in the vicinity. She may place them together, or leave them in different trees while she forages alone. Infant *G. senegalensis* spend the majority of the night by themselves. At first they do not move in the absence of the mother, but she visits them regularly to groom or suckle and she frequently carries them from place to place. Gradually they explore within the tree in which they are left and the frequency of the female's visits decreases accordingly. After approximately 1 month, the young are able to move between trees of their own accord. They then follow the mother away from the sleeping place, maintaining contact through vocalizations, until they reach a well-used group of trees where they spend the rest of the night alone. The mother never fails to retrieve the infants before dawn and she sleeps with them in a huddle during the day.

After 6 weeks, the youngsters leave the sleeping place without the mother and begin independent movements over an ever-increasing range which coincides with their developing physical abilities. They may continue to join the mother in the morning before moving to a sleeping site, and the use of new sleeping trees in different parts of the range no doubt ensures that the movements of the young become more widespread. Sleeping contact with the mother decreases with time, but the subsequent dispersion of maturing individuals remains to be clarified.

Males play no part in the rearing of the young, and infants are at first protected from their attentions by the mother. The relationship between males and juveniles nevertheless remains amicable and a male will occasionally sleep with the maternal group. Small sleeping groups comprising up to seven individuals have been found in all regions where this species has been studied (Sauer and Sauer, 1963; Buettner-Janusch, 1964; Haddow and Ellice, 1964; Bearder and Doyle, 1974). Fully adult males are rarely if ever found in the same group and they frequently sleep alone (Haddow and Ellice, 1964). Sleeping groups split up soon after dark and each animal will spend, on average, 70% of the night alone (Bearder and Doyle, 1974).

3.7.2.2. *Galago crassicaudatus.* One to three infants (generally twins) are born to *G.c. umbrosus*. They are hidden in a leafy nest or tangle of creepers for approximately the first two weeks of life, although they may be transported separately in the mother's mouth from one such place to another. Subsequently, even when the infants are too small to follow the mother (± 25 days), she will carry them with her at night in her mouth or clinging to the fur of her back, or both. The young are able to follow the mother for short distances and to lick gum at an early age (± 30 days), but they continue to be carried and suckled until approximately 10 weeks old. Only occasionally are they left alone. This close association with the mother continues until the offspring are at least subadult (± 300 days), after which they often move alone at night.

Within the isolated population of the main study area, an adult male was seen in the company of a single maternal group during 20% of a total observation period of 411 hours, whereas the female and her infants were together for 90% of that time. Likewise, the male slept with the group, or in an adjacent tree, on 32% of the days on which these animals were found, while the female invariably slept with at least one of her offspring. Sleeping groups in this species are comparable to those of *G. senegalensis*, usually numbering up to 6 individuals (A. Jolly, unpublished paper, 1966; Bearder and Doyle, 1974), although as many as 9 *G.c. panganiensis* have been observed leaving a single sleeping place in East Africa (Haddow and Ellice, 1964).

It is in the structure and functioning of the maternal group at night that *G. crassicaudatus* show important differences from the other lorisids. It has been mentioned that after a period of oral transport alone, the infant *G. crassicaudatus* may also be carried while gripping onto the fur of the mother's back. Riding on the mother was originally reported for this species by Haagner (1920) and Shortridge (1934), although subsequently it was not generally accepted. Buettner-Janusch (1964), for example, considered it unlikely that the infants would ride while their mothers leaped through the trees, despite the fact that he observed infant *G.c. argentatus* clinging to the mother in captivity. In fact, wild *G. crassicaudatus* seldom leap through the trees, although the females may indeed make small jumps while carrying their infants on the fur.

With the exception of *Galago elegantulus*, fur-clinging in *G. crassicaudatus* seems to be unique within the subfamily Galaginae, although it is characteristic among the Lorisinae. Lorisines, however, normally have a single infant which (at least initially) clings to the mother's belly and is not picked up or carried in the mouth. *Galago crassicaudatus*, on the other hand, usually have two infants and sometimes even three, which cling to the fur on the mother's back or are carried in her mouth with approximately equal frequency. Infants of this species have not been observed to cling to their mother's belly in the wild. It therefore seems likely that this method of infant transport has developed from the pattern of oral transport observed with other galagine species. This development may well represent an adaptation linked to slower movement, in association with the large adult body size (*G. crassicaudatus* infants are relatively tiny in relation to the mother's body size, as compared to those of *G. senegalensis*).

4. GENERAL DISCUSSION

Taken as a group, the lorisids occupy a vast range of distribution, covering a wide variety of different forest conditions. The ecological niches involved, corresponding to "small climbing and leaping arboreal and nocturnal mammals with a basically insectivorous diet," are occupied by mammals of some kind in virtually all of the intertropical regions of the world. In the Old World, the lorisids occupy these niches; in Madagascar, it is the cheirogaleids; in South America it is the didelphid marsupials; and in Australasia it is the phalangerid marsupials.

The absence of galagines from Asia poses a problem that is discussed in Section 1 of this chapter. As far as the two lorisine species of this area are concerned, apart from a short field study of *Loris tardigradus* conducted in Ceylon (Petter and Hladik, 1970), little is known of their ecology. Morphologically, *Nycticebus coucang* (the slow loris) is very similar to the potto, while *Loris tardigradus* (the slender loris) resembles the angwantibo. On the other hand, chromosomal characters tend to group the slender loris with the potto and the slow loris with the angwantibo (de Boer, 1973b). It seems likely, in fact, that the slender forms (angwantibo and slender loris) and the robust forms represent individual adaptations acquired secondarily in correspondence to the specific ecological niches exploited. The slender type occupies fine, tangled lianes, while the robust type occurs on branches and in dense foliage. Indeed, among the South African galagines, it would seem that *Galago crassicaudatus* have similarly evolved toward an ecological niche in some ways paralleling that of the robust-bodied lorisines.

The slow-climbing habit characteristic of the lorisines, and to some extent of *G. crassicaudatus*, can be considered as a cryptic strategy, which is adapted for avoidance of the attention of predators (see Charles-Dominique, 1971b). This strategy, which is diametrically opposed to that adopted by the majority of the galagine species (excluding *G. crassicaudatus*), has had far-reaching effects on the behavior of lorisine species:

1. Specialization for capture of relatively unpalatable prey (noxious or irritant forms) which are easily detected with a minimum of locomotor activity.
2. Absence of "matriarchies." Each adult female exploits in isolation a home range shared only, to some extent, with a male. The food supply is consequently more concentrated, and locomotor requirements for foraging are hence even further reduced.
3. Vocal exchanges reduced to a strict minimum, with the particular exclusion of alarm calls, thus permitting discreet activity.
4. Social interactions more indirect than with the galagines, with the main emphasis placed on urine-marking and other scent-marking.
5. Transport of the infant on the mother's fur, first ventrally and then dorsally. Baby-parking is practiced, but not so frequently as with many galagine species. Hence, the mother does not need to move very far to retrieve her infant in the morning before returning to the sleeping site.

6. No nests are used. The infant is born at a more advanced stage of development than with the galagines, and this permits more effective clinging to the mother from birth onward. Since the mother is not attached to a particular nest site for sleeping during the day, at dawn she simply moves to the nearest patch of dense vegetation. Once again, locomotion is kept to a minimum.

The evolutionary pathway followed by the lorisines has influenced morphology primarily in the structure of the hands and feet and in the reduction of the tail.

An analogous evolutionary development has occurred among the cheirogaleid lemurs: *Microcebus coquereli* are a rapid leaping form, *Microcebus murinus* are primarily a branch-running form, though leaping also occurs from time to time, and *Cheirogaleus* species show very little leaping, generally adopting the locomotor style of the lorisines. This, of course, is a case of convergence between *Cheirogaleus* and the lorisines, which is found to a lesser degree with *G. crassicaudatus* as well. In contrast to the lesser bushbaby, the thick-tailed bushbaby is linked to relatively dense forest environments, outside of which a cryptic strategy is ineffective.

It is in Africa that most of the lorisid species are found (8 of the 10 species included in the family), and it has been seen how different the ecological conditions of the rain forest are from those of bordering tropical and subtropical zones.

The relatively stable climatic conditions of the rain forest have permitted diversification of the lorisid species which occur there, each adapted to precise dietary features regularly observable throughout the year. This diversity of species, which is also found with numerous other orders of mammals and birds, permits more effective exploitation of the environmental resources. It is, of course, accompanied by reduction in population densities of the individual species. The total biomass of the five sympatric lorisid species in Gabon is approximately equivalent to that of *Galago senegalensis* (or *G. senegalensis* and *G. crassicaudatus* combined in areas where they are sympatric) in the various ecosystems studied in South Africa (see Hladik, Chapter 8). In the relatively dry forest and wooded savanna regions bordering the rain forest, the lorisids (usually represented by only one species in each ecosystem) must adapt themselves to marked annual variations in climate and diet. These ecological conditions limit species diversification and select, for each ecosystem, one (or at the most two) species with lesser dietary specialization but doubtless with better physiological adaptation for such marked environmental fluctuation. It seems likely that the various subspecies occupying the immense range of distribution of these two bushbaby species each represent populations adapted to the peculiarities of various ecosystems. Cytogenetic studies conducted to date (de Boer, 1973a) indicate marked variability at what is regarded as the subspecific level, and this may well be paralleled by behavioral diversity. Both cytogenetic studies (de Boer, 1973a, b) and comparative investigations of vocalizations (see Petter and Charles-Dominique, Chapter 1) indicate that there is a fairly close evolutionary relationship between *G. senegalensis* and *G. crassicaudatus*. In addition, there is general agreement among various authors that these two species are derived from an ancestral bushbaby originating in the rain forest, perhaps already specialized to some extent for relatively dry conditions.

Although some lorisid species remain to be studied in detail (*Loris tardigradus*, *Nycticebus coucang*, *Galago inustus*), seven out of the ten have already been subjected to intensive study in the field in equatorial rain forest and in surrounding dry forest areas of Africa. It is hence possible to review the family Lorisidae as a whole, at least provisionally, to establish a list of essential behavioral characters and to distinguish the following classes of characters: "primitive placental mammal characters"; "primate characters"; "strepsirhine characters"; and "lorisid characters."

4.1. Placental Mammal Characters

4.1.1. Urine-Marking

This mode of social communication is found in rodents, lagomorphs, carnivores, insectivores, tree shrews, numerous artiodactyls, etc. It is doubtless an archaic character related to the considerable development of olfaction among the earliest mammals. Urine-marking is not practiced as a major behavioral feature among the Old World simians (monkeys and apes) or by certain lemur species (lemurids and *Phaner*), but it is important among the remaining prosimian primates and among the New World monkeys. The pattern of urine-washing, a particular form of urine-marking, is found in some lorisid species, in *Microcebus murinus*, and in at least two New World monkey species (*Saimiri sciureus* and *Aotus trivirgatus*). It is discussed in some detail by Andrew and Klopman (1974). Urine-washing has been interpreted in different ways by various authors: as an anti-slip device for better adhesion to branches (Welker, 1973); as a device for leaving urine trails for location within the home range (Eibl-Eibesfeldt, 1953, Sauer and Sauer, 1963); and as a marking system related to arboreal habits permitting maximum dissemination of urine marks (Charles-Dominique, 1977). In a three-dimensional environment, conspecifics have a greater probability of encountering a scent mark laid as a trail than of finding an isolated, heavily marked spot. In view of its probable association with arboreal habits, urine-washing is possibly a "primitive primate character" that has been retained by various species which still practice urine-marking.

4.1.2. Solitary Life

The system of social organization found, with minor variations, in all lorisid species, is based on the relationships between individual home ranges. In all cases there is relatively little overlap between the home ranges of individual males, whereas the home range of a single male may overlap extensively with the range(s) of one or more females. Among females, there is some variation according to the species, passing from relatively little overlap to almost complete overlap of ranges of females belonging to a matriarchy. This pattern of home ranges is found with many different placental mammal species: rodents, insectivores, carnivores, ruminants, pholidotes, edentates, etc., and it is particularly evident in the more primitive families of each mammalian order. Thus, this simple system of social organization probably represents an archaic mammalian feature.

4.1.3. Oral Transport of the Infant

Most placental mammals exhibit well-developed maternal care. Oral transport of infants is usually an integral part, particularly with nest-living species: rodents, carnivores, insectivores (during the first days of life, at least; Fons, 1975), galagines, cheirogaleids, *Lepilemur*, *Hapalemur griseus*, *Varecia variegata* (*Lemur variegatus*) (Petter, 1962b), *Tarsius spectrum* (Niemitz, 1975), *Colobus verus* (Booth, 1957). However, in contrast to terrestrial mammal species, which grasp the infant by the skin of the neck region, the prosimians which exhibit oral transport usually grasp the infant by the flank. With the burden neatly balanced in this way, the mother may carry the infant, with her head raised, through the branches without risk of injuring it while moving. Grasping by the flank might therefore be considered as a secondary adaptation to arboreal life, derived from the more primitive mammalian pattern of transport by grasping the skin of the neck or back. Sprankel (cited by Sauer, 1967) considers oral transport as a recent acquisition of the primates which exhibit the pattern, but Sauer (1967) and Martin (1972b) interpret this as a primitive retention. In view of the widespread occurrence of oral transport of infants among nest-living placental mammals, the view expressed by Sauer and by Martin would seem to be more reasonable.

4.1.4. Nest Building

Construction of a nest is correlated with the stage of development exhibited by the infants at birth. In this respect, the galagines constitute a stage intermediate between nest-building mammals and those which do not use nests. Infant cheirogaleids are born at a stage of development comparable to that of infant galagines, and this fact, in association with a number of other observations, supports the interpretation that galagines and cheirogaleids have retained many primitive characters (Charles-Dominique and Martin, 1970). This question has been considered in detail by Martin (1972b). A large number of arboreal mammal species construct nests: murids, glirids, sciurids, galagines, cheirogaleids, *Daubentonia*, *Varecia*. However, the structure of the rodent nest differs from that of lorisids and lemurs (see Section 4.3.2).

4.1.5. Insectivorous Diet

All lorisid species include insects in their diet, even though the largest-bodied forms also eat large quantities of fruit or gums. In fact, Beerten-Joly *et al*. (1974) have demonstrated that chitinase is just as abundant in the digestive tract of the potto as in that of the mole or hedgehog. Since it is generally accepted that the earliest mammals were at least partly insectivorous, the dietary regime of the lorisids, like that of the cheirogaleids, may be considered a primitive feature (Charles-Dominique and Martin, 1970).

4.2. Primate Characters

4.2.1. Arboreal Life and Prehension

Like a number of other mammal groups, the primates have become specialized for arboreal life, but they have adopted, in association with this, highly developed patterns of hand use. However, the prosimians and simians have followed two different evolutionary pathways with respect to the morphology and neuromuscular control of the hand (Bishop, 1964; see Section 4.3.4).

4.2.2. Reproduction

Two strategies are generally adopted among animals in response to predation pressures. Either there is production of a large number of young in a relatively immature state, with only a few escaping predation to reach adulthood (r-strategy), or there is production of a small number of well-developed offspring which benefit from special forms of protection which shield them fairly effectively from predators (K-strategy). The primates are among those mammals which have adopted the K-strategy, but the galagines (and the cheirogaleids) are relatively less advanced in this respect than are other primates. Although the number of infants in the average litter is small (*Galago demidovii*: 1.2; *G. alleni*: 1.2; *G. elegantulus*: 1.0; *G. senegalensis*: 1.8; *G. crassicaudatus*: 2.0; all lorisines: 1.0), a young bushbaby barely has its eyes open at birth, and independent locomotion in the branches does not occur until the age of 2 to 3 weeks. It may be concluded that the galagines are already embarked upon the path leading to K-strategy as developed in other primate species, but that their system of social organization is incompatible with a state of great independence of the young in the first two weeks of life.

4.2.3. Transport of the Infant Attached to the Mother

This mode of infant transport is a virtually universal feature of the primates; it is lacking only among the Galaginae, with one or two exceptions; viz. *G. crassicaudatus* (Bearder, 1975) and *G. elegantulus* (Charles-Dominique, 1977), and among the cheirogaleids. It should be noted that in those bushbabies which do carry the infant on the fur, such carriage only commences when the infants have achieved independent locomotion. Prior to that, oral transport of the infant occurs. Since the two bushbabies concerned are the two largest forms in the subfamily Galaginae, there is probably an association between this feature and body size. Among the Lorisinae, only transport on the mother's fur (ventrally or dorsally) occurs. As has been pointed out above, this behavior is advantageous for species which are obliged to keep their locomotion to a strict minimum.

4.3. Strepsirhine Characters

4.3.1. Baby Parking

The term baby parking is used for the pattern of deposition of the infant in vegetation during the mother's nocturnal activity period, involving removal of the infant from the nest where a nest is used (Galaginae). This behavioral adaptation permits the mother to keep within a safe distance of the still vulnerable infant, which remains immobile at the site of parking, while she exploits a relatively large home range. When the mother moves on to feed elsewhere, she picks up her infant again and deposits it closer to the new feeding area. If any disturbance occurs in the surroundings, she quickly returns to retrieve the infant. Such baby parking is practiced by the galagines, the lorisines (to a lesser extent), and by several Madagascar lemur species: *Microcebus murinus* (Martin, 1972a); *Lepilemur* spp.; *Hapalemur griseus*; *Varecia variegata* (Petter *et al.*, 1977).

4.3.2. Nests

The structure of the nests built by various galagine species is peculiar to this lorisid subfamily and to certain nocturnal Madagascar lemur species: *Microcebus murinus* (Petter, 1962b; Martin, 1972a, b), *Microcebus coquereli* (Petter *et al.*, 1971), and *Daubentonia madagascariensis* (Petter and Petter-Rousseaux, 1967). In its characteristic form, the nest consists of interwoven twigs and leaves, forming a ball with a lateral opening. The materials are stripped from trees by the mouth and incorporated into the nest as such with the aid of the hands. The technique involved is quite different from that used by various arboreal rodent species.

In the southern part of their range, both *Galago senegalensis* and *G. crassicaudatus* merely construct a flat platform of leaves on which they rest, and protection above is provided by dense foliage or thorns. In some cases (e.g., in the Northern Transvaal), *G. senegalensis* do not even construct a nest and simply sleep in a thorny tree or in a hollow trunk.

4.3.3. The Tooth Scraper

In numerous mammalian groups, the anterior teeth of the lower jaw have been modified to form a comb, and this is doubtless the result of multiple convergence to produce similar devices in different ways. In the strepsirhine prosimians (lemurs and lorisids), the tooth scraper (or toothcomb) is characteristically formed by the lower incisors and canines, and the anterior lower premolars typically have a caniniform aspect. This device, completed by a cornified sublingua which is used to clean it, is used for grooming in all species and in the collection of gums and sometimes fruit fragments by most omnivorous nocturnal species. (Diurnal lemur species may use the tooth scraper for gouging fruit and for breaking away pieces of bark for ingestion.) Buettner-Janusch and Andrew (1962) have interpreted the formation of the tooth scraper as a special evolutionary adaptation of the strepsirhine primates for social behavior, involving social grooming with a comblike action. However, Martin (1972b), drawing upon various observations under natural conditions showing that the tooth scraper plays a role in feeding, regards the primary function of the tooth scraper in evolution as that of assisting feeding (particularly in obtaining gums), with the grooming function following as a secondary adaptation. Whatever the correct interpretation may be, the utilization (and morphology) of the tooth scraper is entirely comparable between the lorisids and those Madagascar lemurs which possess the typical 6-tooth scraper. Hence, there was probably a single evolutionary origin for the tooth scraper in the strepsirhine primates.

4.3.4. Morphology and Neuromuscular Control of the Hand

With the exception of the obvious secondary specializations of the lorisines and of *Daubentonia*, the structure of the hand is essentially the same in lemurs and lorisids. The strepsirhine hand is distinct from that of the simian primates in that the fingers are long and slender and that it is primarily the terminal phalanges of the strepsirhine hand which operate in prehension (Bishop, 1964). The terminal phalange of each digit is broad and flattened, with the skin bearing simple longitudinal dermatoglyphs (in contrast to the whorls found with simian primates). In locomotion, the terminal phalange of each digit is typically applied to the support by an inverse flexion system between the 2nd and 3rd phalanges. When an object is grasped, it is pressed against the palm of the hand by the terminal phalanges of the fingers. Accordingly, neuromuscular control of the hand is quite different from that found in simian primates. For example, in allogrooming among strepsirhine primates the hands are used only to grasp the partner's fur; there is no fine control of individual digits permitting parting of the fur as in simians.

4.3.5. The Toilet Claw

The strepsirhine foot also typically has a peculiar feature: the toilet claw on the second digit. This claw, which is in fact no more than a modified nail, is primarily used in cleaning of the ears and in scratching areas of the fur which are inaccessible to the tongue and tooth scraper. There is no difference between the lorisids and the Madagascar lemurs in the form and use of this toilet claw, though the specializations of *Daubentonia* have necessarily led to its suppression.

4.3.6. Specific Features of the Lorisids

Although possible homologies between the calls of the various lorisid species may be identified, in terms of their structure (see Petter and Charles-Dominique, Chapter 1), it is virtually impossible to draw homologies of structure with the calls of the various Madagascar lemur species. However, the lorisids and the cheirogaleids respond with analogous calls in approximately similar ways when confronted with standard stimulus situations (evocation of alarm calls, infant calls, vocal communication between males and females, etc.). Thus, it would seem that similar ecological conditions have led to the evolution of analogous calls in lorisids and cheirogaleids, whereas calls characteristic of the family Lorisidae have remained essentially identifiable throughout the family despite more recent adaptation to rather different ecological niches within the arboreal environment.

5. SUMMARY AND CONCLUSIONS

If one considers all of these various behavioral features together, it emerges that, of the total of 14, 5 represent primitive mammalian (symplesiomorph) characters, 3 represent primitive primate characters, 5 represent specialized strepsirhine characters, and only 1 is a specific lorisid feature (see also Martin, 1968). If only these behavioral characters are considered, the indication is that the Lorisidae are a relatively primitive group showing retention of many primitive placental mammal characteristics. Characters common with all other primates are relatively few in number, and this indicates relatively early separation from certain other primates (e.g., the simians). On the other hand, there are many features common to the strepsirhine prosimians in general, and the lorisids are distinguished from the Madagascar lemurs by only one conspicuous behavioral category (namely, vocalization).

This brief summary of behavioral characters is in overall accord with the analysis of comparative anatomical features (Charles-Dominique and Martin, 1970; Szalay and Katz, 1973; Hoffstetter, 1974; Szalay, 1975; Cartmill, 1975). Various recent studies have demonstrated the great similarity which exists between the Lorisidae and the Cheirogaleidae. The Cheirogaleidae, in fact, are intermediate between the Afro-Asian lorisids and the other Madagascar lemurs, despite the fact that lorisids and lemurs are generally classified as two quite different groups (e.g., infraorders). However, this intermediate quality has been interpreted in different ways by different authors:

One hypothesis is that the Galaginae and Cheirogaleidae have retained a large number of ancestral strepsirhine characters, and that this explains their close resemblance. According to this hypothesis, the evolution of the other Madagascar lemurs would have taken place from an ancestral form relatively similar to modern cheirogaleids (Charles-Dominique and Martin, 1970; Martin, 1972b, 1975; Charles-Dominique, 1975, 1977; Petter *et al.,* 1977; Hoffstetter, 1977).

An alternative hypothesis is that the lemurids, which are generally more similar to the Eocene Adapidae in certain characters (e.g., in the pattern of carotid circulation in the ear region), gave rise to the Cheirogaleidae, which in turn gave rise to forms which crossed the Mozambique Channel and gave rise to the Lorisidae (Szalay and Katz, 1973; Szalay, 1975).

At the present time, there is insufficient biological evidence, especially in terms of paleontological remains (e.g., of small-bodied adapids and representatives of more recent relatives of the strepsirhines), to reach a definitive resolution of this question. Nevertheless, it is of great importance to reach a consistent interpretation of the radiation of the strepsirhine prosimians if we are to fully understand primate evolution.

ACKNOWLEDGMENTS

Dr. P. Charles-Dominique wishes to thank his friend and colleague Dr. R. D. Martin, who undertook the translation and correction of the first section. The field studies were carried out at the Laboratoire de Primatologie et d'Ecologie Equatoriale (supported by the Centre National de la Recherche Scientifique) and depended upon the kind hospitality of the government authorities of Gabon and upon the valuable collaboration of the laboratory staff, past and present.

The second section is based on work submitted to the University of the Witwatersrand, South Africa, for the degrees of M.Sc. and Ph.D. Dr. S. Bearder wishes to thank his supervisor, Professor G. A. Doyle, for his continual support. He is grateful to many people for their hospitality and enthusiastic interest in his work, especially Mr. and Mrs. E. A. and R. G. Galpin of Naboomspruit, N. Transvaal; Mr. and Mrs. P. G. McNeil, Mr. and Mrs. R. Speedy and Mr. R. McNeil of Ofcolaco, N. E. Transvaal; Mr. and Mrs. J. Wylie of Umtali, Rhodesia, and Mr. and Mrs. R. E. Webster, Natal Parks Board, Zululand. Technical assistance and advice was willingly given by Mr. T. Bekker, Mr. R. Bérard, Mr. B. Cnoops, Mr. R. Lowe, Mr. M. B. Ormerod, Mr. N. Passmore, Mr. A. Renny, Miss J. G. Rouse, and Mrs. D. Swierstra who he sincerely thanks. The research was supported by the National Geographic Society, Washington, D.C., the Human Sciences Research Council of the Republic of South Africa, and the University Council, University of the Witwatersrand.

REFERENCES

Andersson, A. B. (1969). M.Sc. Dissertation, University of the Witwatersrand (unpublished).

Andrew, R. J., and Klopman, R. B. (1974). *In* "Prosimian Biology" (R. D. Martin, G. A. Doyle, and A. C. Walker, eds.), pp. 303–312. Duckworth, London.

Astley-Maberly, C. T. (1967). "The Game Animals of Southern Africa." Nelson, Johannesburg.

Bearder, S. K. (1969). M.Sc. Dissertation, University of the Witwatersrand (unpublished).

Bearder, S. K. (1975). Ph.D. Thesis, University of the Witwatersrand (unpublished).

Bearder, S. K., and Doyle, G. A. (1974). *In* "Prosimian Biology" (R. D. Martin, G. A. Doyle, and A. C. Walker, eds.), pp. 109–130. Duckworth, London.

Beaudenon, P. (1949). *Mammalia* **13**, 76–99.

Beerten-Joly, B., Piavaux, A., and Goffart, M. (1974). *C.R. Seances Soc. Biol. Ses. Fil.* **168**, 140–143.

Bishop, A. (1964). *In* "Evolutionary and Genetic Biology of Primates" (J. Buettner-Janusch, ed.), Vol. 2, pp. 133–225. Academic Press, New York.

Booth, A. H. (1957). *Proc. Zool. Soc. London* **129**, 421–430.

Buettner-Janusch, J. (1964). *Folia Primatol.* **2**, 93–110.

Buettner-Janusch, J., and Andrew, R. J. (1962). *Am. J. Phys. Anthropol.* **20**, 127–129.

Cansdale, G. (1960). "Bushbaby Book." Phoenix House, London.

Carpenter, C. R. (1934). *Comp. Psychol. Monogr.* **10**, 1–168.

Cartmill, M. (1975). *In* "Phylogeny of the Primates: A Multidisciplinary Approach" (W. P. Luckett and F. S. Szalay, eds.), pp. 313–354. Plenum, New York.

Charles-Dominique, P. (1971a). *Thèse de Doctorat d'Etat* CNRS No. A.O. 5816. University of Paris.

Charles-Dominique, P. (1971b). *Biol. Gabonica* **7**, 121–228.

Charles-Dominique, P. (1972). *Z. Tierpsychol., Beih.* **9**, 7–42.

Charles-Dominique, P. (1974). *Mammalia* **38**, 355–379.

Charles-Dominique, P. (1975). *In* "Phylogeny of the Primates: A Multidisciplinary Approach" (W. P. Luckett and F. S. Szalay, eds.), pp. 69–88. Plenum, New York.
Charles-Dominique, P. (1977). "Ecology and Behaviour of Nocturnal Primates." Duckworth, London.
Charles-Dominique, P., and Hladik, C. M. (1971). *Terre Vie* **25**, 3–66.
Charles-Dominique, P., and Martin, R. D. (1970). *Nature (London)* **227**, 257–260.
Dartnall, H. J. A., Arden, G. B., Ikeda, H., Luck, C. P., Rosenburg, M. E., Pedler, C. M. H., and Tansley, K. (1965). *Vision Res.* **5**, 399–424.
de Boer, L. E. M. (1972). *J. Hum. Evol.* **1,** 83–86.
de Boer, L. E. M. (1973a). *Genetica* **44**, 155–193.
de Boer, L. E. M. (1973b). *Genetica* **44**, 330–367.
Doyle, G. A. (1974). *Behav. Nonhum. Primates* **5**, 155–353.
Doyle, G. A., and Bekker, T. (1967). *Folia Primatol* **7**, 161–168.
Doyle, G. A. *et al.* (1977). Calls of the Lesser Bushbaby, *Galago senegalensis moholi,* and the Thick-tailed Bushbaby, *Galago crassicaudatus* spp., in the laboratory and in the field. (unpublished).
Ehrlich, A. (1970). *Behaviour* **37**, 54–63.
Eibl-Eibesfeldt, I. (1953). *Säugetierkd. Mitt.* **1,** 171–173.
Fons, R. (1975). *Thèse d'Université,* University of Paris VI, Paris, France.
Goodman, M. (1975). *In* "Phylogeny of the Primates: A Multidisciplinary Approach" (W. P. Luckett and F. S. Szalay, eds.), pp. 219–248. Plenum, New York.
Haagner, A. K. (1920). "South African Mammals." Witherby, London.
Haddow, A. J., and Ellice, J. M. (1964). *Trans. R. Soc. Trop. Med. Hyg.* **58**, 521–558.
Hall-Craggs, E. C. B. (1965). *J. Zool.* **147**, 20–29.
Hill, W. C. O. (1953). "Primates: Comparative Anatomy and Taxonomy. Vol. I. Strepshirhini." Univ. Press, Edinburgh.
Hoffstetter, R. (1974). *J. Hum. Evol.* **3,** 327–350.
Hoffstetter, R. (1977). *Studia Geologica* **13,** 211–253.
Jewell, P. A., and Oates, J. F. (1969). *Zool. Afr.* **4,** 231–248.
Jolly, A. (1966). Unpublished paper sent to Dr. E. G. F. Sauer.
Jones, C. (1969). *Folia Primatol.* **11,** 255–267.
Lowther, F. de L. (1939). *Zoologica (N.Y.)* **24**, 477–480.
Martin, R. D. (1968). *Man* **3**, 377–401.
Martin, R. D. (1972a). *Z. Tierpsychol., Beih.* **9**, 43–89.
Martin, R. D. (1972b). *Philos. Trans. R. Soc. London, Ser. B* **264**, 295–352.
Martin, R. D. (1975). *In* "Phylogeny of the Primates: A Multidisciplinary Approach" (W. P. Luckett and F. S. Szalay, eds.), pp. 265–297. Plenum, New York.
Moreau, R. E. (1964). *In* "African Ecology and Human Evolution" (C. Howell and F. Bourlière, eds.), pp. 28–42. Methuen, London.
Napier, J. R., and Walker, A. C. (1967). *Folia Primatol.* **6**, 204–219.
Niemitz, C. (1975). *Folia Primatol.* **462**, 1–24.
Petter, J. J. (1962a). *Terre Vie* **4**, 394–416.
Petter, J. J. (1962b). *Mem. Mus. Natl. Hist. Nat., Paris* **27**, 1–146.
Petter, J. J., and Hladik, C. M. (1970). *Mammalia* **3**, 394–409.
Petter, J. J., and Petter-Rousseaux, A. (1967). *In* "Social Communication among Primates" (S. A. Altmann, ed.), pp. 195–205. Univ. of Chicago Press, Chicago, Illinois.
Petter, J. J., Schilling, A., and Pariente, G. (1971). *Terre Vie* **25**, 287–327.
Petter, J. J., Albignac, R., and Rumpler, Y. (1977). "Faune de Madagascar: Lemuriens." ORSTOM–CNRS, Paris.
Rahaman, H., and Parthasarathy, M. D. (1970). *Myforest* **6**, 23–26.
Rahm, U. (1960). *Bull. Inst. Fr. Afr. Noire* **23**, 331–342.
Roberts, A. (1951). "The Mammals of South Africa." Central News Agency, Johannesburg.
Sanderson, I. T. (1957). "The Monkey Kingdom." Doubleday, New York.

Sauer, E. G. F. (1967). *Folia Primatol.* **7**, 127–149.
Sauer, E. G. F., and Sauer, E. M. (1963). *J. S.W. Afri. Sci. Soc.* **16**, 5–35.
Seitz, E. (1969). *Z. Tierpsychol.* **26**, 73–103.
Shortridge, G. C. (1934). "The Mammals of South West Africa," Vol. I. Heinemann, London.
Silver, P. H. (1966). *Vision Res.* **6**, 153–162.
Southern, H. N. (1955). *Sci. Am.* **193**, 88–98.
Szalay, F. S. (1975). *In* "Phylogeny of the Primates: A Multidisciplinary Approach" (W. P. Luckett and F. S. Szalay, eds.), pp. 91–125. Plenum, New York.
Szalay, F. S., and Katz, C. C. (1973). *Folia Primatol.* **19**, 88–103.
Tansley, K. (1965). "Vision in Vertebrates." Chapman & Hall, London.
Tattersall, I., and Schwartz, J. H. (1974). *Anthropol. Pap. Am. Mus. Nat. Hist.* **52**, 139–192.
Turnbull-Kemp, P. (1967). "The Leopard." Howard Timmins, Cape Town.
Vincent, F. (1969). *Thèse de Doctorat d'Etat*, CNRS No. A.O. 575. University of Paris.
Vincent, F. (1972). *Ann. Fac. Sci. Cameroun* **10**, 135–141.
Walker, A. C. (1969). E. Afr. Wildl. J. **7,** 1–5.
Walker, A. C. (1974). *In* "Prosimian Biology" (R. D. Martin, G. A. Doyle, and A. C. Walker, eds.), pp. 435–447. Duckworth, London.
Walker, E. P. (1954). "The Monkey Book." Macmillan, New York.
Welker, C. (1973). *Folia Primatol.* **20**, 429–452.

Chapter 14

Outline of the Behavior of *Tarsius bancanus*

CARSTEN NIEMITZ

1. INTRODUCTION

As the title suggests this chapter is intended as a preliminary outline of the behavior of only one of the three tarsier species, *Tarsius bancanus* (Horsfield's tarsier). But in order to provide a more general framework for the purposes of this book, it will also deal briefly with the behavior of the other two species, *T. syrichta* (Philippine tarsier) and *T. spectrum* (spectral tarsier), as a review of the somewhat

ISBN 0-12-222150-8

scanty literature. Most previous studies have been made on preserved specimens and have been morphologically oriented, for tarsiers are rarely kept in zoos.

Apart from preliminary field notes recorded by Fogden (1974), giving a very interesting first account of various aspects of the biology of tarsiers in the wild, there is no published report on the behavior of tarsiers in their natural habitat. A few earlier authors reported observations based on statements from local villagers. Only 15 research workers have thus far published observations on captive tarsiers. The earliest author to publish information on the behavior of these rare animals deserves special mention. When Cuming recorded his notes in 1838 there was very little interest in the animal itself, least of all in behavioral considerations. It was not until 86 years later that Le Gros Clark (1924) published his findings on *Tarsius bancanus* in Sarawak. Subsequently Wharton (1950) published some notes from a collecting trip to the Philippines. A modest contribution can now be added to present knowledge of the behavior of *Tarsius* from results gained in the field and under seminatural conditions in Sarawak (East Malaysia, Borneo) in the period 1971–1973. While the main ethological findings will be reported below, the actual methods used during the fieldwork and quantitative behavioral and ecological data will be published elsewhere in more detail than is possible here.

2. PHYLOGENY AND DISTRIBUTION

The current distribution of the genus extends from 99° east longitude on Sumatra to 126° east longitude on Mindanao (Philippines), and from 13° north latitude on Samar (also a Philippine island) to probably almost 11° south of the equator on the small island of Sawu between Timor and Sumba of the Lesser Sunda Islands. On the Indo-Malayan archipelago the tarsiers have survived in a geographically restricted tropical relict distribution, while all other descendants of the Tarsiiformes, formerly inhabiting North America, Europe, and Asia in the Early Tertiary, became extinct. The disappearance of descendants of the Eocene tarsiers is probably associated with their specialized adaptation to nocturnal life and their vertical clinging and leaping habit, which restrict them to an ecological niche where they were, and still are, almost without competition, but which, on the other hand, confined their evolutionary development to a narrow phylogenetic pathway.

Hill (1953b) concluded, on the basis of cutaneous specializations of the tail: "The available evidence would seem to suggest at least diphyletic origin of the tarsiers, one line culminating in a species with naked ridged tails and the other retaining the hairs in modified, specialised form, but lacking any epithelial specialisations." *Tarsius spectrum* is relatively the least specialized representative of the genus. It has scale-like areolae (Hill, 1953b) which recall the scales of reptiles or Edentata, and which the present author did not find in adult *T. bancanus* and *T. syrichta*. But Sprankel (1965) reported that such conditions are recapitulated in the ontogeny of *T. bancanus*, and this has been confirmed by the author's observations. Kiesel (1968) added: "The characters of the caudal integument of all three species of tarsier seem to be based on a common pattern, to which all cutaneous derivates agree" (author's translation). In other respects, such as in craniometrical proportions and in the musculature of the hind foot, the three species are so extremely similar that these conditions cannot be based on mere convergence. Hofer (1962) stated: "The trend of evolution leading to a tarsier-like type has been followed among Anaptomorphidae at least twice: that is to say, among Anaptomorphinae and among Pseudolorisinae" (author's translation). Although the latter subfamily, the Pseudolorisinae from Middle to Upper Eocene of Europe, does in fact show far greater similarities to *Tarsius*, no direct line can be drawn between either the Anaptomorphinae or the Pseudolorisinae and the recent tarsiers. Though fossil evidence for their origin is hence lacking, a monophyletic origin of all living tarsiers is clearly indicated, the implication being that the recent tarsiers have evolved from a rather small genetic pool, probably on Celebes. This hypothesis is also strongly supported by the comparative studies of the functional morphology of the hind foot of the Philippine tarsier and of *T. spectrum*, which demonstrate a clear geographical succession in the elongation of the hind foot from Sawu in the south to Samar in the north. Comparable results were found for the lengths of the 3rd digit of the hand, which has undergone specializations in *T. syrichta* for behavioral reasons (Niemitz, 1977).

3. METHODS

To obtain data from the wild tarsiers were trapped with 6–50 nylon mist nets used simultaneously in the manner used to catch birds for ringing. The nets used were 6–12 m long and 2.5–3.5 m high. Some of the nets were joined together in order to cover longer stretches. The trapping sites were established in secondary and primary jungle, in shrub vegetation in more or less hilly areas some 30 km inland, and also in mangrove vegetation and primary hill forest near the coast. Captured tarsiers were weighed and measured (Niemitz, 1977), ringed around the heel with an 8-mm aluminum ring, and released immediately at the site of capture. It was hoped by this means to throw light on factors such as territoriality, vertical versus horizontal distribution as well as the size of the tree tunks preferred in leaping from one vertical support to another.

In addition, a cage of roughly 7.5 × 12 × 3.5 m was constructed in vegetation on a borderline between secondary and primary jungle. The enclosed area amounted to 90 m^2 with a total capacity of 320 m^3 (Niemitz, 1972). This cage (Fig. 1) contained the trunks of 3 big trees each with a diameter of about 60 cm, 6 trunks between 8 and 14 cm in diameter, and more than 100 tree trunks between 1 and 8 cm in diameter. Trees higher than 3.5 m (about 30) emerged through the roof of the cage so that the vegetation was left undisturbed as far as possible. To achieve this, the wire netting of 2.4-cm mesh size was bound around all emergent tree trunks.

Fig. 1. Site of the cage for seminatural observation of *Tarsius bancanus*.

The cage contained a small artificial pool, a termite heap, some naturally growing bamboo, vines, and a great variety of plants. The mesh size of the fence and roof allowed a great variety of invertebrates and vertebrates to enter and leave the enclosure. Invertebrates included all kinds of leeches, snails, myriapods, chelicerates, crabs, and (especially) insects, while vertebrates included frogs, a few lizards, snakes, some tiny nectarine birds, and mice. Bigger animals such as a tortoise, an owl, a civet cat, and a slow loris (*Nycticebus coucang*) were introduced on occasion for experimental purposes. Inside this enclosure a pair of tarsiers was observed for uninterrupted periods of 72 hours, the observers changing about every 3.5 hours during the day and every 2.5 hours at night. (Variations due to subjective error of the different observers will be calculated in the course of subsequent quantitative behavioral analysis.) The behavioral records were taken on form sheets and comprised different types of locomotor activities, feeding and drinking behavior, defecation and urination, marking and comfort behavior, social and other interactions which will be discussed below. At night an infrared night-viewing device was used at first and later on torchlight which, following an habituation period of several days, did not noticeably disturb the animals (Niemitz, 1973c).

4. THE ECOSYSTEM

4.1. Habitat

Cuming (1838) stated that *T. syrichta* from Bohol and Mindanao were found "living under the roots of trees, particularly the large bamboo of these islands." A more detailed description is given by Cook (1939): "One was seen at the tip of a stalk of tibgao, a tall, strong grass . . . others were captured in vines, hemp plantings and in underbrush . . . only one was caught in a tree." When Fulton (1939) later tried to catch *T. syrichta* on Bohol, at first in vain and then successfully, he found them "in the bamboo near the seashore. It seemed that we had made the fatal error of hunting *Tarsius* in the inland jungle." A slightly different report is given by Lewis (1939) who stated: " . . . the only way to catch them was to locate a hollow tree or crevice and smoke or cut them out." These descriptions indicate a tendency of this species to look for a hollow retreat, near or even on the ground, and to move around low in the forest. They also suggest that *Tarsius syrichta* are at least more abundant in coastal areas than in the inland jungle.

A different impression is given by reports of authors who trapped *T. bancanus* in the field. Le Gros Clark (1924) wrote of the tarsiers from Borneo: "They were all found in jungle of secondary type or in recently cleared primary jungle." This statement is fully supported by Davis (1962), Harrisson (1963), and Fogden (1974). Only K. Frogner (personal communication) has stated that they occurred quite commonly in primary forest at Nanga Tekalit (Ulu Rajang = Rajang river uplands of Borneo). Most reports similarly note for this Bornean species that they are more abundant in coastal areas, but this may be due partly to the fact that most attempts to catch them were made in lowland forests near the coast. Nevertheless, the author trapped *T. bancanus borneanus* in Sarawak primarily in secondary growth, and less often in primary forest. Since it was possible to detect the smell of their urine, deposited for marking purposes (see below), in primary jungle just as often as in secondary forest, it would appear that the Bornean species need larger territories in old, established forests than in secondary growth. Fogden (1974) suggested: "It seems probable that primary forest is only a marginal habitat for tarsiers, and it is utilised mainly by young males which are unable to compete with mature adults for space in secondary forests." Now, for the reasons mentioned above, together with Dr. Frogner's report from Ulu Rajang, it seems more likely that Horsfield's tarsier is in fact not rarer but simply less densely populated in primary vegetation.

In Borneo, *Tarsius* are not restricted to the jungle but also occur in shrubs in coastal areas and near plantations. Their unmistakable scent was detected on one occasion in mangrove vegetation in Bako National Park north of Kuching (the capital of Sarawak), but the author failed to trap any tarsiers there. As is suggested by body proportions of known specimens, there might be a different ecological variety of the Bornean tarsier subspecies inhabiting shrubs and undergrowth near the coast, at least around Kuching. The measurements of Elliot's "*Tarsius saltator*" (1910) (which is neither a distinct species nor a significant subspecies; Niemitz, 1977) from the nearby island Serasen (= Sirhassen) correspond clearly to this coastal type of *T.b. borneanus*.

The type of vegetation inhabited by *T. spectrum* on Celebes is completely unknown, but from aspects of a functional morphology (Niemitz, 1977), *T. spectrum* can be expected to be less specialized for leaping between vertical supports, to climb about more than the other two species rather than jump, and to come to the ground more often than the Philippine and Horsfield's tarsiers do. The habitat of *T. spectrum* is therefore probably more diversified than that of the other tarsiers.

At the site where the author trapped *T. bancanus* with nets in Sarawak, they were caught at an average height of 91 cm above the ground ($n = 52$; S.D. = ± 64 cm). But

Fig. 2. Habitat of *Tarsius bancanus* in secondary vegetation in the Bornean jungle.

merely stating figures for the mean and standard deviation gives a misleading impression of the distribution. About two-thirds of these 52 individuals were trapped between 20 cm and 1 m above the ground; only the remaining third higher than that. Yet some of the latter were trapped as high as 2.8 m. Since some of the nets used to catch the tarsiers were 3.5 m high, there is little likelihood that many individuals move about higher in the trees for any length of time. They go above 2 m only if they feel insecure or endangered, and only a very few minutes before going to sleep in the morning when the sun is already well up. At such times they are more likely to see a net and to avoid it. Fogden (1974) wrote: "... it is probably significant that they were at an average height of nearly 3 m, which is higher than a mist net." But since the vertical distribution observed in the enclosure under seminatural conditions fully agrees with the trapping data from the present study, involving more than 90 individuals, it seems likely that the tarsiers observed by Fogden in the wild were in a state of insecurity. Tarsiers which were trapped in the mist nets and later released at night leaped away at about 1.5 m above ground level to disappear in the darkness. If released during the daytime, however, they swiftly moved up to 4 m or higher, probably for the reason outlined above.

The average distance between the vertical tree trunks in the secondary habitat concerned was about 60 cm (Fig. 2). Yet these vertical clingers and leapers usually cover distances of 100–120 cm in single leaps when moving around in their home ranges. Of 44 measured leaps recorded when catching tarsiers, 14 (approx. 32%) were 1.8 m or longer. Only on rare occasions were tarsiers seen to jump farther than 4 m on a horizontal plane. One leap covered a distance of 5.6 m, and on another occasion a highly threatened individual jumped about 6.3 m, in this case losing considerable height between takeoff and landing. These observations certainly

Trees		
Fagrea racemosa	Tembusu gajan	Type of shrub
Cratoxylum cochinchinense	Patok tilan	Hardwood tree
Ficus bruneiensis	Tempan	Wild fig tree
Saurauja glabra	Mata ikan	Wild fruit tree
Alstonia angustifolia	Mergalang	Tree
Macaranga javanica	Benua	Euphorbia-type tree
Macaranga gigantea	Merkubong	Euphorbia-type tree
Other plants		
Selaginella lobbii		Type of fern
Lycopodium cernum		
Nepenthes rafflesiana		Pitcher plant types —woody vines or climbers
Nepenthes ampullaria		
Nepenthes gracilis		
Blechnum orientale (which occurred only in clearings and less dense vegetation)		Perennial fern

demonstrate that a tarsier is able in some leaps to cover roughly 45 times the average head-and-body-length of about 13 cm, and can effect a leap of at least 30 times head-and-body-length without losing height. The locomotion of *T. bancanus* was described and analyzed by Niemitz (1974), Niemitz and Niemitz (1974), and by Preuschoft *et al*. (in press).

A list of the Latin names of plants typical of the habitat, together with their common or descriptive names where known, providing a brief description of the vegetation is shown on p. 637.

4.2. Synecological Relationships

A consideration of the animals with which *Tarsius* share their living space, and which have an influence on their life or survival, is important for understanding their ecology. Starting with the predators there is only one comment to be found in the literature with respect to *T. syrichta*. Hill *et al*. (1951) introduced a snake in a glass container into a cage where tarsiers of this species were kept. They found that "the tarsiers did not appear fearful of it, maintaining regular feeding habits and coming down to the floor of the cage in normal fashion." The present author confirmed this observation for *T. bancanus*, which (under seminatural conditions) did not usually respond in any way to snakes in their environment. In one case, however, a tarsier caught a poisonous snake (*Maticora intestinalis*), which was about 30 cm long, and ate part of it, starting with the head, as is usual with animal prey, biting it behind the occiput (Fig. 3). As the snake started to coil, the tarsier had to wrestle with it for more than 15 seconds and then jumped to a perch to eat it, with the snake still moving vigorously in its hand. Capture of such prey species required much skill and certainty of immediate killing, especially in the darkness of the jungle night, since the snake is highly neurotoxic (for further information, see Niemitz, 1973a).

In general, it seems that Horsfield's tarsiers do not appear to recognize potential predators. An owl shown to a tarsier in the jungle under seminatural conditions did not evoke any special response, nor did a young civet cat. One slow loris (*Nycticebus coucang*) released in the big cage simply caused the tarsier to leap away one or two jumps when it came too near. The latter obviously forgot about the incident at once, as could easily be seen when the loris followed the tarsier. The loris, which was very hungry, actually tried to catch the tarsier, apparently in order to eat it. While the latter was noisily absorbed in eating a beetle, the slow loris crept down the thin tree trunk so stealthily that the author was forced to intercede at the very last second by vigorously kicking the tree trunk, causing the tarsier to fall to the ground.

Apart from big owls, snakes, possibly civets and lorises, parasites are a natural source of danger to the tarsiers. Hill *et al*. (1951) give an extensive account of the endoparasites they found in *T. syrichta*. Besides the intestinal protozoans *Chilomastix tarsii* and *Eimeria* sp., they reported *Trichomonas* of the *hominis* type in the female tarsier as well as *Trichomonas* (*fetus* type) from its vagina. The intestinal

Fig. 3. A *Tarsius bancanus* from Borneo eating a neurotoxic snake (*Maticora intestinalis*).

parasites they found in *T. syrichta* are: Cestodes: *Hymenolepis* (of the *nana* group); Nematodes: *Trichuris* sp., *Enterobius* (*Oxyuris*) sp., and hookworm, nearer *Necator* than *Ancylostoma*.

In *T. bancanus* from Borneo the following intestinal parasites were discovered: Cestodes: *Hymenolepis* (of the *nana* group); Nematodes: hookworm larva (not yet identified), hookworm ova (not yet identified), *Trichuris* ova (not yet identified), *Oxyuris* sp.

Since the material concerned was recovered and preserved, it is hoped to be able to recover the cestodes and the adult hookworms of *Trichuris* specimens themselves. Creplin (1846) had found and named *Filaria laevis* in the subcutaneous tissue and fat of the interscapular region of *T. spectrum*. Catchpole and Fulton (1943) sacrificed their *T. syrichta* since it suffered from a severe filariasis. To the above list of parasites of *T. bancanus* must be added *Filaria* sp. which was found by the author in the same region as described by Creplin. The similarity of the range of parasites in the various tarsier species yields another argument for the monophyletic origin of the recent species proposed above. With respect to ectoparasites, mites occur on *Tarsius* from Borneo inside the outer ear, on the throat, near the epigastric gland, and on the scrotal skin. But they cannot be described in more detail, as the individuals examined in the Kuching area were all free of them. Only once was a tick (*Ixodes* sp.) found on an immature female, which was also heavily infested with intestinal parasites and appeared to be rather weak. Identification of these parasites (pending) might further elucidate the phylogenetic relationships of the recent tarsier species.

The next category of animals to be considered consists of those which live in competition with the tarsiers. Nothing has yet been published in the literature on this topic. *Tarsius* show striking convergence with owls: big eyes lacking a *tapetum lucidum* and a head which can rotate through 180° (Fig. 4). Three other sim-

Fig. 4. A *Tarsius bancanus* in its typical position before leaping. Its head is turned backwards toward the tree trunk to which it is about to jump.

ilarities are conspicuous: (1) the general correspondence in the lists of animals eaten between *Tarsius* and small owls in the same habitat; (2) completely noiseless locomotion; and (3) ambush-type predation by moving about at nighttime above ground level. These similarities suggest that small owls represent the only potential source of competition in relatively sparse vegetation. Where the undergrowth is too dense for flying, the tarsier probably replaces the owl completely, occupying the ecological niche virtually without competition.

4.3. Diet and Feeding Behavior

The Bornean tarsier is insectivorous/carnivorous, and it is generally agreed in the literature that the diet is not supplemented by any plant material. In the course of more than 1000 hours of observations of tarsiers under seminatural conditions in the present study they were not seen to eat any plant products offered, though gums, ferns, mushrooms, and many monocotyledonous and dicotyledonous plants were present. Under seminatural conditions one female on Borneo was once observed to bite on the mycelium of a fungus, but apparently did not devour any of it. A very few asci of *Ascomycetales* were discovered in the tarsier feces, but this does not mean that tarsiers eat fungus. Since fungi occur simply everywhere, especially in this humid type of habitat, it would have been peculiar if they had not been found in the feces. They might in any case have originated from the intestines of a prey animal which had been eaten. H. Sprankel conducted experiments (unpublished) in which he offered specially selected plants to captive tarsiers. They bit into certain leaves but never ate any part of them. There appears to be little doubt then that *Tarsius bancanus* do not include any vegetable matter in their diet.

The main part of the diet, in terms of numbers of items eaten, certainly consists of insects. All kinds of beetles and grasshoppers serve as food, even if they are able to bite strongly. The same is true for cockroaches. *Tarsius b. borneanus* also eat butterflies, moths, praying mantis and, occasionally, ants. Phasmids are eagerly devoured if they are not too spiny, and cicadas are consumed without hesitation unless they are protected by a pungent odor. All relevant observations, including those below, were conducted under seminatural conditions in the jungle, and included observations on how strongly tarsiers responded to a particular stimulus after having eaten other different kinds of food to satiation prior to the test.

Tarsius bancanus also eat vertebrates. The snake *Maticora intestinalis* has already been mentioned above. The tarsiers observed also caught birds in accordance with Harrisson's observations (1963). They killed and ate not only spider-hunters, warblers, and the fast-flying kingfishers, but even pittas. One large pitta, which was much bigger than the tarsier itself, was killed by an adult female. It took the tarsier more than an hour to crush the skull and eat the head and brain. With striking regularity, the tarsiers ate all the feathers, except those which accidentally fell to the ground, cracked and ate the beak, and sometimes also the feet of the bird. If they were not hungry enough to eat the entire bird (sample weight 13 gm), they invariably started eating the head first and eventually dropped the remainder before finishing.

At this point, the predation pattern of tarsiers should be described. Bornean tarsiers, as observed and filmed in seminatural conditions (Niemitz, 1973c), locate their prey primarily by sound and only secondarily by sight, before approaching and leaping to attack. While killing the animal with a few powerful bites, gustatory control is exercised. *Tarsius bancanus* catch their prey with their eyes closed, usually grasping the prey with one hand and fixing it with the other (H. Sprankel, personal communication; C. Niemitz, personal observation). Variations in this pattern are mainly dependent on the size of the victim (Niemitz, 1973b). The only mammals which some of the tarsiers, at the site at Semongok, caught and ate were bats. But some of them did not like this food and even refused to take them when offered. In general, there seems to be considerable individual variation in the food accepted. The bats eaten were a tiny horseshoe bat (*Taphozus* sp.) and especially two very common fruitbats: *Cynopterus brachyotis* and *Balionycteris maculata*. Mice which lived in, or sometimes entered, the cage were occasionally watched but never caught. A small fish, of about 4 cm total length, which had been caught in the stream nearby and introduced into the small pool in the tarsiers' enclosure, was looked at once in a while but went untouched for more than 4 weeks, though a tarsier drank from the pool almost every night. When this fish was eventually fed to a tarsier, however, it was eaten completely. This is interesting in conjunction with Cook's observation (1939) on *T. syrichta*: "When live shrimp and fish were put in the cage in water the tarsiers showed no hesitation in reaching for them in the water, or scooping them out first and then seizing them." Small freshwater crabs often come to dry land in the Bornean jungle as a result of floods. A tarsier in semicaptivity, on seeing such a crab, jumped closer and looked at it, but neither tried to kill it nor succeeded in ignoring it completely. Possibly this tarsier was inhibited from approaching closer as a result of an unhappy previous experience.

Toads, frogs, and snails are not included in the range of animals which are eaten. A snail which was touched accidentally by a tarsier produced long and intensive reflex shaking "in disgust," a basic reflex common to most mammals, including man.

Fulton (1939) and Ulmer (1963) both report for *T. syrichta* that they eat about one-tenth of their body weight per night. In the present study it was not possible to determine exactly how much *T. bancanus* eat in grams per night, but the figure quoted is probably approximately correct for this species in captivity as well. It is to some extent doubtful whether insects form the main mass of food consumed. All vertebrate prey together might perhaps weigh as much as all the invertebrates together, though the latter are of course eaten in much greater number. On the other hand, it was not possible to confirm Fogden's statement that, in addition to insects, Horsfield's tarsier eats "other small animals in the wild, particularly lizards." One tarsier was once seen to catch a flying lizard (*Draco draco*) and to eat its tiny head. It then dropped the carcase and shortly afterward captured a beetle, indicating that at least some individuals prefer insects to lizards. Others clearly preferred birds to lizards.

Tarsius bancanus from Borneo, under seminatural conditions, live exclusively by predation and take only live food. If a Bornean tarsier happens to drop food which it has already killed, it will not climb down to retrieve and finish eating it. However, an adult male in captivity, which was sick and ate pathologically large amounts, learned to take suitable dead animal food in the following stages: (1) during the first few days, it let insects and pieces of bird drop after biting them in defense when offered in the hand of the keeper; (2) from about day 5 until about day 10, it took the food it had bitten in defense, if the hand presenting the food was withdrawn fast enough; (3) from then until about 2 1/2–3 weeks it leaned toward the approaching food, sniffed it, and then took it; (4) later on, it took the animal food with one or both hands, climbed away, and then ate it after sniffing it.

Healthy tarsiers, caught in the wild, often did not drop beetles which were given to them when they were released. They kept them and sometimes started eating before they were out of sight. This, when added to the above statements, may give some indication of the disposition for learning in adult *T. bancanus*. Cook (1939) reported that *T. syrichta* ate the eggs of common house geckos. *Tarsius syrichta* apparently readily accept dead food as Ulmer (1963) indicated: "Gradually, I swung them over from living food to dead natural food." He was referring to deep-frozen anoles (*Anolis carolinensis*) which he had warmed up before offering them to the tarsiers. He agrees with Wharton (1950) "that lizards and insects were favored foods" for the Philippine species. But the Philippine individuals which Hill *et al*. (1951) kept did not feed readily on lizards and mice. H. Sprankel (personal communication) sometimes gave neonate mice to his captive *T. bancanus*, which they did not particularly like.

5. DIURNAL RESTING AND NOCTURNAL ACTIVITY (CIRCADIAN RHYTHMS)

Though several authors have written in varying detail about the activity rhythms of *Tarsius*, only Harrisson (1963) published quantified data about actual locomotor activity of captive *T. bancanus*. Sprankel is currently preparing a detailed comparative account of *Tarsius syrichta* and *T. bancanus* which he maintained in Frankfurt. In the present study, when observing Bornean tarsiers under seminatural conditions, it seemed best to take locomotion as a primary indicator of activity. Tarsiers appear unable to move their eyeballs sufficiently within their orbits, and move their heads virtually each time they change the direction of their gaze. Such movements were counted for periods of 2-minute intervals, since this provided further information about the general activity of any animal observed. Times of feeding, autogrooming, and eliminatory behavior were also recorded, for locomotion is interrupted during such activities. Although most of the records have not yet been evaluated, the following statement can be made about the typical distribution of activity in *T. bancanus* over the course of 24 hours under seminatural conditions: The tarsiers on Borneo usually wake up shortly before or at sunset, i.e., between 1745 and 1830 hours and in most cases at 1800–1815 hours. Since the animals were actually watched waking up on about 40 occasions, it is fairly certain that any tarsiers which were encountered leaping about earlier in the evening had most probably been disturbed. After waking, each tarsier would rest for 10–20 minutes before sliding down from its sleeping perch and starting to leap. This resting period is rarely shorter and sometimes longer than the range indicated above. The first locomotor activity was usually recorded at between 1830 and 1900 hours. With respect to light intensity, this means that locomotion begins at about the time a human being has some trouble writing without artificial light, or even in almost complete darkness. The overall locomotor activity during the course of the whole night was astonishingly constant, ranging mostly between 650 and 850 leaps per night in an adult female. Harrisson (1963) reported a nightly average for her female of about 1500 leaps in captivity, but she writes: "The length of the average jumps in the type of cage described was roughly 50 cm." As the average length of leaps under seminatural conditions was considerable greater than this (see above), the two sets of figures correspond satisfactorily. Since the total time of nocturnal resting and the distribution of the leaps during the course of the night were not apparently associated with the total number of leaps, and since the length of the average leap does seem to exert an influence on the total number of leaps, the total energy used for nocturnal locomotor activities could be the limiting factor involved.

Bouts of leaping during the night occurred at intervals of approximately 1 hour on average, alternating with periods of rest. Though feeding takes up a considerable amount of time during the night and is often followed by bouts of resting, these intervals are clearly perceptible to the observer, provided that observations are continued over the whole night. Time spent feeding may amount to a total of 2 hours or more during the night, but usually does not last longer than about 1 1/2 hours altogether. Periods of rest after feeding are mainly dependent on the mass eaten. While a small beetle has no effect at all on the number of leaps during immediately subsequent observation periods, a warbler of more than 10 gm certainly decreases the rate of leaps counted for the next 4 hours at least. In most cases, drinking under seminatural conditions took place before 2200 hours (for more information, see Niemitz, 1974).

From the records which have been evaluated to date, distinct peaks of locomotor activity cannot be identified during the first three-quarters of the night. But in the morning period from 0400 to 0630 hours there is a clear average increase in the rate of leaps which sometimes cease only a few minutes before the tarsier begins to look for a sleeping place (see Fig. 5). The same results are obtained by compiling the changes of gaze direction within 2 minutes (Fig. 6). The peak before dawn seems to be highly dependent on the overall quantity of locomotor activity, i.e., upon the total energy spent during the previous part of the night. If the tarsier did not leap

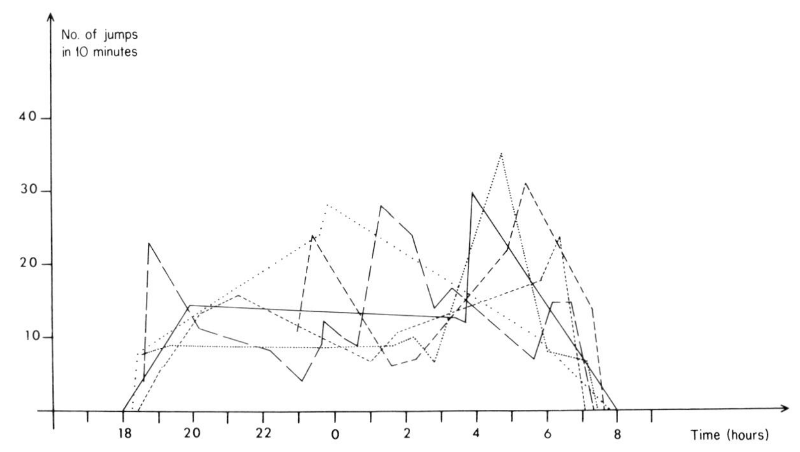

Fig. 5. Overall locomotor activity in *T. bancanus* in seminatural conditions. Each line indicates observational samples for a different night. The observational samples show the variable pattern of activity, with a general tendency toward a peak between 0400 and 0630 hours.

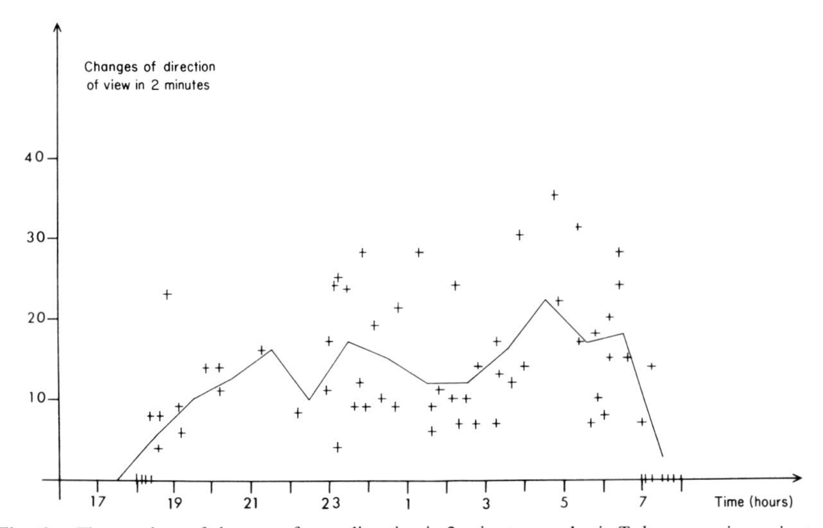

Fig. 6. The numbers of changes of gaze direction in 2-minute samples in *T. bancanus* in seminatural conditions. The observational samples (for approximately 10 nights) indicate only one cycle of activity, with a peak between 0400 and 0630 hours. The continuous line indicates the hourly average of the samples.

much in the earlier parts of the night, the peak was higher than under different conditions.

6. TERRITORIALITY

6.1. Territorial Marking

The terms *territorial* and *territoriality* are abstractions and do not mean, of course, that the related behaviors are necessarily restricted to the territory or the home range, but rather that all social or nonsocial activities relate to the presence of a group of particular individuals in a more-or-less definable area. This would include activities like repelling an intruder and reassuring the occupant.

While nothing is known about territorial behavior in *T. spectrum*, a few details have been provided for the Philippine species. Ulmer (1963) reported, of his captive *T. syrichta*, "The urine hardened in a shiny, orange-yellow layer on the branches, which was difficult to wash off." This observation suggests to some extent that Ulmer's tarsiers more or less habitually urinated on the same spot. Marking behavior is also indicated by the presence of an epigastric gland in male *T. syrichta* (Hill *et al.*, 1951) and in both sexes of *T. bancanus* (Sprankel, 1971). The phylogenetically older and more conservative way of marking (viz. urination) was described by Harrisson (1963), in *T. bancanus* under captive conditions outdoors, as follows:

(i) urinating on specific or prominent perches within the territory;
(ii) rubbing of toe-pads in the urine, thus distributing it over these perches

She also wrote that this method of marking is performed by both sexes and estimated that "thirty percent or more of the tarsier's overall mobility lies outside hunting or social activities and can at this stage only be determined as of a 'general territorial nature'." Fogden (1974) stated: "Marking with urine is frequent, but it is not known whether it has a territorial function."

When walking through the Sarawakian jungle, even people lacking a very sensitive sense of smell are able to detect the "distinct and penetrating odor" which Harrisson (1962) noted with her captive tarsiers. Both male and female *T. bancanus* have certain places for urination, some of which are most probably used only once, others only occasionally, and still others which are used very regularly, i.e., for more than 9 months (this time span being limited by the total duration of field observations in the present study). It is therefore possible that Bornean tarsiers may use marking sites within their territories or home ranges for years. In this context, the continued stability of ranging of one male *T. bancanus*, which Fogden had trapped in 1965 and 1966 (see Fogden, 1974) is confirmed. He ringed this individual, and noted further the number of the net in which he had caught the animal and its weight. In 1972 the same animal was trapped in the present study and it was possible to reconstruct, from the rusty metal number tag on the nets, some of which were still clinging to the trees in the jungle, that this old male had been recaptured less then 30 m away from one of the places where Fogden had previously caught it. This suggests that this particular tarsier's territory (or home range) had not changed considerably in the course of almost seven years. The probability that this individual may have returned after a long absence is remote. As the individual concerned was already fully grown when first captured, it was at least 8 years old in 1972. This is the only record of the minimum life span of a tarsier in the wild.

The regularity of marking certain sites with urine in this same species was observed by H. Sprankel (personal communication) with captive specimens. In the present study, apart from places where both male and female urinated in the enclosed area, there were tree trunks used exclusively by one individual as urination sites. Apart from urinating into water, which did not happen often in seminatural conditions, *T. bancanus* were very seldom ever observed to urinate on to the ground but rather against a more or less vertical substrate. Rubbing of the extremities in the urine was seldom observed and may perhaps be restricted to the male sex. Self-impregnation of the plantar skin with urine, so that the feet can be used as marking pads, has not been confirmed.

Communication by means of olfactory signals is very complex in tarsiers. Females, in particular, rub their external genitalia on tree trunks, most frequently when in estrus. The secretion of the circumanal glands is rubbed off at the same time. Together with urine, vulval smears deposited on tree trunks and the product of the circumanal glands represent three different types of olfactory communication, one with probable territorial significance (or having a social function; see below), the second with at least a sexual significance, and the third with an, as yet, unknown function. A further form of olfactory communication in *Tarsius*, employed more often by the male than the female, is the production of a secretion by the epigastric gland, a cutaneous structure comparable to the jugulosternal scent gland (Sprankel, 1961) in *Tupaia glis*. According to H. Sprankel (personal communication) it clearly has a territorial role, in addition to that of the urine, and in captivity is applied to inclined surfaces, mostly within the core territory rather than in the general living space, i.e., in the center of an occupied area and not at its periphery. To what extent it is of social importance, and to what extent, in this respect, territoriality and social organization are actually identical, or involve at least identical behavioral gestures, cannot yet be definitely decided.

Yet another form of olfactory communication in tarsiers may have a more obvious function despite its being discovered last as a definite form of marking behavior. Sprankel (1971), who studied the morphological structure involved and named it the circumoral gland, observed that both sexes of *T. bancanus* would rub their faces on vertical trunks in captivity. They performed this behavior throughout the year (independently of the reproductive cycle). This marking on the substrate was also observed and described as a form of "comfort behavior" by Polyak (1957), who provides a very good illustration (drawn by Bohlman). This behavior was also reported and similarly interpreted by Hill (1955). Moreover, *T. bancanus* not only deposit this scent on surfaces, but also apply it to their own tail tufts. Hill *et al.* (1951) observed a female *T. syrichta* rubbing her mouth in the fur of one male so persistently that the spot on the male's back gradually became more and more bald. This behavior may be explained as a means of transmitting social signals, such as individual identity or status, rather than as a form of marking, and will be dealt with in the next section. Whether or not the feces have any olfactory function cannot yet be stated, although there are some indications that they have no such communicatory function at all.

The various patterns of olfactory marking and the function of scent as a channel for intraspecific communication clearly show that the extreme development of the visual sense does not necessarily mean that the role of olfaction has decreased, as has been stated by several authors (Starck, 1954; Hofer, 1962). On the contrary, the olfactory channel could well have retained its status or evolved still further into an intricate, complex system for very important basic intraspecific signals. Although *Tarsius* is a microsmatic mammal, with reduced olfactory bulbs, some of the morphological reductions (e.g., the reduction of the number of ethmoturbinal bones) may well be a consequence of the enlargement of the eyes (cf. Spatz, 1968), and do not provide direct evidence of the importance of olfaction in communication in tarsiers.

6.2. Territorial Range

Fogden (1974) suggested a social organization and a system of overlapping ranges in *T. bancanus* which "would have points of similarity with that found in *Galago demidovii*" (see Charles-Dominique, 1972). Following H. Sprankel's observations on pair-bonding in *T. bancanus* in captivity (personal communication), together with a preliminary evaluation of recaptured tarsiers in the present study, which agree fairly well, the present author tentatively concludes that there is typically one pair per home range, possibly accompanied by an infant or juvenile, inhabiting an area of approximately 9000 to 16,000 m^2; hence a population density of 80 tarsiers per square km can be calculated, assuming that there are no areas really free of these animals. Since this is hardly likely in practice, a somewhat smaller estimate would be more accurate. Certain doubts about Fogden's interpretations arise, mainly because of his method of identifying free-moving individuals in the wild (40% of his records; 28 of 71 cases), by looking at a combination of certain individual features, such as characteristic features of the ears, tiny tips of fingers or toes missing, and the side on which the animal carried its ring. Accurate identification, on the basis of this combination of characteristics, of such a small animal, at night, by torchlight and in more or less dense vegetation, is highly questionable. Since Fogden's estimates of home ranges are based on an undisclosed number of captured animals and sight records, they cannot be regarded as definitive.

7. SOCIAL BEHAVIOR

7.1. Social Groupings

In *T. bancanus* the pair seems to be the rule as a social unit. The male and female inhabiting, marking, and defending a territory can be involved in serious fights with intruders of either sex. Under seminatural conditions one female was unrestrainedly aggressive toward a subadult male and killed it on their third or fourth encounter in the cage, before the observer could intervene. After the first encounter the young male, under natural conditions, would probably have been able to avoid a subsequent fatal encounter. Marking behavior as well as visual and vocal communication across the borders of the territories should be effective enough to prevent any serious agonistic clashes.

Both individuals of a pair inhabiting the same area exhibit a strong pair-bond. When an old captive male *T. bancanus* in Frankfurt died of pneumonia, virtually all features of territorial behavior in the female simply collapsed. Moreover, her pattern of activity became completely disturbed so that she exhibited some locomotor activity during the daytime and sometimes was very tired at night. Such obviously abnormal behavior, following the loss of her male partner, could not have been predicted merely from superficial observation. More precise study was necessary to establish the social behavior of the pair. Tarsiers communicate mainly in the visual channel by means of facial expressions which are barely noticeable to the unaided eye of the observer. Vocal signals seem, at first, to play a rather subordinate role in communication between individuals of a pair, which usually keep an individual distance of at least 1 m. Physical contact, which was rare under seminatural conditions, usually had sexual components. However, purely social elements could also be observed in such encounters. For example, the female sometimes grasped the male's tail, and both animals sometimes perched motionless together in this fashion for several minutes.

Wharton (1950), Ulmer (1963), Evans (1967), and others found *T. syrichta* "to huddle up in groups" or even "in one another's arms" when being transported and to some extent when sleeping during the daytime. While Wharton (1950) reported one case of four *T. syrichta* found together in a hollow tree in the wild, *T. bancanus* have never been reported together in groups of more than two. From photographs taken by H. Sprankel (personal communication), Hill *et al.* (1951), and Ulmer (1963), it can be established that adult pairs of *T. syrichta* do not keep an individual distance in captivity at all and even maintain physical contact by intertwining their tails. This was never observed in captive *T. bancanus* by Sprankel during year-long observations, nor in seminatural conditions during the present study on the Bornean species. Intensive mutual marking of adult animals, which Hill *et al.* (1951) observed in captive *T. syrichta*, has not been reported for *T. bancanus*. Despite a paucity of data from the field, reports on captive tarsiers and observations during the present study under seminatural conditions suggest that the social structure in *T. bancanus* is probably highly specialized. *Tarsius syrichta* show much more complex social interactions, since this species apparently has the ability to form groups. Agonistic behavior in *T. syrichta* has only been reported by one author (Wharton, 1950), in densely crowded cages. Hill (1955) characterizes this species as "gentle." At present no information is available on the social organization of *T. spectrum*.

Fogden (1974) found a few "immature male tarsiers" in Sarawak, mainly in primary forest. He assessed their ages from recorded body weights and estimated size of their testes. Fogden reported this as a "possible point of similarity with *Galago demidovii*, for such males resemble the 'vagabond' males of that species" (cf. Charles-Dominique, 1972). Out of 31 individual males the present author caught 6 such lightweight tarsiers with smaller testes. However, 4 of these animals were found in secondary jungle, 1 at the border between secondary and primary vegetation and only 1 in old established forest. Fogden's conclusions that such immature males inhabit a marginal habitat is not confirmed.

When comparisons were conducted not only on the lengths of the testes of such males and their body weights, but also on their body length, length of tail, length of hind foot, and other measurements, these male tarsiers proved to be skeletally as fully grown as the other heavier males with larger testes. In fact the lengths of their tails indicated that they were almost significantly bigger than the latter ($p = 0.056$; Niemitz, 1977). Fogden was, therefore, correct in identifying a particular category of males, now determined as a sexually inactive category of adult males and hence representing a socially distinct part of the male population of the Sarawak tarsiers. The figure of about 20% of the entire male population agrees satisfactorily with Fogden's figures. Although data from the present study have not yet been evaluated sufficiently to allow determination of territoriality in these individuals, the data do show them to be less sedentary than the other potentially sexually active males. Definitive statements, however, cannot yet be made.

7.2. Maternal Behavior and Infant Development

Mother–infant relationships represent an important aspect of social behavior and the ontogenetic key to social integration. Here again information is completely lacking for *T. spectrum*. With respect to *T. syrichta* we are much better informed (Cuming, 1838)

> . . . one morning I was agreeably surprised that she had brought forth. The young appeared to be rather weak but a perfect resemblance to its parent: the eyes were open and [it was] covered with hair; it soon gathered strength and was constantly sucking betwixt the parent's legs, and so well covered by its mother that I could seldom see anything of it but its tail; on the second day it began to creep about the cage with apparent strength, and even climb up to the top by the rods of which the cage was composed. Upon persons wishing to see the young one when covered by the mother, we had to disturb her, upon which the dam would take the young one in its mouth, in the same manner as a cat, and carry it about for some time . . .

Several authors describe the neonate *T. syrichta* to be in an advanced state (Wharton, 1950; Ulmer, 1963; Schreiber, 1968; Hill, 1973; F. Podolschack, personal communication). All criteria for this assessment (weight, measurements, open eyes, full fur, some teeth already erupted, readiness to climb about or at least to hold on tightly) fully agree with the observations of Le Gros Clark (1924), H. Sprankel (personal communication), and with the author's own investigations on living and preserved specimens of *T. bancanus* (Niemitz and Sprankel, 1974). From the reports available it seems unlikely (though it cannot be definitely ruled out) that *T. syrichta* construct nests or make use of holes in trees to care for the newborn. It is somewhat surprising that this has not yet been reliably determined, since at least nine births of full-term Philippine tarsiers in well-equipped modern zoos have been reported by now (Frankfurt, 3; San Diego, 3; Philadelphia, 2; Chicago, 1).

With respect to the Bornean tariser only one successful birth is known at present (H. Sprankel, personal communication). The present author had the fortunate opportunity to observe the behavioral development of *T. bancanus* in two different stages of growth over several weeks (Niemitz, 1974). In the course of these observations it became evident that there is no nest at all for young *T. bancanus*. One of the infants observed still had one ear as yet unfolded. From comparison with the findings of other authors a maximum age of 4 days could be estimated. Its age was taken to be 2 days; thus all statements of its age are to be considered as involving a possible error of ± 2 days.

At the age of 2 days the *T. bancanus* infant did not react to events happening further away than 0.5 m. The main sense was the olfactory one. It was more active at night than in daytime. It would move about quadrupedally, showed interest only in liquid food, and performed some autogrooming, licking its arms, hands, and thighs. At 3 days it made an unsuccessful attempt to jump; one day later it accepted its first solid food. On the same day it made a threatening gesture. By 5 days it showed visual concentration for more than 1 minute and for the first time focused on an object 30 cm away. It no longer awoke completely in the daytime and for the first time combed its hair with its teeth. From 8 days onward it did not sleep at all at nighttime. At 11 days it scratched itself for the first time with its toilet claws. By 13 days it bit for the first time as a protective response and its first leap was recorded. By 14 days orientation was mainly visual. The infant also reacted to events more than 4 m away and showed a strong preference for leaping rather than climbing. At 3 weeks of age no predatory behavior could yet be observed and it was still dependent on its mother's milk.

This individual record is not complete without noting that the infant was adopted by a lactating female at 4 days. The manner of adoption revealed that, while the infant was ready to accept any tarsier female as a mother, the female obviously noticed by olfactory means that the infant was a stranger. Whereas tactile and olfactory signals were more important for the young, visual and acoustic signals were equally important for the mother: the young tarsier uttered a continuous flow of soft, high-pitched calls as a unidirectional signal. Jacobs (in Sprankel, 1965) made an interesting observation with *T. syrichta*. On Mindanao he obtained several tarsier mothers together with their infants. The mothers sometimes lost their infants in the cage after being captured. When the infants uttered distress calls, the mothers began to retrieve them, apparently at random, leading to anomalous situations where, for instance, one female tried to retrieve no less than five infants. This automatic response suggests that the social structure of *T. syrichta* does not permit such situations to arise under natural conditions. Mothers in the wild, each with a single infant, may not live close enough for such situations to develop, perhaps dependent on the season. It may also be concluded that there is either a sensory hierarchy (the distress call of the infant takes priority over olfactory signals) or the social structure of the Philippine tarsier does not provide a basis for mothers to recognize their infants individually. Further studies on captive *T. syrichta* should provide more information on communicatory mechanisms in this species.

When the adult *T. bancanus* female adopted the infant tarsier, in the case described above, the infant, apart from uttering whistles, always grasped its foster mother's fur and impregnated her with the secretion of its circumoral gland (Sprankel, 1971). This provides further support for the social role of the circumoral gland secretion (see Fig. 7), in addition to Sprankel's observations on a caged pair (see above). Characteristically the female first licked the infant completely wet when adopting it, lending still further support for this social explanation.

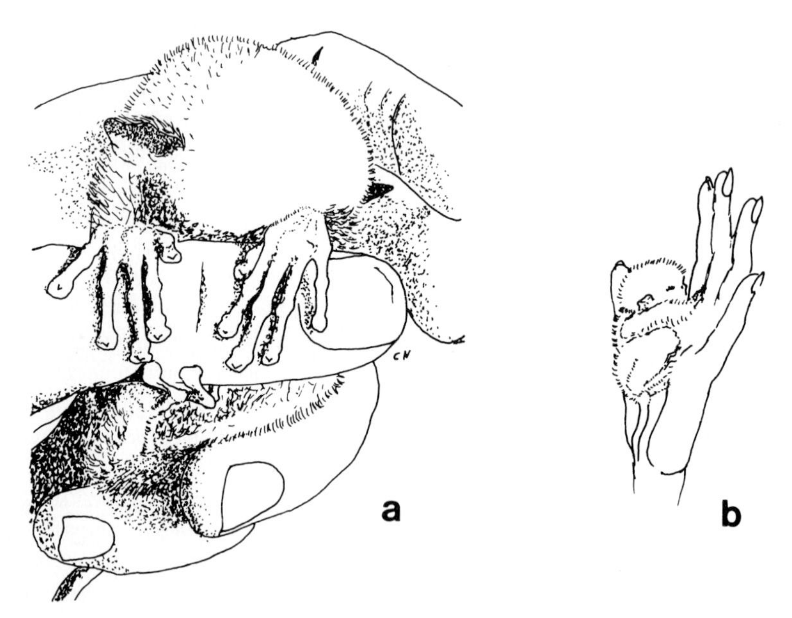

Fig. 7. Infant *T. bancanus* (a few days old) wiping its mouth and face on the fingers of the keeper (a). The young animal also sniffed at the hand of the keeper after depositing the secretion of the circumoral gland there (b). (Drawn from photographs.)

Judging from its greater weight, the second young Bornean tarsier observed was approximately 3 weeks old when captured. For the following account of its behavioral development the reader may assume that, in this case, the range of possible error in estimating the infant's age may well be greater than with the suckling infant. Between the age of 3 and 4 weeks this infant produced greeting sounds toward the keeper, not merely on seeing him but also on hearing him. At this age the infant performed its first proper and balanced landing when jumping to the ground, and also made its first leaps from one vertical trunk to another. After 5 days of observation it also caught an insect larva independently for the first time. Several repetitions of this behavior were observed only 5 days later at the age of 4 1/2 weeks. At this time the infant made strenuous attempts to make physical contact with an adult male tarsier in the keeper's care. The adult tarsier, which was ill, did not permit this though it behaved amicably toward the infant. The adult tried by means of various signals and threat to maintain distance from the juvenile, even preferring to escape from it, being apparently inhibited from attacking it. At 5 1/2 weeks, the first real agonistic behavior appeared. At the same time it started play behavior. At almost 8 weeks it drank water from the pot for the first time (for more information, see Niemitz, 1974).

8. VOCAL COMMUNICATION

While intraspecific communication among tarsiers in the olfactory, tactile, and visual channels has been dealt with in the preceding sections, vocal communication should be considered as a special category, for several authors have particularly noted the "high-pitched," "birdlike," "batlike" and "reedlike" calls. Wharton (1950) gave a classification of the "main calls" produced by *T. syrichta*:

> 1. A loud piercing alarm call of a single note. The writer has heard it while walking down a trail through the habitat, during fights and from deserted babies, often of a very small size.
> 2. A chirping locust-like communication when several tarsiers are within calling distance in the trees.
> 3. A soft sweet bird-like trill, heard when the animal is quiet and apparently contented. It may be heard shortly after dusk in caged animals that have been fed an hour or so before.

The first type given by Wharton had already been reported by Cuming (1838), who described it as a "sharp shrill call and only once." But the latter author stresses that the Philippine tarsier "seldom makes any kind of noise." Cook (1939) wrote about a "twittering noise, rather like the note of a bird" which the Philippine species produce, while Lewis (1939) stated: "One of the animals had a very high-pitched bat-like squeak, around 10,000 cycles per second and up to the limit of my hearing." It is primarily Wharton's report which gives an indication that vocal communication in *T. syrichta* might be much more complex than most authors suspect, mainly because he not only classified three types of calls but also related some of them to various different situations; for example, agonistic calls as opposed to calls having another social function within a group, the former being divergent or dispersion (''diffug''; Tembrock, 1971) signals with distance-increasing characteristics, and the latter signals having distance-maintaining characteristics (''stationär''; Tembrock, 1971). Tembrock (1971) draws attention to a third type of call, an affinity call (''affin'') which has distance-reducing characteristics.

Our present knowledge is similarity limited with respect to *T. bancanus*. Le Gros Clark (1924), before describing the call of an abandoned baby, generally noted: ''With regard to noises, the tarsier is ordinarily silent on every occasion observed.'' This was confirmed by Harrisson (1962) stating: '' . . . our Kuching animals have not been much heard. Calling may only be sexual.'' Hill *et al*. (1951) also referred calls mainly to sexual activities. Harrisson (1963) provided further information when noting: ''Both male and female occasionally call during 'sexual activity' with a piercing-twittering 'chit-chit', slightly higher than a squirrel's. These calls stand out because the tarsier is otherwise silent (except in distress).''

In an attempt to clarify the confusing comparisons, the writer tried to record the calls uttered by *T. bancanus* on tape. This is a difficult task, especially under seminatural conditions with a great deal of ambient noise and a tape recorder limited to an upper range of 16 kc. Recordings of captive animals which were taken at the studios of Radio Malaysia Kuching have, of course, been much easier to reproduce. Despite the lack of technical sophistication a preliminary account may be given based on sonagraphic analyses of the calls up to 12.4 kc, correlated with behavioral records.

It is not surprising that most earlier observers recognized, if at all, only a single type of call. Most calls of the tarsier seem to contain frequencies beyond the human range. Table I and Figs. 8–12 give a preliminary account. By means of sonagraphic diagrams, prepared at the Phonetic Institute of the Linguistic Department at Marburg University (West Germany), the author can so far describe at least seven different types of sounds produced by *T. bancanus*. One twittering type of call produced by a male infant a few days old forms a design of ''patches'' on the sonagram. It is uttered when the baby is feeling cold and appears to be anxious, and it is so faint that it may not necessarily have the function of calling the mother. It shows a rather complex composition of temporally quite isolated portions, the highest of which supersede the range of frequencies analyzed.

The distress call, colon and apostrophe, of the baby is much more simply composed and easily comprehensible in functional terms (Fig. 8). It contains essentially three portions, two very short and faint components and one higher, longer, and louder part, the latter probably yielding the decisive information: it is a distant call for the mother.

Experimentally induced pain (pinching) led to the record of a wave which lies at a distinct frequency of 8500–10,500 cycles per second (Fig. 9). The same faintly perceptible high note of an adult female just entered the scale of analysis with its lowest frequency of 11,500 cycles per second.

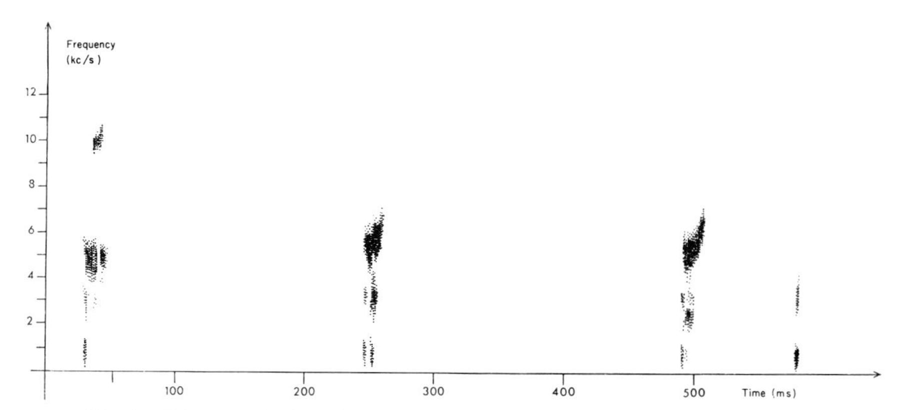

Fig. 8. Distress call of a *T. bancanus* infant when left alone (colon and apostrophe).

The following pattern of sounds (hysteresis, Fig. 10) is named after the shape of its main portion in the sonagram, resembling the hysteretic loop of a magnetic spool. Table I includes reference to a relatively simple modification of this type of call, the behavioral correlation of which is not clearly evident from the corresponding behavioral record. A much more complex example was selected for Fig. 10. A resident female *T. bancanus*, which was pregnant at the time, reacted aggressively

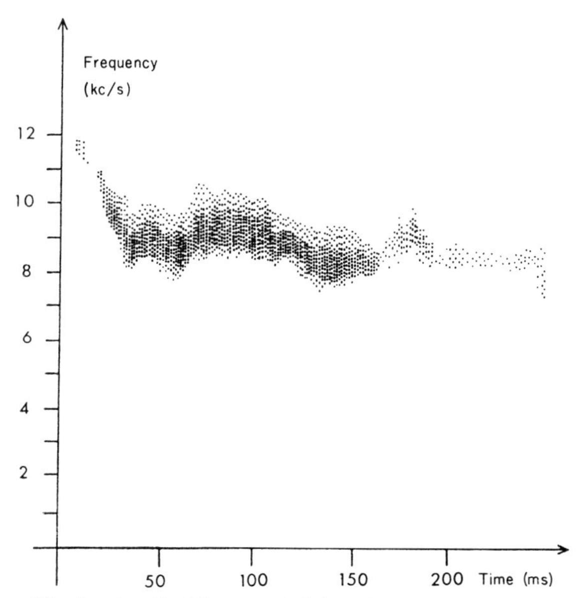

Fig. 9. A call of the same individual in response to pain (wave).

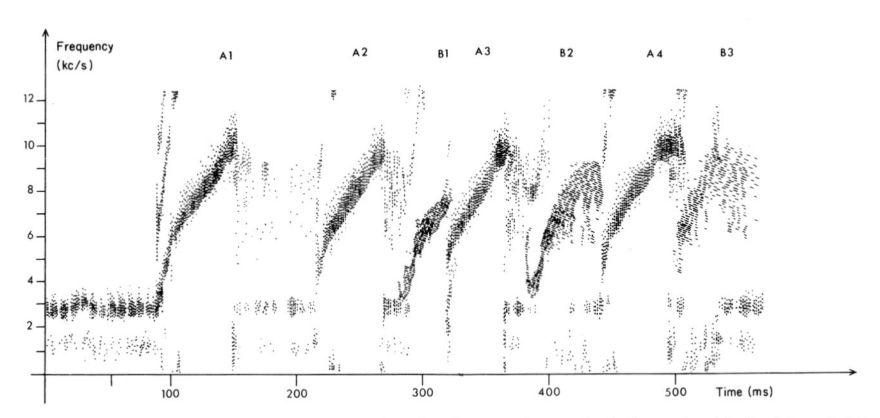

Fig. 10. The calls of two *T. bancanus* females (hysteresis, particularly noticeable in A1 and A3; see text and Table I).

toward a second intruding adult female. A serious fight followed during which the subordinate intruder uttered this hysteresis which is particularly typical in signals A1 and A3. The aggressive dominant resident uttered a slightly lower, but quite similar, call forming an archlike structure in the graph, also beginning at a low frequency and ending at a higher one. This type might actually belong to the same group of whistles, since there are all kinds of transitions between hysteresis and arches. In the case described we have a repetition of very similar sounds in rapid sequence in both animals. Since a subordinate adult male produced whistles similar to the sounds of the aggressive female in Fig. 10, it seems unlikely that the differences in the voices shown indicate the difference in dominance status between the fighting individuals.

A whistle of protest, the so-called clavicula, was recorded in situations where (in anthropomorphic terms) the tarsier was frustrated or annoyed. It also has a rising segment with increasing frequency, but it is easily distinguishable from the preceding hysteresis. Lasting about 140 ms in one case, it is intermediate between the other signals.

Finally, the author recorded two types of sounds to which no biological significance can be attributed as yet (Figs. 11 and 12). The X and the grate comprise frequencies occupying the whole scale of analysis. The first one lasted a total of about 25 ms, while the grate, which was also produced by an adult male, covered a timespan of roughly 75 ms. Since much vocal communication, e.g., during precopulatory situations, was not recorded, one may expect the variety of sounds in *T. bancanus* to be still more complex than shown here. *Tarsius* might also communicate, at least partially, in ultrasonic ranges.

TABLE I

Examples of Sounds and Calls Produced by *T. bancanus*[a]

Designation	Sex and age	Status	Behavioral correlation	Syntax: portions of spectrum, range of frequencies, time
"Patches"	Male, infant	—	Uneasiness, feeling cold and alone; no distant call; perhaps no pragmatics	500–1900 cps, 8 ms 2500–4900 cps, 10 ms 5000–8100 cps, 30 ms 6200–7300 cps, 8 ms
Colon and apostrophe	Male, infant	Affinity	Uneasiness, feeling alone; distant call to mother in distress	500–1900 cps, 2– 4 ms 2500–4400 cps, 2– 4 ms 4900–7800 cps, 6–18 ms
Wave	Male, infant	Dispersion	Feeling pain	8500–10,500 cps, ca. 230 ms
Hysteresis with initial stroke without chirping	Male, infant Female, adult Male, adult	? Dispersion Dispersion	Agonistic correlation up to open fight; partly questionable	500–3500 cps, 40 ms 3100–12,400 cps, 40 ms 5800–12,400 cps, 40 ms
Clavicula with crossed pattern	Male, adult	Dispersion	Protest, "vexation," "frest"	4100–9500 cps, 140 ms plus at least eight more portions within the spectrum, forming a criss-crossed pattern
X	Male, adult	?	?	At least 2 crossed portions within the spectrum: 500–12,400 cps, ca. 25 ms 125–12,400 cps, ca. 25 ms
Grate	Male, adult	?	?	125–12,400 cps, noise with all frequencies, ca. 75 ms.

[a] See Figs. 8–12.

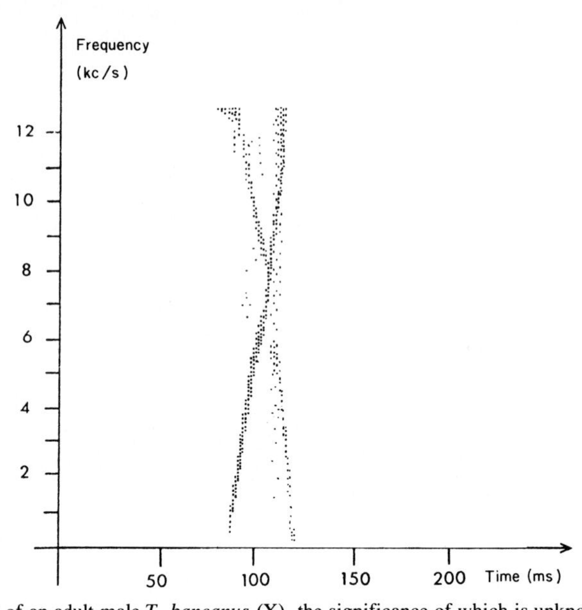

Fig. 11. Call of an adult male *T. bancanus* (X), the significance of which is unknown (see text and Table I).

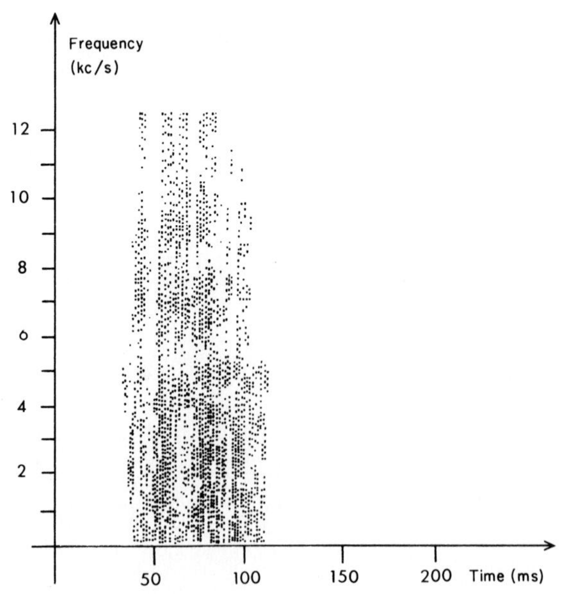

Fig. 12. Noise of an adult male *T. bancanus* (grate), the significance of which is unknown (see text and Table I).

9. DISCUSSION

Following comparison of the three recent species of the genus *Tarsius*, it can be reliably assumed that they are of monophyletic origin, the antecedants of *T. spectrum* forming the basic genetic pool for the evolution of the recent tarsiers. *Tarsius syrichta* are less extremely adapted than *T. bancanus* to the ecological niche the tarsiers occupy. Various facts indicate that the basic behavioral, morphological, and ecological specializations of the genus are to be found in *T. spectrum*.

The range of animals eaten includes insects, crabs, snakes, lizards, birds, and bats. Dead or motionless food is probably not touched by *T. bancanus* in the wild, as indicated by observations under seminatural conditions, but it is accepted in captivity by *T. syrichta* in addition to living animals. The species of parasites found in *T. bancanus* are very similar to those found in *T. syrichta*.

Ecologically, at least, *T. bancanus* essentially occupy the niche of an owl in dense vegetation, as is indicated by morphological and behavioral studies. The activity rhythm seems to be dependent on a circadian oscillator. Overall locomotor acitivity per night might be a function of a genetically determined amount of energy spent per active period.

Tarsiers mark their territories with urine, with the secretion of the epigastric gland, and with the secretion of the circumoral gland. The latter product is also utilized to convey social signals between sexual partners and between mother and infant. In *T. bancanus* a territorial (= sexual) pair seems to be the typical unit, compared to *T. syrichta* which typically form larger social groups. For this reason, the Philippine tarsier probably possesses more ''complex'' means of communication and should, therefore, be more highly evolved than *T. bancanus*. The postnatal behavioral development of *T. bancanus* reflects the different specializations of this species, particularly for vertical clinging and leaping.

Vocal communication in tarsiers is much more complex than previously assumed. The evolution of the visual sense in *Tarsius* did not suppress communication in the olfactory and in the acoustic channel.

ACKNOWLEDGMENTS

I would like to express my gratitude to the Curators of the Sarawak Museum, Mr. Benedict Sandin and Mr. Lucas Chin at Kuching, who made this study possible. I am deeply indebted to Dr. J. R. Anderson and the forest botanist Mr. Paul Chai for their excellent help during the fieldwork and for identifying tree and plant species. I would also like to thank Dr. Fred Chou (Timber Research Institute, Kuching), who was so kind as to let me use his photomicroscope for parasitological investigations and Dr. and Mrs. Kok for allowing me to work in their laboratory. Additional thanks go to Enchik Hossen Haji Usop of Radio Malaysia Timor, who placed a studio at my disposal, to the Chief Game Warden in Kuching, Enchik, L. S. V. Murthy, and also to the staff of the Phonetic Institute at the University of Marburg for providing the sonagrams evaluated in this publication. Finally I would like to gratefully acknowledge the financial support of the Deutsche Forschungsgemeinschaft (Grant Sp. 28/6 - 28/9) provided through Prof. Dr. Heinrich Sprankel, my teacher, to whom this chapter is respectfully dedicated.

REFERENCES

Catchpole, H. R., and Fulton, J. F. (1939). *Nature (London)* **144**, 514.

Catchpole, H. R., and Fulton, J. F. (1943). *J. Mammal.* **24**, 90–93.

Charles-Dominique, P. (1972). *Z. Tierpsychol., Beih.* **9**, 7–41.

Cook, N. (1939). *J. Mammal.* **20**, 173–177.

Creplin, M. (1846). *In* "Beiträge zur näheren Kenntnis der Gattung *Tarsius*" (H. Burmeister, ed.), pp. 131–136. Reimer, Berlin.

Cuming, H. (1838). *Proc. Zool. Soc. London* **6**, 67–68.

Davis, D. D. (1962). *Bull. Natl. Mus. Singapore* **31,** 1–129.

Elliot, D. G. (1910). *Bull. Amer. Mus. Nat. Hist.* **28**, 151–154.

Evans, C. S. (1967). *Int. Zoo Yearb.* **7**, 201–202.

Fogden, M. P. L. (1974). *In* "Prosimian Biology" (R. D. Martin, G. A. Doyle, and A. C. Walker, eds.), pp. 151–165. Duckworth, London.

Fulton, J. F. (1939). *Yale J. Biol. Med.* **11**, 561–573.

Harrisson, B. (1962). *Malay. Nat. J.* **16**, 197–204.

Harrisson, B. (1963). *Malay. Nat. J.* **17**, 218–231.

Hill, C. A. (1973). *Lab. Primate Newsl.* **12,** 15.

Hill, W. C. O. (1953a). *Proc. Zool. Soc. London* **123**, 1–16.

Hill, W. C. O. (1953b). *Proc. Zool. Soc. London* **123**, 17–26.

Hill, W. C. O. (1955). "Primates: Comparative Anatomy and Taxonomy. II. Haplorhini: Tarsioidea." Edinburgh Univ. Press, Edinburgh.

Hill, W. C. O., Porter, A., and Southwick, M. D. (1951). *Proc. Zool. Soc. London* **122,** Part 1, 79–119.

Hofer, H. O. (1962). *Bibl. Primatol.* **1**, 1–31.

Kiesel, U. (1968). *Folia Primatol.* **9**, 182–215.

Le Gros Clark, W. E. (1924). *Proc. Zool. Soc. London* **94,** 216–223.

Lewis, G. C. (1939). *J. Mammal.* **20**, 57–61.

Niemitz, C. (1972). *Sarawak Mus. J.* **20**, 329–337.

Niemitz, C. (1973a). *Borneo Res. Bull.* **5.2**, 61–63.

Niemitz, C. (1973b). *Lab. Primate Newsl.* **12**, 18–19.

Niemitz, C. (1973c). Film (unpublished).

Niemitz, C. (1974). *Folia Primatol.* **21**, 250–276.

Niemitz, C. (1977). *Cour. Forsch.-Inst. Senckenberg Frankfurt* **25,** 1–161.

Niemitz, C., and Niemitz, I. (1974). Film (unpublished).

Niemitz, C., and Sprankel, H. (1974). *Z. Morphol. Tiere* **79**, 155–163.

Polyak, S. (1957). "The Vertebrate Visual System." Univ. of Chicago Press, Chicago, Illinois.

Preuschoft, H., Fritz, M., and Niemitz, C. (in press). *In* "Environment, Behavior and Morphology: Dynamic Interactions in Primates" (M. E. Morbeck and H. Preuschoft, eds.), Fischer, New York.

Schreiber, G. R. (1968). *Int. Zoo Yearb.* **8**, 114–115.

Spatz, W. B. (1968). *Folia Primatol.* **9**, 22–40.

Sprankel, H. (1961). *Verh. Dtsch. Zool. Ges.,* 198–206.

Sprankel, H. (1965). *Folia Primatol.* **3**, 153–188.

Sprankel, H. (1971). *Proc. Int. Congr. Primatol., 3rd, 1970* Vol. 1, 189–197.

Starck, D. (1954). *Z. Wiss. Zool.* **157**, 169–219.

Tembrock, G. (1971). "Biokommunikation. Informationsübertragung im tierischen Bereich," Part II. Akademie-Verlag, Berlin.

Ulmer, F. (1963). *Zool. Gart. Leipzig (N.S.)* **27,** 71–84.

Wharton, C. H. (1950). *J. Mammal.* **31**, 260–268.

Index

Specific characteristics are listed under each genus.

B

C

I

L